Theory of Approximation of Functions of a Real Variable

A.F. TIMAN

TRANSLATED BY

J. BERRY

ENGLISH TRANSLATION EDITED
AND
EDITORIAL PREFACE BY

J. COSSAR

DOVER PUBLICATIONS, INC.
New York

Bibliographical Note

This Dover edition, first published in 1994, is an unabridged republication of the work first published by the Pergamon Press Ltd., Oxford, England, in 1963, and distributed in the United States by The Macmillan Company. The original Russian edition, *Teoriya priblizheniya funktsii deistvitel'nogo peremennogo,* was published by Fizmatgiz, Moscow, in 1960.

Library of Congress Cataloging-in-Publication Data

Timan, A. F. (Aleksandr Filippovich)
[Teorii͡a priblizhenii͡a funkt͡siĭ deĭstvitel 'nogo peremennogo. English]
Theory of approximation of functions of a real variable / A.F. Timan ; translated by J. Berry ; English translation edited and editorial preface by J. Cossar.
p. cm.
Originally published: Oxford, England : Pergammon Press, 1963.
Includes bibliographical references and index.
ISBN 0-486-67830-X
1. Functions of real variables. 2. Approximation theory.
I. Title.
QA331.5.T513 1994
515'.8—dc20 93-33695
CIP

Manufactured in the United States of America
Dover Publications, Inc., 31 East 2nd Street, Mineola, N.Y. 11501

CONTENTS

EDITORIAL PREFACE

THE theory of approximation of functions of a real variable had its origin in the work of Chebyshev (Tchebycheff) and the well-known theorem of Weierstrass on the approximations of continuous functions by polynomials. In the earlier part of this century the theory was developed by, among others, De la Vallée Poussin, Dunham Jackson and S. Bernstein, the last of whom contributed to the subject over a period of more that forty years. Many Russian mathematicians have worked in this field in the last twenty years, and the books of Achiezer (Theory of Approximation 1947) and Natanson (The Constructive Theory of Functions 1949) are well known in translation.** The present book, as well as dealing with the classical theory, gives an account mainly of recent Russian work. About half the papers cited in the bibliography (to which reference is made throughout the book) were published between 1950 and 1959.

PREFATORY NOTE

THIS monograph discusses a number of fundamental parts of the modern theory of approximation of functions of a real variable. The material is grouped round the problem of the connection between the best approximation of functions to their structural properties.

The book is intended for post-graduate students and for mathematical students taking advanced courses; it will also be of interest to workers in the field of the theory of functions.

** See the bibliography.

FOREWORD

THE theory of approximation of functions is now an extremely extensive branch of mathematical analysis.

The present monograph is devoted to a systematic exposition of those sections in which the relationship between the various structural properties of real functions and the character of possible approximations to them by polynomials and other functions of simple construction is investigated.

The basis of this investigation is the classical approximation theorem of Weierstrass, P. L. Chebyshev's concept of the best approximation and the converse theorem of S. N. Bernstein on the existence of a function with a given sequence of best approximations.

The extension and refinement of these two fundamental theorems, and the study of other related questions connected with the concept of best approximation, and related with properties of the classes of functions studied and of the apparatus of approximation, constitute the main content of the book.

Because of this no space has been found for such parts of the constructive theory of functions as the general theory of orthogonal polynomials, the problem of moments and the theory of quadratures, each of which has its special features, or for the theory of trigonometric series which requires a separate treatment.

This book originated from lectures which the author has been giving to senior mathematical students and to post-graduate students of the faculty of mathematics and physics of the Dnepropetrovskii University for a number of years. It consists of eight chapters closely connected with one another. Each chapter concludes with a section containing various problems and theorems which supplement the main text.

In addition to the usual course of analysis, the reader will require to have a knowledge of the fundamentals of the theory of functions of a real variable with elementary functional analysis and of the theory of functions of a complex variable as contained in the usual syllabuses. At the end of the book there is a brief summary of the type of knowledge required.

The sections are designated by two numbers, the first of which indicates the chapter and the second the number of the section within the chapter.

Within each section its individual parts have been numbered on the same principle, theorems, lemmas, corollaries and remarks being distinguished.

The numbers in square brackets are references to the bibliography.

I am grateful to the editor of the book, V. S. Videnskii, who carefully read the manuscript and made a number of suggestions which greatly improve the book, and also to I. M. Ganzburg for his help in reading the proofs.

A. TIMAN

CHAPTER I

WEIERSTRASS'S THEOREM

1.1. Approximation of continuous functions by polynomials on a finite segment. The fundamental theorem

The simplest real functions of the variable x are functions of the form

$$P_n(x) = c_0 + c_1 x + \dots + c_n x^n \tag{1}$$

($c_0, c_1, \dots, c_n$ being constants), known as algebraic (ordinary) polynomials which are the result of applying to x a finite number of times the fundamental arithmetical operations of multiplication and addition in the field of real numbers. By the degree (order) of a polynomial is meant the index of the highest power of x which occurs in it.

The basis of the theory of approximation of functions of a real variable is a theorem discovered by Weierstrass which is of great importance in the development of the whole of mathematical analysis. For continuous functions of a single real variable defined on the finite segment $[a, b]$ it asserts: *for any function $f(x)$ continuous on $[a, b]$, there exists a sequence of ordinary polynomials which converges uniformly to $f(x)$ on $[a, b]$.*

This remarkable constructive and characteristic property of continuous functions applies also to all functions of many variables which are continuous in the closed bounded region $\bar{G}$ of a given multi-dimensional space. If $f(x_1, x_2, \dots, x_m)$ is such a function, there exists a sequence of ordinary polynomials

$$P_{n_1, n_2, \dots, n_m}(x_1, x_2, \dots, x_m) = \sum_{k_1=0}^{n_1} \sum_{k_2=0}^{n_2} \dots \sum_{k_m=0}^{n_m} c_{k_1, k_2, \dots, k_m} x_1^{k_1} \dots x_m^{k_m}, \tag{2}$$

which converges uniformly** to it in $\bar{G}$.

** Such a statement would not be true for arbitrary functions of a complex variable continuous in a bounded closed region. This follows from well-known properties of uniformly convergent series of analytic functions. It is also appropriate to mention here that the stated property of continuous functions of a real variable defined on a finite segment is possessed by all functions of a complex variable continuous on a bounded closed connected set not possessing interior points and not dividing the plane. The following remarkable theorem of M. A. Lavrent'ev [1] is true. If a bounded closed connected set Q in the plane does not possess interior points and the set complementary to it is connected, then for any function $f(z)$ continuous on Q there exists a sequence of polynomials $P_n(z)$ of the form 1.1(1) which converges uniformly to $f(z)$. A new interesting proof of this theorem has been given by S. N. Mergelyan [1].

There exist a number of proofs of this fundamental theorem**.

Many of them, including the proof of Weierstrass himself***, reduce to the direct construction for each function $f(x)$ of the corresponding sequence of polynomials 1.1(1). For $a = 0$, $b = 1$ the sequence of polynomials

$$B_n(x) = \sum_{k=0}^{n} f\left(\frac{k}{n}\right)\binom{n}{k} x^k(1-x)^{n-k}, \tag{3}$$

indicated by S. N. Bernstein [2] is very simple and elegant in its construction.

Theorem 1.1 occupies a central position in the present chapter. All the other propositions depend on it as they are extensions, generalisations or developments in various directions. It was the initial step which gave rise to the formulation of problems connected with the concepts of closedness and completeness.

1.2. Proof of Weierstrass's theorem

We will give here a proof close in idea to that of Lebesgue**** and applicable to a more general case†.

1.2.1. Making use of the uniform convergence of the binomial series on the closed segment [0,1], we note that *if the assertion of the theorem is true for the function $f(x)$, it is true also for the function $|f(x)|$.* In fact, if $|f(x)| \leqslant K$, then

$$\begin{aligned} |f(x)| &= K\sqrt{1-\left[1-\frac{\{f(x)\}^2}{K^2}\right]} \\ &= K\left\{1-\frac{1}{2}\left[1-\frac{f^2(x)}{K^2}\right]-\sum_{n=2}^{\infty}\frac{1.3\ldots(2n-3)}{2.4\ldots 2n}\left[1-\frac{f^2(x)}{K^2}\right]^n\right\}, \end{aligned} \tag{1}$$

and for any $\varepsilon > 0$ we have, for all sufficiently great values of m,

$$\max_{a \leqslant x \leqslant b} \big||f(x)| - S_m(x)\big| < \varepsilon,$$

where $S_m(x)$ is the partial sum of order m of the series 1.2(1). Since we

** Some of them can be found in the monograph of V. L. Goncharov [1] (see also I. P. Natanson [2]). For functions of many variables see C. J. de la Vallée Poussin [4].

*** See Weierstrass [1]. This proof was based on the consideration of an integral in the theory of heat.

**** Lebesgue [2]. See, for example V. L. Goncharov [1].

† Stone [1]. See I. M. Gel'fand, D. A. Raikov, G. E. Shilov [1].

are assuming that the theorem holds for $f(x)$, there is an algebraic polynomial $P_n(x)$ of sufficiently high degree n for which

$$\max_{a \leqslant x \leqslant b} |S_m(x) - P_n(x)| < \varepsilon .$$

Consequently, $|f(x)|$ is the limit of a uniformly convergent sequence of polynomials.

1.2.2. *If the assertion of the theorem is true for the functions* $f(x)$ *and* $g(x)$, *it is true also for the functions* $\phi(x) = \max\{f(x), g(x)\}$, *and* $\psi(x) = \min\{f(x), g(x)\}$. This follows by representing them as

$$\phi(x) = \frac{f(x)+g(x)+|f(x)-g(x)|}{2},$$

$$\psi(x) = \frac{f(x)+g(x)-|f(x)-g(x)|}{2}.$$

1.2.3. *For any two points* $x_1, x_2 \in [a, b]$ *we can find a non-negative continuous function* $f^*(x)$, *not exceeding unity, with the properties*: (1) $f^*(x_1) = 1$; (2) $f^*(x) = 0$ *everywhere in a certain neighbourhood of the point* x_2; (3) *the assertion of theorem* 1.1 *holds for* $f^*(x)$. Let us take an arbitrary polynomial $P(x)$ for which $P(x_1) \neq P(x_2)$. The function

$$\Phi(x) = \frac{P(x)-P(x_2)}{P(x_1)-P(x_2)}$$

possesses the properties $\Phi(x_1) = 1$, $\Phi(x_2) = 0$. In virtue of continuity, for any positive $\varepsilon < 1$ a neighbourhood of the point x_2 can be found such that everywhere within it $\Phi(x) < \varepsilon$. If we also consider the function

$$\Psi(x) = \frac{\max[\Phi(x)-\varepsilon, 0]}{1-\varepsilon},$$

it follows from 1.2.1 and 1.2.2. that the function $f^*(x) = \min[\Psi(x), 1]$ is the one required.

1.2.4. *If* M *is an arbitrary closed set of the segment* $[a, b]$ *and* x_1 *is a point of this segment which does not belong to* M, *there exists a non-negative function* $f(x)$ *continuous on* $[a, b]$ *and not exceeding unity, with the properties*: (1) $f(x_1) = 1$; (2) $f(x) = 0$ *on* M; (3) *the assertion of theorem* 1.1 *holds for* $f(x)$.

For x_1 and each point $x_2 \in M$ it is possible to find a neighbourhood of x_2 and a function which satisfies the requirements of section 1.2.3. By the Borel-Lebesgue lemma it is possible to select from all such neighbourhoods a finite number which cover M. The system of functions

$f_1(x), \ldots, f_m(x)$ corresponding to these neighbourhoods leads to the function

$$f(x) = \min [f_1(x), \ldots, f_m(x)],$$

which is the one required.

1.2.5. *If M and Q are two non-intersecting closed sets of the segment $[a, b]$, there exists a non-negative function $f(x)$, not exceeding unity and continuous on $[a, b]$, with the properties*: (1) $f(x) = 1$ *everywhere on* Q; (2) $f(x) = 0$ *everywhere on* M; (3) *the assertion of theorem* 1.1 *holds for the function* $f(x)$. For every point $x_1 \in Q$ and for the set M it is possible to find a function $F(x)$ which satisfies the requirements of section 1.2.4. In virtue of its continuity, for any positive $\varepsilon < 1$ a neighbourhood of the point x_1 can be found such that everywhere in it $F(x) > 1-\varepsilon > 0$. The function

$$\Phi(x) = \frac{\min [F(x), 1-\varepsilon]}{1-\varepsilon}$$

does not exceed unity anywhere in $[a, b]$, is equal to zero on the set M and equal to unity in this neighbourhood of x_1. It is possible by the Borel-Lebesgue lemma to select from all such neighbourhoods a finite system which covers Q. The system of functions $f_1(x), \ldots, f_m(x)$ which corresponds to these neighbourhoods leads to the function

$$f(x) = \max [f_1(x), \ldots, f_m(x)],$$

which is the one sought.

1.2.6. *The theorem holds for every function $f(x)$ continuous on $[a, b]$.* Without loss of generality, we may assume that $\min\limits_{x \in [a, b]} f(x) = 0$, $\max\limits_{x \in [a, b]} f(x) = 1$. For any fixed n let us denote by M_k the set of those values of x at which $f(x) \leqslant k/n$, and by Q_k the set of those values of x at which $f(x) \geqslant (k+1)/n$ $(k = 0, 1, 2, 3, \ldots, n-1)$. By the continuity of $f(x)$, all these sets are closed and $M_k \subset M_{k+1}$, $Q_{k+1} \subset Q_k$. As M_k and Q_k do not intersect, there exists a function $f_k(x)$ which satisfies the requirements of section 1.2.5: $f_k(x) = 0$ on M_k and $f_k(x) = 1$ on Q_k. Hence for the function

$$F_n(x) = \frac{1}{n} \sum_{k=0}^{n-1} f_k(x) \tag{2}$$

the theorem holds, and as

$$|f(x) - F_n(x)| \leqslant \frac{1}{n}$$

it follows from the arbitrariness of n that the theorem is true also for $f(x)$.

1.3. Generalisation and some particular cases. Periodic functions. Functions of many variables

From the above proof it is seen that, apart from the reality and continuity of the function $f(x)$, use is made only of the following facts: (1) the set of all the polynomials 1.1(1) forms an algebraic ring**, which includes all constants (sections 1.2.1; 1.2.3; 1.2.5); (2) for any two distinct points x_1 and x_2 of the segment $[a, b]$ there exists in the set of all polynomials 1.1(1) a polynomial which assumes distinct values at these points (section 1.2.3); (3) for any closed set of the segment $[a, b]$ the Borel-Lebesgue covering lemma holds (sections 1.2.4; 1.2.5). Hence if in place of the segment $[a, b]$ there is considered *an arbitrary bicompact topological space G of points x, in place of the set of all polynomials* 1.1(1) *an arbitrary family Π of real continuous functions specified on it which contains all constants and is an algebraic ring, in which for any two points $x_1 \neq x_2$ of G there exists a function which assumes distinct values at these points, it can be asserted that every function $f(x)$ continuous on G is the limit of a sequence of functions of Π which converges uniformly to it.* In other words, in the sense of uniform convergence the family Π is everywhere dense in the class of all functions continuous on G. The generalisation of theorem 1.1 was obtained in this form by Stone***.

1.3.1. In particular, if Π is the set of all trigonometric polynomials

$$T_n(x) = \sum_{k=0}^{n} (a_k \cos kx + b_k \sin kx) \tag{1}$$

(a_k, b_k being real numbers), and the length of the segment $[a, b]$ does not exceed 2π****, all the conditions of theorem 1.3 are satisfied, and we obtain Weierstrass's theorem that *every continuous periodic function of period 2π is the limit of some sequence of polynomials* 1.3(1) *which converges uniformly to it.*

1.3.2. If Π is taken as the set of all the polynomials 1.1(2), and G as any closed bounded region of an m-dimensional space with variables x_1, x_2, x_3, ..., x_m, we obtain Weierstrass's theorem for continuous functions of many variables mentioned in section 1.1.

** The family Π of functions $P(x)$ is said to be an algebraic ring if for any two functions of it their sum and product also belongs to it.

*** Stone [1].

**** A segment $[a, b]$ of length 2π is identified with the circumference of a circle of unit radius.

1.3.3. Considering the set of all proper rational functions continuous on $[a, b]$ (see 1.6(1)), we conclude that *any function continuous on* $[a, b]$ *is the limit of a uniformly convergent sequence of such rational functions.*

1.4. Mean approximation of integrable functions by polynomials

Theorem 1.1 permits the constructive definition of the class C of all functions continuous on the finite segment $[a, b]$ as the class of those functions and only those which are the limits of uniformly convergent sequences of polynomials 1.1.(1) i.e. sequences of polynomials $P_n(x)$ for which

$$\|P_n(x)-P_m(x)\|_C = \max_{x\in[a,b]} |P_n(x)-P_m(x)| \to 0 \qquad (m, n \to \infty). \tag{1}$$

Certain other classes of real functions defined on the finite segment $[a, b]$ also admit of a similar constructive definition. Thus, for example, *the class* L_q $(q > 0)$ *of all real functions* $f(x)$ *measurable on* $[a, b]$, *the* q-th *power of the modulus of which is Lebesgue-integrable on this segment, can be defined as the class of those functions and only those which are the limits of sequences of algebraic polynomials* 1.1(1) *which converge in mean, i.e sequences of polynomials* $P_n(x)$ *for which*

$$\int_a^b |P_n(x)-P_m(x)|^q \, dx \to 0 \qquad (m, n \to \infty). \tag{2}$$

On the one hand, every sequence of polynomials $P_n(x)$ which satisfies the condition 1.4(2) corresponds, apart possibly from a set of measure zero, to one and only one function $f(x) \in L_q$ such that

$$\int_a^b |f(x)-P_n(x)|^q \, dx \to 0 \qquad (n \to \infty). \tag{3}$$

This follows from the well-known property of the completeness of the space L_q. On the other hand, if $f(x) \in L_q$, then for any $\varepsilon > 0$ it is possible to find a number $\delta > 0$, such that whatever may be the measurable set $M \subset [a, b]$ with measure less than δ,

$$\int_M |f(x)|^q \, dx < \varepsilon.$$

By the C-property of measurable functions (N. N. Luzin's theorem) there exists a closed set $Q \subset [a, b]$, with measure $\mu Q > b-a-\delta$, on which $f(x)$ is continuous. Then if

$$f_\delta(x) = \begin{cases} f(x), & x \in Q, \\ 0, & x \,\overline{\in}\, Q, \end{cases}$$

we have

$$\int_a^b |f(x)-f_\delta(x)|^q \, dx < \varepsilon.$$

The function $f_\delta(x)$, generally speaking, may turn out to be discontinuous on $[a, b]$, but, considering the intervals adjacent to the set Q, a function $\phi_\delta(x)$ can be found which is continuous everywhere on this segment and for which

$$\int_a^b |f_\delta(x)-\phi_\delta(x)|^q \, dx < \varepsilon.$$

By theorem 1.1, for this function a polynomial $P_n^*(x)$ can be found such that

$$\max_{x \ [a,b]} |\phi_\delta(x)-P_n^*(x)| < \varepsilon,$$

and then *a fortiori*

$$\int_a^b |\varphi_\delta(x)-P_n^*(x)|^q \, dx < \varepsilon^q(b-a).$$

Applying Minkowski's inequality when $q > 1$, or the inequality $|\alpha+\beta|^q \leqslant |\alpha|^q+|\beta|^q$ when $0 < q \leqslant 1$, we see that for any $\varepsilon > 0$ there exists a polynomial $P_n(x)$ possessing the property

$$\int_a^b |f(x)-P_n(x)|^q \, dx < \varepsilon.$$

Consequently, to every function $f(x) \in L_q$ there corresponds some sequence of polynomials $P_n(x)$ which satisfies the relation 1.4(3).

1.4.1. From theorem 1.3.1 it follows that, in the case where the length of the segment $[a, b]$ does not exceed 2π, the assertion and all the reasoning of section 1.4 remains valid if instead of a sequence of ordinary polynomials 1.1(1) a sequence of trigonometric polynomials 1.3(1) is considered.

1.4.2. The proof of theorem 1.4 is based on the completeness of the space L_q, the absolute continuity of the Lebesgue integral, the C-property of measurable functions and on theorem 1.1. These properties persist for measurable functions of several variables defined in a closed bounded region $\bar{G}$ of the multi-dimensional space considered. Hence theorem 1.4 will hold for classes $L_q (q > 0)$ of functions, the q-th power of the modulus of which is integrable on $\bar{G}$, if in place of the polynomials 1.1(1) we consider in them the corresponding polynomials 1.1(2)**.

** In the construction of a function similar to $\phi_\delta(x)$ in 1.4, use can be made of the heorem on the continuation of a continuous function.

1.4.3. It is obvious that in virtue of 1.3.3 it would be possible in theorem 1.4 to replace the polynomials 1.1(1) by proper** rational functions continuous on $[a, b]$.

1.5. Some stronger forms of Weierstrass's theorem

Theorem 1.1 and hence also 1.4 remains valid if instead of the whole set of polynomials 1.1(1) there is considered any part of this set which forms a ring containing all constants and at least one polynomial which varies monotonically on $[a, b]$.

Thus not only is the set of *all* the polynomials 1.1(1) everywhere dense in the space C of all functions continuous on a finite segment (and also in the space L_q), but also certain parts of it. This remark applied to the segment $[a, b]$ $(a \geqslant 0, b > 0)$ is made more precise by the following more general assertion.

1.5.1. *If the infinite sequence of non-negative numbers*

$$n_0 = 0 < n_1 < n_2 < n_3 < \cdots < n_k < \cdots$$

is such that the series

$$\sum_{k=1}^{\infty} \frac{1}{n_k} \tag{1}$$

diverges, then the set of all polynomials of the form

$$P_{n_k}(x) = \sum_{\nu=0}^{k} c_\nu x^{n_\nu} \tag{2}$$

(*where the c_ν are numerical constants*) *is everywhere dense in the space C of all functions*** continuous on* $[a, b]$. The convergence of the series 1.5(2) is not only sufficient but also necessary for the validity of the conclusion of this theorem.

We omit the proof of this interesting stronger form of theorem 1.1, which is given in monographs on the theory of approximation. The reader who wishes to become acquainted with it is referred to these works****.

1.5.2. Among the polynomials 1.1(1) those with integral coefficients may be considered as constructively the simplest. The set of all such poly-

---

** See p. 11 (Translation editor)

*** The indices n_k may also be nonintegral numbers, and may also form a bounded sequence. This proposition, which shows that for theorem 1.1 it is not the arithmetical nature of the indices which is important but only the character of their increase was first advanced in the form of a hypothesis by S. N. Bernstein at an international congress of mathematicians in 1912 (Collected works, Vol.1, page 122) and was proved by Müntz [1] two years later.

**** See N. I. Akhiezer [3], I. P. Natanson [2].

nomials is an algebraic ring. However, since it does not include all constants the remark 1.5 does not apply to it. Nevertheless it is not difficult to see that *when there are no integers on the segment* $[a, b]$ *theorem* 1.1 (*and* 1.4) *remains valid in this case if only the polynomials* 1.1(1) *with integral coefficients are considered***. In fact, to become convinced of the possibility of this it is sufficient, for any constant C, merely to find a sequence of polynomials 1.1(1) with integral coefficients which converges uniformly to it. For this purpose let us consider the identity***

$$C \equiv \sum_{k=0}^{n} C\binom{n}{k}(x-[a]^k(1+[a]-x)^{n-k}.$$

If $A_k^{(n)} = \left[C\binom{n}{k}\right]$ and $\alpha = \max\{b-[a], 1+[a]-a\}$, then

$$C = \sum_{k=0}^{n} A_k^{(n)}(x-[a]^k(1+[a]-x)^{n-k}+O\{(n+1)\alpha^n\}$$

uniformly for all $x \in [a, b]$. Consequently, when $n \to \infty$ the relation

$$C = \sum_{k=0}^{n} A_k^{(n)}(x-[a])^k(1+[a]-x)^{n-k}+o(1),$$

is satisfied uniformly on $[a, b]$, from which it also follows that in the case considered *every function continuous on* $[a, b]$ *is the limit of a sequence of ordinary polynomials with integral coefficients which converges to it uniformly*. It is obvious that the condition imposed here on the segment $[a, b]$ is necessary for the validity of the given theorem. If the segment

** The corresponding result for functions of many variables was obtained in a paper by G. A. Zhirnova [1].

It can be shown (S. N. Bernstein [9]), that for functions $f(x)$ for which $f(a)$ and $f(b)$ are integers, the given assertion is also valid in the case where the segment $[a, b]$ is replaced by the interval (a, b), i.e. it remains valid if we consider a to be integral, and $b = a+1$. Without loss of generality we can put $a = 0$, $f(0) = f(1) = 0$ and assume that the sequence of polynomials

$$P_n(x) = P_n(f;x) = \sum_{k=0}^{n} c_k x^k = \sum_{\nu=0}^{n} d_\nu x^\nu(1-x)^{n-\nu},$$

which by theorem 1.1 converges uniformly to $f(x)$ on $[0, 1]$, also satisfies the condition $P_n(0) = P_n(1) = 0$. Then if

$$Q_n(x) = \sum_{\nu=1}^{n-1} [d_\nu] x^\nu(1-x)^{n-\nu},$$

it follows that

$$|Q_n(x)-P_n(x)| \leqslant \frac{1}{n}\sum_{\nu=1}^{n-1}\binom{n}{\nu}x^\nu(1-x)^{n-\nu} \leqslant \frac{1}{n},$$

so that $\max\limits_{0\leqslant x\leqslant 1}|f(x)-Q_n(x)| \to 0$ when $n \to \infty$.

*** As usual, $[x]$ denotes the integral part of the number x; $O(1)$, and $o(1)$ denote bounded and infinitesimal quantities respectively.

$[a, b]$ contained the integer m, then for any function $f(x)$ for which $f(m)$ is not an integer we should have, for some $\varrho > 0$,

$$\max_{x \in [a,b]} |f(x) - P_n(x)| \geq \varrho > 0,$$

whatever the algebraic polynomial $P_n(x)$ with integral coefficients might be. In addition, *in the case where* $b - a \geq 4$, *no continuous function other than a polynomial* 1.1(1) *with integral coefficients could be the limit of a sequence of such polynomials which converges to it uniformly.* This follows from the inequality

$$\max_{a \leq x \leq b} \left| x^n + \sum_{k=0}^{n-1} c_k x^k \right| \geq 2\left(\tfrac{1}{4}(b-a)\right)^n,$$

which will be proved in 2.9.1. The last statement remains true also in the case where approximation in the mean is considered (see 1.4, $q \geq 1$). By Hölder's inequality this follows from the relation

$$\int_a^b \left| x^n + \sum_{k=0}^{n-1} c_k x^k \right| dx \geq (b-a)\left(\tfrac{1}{4}(b-a)\right)^n,$$

which will be proved in section 2.9.31.

Moreover, on segments of length less than four, any continuous function can be uniformly approximated by a sequence of the polynomials 1.1(1), which have only a finite number (depending on the length of the segment) of non-integral coefficients**. It can be shown*** that for the integral approximations considered in 1.4 it would be sufficient to restrict ourselves in this case to a set of ordinary polynomials having only integral coefficients.

Thus the possibility of approximating uniformly with any degree of precision to a function $f(x)$ continuous on $[a, b]$ by polynomials 1.1(1) with integral coefficients depends on the arithmetical nature of this segment, and also on certain arithmetical properties of the function itself.

1.6. Approximation of continuous functions on an infinite interval. Uniform approximation by rational functions

In the proof of theorem 1.1 an essential part is played by the condition that the segment $[a, b]$ is finite (see 1.3), as a result of which the continuous function considered on it is always bounded and uniformly contin-

** This can be proved by using a theorem of Fekete [1], by which there exists a non-trivial polynomial 1.1(1) with integral coefficients satisfying the inequality

$$\max_{a \leq x \leq b} |P_n(x)| < 2(n+1)\left(\tfrac{1}{4}(b-a)\right)^{\frac{n}{2}}.$$

*** See Aparisio [1], A. O. Gel'fond [3].

uous. This cannot be said of continuous functions defined on an infinite interval. In this case such functions are frequently unbounded, or if bounded may not be uniformly continuous. Apart from this, since no sequence of the polynomials 1.1(1) (with the exception of a stationary sequence) can be uniformly convergent in an infinite interval, not a single function other than a polynomial 1.1(1) can be the limit in an infinite interval of a uniformly convergent sequence of such polynomials.

1.6.1. The position is quite different if, instead of the polynomials 1.1(1) as approximation functions continuous on the infinite interval, use is made of the rational functions

$$R_n(x) = \frac{P_n(x)}{Q_n(x)} = \frac{a_0+a_1x+\cdots+a_nx^n}{b_0+b_1x+\cdots+b_nx^n} \tag{1}$$

($a_0, a_1, \ldots, a_n; b_0, b_1, \ldots, b_n$ being numerical constants), which are the result of applying to the argument x, besides the operations of multiplication and addition in the field of real numbers, an additional operation—division.

It may be supposed that the infinite interval is the whole real axis $-\infty < x < \infty$, and it is also possible to consider only the set of those rational functions 1.6(1) which are continuous everywhere on it and are proper ($Q_n(x)$ has no real roots and $b_n \neq 0$). If a certain sequence of such rational functions $R_n(x)$ converges uniformly on the whole real axis, the limit function $f(x)$ is continuous and bounded there, and has unique limits as $x \to \pm\infty$. The converse proposition is also valid and has for sets of these functions the same significance as theorem 1.1. *For any function $f(x)$ continuous on the whole real axis and having the finite limit* $\lim\limits_{x\to\pm\infty} f(x) = c$, *there exists a sequence of the continuous rational functions* 1.6(1) *such that*

$$\sup_{-\infty<x<\infty} |f(x)-R_n(x)| \to 0 \quad (n \to \infty). \tag{2}$$

This assertion can be obtained as a consequence of theorem 1.3.

If we put

$$f(\infty) = \lim_{x\to\pm\infty} f(x),$$

the function $f(x)$ will be continuous on the whole real axis including also the point at infinity. In this case by a neighbourhood of the point at infinity we understand that part of the x-axis which is situated outside some finite segment.

The real axis generalised in this way is a bicompact space on which all the functions considered here are continuous, and the family of proper rational functions 1.6(1) continuous on $(-\infty, \infty)$ forms an algebraic ring which satisfies the conditions of 1.3.

1.6.11. Theorem 1.6.1 remains valid if instead of the set of all proper rational functions 1.6(1) continuous on $(-\infty, \infty)$, we consider on it any part of this set which forms a ring containing all constants and for any two points $x_1 \neq x_2$ a rational function $R(x)$, which assumes unequal values at them, is different from $R(\infty)$**.

1.6.12. Remark 1.6.11 shows that in the class of functions considered in theorem 1.6.1 not only is the whole set of rational functions 1.6(1) everywhere dense in the sense of uniform convergence, but so also are certain parts of it. Among them we note one important class of such rational functions indicated by S. N. Bernstein [15] and characterised by a given sequence of denominators $Q_n(x)$. *If* $z_k = a_k + ib_k$ $(k = 1, 2, 3, \ldots, n)$ *are the n complex zeros of the polynomial* $Q_n(x)$ *and* $r_k = |z_k|$, $\theta_k = \arg z_k$, *then in the case when the sequence of sums*

$$\sigma_n = \sum_{k=1}^{n} \min\left(\frac{1}{r_k}, r_k\right) |\sin \theta_k| \tag{3}$$

is not bounded for every function $f(x)$ *of theorem* 1.6.1 *there exists a sequence of ordinary polynomials* $P_n(x)$ *such that the corresponding rational functions* 1.6(1) *satisfy the relation* 1.6(2) *of this theorem***.

In particular the denominators of such rational functions may consist of the sequence of polynomials $Q_{2rn}(x) = (1+x^{2r})^n$ $(r > 0$ integral), which has unbounded sums σ_n.

We will not give the proof of this proposition, as it requires special considerations beyond the scope of this chapter. We only mention, as proved by S. N. Bernstein***, that for the validity of the conclusion of the given theorem the condition of the unboundedness of the sequence σ_n is not only sufficient, but also necessary.

1.7. Uniform approximation on the whole real axis by integral functions of finite degree

We see that the class of functions continuous on the whole real axis, being the limits of uniformly convergent sequences of the rational functions 1.6(1), contains only functions the graphs of which have asymptotes parallel to the x-axis. If, however, a function $f(x)$ bounded and continuous on $(-\infty, \infty)$ does not possess this property, then the rational functions 1.6(1), as also the polynomials 1.1(1), become unsuitable as apparatus of approximation and the question of the choice of a new constructive element

** For the generalisation of theorems of the type 1.3 see A. F. Timan [14].

*** See S. N. Bernstein [15], Chap. III, § 4, theorem 1.

arises. At first glance this question is of interest for classes of those functions continuous on the whole real axis which, as in the case of a finite segment, are bounded and uniformly continuous there. Here belong all periodic continuous functions, all the functions considered in section 1.6.1, all functions almost periodic in Bohr's sense**, and many others. In those cases where $f(x)$ is a periodic function, the question is solved by theorem 1.3.1. It appears that if in place of the set 1.3(1) of trigonometric polynomials considered there, which represent the values on the real axis of integral functions of exponential type bounded on it, there are used as constructive elements all such functions, i.e. all possible integral functions

$$G(x) = \sum_{k=0}^{\infty} c_k x^k$$

which like the polynomials 1.3(1) satisfy the inequality

$$|G(x+iy)| \leqslant Ae^{\sigma|y|} \tag{1}$$

for some $A > 0$ and $\sigma > 0$, then it is possible, just as in theorem 1.1, to determine the class of all functions bounded and uniformly continuous on the interval $(-\infty, \infty)$.

The bounded functions $G(x)$ considered are frequently called integral functions of finite degree*** and the lower bound of all the σ in the inequality 1.7(1) the degree of $G(x)$****. One of the most important properties of these functions is their uniform continuity on the whole real axis (see 4.8.2). In virtue of this, if a certain sequence $G_n(x)$ of such functions is uniformly bounded on the whole infinite interval $(-\infty, \infty)$, the limit function $f(x)$ is bounded and uniformly continuous there. Also valid is the converse statement†, according to which *the class of functions which are bounded and uniformly continuous on the whole real axis consists only of those functions which are the limits of sequences of bounded integral functions of finite degree which converge uniformly there, i.e. sequences $g_n(x)$ for which*

$$\sup_{-\infty<x<\infty} |g_n(x)-g_m(x)| \to 0 \quad (m, n \to \infty).$$

** See B. M. Levitan [2].

*** See sections 4.3 and 9.38 for an explanation of the term "degree" as used by Russian writers Cf. Boas [3].

**** The application of the theorem of Phragmén and Lindelöf to the functions $G(z)e^{i\sigma z}$ and $G(z)e^{-i\sigma z}$, considered in the upper and lower half planes respectively, shows that in the inequality 1.7(1) we can put $A = \sup_{-\infty<x<\infty} |G(x)|$. We shall frequently make use of this property.

† S. N. Bernstein [23].

1.7.1. We have to prove that *for any function $f(x)$ bounded and uniformly continuous on the whole real axis there exists a sequence of integral functions $g_n(x)$ of degrees σ_n such that*

$$\sup_{-\infty<x<\infty} |f(x)-g_n(x)| \to 0 \quad \textit{as} \quad \sigma_n \to \infty. \tag{2}$$

Let us note that for the functions considered here the assertions 1.2.1, 1.2.2 are valid, and let us prove that if M and Q are two non-intersecting closed sets of the real axis, the distance between which is positive $\big(\varrho(M,Q)=\varrho>0\big)$, then there exists a non-negative function $\Phi^*(x)$, not exceeding unity, with the properties: (1) $\Phi^*(x)=0$ everywhere on M; (2) $\Phi^*(x)=1$ everywhere on Q; (3) the assertion 1.7.1 holds for $\Phi^*(x)$.

For this purpose let us consider the function

$$g(x) = \frac{4}{\pi}\int_0^x \frac{\sin^2 \frac{t}{2}}{t^2}\,dt. \tag{3}$$

It is an integral function of the first degree, bounded on the real axis $\big(|g(x)|<1\big)$, odd and monotonically increasing. Hence

$$1-|g(x)| = O\left(\frac{1}{1+|x|}\right),$$

so that the sequence of integral functions $h_{2n}(x)$, of degree $\leqslant 2n$, consisting of the partial sums of the corresponding binomial series 1.2(1), satisfies the relation

$$|g(x)-h_{2n}(x)| = O\left(\frac{\varepsilon_n}{1+|x|^2}\right) \qquad (\varepsilon_n \to 0;\ n\to\infty)$$

uniformly with respect to x and n. Let (a_ν, b_ν) $(\nu = 0, \pm 1, \pm 2, \ldots)$ be a sequence of non-intersecting intervals adjacent to M, having a length $2\varrho_\nu \geqslant 2\varrho > 0$ and let $x_\nu = \frac{1}{2}(a_\nu+b_\nu)$. The functions

$$\phi_\nu(x) = \max\{g(x-a_\nu), 0\},$$
$$\psi_\nu(x) = \max\{g(b_\nu-x), 0\},$$

are equal to zero for $x\leqslant a_\nu$ and $x \geqslant b_\nu$ respectively, and as $n\to\infty$ they satisfy the relations

$$\left.\begin{aligned} \left|\phi_\nu(x) - \frac{g(x-a_\nu)+h_{2n}(x-a_\nu)}{2}\right| &= O\left(\frac{\varepsilon_n}{1+|x-a_\nu|^2}\right) \\ \left|\psi_\nu(x) - \frac{g(b_\nu-x)+h_{2n}(b_\nu-x)}{2}\right| &= O\left(\frac{\varepsilon_n}{1+|x-b_\nu|^2}\right) \end{aligned}\right\} \quad (\varepsilon_n\to 0;\ n\to\infty),$$

uniformly with respect to all x and ν. Consequently, for the product $\Phi_\nu(x) = \phi_\nu(x)\psi_\nu(x)$ there exists a sequence of integral functions $G_\nu(x)$ of degree $\leqslant 4n$ such that, as $n \to \infty$,

$$\Phi_\nu(x) - G_\nu(x) = o\left(\frac{1}{1+|x-a_\nu|^2} + \frac{1}{1+|x-b_\nu|^2}\right).$$

Thus for all x the series

$$\sum_{\nu=-\infty}^{\infty} [\Phi_\nu(x) - G_\nu(x)] = o(1) \quad (n \to \infty) \tag{4}$$

converges, the relation 1.7(4) being satisfied uniformly in x. But as $\Phi_\nu(x) = 0$ for all $x \bar{\in} (a_\nu, b_\nu)$, the series

$$\Phi(x) = \sum_{\nu=-\infty}^{\infty} \Phi_\nu(x), \tag{5}$$

converges and consequently so also does the series

$$G(x) = \sum_{\nu=-\infty}^{\infty} G_\nu(x). \tag{6}$$

It is obvious that the partial sums of the series 1.7(5), and hence also the partial sums of the series 1.7(6), are uniformly bounded on the whole real axis. Hence by the inequality 1.7(1)** the series

$$\sum_{\nu=-\infty}^{\infty} G_\nu(x+iy)$$

has uniformly bounded partial sums in any finite part of the complex plane. Hence by the well-known theorem of Vitali it follows that $G(x)$ is an integral function of degree not greater than $4n$.

As the set M does not contain points of any of the intervals (a_ν, b_ν), the function $\Phi(x)$ is zero everywhere on M and is the limit of a sequence of integral functions of finite degree which converges to it uniformly. Considering the sequence of segments $[c_\nu, d_\nu]$, which contain the set Q and are such that $x_\nu = \frac{1}{2}(c_\nu + d_\nu)$, $d_\nu - c_\nu = b_\nu - a_\nu - 2\varrho$, it is easy to find a non-negative continuous function $\Phi^*(x)$, not exceeding unity, with the properties: (1) $\Phi^*(x) = 1$ on all these segments; (2) $\Phi^*(x) = 0$ on M; (3) the assertion of 1.7.1 is true for $\Phi^*(x)$. For this it is sufficient to choose $\varepsilon \in (0, 1)$ such that the function $\Phi(x)$ satisfies the inequality $\Phi(x) > \varepsilon$ for all $x \in (c_\nu, d_\nu)$. Then the function

$$\Phi^*(x) = \frac{\min\,[\Phi(x), \varepsilon]}{\varepsilon}$$

will be the one required.

** See the footnote**** on page 13.

It is now not difficult to prove 1.7.1. Exactly as in 1.2.6 we may suppose without loss of generality that $\inf_{-\infty<x<\infty} f(x) = 0$, $\sup_{-\infty<x<\infty} f(x) = 1$. Let us consider the sets M_k and Q_k defined there for some fixed n. By the uniform continuity of the functions $f(x)$ considered, the quantity

$$\sup_{|x'-x''|\leqslant\delta} |f(x')-f(x'')| = \omega(\delta)$$

is finite and decreases to zero monotonically with δ. If δ_n is such that $\omega(\delta_n) = 1/n$, then for any $k = 0, 1, 2, \ldots, n$ the distance between the sets M_k and Q_k is not less than δ_n, and consequently the adjacent intervals of the set M_k containing points of the set Q_k have length not less than $2\delta_n$.

Hence the assertion proved in the given section may be applied to the sets M_k and Q_k, in virtue of which there exists a non-negative continuous function $f_k(x)$, not exceeding unity, with the properties: (1) $f_k(x) = 0$ on M_k; (2) $f_k(x) = 1$ on Q_k; (3) the assertion of 1.7.1 is valid for $f_k(x)$.

The remainder of the proof is completed exactly as in section 1.2.6.

1.8. On weighted uniform approximation of continuous functions on the whole real axis

Wishing to preserve as constructive element the polynomials 1.1(1) also in the case where the continuous functions considered are defined on the whole real axis, we arrive at the necessity of introducing what is known as a weighted approximation. This means that we consider some function $\phi(x) \geqslant a > 0$ $(-\infty < x < \infty)$ possessing the property

$$\lim_{x\to\pm\infty} \frac{x^n}{\phi(x)} = 0$$

for any $n = 0, 1, 2, \ldots$, and investigate the question of the existence, for every function $f(x)$ continuous on the interval $(-\infty, \infty)$ and satisfying the condition

$$\lim_{x\to\pm\infty} \frac{f(x)}{\phi(x)} = 0, \tag{1}$$

of a sequence of polynomials 1.1(1) such that

$$\sup_{-\infty<x<\infty} \frac{|f(x)-P_n(x)|}{\phi(x)} \to 0 \quad (n\to\infty).$$

The functions $\phi(x)$ for which the answer to this question is in the affirmative are termed weight functions. It can be shown** that *for any weight function* $\varphi(x)$,

$$\int_{-\infty}^{\infty} \frac{\ln \phi(x)}{1+x^2}\, dx = \infty . \tag{2}$$

1.8.1. In certain cases the condition 1.8(2) is not only necessary but also sufficient for $\phi(x)$ to be a weight function. *The transcendental integral function*

$$\phi(x) = 1 + \sum_{k=1}^{\infty} c_k x^{2k} \tag{3}$$

with coefficients $c_k \geqslant 0$ *is a weight function if and only if* 1.8(2) *holds. The condition* 1.8(2) *is satisfied in all those cases where the series* $\sum_{k=1}^{\infty} \sqrt[2k]{c_k}$ *diverges. Hence it follows that any continuous function* $\Phi(x) \geqslant 1$ *for which*

$$\sum_{k=1}^{\infty} \min_{-\infty<x<\infty} \frac{\sqrt[k]{\Phi(x)}}{|x|} = \infty \tag{4}$$

will be a weight function, as for

$$c_k = \frac{1}{2k^2} \left\{ \min_{-\infty<x<\infty} \frac{\sqrt[2k]{\Phi(x)}}{|x|} \right\}^{2k}$$

the inequality $\phi(x) \leqslant 2\Phi(x)$*** holds between the functions 1.8(3) and $\Phi(x)$.

1.8.2. In the case *where* $\Phi(x) = e^{p(x)} > 1$ *is an even function and* $xp'(x)$ *increases to infinity with* $|x|$ *in the strict sense, condition* 1.8(4) *is equivalent to the condition* 1.8(2), *where* $\phi(x) = \Phi(x)$.

Since when 1.8(4) is satisfied the function $\Phi(x)$ is a weight function it follows that 1.8(2) holds for it. Considering the function $n(x) = xp'(x) = x\dfrac{\Phi'(x)}{\Phi(x)}$, which by supposition increases monotonically to infinity with $|x|$, we note that for all sufficiently great values of $n (n \geqslant N)$ the equa-

** N. I. Akhiezer and K. I. Babenko [1].

*** If $\lim_{x\to\pm\infty} f(x)/\Phi(x) = 0$, then for any $\varepsilon > 0$ there can be found an x_0 so great that for $|x| \geqslant x_0$ the inequality $|f(x)| < \frac{1}{4}\varepsilon\Phi(x)$ is satisfied. In addition, for a certain $\delta > 0$ the inequalities $|f(-x_0)| < \frac{1}{4}\varepsilon\,\Phi(x)$ and $|f(x_0)| < \frac{1}{4}\varepsilon\,\Phi(x)$ will hold for values of x on the segments $[-x_0-\delta, -x_0]$ and $[x_0, x_0+\delta]$ respectively. Hence the function $F(x)$ which is identical with $f(x)$ on $[-x_0, x_0]$, is equal to zero for $|x| \geqslant x_0+\delta$ and is linear on the segments $[-x_0-\delta, -x_0]$, $[x_0, x_0+\delta]$, satisfies the condition $|f(x)-F(x)| < \frac{1}{2}\varepsilon\,\Phi(x)$ for all values of x. If $\Phi(x) \geqslant \frac{1}{2}\phi(x)$, and the function $\phi(x)$ is a weight function, there exists an algebraic polynomial $P_n(x)$ such that $|F(x)-P_n(x)| < \frac{1}{4}\varepsilon\,\phi(x) \leqslant \frac{1}{2}\varepsilon\,\Phi(x)$ for all values of x. Thus for all x the inequality $|f(x)-P_n(x)| < \varepsilon\Phi(x)$ holds, i.e. in the given case $\Phi(x)$ is a weight function.

tion $n(x) = n$ has one and only one solution $x = x(n)$, at which the function $\frac{1}{t}\sqrt[n]{\Phi(t)}$ attains a minimum in the interval $(0, \infty)$, and conversely to every $x > 0$ there corresponds a certain $n = n(x)$ such that at the point x this function attains a minimum. Since for $n \geqslant N$ we have $\frac{1}{t}\sqrt{\Phi(t)} \geqslant \frac{1}{x(n)}$, it follows that the convergence of the series 1.8(4) implies the convergence of the series $\sum_{n=N}^{\infty} \frac{1}{x(n)}$, and consequently also the boundedness of the integral

$$\int_1^R \frac{n'(x)}{x} \mathrm{d}x = \int_1^R \frac{p'(x)}{x} \mathrm{d}x + p'(R) - p'(1) > \int_1^R \frac{p(x)}{x^2} \mathrm{d}x - p(1) - p' \tag{1}$$

Hence if the series 1.8(4) converges it follows that $\int_1^\infty \frac{p(x)}{x^2} \mathrm{d}x < \infty$, i.e. in the given case 1.8(4) follows from 1.8(2) for $\varphi(x) = \Phi(x)$.

The last remark shows that such functions as

$$e^{a|x|} \ (a > 0), \qquad e^{\frac{|x|}{\ln(|x|+1)}}, \qquad e^{\frac{|x|}{\ln[\ln(|x|+1)+1]}}$$

are weight functions.

1.8.3. *The integral functions* 1.8(3) *satisfy the condition of theorem* 1.8.2. *Hence the equivalence* 1.8(2) and 1.8(4) where $\Phi(x) = \phi(x)$, *holds for all the integral functions* 1.8(3) ($c_k \geqslant 0$), *i.e. in the class of such functions the condition* 1.8(4) *is not only sufficient but also necessary for* $\Phi(x)$ *to be a weight function.*

1.8.4. Theorems 1.8.1–1.8.3 were established by S. N. Bernstein**. To him is due a series of other criteria for weight functions. Thus, for example, *every even function* $\phi(x)$ *which is a non-decreasing function* $|x|$ *and which satisfies the inequalities* $\phi(|x|) \geqslant Q_{2n}(x)$, *where* $Q_{2n}(x)$ *is a sequence of even non-negative polynomials of degree* $2n$ *with unbounded sums* 1.6(3), *is a weight function.*

1.8.5. In what follows, the results obtained above provide a basis for the clarification of necessary and sufficient conditions which must be satisfied by weight functions in the general case. *The function* $\phi(x)$ *is a weight function if and only if*

$$\sup_{P_n(x) \in \mathfrak{M}_\phi} \int_{-\infty}^{\infty} \frac{\ln|P_n(x)|}{1+x^2} \mathrm{d}x = \infty. \tag{5}$$

** S. N. Bernstein [5, 6, 41].

where $\mathfrak{M}_\phi$ *is the set of all algebraic polynomials* $P_n(x)$ *which satisfy the inequality* $|P_n(x)| \leqslant (1+|x|)\phi(x)$.

Unfortunately, the limits of the present chapter do not permit us to describe these important investigations**.

1.8.6. In conclusion let us note that *in the case where the weight function* $\phi(x) \geqslant a > 0$ *assumes a finite value for some real value* c, *the function* $\dfrac{\phi x}{|x-c|}$ *is also a weight function.* In order to verify this, let us consider an arbitrary continuous function $f(x)$ for which

$$\lim_{x\to\pm\infty} \frac{(x-c)f(x)}{\phi(x)} = 0.$$

Since $\phi(x)$ is a weight function, for any $\varepsilon > 0$ a polynomial $P(x)$ can be found which satisfies the inequality

$$|(x-c)f(x)-P(x)| < \varepsilon\phi(x)$$

on the whole axis $-\infty < x < \infty$

Let

$$R(x) = P(x)-P(c).$$

The polynomial $R(x)$ possesses the property that $R(c) = 0$ and on the whole real axis

$$|(x-c)f(x)-R(x)| < \varepsilon\phi(x)+|P(c)| < \varepsilon\phi(x)+\varepsilon\phi(c)$$
$$= \varepsilon\phi(x)\left\{1+\frac{\phi(c)}{\phi(x)}\right\}.$$

Since $\inf\limits_{-\infty<x<\infty} \phi(x) > 0$ and $\phi(c) < \infty$, it follows that for all values of x we have $\phi(c)/\phi(x) \leqslant M$, where M is some constant. Hence for any x

$$\left|f(x)-\frac{R(x)}{x-c}\right| < \varepsilon\frac{\phi(x)}{|x-c|}(M+1),$$

and consequently the quotient $\phi(x)/(x-c)$ is a weight function.

1.9. On mean approximation of integrable functions on an infinite interval

In section 1.4 there was given a constructive definition using the polynomials 1.1(1) of the class of functions integrable on a finite segment. In this case the condition of the finiteness of the interval (a, b) played an important part. Here we will make a few remarks about the case where the interval considered is unbounded.

** For further information the review papers of N. I. Akhiezer [6] and S. N. Mergelyan [2], which contain proofs of the propositions indicated, can be recommended.

1.9.1. As in section 1.6.1, we choose as constructive elements in this case also the continuous rational functions 1.6(1). We shall take the infinite interval as the whole real axis $(-\infty, \infty)$ and consider the class $L_q (q > 0)$ of all measurable functions $f(x)$ the q-th power of the modulus of which is Lebesgue-integrable over any finite segment and

$$\int_{-\infty}^{\infty} |f(x)|^q \, dx < \infty .$$

The class L_q consists of those functions and only those which are the limits of sequences, convergent in mean, of the rational functions 1.6(1), *i.e. sequences of rational functions such that*

$$\int_{-\infty}^{\infty} |R_n(x) - R_m(x)|^q \, dx \to 0 \qquad (m, n \to \infty),$$

$$\int_{-\infty}^{\infty} |R_n(x)|^q \, dx < \infty .$$

On the one hand this assertion is a consequence of the property of the completeness of the class L_q mentioned already in section 1.4. In order to prove the second part, it is obviously sufficient to suppose that $f(x)$ is continuous and that for some $x_0 > 0$ we have $f(x) = 0$ for all values of x for which $|x| \geqslant x_0$. It is necessary to prove that there exists a sequence of rational fractions $R_n(x)$ which has the property

$$\int_{-\infty}^{\infty} |f(x) - R_n(x)|^q \, dx \to 0 \qquad (n \to \infty). \tag{1}$$

Let us consider the function

$$R_{2m}(x) = R_{2m}(x; \alpha, \beta) = \frac{\alpha}{1+\beta x^{2m}} - 1$$
$$(2 > \alpha > 1, \; \beta > 0, \; m > 0 \text{ an integer}). \tag{2}$$

This function is positive in a neighbourhood of the point $x = 0$ and negative everywhere outside this neighbourhood, which by the choice of the coefficient $\beta > 0$ can be made as small as desired. We see that the sequence of rational functions $S_{4mn}(x)$, as they are the partial sums of the binomial series 1.2(1) for $|R_{2m}(x)|$, satisfies the relation

$$\Big| |R_{2m}(x)| - S_{4mn}(x) \Big| = O\left(\frac{\varepsilon_n}{1+\beta |x|^{2m}}\right) \qquad (\varepsilon_n \to 0; \; n \to \infty)$$

uniformly with respect to all x and n.

Now let M and Q be two closed non-intersecting sets, one of which (say Q) is bounded. Obviously there exists a finite system of intervals (a_ν, b_ν) ($\nu = 1, 2, 3, \ldots, r$) not containing points of M, and a system of segments $[c_\nu, d_\nu] \subset (a_\nu, b_\nu)$ concentric with them which contains the whole of the set Q. The rational functions $R_{2m}(x-x_\nu) = R_{2m}(x-x_\nu;\ \alpha_\nu, \beta_\nu)$, where $x_\nu = \frac{1}{2}(\alpha_\nu+\beta_\nu)$, and m, α_ν and β_ν are chosen in such a way that the neighbourhood of x_ν in which $R_{2m}(x-x_\nu) > 0$ $(R_{2m}(x-x_\nu) \geqslant \frac{1}{2})$ coincides with the interval (a_ν, b_ν) (with the segment $[c_\nu, d_\nu]$)**, satisfy the relation

$$\Big|\,|R_{2m}(x-x_\nu)|-S_{4mn}(x-x_\nu)\Big| = O\left(\frac{\varepsilon_n}{1+\beta_\nu|x-x_\nu|^{2m}}\right) \qquad (\varepsilon_n \to 0;\ n\to\infty)$$

uniformly with respect to all x and n. Hence for the functions

$$\phi_\nu(x) = \max\{R_{2m}(x-x_\nu), 0\},$$

which are equal to zero everywhere outside the interval (a_ν, b_ν), we have

$$\phi_\nu(x)-\tfrac{1}{2}[R_{2m}(x-x_\nu)+S_{4mn}(x-x_\nu)] = o\left(\frac{1}{1+\beta_\nu|x-x_\nu|^{2m}}\right) \qquad (n\to\infty).$$

Consequently, since $|\phi_\nu(x)-\frac{1}{2}| \leqslant \frac{1}{2}$ on the whole axis and $|\phi_\nu(x)-\frac{1}{2}| = \frac{1}{2}$ outside the interval (a_ν, b_v), it follows that for the function $\psi_\nu(x) = 2\min\{\phi_\nu(x), \frac{1}{2}\}$ there exists a sequence of rational functions $\sigma_{k_n}(x)$ for which

$$\psi_\nu(x)-\sigma_{k_n}(x) = o\left(\frac{1}{1+\beta_\nu|x-x_\nu|^{2m}}\right) \qquad (n\to\infty).$$

The function

$$\psi(x) = \sum_{n=1}^{r}\psi_\nu(x), \tag{3}$$

which is non-negative and does not exceed unity, possesses the properties: (1) $\psi(x) = 0$ everywhere on M; (2) $\psi(x) = 1$ everywhere on Q; (3) for

** For what follows it is sufficient for m to be chosen in such a way that

$$\left[\max_{1\leqslant\nu\leqslant r}\left(\frac{d_\nu-c_\nu}{b_\nu-a_\nu}\right)\right]^{2m} < \frac{1}{3}.$$

some $x_0 > 0$, $\psi(x) = 0$ wherever $|x| \geqslant x_0$. Also a sequence of rational functions $S_{i_n}(x)$ can be found such that

$$\psi(x) - S_{i_n}(x) = o\left(\frac{1}{1+|x|^{2m}}\right) \qquad (n \to \infty),$$

uniformly for all x and n.

Since $f(x)$ is continuous and equal to zero for values of x sufficiently great in absolute magnitude, we may suppose without loss of generality, exactly as in section 1.2.6, that $f(x) \geqslant 0$ and $\max\limits_{-\infty<x<\infty} f(x) = 1$.

Let us consider the sets M_k and Q_k defined for some fixed n. Let $f_k(x)$ be the corresponding function 1.9(3). Then for the function 1.2(2), which in our case possesses for some $x_0 > 0$ the property $F_n(x) = 0$ wherever $|x| \geqslant x_0$, we can find a sequence of rational functions $D_{m_\nu}(x) \in L_q$ such that

$$F_n(x) - D_{m_\nu}(x) = o\left(\frac{1}{1+|x|^{2m}}\right)$$

and hence

$$\int_{-\infty}^{\infty} |F_n(x) - D_{m_\nu}(x)|^q \, dx = o\,(1) \qquad (\nu \to \infty).$$

But for sufficiently great $x_0 > 0$

$$\int_{-\infty}^{\infty} |f(x) - F_n(x)|^q \, dx = \int_{-x_0}^{x_0} |F_n(x) - f(x)|^q \, dx \leqslant \frac{2x_0}{n^q}.$$

Hence the correctness of the assertion of 1.9.1 follows.

1.9.2. In addition to the ring of all proper rational functions, some other important classes of functions which are everywhere dense in the sense of integral convergence on the space L_q can be considered. In this connection let us cite the following theorem of N. Wiener [1] which applies to the given problem in the case where $q = 1$. *Let E be some set of functions $f(x)$ of the space L which are integrable on the whole real axis. The set of all possible finite linear combinations*

$$\sum_{k=1}^{n} \lambda_k^{(n)} f_k(x + t_k^{(n)}),$$

where $f_k(x) \in E$, will be everywhere dense in the sense of integral convergence *in the space L if and only if there does not exist a single value of* x_0 *for which*

$$\int_{-\infty}^{\infty} f(t)\,\mathrm{e}^{-ix_0t}\,\mathrm{d}t = 0$$

simultaneously for all the $f(x)$ *of E.*

The reader can find the details of the proof in the monograph by N. I. Akhiezer [3].

1.9.3. In the study of functions integrable over the whole real axis it is possible just as in section 1.7 to make use of integral functions of exponential type in place of the rational functions 1.6(1). *The class* L_q *defined in* 1.9.1 *consist only of those functions which are the limits of convergent-in-mean sequences of integral functions satisfying the condition* 1.7(1).

1.9 4. If for the characterisation of functions integrable over the infinite interval (a, b) we wish to retain as constructive element the ordinary polynomials 1.1(1), we arrive at the necessity of introducing weight functions just as in the case of uniform approximation (section 1.8). In the given case the non-negative function $\phi(x)$, for which all the integrals

$$\int_a^b \frac{x^n}{\phi(x)}\,\mathrm{d}x$$

are finite for any natural number n, is said to be a weight function if to every measurable function $f(x)$ for which

$$\int_a^b \frac{|f(x)|^q}{\phi(x)}\,\mathrm{d}x < \infty \tag{4}$$

there corresponds a sequence of ordinary polynomials $P_n(x)$ satisfying the relation

$$\int_a^b \frac{|f(x)-P_n(x)|^q}{\phi(x)}\,\mathrm{d}x \to 0 \qquad n \to \infty).$$

Without considering in the general case the question of conditions subject to which the function $\phi(x)$ is a weight function**, we mention only two particular results which correspond to the case $q = 2$, i.e. to mean

** See the paper by M. M. Dzhrbashyan [1], which contains the corresponding results for the case $q = 2$.

square approximation**. The first is that *for any* $\alpha \geqslant 0$ *the function* $\phi(x) = |x|^{-\alpha} e^{-x^2}$ *is a weight function for mean square approximation over the whole real axis* (see 2.2). The second, which follows immediately from this on making the substitution $x^2 = t$, *is that the function* $x^{-\alpha} e^x$ *is a weight function for mean square approximation over the range* $[0, \infty]$.

1.10. Various problems and theorems

1. There exists a series of algebraic polynomials $\sum_{n=1}^{\infty} P_n(x)$ which possesses the property that for any finite segment $[a, b]$ and every function $f(x)$ continuous on it, it is possible without changing the order to group its terms in such a way that the new series

$$\sum_{k=0}^{\infty} [P_{n_{k+1}}(x) + \dots + P_{n_{k+1}}(x)]$$

converges uniformly to $f(x)$ on $[a, b]$.

Consider the set of all polynomials with rational coefficients.

2. For any finite segment $[a, b]$ there exists a sequence of functions $\phi_n(x)$ $(n = 0, 1, 2, 3, \dots)$ continuous on it which are basic, i.e. are such that any function $f(x)$ continuous on $[a, b]$ is uniquely representable as the sum of a uniformly convergent series $\sum_{n=0}^{\infty} c_n \phi_n(x)$.

J. Schauder. See L. A. Lyusternik and V. I. Sobolev [1].

3. For any finite segment $[a, b]$ there exists a sequence of algebraic polynomials $P_n(x)$ $(n=0, 1, 2, 3, \dots)$ which is basic.

(This stronger form of the theorem of Weierstrass follows from 1.1, 1.10.2 and a general theorem of functional analysis. See M. G. Krein, D. P. Mil'man and M. A. Rutman [1]).

4. Let the sequence of distinct numbers $a_1, a_2, a_3, \dots$ be such that $\left| a_k - \sqrt{a_k^2 - 1} \right| \neq 1$ and the series

$$\sum_{k=1}^{\infty} \left(1 - \left| a_k - \sqrt{a_k^2 - 1} \right| \right) \tag{1}$$

diverges. Then for any function $f(x)$ continuous on $[-1, 1]$ there can be found a sequence of rational functions

$$R_n(x) = \sum_{k=1}^{n} \frac{\lambda_k}{x - a_k}$$

possessing the property

$$\max_{-1 \leqslant x \leqslant 1} |f(x) - R_n(x)| \to 0 \quad (n \to \infty).$$

Divergence of the series 1.10(1) is also a necessary condition.

See N. I. Akhiezer [3].

** V. A. Steklov [1, 2].

5. A theorem similar to 1.6.1 holds also for functions of many variables. For any bounded continuous function $f(x_1, \ldots, x_m)$ defined throughout the whole m-dimensional space with variables $x_1, \ldots, x_m$, subject to the condition that $f(x_1, \ldots, x_m) \to c$ as $x_1^2 + \ldots + x_m^2 \to \infty$, there exists a sequence of continuous rational functions

$$R_{n_1}, \ldots, {}_{n_m}(x_1, \ldots, x_m) = \frac{\sum\limits_{k_1=0}^{n_1} \ldots \sum\limits_{k_m=0}^{n_m} a_{k_1}, \ldots, {}_{k_m} x_1^{k_1} \ldots x_m^{k_m}}{\sum\limits_{k_1=0}^{n_1} \ldots \sum\limits_{k_m=0}^{n_m} b_{k_1}, \ldots, {}_{k_m} x_1^{k_1} \ldots x_m^{k_m}}$$

which possess the property that

$$\sup_{-\infty < x_1, \ldots, x_m < \infty} |f(x_1, \ldots x_m) - R_{n_1}, \ldots {}_{n_m}(x_1, \ldots, x_m)| \to 0$$

when $n_1, \ldots, n_m$ increase without limit.

6. If $1 \leqslant a < \sqrt{2}$, then for any function $f(x)$ continuous on $[-a, a]$ for which $f(-1)$, $f(0)$ and $\frac{1}{2}[f(-1)+f(1)]$ are integers there exists a sequence $P_n(x)$ of the polynomials 1.1(1) with integral coefficients such that $\max\limits_{|x| \leqslant a} |f(x) - P_n(x)| \to 0$ as $n \to \infty$.

(Consider separately the cases when $f(x) = f(-x)$ and $f(x) = -f(-x)$ and make use of the footnote* on page 9).

CHAPTER II

THE BEST APPROXIMATION

2.1. The best approximation by polynomials on a finite segment

The value of theorem 1.1 and of the other propositions of the preceding chapter consists first and foremost in the fact that they involve a constructive approach based on properties of the simplest functions (the polynomials 1.1(1), the polynomials 1.3(1), the rational functions 1.6(1), the integral functions of finite degree) to the fundamental concepts of analysis (continuity, integrability). However, for a deeper study of the most important structural characteristics of functions these general theorems will have to be generalised and made more precise in what follows. This extension and refinement is possible as a result of the concept of *the best approximation* introduced by P. L. Chebyshev, which chronologically preceded the discovery of theorem 1.1 and forms the basis of the modern constructive theory of functions. It is associated with the following theorem. *For every bounded measurable function* $f(t)$, *defined on the finite segment* $[a, b]$, *and for any natural number n there exists an ordinary polynomial*

$$P_n(f; t) = C_0^{(0)} + C_1^{(0)} t + \cdots + C_n^{(0)} t^n \tag{1}$$

of degree not greater than n which deviates least from it on this segment, i.e. among all the polynomials 1.1(1), *which are of degree not greater than n, it gives the least value of the deviation***

$$\operatorname*{vrai\,sup}_{a \leqslant t \leqslant b} |f(t) - P_n(t)|. \tag{2}$$

The quantity

$$E_n(f) = E_n(f; a, b) = \operatorname*{vrai\,sup}_{a \leqslant t \leqslant b} |f(t) - P_n(f; t)| = \inf_{P_n(t)} \operatorname*{vrai\,sup}_{a \leqslant t \leqslant b} |f(t) - P_n(t)|, \tag{3}$$

which gives a measure of the deviation of $f(t)$ from the polynomial 2.1(1) corresponding to it, has been given the title of the best approximation of

** The essential upper bound $\operatorname*{vrai\,sup}_{t} f(t)$ is the lower bound of all the numbers M, for which $f(t) > M$ on a set of measure zero.

order n of this function**. Thus theorem 1.1 amounts to the fact that *from the continuity of the function $f(t)$ on $[a, b]$ there follows the relation $E_n(f) \to 0$ as $n \to \infty$.*

2.1.1. For a proof of theorem 2.1 we observe first that *for any fixed m the limit of a sequence of polynomials uniformly convergent on $[a, b]$ of the form*

$$\sum_{\nu=0}^{m} c_\nu^{(k)} t^{r_\nu} \qquad (k = 1, 2, 3, \ldots), \tag{4}$$

where r_ν $(\nu = 0, 1, 2, 3, \ldots, m)$ are $m+1$ given non-negative numbers no two of which are equal, is itself a polynomial of the same type. This is obvious when $m = 0$. Moreover if the given assertion is true for $m = n-1$, it is true also for $m = n$. In fact, given that the sequence 2.1(4) converges uniformly, it follows that it is uniformly bounded, and hence in turn there follows the boundedness of the system of coefficients $c_\nu^{(k)} (\nu = 0, 1, \ldots, m)$. If the sequence $c_\nu^{(k)}$ were unbounded for some value ν and if as $i \to \infty$ we had $c_\nu^{(k_i)} \to \infty$, then the sequence

$$-\sum_{\substack{j=0 \\ j \neq \nu}}^{m} \frac{c_j^{(k_i)}}{c_\nu^{(k_i)}} t^{r_j}.$$

would converge uniformly to the function t^{r_ν}, which is not a linear combination of the form

$$\sum_{\substack{j=0 \\ j \neq \nu}}^{m} \lambda_j t^{r_j}.$$

This is impossible by supposition.

By the Bolzano-Weierstrass principle, for bounded sequences $c_\nu^{(k)}$ $(\nu = 0, 1, \ldots, m)$ it is possible to find a subsequence of the natural numbers $\{k_l\}$ $(l = 0, 1, 2, \ldots)$ such that $c_\nu^{(k_l)} \to c_\nu^{(0)}$ $(\nu = 0, 1, 2, \ldots, m)$, as $l \to \infty$. Hence the uniformly convergent sequence 2.1(4) converges to the polynomial

$$\sum_{\nu=0}^{m} c_\nu^{(0)} t^{r_\nu},$$

as it contains a sequence possessing this property.

2.1.2. Now let

$$P_{n,k}(t) = \sum_{\nu=0}^{n} c_\nu^{(k)} t^\nu \qquad (k = 1, 2, 3, \ldots)$$

** Similarly it would be possible to define the best approximation of given order of the function $f(t)$ considered in 1.5.2 by algebraic polynomials with integral coefficients.

be a certain sequence of polynomials for which

$$\lim_{k\to\infty} \operatorname*{vrai\,sup}_{a\leqslant t\leqslant b}|f(t)-P_{n,k}(t)| = \inf_{P_n(t)} \operatorname*{vrai\,sup}_{a\leqslant t\leqslant b}|f(t)-P_n(t)|. \tag{5}$$

This sequence is bounded, that is there exists a constant $M > 0$ such that

$$\max_{a\leqslant t\leqslant b}|P_{n,k}(t)| \leqslant M \quad (k = 1, 2, \ldots). \tag{6}$$

Hence as already noted in 2.1.1 there follows the boundedness of the system of coefficients $c_\nu^{(k)}$ $(\nu = 0, 1, 2, 3, \ldots, n)$ and the existence of the subsequence of natural numbers $\{k_l\}$ $(l = 1, 2, 3, \ldots)$ for which $c_\nu^{(k_l)} \to c_\nu^{(0)} (\nu = 0, 1, 2, 3, \ldots, n)$. The subsequence $P_{n,k_l}(t)$ converges uniformly to the polynomial $P_{n,0}(t) = P_n(f; t)$, which in virtue of 2.1(5) satisfies the inequality 2.1(3) and is hence the polynomial which deviates least from the function $f(t)$ on $[a, b]$. It is easy to see that if $a+b = 0$ and $f(t)$ is an even (odd) function, the polynomial $P_n(f; t)$ can be taken to be even (odd).

2.2. Generalisation to a linear normed space and some particular cases

From the above proof of theorem 2.1 it is obvious that it is possible to replace in it the space of all the functions $f(t)$ measurable on $[a, b]$, for which

$$\|f\| = \operatorname*{vrai\,sup}_{a\leqslant t\leqslant b}|f(t)| < \infty,$$

by any linear normed space F, where the distance of the element x from the element y is measured by the quantity $\|x-y\|$, and the system of functions $1, t, \ldots, t^n$ by an arbitrary linearly independent system of elements $x_0, x_1, \ldots, x_n$ of F, and the uniform convergence of the sequence of polynomials 2.1(4) by the convergence with respect to the norm of the sequence of sums $\sum_{\nu=0}^{n} c_\nu^{(k)} x_\nu (k = 1, 2, \ldots)$. Hence *whatever the linear normed space F may be, for every $x \in F$ there exists a polynomial*

$$\sum_{\nu=0}^{n} c_\nu^{(0)} x_\nu \tag{1}$$

which deviates least from x in norm, i.e. it is such that among all polynomials of the form

$$\sum_{k=0}^{n} c_k x_k \tag{2}$$

it provides the least value for the deviation

$$\left\|x - \sum_{k=0}^{n} c_k x_k\right\|. \tag{3}$$

We thus arrive at a generalisation of the concept of the best approximation, *defining as the best approximation of an element* $x \in F$ *by the polynomials* 2.2 (2) *the quantity*

$$E_n(x) = \left\| x - \sum_{k=0}^{n} c_k^{(0)} x_k \right\| = \inf_{c_k} \left\| x - \sum_{k=0}^{n} c_k x_k \right\|. \tag{4}$$

The sequence of linearly independent elements** $x_0, x_1, \ldots, x_n, \ldots$ is by definition closed in the space F if for any $x \in F$ $E_n(x) \to 0$ as $n \to \infty$.

Using the concept of a linear functional it is possible to indicate general criteria for a system of elements in an arbitrary linear normed space F to be closed. *A sequence* $x_0, x_1, \ldots, x_n, \ldots \in F$ *is closed in* F *if and only if there do not exist linear functionals* $\Phi(x) \not\equiv 0$ *which satisfy the condition* $\Phi(x_k) = 0$ $(k = 0, 1, 2, \ldots)$***. If the system $\{x_k\}$ is closed and $\Phi(x_k) = 0$ $(k = 0, 1, 2, \ldots)$, then whatever $n \geqslant 0$ may be, for any $x \in F$

$$|\Phi(x)| = \left|\Phi\left(x - \sum_{k=0}^{n} c_k^{(0)} x_k\right)\right| \leqslant \|\Phi\|_F E_n(x),$$

i.e. $\Phi(x) = 0$. On the other hand if the sequence $\{x_k\}$ is not closed then, for a certain $x \in F$, we have $\lim_{n\to\infty} E_n(x) = d > 0$. In the linear space F_0 which consists of all possible elements y of the form $y = cx + \sum_{k=0}^{n} c_k x_k$ we can consider the linear functional $\Phi_0(y)$, putting $\Phi_0(y) = c$. It is obvious that since for all $y \in F_0$

$$|\Phi_0(y)|\, d = |c|\, d \leqslant \left\| cx + \sum_{k=0}^{n} c_k x_k \right\| = \|y\|,$$

we have $\|\Phi_0\|_{F_0} \leqslant 1/d$, and as for any $n \geqslant 0$

$$1 = |\Phi_0(x)| = \left|\Phi_0\left(x - \sum_{k=0}^{n} c_k^{(0)} x_k\right)\right| \leqslant \|\Phi_0\|_{F_0} E_n(x),$$

it follows that $\|\Phi_0\|_{F_0} \geqslant 1/d$ and consequently $\|\Phi_0\|_{F_0} = 1/d$. If now, using the well known Hahn–Banach theorem, the functional Φ_0 is extended to the whole space F with preservation of its norm, we obtain a non-trivial linear functional which vanishes for all the elements x_k of the given system.

We illustrate the application of this criterion to the following example. Let F be the Hilbert space of all real measurable functions $f(t)$ defined on $(-\infty, \infty)$ and such that $\|f\| = \{ \int_{-\infty}^{\infty} |f(t)|^2 |t|^\alpha e^{-t^2}\, dt\}^{1/2} < \infty$. The

---

** That is, a sequence for which for any natural number n the system $x_0, x_1, \ldots, x_n$ is linearly independent.

*** S. Banach [1].

system of functions $1, t, \ldots, t^n \ldots$, which belong to the given space, is closed in it. In this space every linear functional Φ is expressed in the form of a scalar product

$$\Phi(f) = \int_{-\infty}^{\infty} \phi(t) f(t) |t|^\alpha e^{-t^2} dt,$$

where $\phi(t) \in F$. Let us consider a certain functional possessing the property that

$$\Phi(t^k) = \int_{-\infty}^{\infty} \phi(t) t^k |t|^\alpha e^{-t^2} dt = 0 \quad (k = 0, 1, 2, \ldots),$$

and a function

$$g(z) = \int_{-\infty}^{\infty} \phi(t) |t|^\alpha e^{-t^2} e^{itz} dt,$$

where z is an arbitrary complex number. As $g(z)$ is analytic for any z and $g^{(k)}(0) = 0$ $(k = 0, 1, 2, 3, \ldots)$ it follows that $g(z) \equiv 0$. In Chapter III it will be proved (see 3.11.21), that this is possible if and only if $\phi(t)$ is zero almost everywhere. Thus every linear functional Φ which vanishes for the functions $\{t^k\}$ $(k = 0, 1, 2, \ldots)$ is identically zero throughout the whole space F.

2.2.1. Let F be the space of measurable functions $f(t)$ of period 2π for which $\|f\| = \operatorname{vrai\,sup}_t |f(t)| < \infty$, and as the linearly independent system of functions to be considered let us take the trigonometric sequence $1, \sin t, \cos t, \sin 2t, \cos 2t, \sin 3t, \cos 3t, \ldots$.

In virtue of theorem 2.2, *for every such function $f(t)$ and any integer $n \geq 0$ there exists a polynomial* 1.3(1), *which deviates less from it than all the other trigonometric polynomials of order not higher than n.*

In the given case the best approximation of order n to the function $f(t)$ is the quantity

$$E_n^*(f) = \inf_{a_k, b_k} \operatorname{vrai\,sup}_t \left| f(t) - \sum_{k=0}^{n} (a_k \cos kt + b_k \sin kt) \right|. \tag{5}$$

Theorem 1.3.1 amounts to the fact that *the continuity of the periodic function $f(t)$ with period 2π* implies that $E_n^*(f) \to 0$ as $n \to \infty$. Let us note that, as in the case of a finite segment (see 2.1.2), in the periodic case considered here evenness of the trigonometric polynomial 1.3(1) which deviates least from $f(t)$ may always accompany evenness of this function.

2.2.2. If F is the space L_q of all the functions measurable on $[a, b]$ (measurable of period 2π), the q-th power ($q \geqslant 1$) of the modulus of which is integrable on this segment (on $[0,2\pi]$), where

$$\|f\| = \left(\int_a^b |f(t)|^q \, dt\right)^{\frac{1}{q}}$$

$$\left(\text{or} \quad \|f\| = \left(\int_0^{2\pi} |f(t)|^q \, dt\right)^{\frac{1}{q}} \text{ respectively}\right),$$

and if as the linearly independent system of functions we have the sequence $1, t, t^2, \ldots, t^n, \ldots$ (the sequence $1, \sin t, \cos t, \sin 2t, \cos 2t, \ldots$), then, in virtue of theorem 2.2, for every function $f(t) \in L_q$ and any integer $n \geqslant 0$ there exists an ordinary polynomial 1.1(1) (a trigonometric polynomial 1.3(1)), which deviates from it by less than all other such polynomials on $[a, b]$ (trigonometric polynomials on $[0, 2\pi]$). The best integral approximation of order n to the function $f(t)$ by the polynomials 1.1(1) on $[a, b]$ (by the polynomials 1.3(1) on $[0, 2\pi]$) is then the quantity

$$E_n(f)_{L_q} = E_n(f; a, b)_{L_q} = \inf_{c_k} \left(\int_a^b \left|f(t) - \sum_{k=0}^{n} c_k t^k\right|^q dt\right)^{\frac{1}{q}}, \tag{6}$$

or

$$E_n^*(f)_{L_q} = \inf_{a_k, b_k} \left(\int_0^{2\pi} \left|f(t) - \sum_{k=0}^{n} (a_k \cos kt + b_k \sin kt)\right|^q dt\right)^{\frac{1}{q}}.$$

The last remark in sections 2.1.2 and 2.2.1 applies in this case also.

Theorem 1.4 (or 1.4.1) expresses the fact that *if the function $f(t)$ belongs to the space L_q, $q < \infty$, then $E_n(f)_{L_q} \to 0$ as $n \to \infty$ (or in the periodic case $E_n^*(f)_{L_q} \to 0$ as $n \to \infty$).*

2.2.3. Let us consider for a given non-decreasing bounded function $\varrho(t)$ the space $L_q(\varrho)$ of all the functions $f(t)$ ϱ-measurable** on $[a, b]$ for which there exists the finite integral

$$\left(\int_a^b |f(t)|^q \, d\varrho(t)\right)^{\frac{1}{q}} \quad (q \geqslant 1),$$

which here expresses $\|f(t)\|$. *For every function $f(t) \in L_q(\varrho)$ and any integral $n \geqslant 0$ there exists an ordinary polynomial* 1.1(1), *which deviates from it*

** For functions measurable in the Lebesgue-Stieltjes sense, see V. I. Smirnov [1], vol. V.

by less than all other such polynomials on $[a, b]$. The best approximation of order n of the function $f(t)$ is now the quantity

$$E_n(f)_{L_q}(\varrho) = \inf_{c_k} \left(\int_a^b \left| f(t) - \sum_{k=0}^{n} c_k t^k \right|^q \mathrm{d}\varrho(t) \right)^{\frac{1}{q}}.$$

2.2.4. Let F be the space $L_q (1 \leqslant q \leqslant \infty)$ of all measurable functions $f(t_1, \ldots, t_m)$ defined on the closed bounded region $\bar{G}$ of the m-dimensional space of the variables $t_k (1 \leqslant k \leqslant m)$, the q-th power of the modulus of which is integrable on G, where

$$\|f\| = \left(\int \underset{G}{\cdots} \int |f(t_1, \ldots, t_m)|^q \, \mathrm{d}t_1, \ldots \mathrm{d}t_m) \right)^{\frac{1}{q}}.$$

The case $q = \infty$ corresponds to the space of all functions measurable on $\bar{G}$ which have there a finite essential upper bound, and in this case we take (see 2.13.2)

$$\|f\| = \operatorname*{vrai\,sup}_{\bar{G}} |f(t_1, \ldots, t_m)| .$$

The system of functions $t_1^{k_1} t_2^{k_2} t_3^{k_3} \ldots t_m^{k_m}$ where $(k_1, k_2, k_3, \ldots, k_m)$ are any non-negative integers) is linearly independent. Thus, applying theorem 2.2, we find that *for any function* $f(t_1, \ldots, t_m)$ *considered there exists a polynomial* 1.1(2) *which for fixed values* $n_1, \ldots, n_n$ *deviates less from it than all other polynomials of the same type.*

The best approximation in the given case is the quantity

$$E_{n_1, \ldots, n_m}(f)_{L_q} = \inf_{c_{k_1, \ldots, k_m}} \left(\int \underset{G}{\cdots} \int \left| f(t_1, \ldots, t_m) - \right.\right.$$

$$\left.\left. - \sum_{k_1=0}^{n_1} \cdots \sum_{k_m=0}^{n_m} c_{k_1, \ldots, k_m} t_1^{k_1} \ldots t_m^{k_m} \right|^q \mathrm{d}t_1 \ldots \mathrm{d}t_m \right)^{1/q}. \quad (7)$$

Considering the space F of all measurable functions $f(t_1, \ldots, t_m)$ of period 2π in each of the variables, which belong to a space L_q on a multi-dimensional cube, $G, 0 \leqslant t_1, \ldots, t_m \leqslant 2\pi$, and also the corresponding linearly independent system of trigonometric functions, we find that there exists an m-dimensional trigonometric polynomial $T_{n_1, \ldots, n_m}(t_1, \ldots, t_m)$ of least deviation and arrive at the similarly defined best approximation $E^*_{n_1, \ldots, n_m}(f)_{L_q}$ of given orders $n_1, \ldots, n_m$.

In virtue of 1.4.2 and 1.3.2 it follows from the fact that the function $f(t_1, \ldots, t_m)$ belongs to the space L_q (for $q = \infty$ the space C) that $E_{n_1, \ldots, n_m}(f)_{L_q} \to 0$ when $n_1, \ldots, n_m$ simultaneously tend to ∞.

2.2.5. If $f(t_1, \ldots, t_m)$ belongs to L_q on G, and $r < m$, then as a function of $t_1, \ldots, t_r$, for all $t_{r+1}, \ldots, t_m$, with the exception possibly of a set of measure zero in the $(m-r)$-dimensional subspace of these variables, f also belongs to L_q on a set $G_{t_{r+1},\ldots,t_m}$ of points of G which have the fixed coordinates $t_{r+1}, \ldots, t_m$; this is a consequence of Fubini's theorem. For all those values of $t_{r+1}, \ldots, t_m$ for which

$$\int_{G_{t_{r+1},\ldots,t_m}} \cdots \int |f(t_1, \ldots, t_m)|^q \, dt_1 \ldots dt_r < \infty$$

it is possible to consider the best approximation

$$E_{n_1,\ldots,n_r}[f;(t_{r+1}, \ldots, t_m)]_{L_q}$$
$$= \inf_{a_{k_1,\ldots,k_r}} \left[\int_{G_{t_{r+1},\ldots,t_m}} \cdots \int \left| f(t_1,\ldots,t_m) - \sum_{k_1=0}^{n_1} \cdots \sum_{k_r=0}^{n_r} a_{k_1,\ldots,k_r} t_1^{k_1} \ldots t_r^{k_r} \right|^q dt_1 \ldots dt_r \right]^{1/q}$$

by the polynomials 1.1(2) with respect to the variables $t_1, \ldots, t_r$, which for $r = m$ is identical with 2.2(7). If G_{m-r} is the projection of G on the subspace of the variables $t_{r+1}, \ldots, t_m$, then it is natural to consider the quantity

$$E_{n_1,\ldots,n_r\infty}(f)_{L_q} = \left[\int_{G_{m-r}} \cdots \int \{E_{n_1,\ldots,n_r}[f;(t_{r+1}, \ldots, t_m)]_{L_q}\}^q \, dt_{r+1} \ldots dt_m \right]^{1/q} \quad (8)$$

as characterising the best approximation to the function $f(t_1, \ldots, t_m)$ as a function of the r variables just chosen. Let us note that the quantity 2.2(8), which we term the partial best approximation, is identical with the best approximation of the function $f(t_1, \ldots, t_m)$ in the metric L_q by sums of the form

$$\sum_{k_1=0}^{n_1} \cdots \sum_{k_r=0}^{n_r} \phi_{k_1,\ldots,k_r}(t_{r+1}, \ldots, t_m) t_1^{k_1} \ldots t_r^{k_r},$$

i.e. with the lower bound

$$\inf_{\phi_{k_1,\ldots,k_r}} \left[\int_G \cdots \int \left| f(t_1, \ldots, t_m - \right.\right.$$
$$\left.\left. - \sum_{k_1=0}^{n_1} \cdots \sum_{k_r=0}^{n_r} \phi_{k_1,\ldots,k_r}(t_{r+1}, \ldots, t_m) t_1^{k_1} \ldots t_r^{k_r} \right|^q dt_1 \ldots dt_m \right]^{1/q} \quad (9)$$

with respect to all possible functions $\phi_{k_1,\ldots,k_r}(t_{r+1}, \ldots, t_m)$ which belong to the space L_q on G_{m-r}. This follows on the one hand from the obvious inequality

$$\left[\int_G \cdots \int \left| f(t_1, \ldots, t_m) - \right.\right.$$
$$\left.\left. - \sum_{k_1=0}^{n_1} \cdots \sum_{k_r=0}^{n_r} \phi_{k_1,\ldots,k_r}(t_{r+1}, \ldots, t_m) t_1^{k_1} \ldots t_r^{k_r} \right|^q dt_1, \ldots dt_m \right]^{1/q} \geqslant E_{n_1,\ldots,n_r\infty}(f)_{L_q}$$

and on the other hand from the fact that there exists a system of functions $\phi_{k_1,\dots,k_r}(t_{r+1},\dots,t_m) \in L_q$ on G_{m-r} such that (see 2.13.1)

$$\Big[\int\cdots\int\limits_{G_{t_{r+1},\dots,t_m}} \Big|f(t_1,\dots,t_m)-\sum_{k_1=0}^{n_1}\cdots\sum_{k_r=0}^{n_r}\phi_{k_1,\dots,k_r}(t_{r+1},\dots,t_m)\times$$

$$\times t_1^{k_1}\dots t_r^{k_r}\Big|^q dt_1\dots dt_r\Big]^{1/q} = E_{n_1,\dots,n_r}[f;(t_{r+1},\dots,t_m)]_{L_q}.$$

From the identity of 2.2(8) with the lower bound 2.2(9) there follows directly the inequality

$$E_{n_1,\dots,n_m}(f)_{L_q} \geqslant E_{n_1,\dots,n_r,\infty}(f)_{L_q} \qquad (r=1,2,\dots,m). \tag{10}$$

2.2.6. If we consider functions $f(t_1,\dots,t_m)$, which are of period 2π in each of the variables, it is possible as in 2.2.5 to speak of their best approximation $E^*_{n_1,\dots,n_m}(f)_{L_q}$ by trigonometric polynomials of orders $n_1,\dots,n_m$ in the corresponding variables $t_1,\dots,t_m$:

$$E^*_{n_1,\dots,n_m}(f)_{L_q} = \inf_{T_{n_1,\dots,n_m}} \Big[\int_0^{2\pi}\cdots\int_0^{2\pi} |f(t_1,\dots,t_m)-$$

$$-T_{n_1,\dots,n_m}(t_1,\dots,t_m)|^q dt_1\dots dt_m\Big]^{1/q}, \tag{11}$$

and also of the partial best approximations $E^*_{n_1,\dots,n_r,\infty}(f)_{L_q}$ defined as in 2.2(8), with only this difference, that instead of algebraic polynomials in $t_1,\dots,t_r$ we now consider multiple trigonometric polynomials with respect to these variables and as the domain G we have an m-dimensional cube of periods. Here, as in 2.2.5, we have the inequality

$$E^*_{n_1,\dots,n_m}(f)_{L_q} \geqslant E^*_{n_1,\dots,n_r,\infty}(f)_{L_q}. \tag{12}$$

Let us note that in some cases it is possible to obtain estimates which in a certain sense satisfy 2.2(12) and indicate a closer connection between the corresponding partial best approximations and the complete best approximation 2.2(11). Using the properties of the partial sums of a Fourier series (see 3.12.22), it can be shown that for every finite $q > 1$

$$E^*_{n_1,\dots,n_m}(f)_{L_q} \leqslant C_{q,m}\sum_{k=1}^{m} E^*_{n_k,\infty}(f)_{L_q}, \tag{13}$$

where $C_{q,m}$ is a constant depending only on m and q.

For $q=1$ and $q=\infty$ we can obtain by the same method similar inequalities**, the right-hand sides of which contain unbounded factors of the type $\ln n_k$.

** For the case $q=\infty$, $m=2$ see S. N. Bernstein [42]. For the general case $q \geqslant 1$ see M. F. Timan [1]; see also 8.7.21.

2.2.7. In the determination of the best approximation of functions it would be possible in place of the segments considered (closed bounded regions) to speak of arbitrary measurable sets contained in them.

2.2.8. The polynomial 2.2(1) which deviates least from an element $x \in F$ is not always unique. Thus, for example, let F be the space M of all functions $f(t)$ bounded and measurable on $[a, b]$, considered in 2.1. If the function $f(t) \equiv 1$ is taken, then for $n = 0$ its best approximation $E_0(f)$ by the polynomials 2.2(2), where $x_0 = x_0(t) = t-a$, is obviously equal to unity, and functions of the form $c_0 x_0(t) = c_0(t-a)$ for which

$$\max_{a \leqslant t \leqslant b} |f(t) - c_0 x_0(t)| = E_0(f) = 1$$

are infinite in number.

For every element $x \in F$ the set of all polynomials 2.2(1) of given order n which deviate from it least is some convex**, closed and bounded set of the space considered. In the study of problems of the theory of the best approximations in a given space F, those linearly independent systems of elements $\{x_k\}$ are first selected for which for any $x \in F$ the polynomial of least deviation from it is unique. Such linearly independent systems are called *Chebyshev* systems.

There are spaces in which any linearly independent system of elements is a Chebyshev system. It is easy to show that every strictly normed space is such a space; a strictly normed space is a linear normed space F in which the norm, in addition to the usual three properties ($\|x\| = 0$ if and only if x is the null element of $F, \|\lambda x\| = |\lambda| \, \|x\|$, $\|x+y\| \leqslant \|x\| + \|y\|$), also possesses the property that $\|x+y\| = \|x\| + \|y\|$ only in the case where $y = \lambda x$ ($\lambda \geqslant 0$). Thus, for example, from the corresponding property of Minkowski's inequality for integrals it follows that for any $q > 1$ the space L_q considered in 2.2.2, and also $L_q(\varrho)$ (see 2.2.3) are strictly normed, and hence in these cases every linearly independent system is a Chebyshev system. As the example given above shows, this assertion would no longer be true for the space M (the case $q = \infty$). It would be false also in the case $q = 1$, as is easily seen by considering a similar example (see 2.1.34). Therefore in these two cases there arises the question of criteria for Chebyshev systems of functions.

2.3. On the uniqueness of the polynomial giving the best uniform approximation. Haar's theorem***

In considering the first of the two cases indicated in section 2.2.8 (approximation in L_∞) we shall in what follows restrict ourselves to approx-

** The set Q is convex if from $x, y \in Q$ it follows that $\alpha x + (1-\alpha) y \in Q$ for $0 \leqslant \alpha \leqslant 1$.

*** Haar [2]. The proof given was proposed by N. I. Akhiezer [3].

imation in the space C of functions continuous on $[a, b]$, where $\|f\| = \max_{a \leqslant t \leqslant b} |f(t)|$. The exhaustive criterion for Chebyshev systems of such functions is due to Haar: *a system of linearly independent functions* $\{\phi_k(t)\}$ $(0 \leqslant k \leqslant n)$ *continuous on* $[a, b]$ *is a Chebyshev system in the space* C *if and only if every nontrivial polynomial*

$$P_n(t) = \sum_{k=0}^{n} c_k \phi_k(t) \tag{1}$$

has not more than n distinct zeros on this segment.

2.3.1. This condition of Haar for the system of functions $\{\phi_k(t)\}$ $(0 \leqslant k \leqslant n)$ is obviously equivalent to the condition that for any $n+1$ distinct points $t_0, t_1, \ldots, t_n$ of the segment $[a, b]$ the determinant $D\begin{pmatrix} \phi_0 \phi_1 \ldots \phi_n \\ t_0 t_1 \ldots t_n \end{pmatrix}$ of the matrix $\{\phi_k(t_i)\}$ is different from zero. Let a linearly independent system of continuous functions $\{\phi_k(t)\}$ satisfy this condition. We show that *if the polynomial*

$$P_n(f; t) = \sum_{k=0}^{n} c_k^{(0)} \phi_k(t)$$

among all the polynomials 2.3 (1) *deviates least on* $[a, b]$ *from the continuous function* $f(t)$, *then there exist at least* $n+2$ *distinct points* $t_0, t_1, \ldots, t_{n+1}$ *of the given segment at which*

$$|f(t_k) - P_n(f; t_k)| = \max_{a \leqslant t \leqslant b} |f(t) - P_n(f; t)| = E_n(f).$$

Assuming the contrary, i.e. that there are altogether only $m+1 \leqslant n+1$ points $t_0, t_1, \ldots, t_m$ at which the maximum deviation $|f(t) - P_n(f; t)|$ is attained, we can determine from the system of equations

$$\sum_{k=0}^{n} c_k \phi_k(t_i) = f(t_i) - P_n(f; t_i) \qquad (i = 0, 1, 2, \ldots, m)$$

the polynomial 2.3 (1) which assumes at these points the same value as the difference $f(t) - P_n(f; t)$. As there are no other points of maximum deviation in modulus of this difference on $[a, b]$, and the polynomial constructed by us is continuous and bounded there, it follows that for sufficiently small $\varepsilon > 0$ the deviation of the polynomial $P_n(f; t) + \varepsilon P_n(t)$ from $f(t)$ will be less than $E_n(f)$ on this segment, which is impossible.

If the function $f(t)$ had two different polynomials of least deviation, $P_n(f; t)$ and $Q_n(f; t)$, the same property would be possessed also by the polynomial $\frac{1}{2}[P_n(f; t) + Q_n(f; t)]$, and on $[a, b]$ there would exist $n+2$ distinct points $t_0, t_1, \ldots t_{n+1}$ where

$$|f(t_k) - \tfrac{1}{2}[P_n(f; t_k) + Q_n(f; t_k)]| = E_n(f).$$

But as $P_n(f; t)$ and $Q_n(f; t)$ are polynomials of least deviation, all these points must be zeros of the difference $P_n(f; t) - Q_n(f; t)$, which by supposition is possible only when it is identically equal to zero. If the system $\{\phi_k(t)\}$ $(0 \leqslant k \leqslant n)$ does not satisfy Haar's condition, $n+1$ points can be found in $[a, b]$ for which the determinant

$$D\begin{pmatrix} \phi_0 \phi_1 \dots \phi_n \\ t_0 \, t_1 \dots t_n \end{pmatrix}$$

of the matrix $\{\phi_k(t_i)\}$ vanishes. Hence there exists a system of numbers $\{\lambda_\nu\}$ $(\nu = 0, 1, \dots, n)$ among which at least one is different from zero and such that

$$\sum_{\nu=0}^{n} \lambda_\nu P_n(t_\nu) = 0$$

whatever the polynomial 2.3(1) may be. Now let $P_n^*(t)$ be a nontrivial polynomial 2.3(1) with zeros at all the points indicated, and $f(t)$ an arbitrary continuous function with modulus not exceeding unity and such that $f(t_k) = \operatorname{sign} \lambda_k$ $(0 \leqslant k \leqslant n)$. If $\varepsilon > 0$ is taken sufficiently small, the best approximation of the function $F(t) = f(t)[1-\varepsilon|P_n^*(t)|]$ by the polynomials 2.3(1) will be equal to unity**. On the other hand, as is easily seen, for any δ less than unity in absolute magnitude

$$\max_{a \leqslant t \leqslant b} |F(t) - \varepsilon\delta P_n^*(t)| = 1,$$

and hence the system of functions considered is not a Chebyshev system.

2.3.2. It follows from theorem 2.3 that:

(1) The system $\phi_k(t) = t^k$ $(k = 0, 1, \dots, n)$ is a Chebyshev system in the space C of functions continuous on the segment $[a, b]$.

(2) The system of functions $\phi_k(t) = \cos kt$ $(k = 0, 1, \dots, n)$ is a Chebyshev system in the space of functions continuous on $[0, 2\pi]$.

(3) The set of functions $\phi_k(t) = \cos kt$ $(k = 0, 1, \dots, n)$, $\psi_k(t) = \sin kt$ $(k = 1, 2, \dots, n)$ is a Chebyshev system in the space C^* of functions which are continuous and periodic with period 2π.

(4) For any continuous function $\varrho(t)$ without zeros on $[a, b]$, the system of functions $\phi_k(t) = \varrho(t)t^k$ $(k = 0, 1, \dots, n)$ is a Chebyshev system in the space of functions continuous on this segment.

(5) For any function $f(t)$ with a continuous derivative of the n-th order, which has no zeros in (a, b), the set of functions $\phi_k(t) = t_k$ $(k = 0, 1, \dots, n-1)$ together with the function $f(t)$ forms a Chebyshev system in the space of functions continuous on this segment.

** Consider the difference $F(t) - P_n(F; t)$ at the points t_ν $(\nu = 0, 1, \dots, n)$.

2.3.3. In the proof of theorem 2.3 use was made only of the general properties of functions continuous on a closed bounded set. Hence *theorems* 2.3 *and* 2.3.1 *remain valid if in place of the class of functions continuous on a finite segment we consider the class of real functions continuous on some compact metric space (which contains at least* $n+1$ *points***), and in particular the class of real continuous functions of points of a bounded closed region $\overline{G}$ of a given multidimensional space.

2.4. On the uniqueness of the polynomial giving the best integral approximation

Let us pause now on the second of the cases (approximation in the metric L) indicated at the end of section 2.2.8, and mention a theorem of Jackson [2] which deals with an interesting question***. *If* $f(t)$ *is continuous on* $[a, b]$ *and the linearly independent system of functions* $\{\phi_k(t)\}$ $(k = 0, 1, 2, \ldots, n)$ *satisfies Haar's condition (see* 2.3) *on this segment, then there exists only one polynomial* 2.3(1) *which deviates least in the mean from* $f(t)$ *on this segment, i.e. only one polynomial*

$$P_n(f; t) = \sum_{k=0}^{n} c_k^{(0)} \phi_k(t) \tag{1}$$

for which

$$\int_a^b | f(t) - P_n(f; t) | \, dt = E_n(f)_L. \tag{2}$$

First of all we notice that as on two distinct polynomials 2.3(1) can assume identical values on a set consisting of more than n points of the segment $[a, b]$, it follows that the set of polynomials 2.4(1), which satisfy the condition 2.4(2) and coincide with $f(t)$ on a set of positive measure cannot be more than countable****. Moreover, in view of the fact that for given n the set of polynomials of least deviation is convex, it follows that of it contains more than one polynomial it has the power of the continuum. Hence if theorem 2.4 were untrue there could be found at least two different polynomials $P_n(f; t)$ and $Q_n(f; t)$, which satisfy con-

** For the extension to the case of complex functions see A. N. Kolmogorow [5].

*** The proof derived below is due to M. G. Krein [1]. Another proof of this theorem was given in the monograph of N. I. Akhiezer [3]. A further extension of this theorem has been obtained in the works of S. Ya. Khavinson [1, 2].

**** For any $\varepsilon > 0$ there can exist only a finite number of polynomials 2.3(1) which coincide with the function $f(t)$ on a set of measure $\geqslant \varepsilon$.

dition 2.4(2) and are such that the differences $\alpha(t) = f(t) - P_n(f; t)$ and $\beta(t) = f(t) - Q_n(f; t)$ are almost everywhere different from zero. Since

$$\int_a^b |\alpha(t)+\beta(t)|\,dt = \int_a^b |\alpha(t)|\,dt + \int_a^b |\beta(t)|\,dt,$$

we should then have sign $\alpha(t) =$ sign $\beta(t)$. Let us also verify the validity of the system of equations

$$\int_a^b \phi_i(t) \operatorname{sign} \alpha(t)\,dt = 0 \qquad (i = 0, 1, 2, \dots, n). \tag{3}$$

In fact, if for some k $(0 \leqslant k \leqslant n)$

$$\int_a^b \phi_k(t) \operatorname{sign}[\alpha(t)\,dt] = \varrho \neq 0,$$

it would be possible to find a number ε so small in absolute value that

$$\varepsilon \int_a^b \phi_k(t) \operatorname{sign}[\alpha(t) - \varepsilon\phi_k(t)]\,dt > 0.$$

But then

$$\int_a^b |f(t) - P_n(f; t)|\,dt = \int_a^b |\alpha(t)|\,dt = \int_a^b \alpha(t) \operatorname{sign} \alpha(t)\,dt$$

$$\geqslant \int_a^b \alpha(t) \operatorname{sign}[\alpha(t) - \varepsilon\phi_k(t)]\,dt$$

$$= \int_a^b |\alpha(t) - \varepsilon\phi_k(t)|\,dt + \varepsilon \int_a^b \phi_k(t) \operatorname{sign}[\alpha(t) - \varepsilon\phi_k(t)]\,dt,$$

i.e.

$$\int_a^b |f(t) - P_n(f; t)|\,dt > \int_a^b |f(t) - P_n(f; t) - \varepsilon\phi_k(t)|\,dt,$$

and $P_n(f; t)$ would not be the polynomial of least deviation.

We see that the points of the interval (a, b) at which the differences $\alpha(t)$ and $\beta(t)$ change sign coincide. As at these points $P_n(f; t) = Q_n(f; t)$ there are not more than n of them. Let them be the points $a < t_0 < t_1 < \cdots < t_m < b$ $(m < n)$. Let us consider an arbitrary null sequence $\{\varepsilon_k\}$ of positive numbers ε_k and for sufficiently great values of k let us choose points $t_{m+1} < t_{m+2} < \cdots < t_{n-1}$ such that $t_m < t_{m+1}$ $t_{m+2r} = t_{m+2r-1} + \varepsilon_k$

$\left(1 \leqslant r \leqslant [\frac{1}{2}(n-m)]\right)$, and when $\frac{1}{2}(n-m)$ is an integer let us put $t_{n-1} = b$. If the numbers $A_k \neq 0$ are chosen so that

$$\left| A_k D \begin{pmatrix} \phi_0 \phi_1 \dots \phi_{n-1} & \phi_n \\ t_0 \; t_1, \dots t_{n-1} & t \end{pmatrix} \right| = \left| \sum_{\nu=0}^{n} c_\nu^{(k)} \phi_\nu(t) \right| \leqslant 1 \qquad (a \leqslant t \leqslant b),$$

then just as in sections 2.1.1, 2.1.2, we see that it is possible to find a sequence of natural numbers $\{k_l\}$ $(l = 1, 2, \dots)$ for which the corresponding sequence of polynomials $\sum_{\nu=0}^{n} c_\nu^{(k_l)} \phi_\nu(t)$ converges uniformly to some polynomial $\sum_{\nu=0}^{n} c_\nu \phi_\nu(t)$. But considering the determinant $D\begin{pmatrix} \phi_0 \phi_1 \dots \phi_{n-1} \phi_n \\ t_0 \; t_1 \dots t_{n-1} \; t_n \end{pmatrix}$, it is easy to see that the polynomials $\sum_{\nu=0}^{n} c_\nu^{(k)} \phi_\nu(t)$ change sign at the points $t_0, t_1, \dots, t_{n-1}$ and only at these points. Consequently, the limit polynomial $\sum_{\nu=0}^{n} c_\nu \phi_\nu(t)$ will change sign within (a, b) only at the points $t_0, t_1, \dots, t_{n-1}$. Hence it follows that

$$\int_a^b \left\{ \sum_{\nu=0}^{n} c_\nu \phi_\nu(t) \right\} \operatorname{sign} d(t) \, dt \neq 0,$$

and this contradicts the system of equations 2.4(3). This contradiction assures us of the uniqueness of the polynomial of least deviation in the case considered.

2.5. A sequence of best approximations. The fundamental characteristic property

As has already been mentioned in section 2.2, we arrive at the concept of the best approximation in all those cases where we consider an arbitrary linear normed space F and a certain linearly independent set of elements in it. It is obvious that every sequence of linearly independent elements $x_0, x_1, x_2, \dots, x_n, \dots$ closed in F gives rise for any $x \in F$ to a non-increasing null sequence of numbers (see 2.2(4)).

$$E_0(x) \geqslant E_1(x) \geqslant E_2(x) \geqslant E_3(x) \geqslant \dots \geqslant E_n(x) \geqslant \dots,$$

which are the corresponding best approximations of the element x. It turns out that this property of the sequence of best approximations characterises it completely**. *If an infinite sequence of linearly independent elements*

** The theorem is due to S. N. Bernstein; see [17], where the space C of functions continuous on $[a, b]$ (the case of uniform approximation) is considered.

$x_0, x_1, \ldots, x_n, \ldots$ *is closed in* F, *and the space* F *is complete***, *then for every sequence of numbers*

$$E_0 \geqslant E_1 \geqslant E_2 \geqslant \ldots \geqslant E_n \geqslant \ldots \tag{1}$$

monotonically decreasing to zero, an element $x \in F$ *can be found such that for any* $n = 0, 1, 2, \ldots$

$$E_n(x) = E_n.$$

In the proof we use the following obvious properties of the best approximation:

(1) $E_n(\lambda x) = |\lambda| E_n(x)$.

(2) $E_n(x+y) \leqslant E_n(x)+E_n(y)$.

(3) The quantity $E_n(x+\lambda y)$ varies continuously with λ and as a function of λ is convex.

(4) If $E_n(y) > 0$, then $E_n(x+\lambda y) \to \infty$ as $\lambda \to \pm\infty$.

(5) $E_n(x+\sum_{k=0}^{n} c_k x_k) = E_n(x)$.

From these properties it follows that for any x and for any $A \geqslant E_{n+1}(x)$ there exists a number c for which

$$E_n(x+c x_{n+1}) = A. \tag{2}$$

In fact, if $A = E_{n+1}(x)$ and $E_{n+1}(x) = \left\| x - \sum_{k=0}^{n+1} c_k^{(0)} x_k \right\|$, then

$$E_n(x-c_{n+1}^{(0)} x_{n+1}) \leqslant E_{n+1}(x) = E_{n+1}(x-c_{n+1}^{(0)} x_{n+1}) \leqslant E_n(x-c_{n+1}^{(0)} x_{n+1}).$$

If, however, $A > E_{n+1}(x)$, then from the inequality

$$E_n(x-c_{n+1}^{(0)} x_{n+1}) < A < E_n(x+\lambda_0 x_{n+1}),$$

in which λ_0 is a sufficiently great number, it follows in virtue of the continuous variation of $E_n(x+\lambda x_{n+1})$ with λ that there exists a constant c satisfying 2.5(2)***. Using this remark, we see that for any n there exist polynomials $\sum_{\nu=0}^{n+1} c_\nu^{(n)} x_\nu$ satisfying the set of equations

$$\left\| \sum_{\nu=0}^{n+1} c_\nu^{(n)} x_\nu \right\| = E_0, \quad E_k\left(\sum_{\nu=0}^{n+1} c_\nu^{(n)} x_\nu \right) = E_k \quad (k = 0, 1, 2, \ldots, n). \tag{3}$$

** That is, every sequence $x^{(n)} \in F$ which possesses the property $\|x^{(n)} - x^{(m)}\| \to 0$ $(m, n \to \infty)$, converges in norm to some element x of $F(\|x - x^{(n)}\| \to 0, n \to \infty)$.

*** In this case there are two such value of c.

In fact, since $E_{n+1}(\theta) = 0$ for the null element θ of the space F, it follows as has just been mentioned, that for some value c_{n+1}

$$E_n(c_{n+1}x_{n+1}) = E_n.$$

On the same grounds, in virtue of the inequalities 2.5(1), for some value c_n we have

$$E_{n-1}(c_n x_n + c_{n+1}x_{n+1}) = E_{n-1}.$$

Then

$$E_n(c_n x_n + c_{n+1}x_{n+1}) = E_n.$$

Furthermore let us choose the constant c_{n-1} so that

$$E_{n-2}(c_{n-1}x_{n-1} + c_n x_n + c_{n+1}x_{n+1}) = E_{n-2}.$$

Then

$$E_{n-1}(c_{n-1}x_{n-1} + c_n x_n + c_{n+1}x_{n+1}) = E_{n-1},$$

$$E_n(c_{n-1}x_{n-1} + c_n x_n + c_{n+1}x_{n+1}) = E_n.$$

Continuing this process, we obtain a polynomial $\sum_{\nu=1}^{n+1} c_\nu x_\nu$ which satisfies the system of equations

$$E_k\Big(\sum_{\nu=1}^{n+1} c_\nu x_\nu\Big) = E_k \quad (k = 0, 1, 2, \ldots, n).$$

If c_0 is chosen so that

$$\Big\| \sum_{\nu=0}^{n+1} c_\nu x_\nu \Big\| = E_0,$$

we obtain the polynomial required.

Let $\sum_{\nu=0}^{k} a_{\nu,k}^{(n)} x_\nu (k = 0, 1, \ldots)$ be the polynomial of order k which deviates least from our polynomial $\sum_{\nu=0}^{n+1} c_\nu^{(n)} x_\nu$. It is obvious that

$$\Big\| \sum_{\nu=0}^{k} a_{\nu,k}^{(n)} x_\nu \Big\| \leqslant 2E_0.$$

For every fixed value of k, applying the Bolzano-Weierstrass principle and then the well-known diagonal process of Cantor, we find for all k a single sequence of natural numbers n_l which possess the property that for any k the sequence $\sum_{\nu=0}^{k} a_{\nu,k}^{(n_l)} x_\nu$ converges (with respect to the norm in F) to a certain polynomial $\sum_{\nu=0}^{k} a_{\nu,k} x_\nu$, i.e.

$$\left\| \sum_{\nu=0}^{k} a_{\nu,k}^{(n_l)} x_\nu - \sum_{\nu=0}^{k} a_{\nu,k} x_\nu \right\| \to 0 \quad (n_l \to \infty). \tag{4}$$

Consequently, for sufficiently great n_l the inequality

$$\left\| \sum_{\nu=0}^{k} a_{\nu,k}^{(n_l)} x_\nu - \sum_{\nu=0}^{k} a_{\nu,k} x_\nu \right\| < E_k$$

holds and hence

$$\left\| \sum_{\nu=0}^{k} a_{\nu,k} x_\nu - \sum_{\nu=0}^{n_l+1} c_\nu^{(n_l)} x_\nu \right\| < 2E_k.$$

Therefore for sufficiently great p and q

$$\left\| \sum_{\nu=0}^{n_p+1} c_\nu^{(n_p)} x_\nu - \sum_{\nu=0}^{n_q+1} c_\nu^{(n_q)} x_\nu \right\| < 4E_k.$$

Hence in virtue of the completeness of the space F and the condition that $E_k \to 0$ as $k \to \infty$ it follows that the sequence of polynomials $\sum_{\nu=1}^{n_l+1} c_\nu^{(n_l)} x_\nu$ converges to some element x of F. But as

$$\left\| \sum_{\nu=0}^{n_l+1} c_\nu^{(n_l)} x_\nu - \sum_{\nu=0}^{k} a_{\nu,k}^{(n_l)} x_\nu \right\| = E_k,$$

it follows from 2.5(4) that

$$\left\| x - \sum_{\nu=0}^{k} a_{\nu,k} x_\nu \right\| = E_k$$

and hence $E_k(x) \leqslant E_k$. If a polynomial $\sum_{\nu=0}^{k} b_\nu x_\nu$ were found for which $\left\| x - \sum_{\nu=0}^{k} b_\nu x_\nu \right\| < E_k$, then for a sufficiently great value of n_l we should have

$$\left\| \sum_{\nu=0}^{n_l+1} c_\nu^{(n_l)} x_\nu - \sum_{\nu=0}^{k} b_\nu x_\nu \right\| < E_k,$$

which for $n_l > k$ would contradict the equality $E_k \left(\sum_{\nu=0}^{n_l+1} c_\nu^{(n_l)} x_\nu \right) = E_k$, This means that $E_k(x) = E_k$ for any value $k = 0, 1, 2, \ldots$.

2.5.1. Theorem 2.5 shows that for any closed linearly independent sequence of elements of F, an element x can always be found of which the best approximation with respect to the given system tends to zero as slowly as desired as $n \to \infty$. Hence if W is some bounded set in F, the upper bound $\sup_{x\in W} E_n(x)$ as n increases without limit may not tend to zero. In the case when $\sup_{x\in W} E_n(x) \to 0 \ (n \to \infty)$, the set W consists of elements for which the rapidity of decrease to zero of the best approximations has a certain single characteristic common to all of them. Such sets W are of particular interest. This is reflected in the further development and refinement of the results of the preceding chapter in our investigations. We shall not go into these investigations here as Chapters V and VI are specially devoted to them.

The following theorem will give an idea of the structure of sets of this type. *A bounded set W of the complete space F possesses the property $\sup_{x\in W} E_n(x) \to 0 \ (n \to \infty)$ if and only if it is compact***.

This assertion is a consequence of the well-known theorem of Hausdorff by which for compactness of the set W of the complete space F the existence of a finite ε-net*** is necessary and sufficient, whatever $\varepsilon > 0$ may be. If W is compact and, for a given $\varepsilon > 0$, $z_1, z_2, \dots, z_m$ is a finite ε-net corresponding to it, then since $E_n(z_k) \to 0$ as $n \to \infty \ (1 \leqslant k \leqslant m)$, we have for all $x \in W$, beginning with a certain value n,

$$E_n(x) \leqslant E_n(z_k)+E_n(x-z_k) < 2\varepsilon,$$

and consequently $\sup_{x\in W} E_n(x) \leqslant 2\varepsilon$. Conversely, if $\sup_{x\in W} E_n(x) \to 0$ as $n \to \infty$, it follows that for any $\varepsilon > 0$ we have for sufficiently great n and for all $x \in W$

$$E_n(x) < \tfrac{1}{2}\varepsilon.$$

This means that the set of polynomials $\sum_{\nu=0}^{n} c_\nu x_\nu$ forms an $\frac{1}{2}\,\varepsilon$-net for W. The set of these polynomials is bounded together with the set W. As n is fixed, the given set of polynomials is compact in virtue of the Bolzano–Weierstrass principle and hence in F there exists for it a finite $\frac{1}{2}\,\varepsilon$-net $Q_{\varepsilon/2}$. It is obvious that $Q_{\varepsilon/2}$ can be used as a finite ε-net for W. Since the space F is complete, the set W is compact.

** That is, any sequence of elements of W contains a convergent subsequence.

*** That is, a finite set $Q_\varepsilon \subset F$, the distance of which from any point $x \in W$ does not exceed ε.

In chapter III we shall consider some structural characteristics of the most frequently encountered compact classes of functions for the spaces indicated in sections 2.2.1–2.2.4.

2.6. The best approximation on an infinite interval

In the case where the function $f(t)$ is defined on the whole real axis, it is possible, just as in sections 2.1 and 2.2.2, to consider its best approximation by the continuous rational functions 1.6(1), by integral functions of exponential type or the weighted best approximation by the polynomials 1.1(1).

2.6.1. If $f(t)$ is a function which is measurable and bounded on $(-\infty, \infty)$, then, for any even non-negative number n, among the rational functions 1.6(1) there is a continuous rational function

$$R_n(f;\, t) = \frac{a_0^{(0)} + a_1^{(0)} t + \dots + a_n^{(0)} t^n}{b_0^{(0)} + b_1^{(0)} t + \dots + b_n^{(0)} t^n},$$

which deviates least from $f(t)$, i.e.

$$\operatorname*{vrai\,sup}_{-\infty t<\infty} |f(t) - R_n(f;\, t)| = \inf_{R_n(t)} \operatorname*{vrai\,sup}_{-\infty t>\infty} |f(t) - R_n(t)|. \tag{1}$$

It is obvious that the sequence

$$R_{n,k}(t) = \frac{\sum_{\nu=0}^{n} a_\nu^{(k)} t^\nu}{\sum_{\nu=0}^{n} b_\nu^{(k)} t^\nu},$$

for which

$$\inf_{R_n(t)} \operatorname*{vrai\,sup}_{-\infty<t<\infty} |f(t) - R_n(t)| = \lim_{k\to\infty} \operatorname*{vrai\,sup}_{-\infty<t<\infty} |f(t) - R_{n,k}(t)| \tag{2}$$

in uniformly bounded on $(-\infty, \infty)$. Hence we may suppose that $|b_\nu^{(k)}| \leqslant 1$ $(0 \leqslant \nu \leqslant n,\ k = 1, 2, \dots)$. Just as in section 2.1.1 we verify that the sequences of coefficients $\{a_\nu^{(k)}\}$ are bounded, and choose a sequence of the natural numbers $(l = 1, 2, \dots)$, for which

$$a_\nu^{(k_l)} \to a_\nu^{(0)},\ b_\nu^{(k_l)} \to b_\nu^{(0)} \qquad (\nu = 0, 1, \dots, n)$$

as $l \to \infty$. The sequence $R_{n,k_l}(t)$ converges to a certain rational function $R_{n,0}(t) = R_n(f;\, t)$, which in virtue of 2.6(2) satisfies 2.6(1) everywhere except possibly at a finite number of points.

Similar considerations show that if $L_q (q > 0)$ is the space of all measurable functions $f(t)$, the q-th power of the modulus of which is integrable over any finite interval and

$$\int_{-\infty}^{\infty} |f(t)|^q \,dt < \infty,$$

then for a fixed even n there exists among all the continuous rational functions $R_n(t) \in L_q$ a rational function $R_n(f; t)$ which deviates least from $f(t)$ on $(-\infty, \infty)$, i.e. it is such that

$$\int_{-\infty}^{\infty} |f(t) - R_n(f; t)|^q \,dt = \inf_{R_n(t)} \int_{-\infty}^{\infty} |f(t) - R_n(t)|^q \,dt. \tag{3}$$

The magnitude of the best approximation $\varrho_n(f)$ $\left(\text{or } \varrho_n(f)_{L_q}\right)$ for a given fixed n is a measure of the deviation of the function from the set of rational functions 1.6(1) and is determined from 2.6(2) (or 2.6(3)). Theorem 1.6.1 thus amounts to the statement that if $f(x)$ is continuous and $\lim\limits_{x\to\pm\infty} f(x) = c$, then $\varrho_n(f) \to 0$ as $n \to \infty$.

2.6.2**. *If $f(t)$ is a function which is measurable and bounded on $(-\infty, \infty)$, then for any $\sigma \geqslant 0$ there exists among the integral functions $g_\sigma(t)$ of degree not greater than σ*** a* i.e. *function $g_\sigma(f; t)$ which deviates least from $f(t)$, i.e. it is such that*

$$\operatorname*{vrai\,sup}_{-\infty<t<\infty} |f(t) - g_\sigma(f; t)| = \inf_{g_\sigma(t)} \operatorname*{vrai\,sup}_{-\infty<t<\infty} |f(t) - g_\sigma(t)|. \tag{4}$$

In fact, the sequence of functions $g_{\sigma,k}(t)$ for which

$$\inf_{g_\sigma(t)} \operatorname*{vrai\,sup}_{-\infty<t<\infty} |f(t) - g_\sigma(t)| = \lim_{k\to\infty} \operatorname*{vrai\,sup}_{-\infty<t<\infty} |f(t) - g_{\sigma,k}(t)| \tag{4'}$$

is uniformly bounded on the whole real axis. Hence applying the Bolzano–Weierstrass principle and the well-known diagonal process of Cantor, we can find a sequence of natural numbers $\{k_l\}$ $(l = 1, 2, \ldots)$ which possess the property that

$$g_{\sigma,k_l}(t) \to g_{\sigma,0}(t)$$

** S. Bernstein [18].

*** That is, any integral functions $g_\sigma(x) = \sum\limits_{\nu=0}^{\infty} c_\nu x^\nu$ the coefficients of which satisfy the inequality $\overline{\lim\limits_{k\to\infty}} \sqrt[k]{k!\,|c_k|} \leqslant \sigma$. In the present theorem it is sufficient to consider only those among them which are uniformly bounded on the whole real axis, i.e. satisfy the inequality 1.7(1). (See 4.3.1). As will be shown in section 4.8.2, for these functions

$$k!\,|c_k| \leqslant \sigma^k \sup_{-\infty<x<\infty} |g_\sigma(x)| \quad (k = 1, 2, \ldots).$$

for all rational values of t. But by the inequality** 1.7(1) the sequence $g_{\sigma, h_i}(t+iy)$ is uniformly bounded in any finite part of the complex plane. Hence as it converges at the rational points of the real axis, in virtue of Vitali's theorem it converges, and moreover uniformly, in every bounded domain of the complex plane. Hence the limit function $g_{\sigma, 0}(t+iy)$ is an integral function, and as it satisfies the inequality 1.7(1) it follows that $g_{\sigma,0}(t) = g_\sigma(f; t)$ is an integral function of degree not higher than σ, bounded on the real axis. In consequence of the relation 2.6(4′) it satisfies 2.6(4) and hence deviates least from $f(t)$ on the whole axis. The lower bound 2.6(4) is called the best approximation of the function $f(t)$ by integral functions of degree not higher than σ and is denoted*** by $A_\sigma(f)$. Theorem 1.7 expresses the fact that for the uniform continuity of a function $f(t)$ bounded on the whole real axis it is necessary that $A_\sigma(f) \to 0$ as $\sigma \to \infty$.

2.6.21. Theorem 2.6.2 can be supplemented as follows****. *If $f(t)$ is a measurable function of period 2π, then among the functions $g_\sigma(f; t)$ which deviate least from $f(t)$ on $(-\infty, \infty)$ there is a trigonometric polynomial* 1.3(1) *of order $\leqslant [\sigma]$ and*

$$A_\sigma(f) = E^*_{[\sigma]}(f), \tag{5}$$

where $E^*_{[\sigma]}(f)$ is determined from 2.2(5).

As for any integral k

$$f(t)-g_\sigma(f; t+2k\pi) = f(t+2k\pi)-g_\sigma(f; t+2k\pi),$$

the functions

$$G_n(f; t) = \frac{1}{2n+1} \sum_{k=-n}^{n} g_\sigma(f; t+2k\pi)$$

are integral functions of degrees $\leqslant \sigma$, bounded on the real axis and possessing the property

$$\operatorname*{vrai\,sup}_{-\infty<t<\infty} |f(t)-G_n(f; t)| = A_\sigma(f). \tag{6}$$

Just as in section 2.6.2, we can select from the sequence $G_n(f; t)$ a subsequence $G_{n_l}(f; t)$ which in any finite part of the complex plane converges uniformly to some integral function $G(t)$ of degree not higher than σ. Since when n increases without limit $G_n(f; t+2\pi)-G_n(f; t)$ tends to zero uniformly for all $t \in (-\infty, \infty)$ it follows that $G(t)$ is of period 2π.

** See the footnote**** on page 13.

*** In those cases where this definition still has a meaning it is retained also for unbounded functions $f(t)$.

**** S. N. Bernstein [23].

In chapter IV (see 4.10.2) it will be shown that every integral function of degree not higher than σ which is of period 2π is a trigonometric polynomial of order not higher than $[\sigma]$. Hence 2.6(5) follows from 2.6(6). Here $G(t)$ is the trigonometric polynomial of least deviation indicated in the theorem.

We remark that since for any function $f(t)$ which is bounded and uniformly continuous on the whole real axis the best approximation $A_\sigma(f) = A(\sigma)$, as a function of σ defined on the semi-axis $0 \leqslant \sigma < \infty$, tends to zero monotonically, it follows from theorem 2.5 that to every such function $f(t)$ there corresponds some function $f^*(t)$ which is continuous and of period 2π, and for all $\sigma \geqslant 0$ possesses the property

$$A_\sigma(f^*) = A_{[\sigma]}(f) = E^*_{[\sigma]}(f^*). \tag{7}$$

2.6.22. There exists a close connection between the best approximation $A_\sigma(f)$ on the whole real axis and the best approximation $E_n(f; a, b)$ considered in section 2.1. *For any measurable function $f(x)$ bounded on $(-\infty, \infty)$ the following relations hold***:

$$\left.\begin{aligned} \lim_{n\to\infty} E_n\left(f; \frac{n}{\sigma+0}\right) &= \lim_{\varepsilon\to 0} \varlimsup_{n\to\infty} E_n\left(f; -\frac{n}{\sigma+\varepsilon}, \frac{n}{\sigma+\varepsilon}\right) = A_\sigma(f), \\ \lim_{n\to\infty} E_n\left(f; \frac{n}{\sigma-0}\right) &= \lim_{\varepsilon\to 0} \varlimsup_{n\to\infty} E_n\left(f; -\frac{n}{\sigma-\varepsilon}, \frac{n}{\sigma-\varepsilon}\right) \\ &= \lim_{\varepsilon\to 0} A_{\sigma-\varepsilon}(f) = A_{\sigma-0}(f) \quad (\varepsilon > 0). \end{aligned}\right\} \tag{8}$$

We remark that as the limit of the sequence of integral functions $g_{\sigma+\varepsilon_n}(x)$ $(\varepsilon_n > 0)$ of degree $\leqslant \sigma + \varepsilon_n$, which converges as $\varepsilon_n \to 0$, satisfies condition 1.7(1) and the condition $\sup\limits_{n} \sup\limits_{-\infty<x<\infty} |g_{\sigma+\varepsilon_n}(x)| < \infty$, it is an integral function of degree $\leqslant \sigma$, so that the non-increasing function $A(\sigma) = A_\sigma(f)$ is always continuous on the right*** for any value $\sigma \geqslant 0$.

** S. N. Bernstein [26].

*** The functions $g_{\sigma+\varepsilon}(x) = g_{\sigma+\varepsilon}(f; x)$ for which

$$A_{\sigma+\varepsilon}(f) = \operatorname*{vrai\,sup}_{-\infty<x<\infty} |f(x) - g_{\sigma+\varepsilon}(f; x)|,$$

satisfy the condition $\sup\limits_{-\infty<x<\infty} |g_{\sigma+\varepsilon}(x)| \leqslant \operatorname*{vrai\,sup}\limits_{-\infty<x<\infty} |f(x)|$ and hence (see 4.3.1) $\sup\limits_{-\infty<x<\infty} |g_{\sigma+\varepsilon}(x+iy)| \leqslant 2 \operatorname*{vrai\,sup}\limits_{-\infty<t<\infty} |f(t)| \cdot e^{(\sigma+\varepsilon)|y|}$. Consequently, using Vitali's theorem, it is possible to select a sequence of numbers $\varepsilon_n \to 0$ such that the sequence of functions $g_{\sigma+\varepsilon_n}(x)$, converges everywhere to some integral function $g_\sigma(x)$, which is of finite degree $\leqslant \sigma$. Then

$$A_\sigma(f) \leqslant \operatorname*{vrai\,sup}_{-\infty<x<\infty} |f(x) - g_\sigma(x)| \leqslant \lim_{\varepsilon\to 0} A_{\sigma+\varepsilon}(f) \leqslant A_\sigma(f).$$

Simple examples show that generally speaking the function $A(\sigma)$, may happen to be discontinuous. Thus, for example, it follows from theorem 2.6.21 that if

$$f(x) = \sum_{k=1}^{m} (a_k \cos kx + b_k \sin kx) \not\equiv 0,$$

then the best approximation $A_\sigma(f) = A(\sigma)$ is a step function which has a jump at least at one integral-valued point. For $A(\sigma)$ to be continuous at a given point σ it is necessary and sufficient, as follows from 2.6(8), that

$$\lim_{n\to\infty} E_n\left(f; \frac{n}{\sigma+0}\right) = \lim_{n\to\infty} E_n\left(f; \frac{n}{\sigma-0}\right).$$

The last condition entails the relation

$$\lim_{n\to\infty} E_n\left(f; -\frac{n}{\sigma}, \frac{n}{\sigma}\right) = A_\sigma(f).$$

Let $g_\sigma(f; x)$ and $P_n(f; x)$ be respectively an integral function of degree $\leqslant \sigma$ and a polynomial 2.1(1) of degree $\leqslant n$, for which

$$\operatorname*{vrai\,sup}_{-\infty<x<\infty} |f(x) - g_\sigma(f; x)| = A_\sigma(f),$$

$$\operatorname*{vrai\,sup}_{|x|\leqslant n/\sigma} |f(x) - P_n(f; x)| = E_n\left(f; -\frac{n}{\sigma}, \frac{n}{\sigma}\right).$$

In 2.9.12 we shall show that the coefficients c_k of any algebraic polynomial $Q_n(t) = \sum_{k=0}^{n} c_k t^k$ satisfy the inequalities

$$|c_k| \leqslant \frac{n^k}{k!} \max_{|t|\leqslant 1} |Q_n(t)|. \tag{9}$$

Moreover, it is established in chapter V (see 5.4.4) that if $g_\sigma(x)$ is an integral function of degree $\leqslant \sigma$ which satisfies condition 1.7(1), then for every fixed $p > \sigma$

$$\lim_{n\to\infty} E_n\left(g_\sigma; -\frac{n}{p}, \frac{n}{p}\right) = 0. \tag{10}$$

From the inequality 2.6(9) it follows that the coefficients $c_k^{(n)}$ of the polynomials $P_n(f; x)$ are governed by the inequalities

$$|c_k^{(n)}| \leqslant \frac{\sigma^k}{k!} 2 \operatorname*{vrai\,sup}_{-\infty<x<\infty} |f(x)|. \tag{11}$$

If we denote by $A_\sigma(f; n/\sigma)$ the lower bound

$$\inf_{g_\sigma(x)} \operatorname*{vrai\,sup}_{|x|\leqslant n/\sigma} |f(x) - g_\sigma(x)|$$

for all integral functions the Taylor coefficients of which satisfy condition 2.6(11), then we obviously have the inequality

$$E_n\left(f; -\frac{n}{\sigma}, \frac{n}{\sigma}\right) \geqslant A_\sigma\left(f; \frac{n}{\sigma}\right)$$

and consequently**,

$$\varliminf_{n\to\infty} E_n\left(f; -\frac{n}{\sigma}, \frac{n}{\sigma}\right) \geqslant \lim_{n\to\infty} A_\sigma\left(f; \frac{n}{\sigma}\right).$$

Using the fact that all integral functions which satisfy condition 2.6(11) are uniformly continuous and uniformly bounded in any finite part of the plane, we see that***

$$\lim_{n\to\infty} A_\sigma\left(f; \frac{n}{\sigma}\right) = A_\sigma(f)$$

and hence

$$\varliminf_{n\to\infty} E_n\left(f; -\frac{n}{\sigma}, \frac{n}{\sigma}\right) \geqslant A_\sigma(f). \tag{12}$$

On the other hand for $p > \sigma$, in virtue of 2.6(10) with $g_\sigma(x) = g_\sigma(f; x)$,

$$\varlimsup_{n\to\infty} E_n\left(f; -\frac{n}{p}, \frac{n}{p}\right) \leqslant \varlimsup_{n\to\infty} E_n\left(f - g_\sigma; -\frac{n}{p}, \frac{n}{p}\right) \leqslant A_\sigma(f),$$

whence as $p \to \sigma$

$$\varlimsup_{n\to\infty} E_n\left(f; \frac{n}{\sigma+0}\right) \leqslant A_\sigma(f) \tag{13}$$

** The sequence of numbers $A_\sigma(f; n/\sigma)$ does not decrease as n increases, and since $g_\sigma(f; x)$ satisfies condition 2.6(11) (see the footnote ** on page 46), it follows that $A_\sigma(f; n/\sigma) \leqslant A_\sigma(f)$.

*** If $g_{\sigma,n}(x) = \sum_{k=0}^{\infty} c_k^{(n)} x^k$ is an integral function the Taylor coefficients of which satisfy condition 2.6(11), and

$$\operatorname*{vrai\,sup}_{|x|\leqslant n/\sigma} |f(x) - g_{\sigma,n}(x)| = A_\sigma(f; n/\sigma),$$

it follows that

$$|g_{\sigma,n}(x+iy)| \leqslant 2 \operatorname*{vrai\,sup}_{-\infty<t<\infty} |f(t)|\, e^{\sigma\sqrt{x^2+y^2}}.$$

Using Vitali's theorem, it is possible to select a subsequence n_k of the natural numbers such that the sequence of functions $g_{\sigma,n_k}(x)$ converges everywhere to some integral function $g_\sigma(x)$ which is of finite degree $\leqslant \sigma$. Then for all real values of x

$$|f(x) - g_\sigma(x)| = \lim_{n_k\to\infty} |f(x) - g_{\sigma,n_k}(x)| \leqslant \lim_{n_k\to\infty} A_\sigma\left(f; \frac{n_k}{\sigma}\right) \leqslant A_\sigma(f).$$

and similarly

$$\overline{\lim_{n\to\infty}}\, E_n\left(f; -\frac{n}{\sigma}, \frac{n}{\sigma}\right) \leqslant A_{\sigma-0}(f). \tag{14}$$

From 2.6(12) and 2.6(13) the first of the relations 2.6(8) follows. The second of these relations at the points of discontinuity of the functions $A(\sigma) = A_\sigma(f)$ follows from 2.6(12) and 2.6(14). In order to prove that it holds at the remaining points it is sufficient to select a sequence of values $\sigma_k \to \sigma\,(\sigma_k < \sigma)$ such that the function $A(\sigma)$ is continuous at the points σ_k and pass to the limit as $k \to \infty$ in the equation

$$\lim_{n\to\infty} E_n\left(f; -\frac{n}{\sigma_k}, \frac{n}{\sigma_k}\right) = A_{\sigma_k}(f).$$

Let us note here that, as was shown by S. N. Bernstein [27], the relation 2.6(8) holds for all $\sigma \geqslant 0$ for a wider class of functions $f(x)$, subject only to a certain restriction on the increase of the function as $|x| \to \infty$, and also on the passage from uniform approximation to mean approximation. For sufficiently great values of σ this relation is satisfied for all functions $f(x)$ for which, from some point onwards, $A_\sigma(f) < \infty$ (see 5.4.6).

2.6.3. An assertion similar to the one derived in section 2.6.2 holds also for passage to an integral metric. If $f(t)$ is a measurable function of the space L_q, defined in 2.6.1, then for any $\sigma \geqslant 0$ there exists among the integral functions $g_\sigma(t) \in L_q$ of degree not higher than σ a function $g_\sigma(f; t)$ such that

$$\left(\int_{-\infty}^{\infty} |f(t) - g_\sigma(f; t)|^q \, dt\right)^{\frac{1}{q}} = \inf_{g_\sigma(t)} \left(\int_{-\infty}^{\infty} |f(t) - g_\sigma(x)|^q \, dt\right)^{\frac{1}{q}}. \tag{15}$$

The quantity $A_\sigma(f)\,L_q$, which is equal to the lower bound of 2.6(15), is known as the best mean approximation of the function $f(t)$ by integral functions of degree not higher than σ.

In order to verify that a function $g_\sigma(f; t)$ satisfying 2.6(15) exists, a sequence of integral functions $g_{\sigma,k}(t)$ is considered for which

$$\inf_{g_\sigma(t)} \int_{-\infty}^{\infty} |f(t) - g_\sigma(t)|^q \, dt = \lim_{k\to\infty} \int_{-\infty}^{\infty} |f(t) - g_{\sigma,k}(t)|^q \, dt.$$

It is obvious that for this sequence there exists a positive constant C such that

$$\int_{-\infty}^{\infty} |g_{\sigma,k}(t)|^q \, dt \leqslant C \quad (k = 1, 2, \ldots). \tag{16}$$

In chap. IV (see 4.9.53) we establish the inequality

$$\sup_{-\infty<x<\infty} |g_\sigma(x)| < 2\sigma^{\frac{1}{q}} \left(\int_{-\infty}^{\infty} |g_\sigma(t)|^q \,dt \right)^{\frac{1}{q}},$$

which holds in the class of all integral functions $g_\sigma(t) \in L_q$ of degree $\leqslant \sigma$. From this and from 2.6(16) it follows that the sequence $g_{\sigma,k}(x)$ is uniformly bounded on the whole real axis. Consequently, just as in section 2.6.2, it is possible to find a subsequence $g_{\sigma,k_l}(x)$ which converges uniformly in any bounded domain to a certain integral function $g_{\sigma,0}(t) = g_\sigma(f;t)$ of degree $\leqslant \sigma$ and is such that

$$\left(\int_{-\infty}^{\infty} |f(t) - g_\sigma(f;t)|^q \,dt \right)^{\frac{1}{q}} = A_\sigma(f)_{L_q}.$$

2.6.4. Theorem 2.2 is applicable to weighted approximation on the whole real axis (see sections 1.8 and 1.9.4), as in this case we have the space C_ϕ (the space $L_{q,\phi}$) of continuous (integrable with weighting) functions $f(x)$ satisfying condition 1.8(1) (condition 1.9(4)), where $\|f\|$ is defined as $\sup_{-\infty<x<\infty} \left| \frac{f(x)}{\phi(x)} \right|$ or as $\left(\int_{-\infty}^{\infty} |f(x)|^q/\phi(x)\,dx \right)^{\frac{1}{q}}$ respectively. *Therefore for any natural number n there exists for every function $f \in C_\phi$ ($f \in L_{q,\phi}$) a polynomial* 2.1(1), *which deviates least from it i.e. is such that*

$$\sup_{-\infty<x<\infty} \left| \frac{f(x) - P_n(f;x)}{\phi(x)} \right| = \inf_{c_k} \sup_{-\infty<x<\infty} \left| \frac{f(x) - \sum_{k=0}^{n} c_k x^k}{\phi(x)} \right|,$$

or

$$\left(\int_{-\infty}^{\infty} |f(x) - P_n(f;x)|^q/\phi(x)\,dx \right)^{\frac{1}{q}} = \inf_{c_k} \left(\int_{-\infty}^{\infty} |f(x) - \sum_{k=0}^{n} c_k x^k|^q/\phi(x)\,dx \right)^{\frac{1}{q}}$$

respectively.

2.7. Properties of best polynomial approximations. A theorem of Chebyshev

In considering questions connected with the concept of a best approximation, there arises the problem of finding criteria characterising polynomials (rational functions, integral functions) which realise this approximation, and also the related problem of finding methods of determining the magnitude of the corresponding least deviation.

2.7.1. Let us commence the study of this problem in its classical formulation, that is for the case considered in section 2.1. Let the function $f(x)$ be continuous on $[a, b]$ and let $P_n(f;x)$ be the polynomial 2.1(1), which deviates least from it on the given segment, i.e. the polynomial for which $\max_{a \leqslant x \leqslant b} |f(x)-P_n(f; x)| = E_n(f)$. It is not difficult to see that there does not exist an algebraic polynomial $Q_n(x)$, of degree not greater than n, for which for some $\varepsilon > 0$

$$Q_n(x)\{f(x)-P_n(f; x)\} > 0$$

at all the points x of the set M where

$$E_n(f)-\varepsilon \leqslant |f(x)-P_n(f; x)| \leqslant E_n(f).$$

In fact, if such a polynomial were to exist, then, as $\min_{x \in M} |Q_n(x)| > 0$, for a sufficiently small value of $\lambda > 0$ and for some positive δ the inequality

$$|f(x)-[P_n(f; x)+\lambda Q_n(x)]| < E_n(f)-\delta$$

would be satisfied at all points x of the segment $[a, b]$, which is impossible. This circumstance completely characterises best polynomial approximations. In fact, if $P_n(x)$ is not the best polynomial, then for sufficiently small $\varepsilon > 0$ there can be found an ordinary polynomial $Q_n(x)$ for which

$$Q_n(x)\{f(x)-P_n(x)\} > 0 \tag{1}$$

at all points x of the set M where

$$\max_{a \leqslant t \leqslant b} |f(t)-P_n(t)|-\varepsilon \leqslant |f(x)-P_n x)|. \tag{2}$$

It is easy to see that for

$$\varepsilon < \max_{a \leqslant t \leqslant b} |f(t)-P_n(t)|-E_n(f)$$

for example, such a polynomial is $Q_n(x) = P_n(f; x)-P_n(x)$.

Consequently, *for a given n the polynomial $P_n(x)$ deviates least from the continuous function $f(x)$ on $[a, b]$ if and only if for an arbitrarily small $\varepsilon > 0$ it is impossible to find a polynomial $Q_n(x)$ of degree $\leqslant n$ which satisfies* 2.7(1) *at all the points of this segment at which the inequality* 2.7(2) *holds.*

Thus *$P_n(x)$ is the best uniform polynomial approximation of degree n to the function $f(x)$ on $[a, b]$ if there does not exist a polynomial $Q_n(x)$ which satisfies the inequality* 2.7(1) *at all points of the given segment at which the maximum of the modulus of the difference $f(x)-P_n(x)$ is attained.*

2.7.11. Theorem 2.7.1 can also be formulated in the following equivalent form**. *For $P_n(x)$ to be the polynomial of least deviation from the*

** In this form it was indicated by A.N. Kolmogorov [5] for functions of a complex variable. This criterion admits of far reaching generalisation to cases of approximation in abstract spaces. For results in this field see S. I. Zukhovitskii, and M. G. Krein [1], S. I. Zukhovitskii [2, 3], S. I. Zukhovitskii and S. B. Stechkin [1].

continuous function $f(x)$ on $[a, b]$, it is necessary and sufficient that for every polynomial $Q_n(x)$ of degree $\leqslant n$ the inequality

$$\max_{x \in M_n} Q_n(x)\{f(x)-P_n(x)\} \geqslant 0 \tag{3}$$

should be satisfied, where M_n is the set of those $x \in [a, b]$ at which the maximum modulus of the difference $f(x) - P_n(x)$ is attained.

In fact, if condition 2.7(3) is not satisfied, there is a polynomial $Q_n(x)$ such that at every point $x \in M_n$ we have $Q_n(x)\{f(x)-P_n(x)\} > 0$. From considerations of continuity it follows that this inequality holds everywhere on a certain open set $U \supset M_n$. If V is the complement of U, then $\max_{x \in V} |f(x)-P_n(x)| = L < E_n(f)$. If $0 < \varepsilon < E_n(f)-L$, the set of points where $|f(x)-P_n(x)| \geqslant E_n(f)-\varepsilon$ is contained in U, i.e. at all its points $Q_n(x)\{f(x)-P_n(x)\} > 0$. By theorem 2.7.1 this is impossible.

2.7.2. The last proposition in section 2.7.1 leads to a remarkable theorem due to P. L. Chebyshev** which gives an elegant algebraic criterion for polynomials of least deviation. It follows directly from section 2.7.1 that if the difference $f(x)-P_n(x)$ attains a maximum of its modulus on $[a, b]$ at least at $n+2$ points with successive changes of sign, then the ordinary polynomial $P_n(x)$ among all the polynomials 1.1(1) of degree $\leqslant n$ deviates least from $f(x)$ on the given segment. The theorem of P. L. Chebyshev establishes the validity also of the converse assertion. *In order that the ordinary polynomial $P_n(x)$ among all the polynomials* 1.1(1) *of degree $\leqslant n$ should deviate least on $[a, b]$ from the continuous function $f(x)$, it is necessary and sufficient that the difference $f(x)-P_n(x)$ should attain the maximum of its modulus on this segment at least at $n+2$ points with successive changes of sign.*

It now only remains for us to verify the necessity of the given condition. Suppose that values of x at which the difference $f(x)-P_n(x)$ attains on $[a, b]$ maxima of its moduls K with successive changes of sign be not more than $n+1$ in number. If these values are the points $a \leqslant x_0 < x_1 < \ldots < x_m \leqslant b$, it is possible to choose a system of values

$$a_1, a_2, \ldots, a_m \qquad (a \leqslant x_0 < a_1 < x_1 < a_2 < x_2 < a_3 < \ldots < x_{m-1} < a_m < x_m \leqslant b)$$

separating them, such that for some $\varepsilon > 0$ on each of the segments $[a, a_1]$, $[a_1, a_2], \ldots, [a_m, b]$ the difference considered alternately satisfies one of the inequalities

$$-K+\varepsilon < f(x)-P_n(x) \leqslant K, \qquad -K \leqslant f(x)-P_n(x) < K-\varepsilon,$$

** P. L. Chebyshev [2].

and vanishes at the points $a_1, a_2, a_3, \ldots, a_m$. In the right-hand neighbourhood of the point a_m, at which

$$-K+\varepsilon < f(x) - P_n(x) < K-\varepsilon,$$

let us in addition choose any $2r = 2[\frac{1}{2}(n-m)]$ points $a_{m+1} < a_{m+2} < \ldots < a_{m+2r}$.

It is obvious that the polynomial $Q_n(x) = D\begin{pmatrix} \phi_0\phi_1\ldots\phi_n \\ xa_1\ldots a_n \end{pmatrix}$ (see section 2.3.1, $\phi_\nu(t) = t^\nu$; $\nu = 0, 1, \ldots n$), in which $a_n = b$, if $m+2r = n-1$ satisfies 2.7(1) apart from sign at all those points at which the inequality 2.7(2) is satisfied, with the exception, possibly, of the point $x = b$ (in the case where $n-m$ is odd). If in the last case $b \in M$ (see section 2.7.1), let us choose a polynomial $Q_n^*(x)$ for which $Q_n^*(b)\{f(b)-P_n(b)\} > 0$. It follows from considerations of continuity that for sufficiently small positive λ the polynomial $Q_n(x) + \lambda Q_n^*(x)$, apart from the sign of $Q_n(x)$, will satisfy 2.7(1) everywhere on $[a, b]$ where the inequality 2.7(2) holds. In virtue of theorem 2.7.1 it follows from this that $P_n(x)$ is not the polynomial of best approximation to the function $f(x)$ on $[a, b]$. This assures us that $m \geqslant n+1$.

2.7.21. If in theorem 2.7.2, instead of the whole set of polynomials 1.1(1), we were to consider only polynomials with integral coefficients, the condition of this theorem would cease to be necessary. To be convinced of this, it is sufficient to consider for the function $f(x) = x\,(1-x)$ its best approximation of order 0 or 1.

2.7.22. Denoting by $M_n(f)$ the set defined in 2.7.11 in the case where $P_n(x) = P_n(f; x)$ is the best polynomial approximation, it is easy to see that for every fixed value n and for any $l (0 \leqslant l < b-a)$ there is always a function $f_n(x)$ continuous on $[a, b]$ for which** mes $M_n(f_n) = l$. If instead of the segment $[a, b]$ we consider the interval (a, b), it is immediately possible to indicate a function $f(x)$ continuous on (a, b), for which mes $M_n(f) = l$ for all $n \geqslant 0$. However, such functions do not exist*** among functions continuous on the closed segment $[a, b]$. *In the last case we always have* mes $M_n(f) \to 0$ as $n \to \infty$. In fact, if $k \neq n$, then $M_k(f)$ and $M_n(f)$ either coincide or have at most a finite number of common points. Hence for non-identical $M_k(f)$ and $M_n(f)$

$$\text{mes}\,[M_k(f)+M_n(f)] = \text{mes}\,M_k(f)+\text{mes}\,M_n(f).$$

If $M_k(f)$ and $M_n(f)$ are identical, then $P_n(f; x) \equiv P_k(f; x)$. But if we suppose that the function $f(x)$ is not itself a polynomial, it follows from

** mes M denotes the measure of the set M.

*** In this connection see Y. A. Brudnyi and I. E. Gopengauz [1].

theorem 1.1 that the degree of the polynomial $P_n(f; x)$ increases without limit. Hence, if in the sequence of sets $M_{n_i}(f)$ no two are identical, we have

$$\sum_{i=1}^{\infty} \text{mes}\, M_{n_i}(f) = \text{mes} \sum_{i=1}^{\infty} M_{n_i}(f) \leqslant b-a,$$

so that mes $M_{n_i}(f) \to 0$ as $n_i \to \infty$.

2.7.3. It is obvious from the proof in section 2.7.2 that the algebraic polynomials 1.1(1) which occur in it may be replaced by the polynomials 2.3(1) with respect to any linearly independent system of continuous functions $\{\phi_k(x)\}$ $(0 \leqslant k \leqslant n)$, which satisfy Haar's condition on $[a, b]$, (see section 2.3).

We note that here it does not matter whether we speak of the best approximation of the function $f(x)$ on the whole segment $[a, b]$ or on some closed set contained in it. Thus *if $P_n(x)$ is a polynomial* 2.3(1) *with respect to some set of continuous functions $\{\phi_k(x)\}$ $(0 \leqslant k \leqslant n)$ which form a Haar system on $[a, b]$, $f(x)$ is continuous and Q is an arbitrary closed set of this segment, then among all the polynomials* 2.3(1) *$P_n(x)$ will deviate least from $f(x)$ in the space C on Q if and only if $f(x)-P_n(x)$ attains a maximum of its modulus on Q, with consecutive changes of sign, at least at $n+2$ points of the given set***.

This more general formulation of Chebyshev's theorem is due to S. N. Bernstein [15].

2.7.31. In the case where the set Q consists in all of only the $n+2$ distinct points $x_0, x_1, \ldots, x_{n+1}$, the problem of the determination of the polynomial $P_n(x)$ which deviates least from $f(x)$ on Q, and of the corresponding best approximation $E_n(f)$, reduces by the above to the solution of the system of $n+2$ linear equations

$$\sum_{k=0}^{n} c_k \phi_k(x_\nu)+(-1)^\nu E_n(f) = f(x_\nu) \quad (\nu = 0, 1, \ldots, n+1). \tag{4}$$

It is immediately clear that if Q consists of $m \leqslant n+1$ points, then $E_n(f) = 0$ and for $m \leqslant n$ the polynomials of best approximation are infinitely many. The case where $m > n+2$ will be considered below (see section 2.10.1).

2.7.32. With respect to the system 2.3.2, (3) consisting of $2n+1$ functions, theorem 2.7.3 shows *that among all the polynomials* 1.3(1) *the*

** Note here that if $f(x)$ vanishes at either of the ends of the segment, then all the functions $\phi_k(x)$ $(k = 0, 1, \ldots, n)$ can assume the value zero there simultaneously, and consequently, in addition to the n distinct zeros within $[a, b]$, the polynomials 2.3(1) can also have a zero at the corresponding end of the segment.

trigonometric polynomial $T_n(x)$ of order $\leqslant n$ deviates least from the function $f(x)$ which is continuous and of period 2π if and only if the difference $f(x)-T_n(x)$ attains on $[0, 2\pi]$ a maximum of its modulus, at least at $2n+2$ distinct points with consecutive changes of sign.

2.7.33. With respect to the system 2.3.2, (4) theorem 2.7.3 shows that *among all the algebraic polynomials* 1.1(1) *the polynomial $P_n(f; x)$ deviates least with weight $\varrho(x) > 0$ on $[a, b]$ from the continuous function $f(x)$, i.e. it realises the least value for* $\max\limits_{a \leqslant t \leqslant b} |f(t)-P_n(t)|\varrho(t)$, *only when the weighted difference $\varrho(x)[f(x)-P_n(f; x)]$ attains on this segment a maximum of its modulus at least at $n+2$ distinct points with consecutive changes of sign.*

2.7.4. The proof of theorem 2.7.11 (and 2.7.1) rests only on the general properties of functions continuous on a bounded closed set (the set of those x where $f(x) \geqslant c$ is closed, and if $f(x) \neq 0$ on the closed set M, then $\inf\limits_{x \in M} |f(x)| > 0$). Hence the theorem remains valid if instead of functions continuous on a segment we consider continuous functions of a point x of an arbitrary compact metric space, and instead of the polynomials 1.1(1) we consider linear combinations of some linearly independent system of continuous functions. *In particular, if $P_{n_1, \ldots, n_m}(x_1, \ldots, x_m)$ is some polynomial* 1.1(2), *it will deviate least from the continuous function $f(x_1, \ldots, x_m)$ in the bounded closed region $\overline{G}$ of the space of the variables $x_1, \ldots, x_m$ if and only if for any polynomial $Q_{n_1, \ldots, n_m}(x_1, \ldots, x_m)$ of degree $\leqslant n_1, n_2, \ldots, n_m$ the inequality*

$$\max_{x_1, \ldots, x_m \in M} Q_{n_1, \ldots, n_m}(x_1, \ldots, x_m)\{f(x_1, \ldots, x_m)- \\ -P_{n_1, \ldots, n_m}(x_1, \ldots, x_m)\} \geqslant 0 \qquad (5)$$

holds, where M is the set of all those points of $\overline{G}$ at which

$$|f(x_1, \ldots, x_m)-P_{n_1, \ldots, n_m}(x_1, \ldots, x_m)| \\ = \max_{\overline{G}} |f(t_1, \ldots, t_m)-P_{n_1, \ldots, n_m}(t_1, \ldots, t)_m| \qquad (6)$$

It is also possible to give an algebraic criterion of the type of Chebyshev's theorem** (or of theorem 2.7.3) for the case of functions of many variables. However, it is considerably more complicated and we will not derive it here.

** See E. Ya. Remez [1, 2]. Further investigation in this direction is due to S. I. Zukhovitskii [1].

2.8. On some properties of polynomials of best integral approximations

Let us consider the problem studied in section 2.7 with reference to the best integral approximation determined in section 2.2.2.

2.8.1. Let us commence with the case where $q = 1$. If on the given segment $[a, b]$ we have $f(x) \in L$, $\phi_k(x) \in L$ $(k = 0, 1, \ldots, n)$, and

$$P_n(x) = \sum_{k=0}^{n} c_k^{(0)} \phi_k(x) \tag{1}$$

is a polynomial which possesses the property that for any polynomial

$$Q_n(x) = \sum_{k=0}^{n} c_k \phi_k(x) \tag{2}$$

we always have

$$\int_a^b Q_n(x) \operatorname{sign}[f(x) - P_n(x)] dx = 0, \tag{3}$$

then from the inequality

$$\int_a^b |f(x) - Q_n(x)| dx \geqslant \int_a^b [f(x) - Q_n(x)] \operatorname{sign}[f(x) - P_n(x)] dx$$

$$= \int_a^b |f(x) - P_n(x)| dx$$

it follows that

$$\int_a^b |f(x) - P_n(x)| dx = \inf_{c_k} \int_a^b |f(x) - \sum_{k=0}^{n} c_k \phi_k(x)| dx. \tag{4}$$

Conversely, the considerations of section 2.4 show that if $P_n(x)$ satisfies 2.8(4) and the difference $\alpha(x) = f(x) - P_n(x)$ is different from zero almost everywhere, then 2.8(3) holds for any polynomial 2.8(2). Thus** *for the polynomial* 2.8(1) *to deviate least from $f(x)$ in the metric L it is sufficient, and in the case where the difference $f(x) - P_n(x)$ is different from zero almost everywhere on $[a, b]$ it is also necessary, that condition* 2.8(3) *be satisfied.*

Simple examples show that condition 2.8(3) cannot be satisfied without the condition that $f(x) - P_n(x) \neq 0$ almost everywhere (see 2.13.4).

2.8.11. Applying the preceding proposition to the case already investigated in section 2.4, where $f(x)$ is continuous and $\{\phi_k(x)\}$ $(k = 0, 1, \ldots, n)$ is a linearly independent system of continuous functions which satisfy

** S. M. Nikol'skii [3].

Haar's condition on $[a, b]$ (see 2.3), we obtain the following criterion due to A. A. Markov**. *If some continuous function $\phi(x)$ changes sign at $n+1$ different points $x_0 < x_1 < \ldots < x_n$ within a given segment and*

$$\int_a^b \phi_k(x) \operatorname{sign} \phi(x)\, dx = 0 \qquad (k = 0, 1, \ldots, n), \tag{5}$$

then the polynomial $P_n(x)$ for which the difference $f(x)-P_n(x)$ changes sign at these points, and only at these points, deviates least from $f(x)$ in the metric L, i.e. it satisfies the relation 2.8(4), *and*

$$E_n(f)_L = \int_a^b |f(x)-P_n(x)|\, dx = \left| \int_a^b f(x) \operatorname{sign} \phi(x)\, dx \right|. \tag{6}$$

As is seen from the proof of theorem 2.4, if $P_n(f; x)$ is the best polynomial approximation in L and the difference $f(x)-P_n(f; x)$ is different from zero almost everywhere, then it is bound always to change sign at least at $n+1$ points within (a, b).

For separate concrete systems which satisfy Haar's condition, it is sometimes not difficult to discover a function $\phi(x)$ with the properties indicated. Thus, for example, from the identity

$$\int_0^{2\pi} e^{imx} \operatorname{sign} \begin{cases} \cos(n+1)x \\ \sin(n+1)x \end{cases} dx = -e^{i\frac{m}{n+1}\pi} \int_0^{2\pi} e^{imx} \operatorname{sign} \begin{cases} \cos(n+1)x \\ \sin(n+1)x \end{cases} dx \tag{7}$$

it is immediately clear that for the system 2.3.2, (2) it is possible to choose as $\phi(x)$ the function $\cos(n+1)x$, and for the system 2.3.2, (3) the function $\cos(n+1)x$ or $\sin(n+1)x$.

We also note that as

$$\int_{-1}^{1} x^k \operatorname{sign} \sin(n+2) \arccos x\, dx = \int_0^{\pi} \sin t \cos^k t \operatorname{sign} \sin(n+2)t\, dt$$

and $\cos^k t$ is an even trigonometric polynomial 1.3(1) of order k, we have, in virtue of the same identity 2.8(7),

$$\int_{-1}^{1} x^k \operatorname{sign} \sin(n+2) \arccos x\, dx = 0, \qquad 0 \leqslant k \leqslant n, \tag{8}$$

i.e. for the system 2.3.2 (1) for $a = -1$, $b = 1$, it is possible to take as $\phi(x)$ the function $\sin(n+2) \arccos x$, which changes sign within $(-1,1)$ at the points $x_\nu = \cos \dfrac{\nu\pi}{n+2}$ $(\nu = 1, 2, \ldots, n+1)$.

** A. A. Markov [1].

2.8.12. From the proof of theorem 2.8.1 it is seen that it remains valid for the best approximation defined in section 2.2.4, i.e. if instead of the function $f(x)$ defined on the segment $[a, b]$ we speak of functions Lebesgue-integrable in a closed bounded region $\bar{G}$ of a given multidimensional space, and instead of the polynomials 2.8(2) we speak of the polynomials 1.1(2), and the condition 2.8(3) is replaced by a similar condition with the integration over $\bar{G}$.

2.8.13. It is also obvious that the given theorem remains valid with respect to the best approximations defined in section 2.2.3 with only this change, that in the given case it is necessary to assume equal to zero not the ordinary Lebesgue measure of those x for which $f(x)-P_n(x) = 0$, but the corresponding Lebesgue-Stieltjes measure.

2.8.2. Let us now pass to polynomials of best power approximation ($q > 1$) and let us consider specially the commonest case, where $q = 2$ (the best mean square approximation) (see 2.2.2).

If on the segment $[a, b]$ there is defined a linearly independent system of functions $\phi_k(x) \in L_2$ ($k = 0, 1, \ldots, n$), then *the polynomial*

$$P_n(f; x) = \sum_{k=0}^{n} c_k^{(0)} \phi_k(x) \tag{9}$$

deviates least on $[a, b]$ from the function $f(x) \in L_2$ in the sense of the metric L_2 if and only if for any polynomial

$$Q_n(x) = \sum_{k=0}^{n} c_k \phi_k(x) \tag{10}$$

the equality

$$\int_a^b Q_n(x)[f(x)-P_n(f; x)]\,\mathrm{d}x = 0 \tag{11}$$

is satisfied.

In fact, if for a certain polynomial $Q_n(x)$ the integral 2.8(11) were to assume a value $\mathrm{d} \neq 0$, then for

$$\lambda = \frac{d}{\int_a^b Q_n^2(x)\,\mathrm{d}x}$$

we should have

$$\int_a^b [f(x)-P_n(f; x)-\lambda Q_n(x)]^2\,\mathrm{d}x < \int_a^b [f(x)-P_n(f; x)]^2\,\mathrm{d}x,$$

and this is impossible. The converse assertion follows from the identity

$$\int_a^b |f(x)-Q_n(x)|^2\,dx = \int_a^b |f(x)-P_n(f;x)|^2\,dx + \int_a^b |P_n(f;x)-Q_n(x)|^2\,dx.$$

2.8.21. The condition 2.8(11) is equivalent to the system of equations

$$\int_a^b \phi_k(x)\,P_n(f;x)\,dx = \int_a^b f(x)\,\phi_k(x)\,dx \qquad (k=0,1,\dots,n) \tag{12}$$

for the determination of the coefficients of the polynomial 2.8(9). If with this system there is also associated the equation**

$$\int_a^b f(x)\,P_n(f;x)\,dx = \int_a^b f^2(x)\,dx - E_n^2(f)_{L_2},$$

which follows from 2.8(11), we find from the condition of compatibility that

$$E_n^2(f)_{L_2} = \frac{\Gamma(f,\phi_0,\phi_1,\dots,\phi_n)}{\Gamma(\phi_0,\phi_1,\dots,\phi_n)}, \tag{13}$$

where $\Gamma(\phi_0,\phi_1,\dots,\phi_n)$ is Gram's determinant, i.e. the determinant of the numbers

$$\gamma_{\mu\nu} = \int_a^b \phi_\mu(x)\,\phi_\nu(x)\,dx \qquad (\mu,\nu=0,1,\dots,n).$$

In the case *where the system* $\{\phi_k(x)\}$ is *orthonormal*, i.e. $\gamma_{\mu\nu}=0$ if $\mu\neq\nu$ and $\gamma_{\nu\nu}=1$ $(\mu,\nu=0,1,\dots,n)$, it follows from 2.8(12) or directly from the identity

$$\int_a^b \Big|f(x)-\sum_{k=0}^n c_k^{(0)}\phi_k(x)\Big|^2 dx$$

$$= \int_a^b f^2(x)\,dx - \sum_{k=0}^n \Big(\int_a^b f(x)\,\phi_k(x)\,dx\Big)^2 + \sum_{k=0}^n \Big[c_k^{(0)} - \int_a^b f(x)\,\phi_k(x)\,dx\Big]^2$$

that the coefficients of the best mean square polynomial approximation to the function $f(x)$ *are determined by the equations*

$$c_k^{(0)} = \int_a^b f(x)\,\phi_k(x)\,dx.$$

** Where

$$E_n(f)_{L_2} = \inf_{c_k}\Big\{\int_a^b \Big|f(x)-\sum_{k=0}^n c_k\phi_k(x)\Big|^2 dx\Big\}^{1/2}.$$

If this condition is not satisfied, it is then convenient in place of the system $\{\phi_k(x)\}$ to introduce a new system of functions

$$\psi_k(x) = \frac{1}{\sqrt{\Gamma(\phi_0, \dots, \phi_{k-1})\Gamma(\phi_0, \dots, \phi_k)}} \begin{vmatrix} \gamma_{00} \cdots \gamma_{0,k-1} \, \phi_0(x) \\ \gamma_{10} \cdots \gamma_{1,k-1} \, \phi_1(x) \\ \dots \dots \dots \quad \dots \\ \gamma_{k,0} \cdots \gamma_{k,k-1} \, \phi_k(x) \end{vmatrix}, \tag{14}$$

which are orthonormal to begin with and possess the property that the set of all possible polynomials $\sum_{k=0}^{n} c_k \psi_k(x)$ is identical with the set of all the polynomials 2.8(10).

2.8.22. We will give some important examples of orthonormal systems of functions (see also 2.13.29):

(1) The system

$$\frac{1}{\sqrt{\pi}}, \quad \sqrt{\frac{2}{\pi}} \cos kx \quad (k = 1, 2, \dots)$$

is orthonormal on $[0,2\pi]$.

(2) The system

$$\sqrt{\frac{2}{\pi}} \sin kx \quad (k = 1, 2, \dots)$$

is orthonormal on the same segment.

(3) The system of functions

$$\hat{T}_0(x) = \sqrt{\frac{1}{\pi}}, \quad \hat{T}_k(x) = \sqrt{\frac{2}{\pi}} \cos k \arccos x \quad (k = 1, 2, \dots)$$

is orthonormal with weight $\frac{1}{\sqrt{1-x^2}}$ on $[-1,1]$, i.e.

$$\int_{-1}^{1} \hat{T}_m(x)\hat{T}_n(x) \frac{1}{\sqrt{1-x^2}} \, dx = \begin{cases} 1, & m = n; \\ 0, & m \neq n. \end{cases}$$

2.8.23. The statements in sections 2.82 and 2.8.21 remain valid if instead of ordinary mean square best approximation we consider what is known as the best weighted mean square approximation, defined in section 2.2.3 ($q = 2$), where the algebraic polynomials 1.1(1) are replaced by the polynomials 2.8(10) with reference to an arbitrary linearly independent system of functions $\phi_k(x) \in L_2(\varrho)$. Here we must have everywhere, instead of ordinary Lebesgue integrals, the corresponding Lebesgue-Stieltjes integrals.

2.8.24. In particular, if $\{\phi_k(x)\}$ is the trigonometric system 2.3.2, (3) and $\varrho_m(x)$ is a step function which assumes the values $x_\nu = 2\nu\pi/m$

for $2(\nu-1)\pi/m < x \leqslant 2\nu\pi/m$ $(\nu = 0, 1, \ldots, m)$, then from the identity

$$\sum_{k=0}^{m-1} e^{2ikr\frac{\pi}{m}} = 0 \qquad (m > r)$$

for $m \geqslant 2n+1$ there follows its orthogonality with weight $\varrho_m(x)$ on $[0, 2\pi]$. Since in this case

$$\int_0^{2\pi} \sin^2 kx \, d\varrho_m(x) = \int_0^{2\pi} \cos^2 kx \, d\varrho_m(x) = \pi$$

for $0 \leqslant k \leqslant n$, the coefficients of the trigonometric polynomial

$$T_n(f; x) = \frac{a_0}{2} + \sum_{k=1}^{n} (a_k \cos kx + b_k \sin kx)$$

of the corresponding best weighted mean square approximation to the function $f(x)$ are determined by the formulae

$$\begin{aligned} a_k &= \frac{1}{\pi} \int_0^{2\pi} f(t) \cos kt \, d\varrho_m(t) = \frac{2}{m} \sum_{k=0}^{m-1} f(x_\nu) \cos kx_\nu, \\ b_k &= \frac{1}{\pi} \int_0^{2\pi} f(t) \sin kt \, d\varrho_m(t) = \frac{2}{m} \sum_{k=0}^{m-1} f(x_\nu) \sin kx_\nu, \end{aligned} \tag{15}$$

whence it follows that

$$T_n(f; x) = \frac{1}{\pi} \int_0^{2\pi} f(t) \left[\frac{1}{2} + \sum_{k=1}^{n} \cos k(t-x)\right] d\varrho_m(t)$$

$$= \frac{1}{2\pi} \int_0^{2\pi} f(t) \frac{\sin \frac{2n+1}{2}(t-x)}{\sin \frac{t-x}{2}} d\varrho_m(t) = \frac{1}{m} \sum_{\nu=0}^{m-1} f(x_\nu) \frac{\sin \frac{2n+1}{2}(x_\nu - x)}{\sin \frac{x_\nu - x}{2}}. \tag{16}$$

Since the weighted mean square approximation considered depends only on the values of the function $f(x)$ at the given equidistant points x_ν $(\nu = 0, 1, \ldots, m-1)$, the polynomial 2.8(16) is the trigonometric polynomial of order $\leqslant n$ which deviates least from $f(x)$ in the mean square sense at this system of points, i.e.

$$\sum_{\nu=0}^{m-1} [f(x_\nu) - T_n(f; x_\nu)]^2 = \inf_{T_n(x)} \sum_{\nu=0}^{m-1} [f(x_\nu) - T_n(x_\nu)]^2.$$

2.8.25. For other values of the index q effective criteria for the best polynomial approximation, similar to those indicated in sections 2.8.11 ($q = 1$), 2.8.21 ($q = 2$) and 2.7.3 ($q = \infty$), are unknown.

At the same time theorems 2.8.1 ($q = 1$) and 2.8.2 ($q = 2$) are easily generalised to arbitrary power approximations**.

Whatever the finite $q \geqslant 1$ may be, in order that the polynomial

$$P_n(x) = \sum_{k=0}^{n} c_k^{(0)} \phi_k(x)$$

in the linearly independent system of functions $\phi_k(x) \in L_q$ $(k = 0, 1, \ldots, n)$ defined on the segment $[a, b]$ should deviate least in the metric L_q on $[a, b]$ from the function $f(x) \in L_q$, it is sufficient and (for $q = 1$ in the case where the difference $f(x) - P_n(x)$ is different from zero almost everywhere on $[a, b]$) it is also necessary that for any polynomial

$$Q_n(x) = \sum_{k=0}^{n} c_k \phi_k(x)$$

the equality

$$\int_a^b Q_n(x) |f(x)-P_n(x)|^{q-1} \operatorname{sign} \{f(x)-P_n(x)\} \, dx = 0 \tag{17}$$

should hold.

In fact if condition 2.8(17) is satisfied, then since

$$\begin{aligned}\int_a^b |f(x)-P_n(x)|^q \, dx \\ &= \int_a^b [f(x)-P_n(x)] |f(x)-P_n(x)|^{q-1} \operatorname{sign} [f(x) - P_n(x)] \, dx \\ &= \int_a^b [f(x)-Q_n(x)] |f(x)-P_n(x)|^{q-1} \operatorname{sign} [f(x)-P_n(x)] \, dx \\ &\leqslant \int_a^b |f(x)-Q_n(x)| \, |f(x)-P_n(x)|^{q-1} \, dx \\ &\leqslant \left\{\int_a^b |f(x)-Q_n(x)|^q \, dx\right\}^{\frac{1}{q}} \left\{\int_a^b |f(x)-P_n(x)|^q \, dx\right\}^{1-\frac{1}{q}},\end{aligned}$$

it follows that

$$\int_a^b |f(x)-P_n(x)|^q \, dx \leqslant \int_a^b |f(x)-Q_n(x)|^q \, dx,$$

** These theorems can also be generalised to the case of best approximations in any linear normed space (see 2.2). See Zinger [1].

whatever the polynomial $Q_n(x)$ may be, i.e. $P_n(x)$ is the best polynomial approximation to the function $f(x)$ in the metric L_q. We shall now show that if $P_n(x)$ is the best polynomial approximation to the function $f(x)$ in the metric L_q, then (for $q = 1$ in the case where the difference $f(x)-P_n(x)$ is different from zero almost everywhere on $[a, b]$)

$$\int_a^b \phi_k(x)\,|f(x)-P_n(x)|^{q-1}\,\mathrm{sign}\,[f(x)-P_n(x)]\,\mathrm{d}x = 0$$

for any $k = 0, 1, \ldots, n$.
If it should happen that for some k

$$\int_a^b \phi_k(x)\,|f(x)-P_n(x)|^{q-1}\,\mathrm{sign}\,[f(x)-P_n(x)]\,\mathrm{d}x = \delta \neq 0,$$

then it would be possible to find a number ε so small in absolute magnitude that

$$\varepsilon \int_a^b \phi_k(x)\,|f(x)-P_n(x)-\varepsilon\phi_k(x)|^{q-1}\,\mathrm{sign}\,[f(x)-P_n(x)-\varepsilon\phi_k(x)]\,\mathrm{d}x > 0.$$

But then

$$\begin{aligned}
&\int_a^b |f(x)-P_n(x)-\varepsilon\phi_k(x)|^q\,\mathrm{d}x \\
&\quad= \int_a^b [f(x)-P_n(x)]\,|f(x)-P_n(x)-\varepsilon\phi_k(x)|^{q-1}\,\mathrm{sign}\,[f(x)-P_n(x)- \\
&\quad\quad -\varepsilon\phi_k(x)]\,\mathrm{d}x - \varepsilon\int_a^b \phi_k(x)\,|f(x)-P_n(x)-\varepsilon\phi_k(x)|^{q-1}\,\mathrm{sign}[f(x)- \\
&\quad\quad -P_n(x)-\varepsilon\phi_k(x)]\,\mathrm{d}x < \\
&\quad< \int_a^b |f(x)-P_n(x)|\,|f(x)-P_n(x)-\varepsilon\phi_k(x)|^{q-1}\,\mathrm{d}x \leqslant \\
&\quad\leqslant \left\{\int_a^b |f(x)-P_n(x)|^q\,\mathrm{d}x\right\}^{\frac{1}{q}} \left\{\int_a^b |f(x)-P_n(x)-\varepsilon\phi_k(x)|^q\,\mathrm{d}x\right\}^{1-\frac{1}{q}}.
\end{aligned}$$

Consequently

$$\int_a^b |f(x)-P_n(x)-\varepsilon\phi_k(x)|^q\,\mathrm{d}x < \int_a^b |f(x)-P_n(x)|^q\,\mathrm{d}x,$$

and we arrive at a contradiction on the assumption concerning the polynomial $P_n(x)$.

Among other results related to this we also note the theorem that *the algebraic polynomial $P_{n,q}(f; x)$ of degree $\leqslant n$ which deviates least* *on the finite segment* $[a, b]$ *from the given continuous function* $f(x)$ *in the metric* L_q, *tends as* $q \to \infty$ *to the ordinary polynomial* $P_n(f; x)$ *which gives the best uniform approximation to the same function on this segment***.

To prove this theorem it is sufficient to note that, from considerations similar to those given in section 2.1.1, the coefficients $c_k^{(q)}$ of the polynomials $P_{n,q}(f; x) = \sum_{k=0}^{n} c_k^{(q)} x^k$ are uniformly bounded with respect to q, and hence from any sequence $q_\nu \to \infty$ it is possible to select a subsequence q_{ν_l} such that for $k = 1, 2, \ldots, n$ the sequence $c_k^{(q_{\nu_l})}$ converges to some finite number c_k. In addition, for any $\varepsilon > 0$ it is possible to find an N such that for $q \geqslant N$ (see 2.13.2)

$$\left\{\int_a^b |f(x)-P_{n,q}(f; x)|^q \,dx\right\}^{\frac{1}{q}} \leqslant \left\{\int_a^b |f(x)-P_n(f; x)|^q \,dx\right\}^{\frac{1}{q}} \leqslant E_n(f; a, b)+\varepsilon.$$

But since the subsequence of polynomials $P_{n,q_{\nu_l}}(f; x)$ converges uniformly on $[a, b]$ to the polynomial $P_n(x) = \sum_{k=0}^{n} c_k x^k$, it follows that for sufficiently great values of q_{ν_l}

$$(1-\varepsilon) \max_{a \leqslant x \leqslant b} |f(x)-P_n(x)| \leqslant \left\{\int_a^b |f(x)-P_n(x)|^{q_{\nu_l}} \,dx\right\}^{\frac{1}{q_{\nu_l}}}$$

$$\leqslant E_n(f; a, b)+\varepsilon+\left\{\int_a^b |P_n(x)-P_{n,q_{\nu_l}}(f; x)|^{q_{\nu_l}} \,dx\right\}^{\frac{1}{q_{\nu_l}}}$$

$$\leqslant E_n(f; a, b)+2\varepsilon$$

i.e.

$$\max_{a \leqslant x \leqslant b} |f(x)-P_n(x)| = E_n(f; a, b), \qquad \text{or} \qquad P_n(x) = P_n(f; x).$$

2.9. Polynomials which deviate least from zero

P. L. Chebyshev arrived at theorem 2.7.2 as a result of considering the following particular problem, which arose in connection with certain of his investigations in the theory of mechanisms***.

2.9.1. This problem, known as the problem of polynomials which deviate least from zero, consists of finding the algebraic polynomial

$$P_n(x) = x^n+c_1x^{n-1}+\cdots+c_n \tag{1}$$

** G. Pólya [1].
*** P. L. Chebyshev [1].

for which on the given segment $[a, b]$ the quantity $\max_{a \leqslant x \leqslant b} |P_n(x)|$ assumes the least possible value, or what is the same thing, of finding a polynomial 2.1(1) of degree $\leqslant n-1$ which deviates least from the function $f(x) = x^n$ on $[a, b]$. Its solution is easily obtained from the general theorem 2.7.2, from which it follows that such a polynomial attains a maximum of its modulus on $[a, b]$ at the ends of this segment and exactly $n-1$ points within it. If for definiteness we put $a = -1$, $b = 1$ and introduce the notation $L = \max_{|x| \leqslant 1} |P_n(x)|$, it follows from this that the polynomials $(1-x^2)P_n'^2(x)$ and $L^2 - P_n^2(x)$ have exactly the same zeros. Hence $y = P_n(x)$ satisfies the differential equation

$$(1-x^2)y'^2 = n^2(L^2 - y^2), \tag{2}$$

the general solution of which is of the form

$$y = \pm L \cos(n \arccos x + C). \tag{3}$$

But since it follows from the identity

$$2 \cos nt = (\cos t + i \sin t)^n + (\cos t - i \sin t)^n \tag{4}$$

that

$$\cos n \arccos x = \tfrac{1}{2}[(x+\sqrt{x^2-1})^n + (x-\sqrt{x^2-1})^n] \tag{5}$$

is a polynomial and hence $\sin n \arccos x$ is not a polynomial, among all the solutions of 2.10 (3) only that solution will be a polynomial which corresponds to the value $C = 0$. The quantity L is determined by the condition that the coefficient of x^n in the polynomial $P_n(x)$ is equal to unity. From 2.9(5) we conclude that $L = \dfrac{1}{2^{n-1}}$ and thus

$$P_n(x) = \frac{1}{2^{n-1}} \cos n \arccos x. \tag{6}$$

The polynomials 2.9(6) are known as Chebyshev polynomials. They attain maxima of their modulus on the segment $[-1, 1]$ with successive changes of sign at the points $x_k = \cos\dfrac{k\pi}{n}$ $(k = 0, 1, \ldots, n)$. The case of an arbitrary segment $[a, b]$ is reduced to the case considered by a linear transformation. Thus *the algebraic polynomial* 2.9(1) *which deviates least from zero on* $[a, b]$ *is of the form*

$$P_n(x) = 2\left(\frac{b-a}{4}\right)^n \cos n \arccos \frac{2x-a-b}{b-a}. \tag{7}$$

2.9.11. From theorem 2.7.2 and Descartes's theorem on the positive zeros of algebraic polynomials it follows that in the case where $ab > 0$, the polynomial 2.9(7), apart from a constant factor, deviates least from zero on $[a, b]$ among the polynomials 1.1(1) with a given coefficient for an arbitrary degree $k (0 \leqslant k \leqslant n)$** (for $ab = 0$ the case $k = 0$ is excluded). Hence it follows that *if on* $[a, b]$ $(ab > 0)$ *the polynomial* 1.1(1) *satisfies the inequality* $|P_n(x)| \leqslant L$, *then its coefficients* $c_k (0 \leqslant k \leqslant n)$ *satisfy the inequalities*

$$|c_k| \leqslant |a_k|,$$

where a_k *are the coefficients of the polynomial*

$$L \cos n \arccos \frac{2x-a-b}{b-a} = \sum_{k=0}^{n} a_k x^k. \tag{8}$$

Since $c_0 = P_n(0)$, after the substitution $x = t-c$ we immediately conclude that *if the polynomial* 1.1(1) *satisfies the inequality* $|P_n(x)| \leqslant L$ *on the segment* $[a, b]$ *then at any point outside the segment we have*

$$|P_n(x)| \leqslant L \left| \cos n \arccos \frac{2x-a-b}{b-a} \right|. \tag{9}$$

2.9.12. Let us note that for even (cdd) values of n the polynomial 2.9(5) contains only even (cdd) powers of x. Hence from the last remark of section 2.1.2 and the identities

$$\cos 2n \arccos x = \sum_{k=J}^{n} (-1)^{n-k} 2^{2k} \frac{n}{n+k} \binom{n+k}{2k} x^{2k},$$

$$\cos (2n+1) \arccos x = \sum_{k=0}^{n} (-1)^{n-k} 2^{2k} \frac{2n+1}{2k+1} \binom{n+k}{2k} x^{2k+1}$$

and in virtue of theorem 2.7.3 it follows that *if for all x on the segment* $[-1,1]$

$$\left| \sum_{k=0}^{2n+1} c_k x^k \right| \leqslant 1$$

then

$$|c_{2\nu}| \leqslant 2^{2\nu} \frac{n}{n+\nu} \binom{n+\nu}{2\nu}, \qquad |c_{2\nu+1}| \leqslant 2^{2\nu} \frac{2n+1}{2\nu+1} \binom{n+\nu}{2\nu}$$
$$(\nu = 0, 1, \ldots, n)***.$$

** In this case the system of functions $1, x, \ldots, x^{k-1}, x^{k+1}, \ldots, x^n$ satisfies the conditions of Haar's theorem (section 2.3) on $[a, b]$.

*** V. A. Markov [1]. From this there follows the estimate 2.6(9).

In particular, *the inequality***

$$\max_{-1\leqslant x\leqslant 1}\left|\sum_{k=0}^{2n+1} c_k x^k\right| \geqslant \frac{|c_1|}{2n+1},$$

is always satisfied, and this should be noted together with the inequality

$$\max_{-1\leqslant x\leqslant 1}\left|\sum_{k=0}^{n} c_k x^k\right| \geqslant \frac{|c_n|}{2^{n-1}}$$

obtained in section 2.9.1, and the trivial inequality

$$\max_{-1\leqslant x\leqslant 1}\left|\sum_{k=0}^{n} c_k x^k\right| \geqslant |c_0|.$$

2.9.13. Theorem 2.9.1 was supplemented by G. Pólya***, who considered the problem of the least deviation from zero of the polynomials 2.9(1) on an arbitrary closed set of the real axis. *The maximum modulus of any polynomial* 2.9(1) *on the closed set* Q, *the measure of which is equal to* $2h$, *is always not less than* $2\left(\frac{h}{2}\right)^n$, *and can be equal to this number only in the case where* Q *is a segment of the same length.* So far as concerns the polynomials 2.9(1) with least possible deviation form zero on a given closed set, the problem of finding their general form is very much more complex than problem 2.9.1. In particular cases where Q consists of two intersecting sets, it was investigated in detail in the work of N. I. Akhiezer [1].

2.9.14. It is also necessary to point out some other interesting generalisations of this problem which are due to E. I. Zolotarev [1], who gave a solution of the problem of algebraic polynomials which deviate least from zero when the coefficients of the two highest powers are fixed. These investigations were significantly advanced in various directions by N. I. Akhiezer [1], who in particular solved the same problem for the case where the coefficients of the three highest powers are given.

2.9.15. It is seen from 2.9(5) and 2.9(7) that for $a = -2$, $b = 2$ all coefficients of the polynomial 2.9(7) are integers. Hence for the given segment it also deviates least from zero in the class of all polynomials 2.9(1) and, consequently, in the class of all algebraic polynomials of degree $\leqslant n$, which have only integral coefficients. For other segments $[a, b]$ the problem of finding non-trivial algebraic polynomials with integral coeffi-

---

** For a similar result for the upper bound of transcendental integral functions of exponential type on the whole real axis see section 4.7.1.

*** See S. N. Bernstein [15].

cients, deviating least from zero, requires investigations of a more specialised character which would take us outside the framework of the theory given in section 2.7.

2.9.2. The problem similar to problem 2.9.1 for the polynomials 1.3(1) reduces to the finding of the trigonometric polynomial

$$T_n(x) = a\cos nx + b\sin nx + \sum_{k=0}^{n-1} (a_\kappa \cos kx + b_k \sin kx), \tag{10}$$

for which for fixed coefficients b and a the quantity $\max\limits_x |T_n(x)|$ assumes its least possible value, or what is the same thing, the finding of the trigonometric polynomial of order $\leqslant n-1$ which deviates least from the function $f(x) = \sqrt{a^2+b^2}\cos(nx+\theta)$. Its solution follows directly from theorem 2.7.32 if we note that the function $f(x) = \sqrt{a^2+b^2}\cos(nx+\theta)$ attains maxima of its modulus on $[0,2\pi]$ at $2n$ equally spaced points with successive changes of sign. Thus among the polynomials 2.9(10) *with fixed coefficients of* $\cos nx$ *and* $\sin nx$ *the trigonometric polynomial*

$$T_n(x) = a\cos nx + b\sin nx \tag{11}$$

deviates least from zero.

2.9.21. Let us note that a problem similar to problem 2.9.11, that is the problem of the determination of the trigonometric polynomial of order $\leqslant n$ with given coefficients of $\cos kx$ and $\sin kx$, which deviates least from zero for $0 < k < n$ is rather complicated.

At the same time for sufficiently great n an approximate solution can be obtained** (see 2.13.24).

2.9.3. Let us consider the problems of sections 2.9.1 and 2.9.2 for an integral metric.

2.9.31. Using the proposition of section 2.8.11 and equation 2.8(8), it is easy to find the polynomial 2.9(1) which on [—1,1] deviates least from zero in the metric L, that is, for which

$$\int_{-1}^{1} |P_n(x)|\,dx$$

assumes the least possible value. Such a polynomial is the difference between the function $f(x) = x^n$ and the polynomial of degree $\leqslant n-1$ which deviates from it least in the metric L. If this difference changes sign at the points $x_\nu = \cos\dfrac{\nu\pi}{n+1}$ $(\nu = 1, \dots, n)$, then by virtue of a theorem of

** See S. N. Bernstein [15].

A. A. Markov (see section 2.8.11) it also gives us the required polynomial. But the polynomial 2.9.(1) which changes sign at the given points, as is easily seen, is the polynomial

$$P_n(x) = \frac{1}{2^n} \frac{\sin(n+1)\arccos x}{\sqrt{1-x^2}}, \tag{12}$$

which apart from a constant factor is identical with the derivative of Chebyshev's polynomial $\cos(n+1)\arccos x$. It also gives a solution of the problem considered. For this,

$$\int_{-1}^{1} |P_n(x)|\,dx = \frac{1}{2^n}\int_0^{\pi} |\sin(n+1)t|\,dt = \frac{1}{2^{n-1}}. \tag{13}$$

Reasoning in exactly the same way, it is possible to find the best approximation of any function $f(x)$ which forms part of the Chebyshev system 2.3.2, (5). For the segment [—1,1] it is calculated by the formula

$$E_{n-1}(f)_L = \left| \int_{-1}^{1} f(x) \operatorname{sign} \sin(n+1)\arccos x\,dx \right|. \tag{14}$$

The passage to any segment $[a, b]$ is carried out exactly as in section 2.9.1.

2.9.32. Let us consider the same problem for the case $q = 2$.

From section 2.8.2 it follows that the problem of finding the polynomial 2.9(1) which deviates least from zero in the metric L_q reduces to finding the function $\psi_n(x)$ of 2.8(14), where $\phi_\nu(x) = x^\nu$ $(\nu = 0, 1, \ldots, n)$, which in the given case is a polynomial of the ν-th degree. To solve this problem this polynomial is integrated n times in succession and the arbitrary constants are chosen so that at the left-hand end of the given segment the new polynomial $Q_{2n}(x)$, and all its derivatives up to the $(n-1)$-th order inclusive, vanish. But since

$$\int_a^b Q_{2n}^{(n)}(x)x^k\,dx = \int_a^b \psi_n(x)x^k\,dx = 0 \qquad (k = 0, 1, \ldots, n-1)$$

we see, on putting successively $k = 0, 1, \ldots, n-1$, that such polynomials possess these same properties at the right-hand end of the segment also, so that $Q_{2n}(x) = A(x-a)^n(x-b)^n$, and consequently *the polynomial* 2.9(1) *which deviates least from zero on* $[a, b]$ *in the mean square sense is the polynomial*

$$P_n(x) = \frac{n!}{(2n)!} \cdot \frac{d^n[(x-a)^n(x-b)^n]}{dx^n}, \tag{15}$$

known as Legendre's polynomial. Formula 2.9(15) is known as Rodrigues's formula For $a = -1$, $b = 1$ direct calculation (integration by parts) shows that

$$\left\{ \int_{-1}^{1} |P_n(x)|^2 \, dx \right\}^{1/2} = \sqrt{\frac{2}{2n+1}} \, 2^n \frac{(n!)^2}{(2n)!} . \tag{16}$$

For comparison with 2.9(6) and 2.9(13) we note that the right-hand side of 2.9(16) for integral n is contained between $\frac{1}{2^{n-1}\sqrt{2}}$ and $\frac{\sqrt{2}}{2^{n-1}}$,

2.9.33. We see from sections 2.9.1, 2.9.31, 2.9.32 that when the metric is changed the algebraic polynomial 2.9(1) which deviates least from zero is also changed**. It is curious to note that the trigonometric polynomials 2.9(10) are not connected with the metric in this sense. For any $q \geqslant 1$ the polynomial 2.9(11) deviates least from zero on $[0,2\pi]$ in the metric L_q among all the trigonometric polynomials 2.9(10) with the given coefficients b and a***. In fact, for $1 \leqslant q < \infty$ the polynomial $T_n(x)$ which deviates least from zero is unique (see sections 2.2.8 and 2.4). Consequently, as

$$\int_0^{2\pi} |T_n(x)|^q \, dx = \int_0^{2\pi} \left| T_n\left(x + \frac{\pi}{n}\right) \right|^q dx = \int_0^{2\pi} \left| T_n\left(x + \frac{2\pi}{n}\right) \right|^q dx ,$$

it follows that $T_n\left(x+\frac{2\pi}{n}\right) \equiv T_n(x) \equiv -T_n\left(x+\frac{\pi}{n}\right)$. Hence it follows that in the given case this polynomial is 2.9(11).

2.10. Estimation of the best approximation. De la Vallée Poussin's theorem

In the study of the problem of the magnitude of the best approximation for continuous functions, besides theorem 2.7.2 the following proposition is also of great importance****. *If for the function $f(x)$ defined on the set $Q \in [a, b]$ and for some polynomial $P_n(x) = \sum_{k=0}^{n} c_k \phi_k(x)$,* where $\{\phi_k(x)\}$ *$(k = 0, 1, \dots, n)$ is a linearly independent system of continuous functions which satisfy Haar's condition (see 2.3) on $[a, b]$, the difference $f(x) - P_n(x)$ assumes values $a_0, a_1, \dots, a_{n+1}$ different from zero, with alternating signs, at*

** It is easy to see from what was said in section 2.8.25 that for any $q \geqslant 1$ such a polynomial has n distinct real zeros within the interval considered.

*** S. N. Bernstein [16].

**** De la Vallée Poussin [2]. See also S. N. Bernstein [15].

the $n+2$ *distinct points* $x_0, < x_1 < \ldots < x_{n+1}$ *of the set* Q, *then the best uniform approximation of* $f(x)$ *on* Q *by such polynomials satisfies the inequality*

$$E_n(f) = \inf_{c_k} \sup_{x \in Q} \left| f(x) - \sum_{k=0}^{n} c_k \phi_k(x) \right| \geqslant \mu = \min \{ |a_0|, |a_1|, \ldots, |a_{n+1}| \}.$$

In fact, if for some polynomial

$$P_n(f; x) = \sum_{k=0}^{n} c_k^{(0)} \phi_k(x)$$

$$|f(x) - P_n(f; x)| < \mu,$$

at all the points $x \in Q$, then at the $n+2$ given points the difference $P_n(x) - P_n(f; x)$ assumes values with alternating signs and as a result of this vanishes at least at $n+1$ points of the segment $[a, b]$, which is impossible.

2.10.1. In section 2.7.31 the problem was considered of the best approximation of order n of the function $f(x)$ on the set Q consisting of $m \leqslant n+2$ points. For finite $m > n+2$ this problem reduces to the solution of $\binom{m}{n+2}$ systems of $n+2$ equations of the type 2.7(4), in each of which the term $(-1)^\nu E_n(f)$ is replaced by the quantity $(-1)^\nu E_n(f; x_0; x_1, \ldots; x_{n+1})$, which is the best approximation of the function $f(x)$ on the corresponding part of Q, consisting of the points $x_0, x_1, \ldots, x_{n+1}$. It follows from section 2.7.3 that the best approximation $E_n(f)$ on the whole set Q is one of the numbers $E_n(f; x_0; x_1; \ldots; x_{n+1})$, and in virtue of 2.10 it is equal to the best approximation $E_n(f; x_0; x_1; \ldots; x_{n+1})$ corresponding to that set of $n+2$ points for which $E_n(f; x_0; x_1; \ldots; x_{n+1})$ is maximal**. It is obvious that in the general case also, *when* Q *is an arbitrary closed set in which there are not less than* $n+2$ *points,*

$$E_n(f) = \max_{x_0, x_1, \ldots, x_{n+1} \in Q} E_n(f; x_0; x_1; \ldots; x_{n+1}). \tag{1}$$

2.10.2. Let us examine a simple case of the calculation of the quantity $E_n(f)$ which does not require us to consider the system of equations 2.7(4). *If* Q *consists of the* $2n+2$ *points* $\dfrac{2\nu\pi}{2n+2}$ $(\nu = 0, 1, \ldots, 2n+1)$, *then the trigonometric polynomial of order* $\leqslant n$ *which gives the best uniform approx-*

** De la Vallée Poussin [3].

imation to the function $f(x)$ on Q is the polynomial 2.8(16), *i.e. the best mean square polynomial approximation for this system of points***. In fact, in the given case

$$T_n(f;x_k) = \frac{1}{2n+2}\sum_{\substack{\nu=0\\ \nu\neq k}}^{2n+1}(-1)^{\nu-k+1}f(x_\nu)+\frac{2n+1}{2n+2}f(x_k)$$

and hence

$$f(x_k)-T_n(f;x_k) = \frac{(-1)^k}{2n+2}\sum_{\nu=0}^{2n+1}(-1)^\nu f(x_\nu),$$

i.e. the difference $f(x)-T_n(f;x)$ assumes values identical in absolute magnitude at the $2n+2$ points $x_k (k = 0, 1, \ldots, 2n+1)$ with consecutive changes of sign, and hence in virtue of 2.7.32 the statement follows.

2.10.3. Theorem 2.10 (see also 2.10(1)) is a particular case of a more general assertion, which enables us to estimate from below the best approximation 2.2(4) in any linear normed space, and by which** the quantity $E_n(x)$ is the greatest of the values of all possible linear functionals $\Phi(x)$ in the given space for which $\Phi(x_k) = 0$ $(k = 0, 1, \ldots, n)$, $\|\Phi\| = 1$. If Φ is such a functional and $z_n = \sum_{k=0}^{n} c_k^{(0)}x_k \neq 0$ (see 2.2(4)), then $|\Phi(x)| = |\Phi(z_n)| \leqslant \|z_n\| = E_n(x)$. On the other hand, for a linear functional Φ defined on the $(n+2)$-dimensional space of elements $cz_n+\sum_{k=0}^{n} c_k x_k$ so that $\Phi\left(cz_n+\sum_{k=0}^{n} c_k x_k\right) = c\|z_n\|$ and then extended to the whole space with preservation of the norm, we have $\|\Phi\| = 1$ and $E_n(x) = \|z_n\| = \Phi\left(x-\sum_{k=0}^{n} c_k^{(0)}x_k\right) = \Phi(x)$.

Using the integral representation of linear functionals in the spaces considered, it is possible to obtain for them various estimates of the type

** See V. B. Gurevich [1], where it is proved that for the given system of equally spaced points the best uniform approximation by trigonometric polynomials is identical with the best mean approximation for any $q \geqslant 1$. This result was generalised by V. S. Videnskii [6], who considered the case of approximation on an arbitrary system of points in the complex plane.

*** S. M. Nikol'skii [3].

type of theorem 2.10 corresponding to the space of continuous functions (see, for example, 2.13.34)**.

2.11. Some cases of exact solution of the problem of the best approximation

In section 2.8.2 a simple method was indicated for the solution of the problem of the best mean square approximation. Of course algorithms do not exist*** for the exact solution of such a problem for the other methods of approximation (uniform approximation, approximation in the mean ($q = 2$)) which we considered. Hence the finding of the exact magnitude of the best approximation, and similarly the actual discovery of the polynomial which deviates least from the given function, constitutes the object of a special investigation in each concrete case. There are comparatively few examples of functions for which such an investigation can be reduced to the direct application of theorems 2.7.2, 2.8.11 and 2.10. We will consider some of them here (see also 2.13.31–33).

2.11.1. Let us find the best uniform approximation by algebraic polynomials to the function $f(x) = \dfrac{1}{x-a}$ ($|a| > 1$) on [—1,1]. If $P_n(x)$ is

---

** We obtain theorem 2.10 from 2.10.3 if in the space C of functions $f(x)$ continuous on $[a, b]$ we consider the linear functional $\Phi(f) = \int_a^b f(x) dg_n(x)$, where $g_n(x)$ is a step function with jumps d_k at the points $x_k (k = 0, 1, \ldots, n+1)$, which satisfy the system of equations

$$\Phi(\phi_\nu) = \sum_{k=0}^{n+1} d_k \phi_\nu(x_k) = 0 \quad (\nu = 0, 1, \ldots, n), \quad \sum_{k=0}^{n+1} (-1)^k d_k = 1,$$

where $\phi_0(x), \phi_1(x), \ldots, \phi_n(x)$ are the functions considered in section 2.10. The determinant of this system is different from zero (see section 2.3), and the corresponding values of d_k are such that $\|\Phi\| = \operatorname*{var}_{a \leqslant t \leqslant b} g_n(x) = \sum_{k=0}^{n+1} (-1)^k d_k = 1$. Hence for the function $f(x)$ considered in section 2.10

$$E_n(f) \geqslant |\Phi(f)| = \left| \int_a^b \{f(x) - P_n(x)\} dg_n(x) \right|$$

$$= \left| \sum_{k=0}^{n+1} \{f(x_k) - P_n(x_k)\} d_k \right| \geqslant \mu \sum_{k=0}^{n+1} (-1)^k d_k = \mu.$$

*** Some methods are known for the approximate solution of such problems. In a number of cases they were pointed out as early as P. L. Chebyshev [1] and S. N. Bernstein [15]. Further investigations are due to E. Ya. Remez [1] and S. I. Zukhovitskii [4].

the polynomial 1.1(1) which deviates least from $\dfrac{1}{x-a}$ on this segment, then in virtue of theorem 2.7.2 the difference

$$y = \frac{1}{x-a} - P_n(x) \tag{1}$$

attains a maximum of its modulus $L = E_n\left(\dfrac{1}{x-a}\right)$ with consecutive changes of sign at $n+2$ points of the segment $[-1,1]$. As $y^{(n+1)}(x)$ does not vanish in $[-1,1]$, n of these points are located within the given segment, and the two remaining points coincide with its ends. If we denote by b that root of the equation $y' = 0$, which lies outside $[-1,1]$, then the polynomials $(1-x^2)\,y'^2\,(x-a)^4$ and $(L^2-y^2)\,(x-b)^2\,(x-a)^2$ have the same zeros. Hence the difference 2.11(1) satisfies the differential equation

$$y'^2(1-x^2)(x-a)^2 = n^2(L^2-y^2)(x-b)^2, \tag{2}$$

the general solution of which is of the form

$$y(x) = L\cos[n \arccos x + \phi(x) + c], \tag{3}$$

where

$$\phi(x) = n(a-b)(a^2-1)^{-\frac{1}{2}} \arccos \frac{ax-1}{x-a}.$$

So that function $y(x)$ may assume the value $\pm L$ at $n+2$ points of the segment $[-1,1]$, it is necessary that as x varies from -1 to $+1$ the function $n \arccos x + \phi(x)$ should vary along a segment of length $(n+1)\pi$. Hence it follows that

$$b = a \pm \frac{\sqrt{a^2-1}}{n}.$$

Thus, taking account of 2.9(5), we obtain

$$y = \pm\frac{L}{2}\left\{\frac{(x+\sqrt{x^2-1})^n[ax-1+\sqrt{(a^2-1)(x^2-1)}]}{x-a} + \right.$$
$$\left. + \frac{(x-\sqrt{x^2-1})^n[ax-1-\sqrt{(a^2-1)(x^2-1)}]}{x-a}\right\}.$$

Division by a negative power of $\dfrac{1}{x-a}$ shows that**

$$L = E_n\left(\frac{1}{x-a}\right) = \frac{1}{(a^2-1)(a+\sqrt{a^2-1})^n} \tag{4}$$

and

$$P_n(x) = \frac{1}{x-a} - \frac{\cos[n \arccos x + \phi(x)]}{(a^2-1)(a+\sqrt{a^2-1})^n}. \tag{5}$$

** S. N. Bernstein [15].

The given method of solving the problem of the best approximation, which has been used already in section 2.9.1, can also be applied in some other cases (see 2.13.31).

2.11.2. Let us consider a continuous function $f(x)$ of period 2π with the Fourier series

$$f(x) = \sum_{k=0}^{\infty} a_k \cos n_k x \tag{6}$$

subject to the conditions that $a_k > 0$ and $\dfrac{n_{k+1}}{n_k} = 2p_k + 1$, where p_k is a positive integer**. It is obvious that the difference attains on $[0,2\pi]$ maxima of its modulus, equal to $\sum\limits_{k=\nu+1}^{\infty} a_k$, with consecutive changes of sign at the $2n_{\nu+1}$ points $x_1 = \dfrac{i\pi}{n_{\nu+1}}$ $(i = 0, 1, \ldots, 2n_{\nu+1}-1)$. Hence by theorem 2.7.32, *for any integer n the trigonometric polynomial of order $\leqslant n$ which deviates least from $f(x)$ is the polynomial*

$$T_n(f; x) = \sum_{k=0}^{\nu} a_k \cos n_k x \qquad (n_\nu \leqslant n < n_{\nu+1}) \tag{7}$$

and

$$E_n^*(f) = \sum_{k=\nu+1}^{\infty} a_k. \tag{8}$$

2.11.21. The function 2.11(6) possesses the remarkable property that for any $n > 0$ its trigonometric polynomials of best uniform approximation of order $\leqslant n$ are the corresponding partial sums of the Fourier series, i.e. polynomials of best mean square approximation (see section 2.8.2). It turns out that other even functions of period 2π with positive Fourier coefficients and possessing this property do not exist***. *If for any n the trigonometric polynomial of best uniform approximation of order $\leqslant n$ to the function $f(x)$ with Fourier series* 2.11(6), *in which $a_k > 0$* $(k = 0, 1, 2, \ldots)$, *is identical with the partial sum of this series of the same order, then $\dfrac{n_{k+1}}{n_k} = 2p_k + 1$, where p_k is a positive integer.*

** With this condition the series 2.11(6) always converges absolutely. See A. Zygmund [1], Chap. VI.

$$f(x) - \sum_{k=0}^{\nu} a_k \cos n_k x = \sum_{k=\nu+1}^{\infty} a_k \cos n_k x$$

*** S. N. Bernstein [15].

In fact, since

$$|f(x)-T_{n_{k-1}}(f;x)|\leqslant E^*_{n_{k-1}}(f),\quad |f(x)-T_{n_{k+1}}(f;x)|\leqslant E^*_{n_{k+1}}(f)$$

and

$$\max_x |a_{k+1}\cos n_{k+1}x+a_k\cos n_k x| = a_k+a_{k+1} = E^*_{n_{k-1}}(f)-E^*_{n_{k+1}}(f)>0,$$

among trigonometric polynomials of order $\leqslant n_k-1$ the polynomial $T_{n_{k-1}}(f;x)$ is the one which deviates least from $T_{n_{k+1}}(f;x)$. If this were not so, the inequality

$$\max_x |T_{n_{k+1}}(f;x)-T^*_{n_k-1}(x)|<a_k+a_{k+1}$$

would hold for some trigonometric polynomial $T^*_{n_k-1}(x)$, from which it would follow that

$$|f(x)-T^*_{n_k-1}(x)|<a_k+a_{k+1}+E^*_{n_{k+1}}(f)=E^*_{n_{k-1}}(f)=E^*_{n_k-1}(f),$$

and this is impossible. Thus the sum $a_k\cos n_k x+a_{k+1}\cos n_{k+1}x$ must by theorem 2.7.32 attain the value a_k+a_{k+1} at $2n_k$ points at least, with consecutive changes of sign on $[0,2\pi)$. But this can only occur when $\cos\frac{n_{k+1}}{n_k}\nu\pi=(-1)^\nu$, i.e. when $\frac{n_{k+1}}{n_k}$ is an odd number.

2.11.3. Let

$$f(x)=\frac{4}{\pi}\sum_{k=0}^{\infty}\frac{\sin\left[(2k+1)(n+1)x-\frac{r\pi}{2}\right]}{(2k+1)^{r+1}}\tag{9}$$

where r is a positive integer. At the $2n+2$ points

$$x_\nu=\frac{\nu\pi}{n+1}+\frac{\pi}{4(n+1)}[1+(-1)^r]\qquad(\nu=0,1,\ldots,2n+1)$$

of the half-interval $[0,2\pi]$ this function assumes with consecutive changes of sign the value

$$K_r=\frac{4}{\pi}\sum_{k=0}^{\infty}\frac{(-1)^{k(r+1)}}{(2k+1)^{r+1}},\tag{10}$$

which, as is not difficult to verify (using Rolle's theorem), represents its maximum modulus.

Hence by theorem 2.7.32, among the trigonometric polynomials of order $\leqslant n$ the polynomial which deviates least from it is identically equal to zero, and $E^*_n(f)=K_r$.

2.11.4. Let us also consider an example of the exact solution of the problem of the best approximation in the metric L and let us find the best approximation in the mean ($q = 1$) by trigonometric polynomials of order $\leqslant n$ to the function**

$$\phi_r(t) = \sum_{k=1}^{\infty} (-1)^{k-1} \frac{\cos\left(kt - \frac{r\pi}{2}\right)}{k^r} \quad (r > 0 \text{ an integer}). \tag{11}$$

For this purpose let us note that whatever the number c_ν may be, the difference

$$\phi_r(t) - \sum_{\nu=0}^{n} c_\nu \cos\left(\nu t - \frac{r\pi}{2}\right)$$

cannot have more than $n+1$ zeros within the interval $(0, \pi)$.

If this were not so, then after r differentiations we should arrive at the polynomial

$$\frac{1}{2} - \sum_{\nu=1}^{n} \nu^r c_\nu \cos \nu t,$$

which in virtue of Rolle's theorem must have in the given interval not less than $n+1$ zeros, and this is impossible.

Hence it follows that the polynomial whose coefficients satisfy the system of equations

$$\phi_r(t_k) - \sum_{\nu=0}^{n} c_\nu \cos\left(\nu t_k - \frac{r\pi}{2}\right) = 0 \quad \left(t_k = \frac{2k+1}{2(n+1)}\pi + \frac{r\pi}{2(n+1)}\right)$$

will change sign where the function $\cos\left[(n+1)t - \frac{r\pi}{2}\right]$ changes sign, and only there, and hence by theorem 2.8.11 it will be the best mean polynomial approximation on $[0, \pi]$. From the last remark of section 2.2.2 it follows that the same polynomial will be best on $[0, 2\pi]$. Thus by theorem 2.8.11

$$E_n^*(\varphi_r)_L = \left|\int_0^{2\pi} \phi_r(t) \operatorname{sign} \cos\left[(n+1)t - \frac{r\pi}{2}\right] dt\right|$$

$$= \frac{4}{\pi} \int_0^{2\pi} \sum_{k=1}^{\infty} (-1)^{k-1} \frac{\cos\left(kt - \frac{r\pi}{2}\right)}{k^r} \sum_{k=0}^{\infty} (-1)^{k(r+1)} \times$$

$$\times \frac{\cos\left[(2k+1)(n+1)t - \frac{r\pi}{2}\right]}{2k+1} dt = \frac{\pi K_r}{(n+1)^r} = E_n^*[\phi_r(t+\pi)]_L. \tag{12}$$

** Favard [1]. The case of fractional r $(0 < r < 1)$ was considered by V. K. Ddzyadyk [2] and Sun Yun-shen [1] $(r > 6)$.

2.11.5. The result of section 2.11.4 can be generalised if in place of the function $\phi_r(t+\pi)$ we consider the function

$$F(x) = \sum_{k=0}^{\infty} a_k \cos\left(kx - \frac{r\pi}{2}\right) \qquad (r \text{ an integer}) \tag{13}$$

with Fourier coefficients satisfying some special conditions; the exac solution of the given problem of the best mean ($q = 1$) approximation to the function $F(x)$ by trigonometric polynomials on the segment $[0, 2\pi]$ then reduces to the direct application of theorem 2.8.11. For definiteness, let us consider the case of even r (for odd r see theorem 2.13.32). By theorem 2.8.11 the solution is found in this case if, for the function 2.11(13), we succeed in finding a polynomial

$$T_n(x) = \frac{a_0^{(0)}}{2} + \sum_{k=1}^{n} a_k^{(0)} \cos kx \tag{14}$$

possessing the property that

$$F(x) - T_n(x) = \cos(n+1)x \cdot f(x), \tag{15}$$

where $f(x)$ is a non-negative integrable function. Let us take an arbitrary convex sequence of numbers

$$\{c_k\} \quad (k = 0, 1, 2, \ldots; \quad c_k \geqslant c_{k+1}, \quad c_k - 2c_{k+1} + c_{k+2} \geqslant 0)$$

which decreases monotonically to zero, and consider the series

$$\frac{c_0}{2} + \sum_{k=1}^{\infty} c_k \cos kx. \tag{16}$$

On applying Abel's transformation twice, it can be seen that for all $x \neq 2\nu\pi$ the given series converges to a negative integrable function** $f(x)$, and consequently is its Fourier series. Then

$$2\cos mx \cdot f(x) = c_0 \cos mx + \sum_{k=1}^{\infty} c_k[\cos(m+k)x + \cos(m-k)x]$$

$$= c_m + \sum_{k=1}^{m-1} (c_{m-k} + c_{m+k}) \cos kx + (c_0 + c_{2m}) \cos mx +$$

$$+ \sum_{k=m+1}^{\infty} (c_{k-m} + c_{k+m}) \cos kx.$$

The numbers c_k ($k = 0, 1, 2, \ldots$) can be determined in such a way that

$$c_{k-m} + c_{k+m} = a_k \qquad (k = m, m+1, m+2, m+3, \ldots).$$

** See Zygmund [1], section 5.12.

From the system of equations

$$c_{2\nu m+k}+c_{2(\nu+1)m+k} = a_{(2\nu+1)m+k} \qquad (k = 0, 1, 2, \ldots)$$

we find

$$c_k = \sum_{\nu=0}^{\infty} (-1)^\nu a_{(2\nu+1)m+k} \qquad (k = 0, 1, 2, \ldots). \tag{17}$$

Hence it follows that if $a_k \geqslant a_{k+1}$, $a_k-2a_{k+1}+a_{k+2} \geqslant 0$, $a_k-3a_{k+1}+ +3a_{k+2}-a_{k+3} \geqslant 0 (k \geqslant 0)$, then the sequence $\{c_k\}$ is positive, decreases monotonically to zero, and is convex. Choosing $m = n+1$ we thus find** that *if the Fourier coefficients of the function* 2.11(13) *form a non-negative triply monotonic sequence, the polynomial*

$$T_n(x) = \sum_{k=0}^{n} a_k \cos kx - c_{n+1} - \sum_{k=1}^{n} (c_{n-k+1} c_{n+k+1}) \cos kx, \tag{18}$$

where the numbers c_k are given by equations 2.11.(17), *satisfies condition* 2.11(15) *for even r and in consequence of this deviates least from* $(-1)^{\frac{r}{2}} F(x)$ *in the metric L on* $[0, 2\pi]$. *By* 2.8(6) *the magnitude of the best approximation is determined by the relation*

$$E_n^*(F)_L = \left| \int_0^{2\pi} F(t) \operatorname{sign} \cos(n+1)t \, dt \right| = 4 \sum_{\nu=0}^{\infty} (-1)^\nu \frac{a_{(2\nu+1)(n+1)}}{2\nu+1}. \tag{19}$$

2.12. On some criteria for the best approximation in an infinite interval

For the best approximations considered in section 2.6 a series of properties can be indicated which are similar to those given in sections 2.7, 2.8 and 2.10.

2.12.1. Let $f(x)$ be a function which is bounded and continuous on the real axis (or semi-axis)***. *Among all the proper continuous rational functions*

$$R_n(x) = \frac{\sum_{k=0}^{n} a_k x^k}{\sum_{k=0}^{n} b_k x^k} \tag{1}$$

** B. Sz. Nagy [1].

*** See N. I. Akhiezer [3].

of degree not higher than n the one which deviates least from $f(x)$ is the irreducible rational function

$$R_m(f;x)=\frac{\sum\limits_{k=0}^{m} a_k^{(0)}x^k}{\sum\limits_{k=0}^{m} b_k^{(0)}x^k} \qquad (m\leqslant n,\ b_m^{(0)}\neq 0) \tag{2}$$

*for which the difference $f(x)-R_m(f;x)$ assumes the value $L=\sup\limits_x |f(x)-R_m(f;x)|$ on this interval with alternating signs, at least at $n+m+2$ points***. The proof of this proposition is in part similar to the proof of theorem 2.7.2, and we will omit it***.

2.12.2. *If for the irreducible rational function* 2.12(2) *the difference $f(x)-R_m(f;x)$ assumes the non-zero values $a_0, a_1, \dots a_{n+m+1}$ with alternating signs at the $n+m+2$ distinct points $x_0<x_1<\dots<x_{n+m+1}$ of the interval considered**, then the best approximation $\varrho_n(f)$ of the function $f(x)$ by the rational functions* 2.12(1) *of degree not greater than n on this interval satisfies the inequality*

$$\varrho_n(f)\geqslant\mu=\min\{|a_0|,|a_1|,\dots,|a_{n+m+1}|\}. \tag{3}$$

This is proved by assuming the contrary (cf. section 2.10).

2.12.3. Let $f(x)$ be continuous and $g_\sigma(f;x)$ an integral function of degree $\leqslant\sigma$, for which $\sup\limits_{-\infty<x<\infty}|f(x)-g_\sigma(f;x)|=A_\sigma(f)$. Reasoning as in section 2.7.1, we note that**** *there does not exist an integral function $\phi_\sigma(x)$ of degree $\leqslant\sigma$, bounded on the whole real axis for which, for some $\varepsilon>0$, we have simultaneously*

$$|\phi_\sigma(x)|\geqslant 1,\quad \phi_\sigma(x)\{f(x)-g_\sigma(f;x)\}>0$$

at all the points x where

$$A_\sigma(f)-\varepsilon\leqslant|f(x)-g_\sigma(f;x)|\leqslant A_\sigma(f).$$

Conversely, if for an integral function $g_\sigma(x)$ of degree $\leqslant\sigma$

$$\sup_{-\infty<x<\infty}|f(x)-g_\sigma(x)|>A_\sigma(f),$$

then for a sufficiently small $\varepsilon>0$ it is possible to find a bounded integral function $\phi_\sigma(x)$ of degree $\leqslant\sigma$ for which

$$|\phi_\sigma(x)|\geqslant 1,\quad \phi_\sigma(x)\{f(x)-g_\sigma(x)\}>0 \tag{4}$$

** In the case considered one of these points may be the point at infinity.

*** See N. I. Akhiezer [3], where there is also a proof of the uniqueness of the best rational function approximation.

**** S. N. Bernstein [24]. Here the function $f(x)$ may also be unbounded on the whole real axis.

wherever

$$|f(x)-g_\sigma(x)| \geqslant \sup_{-\infty<x<\infty} |f(x)-g_\sigma(x)|-\varepsilon.$$

From this there follows the assertion**: *if $g_\sigma(x)$ possesses the property that there does not exist an integral function $\phi_\sigma(x)$ of degree $\leqslant \sigma$ which satisfies the inequality* 2.12(4) *at all the points x where $|f(x)-g_\sigma(x)| = \sup\limits_{-\infty<t<\infty} |f(t)-g_\sigma(t)|$, then $g_\sigma(x) = g_\sigma(f; x)$,* i.e.

$$A_\sigma(f) = \sup_{-\infty<x<\infty} |f(x)-g_\sigma(x)|.$$

2.12.4. It is convenient to formulate the last assertion by introducing he concept of a complete set of degree σ. A monotonic sequence of points, $\{x_k\}$ $(k = \ldots -3, -2, -1, 0, 1, 2, 3, \ldots)$, following S. N. Bernstein***, is said to be complete of degree σ if there does not exist an integral function $\phi_\sigma(x)$ of degree $\leqslant \sigma$, bounded on the whole real axis, for which

$$(-1)^k\phi_\sigma(x_k) \geqslant 1 \quad (k = \ldots -3, -2, -1, 0, 1, 2, 3, \ldots).$$

Thus if the integral function $g_\sigma(x)$ of degree $\leqslant \sigma$ possesses the property that the difference $f(x)-g_\sigma(x)$ assumes the value $\sup\limits_{-\infty<t<\infty} |f(t)-g_\sigma(t)|$ with consecutive changes of sign at the points of some complete set of degree σ, then among all integral functions of degree not higher than σ the function $g_\sigma(x)$ deviates least from $f(x)$ on the whole real axis.

As an example of a complete set of points of degree $\sigma \geqslant 0$ we have the sequence $x_k = \dfrac{k\pi}{[\sigma]+1}+\alpha$ $(k = 0, \pm 1, \pm 2, \ldots)$. In fact, in virtue of theorems 2.6.21 and 2.9.2, a function which is identically zero deviates least on the whole real axis, among integral functions of degree $\leqslant \sigma$, from the function $f(x) = \cos([\sigma]+1)(x-\alpha)$, and hence $A_\sigma(f) = 1$. In chapter IV (see 4.8.2) it is established that any arbitrary integral function of exponential type is also bounded. Hence it follows that if for some function $\phi_\sigma(x)$ of degree $\leqslant \sigma$ bounded on $(-\infty, +\infty)$ the inequality $(-1)^k\phi_\sigma(x_k) \geqslant 1$ $(k = 0, \pm 1, \pm 2, \ldots)$ were to be satisfied, then for sufficiently small $\varepsilon > 0$ we should have****

$$\sup_{-\infty<x<\infty} |\cos([\sigma]+1)(x-\alpha)-\varepsilon\phi_\sigma(x)| < 1,$$

and this is impossible.

** S. N. Bernstein [24].

*** S. N. Bernstein [24]. Other criteria for integral functions which give the best approximation were indicated by N. I. Akhiezer [5].

**** Since $(-1)^k\phi_\sigma(x_k) \geqslant (-1)^k \cos([\sigma]+1)(x_k-\alpha) = 1$, and $\sup|\phi'_\sigma(x)| \leqslant M$ $(-\infty < x < \infty)$, where M is some constant, there exists a $\delta > 0$ $(\delta < \pi/2)$, such that $\operatorname{sign}\phi_\sigma(x) = \operatorname{sign}\cos([\sigma]+1)(x-\alpha)$ for $|x-x_k| \leqslant \delta (k = 0, \pm 1, \pm 2, \ldots)$. For the remaining values of x $|\cos([\sigma]+1)(x_k-\alpha)| \leqslant \cos\delta$.

2.12.5. *If for some integral function $g_\sigma(x)$ of degree $\leqslant \sigma$ the difference $f(x)-g_\sigma(x)$ is bounded on the whole real axis and at the points x_k of a certain complete set of degree σ*

$$(-1)^k[f(x_k)-g_\sigma(x_k)] \geqslant \mu > 0, \tag{5}$$

then $A_\sigma(f) \geqslant \mu$.

If for some $\varepsilon > 0$ we were to have $A_\sigma(f) = \sup\limits_{-\infty<x<\infty} |f(x)-g_\sigma(f;x)| = \mu-\varepsilon$, then the integral function $\phi_\sigma(x) = \dfrac{1}{\varepsilon}[g_\sigma(f;x)-g_\sigma(x)]$ would satisfy the inequalities $(-1)^k\phi_\sigma(x_k) \geqslant 1$ at all the points x_k and the sequence $\{x_k\}$ would not form a complete set.

2.12.6. Let us derive criteria for best integral approximations on an infinite interval, similar to that indicated in section 2.8.25. Let $B_\sigma(L_q)$ $(1 \leqslant q < \infty)$ be the class of all integral functions $g_\sigma(x)$ of degree $\leqslant \sigma$ for which

$$\|g_\sigma(x)\|_{L_q} = \left\{ \int_{-\infty}^{\infty} |g_\sigma(x)|^q \, dx \right\}^{\frac{1}{q}} < \infty,$$

and let $f(x)$ be an arbitrary measurable function with $\|f(x)\|_{L_q} < \infty$.

In order that among all functions of the class $B_\sigma(L_q)$ the function $g_\sigma(f;x) \in B_\sigma(L_q)$ should deviate least in the metric L_q from the function $f(x)$, it is sufficient and (for $q = 1$, in the case where the difference $f(x)-g_\sigma(f;x)$ is different from zero almost everywhere) necessary, that for any function $g_\sigma(x) \in B_\sigma(L_q)$ the equality

$$\int_{-\infty}^{\infty} |f(x)-g_\sigma(f;x)|^{q-1} g_\sigma(x) \operatorname{sign} \{f(x)-g_\sigma(f;x)\} \, dx = 0 \tag{6}$$

should be satisfied.

If the condition 2.12(6) is satisfied, then from the fact that

$$\begin{aligned}
&\int_{-\infty}^{\infty} |f(x)-g_\sigma(f;x)|^q \, dx \\
&\quad = \int_{-\infty}^{\infty} \{f(x)-g_\sigma(f;x)\} \, |f(x)-g_\sigma(f;x)|^{q-1} \operatorname{sign} \{f(x)-g_\sigma(f;x)\} \, dx \\
&\quad = \int_{-\infty}^{\infty} [f(x)-g_\sigma(x)] \, |f(x)-g_\sigma(f;x)|^{q-1} \operatorname{sign}[f(x)-g_\sigma(f;x)] \, dx \leqslant \\
&\quad \leqslant \int_{-\infty}^{\infty} |f(x)-g_\sigma(x)| \, |f(x)-g_\sigma(f;x)|^{q-1} \, dx \leqslant \\
&\quad \leqslant \|f(x)-g_\sigma(x)\|_{L_q} \, \|f(x)-g_\sigma(f;x)\|_{L_q}^{q-1}
\end{aligned}$$

the inequality

$$\|f(x)-g_\sigma(f;x)\|_{L_q} \leqslant \|f(x)-g_\sigma(x)\|_{L_q}$$

follows for any function $g_\sigma(x)$ of $B_\sigma(L_q)$, i.e. $g_\sigma(f;x)$ deviates least from $f(x)$ in the metric L_q.

Let us now verify that if $g_\sigma(f;x)$ is the best integral function approximation, then (for $q=1$, in the case where the difference $f(x)-g_\sigma(f;x)$ is different from zero almost everywhere), 2.12 (6) is satisfied for any function $g_\sigma(x) \in B_\sigma(L_q)$.

In fact, if for some function $g_\sigma(x) \in B_\sigma(L_q)$

$$\int_{-\infty}^{\infty} g_\sigma(x)\,|f(x)-g_\sigma(f;x)|^{q-1} \operatorname{sign}\{f(x)-g_\sigma(f;x)\}\,dx = \varrho \neq 0,$$

then for a sufficiently great $a>0$ we should have

$$\int_{|x|\geqslant a} |g_\sigma(x)|\cdot|f(x)-g_\sigma(f;x)|^{q-1}\,dx < \tfrac{1}{5}[\varrho].$$

Let us choose $\delta>0$ such that for any set $E\subset[-a,a]$ with $\operatorname{mes} E<\delta$

$$\int_E |g_\sigma(x)|\cdot|f(x)-g_\sigma(f;x)|^{q-1}\,dx < \tfrac{1}{5}|\varrho|.$$

In virtue of the C-property of N. N. Luzin it is possible to find in $[-a,a]$ a closed set Q such that $\operatorname{mes} Q > 2a-\delta$ and $f(x)$ is continuous on Q. We may also suppose that $f(x)-g_\sigma(f;x)\neq 0$ on Q. Let $\alpha = \min_{x\in Q}|f(x)-g_\sigma(f;x)|$. Let us choose ε so small that the inequality $|\varepsilon g_\sigma(x)| < \alpha$ is satisfied everywhere on $[-a,a]$. We then have

$$\left|\int_{-a}^{a} g_\sigma(x)\,|f(x)-g_\sigma(f;x)|^{q-1}\operatorname{sign}\{f(x)-g_\sigma(f;x)-\varepsilon g_\sigma(x)\}\,dx\right| > \tfrac{1}{4}|\varrho|.$$

It is obvious that for a value of ε sufficiently small in absolute value the inequality

$$\left|\int_{-a}^{a} g_\sigma(x)\,|f(x)-g_\sigma(f;x)-\varepsilon g_\sigma(x)|^{q-1}\operatorname{sign}\{f(x)- -g_\sigma(f;x)-\varepsilon g_\sigma(x)\}\,dx\right| > \tfrac{1}{5}|\varrho|$$

is also satisfied.

In case of necessity again decreasing the absolute magnitude of ε and choosing its sign appropriately, we obtain

$$\varepsilon\int_{-\infty}^{\infty} g_\sigma(x)\,|f(x)-g_\sigma(f;x)-\varepsilon g_\sigma(x)|^{q-1}\operatorname{sign}\{f(x)- -g_\sigma(f;x)-\varepsilon g_\sigma(x)\}\,dx > 0.$$

But then

$$\int_{-\infty}^{\infty} |f(x)-g_\sigma(f; x)-\varepsilon g_\sigma(x)|^q \, dx$$

$$= \int_{-\infty}^{\infty} [f(x)-g_\sigma(f; x)] \, |f(x)-g_\sigma(f; x)-\varepsilon g_\sigma(x)|^{q-1} \times$$

$$\times \operatorname{sign}\{f(x)-g_\sigma(f; x)-\varepsilon g_\sigma(x)\} \, dx -$$

$$-\varepsilon \int_{-\infty}^{\infty} g_\sigma(x) \, |f(x)-g_\sigma(f; x)-\varepsilon g_\sigma(x)|^{q-1} \times$$

$$\times \operatorname{sign}\{f(x)-g_\sigma(f; x)-\varepsilon g_\sigma(x)\} \, dx <$$

$$< \int_{-\infty}^{\infty} |f(x)-g_\sigma(f; x)| \, |f(x)-g_\sigma(f; x)-\varepsilon g_\sigma(x)|^{q-1} \leqslant$$

$$\leqslant \|f(x)-g_\sigma(f; x)\|_{L_q} \|f(x)-g_\sigma(f; x)-\varepsilon g_\sigma(x)\|_{L_q}^{q-1}.$$

Consequently

$$\|f(x)-g_\sigma(f; x)-\varepsilon g_\sigma(x)\|_{L_q} < \|f(x)-g_\sigma(f; x)\|_{L_q},$$

and we arrive at a contradiction of the assumptions made with respect to $g_\sigma(f; x)$.

2.13. Various problems and theorems

1. Let $x_1, x_2, \ldots, x_n$ be a linearly independent system of elements of the linear normed space F and

$$d_k = \inf_{\lambda_\nu} \|x_k - \sum_{\substack{\nu=1 \\ \nu \neq k}}^{n} \lambda_\nu x_\nu\|.$$

If $\|\sum_{\nu=1}^{n} c_\nu x_\nu\| \leqslant M$ and $d = \min_\nu d_\nu$, then $|c_\nu| \leqslant \frac{M}{d}$.

2. Let G be a certain region in an m-dimensional space with the variables $t_1, t_2, \ldots, t_m$ and $f(t_1, \ldots, t_m)$ a measurable function with finite upper bound. Then

$$\lim_{q\to\infty} \left\{\int \cdots \int_{\bar{G}} |f(t_1, \ldots, t_m)|^q \, dt_1 \ldots dt_m\right\}^{\frac{1}{q}} = \operatorname*{vrai\,sup}_{\bar{G}} |f(t_1, \ldots, t_m)|. \tag{1}$$

3. Let G be a bounded region of an m-dimensional space in the variables $t_1, \ldots, t_m$. The quantity $\left\{\frac{1}{\operatorname{mes} G} \int \cdots \int_G |f(t_1, \ldots, t_m)|^q \, dt_1 \ldots dt_m\right\}^{\frac{1}{q}}$ is a non-decreasing function of q.

Hausdorff [1]. Use Hölder's inequality.

4. Let

$$f(t) = \begin{cases} 1, & 0 \leqslant t \leqslant \frac{2}{3}, \\ 0, & \frac{2}{3} < t \leqslant 1; \end{cases} \quad x_0(t) = \begin{cases} 0, & 0 \leqslant t \leqslant \frac{1}{3}, \\ 1, & \frac{1}{3} < t \leqslant 1. \end{cases}$$

In the given case

$$E_0(f)_L = \inf_c \int_0^1 |f(t) - cx_0(t)| \, dt = \tfrac{2}{3},$$

and there are infinitely many functions of the form $cx_0(t)$ for which $\|f(t) - cx_0(t)\|_L = E_0(f)_L = \frac{2}{3}$ (see section 2.2.8).

5. Let $f(x, y)$ be an arbitrary function, continuous in a given closed bounded region $\overline{G}$. There exist two polynomials $P(x, y) = a_0x + b_0y + c_0$ and $Q(x, y) = a_1x + b_1y + c_1$ such that any polynomial $R(x, y) = ax + by + c$ which, among all other polynomials of not more than the first degree, deviates least from $f(x, y)$ on G can be represented in the form $R(x, y) = \alpha P(x, y) + \beta Q(x, y)$.

M. G. Krein. See section 2.13.6.

6. The assertion of 2.13.5 follows directly from the following generalization of a theorem of Haar (see section 2.3.3).

Let G be a certain compact metric space of points M and $\phi_k(M)$ $(k = 0, 1, \ldots, n)$ a linearly independent system of functions continuous on G. The set of polynomials $P_n(M) = \sum_{\nu=0}^{\infty} c_\nu \phi_\nu(M)$, deviating least from the set of functions $f(M)$ continuous on G (see section 2.2.6) has for all such f not more than $k (0 \leqslant k \leqslant n)$ dimensions if and only if any system of $k+1$ linearly independent polynomials $P_n(M)$ has not more than $n-k$ common roots on G.

G. Sh. Rubinstein [1]. The proof can be derived in a similar way to the proof of theorem 2.3. See S. I. Zukhovitskii [1].

7. Let $x_0, x_1, \ldots x_{n1} \ldots$ be a sequence of linearly independent elements of a linear normed space F and

$$E_0 > E_1 > E_2 > \ldots > E_n > \ldots$$

an arbitrary monotonically decreasing sequence of numbers. For any natural number n there exist 2^{n+1} elements $x \in F$ which possess the property that $E_k(x) = \inf_{c_\nu} \|x - \sum_{\nu=0}^{k} c_\nu x_\nu\| = E_k (k = 0, 1, \ldots, n)$. See 2.5. Use the property 2.5,(3).

8. If $-\infty < a \leqslant c < d < b < \infty$, then there does not exist a continuous function $f(x) \not\equiv$ constant for which $E_n(f; a, b) = E_n(f; c, d)$ for all values $n = 0, 1, 2, \ldots$.

S. N. Bernstein. Consider the case $a = 0$, $b = 1$ and apply theorem 2.7.2.

9. For any algebraic polynomial $P_n(x)$ of the type 2.9(1) we have the inequality

$$\int_{-1}^{1} \frac{P_n^2(x)}{\sqrt{1-x^2}} \, dx \geqslant \frac{\pi}{2^{2n-1}}, \tag{2}$$

which becomes an equality only when $P_n(x) = \dfrac{1}{2^{n-1}} \cos n \arccos x$.

Expand the polynomial $P_n(x)$ in terms of Chebyshev polynomials and use the remark 2.8.22,(3).

10. If x is real and $|x| > 1$, then $\cos n \arccos x = \frac{1}{2}\left(R^n + \frac{1}{R^n}\right)$, where R is the sum of the semi-axes of an ellipse which passes through the point x and has foci at the points $+1$ and -1. See 2.9(5).

11. Let $N_n(\alpha, \beta)$ be the number of zeros of the Chebyshev polynomial $\cos n \arccos x$ which occur in the interval $[\alpha, \beta[$. At every point $x \in (-1, 1)$

$$\sigma(x) = \lim_{\Delta x \to 0} \frac{1}{\Delta x} \lim_{n\to\infty} \frac{N_n(x, x+\Delta x)}{N} = \frac{1}{\pi\sqrt{1-x^2}}. \tag{3}$$

See V. L. Goncharov [1] or I. P. Natanson [2].

12. For any $n \geqslant 0$ the polynomials 2.9(5) and 2.9(12) are connected by the relation

$$\frac{\sin(n+1)\arccos x}{\sqrt{1-x^2}} = \frac{1}{\pi}\int_{1-}^{1} \frac{\cos(n+1)\arccos x - \cos(n+1)\arccos t}{(x-t)\sqrt{1-t^2}}\,dt. \tag{4}$$

See V. L. Goncharov [1], page 154.

13. Let Q be a bounded closed set of points z in the complex plane and

$$d_n(Q) = \min_{c_0, c_1, \ldots, c_{n-1}} \max_{z \in Q} |z^n + c_{n-1}z^{n-1} + \ldots + c_1 z + c_0|. \tag{5}$$

(1) As $n \to \infty$ the limit

$$\lim_{n\to\infty} \sqrt[n]{d_n(Q)} = d(Q) \tag{6}$$

exists.

(2) The number $d(Q)$, known as the transfinite diameter of the set Q, is identical with the limit

$$\lim_{n\to\infty} \left\{ \max_{z_1, \ldots, z_n \in Q} \prod_{\substack{k, i=1 \\ k<i}}^{n} |z_k - z_i| \right\}^{\frac{2}{n(n+1)}} \tag{7}$$

(3) The transfinite diameter of a finite segment is equal to a quarter of its length

(4) The transfinite diameter of a circular area is equal to its radius.

(5) The transfinite diameter of the set Q consisting of the two segments $[-1, -\alpha]$ $[\alpha, 1]$, where $0 < \alpha < 1$, is equal to $\frac{1}{2}\sqrt{1-\alpha^2}$.

For (1)–(4) see Fekete [1]; for (5) see N. I. Akhiezer [3].

14. If a is real and $|a| > 1$, then any polynomial

$$P_n(x) = \sum_{k=0}^{n} c_k x^k,$$

with the property $P_n(a) = 1$ satisfies the inequality

$$\max_{-1 \leqslant x \leqslant 1} |P_n(x)| \geqslant \frac{1}{|\cos n \arccos a|},$$

which becomes an equality if and only if

$$P_n(x) = \frac{\cos n \arccos x}{\cos n \arccos a}. \tag{8}$$

See V. L. Goncharow [1]. Consider the difference between $P_n(x)$ and the polynomial 2.13(8).

15. The sequence of Chebyshev polynomials $T_n(x) = \cos n \arccos x$ satisfies the recurrence formula

$$T_n(x) = 2xT_{n-1}(x) - T_{n-2}(x) \quad (n \geqslant 2). \tag{9}$$

16. Chebyshev polynomials satisfy the differential equation

$$(1-x^2)\, T_n''(x) - xT_n'(x) + n^2 T_n(x) = 0. \tag{10}$$

17. For $x \in [-1, 1]$,

$$\frac{1-tx}{1-2tx+t^2} = \sum_{k=0}^{\infty} t^k T_k(x) \quad (-1 < t < 1). \tag{11}$$

18. The sequence of polynomials 2.9(12) satisfies the recurrence formula

$$P_n(x) = xP_{n-1}(x) - \tfrac{1}{4}P_{n-2}(x).$$

19. The sequence of polynomials 2.9(15) satisfies the recurrence formula

$$P_n(x) = xP_{n-1}(x) - \frac{(n-1)^2}{(2n-1)(2n-3)} P_{2n-2}(x) \tag{12}$$

for $a = -1$, $b = 1$.

See I. P. Natanson [2].

20. The polynomials 2.9(15) satisfy the differential equation

$$(1-x^2)P_n''(x) - 2xP_n'(x) + n(n+1)P_n(x) = 0 \tag{13}$$

for $a = -1$, $b = 1$.

See I. P. Natanson [2].

21. For $a = -1$, $b = 1$ the polynomials 2.9(15) are the coefficients of the expansion

$$\frac{1}{\sqrt{1-2tx+t^2}} = \sum_{n=0}^{\infty} \frac{1.3 \ldots (2n-1)}{n!} P_n(x)t^n \tag{14}$$

and satisfy the inequalities

$$\begin{aligned}
&(1)\quad \frac{1.3 \ldots (2n-1)}{n!} |P_n(x)| \leqslant 1;\\
&(2)\quad \frac{1.3 \ldots (2n-1)}{n!} |P_n(x)| \leqslant \sqrt{\frac{\pi}{2n(1-x^2)}};\\
&(3)\quad \frac{1.3 \ldots (2n-1)}{n!} |P_n'(x)| \leqslant \tfrac{1}{2}n(n+1);\\
&(4)\quad \frac{2n+1}{2n-1} P_{n-1}(x)\, P_{n+1}(x) \leqslant \frac{n+1}{n} P_n^2(x).
\end{aligned} \tag{15}$$

For 2.13(14) and 2.13(15), (1) see I. P. Natanson [2]; for 2.13(15), (2) and (3) see G. Szegö [1]; see also F. Tricomi [1]. The inequality 2.13(15), (4) is due to P. Turán.

22. If $q > 1$ is integral, then for any polynomial

$$P_n(x) = x^n + c_{n-1}x^{n-1} + \ldots + c_1 x + c_0$$

with integral coefficients the inequality

$$\max_{-\frac{1}{q} \leqslant x \leqslant \frac{1}{q}} |P_n(x)| \geqslant \frac{1}{q^n}, \tag{16}$$

holds, and for $P_n(x) = x^n$ this becomes an equality.

Consider $P_n(\pm 1/q)$. From the same considerations it follows that for any n the best uniform approximation of $\frac{1}{1-x}$ by the polynomials $P_n(x)$ on $[-\frac{1}{2}, \frac{1}{2}]$ is exactly equal to $\frac{1}{2n}$ and is realised by the corresponding partial sums of its Taylor series $1+x+x^2+x^3+\ldots$

S. N. Bernstein [9].

23. If $0 < \alpha \leqslant \pi$, then

$$\max_{-\alpha\leqslant t\leqslant\alpha} \left|\cos nt + \sum_{k=0}^{n-1} a_k \cos kt\right| \geqslant \left(\sin\frac{\alpha}{2}\right)^{2n} \tag{17}$$

S. N. Bernstein [15]. Make the substitution $x = \cos t$ and use theorem 2.9.1 for $a = \cos\alpha$, $b = 1$.

24. If $f(x) = \sum_{k=0}^{\infty} a_k \cos kx$, then for any $n \geqslant 0$ we have the precise inequality

$$E_n^*(f) \geqslant \tfrac{1}{4}\pi|a_{n+1}|.$$

Let $0 < k < n$ and

$$M_{k,n} = \min_{a_\nu} \max_{t} \left|\cos kt + \sum_{\substack{\nu=0\\ \nu\neq k}}^{n} a_\nu \cos \nu t\right|.$$

Then $\frac{1}{4}\pi < M_{k,n} < 1$, and for any fixed value of k

$$\lim_{n\to\infty} M_{k,n} = \tfrac{1}{4}\pi.$$

S. N. Bernstein [15]. Consider the function

$$f(t) = \sum_{m=0}^{\infty} (-1)^m \frac{\cos(2m+1)t}{2m+1}.$$

25. If $\beta-\alpha < \pi$, then

$$\max_{\alpha\leqslant x\leqslant\beta} |1+a\cos x+b\sin x| \geqslant \tan^2\frac{\alpha-\beta}{4}. \tag{18}$$

See V. L. Goncharov [1]. Use theorem 2.7.3.

26. If in the interval $-\alpha \leqslant x \leqslant \alpha$, where $0 < \alpha < \pi$, the trigonometric polynomial

$$T_n(x) = \sum_{k=0}^{n} (a_k \cos kx + b_k \sin kx)$$

satisfies the inequality $\max_{-\alpha\leqslant x\leqslant\alpha} |T_n(x)| \leqslant 1$, then for any value of x

$$|T_n(x)| \leqslant \frac{1}{2}\left(\tan^{2n}\tfrac{1}{4}\alpha + \frac{1}{\tan^{2n}\frac{1}{4}\alpha}\right). \tag{19}$$

See V. L. Goncharov [1]. The proof is similar to the proof of theorem 2.9.1.

27. If the algebraic polynomial

$$P_n(x) = \sum_{k=0}^{n} c_k x^k$$

satisfies the inequality $\max_{-1 \leqslant x \leqslant 1} |P_n(x)| \leqslant M$, then for all values of $z = x+iy$ on the ellipse with foci at the points ± 1 and with the sum of the semiaxes equal to $R > 1$,

$$|P_n(z)| \leqslant MR^n. \tag{20}$$

See S. N. Bernstein [15]. Consider the function $\dfrac{P_n(z)}{(z+\sqrt{z^2-1})^n}$ and make use of the principle of the maximum modulus from the theory of analytic functions. For real values of z see 2.9(9) and 2.13.9.

28. For $a = -1$, $b = 1$ the polynomials 2.9(15) can be written in the form

$$\frac{1.3 \ldots (2n-1)}{n!} P_n(x) = \frac{1}{\pi} \int_0^\pi (x+\sqrt{x^2-1}\cos t)^n \, dt \tag{21}$$

and

$$\frac{1.3 \ldots (2n-1)}{n!} P_n(\cos t) = \frac{\sqrt{2}}{\pi} \int_0^t \frac{\cos (n+\frac{1}{2}) n}{\sqrt{\cos n - \cos t}} \, du.$$

If $0 < \varepsilon \leqslant t \leqslant \pi - \varepsilon$, then as $n \to \infty$ we have the relation

$$\frac{1.3 \ldots (2n-1)}{n!} P_n(\cos t) = \sqrt{\frac{2}{n\pi \sin t}} \cos \left[\left(n+\frac{1}{2}\right)t - \frac{\pi}{4}\right] + O\left(n^{-\frac{3}{2}}\right).$$

See I. P. Natanson [2]. For the last two formulae see F. Tricomi [1]. The representation 2.13(21) for Legendre polynomials is known as Laplace's formula.

29. The sequence of polynomials 2.9(15) forms an orthogonal system on the segment $[a, b]$, i.e. for $n \neq m$

$$\int_a^b P_n(x) P_m(x) \, dx = 0. \tag{22}$$

See 2.9.32 and 2.8(14).

30. Whatever the real number c may be, for any $\sigma > 0$

$$A_\sigma\left(\frac{1}{x^2+c^2}\right) = \frac{e^{-c\sigma}}{2c^2}. \tag{23}$$

S. N. Bernstein [29]. Consider the sequence of best approximations $E_{2n}\left(\frac{1}{x^2+c^2}; -1, 1\right)$, use formula 2.11(4) and theorem 2.6.22.

31. The method applied for the solution of the problem of the best approximation considered in section 2.11.1 can be used for the exact solution of such problems in certain other cases. If $a > 1$, then for even n

$$E_n\left(\frac{1}{x^2-a^2}; -1,1\right) = E_{n+1}\left(\frac{1}{x^2-a^2}; -1,1\right) = \frac{1}{2a^2(a^2-1)(a+\sqrt{a^2-1})^n} \tag{24}$$

and for odd n

$$E_n\left(\frac{x}{x^2-a^2}; -1,1\right) = E_{n+1}\left(\frac{x}{x^2-a^2}; -1,1\right) = \frac{1}{2a(a^2-1)(a+\sqrt{a^2-1})^n} \tag{25}$$

S. N. Bernstein [15].

32. If the Fourier coefficients of the function

$$F(x) = \sum_{k=1}^{\infty} b_k \sin kx$$

form a non-negative doubly monotonic sequence of numbers ($b_k \geqslant b_{k+1}$, $b_k - 2b_{k+1} + b_{k+2} \geqslant 0$), then

$$E_n^*(F)_L = \int_0^{2\pi} \left| F(t) - T_n(F;t) \right| \mathrm{d}t = \left| \int_0^{2\pi} F(t) \operatorname{sign} \sin (n+1)t \, \mathrm{d}t \right|$$
$$= 4 \sum_{\nu=0}^{\infty} \frac{b_{(2\nu+1)(n+1)}}{2\nu+1}, \tag{26}$$

where $T_n(F;t) = \sum_{k=1}^{n} c_k \sin kt$ and $c_k = b_k - \sum_{\nu=1}^{\infty} [b_{2\nu(n+1)+k} - b_{2\nu(n+1)-k}]$.

B. Sz. Nagy [1]. See section 2.11.5. The proof is similar.

33. If $a \geqslant 1$, then for any $n \geqslant 0$

$$E_n[\ln (a-x); -1,1]_L = 2 \int_{\alpha}^{\infty} \ln \frac{(t+\sqrt{t^2-1})^{n+2}+1}{(t+\sqrt{t^2-1})^{n+2}-1} \, \mathrm{d}t. \tag{27}$$

Consider the function $\ln \frac{b-x}{b+x}$, where $b > a$, and make use of the last remark of section 2.9.31. The best mean approximation by algebraic polynomials of functions of the form $(a-x)^n$ can be calculated in a similar way. See N. I. Akhiezer [3].

34. If $\left\{ \int_a^b \left| f(t) \right|^q \mathrm{d}t \right\}^{\frac{1}{q}} < \infty$ $(1 \leqslant q \leqslant \infty)$, then for $q < \infty$

$$E_n(f)_{L_q} = \min_{c_k} \left\{ \int_a^b \left| f(t) - \sum_{k=0}^{n} c_k t^k \right|^q \mathrm{d}t \right\}^{\frac{1}{q}} = \sup_{\phi(t)} \int_a^b f(t)\phi(t) \, \mathrm{d}t,$$

where the upper bound extends to all the functions $\phi(t)$ integrable on $[a, b]$ for which

$$\int_a^b t^k \phi(t) \, \mathrm{d}t = 0 \quad (k = 0, 1, \ldots, n) \quad \text{and} \quad \left\{ \int_a^b \left| \phi(t) \right|^{\frac{q}{q-1}} \mathrm{d}t \right\}^{\frac{q-1}{q}} = 1,$$

and for $q = \infty$, if $f(x)$ is continuous,

$$E_n(f) = \min_{c_k} \max_{a \leqslant t \leqslant b} \left| f(t) - \sum_{k=0}^{n} c_k t^k \right| = \sup_{g(t)} \int_a^b f(t) \, \mathrm{d}g(t),$$

where the upper bound extends to all functions $g(t)$ which have on $[a, b]$ bounded variation, equal to unity, and are such that

$$\int_a^b t^k \, \mathrm{d}g(t) = 0 \quad (k = 0, 1, \ldots, n).$$

See section 2.10.3.

CHAPTER III

SOME COMPACT CLASSES OF FUNCTIONS AND THEIR STRUCTURAL CHARACTERISTICS

3.1. Compact classes of functions of a single variable. Criteria of the type of Arzelà's theorem. The concepts of the ε-entropy and ε-capacity of a compact set

In connection with theorem 2.5.1 the question of the internal properties of compact sets W of the functional spaces considered and of corresponding criteria equivalent to the condition

$$\sup_{f \in W} E_n(f) \to 0 \quad \text{as} \quad n \to \infty \tag{1}$$

are of interest.

Such criteria are known for some of the frequently encountered function spaces. The simplest of them are tests of the type of Arzelà's theorem**. This theorem corresponds to the classical case relating to the space C of functions of a single real variable continuous on a finite segment, and is formulated in the following way. *The bounded set W of functions $f(x)$ continuous on $[a, b]$ is compact in the sense of uniform convergence if and only if for any $\varepsilon > 0$ it is possible to find a number $\delta > 0$ which possesses the property, that if $|x_1 - x_2| < \delta$, then all the functions $f(x) \in W$ satisfy the inequality $|f(x_1) - f(x_2)| < \varepsilon$.*

3.1.1. Theorem 3.1 can be conveniently formulated by the use of the concept of the *modulus of continuity*. If the function $f(x)$ is bounded on $[a, b]$, then by its modulus of continuity is understood the function

$$\omega(f; t) = \omega(f; a, b; t) = \sup_{\substack{|x_1 - x_2| \leqslant t \\ x_1, x_2 \in [a, b]}} |f(x_1) - f(x_2)|$$

defined for non-negative values of $t \leqslant b - a$.

A necessary and sufficient condition for the continuity of $f(x)$ is that $\omega(f; t)$ should tend to zero as $t \to 0$. Arzelà's theorem indicates that *the*

** See, for example, L. A. Lyusternik and V. I. Sobolev [1].

bounded set W of functions f(x) continuous on [a, b] is compact in the sense of uniform convergence if and only if

$$\sup_{f\in U} \omega(f;\, t) \to 0 \quad \text{as} \quad f \to 0. \tag{2}$$

Thus conditions 3.1(1) and 3.1(2) are equivalent for the sets considered.

3.1.2. A similar criterion for the space L_q which we studied in section 1.4 is given by the well-known theorem of M. Riesz**, which uses the concept of the *integral modulus of continuity*. If the measurable function $f(x)$ of period $b-a$ belongs to the class L_q on $[a, b]$***, then by its integral modulus of continuity is understood the function

$$\omega(f;\, t)_{L_q} = \sup_{|h|\leqslant t} \left\{ \int_a^b |f(x+h)-f(x)|^q \, dx \right\}^{\frac{1}{q}}.$$

From the theorem proved in section 1.4, it follows that *if* $f(x) \in L_q$ *then* $\omega(f;\, t)_{L_q} \to 0$ *as* $t \to 0$. The above theorem of M. Riesz asserts that *the bounded set W of functions f(x) of* L_q *is compact in this space if and only if*

$$\sup_{f\in W} \omega(f;\, t)_{L_q} \to 0 \quad \text{as} \quad t \to 0 \tag{3}$$

(see also 3.12.2). Condition 3.1(3) here plays the same part as does condition 3.1(2) for a uniform metric and is equivalent to the corresponding condition 3.1(1) in the space L_q.

Condition 3.1(1) formulated in terms of the best approximation is of a constructive character. However, the relations 3.1(2) and 3.1(3) use only the internal properties of the functions forming the set W themselves, and are of a structural character.

In what follows (see chapters V and VI) it will be made clear that the external similarity of condition 3.1(1) with conditions of the type 3.1(2) or 3.1(3) is not accidental but is a reflection of the close relationship which exists between the best approximations of functions and their moduli of continuity.

3.1.3. The upper bounds 3.1(2) and 3.1(3), being purely internal characteristics of the set W, in distinction from the upper bounds 3.1(1), are as a result closely connected with the specific nature of its elements.

Characteristics of compact sets exist which are free from this peculiarity and in a certain sense reflect this or that general property of such sets as a whole.

** See L. A. Lyusternik and V. I. Sobolev [1], and also V. V. Nemytskii [1].

*** In what follows, when $b-a = 2\pi$ this class and the space L_q corresponding to it will be denoted by L_q^*.

If the set W, belonging to a given linear normed space F, is compact, then as has already been mentioned in 2.5.1, in virtue of the well-known theorem of Hausdorff, for any $\varepsilon > 0$ there exists for it a finite ε-net. The minimal number of elements which belong to W and form an ε-net for it is a certain function $N_W^a(\varepsilon)$ which increases with $\frac{1}{\varepsilon}$.

The function $I_W^a(\varepsilon) = \log_2 N_W^a(\varepsilon)$, which can serve as a measure of the "quantity of information" contained in the given definite element $x \in W$ with precision ε**, known as the ε-entropy of the set W, is, like the function $N_W^a(\varepsilon)$, an example of one of the characteristics of the compact set W mentioned above.

We obtain a similar example if we make use of the fact that, thanks to the existence of finite ε-nets, it is possible for any $\varepsilon > 0$ to cover the compact set W by a finite system of sets of diameter $\leqslant \varepsilon$***. The function $I_W^b(\varepsilon) = \log_2 N_W^b(\varepsilon)$, where $N_W^b(\varepsilon)$ is the minimal number of sets of diameter $\leqslant \varepsilon$ necessary to cover W, which in a similar sense can be considered as the ε-entropy of this set, belongs to the same type of characteristic of certain general geometrical properties of W****.

If for any $\varepsilon > 0$ we consider in W, instead of ε-nets, all possible ε-chains, i.e. systems of elements of W of which adjacent pairs are at distance $< \varepsilon$ apart, we arrive at another characteristic of this set. This is known as the ε-capacity of the set W, that is the function $I_W^c(\varepsilon) = \log_2 N_W^c(\varepsilon)$, where N_W^c is the maximal number of elements belonging to W and forming in it an ε-chain.

The following proposition shows, that the ε-capacity of a compact set occupies an intermediate position between the ε-entropy $I_W^a(\varepsilon)$ defined above and the ε-entropy $I_W^b(\varepsilon)$†.

If W is any compact set, then for any $\varepsilon > 0$

$$N_W^a(\varepsilon) \leqslant N_W^c(\varepsilon) \leqslant N_W^b(\varepsilon) \leqslant N_W^a\left(\frac{\varepsilon}{2}\right). \tag{4}$$

The validity of 3.1(4) follows directly from the definition of the functions $N_W^a(\varepsilon)$, $N_W^b(\varepsilon)$ and $N_W^c(\varepsilon)$. We obtain the first of these inequalities if we note that every element of the set W is located in at least one of the

** If the binary system of calculation is used, in order to select one definite element from the N elements $z_1, \ldots, z_N$, or what is the same thing, from the numbers $1, 2, \ldots, N$, it is sufficient to give $[\log_2 N]+1$ binary numbers.

*** By the diameter of a set in a metric space, as usual, is understood the upper bound of the distances between all possible pairs of elements belonging to it.

**** For this see L. S. Pontryagin and L. G. Shnirel'man [1].

† A. N. Kolmogorov [6].

closed spheres of radius ε with centres at the elements of the maximal ε-chain, and consequently the maximal ε-chain is at the same time an ε-net. The second inequality follows from the fact in a set of diameter $\leqslant \varepsilon$ there can be located not more than one element of an ε-chain. Finally, in order to obtain the third inequality it must be noted that every $\frac{\varepsilon}{2}$-net determines a closed set W of spheres of diameter $\leqslant \varepsilon$.

The inequalities 3.3(4) may be useful for obtaining estimates of the increase of ε-entropy or ε-capacity as $\varepsilon \to 0$ in concrete cases (see, for example, section 3.2.7)**. In this connection let us note also that*** *if W is the sum of a finite number of sets $\mathfrak{M}_1, \ldots, \mathfrak{M}_n$, then for $z = a, b, c$*

$$\max_{1 \leqslant k \leqslant n} N^z_{\mathfrak{M}_k}(\varepsilon) \leqslant N^z_W(\varepsilon) \leqslant \sum_{k=1}^{n} N^z_{\mathfrak{M}_k}(\varepsilon). \tag{5}$$

This relation, like 3.3(4), follows directly from the definition of the functions $N^z_W(\varepsilon)$.

3.2. Moduli of continuity and some of their properties

From the definition of the modulus of continuity $\omega(t) = \omega(f; t)$ for continuous functions the following four properties immediately follow:

(1) $\omega\ (0) = 0$.

(2) The function $\omega(t)$ is non-decreasing.

(3) The function $\omega(t)$ is semi-additive****, that is

$$\omega(t_1+t_2) \leqslant \omega(t_1)+\omega(t_2). \tag{1}$$

(4) The function $w(t)$ is continuous on the segment $0 \leqslant t \leqslant b-a$.

The properties (1) and (2) are obvious. If $|x_1-x_2| \leqslant t_1+t_2$ and $x-x_1 = t_1$, and $x_1 \leqslant x \leqslant x_2$, then

$$|f(x_1)-f(x_2)| \leqslant |f(x)-f(x_1)|+|f(x_2)-f(x)| \leqslant \omega(x-x_1)+\omega(x_2-x).$$

As $x_2-x \leqslant t_2$, it follows that

$$|f(x_1)-f(x_2)| \leqslant \omega(t_1)+\omega(t_2).$$

The property (4) follows from (1) and (3) if we take account of the fact that $\omega(t) \to 0$ as $t \to 0$.

** A. N. Kolmogorov [6], who first introduced the concepts of the ε-entropy and ε-capacity of a compact set, also obtained estimates in a number of important cases.

*** A. N. Kolmogorov [6].

**** For more details of this see Hille [1], Chap. VI.

It is not difficult to see that the properties (1)–(4) completely define the function $\omega(t)$ as a modulus of continuity. Every function $\omega(t)$ which possesses these properties is automatically a modulus of continuity**.

3.2.1. It is clear that the properties (1)–(4) remain valid if in place of the ordinary modulus of continuity we consider the integral modulus of continuity $\omega(f; t)L_{L_q}$ for arbitrary $q \geqslant 1$.

3.2.2. From what was said in section 3.2 it follows that in the case where the set W is compact, the upper bounds in the relations 3.1(2) and 3.1(3), represent moduli of continuity. They can also serve as structural characteristics for W, in exactly the same way as the upper bound in 3.1(1), representing in the given case a sequence of best approximations, serves as a constructive characteristic of this set.

3.2.3. It is possible to give a fairly simple sufficient criterion of whether a given function is a modulus of continuity***. *If the function* $\omega(t)$ *is continuous and non-decreasing on* $[0, b-a]$ *and is such that* $\omega(0) = 0$ *and* $\dfrac{\omega(t)}{t}$ *is non-increasing, then it is a modulus of continuity*. It is only necessary to verify that the given function is semi-additive. In fact,

$$\omega(t_1+t_2) = t_1\frac{\omega(t_1+t_2)}{t_1+t_2}+t_2\frac{\omega(t_1+t_2)}{t_1+t_2} \leqslant t_1\frac{\omega(t_1)}{t_1}+t_2\frac{\omega(t_2)}{t_2}$$
$$= \omega(t_1)+\omega(t_2) \qquad (2)$$

It must be noted that for semi-additiveness of the function $\omega(t)$ it is not necessary for $\dfrac{\omega(t)}{t}$ to be a decreasing function. However, this always happens when $\omega(t)$ is for $t > 0$ a function which is non-decreasing and convex upwards, and for which $\omega(0) = 0$. Consequently, any such function is a modulus of continuity.

3.2.4. Moduli of continuity exist which do not satisfy the condition of convexity. We will give an example which is importanti n some other respects also.

Let us consider a perfect Cantor set on [0,1].

The right hand ends of the intervals adjacent to it are numbers of the from $\sum\limits_{k=1}^{n} \dfrac{a_k}{3^k}$, where a_k assumes only the values 2 and 0. For all values of t within or at the ends of these intervals let us put $\omega(t) = \sum\limits_{k=1}^{n} \dfrac{\nu_k}{2^u}$ where

$$\nu_k = \begin{cases} 1 & \text{if} \quad a_k = 2, \\ 0 & \text{if} \quad a_k = 0. \end{cases} \qquad (3)$$

** S. M. Nikol'skii [14]. In what follows such functions $\omega(t)$ will be called moduli of continuity.

*** See, for example, E. Hille [1], Chap. VI.

In addition let us put $\omega(0) = 0$, $\omega(1) = 1$, and at all other points of the segment $[0,1]$ let us define $\omega(t)$ by continuity. It is not difficult to see that the function $\omega(t)$ does not satisfy the condition of convexity and that it has almost everywhere a derivative equal to zero. Let us verify that it is a modulus of continuity, i.e. that it possesses the properties (1)–(4) of section 3.2. It is only necessary to prove that it is semi-additive. For this purpose let us for the present assume that t_1 and t_2 are the right-hand ends of intervals adjacent to the Cantor set and let $t_1 < t_2$. If

$$t_1 = \sum_{k=1}^{n} \frac{\alpha_k}{3^k}, \qquad t_2 = \sum_{k=1}^{m} \frac{\beta_k}{3^k} \qquad (\alpha_k, \beta_k = 0, 2),$$

then

$$\omega(t_1) = \sum_{k=1}^{n} \frac{\nu_k}{2^k}, \qquad \omega(t_2) = \sum_{k=1}^{m} \frac{\mu_k}{2^k} \qquad (\nu_k, \mu_k = 0,\ \ 1).$$

Let us denote by s the natural number which possesses the property that $\alpha_k = \beta_k\,(k \leqslant s)$, $\alpha_{s+1} \neq \beta_{s+1}$, and let l be the least of the indices for which $\beta_l = 0$, $\alpha_l = 2$.

Then the point

$$t_2 - t_1 = \frac{2}{3^{s+1}} + \sum_{k=s+2}^{m} \frac{\beta_k}{3^k} - \sum_{k=s+2}^{n} \frac{\alpha_k}{3^k}$$

will belong to a closed adjacent interval with right-hand end

$$\frac{2}{3^{s+1}} + \sum_{k=s+2}^{l-1} \frac{\beta_k - \alpha_k}{3^k}.$$

Hence

$$\omega(t_2 - t_1) = \frac{1}{2^{s+1}} + \sum_{k=s+2}^{l-1} \frac{\mu_k - \nu_k}{2^k}$$

while

$$\omega(t_2) - \omega(t_1) = \frac{1}{2^{s+1}} + \sum_{k=s+2}^{m} \frac{\mu_k}{2^k} - \sum_{k=s+2}^{n} \frac{\nu_k}{2^k}$$

$$= \omega(t_2 - t_1) - \frac{1}{2^l} + \sum_{k=l+1}^{m} \frac{\mu_k}{2^k} - \sum_{k=l+1}^{n} \frac{\nu_k}{2^k}.$$

Consequently, in the case where t_1 and t_2 are the right-hand ends of intervals adjacent to the Cantor set, we have the inequality

$$\omega(t_2) - \omega(t_1) \leqslant \omega(t_2 - t_1), \tag{4}$$

which is equivalent to 3.2(4). But as the left-hand end of every such adjacent interval is the limit point of some sequence of right-hand ends, it follows in virtue of the continuity of $\omega(t)$ that this inequality holds for any two end points of the Cantor set. If now t_1 and t_2 are two arbitrary points belonging respectively to the closed adjacent intervals U and V, then denoting the right-hand end of the interval U by x, and the left-hand end of the interval V by y, we have

$$\omega(t_2-t_1) \geqslant \omega(y-x) \geqslant \omega(y)-\omega(x) = \omega(t_2)-\omega(t_1).$$

The correctness of the inequality 3.2(4) for other pairs of points follows from considerations of continuity.

It is easy to show that for the modulus of continuity considered here the function $\frac{\omega(t)}{t}$ is not non-increasing**. However, from the inequality

$$\omega(nt) \leqslant n\omega(t) \quad (n \text{ an integer}) \tag{5}$$

it immediately follows that for any modulus of continuity $\omega(t)$

$$\frac{\omega(t_2)}{t_2} \leqslant 2\frac{\omega(t_1)}{t_1} \quad (t_1 < t_2). \tag{6}$$

We have already seen in section 3.2.3, that subject to the condition of convexity of the function $\omega(t)$ the factor 2 on the right-hand side of this inequality can be omitted.

3.2.5. From 3.2.5 it follows immediately that *if* $\omega(t)$ *is a function which is non-decreasing and continuous on* $[0, b-a]$, *and for which* $\omega(0) = 0$, *then the function* $\omega^*(t) = t \inf\limits_{0<x\leqslant t} \frac{\omega(x)}{x}$ *is a modulus of continuity.*

Since the function $\frac{\omega^*(t)}{t}$ is non-increasing it is only necessary to verify that $\omega^*(t)$ is non-decreasing. If $t_1 < t_2$ and

$$\inf_{0<x\leqslant t_2} \frac{\omega(x)}{x} < \inf_{0<x\leqslant t_1} \frac{\omega(x)}{x}$$

then

$$\inf_{0<x\leqslant t_2} \frac{\omega(x)}{x} = \inf_{t_1\leqslant x\leqslant t_2} \frac{\omega(x)}{x},$$

and hence

$$\omega^*(t_2) = \inf_{t_1\leqslant x\leqslant t_2} \frac{t_2}{x}\omega(x) > \omega(t_1) \geqslant t_1 \inf_{0<x\leqslant t_1} \frac{\omega(x)}{x} = \omega^*(t_1).$$

** We note that the given modulus of continuity has the same order of convergence to zero as $t^{\log_2 3}$.

We note that if $\omega(t)$ is a modulus of continuity, then by the inequalities

$$\tfrac{1}{2}\omega(t) \leqslant \omega^*(t) \leqslant \omega(t) \tag{7}$$

the moduli of continuity $\omega(t)$ and $\omega^*(t)$ have the same order of convergence to zero as $t \to 0$. In this case, as mentioned above, for $\omega^*(t)$ the ratio $\frac{\omega^*(t)}{t}$ does not increase.

3.2.6. From the inequality 3.2(6) it follows immediately that *for a function $f(x)$ which is not identically a constant the modulus of continuity $\omega(t) = (f; t)$ satisfies the relation*

$$\lim_{t\to 0} \frac{\omega(t)}{t} > 0. \tag{8}$$

From the remark 3.2.1 it follows, exactly as in 3.2.5, that for every function $f(x) \in L_q$, where $q \geqslant 1$, the inequalities

$$\frac{\omega(f;\, t_2)}{t_2} \leqslant 2\frac{\omega(f;\, t_1)_{L_q}}{t_1}\ (t_1 < t_2); \quad \tfrac{1}{2}\omega(f;\, t)_{L_q} \leqslant \omega^*(t)_{L_q} < \omega(f;\, t)_{L_q},$$

similar to 3.2. (6) and 3.2. (7), are satisfied, where for the function $\omega^*(t)_{L_q}$ the ratio $\frac{\omega^*(t)_{L_q}}{t}$, determined in the same way as in 3.2.5, is non-increasing. Hence with the help of well-known properties of the Lebesgue integral it follows that *for any function $f(x) \in L_q$ which is not equivalent to a constant,*

$$\lim_{t\to 0} \frac{\omega(f;\, t)}{t} > 0. \tag{9}$$

3.2.7. Let us denote by $H_\omega\big(H_\omega(L_q)\big)$ the class of all functions $f(x)$, the modulus of continuity (integral modulus of continuity) of which does not exceed a given modulus of continuity $\omega(t)$. In the case where $\omega(t) = Mt^\alpha (0 < \alpha \leqslant 1)$ (the class of functions, which satisfies a Lipschitz condition of order α with constant M), the notation $H^{(\alpha)}M\big(H^{(\alpha)}M(L_q)\big)$ is sometimes used.

As an example of section 3.1.3. we will derive an estimate of the growth of the ε-entropy or ε-capacity of the set AH_ω of all functions $f(x)$ which belong to the class H_ω on $[0,1]$ and satisfy the inequality $|f(x)| \leqslant A$.

Let $t = \Omega(\tau)$ be the function inverse to the modulus of continuity $\tau = \omega(t)$ considered. Let us divide the segment $[-A, A]$ on the ordinate axis by the points $\frac{\nu\varepsilon}{2}$, where $-\left[\frac{2A}{\varepsilon}\right] \leqslant \nu \leqslant \left[\frac{2A}{\varepsilon}\right]$, and the segment $[0,1]$ on the axis of abscissae by the points $k\delta_0$ where $\delta_0 = \Omega(\varepsilon/2)$, $0 \leqslant k \leqslant [1/\delta_0]$. Through the resulting points let us draw straight lines parallel to the coordinate axes, dividing the rectangle [0,1; —A, A] into rectangles with sides δ_0 and $\varepsilon/2$. From the definition of δ_0 it follows that the graph of any function $f(x) \in AH_\omega$ can belong to not more than two such adjacent rectangles arranged one above the other**. Hence from these pairs of rectangles it is possible to form not more than $2\left(\left[\frac{2A}{\varepsilon}\right]+1\right)3^{\left[\frac{1}{\delta_0}\right]+1}$ systems, each of which may contain a complete graph of some function $f(x) \in AH_\omega$. The family of all functions $f(x)$ of AH_ω which belong to one such system of rectangles forms a certain set of diameter $\leqslant \varepsilon$, and the sum of all these sets constitutes some ε-cover of all the AH_ω. Consequently,

$$N^b_{AH_\omega}(\varepsilon) \leqslant 2\left(\left[\frac{2A}{\varepsilon}\right]+1\right)3^{\left[\frac{1}{\delta_0}\right]+1},$$

i.e.

$$I^b_{AH_\omega}(\varepsilon) \leqslant \left\{\frac{1}{\Omega(\frac{1}{2}\varepsilon)}+1\right\}\log_2 3+\log_2\left(\frac{2A}{\varepsilon}+1\right)+1.$$

Taking account of 3.2(8), we thus see that for all values of ε considered

$$I^b_{AH_\omega}(\varepsilon) \leqslant \frac{C_0}{\Omega\left(\frac{\varepsilon}{2}\right)}, \tag{10}$$

where C_0 is some constant.

Now let $\delta = \Omega(\varepsilon)$,

$$a = \begin{cases} \Omega(A) & \text{if} \quad A < \omega(1), \\ 1 & \text{if} \quad A \geqslant \omega(1), \end{cases}$$

and $n = \left[\frac{a}{\delta}\right]$. Let us consider any system of m $(1 \leqslant m \leqslant n)$ integers $0 \leqslant k_1 < k_2 < \ldots < k_m \leqslant n$ and the corresponding function

$$\omega_{k_1,\ldots,k_m}(u) = \begin{cases} \omega(k_1\delta-u), & 0 \leqslant u \leqslant k_1\delta, \\ \omega(u-k_\nu\delta), & k_\nu\delta \leqslant u \leqslant \frac{1}{2}(k_\nu+k_{\nu+1})\delta, \\ \omega(k_{\nu+1}\delta-u), & \frac{1}{2}(k_\nu+k_{\nu+1})\delta \leqslant u \leqslant k_{\nu+1}\delta, \end{cases} \Bigg\} (1 \leqslant \nu \leqslant k_{m-1}),$$
$$\qquad \begin{cases} \omega(u-k_m\delta), & k_m\delta \leqslant u \leqslant a, \\ 0, & a \leqslant u \leqslant 1. \end{cases}$$

** Similar reasoning was applied by L. A. Lyusternik for the proof of Arzelà's heorem (see I. G. Petrovskii [1]).

It is obvious that all these functions belong to the set AH_ω and form an ε-chain in it (see section 3.1.3). Since the number of such functions is given by

$$\sum_{m=1}^{n}\binom{n+1}{m} = 2^{n+1}-2 > 2^{\frac{a}{\delta}}-2 = 2^{\frac{a}{(\Omega\varepsilon)}}-2,$$

it follows that

$$N^c_{AH_\omega}(\varepsilon) \geqslant 2^{\frac{a}{\Omega(\varepsilon)}}-2,$$

i.e.

$$I^c_{AH_\omega}(\varepsilon) \geqslant \frac{a}{\Omega(\varepsilon)} + \log_2\left\{1-2^{1-\frac{a}{\Omega(\varepsilon)}}\right\}.$$

Since $\lim_{\varepsilon\to 0}\Omega(\varepsilon) = 0$ it follows from this that for all values of ε considered

$$I^c_{AH_\omega}(\varepsilon) \geqslant \frac{C}{\Omega(\varepsilon)}, \tag{11}$$

where C is a certain constant.

From 3.1(4), 3.2(10) and 3.2(11) follow the inequalities

$$\frac{C_0}{\Omega(\frac{1}{4}\varepsilon)} \geqslant I^b_{AH_\omega}\left(\frac{\varepsilon}{2}\right) \geqslant I^a_{AH_\omega}\left(\frac{\varepsilon}{2}\right) \geqslant I^b_{AH_\omega}(\varepsilon) \geqslant I^c_{AH_\omega}(\varepsilon) \geqslant \frac{C}{\Omega(\varepsilon)}, \tag{12}$$

which show that, for those moduli of continuity $\omega(t)$ for which $\Omega(2\varepsilon)$ and $\Omega(\varepsilon)$ are infinitesimals of the same order as $\varepsilon \to 0$, the exact order of increase of the ε-entropy and ε-capacity of the set AH_ω is equal to $\frac{1}{\Omega(\varepsilon)}$. In particular, the ε-entropy (ε-capacity) of the set of all functions $f(x)$ which belong to the class $H^{(\alpha)}M$ on a given finite segment and satisfy the condition $|f(x)| \leqslant A$ is a quantity of the order $\mathrm{e}^{-\frac{1}{\alpha}}$.

3.3. Moduli of smoothness of various orders.

In some cases, for a fuller description of the structural properties of the functions of some compact class we study, in addition to the moduli of continuity of these functions, their moduli of smoothness of various orders, which are defined as follows. If the function $f(x)$ is bounded on $[a, b]$, by its modulus of smoothness of order $k \geqslant 1$ is meant the function

$$\omega_k(f; t) = \omega_k(f; a, b; t) = \sup_{x\in[a,b],\ x+kh\in[a,b],\ |h|\leqslant t} |\Delta_h^k f(x)|,$$

$$\Delta_h^k f(x) = \sum_{\nu=0}^{k}(-1)^{k-\nu}\binom{k}{\nu} f(x+\nu h),$$

defined for non-negative values of $t \leqslant \frac{b-a}{k}$; in the case when $k = 1$, this is the ordinary modulus of continuity.

The integral modulus of smoothness is defined similarly. If the function $f(x)$, measurable and of period $b-a$, belongs to the class L_q on $[a, b]$, then by its integral modulus of smoothness of order $k \geqslant 1$ is meant the function

$$\omega_k(f; t)_{L_q} = \sup_{|h| \leqslant t} \left\{ \int_a^b |\Delta_h^k f(x)|^q \, dx \right\}^{\frac{1}{q}}.$$

The following properties of the modulus of smoothness $\omega_k(t) = \omega_k(f; t)$ follow directly from the definition**:

(1) $\omega_k(0) = 0$.

(2) The function $\omega_k(t)$ is non-decreasing.

(3) If the function $f(x)$ is continuous, then $\varphi_k(t)$ is also continuous on the segment $0 \leqslant t \leqslant \frac{b-a}{k}$.

(4) If the function $f(x)$ has a bounded derivative of the r-th order (r an integer) on $[a, b]$, then for any integer $k \geqslant 0$

$$\omega_{k+r}(f; t) \leqslant t^r \omega_k(f^{(r)}; t). \tag{1}$$

Similarly if $f^{(r-1)}(x)$ is absolutely continuous and $f^{(r)}(x) \in L_q$ ($q \geqslant 1$), then $\omega_{k+r}(f; t)_{L_q} \leqslant t^r \omega_k(f^{(r)}; t)_{L_q}$.

(5) If $n \geqslant 0$ is an integer, then

$$\omega_k(nt) \leqslant n^k \omega_k(t). \tag{2}$$

Similarly for any $q \geqslant 1$ $\omega_k(f; nt)_{L_q} \leqslant n^k \omega_k(f; t)_{L_q}$.

Properties (1) and (2) are obvious, and (3), (4) and (5) follow respectively from the simple inequality

$$|\omega_k(t_2) - \omega_\kappa(t_1)| \leqslant 2^k \omega_1(k|t_2 - t_1|) \tag{3}$$

and the identities

$$\Delta_h^r f(x) = \int_0^h \cdots \int_0^h f^{(r)}(x + t_1 + \cdots + t_r) \, dt_1 \ldots dt_r, \tag{4}$$

$$\Delta_{nh}^r f(x) = \sum_{\nu_1=0}^{n-1} \cdots \sum_{\nu_k=0}^{\nu-1} \Delta_h^k f(x + \nu_1 h + \cdots + \nu_h h). \tag{5}$$

which are easily verified by induction.

** See A. Marchoud [1].

3.3.1. It follows from (2) and (5) that *any modulus of smoothness satisfies the inequality*

$$\frac{\omega_k(t_2)}{t_2^k} \leqslant 2^k \frac{\omega_k(t_1)}{t_1^k} \qquad (t_1 < t_2). \tag{6}$$

a particular case of which is 3.2(6)**. We conclude from this that *in those cases where $\omega_k(t) \not\equiv 0$ we have*

$$\lim_{t\to 0} \frac{\omega_k(t)}{t^k} > 0. \tag{7}$$

The factor 2^k on the right-hand side of 3.3(6) can be omitted if the function $\frac{\omega_k(t)}{t^k}$ does not increase. Just as in section 3.2.5, for every modulus of smoothness $\omega_k(t)$ there can be found a non-decreasing function, for example

$$\omega_k^*(t) = t^k \inf_{0<x\leqslant t} \frac{\omega_k(x)}{x^k}, \tag{8}$$

connected with it by the inequalities

$$\frac{1}{2^k}\omega_k(t) \leqslant \omega_k^*(t) \leqslant \omega_k(t) \tag{9}$$

and such that the ratio $\frac{\omega_k^*(t)}{t^k}$ does not increase.

As a result of 3.3(9) the order of convergence to zero as $t \to 0$ is the same for $\omega_k^*(t)$ as for $\omega_k(t)$.

3.3.2. There is a well-known connection between moduli of smoothness of various orders. In the first place it is obvious that *if $k < \nu$ then*

$$\omega_\nu(f; t) \leqslant 2^{\nu-k}\omega_k(f; t) \tag{10}$$

for any function $f(x)$.

As is demonstrated by the example

$$f(x) = \begin{cases} 0, & x \leqslant 0, \\ x & x \geqslant 0, \end{cases}$$

this inequality cannot be improved as regards order (as $t \to 0$) in the general case. Thus it enables us to estimate the modulus of smoothness of a function from above in terms of its moduli of smoothness of lower orders. On the other hand, the moduli of smoothness of a function can be esti-

** It follows from this, that

$$\omega_k(t_1 - t_2) \leqslant 2^k \left\{ t_1^k \frac{\omega_k(t_1+t_2)}{(t_1+t_2)^k} + t_2^k \frac{\omega_k(t_1+t_2)}{(t_1+t_2)^k} \right\} \leqslant 2^{2k}\{\omega_k(t_1) + \omega_k(t_2)\}.$$

mated from above also in terms of its moduli of smoothness of higher orders. *Whatever the natural number k may be, the inequality***

$$\omega_k(f;t) \leqslant C_k t^k \left\{ \int_t^c \frac{\omega_{k+1}(f;u)}{u^{k+1}}\,du + \sup_x |f(x)| \right\} \tag{11}$$

holds for all sufficiently small values of t, where C_k and c are positive constants, independent of f.

To prove this we note that in virtue of the identity 3.3(5) we have

$$\Delta_{2h}^k f(x) = \sum_{\nu=0}^{k} \binom{k}{\nu} \Delta_h^k f(x+\nu h).$$

Hence

$$\Delta_{2h}^k f(x) - 2^k \Delta_h^k f(x) = \sum_{\nu=0}^{k} \binom{k}{\nu} [\Delta_h^k f(x+\nu h) - \Delta_h^k f(x)]$$

$$= \sum_{\nu=1}^{k} \binom{k}{\nu} \sum_{\mu=0}^{\nu-1} \Delta_h^{k+1} f(x+\mu h),$$

or

$$\Delta_{2h}^k f(x) - 2^k \Delta_h^k f(x) = \sum_{\nu=0}^{k-1} \sum_{\mu=\nu+1}^{k} \binom{k}{\mu} \Delta_h^{k+1} f(x+\nu h).$$

Hence by the condition*** $a \leqslant x+2kh \leqslant b$, $h > 0$, it follows that

$$|\Delta_{2h}^k f(x) - 2^k \Delta_h^k f(x)| \leqslant \omega_{k+1}(f;h) \sum_{\nu=0}^{k-1} \sum_{\mu=\nu+1}^{k} \binom{k}{\mu},$$

i.e.

$$|\Delta_{2h}^k f(x) - 2^k \Delta_h^k f(x)| \leqslant k 2^{k-1} \omega_{k+1}(f;h).$$

Putting here successively $h = 2^m \varepsilon$ $(m = 0, 1, \ldots, r-1)$ and taking account of the fact that $0 < \varepsilon \leqslant t \leqslant \dfrac{b-x}{2^{m+1}k}$, we obtain a system of inequalities of the form

$$|2^{-(m+1)k} \Delta_{2^{m+1}\varepsilon}^k f(x) - 2^{-mk} \Delta_{2^m\varepsilon}^k f(x)| \leqslant \frac{k}{2} 2^{-mk} \omega_{k+1}(f; 2^m t).$$

** A. Marchoud [1]. In virtue of the inequality 3.3(1) we conclude from this that if $f(x)$ has a bounded derivative $f'(x)$, then for sufficiently small t

$$\frac{\omega_k(f;t)}{t^k} \leqslant C_k \left\{ \int_t^c \frac{\omega_k(f;u)}{u^k}\,du + \sup_x |f(x)| \right\}$$

for all $k \geqslant 2$.

*** In the case where the function $f(x)$ is defined on the whole real axis this condition is superfluous.

From these follows the estimate

$$|2^{-rk}\Delta^k_{2^r\varepsilon}f(x)-\Delta^k_\varepsilon f(x)|\leqslant\frac{k}{2}\sum_{m=0}^{r-1}\frac{\omega_{k+1}(f;2^m t)}{2^{mk}},$$

which shows that

$$|\Delta^k_\varepsilon f(x)|\leqslant 2^{-(r-1)k}\sup_u|f(u)|+\frac{k}{2}\sum_{m=0}^{r-1}\frac{\omega_{k+1}(f;2^m t)}{2^{mk}}$$

$$\leqslant 2^{-(r-1)k}\sup_u|f(u)|+k^2t^k\sum_{m=0}^{r-1}\int_{2^m t}^{2^{m+1}t}\frac{\omega_{k+1}(f;u)}{u^{k+1}}\,du.$$

Thus, whatever the sufficiently small positive $t\leqslant\frac{b-x}{2^r k}$ may be, for any non-negative $\varepsilon\leqslant t$

$$|\Delta^k_\varepsilon f(x)|\leqslant 2^{-(r-1)k}\sup_u|f(u)|+k^2t^k\int_t^{2^r t}\frac{\omega_{k+1}(f;u)}{u^{k+1}}\,du.$$

Consequently, this inequality is satisfied uniformly with respect to all $x\in\left[a,\frac{a+b}{2}\right]$ and $0\leqslant\varepsilon\leqslant t\leqslant\frac{b-a}{2^{r+1}k}$. If now in place of $f(x)$ we substitute the function $\phi(x)=f(a+b-x)$, we find that the inequality

$$|\Delta^k_\varepsilon \phi(x)|\leqslant 2^{-(r-1)k}\sup_u|f(u)|+k^2t^k\int_t^{2^r t}\frac{\omega_{k+1}(f;u)}{u^{k+1}}\,du$$

also satisfies the same condition. As a result of this, choosing $r=r(t)$ such that $\frac{b-a}{4k}<2^r t\leqslant\frac{b-a}{2k}$, we obtain the estimate 3.3(11).

Let us note that this inequality can be written in the more general form:

$$\omega_k(f;t)\leqslant C_{k,\nu}t^k\int_t^c\frac{\omega_\nu(f;u)}{u^{k+1}}\,du+O(t^k) \tag{12}$$

where $k<\nu$, and $C_{k,\nu}$ and c are constants which do not depend on f.

The estimate 3.3(12) shows that** $\omega_k(f;t^\nu)=O\{\omega_\nu(f;t^k)\}$ $(k<\nu)$.

** In order to verify this, it is sufficient to divide the range of integration $[t,c]$ in 3.3(12) into two parts by the point $t^{\frac{k}{\nu}}$ and make use of the properties 3.3(2), 3.3(6) and 3.3(7).

3.3.3. As a corollary of the inequality 3.3(12) we note that *if* $\nu > k$, $\alpha \leqslant \nu$ *and* $\omega_\nu(f; t) = O(t^\alpha)$, *then as* $t \to 0$

$$\omega_k(f; t) = \begin{cases} O(t^k), & \text{if} \quad \alpha > k, \\ O(t^k |\ln t|), & \text{if} \quad \alpha = k, \\ O(t^\alpha), & \text{if} \quad \alpha < k. \end{cases} \tag{13}$$

Here the replacement of O by o (as $t \to 0$) in the condition entails the same substitution in the last two cases. In particular, for $k = 1, \nu = 2, 0 < \alpha \leqslant 1$, there follows from this A. Zygmund's theorem** which states that for periodic functions the *conditions*

$$\omega_2(f; h) = O(h^\alpha) \quad \left(\omega_2(f; h) = o(h^\alpha)\right)$$

and

$$\omega_1(f; h) = O(h^\alpha) \quad \left(\omega_1(f; h) = o(h^\alpha)\right)$$

are equivalent when $\alpha < 1$, *and if* $\omega_2(f; h) = O(h)$ $\left(\omega_2(f; h) = o(h)\right)$, *then* $\omega_1(f; h) = O\left(h \ln \frac{1}{h}\right) \left(\omega_1(f; h) = o\left(h \ln \frac{1}{h}\right)\right)$.

Examples exist which show that these estimates cannot be improved as regards their orders. For $\alpha = 1$ in the case of the segment [—1,1] an example of this is given by the function***

$$f(x) = \begin{cases} -n(x-1) - \dfrac{1}{2^n}, & x \in \left[1 - \dfrac{1}{2^n}, 1 - \dfrac{1}{2^{n+1}}\right] \\ 0, & x = 1 \\ f(-x), & -1 \leqslant x \leqslant 0, \end{cases} \tag{14}$$

or for the whole axis $-\infty < x < \infty$

$$f(x) = \sum_{k=1}^{\infty} \frac{\sin kx}{k^2}. \tag{15}$$

As an example of this the function (see 9.16)

$$f(x) = \sum_{k=0}^{\infty} a^k \cos(b^k x) \quad (0 < a < 1, \ b \text{ an integer } > 1), \tag{16}$$

of the type 2.11(6) in which $ab = 1$, is frequently taken.

** A. Zygmund [2]. For arbitrary bounded functions defined on the whole real axis this result was established by S. N. Bernstein (36].

*** The function 3.3(14) possesses the property that for any $x \in [-1, 1]$, $(x) = \sup |\phi(x)|$, where the upper bound extends to all $\phi(x)$ for which $\phi(\pm 1) = 0$, $f\max_{|x| \leqslant 1} |\phi(x)| \leqslant 1$ and $\omega_2(\phi; t; -1, 1) \leqslant 2t$. See A. F. Timan [9].

3.3.31. In the case where $k = 1$, $\nu = 2$ and $0 < \alpha \leqslant 1$, the relation 3.3(13) can be made more precise and we obtain for $\omega_1(f; t)$ a final asymptotic estimate. *If for* $0 < \alpha \leqslant 1$

$$\omega_2(f; t) \leqslant (2t)^\alpha \tag{17}$$

then

$$\omega_1(f; t) \leqslant \begin{cases} \dfrac{1}{\ln 2} t \left|\ln \dfrac{1}{t}\right| + O(t), & \text{when} \quad \alpha = 1, \\ \dfrac{1}{2^{1-\alpha} - 1} t^\alpha + O(t), & \text{when} \quad 0 < \alpha < 1. \end{cases} \tag{18}$$

In the class of all functions defined on a certain segment and satisfying there condition 3.3(17), *this inequality is asymptotically exact as* $t \to 0$. *In this class there are functions (for example the function* 3.3(14), *if* $\alpha = 1$), *for which the left-hand side of* 3.3(18) *is asymptotically equal to the right**).*

It should be observed that in the subclass of all the periodic functions $f(x)$ the constants $\dfrac{1}{\ln 2}$ and $\dfrac{1}{2^{1-\alpha} - 1}$ in equation 3.3(18) can now be reduced***. Thus, for example, *the definitive constant in the first of these inequalities for periodic functions, exactly as for any bounded functions defined on* $(-\infty, \infty)$, *is* $\dfrac{1}{\ln(\sqrt{2} + 1)}$****. The reader will find proofs of the more precise inequalities indicated in the references quoted.

3.3.4. As is not difficult to see the inequality 3.3(12) remains valid if, in the case of functions defined on the whole real axis, the ordinary moduli of smoothness are replaced by the integral moduli of smoothness $\omega_k(f; t)$ and $\omega_\nu(f; t)_{L_q}$.

However, in some cases the inequality 3.3(12), being as regards order (as $t \to \infty$) definitive for uniform metrics, now turns out to be crude for the metric L_q†.

3.3.41. We will illustrate this statement by the example of the space L_2^* when $k = 1$, $\nu = 2$, $\alpha = 1$, and will show that *if the function* $f(x)$ *is periodic and* $\omega_2(f; t)_{L_2^*} = O(t)$ *then*

$$\omega_1(f; t)_{L_2^*} = O\left(t\sqrt{\ln \frac{1}{t}}\right)^{\dagger\dagger}.$$

** A. F. Timan [9], G. N. Sakovich [1].

*** I. E. Gopengauz and A. F. Timan [1], A. F. Timan [10].

**** A. V. Efimov [1] for the periodic case; A. F. Timan [16] for the general case.

† See A. F. Timan and M. F. Timan [1], and also A. Zygmund [4].

†† This statement is a particular case of a more general theorem by which the inequality

$$\omega_k(f; t)_{L_2^*} \leqslant C_k t^k \left[\int_t^1 \frac{\omega_{k+1}^2(f; u)}{u^{2k+1}} du\right]^{\frac{1}{2}} \quad (0 < t \leqslant \tfrac{1}{2}),$$

which follows from theorem 6.1.5, always holds. See A. F. Timan and M. F. Timan [2].

With the same conditions an inequality similar to 3.3(12) gives only the estimate $\omega_1(f;\, t)_{L_2^*} = O\left(t \ln \frac{1}{t}\right)$.

From what was said in section 2.8.2, and Parseval's equality, it follows in virtue of theorem 5.11.1 that if

$$c_k = \frac{1}{2\pi} \int_0^{2\pi} f(t) e^{-ikt} \, dt \qquad (k = 0, \pm 1, \pm 2, \dots)$$

and

$$\omega_2(f;\, t)_{L_2^*} = O(t),$$

then

$$\sum_{|k|>n} |c_k|^2 = O\left(\frac{1}{n^2}\right).$$

Also, the function $f(x+h) - f(x-h)$ has as its complex Fourier coefficients the numbers $2iC_k \sin kh$. Hence

$$\frac{1}{2\pi} \int_0^{2\pi} |f(x+h) - f(x-h)|^2 \, dx = 4 \sum_{k=-\infty}^{\infty} |c_k \sin kh|^2.$$

Choosing $n = \left[\frac{1}{h}\right]$, we find from this

$$\frac{1}{2\pi} \int_0^{2\pi} |f(x+h) - f(x-h)|^2 \, dx = 4 \sum_{k=-n}^{n} |c_k \sin kh|^2 + O(h^2)$$

$$\leqslant h^2 \left\{4 \sum_{k=-n}^{n} k^2 |c_k|^2 + O(1)\right\},$$

or

$$\frac{1}{2\pi} \int_0^{2\pi} |f(x+h) - f(x-h)|^2 \, dx$$

$$\leqslant O(h^2) \left\{\sum_{k=1}^{n} (2k+1) \sum_{|i| \geqslant k} |c_i|^2 + O(1)\right\}$$

$$= O(h^2) \left\{\sum_{k=1}^{n} (2k+1) O\left(\frac{1}{k^2}\right) + O(1)\right\} = O\left(h^2 \ln \frac{1}{h}\right).$$

Consequently,

$$\left\{\int_0^{2\pi} |f(x+h)-f(x-h)|^2\, dx\right\}^{\frac{1}{2}} = O\left(h\sqrt{\ln\frac{1}{h}}\right)$$

and

$$\omega_1(f;\, t)_{L_2} = O\left[t\left(\ln\frac{1}{t}\right)^{\frac{1}{2}}\right].$$

Let us note that the proposition proved can be generalised to any space $L_q^*(1 \leqslant q < \infty)$**. *If* $\omega_2(f;\, t)_{L_q} = O(t)$, *then*

$$\omega_1(f;\, t)_{L_q} = \begin{cases} O\left[t\left(\ln\frac{1}{t}\right)^{\frac{1}{q}}\right] & \textit{when} \quad 1 \leqslant q \leqslant 2, \\ O\left[t\left(\ln\frac{1}{t}\right)^{\frac{1}{2}}\right] & \textit{when} \quad 2 \leqslant q < \infty. \end{cases} \tag{19}$$

It can be shown that the estimate 3.3(19) for the integral modulus of continuity is definitive as regards order (as $t \to 0$) (see 3.12.26)***.

3.3.5. From the inequalities 3.3(10) and 3.3(12) (see also section 3.3.4) follows a new formulation of the criterion of compactness of sets of functions in the space C (or L_q, respectively). *A bounded set W of functions which are continuous (belong to the set L_q) on a finite segment is compact in the sense of uniform convergence (of mean convergence over the period) if and only if, when $k \geqslant 1$,*

$$\sup_{f\in W} \omega_k(f;\, t) \to 0 \quad \textit{as} \quad t \to 0, \tag{20}$$

or

$$\sup_{f\in W} \omega_k(f;\, t)_{L_q} \to 0 \quad \textit{as} \quad t \to 0, \tag{21}$$

respectively.

3.3.1. It must be noted that if the segment considered is infinite, then condition 3.3(20) (or 3.3(21), respectively) entails compactness in the sense of uniform convergence (convergence with respect to the metric L_q) only over any finite range contained in it.

3.4. On compact classes of functions of many variables

Let us now consider the question of the structural characteristics of compact sets of functions of many variables and for definiteness let us restrict ourselves to the case where the number of variables is two.

** A. Zygmund [4].

*** M. F. Timan showed that such estimates also hold for $f(x) \in L_q\ (-\infty,\ \infty)$.

3.4.1. Let W be the set, compact in the sense of uniform convergence, of functions $f(x, y)$ which are continuous in the closed bounded rectangle $\bar{G}$ of the variables x and y. If we consider the modulus of continuity

$$\omega(f; u, v) = \sup_{\substack{|x_1-x_2|\leqslant u,\ |y_1-y_2|\leqslant v,\\ (x_1, y_1)\in \bar{g},\ (x_2 y_2)\in \bar{g}}} |f(x_1, y_1)-f(x_2, y_2)|, \tag{1}$$

it immediately follows from Hausdorff's theorem on finite ε-nets for compact sets (see 2.5.1) that:

$$\sup_{f\in W} \omega(f; u, v) \to 0 \quad \text{as} \quad u, v \to 0. \tag{2}$$

Conversely, if the relation 3.4(2) is satisfied and the set W is bounded, then, reasoning exactly as in the proof of Arzelà's theorem for sets of continuous functions of a single variable**, we see that W is compact in the sense of uniform convergence on $\bar{G}$. Consequently, *the relation* 3.4(2), *and also the relation*

$$\sup_{f\in W} E_{n_1, n_2}(f) \to 0 \quad \textit{as} \quad n_1, n_2 \to \infty, \tag{3}$$

are necessary and sufficient conditions for the compactness of a bounded set W of functions continuous on $\bar{G}$.

3.4.2. It is possible to formulate in a similar way the condition for the compactness of the set W of functions $f(x, y)$, periodic in each of the variables, and belonging to the class L_q on the rectangle of periods $\bar{G}$, if use is made of the integral modulus of continuity

$$\omega(f; u, v)_{L_q} = \sup_{|h|\leqslant u, |\eta|\leqslant v} \left\{\iint_{\bar{G}} |f(x+h, y+\eta)-f(x, y)|^q \, dx \, dy\right\}^{\frac{1}{q}}.$$

Since in virtue of section 1.4.2 $\omega(f; u, v)_{L_q} \to 0$ as $u, v \to 0$ for each of the functions $f(x, y)\in L_q$ on $\bar{G}$, it follows from considerations similar to those applied in section 3.4.1 that *the bounded set W of periodic functions $f(x, y)\in L_q$ is compact in the sense of mean convergence if and only if*

$$\sup_{f\in W} \omega(f; u, v)_{L_q} \to 0 \quad \textit{as} \quad u, v \to 0. \tag{4}$$

3.4.3. From the definition of the modulus of continuity $\omega(u, v) = \omega(f; u, v)$ for continuous functions $f(x, y)$ the following properties immediately follow:

(1) $\omega(0, 0) = 0$.

(2) The function $\omega(u, v)$ is non-decreasing with respect to u and v.

** See V. I. Smirnov [1], Vol. V, sec. 1.4.

(3) The function $\omega(u, v)$ is semi-additive, i.e.

$$\omega(u_1+u_2, v_1+v_2) \leqslant \omega(u_1, v_1)+\omega(u_2, v_2).$$

(4) The function $\omega(u, v)$ is continuous.

The properties enumerated are proved in exactly the same way as were the properties (1)–(4) in section 3.2, and as a whole completely characterise the function $\omega(u,v)$ as a modulus of continuity. If $\omega(u, v)$ possesses all these properties, then it is automatically a modulus of continuity.

The integral modulus of continuity $\omega(f; u, v)_{L_q}$ also possesses the four properties indicated for any $q \geqslant 1$.

3.4.31. The complete modulus of continuity $\omega(f; u, v)$ of the function $f(x, y)$ is connected with its partial moduli of continuity

$$\omega(f; u, 0) = \sup_{y} \sup_{|x_1-x_2| \leqslant u} |f(x_1, y)-f(x_2, y)|.$$

$$\omega(f; 0, v) = \sup_{x} \sup_{|y_1-y_2| \leqslant v} |f(x, y_1)-f(x, y_2)|.$$

by the inequalities

$$\max\{\omega(f; u, 0), \omega(f; 0, v)\} \leqslant \omega(f; u, v) \leqslant (f; u, 0)+\omega(f; 0, v). \quad (5)$$

Hence if $\Omega_1(t)$ and $\Omega_2(t)$ are two simple moduli of continuity, characterised by the properties 3.2, (1)–(4), the class of functions the partial moduli of continuity of which $\omega(f; u, 0)$ and $\omega(f; 0, v)$ do not exceed $\Omega_1(u)$ and $\Omega_2(v)$ respectively, is identical with the class of functions $f(x, y)$ with complete modulus of continuity $\omega(f; u, v)$ which satisfies the inequality

$$\omega(f; u, v) \leqslant \Omega_1(u)+\Omega_2(v). \quad (6)$$

3.4.32. The inequality 3.4(5) remains valid if the ordinary complete and partial moduli of continuity in it are replaced by the corresponding complete integral modulus of continuity $\omega(f; u, v)_{L_q} (q \geqslant 1)$ and the partial integral moduli of continuity

$$\omega(f; u, 0)_{L_q} = \sup_{|h| \leqslant u} \left\{\iint_G | f(x+h, y)-f(x, y)|^q \, dx\, dy\right\}^{\frac{1}{q}},$$

$$\omega(f; 0, v)_{L_q} = \sup_{|h| \leqslant v} \left\{\iint_G |f(x, y+h)-f(x, y)|^q \, dx\, dy\right\}^{\frac{1}{q}}.$$

3.4.32. It follows from what was said in section 3.4.3, that the upper bounds in the relations 3.4(2) and 3.4(4), in the case where the set is compact, represent the moduli of continuity. They can serve as structural characteristics for W, exactly as the simple moduli of continuity do for classes of functions of a single variable. However in virtue of the remark made in 3.4.31 the specification of such a characteristic $\omega(u, v)$ is equivalent to the specification of some pair of simple moduli of continuity $\Omega_1(t)$ and $\Omega_2(t)$. For this reason the class W of all the functions $f(x, y)$, the modulus of continuity (integral modulus of continuity) of which does not exceed the given moduls of continuity $\omega(u, v)$, is indicated by H_{Ω_1,Ω_2} $\left(H_{\Omega_1,\Omega_2}(L_q)\right)$, where

$$\Omega_1(t) = \sup_{f\in W} \omega(f; t, 0) \qquad \Omega_2(t) = \sup_{f\in W} \omega(f; 0, t)$$

$$\left(\Omega_1(t) = \sup_{f\in W} \omega(f; t, 0)_{L_q}, \quad \Omega_2(t) = \sup_{f\in W} \omega(f; 0, t)_{L_q}\right).$$

3.4.34. As in section 3.3, for a more complete description of the structural properties of functions of many variables it is possible to consider, in addition to their moduli of continuity, their moduli of smoothness of various orders. Thus, for example, if $f(x, y)$ is a function continuous on the rectangle G, then its total, mixed and partial moduli of smoothness are naturally defined respectively as the upper bounds

$$\omega_\kappa(f; u, v) = \sup_{|h|\leqslant u,\, |\eta|\leqslant v} |\Delta^k_{h\eta} f(x,y)|$$

$$= \sup_{|h|\leqslant u,\, |\eta|\leqslant v} \left| \sum_{\nu=0}^{k} (-1)^{k-\nu} \binom{k}{\nu} f(x+\nu h, y+\nu\eta) \right|,$$

$$(x, y)\in G, \ (x+ku, y+kv)\in G;$$

$$\omega_{k,l}(f, u, v) = \sup_{|h|\leqslant u,\, |\eta|\leqslant v} |\Delta^k_h \Delta^l_\eta f(x, y|$$

$$= \sup_{|h|\leqslant u,\, |\eta|\leqslant v} \left| \sum_{\nu=0}^{k} \sum_{\mu=0}^{l} (-1)^{k+l-\nu-\mu} \binom{k}{\nu} \binom{l}{\mu} f(x+\nu h, y+\mu\eta) \right|,$$

$$(x, y)\in G, \ (x+ku, y+lv)\in G;$$

$$\omega_k(f; u, 0) = \omega_{k,0}(f; u, v) = \sup_y \sup_{|h|\leqslant u} \left| \sum_{\nu=0}^{k} (-1)^{k-\nu} \binom{k}{\nu} f(x+\nu h, y) \right|,$$

$$(x, y)\in G, \ (x+ku, y)\in G;$$

$$\omega_k(f; 0, v) = \omega_{0,k}(f; u, v) = \sup_x \sup_{|h|\leqslant v} \left| \sum_{\nu=0}^{k} (-1)^{k-\nu} \binom{k}{\nu} f(x, y+\nu h) \right|,$$

$$(x, y)\in G, \ (x, y+kv)\in G.$$

These four functions possess a series of properties similar to those indicated in section 3.3 for simple moduli of smoothness. The nature of their behaviour for sufficiently small values of u and v reflects the various differential properties of the functions studied. Thus, for example, if $\omega_1(f; u, 0) = O(u)$, $\omega_1(f; 0, v) = o(v)$ and $\omega_{1,1}(f; u, v) = o(u+v)$, then the function

$f(x, y)$ is differentiable almost everywhere. By induction with respect to k it is not difficult to show that the inequality

$$\max\{\omega_{\kappa}(f; u, 0),\ \omega_k(f; 0, v)\} \leqslant \omega_{\dot{\kappa}}(f, u, v) \leqslant$$
$$\leqslant \sum_{\nu=0}^{k} \binom{k}{\nu} \omega_{k-\nu, \nu}(f; u, v), \tag{7}$$

which gives an estimate of the complete modulus of smoothness of the function f in terms of its mixed and partial moduli of smoothess of the same order, is always valid.

Hence, using the inequalities 3.3(12) and 3.4(5), it is easy to arrive at a new formulation of the compactness of a set W of functions continuous on a closed bounded rectangle. A bounded set W of functions $f(x, y)$ continuous on G is compact in the sense of uniform convergence if and only if

$$\sup_{f\in W} \omega_k(f; u, v) \to 0 \qquad (u, v \to 0) \tag{8}$$

for $k \geqslant 1$.

3.5. Compact classes of differentiable functions

Let us pass on to the consideration of sets consisting of differentiable functions.

3.5.1. Let the class W consist of the functions $f(x)$, which possess on the segment $[a, b]$ an $(r-1)$-th $(r \geqslant 1)$ absolutely continuous derivative and, consequently, almost everywhere a derivative $f^{(r)}(x)$ of order r. *For W to be compact in the sense of uniform convergence it is sufficient that* $\sup\limits_{f\in W} \max\limits_{x} |f(x)|\ \infty$ *and*

$$\sup_{f\in W}\ \operatorname*{vrai\,sup}_{a\leqslant x\leqslant b} |f^{(r)}(x)| < \infty. \tag{1}$$

If these conditions are satisfied, then the classes $W^{(1)}, W^{(2)}, W^{(3)}, \ldots, W^{(r-1)}$ consisting respectively of the $(r-1)$-th, $(r-2)$-th, $(r-3)$-th,..., *first, derivatives of all the functions of W will also be compact.*

The class W consisting of all functions $f(x)$ possessing on $[a, b]$ an absolutely continuous $(r-1)$-th derivative $f^{(r-1)}(x)$ for which the upper bound in 3.5(1) is equal to M, will be denoted by $W^{(r)}M$ in what follows.

It goes without saying that condition 3.5(1) is not necessary for the compactness of W. It would be possible to require only the satisfaction of the conditions

$$\begin{gathered} \sup_{f\in W} \max_{x} |f(x)| < \infty \\ \sup_{f\in W}\ \operatorname*{vrai\,sup}_{a\leqslant x\leqslant b} |f'(x)| < \infty \end{gathered} \tag{2}$$

and thereby significantly widen the given class of functions (up to the class $W^{(1)}$), without destroying its compactness.

3.5.2. It is also possible to indicate other conditions for the compactness of the sets considered, which are less restrictive than 3.5(1). Among them we note the following. *If the set W consists of the functions $f(x)$ which on the segment $[a, b]$ possess an* $(r-1)$-th *absolutely continuous derivative and almost everywhere a derivative of order $r > 1$ for which*

$$\sup_{f \in W} \int_a^b |f^{(r)}(t)| \, dt < \infty, \tag{3}$$

and if in addition

$$\sup_{f \in W} \max_x |f(x)| < \infty, \tag{4}$$

then W is compact in the sense of uniform convergence.

The validity of this proposition follows directly from the formula

$$f(x) = \frac{1}{(r-1)!} \int_a^x (x-t)^{r-1} f^{(r)}(t) \, dt + \sum_{k=0}^{r-1} c_k x^k. \tag{5}$$

in fact, by 3.5(3)

$$\sup_{f \in W} \left| \int_a^x (x-t)^{r-1} f^{(r)}(t) \, dt \right| \leqslant \sup_{f \in W} \int_a^b |x-t|^{r-1} \, |f^{(r)}(t)| \, dt$$

$$\leqslant (b-a)^{r-1} \sup_{f \in W} \int_a^b |f^{(r)}(t)| \, dt < \infty. \tag{6}$$

Thus for the set of algebraic polynomials on the right-hand side of 3.5(5) we have

$$\sup_{f \in W} \max_{a \leqslant x \leqslant b} \left| \sum_{k=0}^{r-1} c_k x^k \right| < \infty.$$

Consequently, as has already been mentioned in section 2.1.1 (see also 2.13.1), $\sup_{f \in W} |c_k| < \infty$ $(k = 0, 1, \ldots, r-1)$ and the given set of polynomials is compact in the sense of uniform convergence. It only remains to verify the compactness of the family of integrals

$$F_r(x) = \frac{1}{(r-1)!} \int_a^x (x-t)^{r-1} f^{(r)}(t) \, dt$$

$$= \frac{(x-a)^r}{(r-1)!} \int_0^1 (1-t)^{r-1} f^{(r)}[xt + a(1-t)] \, dt \tag{7}$$

on the right-hand side of 3.5(5). For this we have to note that $\sup_{f\in W}\omega(F_r; h)\to 0$ as $h \to 0$, and make use of theorem 3.1.1. From condition 3.5(3) it follows that

$$\sup_{f\in W}\ \sup_{a\leqslant x\leqslant b} |f^{(r-1)}(x)-f^{(r-1)}(a)| < \infty,$$

and hence using Taylor's formula

$$f(x) = \sum_{k=0}^{r-1} \frac{f^{(k)}(a)}{k!}(x-a)^k + \frac{f^{(r-1)}(\xi)-f^{(r-1)}(a)}{(r-1)!}(x-a)^{r-1}$$

we conclude that from condition 3.5(4) it follows that $\sup_{f\in W} |f^{(k)}(a)| < \infty$ $(k = 0, 1, \ldots, r-1)$ and consequently that

$$\sup_{f\ W}\ \sup_{a\leqslant x\leqslant b} |f^{(k)}(x)| < \infty \qquad (k = 0, 1, \ldots, r-1). \tag{8}$$

The compactness of the classes $W^{(k)} (k = 1, 2, \ldots, r-1)$ follows from the last inequalities (see theorem 3.5.1).

The example of the sequence of functions $f_n(x) = x^n$, considered on the segment [0,1], shows that for $r = 1$ theorem 3.5.2 is no longer true. In the case $r = 1$, *as is obvious from the proof, it would be necessary to require in place of* 3.5(3) *that*

$$\sup_{f\in W}\ \sup_{|x_1-x_2|\leqslant h} \int_{x_1}^{x_2} |f'(t)|\,dt \to 0 \quad \textit{as} \quad h\to 0.$$

3.5.21. The assertion 3.5.2 will be true *a fortiori* if we replace 3.5(3) by the condition $f^{(r)}(x)\in L_q$ $(q \geqslant 1)$, on $[a, b]$ and

$$\sup_{f\in W} \int_a^b |f^{(r)}(x)|^q\,dx < \infty \tag{9}$$

when $r > 1$, or

$$\sup_{f\in W}\ \sup_{|x_1-x_2|\leqslant h} \int_{x_1}^{x_2} |f'(x)|^q\,dx \to 0 \quad \text{as} \quad h \to 0$$

when $r = 1$.

3.5.3. The class of functions $f(x)$ which have on $[a, b]$ an $(r-1)$-th absolutely continuous derivative for which 3.5(3) and 3.5(4) hold, is in virtue of 3.5(8) contained in the class $W^{(r-1)}M$ to within a constant M. If we consider a bounded set (i.e. a set satisfying condition 3.5(4)) in the class of all functions of which the modulus of continuity of the $(r-1)$-th derivative $(r \geqslant 1)$ does not exceed the given modulus of continuity $\omega(t)$, then we obtain another example of a compact set contained in $W^{(r-1)}M$.

In what follows we shall denote such a class of functions by $W^{(r-1)}H_\omega$, and, in the case where $\omega(t) = Mt^\alpha$ $(0 < \alpha \leqslant 1)$, by $W^{(r-1)}H^{(\alpha)}M$**. The class $W^{(r-1)}H^{(1)}M$ is identical with $W^{(r)}M$.

In the periodic case the classes $W^{(r)}H_\omega(L_q)$ are similarly defined, consisting of the periodic functions $f(x)$, the $(r-1)$-th derivative of which is absolutely continuous, the r-th derivative belongs to L_q $(q \geqslant 1)$ on a period and has an integral modulus of continuity $\omega(f^{(r)}; t)_{L_q}$ which does not exceed the given modulus of continuity $\omega(t)$. In many problems an important part is played by the class $W^{(r)}H_2^{(\alpha)}M(L_q)$, which consists of the periodic functions which have an $(r-1)$-th $(r \geqslant 1)$ absolutely continuous derivative, an r-th derivative belonging to L_q on a period, and are such that

$$\omega_2(f^{(r)}; t)_{L_q} \leqslant Mt^\alpha \quad (1 \leqslant q \leqslant \infty,\ 0 \leqslant \alpha \leqslant 1).$$

It is sometimes also denoted by $H_2^{(r+\alpha)}M(L_q)$.

3.5.4. The functions $F_r(x)$ defined in 3.5(2) are the r-th integrals of the functions $F(x) = f^{(r)}(x)$: $F_1(x)$ is the integral of $F(x)$ over the interval (a, x) and, for all other integers $k > 1$, $F_k(x)$ is the integral of $F_{k-1}(x)$ over the same interval. Thus the given set W of integrable functions $F(x)$ is mapped into the set $W^{(r)}$ of functions

$$F_r(x) = \frac{1}{\Gamma(r)} \int_a^x (x-t)^{r-1} F(t)\,dt, \tag{10}$$

and depending on the character of the set W we obtain one or other of the classes of functions $W^{(r)}$, indicated in 3.5.1–3.5.3.

Since the functions which form the integrands on the right-hand side of 3.5(10) are integrable*** for any $r > 0$ for almost all x, the concept of an integral of the r-th order of the function $F(x)$ is extended directly to fractional values of r also, if formula 3.5(10) is taken as its definition****. The r-th integral $F_r(x)$ defined in this way is finite for almost all x, being an integrable and, in the case where $r \geqslant 1$, even a continuous function.

By considering various classes W of functions $F(t)$ and giving r also fractional values, we can obtain new compact sets, consisting of the functions $F_r(x)$.

** The class of all the functions of period 2π, which belong to $W^{(r)}H_\omega$ (or to $W^{(r)}H^{(\alpha)}M$ or $W^{(r)}M$ respectively) on $(-\infty, \infty)$, will in what follows be denoted by $W_*^{(r)}H_\omega$ (or by $W_*^{(r)}H^{(\alpha)}M$ or $W_*^{(r)}M$ respectively).

*** Use Fubini's theorem for multiple integrals.

**** This definition of fractional integration was indicated by Riemann [1] and Liouville [1].

3.5.5. In the consideration of periodic functions $F(x)$ it is more convenient to replace the definition of an integral of fractional order given by formula 3.5(10) in 3.5.4, by a definition which is more natural in this case**. Let $F(x)$ be an integrable function of period 2π for which

$$\int_0^{2\pi} F(t)\,dt = 0, \tag{11}$$

and let

$$\sum_{k=-\infty}^{\infty} c_k e^{ikx} \qquad (\bar{c}_k = c_{-k}) \tag{12}$$

be its Fourier–Lebesgue series, i.e.

$$c_k = \frac{1}{2\pi}\int_0^{2\pi} F(t)e^{-ikt}\,dt \qquad (c_0 = 0). \tag{13}$$

For any integer $r \geqslant 0$ the r-th integral $F_r(x)$ of the function $F(x)$ is determined in such a way that, like the function $F(x)$, it is a periodic function. To achieve this, from all the primitive functions of $F(x)$ we choose as the first integral $F_1(x)$ that of which the integral over $(0,2\pi)$ is equal to zero. The remaining successive integrals $F_k(x)$ $(k > 1)$ are determined in exactly the same way, i.e. so that

$$\int_0^{2\pi} F_k(t)\,dt = 0.$$

Since Fourier–Lebesgue series can be integrated term by term, the resulting series being always uniformly convergent, we find that for any integer $r \geqslant 1$

$$F_r(x) = \sum_{\substack{k=-\infty \\ k\neq 0}}^{\infty} \frac{c_k}{(ik)^r} e^{ikx}. \tag{14}$$

Substituting there the values of the c_k from 3.5(13) and interchanging*** the order of the summation with respect to k and integration with respect to t, we find that for all integers $r \geqslant 1$

$$F_r(x) = \frac{1}{\pi}\int_0^{2\pi} F(t)\,D_0^{(r)}(x-t)\,dt, \tag{15}$$

** Weyl [1].

*** The change of the order of summation and integration is possible owing to the fact that for $r \geqslant 1$ the partial sums of the series 3.5(16) are uniformly bounded. (See 4.12.12).

where the kernel

$$D_0^{(r)}(t) = \frac{1}{2} \sum_{\substack{k=-\infty \\ k \neq 0}}^{\infty} \frac{e^{ikt}}{(ik)^r} = \sum_{k=1}^{\infty} \frac{\cos(kt - \frac{1}{2}r\pi)}{k^r}, \tag{16}$$

apart from sign, is identical with the function $\phi_r(t+\pi)$ already considered in section 2.11.4 (see 2.11(11)).

If Abel's transformation is applied, it may be verified that whatever $r > 0$ may be, the trigonometric series 3.5(16) converges on $[0,2\pi)$ for all values of t except perhaps for $t = 0$. A second application of Abel's transformation shows that its sum is an integrable function. Hence from a well-known theorem of the theory of trigonometric series it follows that this series is the Fourier series of its sum $D_0^{(r)}(t)$, which is now defined for all $r > 0$. Since the function which forms the integrand on the right-hand side of formula 3.5(15) is integrable** for any $r > 0$, it is natural to define the integral of the r-th order $F_r(x)$ of the function $F(x)$ in the general case by this formula for almost all x. In exactly the same way, in the definition given in 3.5.5, the r-th integral $F_r(x)$ is finite for any $r > 0$ for almost all values of x, being an integrable, and in the case when $r \geqslant 1$, even a continuous function.

Evaluation of the Fourier coefficients of the function $F_r(x)$ shows that its Fourier series is a trigonometric series with the coefficients $\frac{c_k}{k^r} e^{-\frac{ir\pi}{2}}$. Applying Abel's transformation twice to its partial sum and then using the well-known theorem of the arithmetical means of the partial sums of a Fourier series, we can verify that this series converges almost everywhere and, consequently, has $F_r(x)$ as its sum.

3.5.51. The definition of a fractional integral by formula 3.5(15) in some respects differs little from the definition given in section 3.5.4. This can be verified for the case $0 < r < 1$ by considering the expansion

$$(1-z)^{r-1} = \sum_{k=0}^{\infty} A_k^{-r} z^k, \quad A_k^{-r} = \frac{(1-r)(2-r)(3-r)\dots(k-r)}{k!}, \quad A_0^{-r} = 1,$$

which holds everywhere in the closed circle $|z| \leqslant 1$ with the exception of the point $z = 1$. If we substitute in this $z = e^{it}$ $(t \neq 0)$, we obtain the two identities:

$$\left.\begin{aligned} \sum_{k=0}^{\infty} A_k^{-r} \cos kt &= \cos \frac{r-1}{2}(t-\pi) \cdot \left(2 \sin \frac{t}{2}\right)^{r-1} \\ \sum_{k=1}^{\infty} A_k^{-r} \sin kt &= \sin \frac{r-1}{2}(t-\pi) \cdot \left(2 \sin \frac{t}{2}\right)^{r-1}, \end{aligned}\right\} \quad (0 < t < 2\pi)$$

** This follows from Fubini's theorem on multiple integrals.

from which it follows that

$$\sum_{k=0}^{\infty} A_k^{-r}\cos\left(kt-\frac{r\pi}{2}\right)=\begin{cases} t^{r-1}\sin r\pi+P_r(t) & (0<t<2\pi)\\ Q_r(t) & (-2\pi<t<0),\end{cases}$$

where $P_r(t)$, $Q_r(t)$ are functions analytic in the corresponding intervals, which satisfy in $0<t<2\pi-\delta$ ($\delta>0$), $-2\pi+\delta\leqslant t<0$ respectively the relations $P_r'(t)=O(t^{r-1})$, $Q_r'(t)=O(t^{r-1})$. In order to pass to the function $D_0^{(r)}(t)$, it is necessary to make use of the asymptotic equation

$$\frac{k^{-r}}{A_k^{-r}\Gamma(1-r)}=1+\frac{\alpha_r}{k}+O\left(\frac{1}{k^2}\right)$$

and take account of the fact that**

$$\Gamma(1-r)\sin r\pi=\frac{\pi}{\Gamma(r)}.$$

We then obtain:

$$D_0^{(r)}(t)=\frac{\pi}{\Gamma(r)}K_r(t)+G_r(t), \tag{17}$$

where

$$K_r(t)=\begin{cases} t^{r-1}, & t>0,\\ 0\ , & t\leqslant 0,\end{cases}$$

and $G_r'(t)=O(t^{r-1})$, $|t|\leqslant 2\pi-\sigma$. Thus if $F(t)$ is an integrable function of period 2π then

$$\frac{1}{\pi}\int_0^{2\pi}F(t)D_0^{(r)}(x-t)\,dt=\frac{1}{\Gamma(r)}\int_0^{x}(x-t)^{r-1}F(t)\,dt+\frac{1}{\pi}\int_0^{2\pi}F(t)G_r(x-t)\,dt. \tag{18}$$

Here the kernel $G_r(t)$ possesses in the neighbourhood of the point $t=0$ better differential properties than $D_0^{(r)}(t)$***.

3.5.6. Now that the concept of an r-th integral has been considered for fractional values of r, it is not difficult to define also the operation of fractional differentiation. The function $F(x)$ integrable on the given segment****, is by definition the r-th ($r>0$) derivative $f^{(r)}(x)$ of a function $f(x)$ if almost everywhere $f(x)=F_r(x)$. For the periodic functions considered in section 3.5.5, a somewhat wider definition of a fractional derivative (in Weyl's sense) is sometimes used. In the case where $0<r<1$

** See G. M. Fikhtengol'ts [1], vol. II, chap. XIV, page 493.

*** It can be shown (see 3.12.9), that subject to the condition 3.5(11) the second term on the right-hand side of 3.5(18) is identical with $\frac{1}{\Gamma(r)}\int_{-\infty}^{x}F(t)(x-t)^{r-1}\,dt$ and that the function $G_r(t)$ is analytic in $(-2\pi, 2\pi)$.

**** In the periodic case subject to the condition that the integral of $F(x)$ over a period is equal to zero.

the r-th derivative $f^{(r)}(x)$ of the function $f(x)$ is understood to mean the ordinary derivative of the $(1-r)$-th integral $f_{1-r}(x)$, i.e.

$$f^{(r)}(x) = \frac{\mathrm{d}}{\mathrm{d}x} f_{1-r}(x),$$

if it exists. When $r \geqslant 1$ the r-th derivatives are defined successively thus:

$$f^{(r)}(x) = \left(f^{[r]}(x)\right)^{(r-[r])}. \tag{19}$$

It follows from the representation 3.5(14) that in those cases where the function $f_{1-r}(x)$ $(0 < r < 1)$ is absolutely continuous, the function is the r-th integral of $f^{(r)}(x)$.

In a number of cases, when speaking of the classes of functions indicated in sections 3.5.1–3.5.3, we shall consider r to be any non-negative number.

3.5.7. An idea of the structural properties of functions possessing fractional derivatives is given by the following proposition which we give without proof**.

*If the function $f(x)$ has an absolutely continuous integral $f_{r-1}(x)$ and $|f^{(r)}(x)| \leqslant M$ almost everywhere***, then it satisfies a Lipschitz condition of order r and $\omega(f; t) \leqslant C_r Mt^r$, where the constant C_r does not depend on f.*

The example of the function 3.3(16), in which $a = b^{-r}$ (see 9.16), shows that for $0 < r < 1$ the converse theorem is false. *It can only be asserted that if the function $f(x)$ satisfies a Lipschitz condition of order α $(0<\alpha<1)$, then its $(1-\alpha)$-th integral satisfies the condition of quasismoothness, i.e. $\omega_2(f; t) = O(t)$**** and for any $r < \alpha$ it has an* r-th *derivative $f^{(r)}(x)$ which satisfies a Lipschitz condition of order $\alpha - r$*†.

From this theorem of Hardy and Littlewood it follows directly in virtue of 3.1.1 that any bounded set of classes $W^{(r)}M$ is compact (in the sense of uniform convergence) for all $r > 0$.

3.5.71. In the study of functions defined on a finite segment the following remarks are useful in some cases.

Any function $f(x)$ continuous on a finite segment $[a, b]$ can be con-

---

** Hardy and Littlewood [1]. See also A. Zygmund [1].

*** In this case we write $f(x) \in W^r M$, and, if $f(x)$ is of period 2π, $f(x) \in W_*^{(r)} M$.

**** A. Zygmund [2]. There also the converse statement is established, in virtue of which, if $\omega_r(f; t) = O(t)$, the derivative $f^{(1-\alpha)}(x)$ exists and satisfies a Lipschitz condition of order α.

† Hardy and Littlewood [1].

tinued as a periodic function of period $2(b-a)$ with the same modulus of continuity. We confirm this by considering the periodic function

$$F(x) = \begin{cases} f(x), & a \leqslant x \leqslant b \\ f(2a-x), & 2a-b \leqslant x \leqslant a \end{cases} \tag{20}$$

of period $2(b-a)$.

We note that the difference $\phi(x) = f(x) - \frac{f(b)-f(a)}{b-a}(x-a) - f(a)$ can be continued outside the segment $[a, b]$ to the whole real axis in such a way that its modulus of smoothness $\omega_2(\phi; t)$ is multiplied by a factor of not more than five and the modulus of continuity $\omega_1(\phi; t)$ by a factor of not more than two. For this it is sufficient to continue $\phi(x)$ so as to be odd with respect to the ends of the segment, and then periodically with period $2(b-a)$ on the whole axis.

If $a \leqslant x \leqslant \frac{1}{2}(a+b)$, $x-h < a$ and $x+h \leqslant b$, then

$$\begin{aligned} |\phi(x+h)-2\phi(x)+\phi(x-h)| &= |\phi(x-h)-2\phi(x)-\phi(2a+h-x)) \\ &\leqslant |\phi(h+x)-2\phi(a+h)+\phi(2a+h-x)+ \\ &+2\left|\phi(a+h)-2\phi\left(a+\frac{h}{2}\right)+\phi(a)\right. \\ &+2\left|\phi(x)-2\phi\left(a+\frac{h}{2}\right)+\phi(2a+h-x)\right| \leqslant 5\omega_2(\phi; h). \end{aligned}$$

Similarly for $a \leqslant x \leqslant \frac{1}{2}(a+b)$, $x-h < a$ and $x+h > b$.

3.5.8. Let W consist of functions $f(x, y)$ continuous in the closed rectangle G $(a \leqslant x \leqslant b, c \leqslant y \leqslant d)$, possessing on G for almost all values of y an absolutely continuous (r_1-1)-th partial derivative with respect to x, $\frac{\partial^{r_1-1}f(x, y)}{\partial x^{r_1-1}}$ $(r_1 \geqslant 1$, an integer) and for almost all values of x an absolutely continuous (r_2-1)-th partial derivative with respect to y, $\frac{\partial^{r_2-1}f(x, y)}{\partial y^{r_2-1}}$ $(r_2 \geqslant 1$, an integer). As in 3.5.1 it can be shown that *for $r_1 > 1$, $r_2 > 1$ the set W is compact in the sense of uniform convergence on G if*

$$\sup_{f \in W} \max_{x, y} |f(x, y)| < \infty \tag{21}$$

and if the partial derivatives $\frac{\partial^{r_1}f(x, y)}{\partial x^{r_1}}, \frac{\partial^{r_2}f(x, y)}{\partial y^{r_2}}$ *which exist almost everywhere on G satisfy for some $q \geqslant 1$ the conditions*

$$\left.\begin{aligned}&\sup_{f\in W}\ \operatorname{vrai\,sup}_{y}\left\{\int_a^b\left|\frac{\partial^{r_1}f(x,y)}{\partial x^{r_1}}\right|^q \mathrm{d}x\right\}^{\frac{1}{q}}<\infty,\\&\sup_{f\in W}\ \operatorname{vrai\,sup}_{x}\left\{\int_a^b\left|\frac{\partial^{r_2}f(x,y)}{\partial y^{r_2}}\right|^q \mathrm{d}y\right\}^{\frac{1}{q}}<\infty.\end{aligned}\right\}\qquad(22)$$

In fact, by Taylor's formula

$$\sum_{k=0}^{r_1-1}\frac{1}{k!}\left.\frac{\partial^k f(x,y)}{\partial x^k}\right|_{x=a}(x-a)^k$$
$$=f(x,y)-\frac{1}{(r_1-1)!}\left[\left.\frac{\partial^{r_1-1}f(x,y)}{\partial x^{r_1-1}}\right|_{x=\xi}-\left.\frac{\partial^{r_1-1}f(x,y)}{\partial x^{r_1-1}}\right|_{x=a}\right](x-a)^{r_1-1},$$

$$\sum_{k=0}^{r_2-1}\frac{1}{k!}\left.\frac{\partial^k f(x,y)}{\partial y^k}\right|_{y=c}(y-c)^k$$
$$=f(x,y)-\frac{1}{(r_2-1)!}\left[\left.\frac{\partial^{r_2-1}f(x,y)}{\partial y^{r_2-1}}\right|_{y=\eta}-\left.\frac{\partial^{r_2-1}f(x,y)}{\partial y^{r_2-1}}\right|_{y=c}\right](y-c)^{r_2-1}$$

for almost all y in the first case and for almost all x in the second. The uniform boundedness (in the sense of the essential upper bound), of the sum on the left-hand side follows directly from 3.5(21) and 3.5(22), and hence in turn it follows (see section 2.1.1) that

$$\sup_{f\in W}\ \operatorname{vrai\,sup}_{y}\left|\frac{\partial^k f(x,y)}{\partial x^k}\right|_{x=a}<\infty\qquad(k=1,2,\ldots,r_1-1)$$
$$\sup_{f\in W}\ \operatorname{vrai\,sup}_{x}\left|\frac{\partial^k f(x,y)}{\partial y^k}\right|_{y=c}<\infty\qquad(k=1,2,\ldots,r_2-1).$$

As a result of this, by 3.5(22),

$$\left.\begin{aligned}&\sup_{f\in W}\ \operatorname{vrai\,sup}_{x,y}\left|\frac{\partial^k f(x,y)}{\partial x^k}\right|<\infty\qquad(k=1,2,\ldots,r_1-1),\\&\sup_{f\in W}\ \operatorname{vrai\,sup}_{x,y}\left|\frac{\partial^k f(x,y)}{\partial x^k}\right|<\infty\qquad(k=1,2,\ldots,r_2-1).\end{aligned}\right\}\qquad(23)$$

It remains to refer to the inequality 3.4(5) and theorem 3.4.1. In the case where even one of the two numbers r_1, or r_2 is equal to unity, for compactness of W the corresponding condition of 3.5(22) must be replaced by a stronger one, for example condition 3.5(23) for $k = 1$.

In what follows the set W consisting of all functions $f(x, y)$ for which there exist almost everywhere on G absolutely continuous partial derivatives $\dfrac{\partial^{r_1-1}f(x,y)}{\partial x^{r_1-1}}$, $\dfrac{\partial^{r_2-1}f(x,y)}{dy^{r_2-1}}$, and the upper bounds 3.5(22) do not exceed M_1 and M_2 respectively, will be denoted by $W^{(r_1,r_2)}M_1M_2(L_q)$ It follows from the above that whatever the finite $q \geqslant 1$ may be, any bounded set of this class is for some M_1^* and M_2^* contained in $W^{(r_1-1,r_2-1)}M_1^*M_2^*(L_\infty)$. Here in fact is contained any bounded set of the class $W^{(r_1-1,r_2-1)}H_{\Omega_1\Omega_2}$ of functions $f(x,y)$ the partial moduli of continuity of which, $\omega\left(\dfrac{\partial^{r_1-1}f}{dx^{r_1-1}}; t, 0\right)$ and $\omega\left(\dfrac{\partial^{r_2-1}f}{\partial y^{r_2-1}}; 0, t\right)$ respectively, do not exceed the given moduli of continuity $\Omega_1(t)$ and $\Omega_2(t)$. In the case where $\Omega_1(t) = M_1 t^\alpha$ $(0 < \alpha \leqslant 1)$ and $\Omega_2(t) = M_2 t^\beta (0 < \beta \leqslant 1)$, this class is denoted by $W^{(r_1-1,r_2-1)}H^{(\alpha,\beta}M_1M_2$. In the periodic case classes $W^{(r_1,r_2)}$ $H_{\Omega_1\Omega_2}(L_q)$ are similarly defined, consisting of the periodic functions $f(x,y)$ the partial derivatives of which $\dfrac{\partial^{r_1}f(x,y)}{\partial x^{r_1}}$ and $\dfrac{\partial^{r_2}f(x,y)}{\partial y^{r_2}}$ belong to L_q $(1 \leqslant q < \infty)$ on the rectangle of periods G and possess the integral moduli of continuity $\omega\left(\dfrac{\partial^{r_1}f}{\partial x^{r_1}}; t, 0\right)_{L_q}$, $\omega\left(\dfrac{\partial^{r_2}f}{\partial y^{r_2}}; 0, t\right)_{L_q}$, which respectively do not exceed the moduli of continuity $\Omega_1(t)$ and $\Omega_2(t)$.

Similar classes of functions of a greater number of variables are defined similarly.

In some cases an important part is played in the study of functions of many variables by the classes $H_{2,x_k}^{(r_k)}M_k(L_q)$ $(1 \leqslant q \leqslant \infty$, $r_k = s_k + \alpha_k$, $0 < \alpha_k \leqslant 1$, s_k an integer), which consist of the periodic functions $f(x_1, \dots, x_m)$ possessing for almost all $x_1, \dots, x_{k-1}, x_{k+1}, \dots, x_m$ $(1 \leqslant k \leqslant m)$ an absolutely continuous (with respect to x_k) partial derivative $\dfrac{\partial^{s_k-1}f}{\partial x_k^{s_k-1}}$, a partial derivative $\dfrac{\partial^{s_k}f}{\partial x_k^{s_k}}$, belonging to L_q on an m-dimensional parallelepiped of periods G, and such that

$$\left\{\int \cdots \int_G \left| \frac{\partial^{s_k}f(x_1,\dots,x_{k-1},x_k-h,x_{k+1},\dots,x_m)}{\partial x_k^{s_k}} - 2\frac{\partial^{s_k}f(x_1,\dots,x_m)}{\partial x_k^{s_k}} + \right.\right.$$
$$\left.\left. + \frac{\partial^{s_k}f(x_1,\dots,x_{k-1},x_k+h,x_{k+1},\dots,x_m)}{\partial x_k^{s_k}}\right|^q dx_1 \dots dx_m\right\}^{\frac{1}{q}} \leqslant M_k|h|^{\alpha_k}. \quad (24)$$

The class of functions which belongs simultaneously to all the $H_{2,x_k}^{(r_k)}M_k(L_q)$ $(k = 1, \dots, m)$ is denoted by $H_2^{(r_1,\dots,r_m)}M_1, \dots, M_m(L_q)$.

3.5.81. From what was said in section 3.5.8 it is obvious that for the classes of functions $W^{(r_1,r_2)}M_1, M_2(L_q)$ $(1 \leqslant q \leqslant \infty)$ and $W^{(r_1,r_2)}H_{\Omega_1\Omega_2}$ (if the domain G is bounded) considered there, the relations**

$$AW^{(r_1,r_2)}H_{\Omega_1\Omega_2} \subset AW^{(r_1,r_2)}M_1, M_2(L_\infty) \subset$$
$$\subset AW^{(r_1,r_2)}M_1', M_2'(L_q) \subset W^{(r_1-1,r_2-1)}M_1^*, M_2^*(L_\infty), \quad (25)$$

are satisfied, each of which is an assertion of the inclusion of one of these classes in another. These assertions (see also section 3.5.3) are the simplest theorems of inclusion for classes of functions. Let us indicate here another, deeper, inclusion theorem relating to the classes $H_2^{(r_1,\dots,r_m)}M_1,\dots,M_m(L_q)$ $(1 \leqslant q \leqslant \infty)$ of functions $f(x_1,\dots,x_m)$ defined on the whole of an m-dimensional space $R_m(-\infty < x_1,\dots,x_m < \infty)$ (see section 3.5.8, $G = R_m$)***. *If* $1 \leqslant q \leqslant q' \leqslant \infty$, $r_i > 0$ $(i = 1,\dots,m)$, $\varrho_i = \lambda_m r_i$ $(i = 1,\dots,m)$, *where*

$$\lambda_m = 1 - \left(\frac{1}{q} - \frac{1}{q'}\right)\sum_{i=1}^{m}\frac{1}{r_i} > 0, \quad (26)$$

and $f \in H_2^{(r_1,\dots,r_m)}M_1,\dots,M_n$ *then for some finite* $M_1^*,\dots,M_m^*$, *we have* $f \in H_2^{(\rho_1,\dots,\rho_m)}M_1^*,\dots,M_m^*(L_{q'})$. The validity of this assertion follows from the results of Chaps. V and VI (see section 6.4.4). From these results, both in the periodic case and in the case of the whole of R_m, a more general theorem follows* (see section 6.4.4). *If* $1 \leqslant q \leqslant q' \leqslant \infty$, $r_i > 0$ $(i = 1,\dots,m)$, $1 \leqslant s \leqslant m$ *and* $\varrho_i^{(s)} = \lambda_m^{(s)} r_i$ $(i = 1,\dots,m)$, *where*

$$\lambda_m^{(s)} = 1 - \left(\frac{1}{q} - \frac{1}{q'}\right)\sum_{i=1}^{s}\frac{1}{r_i} - \frac{1}{q}\sum_{i=s+1}^{m}\frac{1}{r_i} > 0, \quad (27)$$

then every function $f(x_1,\dots,x_m) \in H_2^{(r_1,\dots,r_m)}M_1,\dots,M_m(L_q)$ *of the variables* $x_1,\dots,x_s$, *for any fixed values* $x_{s+1},\dots,x_m$, *belongs to the class* $H_2^{(\rho_1^{(s)}),\dots,\rho_s^{(s)}}M_1^*,\dots,M_s^*(L_{q'})$.

3.5.82. The classes of functions defined in section 3.5.8 are characterised by the fact that the functions composing them possess a given number of *simple* (*not mixed*) partial derivatives in each of the variables and by certain properties of these derivatives. In chap. VI (see section 6.3.51) it will be proved that such functions must also have a known number of *mixed* partial derivatives. Thus, for example, *if the function* $f(x, y)$ *defined throughout the whole of the plane* R_2 $(-\infty < x, y < \infty)$ *belongs*

** AW denotes a set of functions $f(x) \in W$ for which $\sup_{x,y} |f(x,y)| \leqslant A$.

*** S. M. Nikol'skii [13].

to the class $W^{(r_1,r_2)}M_1M_2(L_\infty)$ ($r_1>0$, $r_2>0$ *integers*), *then for any two numbers* $k\geqslant 0$ *and* $l\geqslant 0$, *which satisfy the condition*

$$\frac{k}{r_1}+\frac{l}{r_2}<1,$$

its continuous mixed derivative $\dfrac{\partial^{k+l}f(x,y)}{\partial x^k\,\partial y^l}$ *exists.*

3.6. Sets of functions of bounded variation

A bounded set W, consisting of functions $f(x)$ defined of the segment $[a,b]$ and having there a finite total variation $V_a^b(f)$, is not always compact. It is obvious that it is also impossible to guarantee compactness of W in the sense of uniform convergence if, in addition to the condition

$$\sup_{f\in W}\sup_x|f(x)|<\infty, \tag{1}$$

it is also required that

$$\sup_{f\in W}V_a^b(f)<\infty \tag{2}$$

(see the example at the end of section 3.5.2). At the same time, it will follow from what is given below that conditions 3.6(1) and 3.6(2), though not sufficient for the compactness of W in a uniform metric, nevertheless ensure the compactness of this set in a somewhat wider sense.

3.6.1. *If the set W of periodic functions $f(x)$ with period $b-a$ satisfies condition* 3.6(2), *then for all* $t>0$

$$\sup_{f\in W}\sup_t\frac{\omega(f;t)_L}{t}<\infty. \tag{3}$$

The truth of 3.6(3) follows the relations

$$\begin{aligned}\int_a^b|f(x+h)-f(x)|\,\mathrm{d}x&\leqslant\int_a^b V_x^{x+h}(f)\,\mathrm{d}x\\&=\int_{a+h}^{b+h}V_x^a(f)\,\mathrm{d}x-\int_a^b V_a^x(f)\,\mathrm{d}x\\&=\int_b^{b+h}V_a^x(f)\,\mathrm{d}x-\int_a^{a+h}V_a^x(f)\,\mathrm{d}x\leqslant M|h|,\end{aligned}$$

where the constant M does not depend on f, if condition 3.6(2) is satisfied.

Thus if $f(x)$ is of bounded variation on $[a, b]$, then $\omega(f; t)_L = O(t)$.

Hardy and Littlewood [1] have mentioned that the following assertion, which is in a certain sense a converse, is also true. *If* $\omega(f; t)_L = O(t)$, *then almost everywhere on* $[a, b]$ *the function* $f(x)$ *is identical with some function of bounded variation.*

In fact, let $\omega(f; t)_L = Mt$ and**

$$f_n(x) = n\int_0^{\frac{1}{n}} f(x+t)\,dt.$$

Then

$$\int_a^b |f_n(x+h)-f_n(x)|\,dx = n\int_a^b \left|\int_0^{\frac{1}{n}} [f(x+t+h)-f(x+t)]dt\right| dx$$

$$\leqslant n\int_a^b \int_0^{\frac{1}{n}} |f(x+t+h)-f(x+t)|\,dt\,dx$$

$$= n\int_0^{\frac{1}{n}} \int_a^b |f(x+t+h)-f(x+t)|\,dx\,dt$$

$$= n\int_0^{\frac{1}{n}} \int_a^b |f(x+h)-f(x)|\,dx\,dt \leqslant M\,|h|.$$

If $[x_k, x_k+h_k]$ $(k = 1, \ldots, m)$ is a finite system of pairs of non-intersecting intervals, then

$$\sum_{k=1}^{m} |f_n(x_k+h_k)-f_n(x_k)| = \sum_{k=1}^{m} \left|\int_{x_k}^{x_k+h_k} f_n'(x)\,dx\right|$$

$$\leqslant \sum_{k=1}^{m} \int_{x_k}^{x_k+h_k} |f_n'(x)|\,dx \leqslant \int_a^b |f_n'(x)|\,dx.$$

** It follows by Fubini's theorem from the inequality

$$\int_a^b dt \int_a^b |f(x+t)-f(x)|\,dx \leqslant \tfrac{1}{2}M(b^2-a^2),$$

that the function $f(x)$ is integrable over $[a, b]$.

Hence from the inequality

$$\int_a^b |f_n'(x)|\,dx \leqslant \lim_{h\to 0}\int_a^b \left|\frac{f_n(x+h)-f_n(x)}{h}\right| dx \leqslant M.$$

which follows from Fatou's lemma, we have

$$\sum_{k=1}^{m} |f_n(x_k+h_k)-f_n(x_k)| \leqslant M. \tag{4}$$

Let us now make use of the fact that $f_n(x)\to f(x)$ at all points x of the segment $[a, b]$, except perhaps for those of some set Q of measure zero. If the points x_k, x_k+h_k do not belong to Q, then, passing to the limit as $n\to\infty$ in 3.6(4), we obtain

$$\sum_{k=1}^{m} |f(x_k+h_k)-f(x_k)| \leqslant M.$$

The last inequality, true for any finite system of pairs of non-intersecting intervals $(x_k,\ x_k+h_k)$ with ends not belonging to Q, indicates that on the complement to the set Q the function $f(x)$ is of bounded variation, and hence may be represented there as the difference of two bounded, non-negative and non-decreasing functions. Hence at every point of the set Q it is possible to define it as the limit on the right of the values on the complement.

3.6.11. Thus the inequality 3.6(3) represents an exhaustive structural description of the class of functions which satisfy the condition 3.6(2) in the space of functions integrable on $[a, b]$, apart from their values on a set of measure zero. In virtue of theorem 3.1.2 *there follows from this the compactness of any bounded set of functions of this class in the sense of mean convergence* (in the metric L).

3.6.12. The class of all functions $f(x)$ for which the upper bound in 3.6(2) does not exceed M on the segment considered, will be denoted by VM. Let us note that in the inequality 3.6(3) it is impossible to replace the integral modulus of continuity $\omega(f;\, t)_L$ by $\omega(f;\, t)_{L_q}$ when $q>1$. This is easily seen if in the class VM we take, for example, the function

$$f_c(x) = \begin{cases} 0, & a\leqslant x<c, \\ M, & c\leqslant x\leqslant b \end{cases} \quad (a<c<b).$$

The inequalities

$$\sup_{f\in W}\ \sup_{t} \frac{\omega(f;\, t)_{L_q}}{t} < \infty, \tag{5}$$

similar to 3.6(3) for values of q greater than unity, now characterise distinct classes W consisting of absolutely continuous functions (see 3.12.13).

3.6.2. We have seen that the bounded set of classes VM, not compact in the sense of uniform convergence, is compact in the sense of convergence in the metric L.

In addition to this it should be noted that if instead of uniform convergence we consider only convergence at each point, then, as Helly** showed, any set W of the class VM which satisfies condition 3.6(1) is compact in this sense. *From every uniformly bounded sequence of functions* $f_n(x) \in VM$ *it is possible to select a subsequence* $f_{m_k}(x)$ *which converges at every point* x *to some function* $f(x)$ *of bounded variation.*

Since the proof of this important theorem is contained in many textbooks on the theory of functions***, we shall not derive it here. It is not complicated and is based on the possibility of representing any function of bounded variation on $[a, b]$ as the difference of two non-negative, non-decreasing functions, and also on the application of the well-known diagonal process of Cantor.

3.7. Some classes of functions analytic on a finite segment

3.7.1. Let W be a set of functions $f(x)$ analytic on the given segment $[a, b]$. Every function $f(x)$ which occurs in W, being analytic on the given segment, is also analytic in some plane domain G containing it and, consequently, within a certain ellipse with foci at the ends of this segment. The given ellipse can always be chosen so that there is located on it at least one singularity of the analytic continuation of the function $f(x)$. Such an ellipse will be called the ellipse of analyticity of the given function. Thus, for every function $f(x) \in W$ there exists a certain number $\delta > 0$ which possesses the property that, whatever point x of the finite segment $[a, b]$ may be chosen, the function $f(z)$ is analytic within a circle of radius δ with centre at this point. Consequently, for any $x \in [a, b]$ the radius of convergence of the Taylor series

$$\sum_{k=0}^{\infty} \frac{f^{(k)}(x)}{k!}(z-x)^k$$

is not less than δ and hence

$$\overline{\lim_{k\to\infty}} \max_{a\leqslant x\leqslant b} \sqrt[k]{\frac{|f^{(k)}(x)|}{k!}} < \infty. \tag{1}$$

** Helly [1].

*** See I. P. Natanson [3].

If we introduce the notation

$$M_k(f) = \max_{a \leqslant x \leqslant b} |f^{(k)}(x)|$$

and use Stirling's formula

$$k! = k^k e^{-k} \sqrt{2\pi k}\,(1+\varepsilon_k) \qquad (0 < \varepsilon_k \to 0) \tag{2}$$

we find that**, *for any function $f(x)$ analytic on the finite segment $[a, b]$, there exists a finite positive number $M(f)$ such that*

$$\sqrt[k]{M_k(f)} \leqslant kM(f) \tag{3}$$

for all natural numbers k. Conversely, if the function $f(x)$ is differentiable an infinite number of times and the inequality 3.7(3), *holds, where $M(f)$ is a finite number, then it is analytic on $[a, b]$.*

The second part of this assertion follows from the consideration of the remainder term in Taylor's formula.

It follows from the inequality 3.7(1) that an *infinitely differentiable function $f(x)$ is analytic and can be continued analytically throughout the whole of the complex plane, i.e. it is an integral function, if and only if*

$$\lim_{k\to\infty} \frac{\sqrt[k]{M_k(f)}}{k} = 0 \tag{4}$$

Thus, in the class of infinitely differentiable functions the inequality 3.7(3) can be taken as the definition of analyticity of a function on the given segment***, and the relation 3.7(4) can be used as the definition of an integral function.

3.7.2. It follows from the definition of section 3.5.1, that *if the bounded set W of analytic functions $f(x)$ possesses the property*

$$\sup_{f\in W} \sup_{1 \leqslant k < \infty} \frac{\sqrt[k]{M_k(f)}}{k} < \infty, \tag{5}$$

it is compact in the sense that from any sequence $f_n(x)$ belonging to it there can be selected a subsequence $f_{n_i}(x)$ which, together with all its derivatives $f_{n_i}^{(k)}(x)$ $(k = 0, 1, 2, \ldots)$, will converge uniformly on the given finite segment $[a, b]$.

3.7.21. Let us denote by $A_R(a, b)$ the clase of all functions $f(x)$ defined on the finite segment $[a, b]$ which are analytic**** within the ellipse

---

** Pringsheim [1].

*** See also 3.12.12.

**** The analytic functions considered here are assumed to be single-valued. The same remark applies also to section 3.8.

C_R with foci at the ends of this segment and with the sum of its semiaxes equal to R. Using Cauchy's integral formula

$$f^{(k)}(x) = \frac{k!}{2\pi i} \int\limits_{C_{R-\varepsilon}} \frac{f(z)}{(z-x)^{k+1}} \, dz,$$

it is easy to show that if $A_R M(a, b)$ is the set of all functions $f(x) \in A_R(a, b)$ for which $|f(z)| \leqslant M$ everywhere within C_R, then

$$\sup_{f \in A_R M(a,\, b)} \; \sup_{1 \leqslant k \infty} \frac{\sqrt[k]{M_k(f)}}{k} < \infty. \tag{6}$$

Hence, as

$$\sup_{f \in A_R M(a,b)} \; \sup_{a \leqslant x \leqslant b} |f(x)| \leqslant M, \tag{7}$$

the class $A_R M(a, b)$ is compact in the sense indicated in section 3.7.2.

3.7.3. It is well known that in investigations connected with functions which are bounded and analytic in some circle, Cauchy's classical inequality for the coefficients of the Taylor series is of great importance. In the study of functions which are analytic and bounded in an ellipse with foci at the ends of the given finite segment of the real axis, i.e. functions of the classes $A_R M$, a similar place is occupied by the estimates of the coefficients of the corresponding Chebyshev series. If for definiteness we consider the segment $[-1,1]$, these coefficients are defined by the formula

$$c_k = \int\limits_{-1}^{1} \frac{f(x)\hat{T}_k(x)}{\sqrt{1-x^2}} \, dx \qquad (k = 0, 1, 2, \ldots), \tag{8}$$

where $\hat{T}_k(x)$ $(k = 0, 1, 2, \ldots)$ is the orthogonal system of Chebyshev polynomials, indicated previously in section 2.8.22.

For all $k = 0, 1, 2, \ldots$ *the inequality***

$$\sup_{f \in A_R M} |c_k| \leqslant \frac{\sqrt{2\pi}\, M}{R^k}. \tag{9}$$

holds in the class $A_R M$ *of functions defined on the segment* $[-1,1]$. In fact, if we put $x = \cos t$ and make the change of variable $z = e^{it}$, then, denoting by Γ_r the circumference of a circle of radius r with centre at the point $z = 0$, we obtain for $k \geqslant 1$

$$c_k = \frac{1}{\sqrt{2\pi}} \int\limits_{-\pi}^{\pi} f(\cos t) \cos kt \, dt = \frac{1}{2i\sqrt{2\pi}} \int\limits_{\Gamma_1} f\left[\frac{1}{2}\left(z + \frac{1}{2}\right)\right]\left(z^k + \frac{1}{z^k}\right)\frac{dz}{z}. \tag{10}$$

** S. N. Bernstein [15].

We note that Zhukovskii's function

$$w = \frac{1}{2}\left(z + \frac{1}{z}\right)$$

is analytic in the annulus $\frac{1}{R} < |z| < R$ $\left(\text{or } R < |z| < \frac{1}{R}\right)$ and maps it into the interior of the ellipse C_R. Hence in virtue of Cauchy's theorem, we have

$$c_k =$$

$$\frac{1}{2i\sqrt{2\pi}} \int_{\Gamma_{1(R\mp\varepsilon)}} f\left[\left(z+\frac{1}{z}\right)\frac{1}{2}\right] z^{k-1}\,\mathrm{d}z + \frac{1}{2i\sqrt{2\pi}} \int_{\Gamma_{R\mp\varepsilon}} f\left[\frac{1}{2}\left(z+\frac{1}{z}\right)\right] z^{-k-1}\,\mathrm{d}z,$$

where $\varepsilon > 0$ and the sign $+$ or $-$ is taken according as $R < 1$ or $R > 1$. Hence

$$|c_k| \leqslant \frac{\sqrt{2\pi}\, M}{R^k}.$$

3.8. Some classes of functions analytic on the whole real axis

Passing to the consideration of real functions analytic on the whole real axis $-\infty < x < \infty$, we restrict ourselves to the classes $A_{\infty,\delta}$ of bounded functions $f(x)$ which can be continued analytically in some strip of values of the complex variable $z = x+iy$ of width 2δ

$$(|y| < \delta;\ \infty < x < \infty),$$

and those for which, as in the case of a finite segment,

$$\overline{\lim_{k\to\infty}} \frac{e}{k} \sqrt[k]{M_k(f)} \leqslant \frac{1}{\delta}\, M(f) = \sup_{-\infty < x < \infty} |f^{(k)}(x)|.$$

The investigation of such functions is connected with the study of functions harmonic in the unit circle. Hence we will first pause for some properties of harmonic functions.

3.8.1. Let $u(r, \phi)$ be a function which is harmonic in the unit circle $0 \leqslant r < 1$, $-\pi \leqslant \phi \leqslant \pi$, i.e. it is the real part of the sum of a power series

$$f(re^{i\phi}) = \frac{c_0}{2} + \sum_{k=1}^{\infty} c_k z^k = \frac{c_0}{2} + \sum_{k=1}^{\infty} c_k r^k e^{ik\phi} \qquad (c_k = a_k - ib_k) \tag{1}$$

with radius of convergence not less than unity. It is well known that if for some value $\phi = \phi_0$ the series

$$\frac{c_0}{2} + \sum_{k=1}^{\infty} c_k e^{ik\phi_0} \tag{2}$$

converges, then as $r \to 1$ the function $f(re^{i\phi_0})$ has a finite limit equal to the sum of this series. Without this supposition that the series 3.8(2) converges, it is impossible to guarantee the existence of a finite limit of $f(re^{i\phi_0})$ as $r \to 1$. Examples are known of such functions $u(r, \phi)$ harmonic in the circle $|z| < 1$, for which $u(r, \phi) \to \infty$ as $r \to 1$ for almost all values of ϕ**. In this connection the following assertion is true. *If the function $u(r, \phi)$, harmonic in the unit circle, is such that as $r \to 1$*

$$\int_0^{2\pi} |u(r, \phi)|\, d\phi = O(1), \tag{3}$$

then for almost all values of ϕ $u(r, \phi)$ tends to a finite limit as $r \to 1$.

3.8.2***. In order to prove theorem 3.8.1 we will establish, that *the class of functions $u(r, \phi)$, harmonic in the unit circle, for which* 3.8(3) *holds is identical with the class of functions which can be represented in the form*

$$u(r, \phi) = \frac{1}{2\pi} \int_0^{2\pi} \frac{1-r^2}{1-2r\cos(\phi-t)+r^2}\, dF(t), \tag{4}$$

where $F(t)$ is an arbitrary function of bounded variation on $[0,2\pi]$.

The fact that every function $u(r, \phi)$ of the form 3.8(4) is harmonic in the unit circle follows from the equality

$$u(r, \phi) = \frac{a_0}{2} + \sum_{k=1}^{\infty} (a_k \cos k\phi + b_k \sin k\phi) r^k, \tag{5}$$

in which the sequences of coefficients

$$\left.\begin{matrix} a_k \\ b_k \end{matrix}\right\} = \frac{1}{\pi} \int_0^{2\pi} \left.\begin{matrix} \cos kt \\ \sin kt \end{matrix}\right\} dF(t) \tag{6}$$

are bounded. The relation 3.8(3) can be verified directly with the help of 3.8(4). Conversely, let $u(r, \phi)$ be a function harmonic in the unit circle

** The existence of such harmonic functions was first observed by I. I. Privalov and N. N. Luzin [1].

*** A. Zygmund [1].

which satisfies the relation 3.8(3). Then the function $\psi(r, t) = \int\limits_0^t u(r, \phi)\,\mathrm{d}\phi$ is absolutely continuous with respect to t and

$$V_0^{2\pi}(\psi) = \int\limits_0^{2\pi} |u(r, \phi)|\,\mathrm{d}\phi = O(1). \tag{7}$$

Moreover, since

$$\frac{1}{2\pi}\int\limits_0^{2\pi} u(r, \phi)\mathrm{e}^{-ik\phi}\,\mathrm{d}\phi = \frac{1}{2\pi}\psi(r, 2\pi) + \frac{ik}{2\pi}\int\limits_0^{2\pi}\psi(r, t)\mathrm{e}^{-ikt}\,\mathrm{d}t, \tag{8}$$

and $u(r, \phi)$ is the real part of the sum of the series 3.8(1) i.e.

$$u(r, \phi) = \frac{a_0}{2} + \sum_{k=1}^{\infty}(a_k \cos k\phi + b_k \sin k\phi)r^k, \tag{9}$$

we obtain after substitution in 3.8(8) and term by term integration,

$$\frac{1}{2}(a_k - ib_k)r^k = \frac{1}{2\pi}\psi(r, 2\pi) + \frac{ik}{2\pi}\int\limits_0^{2\pi}\psi(r, t)\mathrm{e}^{-ikt}\,\mathrm{d}t.$$

On account of 3.8(7) it is possible to make use of Helly's theorem (see section 3.6.2) and choose the corresponding sequence $r_n \to 1$ in such a way that after passing to the limit we obtain

$$\frac{1}{2}(a_k - ib_k) = \frac{1}{2\pi}\int\limits_0^{2\pi}\mathrm{e}^{-ikt}\,\mathrm{d}F(t),$$

where $F(t) = \lim\limits_{n\to\infty}\psi(r_n, t)$ is some function of bounded variation. Hence

$$a_k = \frac{1}{\pi}\int\limits_0^{2\pi}\cos kt\,\mathrm{d}F(t), \quad b_k = \frac{1}{\pi}\int\limits_0^{2\pi}\sin kt\,\mathrm{d}F(t) \quad (k = 0, 1, 2, \ldots). \tag{10}$$

Substituting the values obtained for the number a_k, b_k in 3.8(9) and making use of the identity

$$P_r(\alpha) = \frac{1-r^2}{1-2r\cos\alpha + r^2} = \frac{1}{2} + \sum_{k=1}^{\infty} r^k \cos k\,\alpha, \tag{11}$$

already employed, we arrive at the representation 3.8(4). We note that without loss of generality (apart from an additive constant in 3.8(4)) it is possible in this representation to put $F(2\pi) = F(0)$ and after integration by parts to replace it by another:

$$u(r, \phi) = -\frac{1}{\pi}\int_0^{2\pi} P_r'(t-\phi)\, F(t)\, \mathrm{d}t. \tag{12}$$

It is also obvious from the proof given here that $u(r, \phi) \geqslant 0$ if and only if $F(t)$ does not decrease on $[0, 2\pi]$.

3.8.3. Theorem 3.8.1 is easily proved by the use of formula 3.8(12). Let us take a value of ϕ for which the derivative $F'(\phi)$ exists and is finite. Then

$$\frac{F(\phi+t)-F(\phi-t)}{2t} \to F'(\phi) \qquad (t \to 0). \tag{13}$$

Since

$$-\frac{1}{\pi}\int_0^{2\pi} \sin t\, P_r'(t)\, \mathrm{d}t = r,$$

then in virtue of the oddness of the kernel $P_r'(t)$

$$|u(r, \phi)-F'(\phi)| = \frac{2}{\pi}\left|\int_0^{\pi}\left\{\frac{F(\phi+t)-F(\phi-t)}{2 \sin t} - \frac{1}{r}F'(\phi)\right\} \sin t\, P_r'(t)\, \mathrm{d}t\right|$$

$$\leqslant -\frac{2}{\pi}\int_0^{\delta}\left|\frac{F(\phi+t)-F(\phi-t)}{2 \sin t} - \frac{1}{r}F'(\phi)\right| \sin t\, P_r'(t)\, \mathrm{d}t -$$

$$-\frac{2}{\pi}\int_{\delta}^{\pi}\left|\frac{F(\phi+t)-F(\phi-t)}{2 \sin t} - \frac{1}{r}F'(\phi)\right| \sin t\, P_r'(t)\, \mathrm{d}t.$$

Taking account of 3.8(13) and also of the fact that $\max_{\delta \leqslant t \leqslant \pi} \sin t\, P_r'(t) \to 0$ $(r \to 1)$ for any $\delta > 0$, we obtain (Fatou**):

$$u(r, \phi) \to F'(\phi) \qquad (r \to 1).$$

From the proof of theorem 3.8.1 it is not difficult to see that the last relation still applies if the points (r, ϕ) tend to the boundary of the circle not along a radius but along any curve orthogonal to the boundary.

3.8.4. We note that if $|u(r, \phi)| \leqslant M$, the function $\psi(r, t) = \int_0^t u(r, \phi)\, \mathrm{d}\phi$, and together with it the function $F(t)$ also, satisfies a Lipschitz condi-

** See A. Zygmund [1].

tion of the first order with constant M and, consequently, is absolutely continuous. Hence if $g(t) = F'(t)$ in the given case formula 3.8(4) assumes the form

$$u(r, \phi) = \frac{1}{2\pi}\int_0^{2\pi} P_r(\phi - t)\, g(t)\, dt. \tag{14}$$

Conversely, if $g(t)$ is a measurable essentially bounded function, then the function $u(r, \phi)$ defined by 3.8(14) is harmonic and bounded in the unit circle. *Thus $u(r, \phi)$ is a function harmonic and bounded in the unit circle if and only if formula* 3.8(14) *holds, where $g(t)$ is a measurable essentially bounded function. Then $u(r, t) \to g(t)$ $(r \to 1)$ for almost all values of t.*

3.8.5. If $f_0(\zeta) = u_0(r, \phi) + iv_0(r, \phi)$ is some function which is analytic and bounded in the unit circle, which assumes real values on the interval $(-1,1)$ of the real axis and has in this circle a bounded real part $u_0(r, \phi)$, then by 3.8(14) Schwarz's formula

$$f_0(\zeta) = \frac{1}{2\pi}\int_{-\pi}^{\pi} \frac{e^{i\phi} + \zeta}{e^{i\phi} - \zeta}\, g_0(\phi)\, d\phi \tag{15}$$

is satisfied, in which $g_0(t) = \lim_{r\to 1} u_0(r, t)$ is, by the principle of symmetry, an even function of period 2π. By the transformation

$$\zeta = \frac{w - i}{w + i}$$

the unit circle of the ζ-plane is mapped onto the upper half-plane $w = \tau + i\delta\, (\delta > 0)$, the segment $[-1,1]$ of the real axis onto the imaginary semi-axis $\tau = 0$, $\delta \geqslant 0$, and the circumference $r = 1$, $-\pi \leqslant \phi \leqslant \pi$, onto the real axis $\delta = 0$, $-\infty < \tau < \infty$. On the imaginary axis $\tau = 0$ formula 3.8(15) assumes the form

$$f_0\left(\frac{\sigma - 1}{\sigma + 1}\right) = \frac{1}{\pi}\int_{-\infty}^{\infty} \frac{\tau\sigma - i}{(r - i\sigma)(1 + \tau^2)}\, g_0\left(\frac{1}{i}\ln\frac{\tau - i}{\tau + i}\right) d\tau$$

$$= \frac{1}{\pi}\int_{-\infty}^{\infty} \frac{\sigma\, d\tau}{\tau^2 + \sigma^2} g_0\left(\frac{1}{i}\ln\frac{\tau - i}{\tau + i}\right) = \frac{2}{\pi}\int_0^{\infty} \frac{\sigma}{\tau^2 + \sigma^2} g_0\left(\frac{1}{i}\ln\frac{\tau - i}{\tau + i}\right) d\tau.$$

Thus if a function $f(w) = u,\ (\tau, \sigma) + iv_1(\tau, \sigma)$ is analytic in the upper half-plane $w = \tau + i\sigma$, has there a bounded real part $u_1(\tau, \sigma)$ and assumes real values on the imaginary axis, then for almost all τ the finite limit $\lim_{\sigma\to 0} u_1(\tau, \sigma) = g_1(\tau)$ exists. Also

$$f_1(i\sigma) = \frac{2}{\pi}\int_0^\infty \frac{\sigma g_1(\tau)}{\tau^2+\sigma^2}\,d\tau. \tag{16}$$

The change of variable $w = ie^{\frac{\pi z}{2\delta}}$ maps the upper half-plane of $w = \tau + i\sigma$ onto the strip $|y| < \delta$ of the plane $z = x + iy$, the imaginary semi-axis $\tau = 0$, $\sigma > 0$ onto the real axis $-\infty < x < \infty$, $y = 0$, the semi-axes $\sigma = 0$, $\tau > 0$ and $\sigma = 0$, $\tau < 0$ respectively onto the straight lines $y = -\delta$, $-\infty < x < \infty$ and $y = \delta$, $-\infty < x < \infty$. On the real axis $y = 0$, formula 3.8(16) then assumes the form

$$f_1(ie^{\frac{\pi x}{2\delta}}) = \frac{1}{\delta}\int_{-\infty}^{\infty} \frac{g_1(e^{\frac{\pi t}{2\delta}})}{e^{\frac{\pi(x-t)}{2\delta}} + e^{\frac{\pi(t-x)}{2\delta}}}\,dt = \frac{1}{2\delta}\int_{-\infty}^{\infty} \frac{g_1(e^{\frac{\pi t}{2\delta}})}{\operatorname{ch}\frac{\pi(x-t)}{2\delta}}\,dt.$$

Consequently**, *if the function* $f(z) = u(x, y) + iv(x, y)$, *analytic in the strip* $|y| < \delta$, *has there a bounded real part* $u(x, y)$ *and assumes real values on the straight line* $y = 0, -\infty < x < \infty$, *then for almost all* x *the finite limit* $\lim_{y\to\delta} U(x, y) = \lim_{y\to-\delta} u(x, y) = g(x)$ exists; and

$$f(x) = \frac{1}{2\delta}\int_{-\infty}^{\infty} \frac{g(t)}{\operatorname{ch}\frac{\pi(x-t)}{2\delta}}\,dt. \tag{17}$$

Conversely, from the fact that the function $\dfrac{1}{\operatorname{ch}\frac{\pi(x+iy)}{2\delta}}$ is analytic in the strip $|y| < \delta$, $-\infty < x < \infty$,
and

$$\operatorname{Re}\frac{1}{\operatorname{ch}\frac{\pi(x+iy)}{2\delta}} > 0, \qquad \frac{1}{2\delta}\int_{-\infty}^{\infty} \operatorname{Re}\frac{1}{\operatorname{ch}\frac{\pi(x+iy)}{2\delta}}\,dx = 1,$$

it follows that *if* $\operatorname{vrai\,sup}_{-\infty<t<\infty} |g(t)| \leqslant M$, *then the function*

$$f(x+iy) = \frac{1}{2\delta}\int_{-\infty}^{\infty} \frac{g(t)}{\operatorname{ch}\frac{\pi(x+iy-t)}{2\delta}}\,dt$$

is analytic in the given strip and $|\operatorname{Re} f(x+iy)| \leqslant M$.

** N. I. Akhiezer [3].

3.8.6. Let us denote by $A_{\infty,\delta}UM$ the class of all analytic functions $f(x)$ defined on the real axis $-\infty < x < \infty$, analytic in the strip of values of the complex variable $z = x+iy$ of width $2\delta > 0$ ($|y| < \delta$, $-\infty < x < \infty$) and such that $|\text{Re} f(z)| \leqslant M$ for all z of this strip. The example of the function

$$f(z) = \frac{1}{2i} \ln \frac{e^{\frac{\pi z}{2\delta}}+i}{1+ie^{\frac{\pi z}{2\delta}}},$$

which belongs to $A_{\infty,\delta}UM$ but is not bounded in the strip considered, shows that the given class is essentially wider than the class $A_{\infty,\delta}M$ of all functions of $A_{\infty,\delta}UM$ for which $|f(x+iy)| \leqslant M$ for $|y| < \delta, -\infty < x < \infty$. By using the integral formula 3.8(17), it is easily verified that the class $A_{\infty,\delta}UM$ is compact, and hence also that the class $A_{\infty,\delta}M$ is compact, in the sense that from any sequence $f_n(x)$ belonging to it there can be selected a subsequence $f_{n_i}(x)$ which, together with all its derivatives $f_{n_i}^{(k)}(x)$ ($k = 0, 1, 2, \ldots$), converges uniformly on any finite segment of the x-axis.

For this it is sufficient to note, by theorem 3.1.1, that all the functions $f(x)$, can be represented in the form of the integral 3.8(17), in which $\text{vrai sup}\,|gt| \leqslant M$, uniformly continuous on any finite segment of the real axis, and that the given integral and also all its derivatives with respect to x converge uniformly on every such segment.

As is shown by the example of the sequence $f_n(x) = \sin \dfrac{x}{\pi}$, in the assertion made above it is impossible to replace uniform convergence on any finite segment by uniform convergence on the whole real x-axis.

3.8.7. It is obvious that any real periodic function $f(x)$, analytic on the real axis, can always be continued analytically in to some strip containing it. If the real part of the analytic continuation of such a function is bounded in the strip, then $f(x)$ is expressed by the integral 3.8(17). In addition, in this case it is possible to indicate another and in some ways a more appropriate representation for $f(x)$ (see below 3.8(20)).

The function $\zeta = e^{iz}$ maps the strip considered onto the circular annulus $e^{-\delta} < |\zeta| < e^{\delta}$. On account of the periodicity of $f(x)$ the function $f\left(\dfrac{1}{i} \ln \zeta\right)$ is single-valued and analytic in the given annulus. Expanding it in a Laurent series in this annulus, we have

$$f(z) = u(x, y)+iv(x, y) = f\left(\frac{1}{i} \ln \zeta\right) = \sum_{k=-\infty}^{\infty} c_k \zeta^k, \tag{18}$$

where for $|y| < \delta$

$$c_k = \frac{1}{2\pi i} \int_{\Gamma_{e^y}} f\left(\frac{1}{i} \ln \zeta\right) \zeta^{-k-1} \, d\zeta = \frac{1}{2\pi i} \int_{\Gamma_{e^{-y}}} f\left(\frac{1}{i} \ln \zeta\right) \zeta^{-k-1} \, d\zeta$$

i.e.

$$c_k = \frac{1}{2\pi} \int_0^{2\pi} f(x-iy) e^{-ky} e^{-ikx} \, dx = \frac{1}{2\pi} \int_0^{2\pi} f(x+iy) e^{ky} e^{-ikx} \, dx.$$

But since the function $f(z)$ assumes real values on the x-axis, we have $f(x-iy) = \overline{f(x+iy)}$. Hence

$$c_k = \frac{1}{\operatorname{ch} ky} \frac{1}{2\pi} \int_0^{2\pi} u(x, y) e^{-ikx} \, dx, \tag{19}$$

or, passing to the limit as $y \to \delta$ and introducing the notation $q = e^{-\delta}$, we obtain

$$c_k = \frac{1}{\pi} \frac{q^k}{1+q^{2k}} \int_0^{2\pi} g(t) e^{-ikt} \, dt.$$

Consequently we have

$$f(x) = \sum_{k=-\infty}^{\infty} c_k e^{ikx} = \frac{1}{2\pi} \int_0^{2\pi} \left\{1 + 4 \sum_{k=1}^{\infty} \frac{q^k}{1+q^{2k}} \cos k(x-t)\right\} g(t) \, dt. \tag{20}$$

In what follows, the class of all functions $f(x)$ which can be represented in the form 3.8(20), where $q(t)$ is any function of period 2π which is measurable, real, and the upper bound of the modulus of which does not exceed M, will be denoted by $A_\delta^* UM$.

It can be proved** that for $|y| < \delta$ and $0 < q = e^{-\delta} < 1$ the real part of the function

$$H(x+iy) = 1 + 4 \sum_{k=1}^{\infty} \frac{q^k}{1+q^{2k}} \cos k(x+iy)` \tag{21}$$

i.e. the sum of the series

$$\operatorname{Re} H(x+iy) = 1 + 4 \sum_{k=1}^{\infty} \frac{\operatorname{ch} ky}{\operatorname{ch} k\delta} \cos kx, \tag{22}$$

is positive. Hence if $f(x) \in A_s^* UM$, the function

$$f(x+iy) = \frac{1}{2i} \int_0^{2\pi} H(x+iy-t) g(t) \, dt \tag{23}$$

** See N. I. Akhiezer [3].

is analytic in the strip $|y| < \delta$, $-\infty < x < \infty$, and has a bounded real part there. Thus *the class* $A_\delta^* UM$ *is identical with the set of all the periodic functions with period* 2π *of* $A_{\infty,\delta} UM$.

3.8.8. The class $A_\delta^* UM$ is the particular case (for $q = \infty$) of the classes $A_\delta^* UM(L_q)$ of real periodic functions $f(x)$, which admit an analytic continuation $f(x+iy) = u(x, y)+iv(x, y)$ into the strip $|y| < \delta$ and are such that

$$\left\{\int_0^{2\pi} |u(x, y)|^q \, dx\right\}^{\frac{1}{q}} \leqslant M \quad (q \geqslant 1). \tag{24}$$

By commencing with 3.8(19), it is not difficult to obtain the representation of all functions of $A_\delta^* UM(L_q)$ for any $q \geqslant 1$. In fact, if

$$\Psi(t, y) = \int_0^t u(x, y) dx,$$

then as in section 3.8.2 we obtain for the coefficients c_k of the series 3.8(18)

$$c_k = \frac{1}{\operatorname{ch} ky} \cdot \frac{1}{2\pi} \int_0^{2\pi} e^{ikx} \, d\Psi(t, y),$$

and after the application of Helly's theorem (see section 3.6.2) and passing to the limit with respect to some sequence $y_n \to \delta$ we find

$$c_k = \frac{1}{\operatorname{ch} k\delta} \cdot \frac{1}{2\pi} \int_0^{2\pi} e^{-ikt} \, dF(t),$$

where $F(t)$ is a function of bounded variation. Consequently,

$$f(x) = \frac{1}{2\pi} \int_0^{2\pi} H(x-t) \, dF(t). \tag{25}$$

If now the function $f(x)$ can be represented by formula 3.8(25), then $f(x+iy)$ is analytic in the strip $|y| < \delta$, $-\infty < x < \infty$, and as a result of the positiveness of the sums of the series 3.8(22) has a real part $U(x, y)$ which there satisfies the condition 3.8(24) for $q = 1$, $M = \operatorname*{var}_{0 \leqslant t \leqslant 2\pi} F(t)$. Therefore *the class of real periodic functions* $f(x)$ *of period* 2π *which admit of the analytic continuation* $f(x+iy) = u(x, y)+iv(x, y)$ *in the strip* $|y| < \delta$ *and satisfy the condition* 3.8(24) *for* $q = 1$, *is identical with the class of functions having the representation* 3.8(25).

In the case when $q > 1$, in virtue of the inequality

$$\int_Q |u(x, y)| \, dx \leqslant (\operatorname{mes} Q)^{\frac{q-1}{q}} \left\{\int_Q |u(x, y)|^q \, dx\right\}^{\frac{1}{q}} \leqslant M(\operatorname{mes} Q)^{\frac{q-1}{q}}$$

the function $\Psi(t, y)$ is absolutely continuous with respect to t, uniformly with respect to y. Hence the limit function $F(t)$ is also absolutely continuous and so formula 3.8(25) assumes the form

$$f(x) = \frac{1}{2\pi} \int_0^{2\pi} H(x-t) g(t) \, dt, \tag{26}$$

where $g(t) = F'(t) = \lim\limits_{y_n \to \delta} u(t, y_n)$. By the well-known lemma of Fatou

$$\left\{\int_0^{2\pi} |g(t)|^q \, dt\right\}^{\frac{1}{q}} \leqslant M. \tag{27}$$

Since $\operatorname{Re} H(x+iy) > 0$ and

$$\frac{1}{2\pi} \int_0^{2\pi} \operatorname{Re} H(x+iy) \, dx = 1$$

we see also that conversely, every function $f(x)$ which can be represented by the integral 3.8(26), in which 3.8(27) holds for $g(t)$, admits of the analytic continuation $f(x+iy) = u(x, y) + iv(x, y)$ in the strip $|y| < \delta$, and in virtue of the inequality

$$\int_0^{\pi} |u(x, y)|^q \, dx = \frac{1}{2\pi} \int_0^{2\pi} \left| \int_0^{2\pi} \operatorname{Re} H(x+iy-t) g(t) \, dt \right|^q dx \leqslant$$

$$\leqslant \frac{1}{2\pi} \int_0^{2\pi} \int_0^{2\pi} \operatorname{Re} H(x+iy-t) \, |g(t)|^q \, dt \, dx = \int_0^{2\pi} |g(t)|^q \, dt$$

satisfies condition 3.8(24).

3.9. On regularly monotonic functions

As is shown by the example

$$f(x) = e^{-\frac{1}{x^2}} \tag{1}$$

a function $f(x)$ may be infinitely differentiable in a given range without being analytic. *However, if a function $f(x)$ is such that each of its deriva-*

tives on $[a, b]$ *retains its sign, then it is analytic within this segment***. The proof is based on the following lemma***. *If at all the points of the interval* (a, b)

$$0 < \phi^{(n+1)}(x) < f^{(n+1)}(x), \tag{2}$$

then for the best approximation of order n by algebraic polynomials on $[a, b]$ *(see* 2.1(3)) *the inequality*

$$E_n(\phi) < E_n(f) \tag{3}$$

is satisfied. In virtue of a theorem of P. L. Chebyshev (see section 2.7.2) the difference $f(x)-P_n(f; x)$ between the function $f(x)$ and the polynomial $P_n(f; x)$ which deviates least from it vanishes at least at $n+1$ points $x_0, x_1, \ldots, x_n$ of the interval (a, b). If the polynomial $Q_n(x)$ satisfies the condition

$$Q_n(x_i) = \phi(x_i) \quad (i = 0, 1, \ldots, n),$$

then the function

$$F(x, t) = [f(x)-P_n(f; x)]\,[\phi(t)-Q_n(t)]-[f(t)-P_n(f; t)]\,[\phi(x)-Q_n(x)]$$

has zeros with respect to the variable t at $t = x$ and $t = x_k$ $(k = 0, 1, \ldots, n)$. Hence, assuming that $x \neq x_k (k = 0, 1, \ldots, n)$, we have by Rolle's theorem at some point $c \in (a, b)$

$$\left.\frac{\partial F^{(n+1)}(x, t)}{\partial t}\right|_{t=c} = 0,$$

i.e.

$$|\phi(x)-Q_n(x)|\, f^{(n+1)}(c) = |f(x)-P_n(f; x)|\, \phi^{(n+1)}(c).$$

But this is compatible with 3.9(2) only when

$$|\phi(x)-Q_n(x)| < |f(x)-P_n(f; x)| \leqslant E_n(f).$$

Consequently 3.9(3) holds.

In particular, choosing $\phi(x) = \dfrac{Mx^{n+1}}{(n+1)!}$ and using theorem 2.9 as a lemma, we find that if on $[a, b]$

$$|f^{(m)}(x)| > M, \tag{4}$$

then

$$M_0 = \max_{a \leqslant x \leqslant b} |f(x)| \geqslant E_{m-1}(f) > \frac{2M}{m!}\left(\frac{b-a}{4}\right)^m. \tag{5}$$

** See S. N. Bernstein [7], where this theorem is proved with less restrictive conditions.

*** S. N. Bernstein [15]. The proof given was proposed by I. V. Tsenov [1].

It is now not difficult to prove the theorem. Since for any n the derivati.e $f^{(n+1)}(x)$ retains its sign on $[a, b]$, it follows that the inequality

$$|f^{(n)}(x)| \geqslant \left| f^{(n)}\left(\frac{a+b}{2}\right) \right|,$$

is satisfied on one of the two segments $\left[a, \frac{a+b}{2}\right], \left[\frac{a+b}{2}, b\right]$. Consequently

$$M_0 = \max_{a \leqslant x \leqslant b} |f(x)| \geqslant \frac{2f^{(n)}\left(\frac{a+b}{2}\right)}{n!} \left(\frac{b-a}{8}\right)^n,$$

or

$$\frac{1}{n!} f^{(n)}\left(\frac{a+b}{2}\right) \leqslant \frac{M_0}{2}\left(\frac{8}{b-a}\right)^n.$$

Hence we conclude that on every segment $[a_1, b_1]$ entirely contained within (a, b) we have

$$\overline{\lim_{n\to\infty}} \frac{1}{n} \sqrt[n]{\max_{a_1 \leqslant x \leqslant b_1} |f^{(n)}(x)|} < \infty.$$

Hence in virtue of theorem 2.7.1 the function $f(x)$ is analytic on any segment $[a_1, b_1] \subset (a, b)$, i.e. everywhere within (a, b).

3.9.1. The functions $f(x)$, which satisfy the condition of theorem 3.9, are said to be *regularly monotonic*. Among them a special place is occupied by *absolutely monotonic functions*, i.e. those which are such that all their derivatives have the same constant sign on $[a, b]$.

In this particular case the validity of theorem 3.9 is almost obvious and follows directly from formula 3.5(5) (see also 3.5(7)). Also, since

$$f(x) - \sum_{k=0}^{n} \frac{f^{(k)}(a)}{k!}(x-a)^k = \frac{(x-a)^{n+1}}{n!} \int_0^1 (1-t)^n f^{(n+1)}[tx+(1-t)a]\,dt \tag{6}$$

it follows that in the case where the function $f(x) \geqslant 0$ is absolutely monotonic for $a \leqslant x < b$

$$0 \leqslant f(x) - \sum_{k=0}^{n} \frac{f^{(k)}(a)}{k!}(x-a)^k \leqslant \left(\frac{x-a}{b-a}\right)^{n+1} f(b),$$

i.e.

$$f(x) = \sum_{k=0}^{\infty} \frac{f^{(k)}(a)}{k!} (x-a)^k, \tag{7}$$

and the radius of convergence of the last series is not less than $b-a$.

If now $f(x)$ is any function analytic in the range $[a, b]$, and the radius of convergence of its Taylor series 3.9(7) is not less than $b-a$, then on grouping together its terms with the same signs we obtain

$$f(x) = f_1(x) - f_2(x),$$

where $f_1(x)$ and $f_2(x)$ are functions absolutely monotonic on $[a, b]$. Consequently, *in order that the function $f(x)$, defined on $[a, b]$, can be expanded in a Taylor series in powers of $x-a$ with radius of convergence not less than $b-a$, it is necessary and sufficient that it should be the difference of two functions which are absolutely monotonic on this segment***.

Thus we see that in the theory of real analytic functions absolutely monotonic functions play a similar part to that which ordinary monotonic functions play for functions of bounded variation.

3.9.2. There also exists a direct connection between absolutely monotonic functions and the class of all monotonic functions. This is obvious from the following proposition***. *The function $f(x)$ defined of the semi-axis $(-\infty, 0]$, is absolutely monotonic there if and only if*

$$f(x) = \int_0^{\infty} e^{tx} \, dF(t), \tag{8}$$

where $F(t)$ is a function which is non-decreasing on $[0, \infty)$. In the first place we note that if $f(x)$ is bounded and absolutely monotonic on $(-\infty, 0]$, then for $n = 1, 2, \ldots$ we have:

$$f^{(n)}(x) \leqslant -\frac{2}{x} \int_x^{\frac{x}{2}} f^{(n)}(t) \, dt = -\frac{2}{x} \left[f^{(n-1)}\left(\frac{x}{2}\right) - f^{(n-1)}(x) \right]$$

$$\leqslant -\frac{2}{x} f^{(n-1)}\left(\frac{x}{2}\right) \leqslant \ldots \leqslant \frac{(-1)^n}{x^{2n}} 2^{n^2} \left[f\left(\frac{x}{2^n}\right) - f\left(\frac{x}{2^{n-1}}\right) \right] = o\left(\frac{1}{x^n}\right)$$

$$(x \to -\infty),$$

and hence

$$\frac{1}{n!} \left| \int_{-\infty}^{0} x^n f^{(n+1)}(x) \, dx \right| = \int_{-\infty}^{0} f'(x) \, dx = f(o) - f(-\infty) < \infty.$$

** S. N. Bernstein [3].

*** S. N. Bernstein [8]. The proof given here was proposed by B. I. Korenblyum [1].

Also, since

$$f(x)-f(-\infty)=\int_{-\infty}^{x} f'(t)\,\mathrm{d}t=\frac{1}{n!}\int_{-\infty}^{x}(x-t)^n f^{(n+1)}(t)\,\mathrm{d}t$$

$$=\frac{(-1)^n}{(n-1)!}\int_{-\infty}^{\frac{x}{n}}\left(1-\frac{x}{nt}\right)^n(nt)f^{(n+1)}(nt)\,\mathrm{d}t,$$

on putting

$$\phi_n(u)=\begin{cases}\left(1-\dfrac{u}{n}\right)^n, & 0\leqslant u\leqslant n,\\ 0, & n\leqslant u,\end{cases}$$

we have

$$f(x)-f(-\infty)=\frac{(-1)^n}{(n-1)!}\int_{-\infty}^{0}\phi_n\left(\frac{x}{t}\right)(nt)^n f^{(n+1)}(nt)\,\mathrm{d}t$$

$$=\int_{-\infty}^{0}\phi_n(xt)\,\mathrm{d}F_n(t),$$

where

$$F_n(t)=\frac{(-1)^{n+1}}{(n-1)!}\int_{-\infty}^{\frac{1}{t}}(nu)^n f^{(n+1)}(nu)\,\mathrm{d}u.$$

It is obvious that $F_n(t)$ is a nondecreasing function on $(-\infty, 0]$, and

$$F_n(0)-F_n(-\infty)=\frac{(-1)^n}{(n-1)!}\int_{-\infty}^{0}(nu)^n f^{(n+1)}(nu)\,\mathrm{d}u$$

$$=\frac{(-1)^n}{n!}\int_{-\infty}^{0}t^n f^{(n+1)}(t)\,\mathrm{d}t<f(0)-f(-\infty).$$

Moreover, as the greatest value of the modulus of the difference $\phi_n(u)-\mathrm{e}^{-u}$ is attained at the point u_n $(0<u_n<n)$ at which

$$\left(1-\frac{u_n}{n}\right)^{n-1}=\mathrm{e}^{-u_n},$$

it follows that for $n\geqslant 2$

$$|\phi_n(u)-\mathrm{e}^{-u}|<|\phi_n(u_n)-\mathrm{e}^{-u_n}|=\frac{u_n}{n}\mathrm{e}^{-u_n}\leqslant\frac{1}{n\mathrm{e}},$$

i.e. $\phi_n(u) \to e^{-u}$ as $n \to \infty$ uniformly on the semi-axis $[0, \infty)$. Applying Helly's theorem (see section 3.6.2) and choosing a subsequence n_k of the natural numbers, after passing to the limit we obtain

$$f(x) = f(-\infty) + \int_{-\infty}^{0} e^{-xt}\, dF^*(t) = f(-\infty) - \int_{0}^{\infty} e^{xt}\, dF^*(-t),$$

where $F^*(t) = \lim_{n_k \to \infty} F_{n_k}(t)$, or

$$f(x) = \int_0^\infty e^{xt}\, dF(t),$$

where $F(t)$ is a function which is non-decreasing and bounded on $[0, \infty)$. Conversely, if the function $f(x)$ is given by formula 3.9(8) on the semi-axis $(-\infty, 0]$, it is immediately clear that it is infinitely differentiable and absolutely monotonic on $(-\infty, 0]$.

3.9.3. It is obvious that if $f(x)$ is absolutely monotonic on $(0, b)$, consecutive derivatives of the function $\phi(x) = f(b-x)$ have opposite signs on this range, and conversely. Thus, the class of functions which are regularly monotonic on $(0, b)$, and for which for all $k = 0, 1, 2, \ldots$

$$f^{(k)}(x) f^{(k+1)}(x) \leqslant 0, \tag{9}$$

becomes, if we replace x by $b-x$, identical with the class of functions absolutely monotonic in this range. An essentially new class, in a certain sense opposite to the last, is formed by functions regularly monotonic on the given segment, the successive derivatives of which, in place of condition 3.9(9), satisfy the condition

$$f^{(k)}(x) f^{(k+2)}(x) \leqslant 0 \tag{10}$$

for all $k = 0, 1, 2, \ldots$ Such functions are known as cyclically monotonic. *Any function cyclically monotonic in the range (a, b) is an integral function of finite order***. Without loss of generality it is possible to put $a = 0$. Since

$$f^{(k)}(x) = f^{(k)}(0) + \int_0^x f^{(k+1)}(t)\, dt$$

we have, supposing that $f(x) \geqslant 0$ and $f'(x) \geqslant 0$,

$$|f^{(2k)}(x)| \geqslant \left| \int_0^x f^{(2k+1)}(t)\, dt \right| \geqslant x\, |f^{(2k+1)}(x)|. \tag{11}$$

** S. N. Bernstein [40].

The function $\phi(x) = f(b-x)$ is also cyclically monotonic in $(0, b)$ and $\phi'(x) \leqslant 0$, $\phi''(x) \leqslant 0$. Hence since

$$\phi^{(2k+1)}(x) = \phi^{(2k+1)}(0) + \int_0^x \phi^{(2k+2)}(t)\,dt,$$

it follows that

$$|\phi^{(2k+1)}(x)| \geqslant \left| \int_0^x \phi^{(2k+2)}(t)\,dt \right| \geqslant x\,|\phi^{(2k+2)}(x)|. \tag{12}$$

Putting $x = \frac{b}{2}$ in 3.9(11) and 3.9(12), we obtain for any $k = 0, 1, 2, \ldots$

$$\left| f^{(k)}\left(\frac{b}{2}\right) \right| \geqslant \frac{b}{2} \left| f^{(k+1)}\left(\frac{b}{2}\right) \right|.$$

Consequently

$$\left| f^{(k)}\left(\frac{b}{2}\right) \right| \leqslant \left(\frac{2}{b}\right)^k \left| f\left(\frac{b}{2}\right) \right|. \tag{13}$$

This means that the radius of convergence of the Taylor series

$$f(x) = \sum_{k=0}^{\infty} \frac{1}{k!} f^{(k)}\left(\frac{b}{2}\right)\left(x - \frac{b}{2}\right)$$

is infinite, i.e. $f(x)$ is an integral function. The inequality 3.9(13) shows that $f(x)$ is an integral function of exponential type of degree σ not greater than $\frac{2}{b-a}$. This result can be improved. S. N. Bernstein [40] proved that in fact *the degree σ of a function cyclically monotonic in the range* (a, b) *is not greater than* $\frac{\pi}{2(b-a)}$. The last inequality is now definitive. For the function $f(x) = \sin\frac{\pi(x-a)}{2(b-a)}$ it is an equality.

As is obvious from the inequality 3.9(13) a function $f(x)$ (which is not linear) can possess the property of cyclical monotonicity only on a finite segment**.

In the class of all integral functions of finite degree, cyclically monotonic functions play a part similar to that played by absolutely monotonic functions in the study of all analytic functions (see section 3.9.1) and by

** In virtue of 3.9(13) as $x \to \infty$ we have $|f^{(k)}(x)| = O(x^{-k+1})$, since $f(x) \geqslant 0$, $f''(x) \leqslant 0$. Consequently, $f''(x) = O\left(\frac{1}{x}\right)$ and hence $f''(x) = 0$.

ordinary monotonic functions in the theory of functions of bounded variation.

On any segment $[a, b]$ *of length less than* $\frac{\pi}{2\sigma}$ *any integral function* $f(x)$ *of degree* σ *is the difference of two functions cyclically monotonic on this segment***. If $p > \sigma$, then $g(x) = f\left(\frac{x}{p}\right)$ is an integral function of degree $\frac{\sigma}{p} < 1$ and consequently*** there exists a finite number M such that, for all $k \geqslant 0$,

$$|g^{(k)}(x)| \leqslant M$$

on any finite segment $[\alpha, \beta] \in [a, b]$.

If we take $\varepsilon > 0$ and choose a constant A such that $M < A \sin \varepsilon$, the function $\phi(x) = A \sin (x-a)$, which is cyclically monotonic on $\left[a+\varepsilon, a+\frac{\pi}{2}-\varepsilon\right]$, will possess the property that everywhere on this segment

$$|\phi^{(k)}(x)| > M \geqslant |g^{(k)}(x)|$$

for any $k \geqslant 0$.

Hence the function $\psi(x) = \phi(x) - g(x)$ is cyclically monotonic on $\left[a+\varepsilon, a+\frac{\pi}{2}-\varepsilon\right]$, and hence on the given segment $g(x)$ is representable as the difference

$$g(x) = \phi(x) - \psi(x).$$

It follows that the function $f(x) = g(px)$ is representable as the difference of two cyclically monotonic functions in the range $\left[\frac{a+\varepsilon}{p}, \frac{a-\varepsilon}{p}+\frac{\pi}{2p}\right]$.

It is obvious from the example of absolutely monotonic and cyclically monotonic functions that changing the law of the alternation of signs of the successive derivatives of regularly monotonic functions has an effect on their analytical nature. In the general case in order to characterize the type of regular monotonicity we may take into consideration the frequency $\nu(f; n)$ of the alternation of signs in the series $f(x)$, $f'(x), \ldots, f^{(n)}(x)$. Absolute monotonicity corresponds to the frequencies

---

** S. N. Bernstein [40].

*** By the inequality

$$\overline{\lim_{\nu \to \infty}} \sqrt[\nu]{|g^{\nu}(0)|} \leqslant \frac{\sigma}{p} < 1.$$

$\nu(f; n) \equiv 0$ and $\nu(f; n) \equiv n$, and cyclical monotonicity to the frequency $\nu(f; n) = \left[\frac{n}{2}\right]$. Thus in a certain sense the two cases of regular monotonicity considered are extremes.

3.10. On quasi-analytic classes of functions

One of the most important properties of the class of analytic functions is the property of uniqueness. A function $f(x)$ which is analytic on a given segment is completely determined there by its values in any interval or by its value and the values of all its derivatives at any point.

3.10.1. However it is easily seen that for the validity of this property of the function $f(x)$ there is in general no need to assume its analyticity on the given segment, i.e. the satisfaction of condition 3.7(3) for all natural numbers k. Thus, for example, it is sufficient merely to assume that

$$\varliminf_{n\to\infty} \frac{1}{n} \sqrt[n]{M_n(f)} < \infty, \tag{1}$$

and it can then be asserted that *if at some point $c \in (a, b)$ the given function and all its derivatives vanish, then $f(x) \equiv 0$ on $[a, b]$.* In fact, as

$$f(x) = \frac{f^{(n)}[c+\theta_n(x-c)]}{n!}(x-c)^n,$$

we see, choosing a value of n from a subsequence n_k for which

$$\lim_{n_k\to\infty} \frac{1}{n_k} \sqrt[n_k]{M_{n_k}(f)} = \varliminf_{n\to\infty} \frac{1}{n} \sqrt[n]{M_n(f)},$$

that for some $M > 0$

$$|f^{(n_k)}[c+\theta_{n_k}(x-c)]| \leqslant M_{n_k}(f) < M^{n_k} n_k!$$

and consequently

$$|f(x)| < (M|x-c|)^{n_k}.$$

Hence if $|x-c| < \frac{1}{M}$ it follows that $f(x) = 0$.

Choosing in place of c the points $c \pm \frac{1}{2M}$ and once more repeating the same reasoning, we obtain $f(x) \equiv 0$ on the whole segment $[a, b]$.

If now $n_1 < n_2 < n_3 < \ldots$ is a given infinite sequence of natural numbers, the class of all infinitely differentiable functions $f(x)$, defined on $[a, b]$ and such that

$$\sqrt[n_k]{M_{n_k}(f)} \leqslant M(f) n_k \quad \big(M(f) \text{ constant}\big), \tag{2}$$

exactly as in the case $n_k = k$ (analytic functions) possess the property that any two functions $f_1(x)$ and $f_2(x)$ belonging to it and identical in an arbitrarily small interval are identical over the whole segment. Classes of functions which possess this property are called *quasi-analytic*.

3.10.2. Thus any sequence of natural numbers n_k which increases without limit defines some quasi-analytic class of functions which satisfies condition 3.10(2) on a given range. We note that *if the sequence n_k is such that the ratio $\frac{n_{k+1}}{n_k}$ remains bounded as $k \to \infty$, then the corresponding quasi-analytic class contains only analytic functions.* In fact, since

$$f(x) = \sum_{k=0}^{n-1} \frac{1}{k!} f^{(k)}(x_0)\,(x-x_0)^k + \\ + \frac{f^{(n)}\,[x_0+\theta\,(x-x_0)]}{n!}(x-x_0)^n (0 < \theta < 1),$$

where $x_0 \in [a, b]$, we have for the polynomial

$$P_{n-1}(x) = \sum_{k=0}^{n-1} \frac{1}{k!} f^{(k)}(x_0)\,(x-x_0)^k$$

the estimate

$$|P_{n-1}(x)| \leqslant M_0(f) + \frac{|x-x_0|^n}{n!} M_n(f). \tag{3}$$

From this it immediately follows with the help of theorem 4.8.8 (see 4.8(48)) that for any $k < n$

$$|f^{(k)}(x_0)| = |P^{(k)}_{n-1}(x_0)| \leqslant \left(\frac{Cn^2}{k}\right)^k \left[M_0(f) + \frac{(b-a)^n}{n!} M_n(f)\right], \tag{4}$$

where C is a constant which does not depend on f, n and k. Since the point x_0 is arbitrary, it follows that

$$M_k(f) \leqslant \left(\frac{Cn^2}{k}\right)^k \left[M_0(f) + \frac{(b-a)^n}{n!} M_n(f)\right].$$

Consequently, if $n_{k-1} \leqslant n < n_k$, then

$$M_n(f) \leqslant \left(\frac{Cn_k^2}{n}\right)^n \left[M_0(f) + \frac{(b-a)^{n_k}}{n_k!} M_{n_k}(f)\right], \tag{5}$$

and in virtue of 3.10(2) with the condition $\frac{n_{k+1}}{n_k} = O(1)$ it follows that

$$\sqrt[n]{M_n(f)} \leqslant C(f)\,n$$

for all natural numbers n, i.e. the function $f(x)$ is analytic on $[a, b]$ (see 3.7(1)).

In order to verify the validity of the given assertion in the case where the range $[a, b]$ is not bounded, we note that for $a = -1$, $b = 1$, $x_0 = 0$ we have from 3.10(3), thanks to theorem 2.9.12 (see also 2.6(9)), the inequality

$$|f^{(k)}(0)| = |P_{n-1}^{(k)}(0)| \leqslant \frac{n^k}{r^k}\left[M_0(f) + \frac{r^n M_n(f)}{n!}\right]$$

for $0 < r \leqslant 1$. Substituting $r = \left[\frac{M_0(f)n!}{M_n(f)}\right]$ if $M_0(f)n! \leqslant M_n(f)$, and $r = \frac{\sqrt[n]{n!}}{n}$ if $M_0(f)n! > M_n(f)$, we obtain (see 3.7(2)) the estimate**

$$|f^{(k)}(0)| \leqslant 2e^k \max\{M_0^{1-\frac{k}{n}}(f) M_n^{\frac{k}{n}}(f), n^k M_0(f)\}. \tag{6}$$

Similarly for $a = -1$, $b = 1$, $x_0 = -1$, $0 < r \leqslant 1$ we obtain from 3.10(4)

$$|f^{(k)}(-1)| = |P_{n-1}^{(k)}(-1)| \leqslant \frac{1}{r^k}\left(\frac{Cn^2}{k}\right)^k \left[M_0(f) + \frac{(2r)^n}{n!} M_n(f)\right],$$

whence, putting $r = \left[\frac{M_0(f)n!}{2^n M_n(f)}\right]^{\frac{1}{n}}$ if $M_0(f)n! \leqslant 2^n M_n(f)$ and $r = \frac{\sqrt[n]{n!}}{n}$ otherwise, we find*:

$$|f^{(k)}(-1)| \leqslant \left(\frac{C_0 n}{k}\right)^k \max\{M^{1-\frac{k}{n}}(f) M_n^{\frac{k}{n}}(f), n^k M_0(f)\}, \tag{7}$$

where C_0 is an absolute constant.

If now $F(x)$ is defined on the whole real axis, then, considering the function $f(x) = F(t+\lambda x)$ $(\lambda > 0,\ -\infty < t < \infty)$ on the segment $-1 \leqslant x \leqslant 1$, we have by the inequality 3.10(6)

$$|F^{(k)}(t)| \leqslant 2e^k \max\left\{M_0^{1-\frac{k}{n}}(F) M_n^{\frac{k}{n}}(F), \left(\frac{n}{\lambda}\right)^k M_0(F)\right\},$$

** Gorny [1]; see also H. Cartan [1].

and in consequence of the arbitrariness of λ

$$M_k(F) \leqslant 2e^k M^{1-\frac{k}{n}}(F) M_n^{\frac{k}{n}}(F)^{**}. \tag{8}$$

Similarly, if $F(x)$ is defined on the semi-axis $[0, \infty)$, then considering the function $f(x) = F[t+\lambda(x+1)]$ and using the inequality 3.10(7) we obtain:

$$M_k(F) \leqslant \left(\frac{C_0 n}{k}\right)^k M_0^{1-\frac{k}{n}}(F) M_n^{\frac{k}{n}}(F). \tag{9}$$

On account of the inequalities 3.10(8) and 3.10(9), in order to prove the analyticity of $F(x)$ it is sufficient, as in the case of a finite segment, to make use of 3.10(2) and $\frac{n_{k+1}}{n_k} = O(1)$.

3.10.3. Denjoy and Carleman were the first to consider the problem of the conditions which must be satisfied by the sequence of numbers $M_n (n = 0, 1, 2, \ldots)$ in order that the class of infinitely differentiable functions $f(x)$ determined by it which possess the property

$$\sqrt[n]{M_n(f)} \leqslant M(f) M_n, \tag{10}$$

should be quasi-analytic in the sense that any two functions $f_1(x)$ and $f_2(x)$ of the class, which together with all their derivatives assume identical values at some single point, should be identically equal over the whole segment***. It turns out**** that the class *of all infinitely differentiable functions* $f(x)$ *which satisfy condition* 3.10(10) *is quasi-analytic in the given sense if and only if the series*

$$\sum_{n=1}^{\infty} \frac{1}{M_n^*}, \tag{11}$$

where $M_n^* = \inf_{k \geqslant n} M_k^{\frac{1}{k}}$, *diverges.* Several different proofs of this important theorem are known†, but an exposition of any of them would take us outside the scope of the present chapter.

** This inequality can be improved. For functions considered on the whole real axis A. N. Kolmogorov [4] obtained the inequality 3.10(8) with an exact constant which in fact is always less than $\frac{\pi}{2}$. In the integral metric L_p a similar inequality was obtained by Stein [1].

*** This problem was stated by Hadamard [1].

**** Carleman [1].

† See, for example, S. Mandelbrojt [2].

The criterion for the function $f(x)$ to belong to the quasi-analytic class of Denjoy and Carleman is the condition

$$\sum_{n=1}^{\infty} \frac{1}{\inf\limits_{k \geqslant n} \sqrt[k]{M_k(f)}} = \infty .$$

In particular, every infinitely differentiable function $f(x)$ for which

$$\sum_{n=1}^{\infty} \frac{1}{\sqrt[n]{M_n(f)}} = \infty \qquad (12)$$

is quasi-analytic in this sense, as also is any function which satisfies the condition 3.10(1)**. In chapter V the fundamental constructive property of such functions is established.

The set of all functions which satisfy condition 3.10(12), as was discovered by Mandelbrojt [1], plays the same part in the class of all infinitely differentiable functions as the set of all monotonic functions does in the class of functions of bounded variation, the absolutely monotonic functions in the theory of analytic functions and the set of cyclically monotonic functions in the class of integral functions of exponential type. *Any infinitely differentiable function is the difference of two functions which are quasi-analytic in the sense of Denjoy (i.e. they satisfy condition* 3.10(12)).

3.10.31. The set W of infinitely differentiable functions $f(x)$ defined on the finite segment $[a, b]$ and satisfying the conditions $\sup\limits_{f \in W} M_0(f) < \infty$

** The fact that functions satisfying condition 3.10(1) are quasi-analytic in the sense of Denjoy and Carleman follows directly from the following simple lemma. *If the terms of the series of numbers* $\sum\limits_{n=1}^{\infty} u_n$ *decrease monotonically and an infinite sequence* $\{n_k\}$ *exists such that* $u_{n_k} \geqslant \frac{1}{n_k}$, *then the given series diverges.* In virtue of our assumptions

$$\sum_{n=1}^{\infty} u_n \geqslant \sum_{n=1}^{n_1} u_n + \sum_{k=1}^{\infty} \frac{n_{k+1}-n_k}{n_{k+1}} .$$

It may be assumed that as $k \to \infty$ the limit of the ratio $\frac{n_k}{n_{k+1}}$ exists and is equal to unity. Otherwise the divergence of the series on the right-hand side is obvious. With this condition, if $\frac{n_k}{n_{k+1}} \geqslant A > 0$ it follows that

$$\sum_{k=1}^{\infty} \frac{n_{k+1}-n_k}{n_{k+1}} \geqslant A \sum_{k=1}^{\infty} \frac{n_{k+1}-n_k}{n_k} > A \int_{n_1}^{\infty} \frac{\mathrm{d}x}{x} = \infty ,$$

i.e. the series $\sum\limits_{n=1}^{\infty} u_n$ diverges.

and 3.10(10), where $M_n < \infty$ $(n = 1, 2, \ldots)$ and $\sup_{f \in W} M(f) < \infty$, is by section 3.51 compact in the same sense as in theorem 3.7.2.

3.10.4. An important generalisation in another direction of quasi-analytic classes of functions which satisfy condition 3.10.(2) was indicated by S. N. Bernstein [7]. It is based on a natural constructive property of such functions (see section 5.8.1). It consists in the fact that for any infinitely differentiable function which satisfies condition 3.10(1) we always have

$$\varlimsup_{n \to \infty} \sqrt[n]{E_n(f; a, b)} < 1. \tag{13}$$

In chapter VI classes of functions $f(x)$ are considered which satisfy condition 3.10(13), and it is proved that whatever may be the sequence of natural numbers $n_1 < n_2 < n_3 < \ldots$ increasing without limit, the class of all continuous functions $f(x)$ for which for some $\varrho < 1$

$$E_{n_k}(f; a, b) \leqslant M(f)\varrho^{n_k} \tag{14}$$

is quasi-analytic**. It is clear that the differential properties of the functions composing such a class depend on the nature of the sequence n_k. In particular, it may happen that functions are quasi-analytic in this sense which are differentiable only a finite number of times or even not differentiable anywhere.

We also note that the set of all continuous functions which satisfy condition 3.10(13) occupies the same position in the class of all continuous functions as functions quasi-analytic in the sense of Denjoy in the class of all infinitely differentiable functions. It will be proved in section 6.6.3, that every function continuous on $[a, b]$ is the difference of two functions quasi-analytic in the sense of S. N. Bernstein***.

3.11. Conjugate classes of functions defined on the whole real axis

Side by side with the classes W of functions $f(x)$ defined on the whole real axis, which we have already considered, the classes W conjugate to them are also frequently studied.

3.11.1. In the case where W is the corresponding class of periodic functions, the class $\tilde{W}$ conjugate to it consists of the functions $\tilde{f}(x)$ known as trigonometrically conjugate to $f(x)$ and defined as follows.

** The classes of functions which satisfy condition 3.10(14), will be called quasi-analytic in the sense of S. N. Bernstein.

*** A. I. Markushevich [1].

If $f(x)$ is a Lebesgue-integrable function of period 2π with Fourier series

$$f(x) \sim \sum_{k=-\infty}^{\infty} c_k e^{ikx} = \frac{a_0}{2} + \sum_{k=1}^{\infty} (a_k \cos kx + b_k \sin kx), \tag{1}$$

then in the case where the conjugate trigonometric series

$$-i \sum_{k=-\infty}^{\infty} \operatorname{sign} k \cdot c_k e^{ikx} = \sum_{k=1}^{\infty} (a_k \sin kx - b_k \cos kx) \tag{2}$$

is the Fourier series of some function $\tilde{f}(x)$, the latter is said to be trigonometrically conjugate to $f(x)$**. This terminology is justified by the fact that if the series 3.1(1) is added to the series 3.11(2) multiplied by i, we obtain the power series

$$\frac{a_0}{2} + 2\sum_{k=1}^{\infty} c_k z^k$$

for $|z| = 1$, the real and imaginary parts of which represent a pair of conjugate harmonic functions in the circle $|z| < 1$.

There also exists another, wider, definition of conjugate functions as the limit

$$\tilde{f}(x) = \lim_{h \to 0} \left\{ -\frac{1}{\pi} \int_h^{\pi} \frac{f(x+t) - f(x-t)}{2 \tan \frac{t}{2}} \, dt \right\}$$

$$= -\frac{1}{\pi} \int_0^{\pi} \frac{f(x+t) - f(x-t)}{2 \tan \frac{t}{2}}, \tag{3}$$

which exists and is finite almost everywhere for any integrable function $f(x)$***. However, a function $\tilde{f}(x)$ defined in this way is not necessarily

** By the Riesz–Fisher theorem a function $\tilde{f}(x)$ conjugate in the given sense is uniquely defined for any function $f(x)$ which is of integrable square. It must also be borne in mind that in the general case, a trigonometric series conjugate to a Fourier–Lebesgue series need not be a Fourier–Lebesgue series. This is easily seen by considering the example of the series

$$\sum_{k=2}^{\infty} \frac{\cos kx}{\ln k},$$

which converges at every point $x \in (0, 2\pi)$ to an integrable function and consequently is the Fourier series of its sum. (See A. Zygmund [1]).

*** This could be verified by using theorem 3.8.1 (see A. Zygmund [1] section 3.45).

Lebesgue-integrable**. In those cases where it is integrable, its Fourier–Lebesgue series is the series 3.11(2), and hence the definition 3.11(3) is identical with the previous one. It follows from this that in those cases where the function $\tilde{f}(x)$ of 3.11(3) is integrable, the dual of 3.11(3)

$$f(x) = \frac{1}{\pi}\int_0^\pi \frac{\tilde{f}(x+t)-\tilde{f}(x-t)}{2\tan\frac{t}{2}}\,\mathrm{d}t \tag{4}$$

holds. These facts are treated in the fundamental monograph of A. Zygmund [1], where the reader will find detailed proofs and examples. We will give another important statement proved there, due to M. Riesz.

If $f(x)\in L_q^*(1<q<\infty)$, *then* $\tilde{f}(x)\in L_q^*$ *also. In this case the inequality*

$$\int_0^{2\pi} |\tilde{f}(x)|^q\,\mathrm{d}x \leqslant M_q^* \int_0^{2\pi} |f(x)|^q\,\mathrm{d}x \tag{5}$$

holds, in which the constant M_q^* *does not depend on* f.

The definition of the conjugate function $\tilde{f}(x)$ by formula 3.11(3), as is not difficult to prove with the help of the identity

$$\frac{t}{2}\cot\frac{t}{2} = 1 + t\sum_{k=1}^{\infty}\left\{\frac{1}{t-2k\pi}+\frac{1}{t+2k\pi}\right\}, \tag{6}$$

is equivalent to

$$\tilde{f}(x) = -\frac{1}{\pi}\int_0^\infty \frac{f(x+t)-f(x-t)}{t}\,\mathrm{d}t. \tag{7}$$

In this form the given formula can be taken as the definition of a conjugate function for every function $f(x)$, which belongs to the class L_q $(1\leqslant q<\infty)$ on $(-\infty, \infty)$.

For arbitrary functions defined on the whole real axis, as in the periodic case, a theorem holds which is similar to the theorem of M. Riesz mentioned above***. *If* $f(x)$ *belongs to the class* L_q $(1<q<\infty)$ *on the whole interval* $(-\infty, \infty)$, *then the conjugate function* $f(x)$, *defined by formula* 3.11(7), *belongs to the same class and*

$$\int_{-\infty}^{\infty} |\tilde{f}(x)|^q\,\mathrm{d}x \leqslant M_q \int_{-\infty}^{\infty} |f(x)|^q\mathrm{d}x. \tag{8}$$

** To verify this, it is sufficient to take the periodic function $f(x)$, equal to zero on $[-\pi, 0]$ and equal to $\left(x\ln^2\frac{x}{2\pi}\right)^{-1}$ on $(0, \pi)$.

*** M. Riesz [1]. See also E. C. Titchmarsh [2].

Here the constant M_q does not depend on f. We note also that for functions $f(x)$ which belong to the class L_q $(1 < q < \infty)$ on the whole interval $(-\infty, \infty)$, as in the periodic case, the formula**

$$f(x) = \frac{1}{\pi}\int_0^\infty \frac{\tilde{f}(x+t)-\tilde{f}(x-t)}{t}\,dt, \tag{10}$$

dual to 3.11(7), holds.

3.11.2. In the consideration of arbitrary functions defined on the whole real axis, it is sometimes more convenient to use another expression which is connected with the Fourier transform of the function $f(x)$ and is similar to the first of the definitions of a conjugate function in terms of a sequence of Fourier coefficients given in section 3.11.1. Let us consider the case of functions $f(x)$ belonging on $(-\infty, \infty)$ to the class L_2, which is most important in many other problems also. Let $f(x)$ be an arbitrary complex function defined on the whole real x-axis and such that

$$\int_{-\infty}^\infty |f(x)|^2\,dx < \infty. \tag{10}$$

3.11.21. The following fundamental theorem is due to Plancherel***. *For any function $f(x)$ which satisfies condition* 3.11(10) *the integral*

$$\frac{1}{\sqrt{2\pi}}\int_{-\infty}^\infty f(t)\frac{e^{-itx}-1}{it}\,dt$$

has almost everywhere a finite derivative

$$F(x) = \frac{1}{\sqrt{2\pi}}\frac{d}{dx}\int_{-\infty}^\infty f(t)\,\frac{e^{-itx}-1}{-it}\,dt \tag{11}$$

for which

$$\int_{-\infty}^\infty |F(x)|^2\,dx = \int_{-\infty}^\infty |f(x)|^2\,dx \tag{12}$$

and

$$f(x) = \frac{1}{\sqrt{2\pi}}\frac{d}{dx}\int_{-\infty}^\infty F(t)\frac{e^{itx}-1}{it}\,dt. \tag{13}$$

** See E. C. Titchmarsh [2].

*** Plancherel [1]. See also E. C. Titchmarsh [2]. The proof given here is due to S. Bochner [1].

Moreover, as $\lambda \to \infty$

$$\int_{-\infty}^{\infty} \left| F(x) - \frac{1}{\sqrt{2\pi}} \int_{-\lambda}^{\lambda} f(t)\, e^{itx}\, dt \right|^2 dx \to 0,$$

$$\int_{-\infty}^{\infty} \left| f(x) - \frac{1}{\sqrt{2\pi}} \int_{-\lambda}^{\lambda} F(t)\, e^{itx}\, dt \right|^2 dx \to 0. \tag{14}$$

The function $F(x)$ is said to be the Fourier transform of $f(x)$.

If $f(x)$ satisfies condition 3.11(10), then from theorem 1.4 there follows the existence of a sequence of functions $f_n(x)$, each of which vanishes outside some finite interval, such that

$$\int_{-\infty}^{\infty} | f(x) - f_n(x) |^2\, dx \to 0 \quad (n \to \infty)$$

and $\omega(f_n; t) \leqslant M_n t$. For the function

$$F_n(x) = \frac{1}{\sqrt{2\pi}} \int_{-\infty}^{\infty} f_n(t)\, e^{-ixt}\, dt$$

and arbitrary $\lambda > 0$

$$\int_{-\lambda}^{\lambda} | F_n(x) |^2\, dx = \frac{1}{2\pi} \int_{-\lambda}^{\lambda} \int_{-\infty}^{\infty} f_n(t)\, e^{-ixt}\, dt \int_{-\infty}^{\infty} \overline{f_n(u)}\, e^{ixu}\, du\, dx$$

$$= \frac{1}{\pi} \int_{-\infty}^{\infty} f_n(t)\, dt \int_{-\infty}^{\infty} \overline{f_n(u+t)} \frac{\sin \lambda u}{u}\, du.$$

But as $\omega(f_n; t) \leqslant M_n t$, it is not difficult to prove that as $\lambda \to \infty$

$$\int_{-\infty}^{\infty} \overline{f_n(t+u)} \frac{\sin \lambda u}{u}\, du \to \overline{f_n(t)}$$

uniformly with respect to all t of any finite segment. Passing to the limit, we obtain

$$\int_{-\infty}^{\infty} | F_n(x) |^2\, dx = \int_{-\infty}^{\infty} | fn(x) |^2\, dx.$$

From the same considerations we conclude that

$$\int_{\infty}^{\infty} | F_n(x) - F_m(x) |^2\, dx = \int_{-\infty}^{\infty} | f_n(x) - f_m(x) |^2\, dx.$$

Hence in virtue of the completeness of the space of functions of integrable square there follows the existence of a function $F(x)$ for which as $n \to \infty$

$$\int_{-\infty}^{\infty} | F(x)-F_n(x)|^2 \, dx \to 0,$$

and hence 3.11(12) holds for $F(x)$.

The validity of 3.11(11) follows from the fact that

$$\int_0^x F_n(y) \, dy = \frac{1}{\sqrt{2\pi}} \int_0^x dy \int_{-\infty}^{\infty} f_n(u) \, e^{-iyu} \, du = \frac{1}{\sqrt{2\pi}} \int_{-\infty}^{\infty} f_n(u) \frac{e^{-iux}-1}{-iu} \, du,$$

after passing to the limit and differentiating with respect to x.

We note that 3.11(12) is equivalent to the equation

$$\int_{-\infty}^{\infty} F(x) \overline{\Phi(x)} \, dx = \int_{-\infty}^{\infty} f(x) \overline{\phi(x)} \, dx \tag{15}$$

for any two functions $f(x)$ and $\phi(x)$ which satisfy the condition 3.11(10), and their Fourier transforms $F(x)$ and $\Phi(x)$. Putting here $\phi(x) = 1$ for $0 \leqslant x \leqslant u$ and $\phi(x) = 0$ for the remaining values of x, we have by formula 3.11(11)

$$\Phi(x) = \frac{1}{\sqrt{2x}} \frac{d}{dx} \int_0^u \frac{e^{-itx}-1}{-it} \, dt = \frac{1}{\sqrt{2\pi}} \frac{e^{-iux}-1}{-ix},$$

i.e.

$$\int_0^u f(x) \, dx = \frac{1}{\sqrt{2\pi}} \int_{-\infty}^{\infty} F(x) \frac{e^{iux}-1}{ix} \, dx,$$

and after differentiation with respect to u we obtain 3.11(13). Putting $\phi(x) = f(x)$ for $|x| \leqslant \lambda$ and $\phi(x) = 0$ for the remaining values of x, we have by formula 3.11(11)

$$\Phi(x) = \frac{1}{\sqrt{2\pi}} \frac{d}{dx} \int_{-\lambda}^{\lambda} f(t) \frac{e^{-itx}-1}{-it} \, dt = \frac{1}{\sqrt{2\pi}} \int_{-\lambda}^{\lambda} f(t) e^{-ixt} \, dt.$$

Consequently, if we take account of the equation

$$\int_{-\infty}^{\infty} |F(x)-\Phi(x)|^2 \, dx = \int_{-\infty}^{\infty} |f(x)-\phi(x)|^2 \, dx,$$

we obtain the first of the relations 3.11(14). The second is established in a similar way.

3.11.22. If in 3.11(15) we replace $\phi(x)$ by $\overline{\phi(u-x)}$ and note that the Fourier transform $\Phi(x)$ is then replaced by $\overline{\Phi(x)}\,\mathrm{e}^{-iux}$, we also obtain the following formula for the convolution of two functions:

$$\int_{-\infty}^{\infty} f(t)\,\phi(x-t)\,\mathrm{d}t = \int_{-\infty}^{\infty} F(t)\Phi(t)\,\mathrm{e}^{itx}\,\mathrm{d}t. \tag{16}$$

In the case where the function $f(x)$ ($F(x)$) in formula 3.11(11) (formula 3.11(13)) is absolutely integrable on $(-\infty, \infty)$, the differentiation can be carried out under the integral sign, and then the Fourier transform 3.11(11) (the transform 3.11(13)) assumes the form

$$F(x) = \frac{1}{\sqrt{2\pi}} \int_{-\infty}^{\infty} f(t)\,\mathrm{e}^{-itx}\,\mathrm{d}t \tag{17}$$

$$\left(\text{or } f(x) = \frac{1}{\sqrt{2\pi}} \int_{-\infty}^{\infty} F(t)\,\mathrm{e}^{itx}\,\mathrm{d}t, \text{ respectively}\right).$$

Formula 3.11(17) defines the Fourier transform for all functions absolutely integrable on $(-\infty, \infty)$. It follows from 3.11(16) that the Fourier transform of the convolution of two functions is identical with the product of their Fourier transforms.

Similarly, applying 3.11(15), we obtain:

$$\int_{-\infty}^{\infty} f(t)\,\phi(t)\,\mathrm{e}^{-itx}\,\mathrm{d}t = \int_{-\infty}^{\infty} F(t)\Phi(x-t)\,\mathrm{d}t, \tag{18}$$

i.e. the Fourier transform of the product of two functions of integrable square is identical with the convolution of their Fourier transforms.

3.11.23. In the case where the function $f(x)$ is even (odd), the Fourier transform 3.11(11) and the inverse transform 3.11(13) are of the form

$$F(x) = \sqrt{\frac{2}{\pi}}\,\frac{\mathrm{d}}{\mathrm{d}x}\int_{0}^{\infty} f(t)\,\frac{\sin tx}{t}\,\mathrm{d}t,$$

$$f(x) = \sqrt{\frac{2}{\pi}}\,\frac{\mathrm{d}}{\mathrm{d}x}\int_{0}^{\infty} F(t)\,\frac{\sin tx}{t}\,\mathrm{d}t.$$

(correspondingly $F(x) = \sqrt{\dfrac{2}{\pi}}\,\dfrac{\mathrm{d}}{\mathrm{d}x}\displaystyle\int_{0}^{\infty} f(t)\,\frac{1-\cos tx}{t}\,\mathrm{d}t,$

$$f(x) = \sqrt{\frac{2}{\pi}}\,\frac{\mathrm{d}}{\mathrm{d}x}\int_{0}^{\infty} F(t)\,\frac{1-\cos tx}{t}\,\mathrm{d}t.$$

The corresponding form of Plancherel's theorem for the class of functions of integrable square on the semiaxis is formulated as in theorem 3.11.2.

3.11.24. The result of the formal operations

$$\frac{1}{\sqrt{2\pi}}\int_{-\infty}^{\infty}\tilde{f}(t)\,e^{-ixt}\,dt = -\frac{1}{\pi\sqrt{2\pi}}\int_{-\infty}^{\infty}\int_{0}^{\infty}\frac{f(t+u)-f(t-u)}{u}\,du\,e^{-itx}\,dt$$

$$= -\frac{1}{\pi\sqrt{2\pi}}\int_{0}^{\infty}\frac{du}{u}\int_{-\infty}^{\infty}[f(t+u)-f(t-u)]\,e^{-itx}\,dt$$

$$= -\frac{1}{\pi}\int_{0}^{\infty}\frac{du}{u}(e^{iux}-e^{-iux})\,F(x)$$

$$= -\frac{2i}{\pi}F(x)\int_{0}^{\infty}\frac{\sin ux}{u}\,du = -iF(x)\operatorname{sign} x$$

indicates the connection between the Fourier transforms of two conjugate functions. In what follows we shall use this relation as the basis of the definition of the function $\tilde{f}(x)$ conjugate to the function $f(x)$, satisfying the condition 3.11(10). Thus in the class L_2, by definition,

$$\tilde{f}(x) = \frac{1}{\sqrt{2\pi}}\frac{d}{dx}\int_{-\infty}^{\infty}\operatorname{sign} t\cdot F(t)\frac{1-e^{ixt}}{t}\,dt. \tag{19}$$

The given definition of a conjugate function is easily generalized to a wider class of functions, defined on the whole real axis, including, in particular, all periodic functions** of integrable square. This class (Wiener's class) consists of all integrable functions $f(x)$ for which the function $\dfrac{|f(x)|^2}{1+x^2}$ is integrable on $(-\infty, \infty)$. If $f(x)$ is such a function and for certain values of the numbers a, b (in general complex) the modulus of the quotient $\dfrac{f(x)-a}{x-b}$ is of integrable square on $(-\infty, \infty)$, and $F^*(x)$ is the Fourier transform of this function, then by formula 3.11(13)

$$f(x) = a+\frac{x-b}{\sqrt{2\pi}}\frac{d}{dx}\int_{-\infty}^{\infty}F^*(t)\frac{e^{itx}-1}{it}\,dt.$$

** See N. I. Akhiezer [3].

From this, by analogy with 3.11(19), the conjugate function is defined by the formula

$$\tilde{f}(x) = \frac{x-b}{\sqrt{2\pi}} \frac{d}{dx} \int_{-\infty}^{\infty} \operatorname{sign} t \cdot F^*(t) \frac{1-e^{ixt}}{t} dt. \tag{20}$$

It would be possible to prove** that as for periodic functions of integrable square, so also for functions the square of the modulus of which is integrable on $(-\infty, \infty)$, the conjugate function defined by equation 3.11(20), differs only by a constant from the corresponding conjugate function defined at the beginning of section 3.11.1 or by equation 3.11(19).

3.11.3. The example

$$f(x) = \sum_{k=1}^{\infty} \frac{\sin kx}{k} \tag{21}$$

show that the function conjugate to a bounded function need not be bounded. Another example

$$f(x) = \sum_{k=2}^{\infty} \frac{\sin kx}{k \ln k} \tag{22}$$

shows that a function conjugate to a continuous (even absolutely continuous) function may be discontinuous and unbounded. Moreover, starting from 3.11(7), it can be verified immediately that in those cases where the continuous function $f(x)$ is periodic or belongs to the class $L_q(1 \leqslant q < \infty)$ on $(-\infty, \infty)$ and

$$\int_0^1 \frac{\omega(f; t)}{t} dt < \infty, \tag{23}$$

the conjugate function $\tilde{f}(x)$ is always continuous. Moreover, the results of chap. V (section 5.1.32) and chap. VI (sections 6.7.1 to 6.7.5) give the following theorem. *If the function $f(x)$ is periodic and for $l > r$* (r an integer)

$$\int_0^1 \frac{\omega_l(f; u)}{u^{r+1}} du < \infty, \tag{24}$$

** See N. I. Akhiezer [3].

then $\tilde{f}(x)$ has a continuous derivative of order r and

$$\omega_k(\tilde{f}^{(r)}; t) \leqslant C_{k,r,l}\left\{t^k \int_t^1 \frac{\omega_l(f; u)}{u^{k+r+1}}\,\mathrm{d}u + \int_0^t \frac{\omega_l(f; u)}{u^{r+1}}\,\mathrm{d}u\right\}, \tag{25}$$

where $C_{k,r,l}$ is a constant which does not depend on f**. In particular, putting $l = r+1$ and $\omega_{r+1}(f; t) = O(t^{r+\alpha})$ $(0 < \alpha < 1)$, we obtain from this that *if $f(x) \in W_*^{(r)}H^{(\alpha)}M (0 < \alpha < 1)$, then $\tilde{f}(x) \in W_*^{(r)}H^{(\alpha)}M_1$****. Thus, *for r an integer and $0 < \alpha < 1$ the classes $W_*^{(r)}H^{(\alpha)}M$ and $\tilde{W}_*^{(r)}H^{(\alpha)}M_1$ are identical apart from the values of the constants M and M_1*. As is shown by the example

$$f(x) = \sum_{k=1}^{\infty} \frac{\cos\left(kx + \dfrac{r\pi}{2}\right)}{k^{r+2}} \tag{26}$$

this assertion would no longer be true for $\alpha = 0$ or $\alpha = 1$.

It follows directly from the last theorem that if the periodic function $f(x)$ of period 2π is infinitely differentiable, then the same property is possessed by $\tilde{f}(x)$ also.

3.12. Various problems and theorems

1. Let $f(x)$ be an integrable periodic function of period 2π and

$$f_{h,r}(x) = \frac{1}{2h}\int_{x-h}^{x+h} f_{h,r-1}(t)\,\mathrm{d}t, \qquad f_{h,0}(x) = f(x) \tag{1}$$

be its r-th Steklov function with increment h.

(1) As $h \to 0$ we have for almost all x

$$f_{h,1}(x) \to f(x). \tag{2}$$

(2) If $\sum_{k=-\infty}^{\infty} c_k e^{ikx}$ is the Fourier series of the function $f(x)$, then for any integer $r \geqslant 1$

$$f_{h,r}(x) = \sum_{k=-\infty}^{\infty} \left(\frac{\sin kh}{kh}\right)^r c_k e^{ikx}. \tag{3}$$

(3) The following inequalities hold:

$$|f(x) - f_{h,r}(x)| \leqslant \int_{-h}^{h} \cdots \int_{-h}^{h} \omega(f; |t_1 + \cdots + t_r|)\,\mathrm{d}t_1 \ldots \mathrm{d}t_r,$$

$$|f(x) - f_{h,1}(x)| \leqslant \tfrac{1}{2}\omega_2(f; h),$$

$$|f(x) - f_{h,2}(x)| \leqslant \tfrac{1}{2}\omega_2(f; 2h),$$

$$|f_{h,r}^{(r)}(x)| \leqslant \frac{1}{(2h)^r}\,\omega_r(f; 2h). \tag{4}$$

** A. Zygmund [5] in the case $r = 0$, $k = 1$, $l = 1$. N. K. Bari and S. V. Stechkin [1] in the general case.

*** For $r = 0$ this was first established by I. I. Privalov [1].

(4) If the function $f(x)$ is differentiable and $|f'(x)| \leqslant 1$, then for $0 \leqslant h \leqslant \frac{\pi}{r}$

$$|f(x)-f_{h,r}(x)| \leqslant \frac{1}{2^{r-1}(r+1)!} A_{r+1}^{(r)} h, \tag{5}$$

where

$$A_m^{(n)} = \sum_{\nu=0}^{\left[\frac{n-1}{2}\right[} (-1)^\nu \binom{n}{\nu} (n-2\nu)^m \qquad (m \geqslant 0,\ n \geqslant 1).$$

The inequality 3.12(5) is precise.

(5) If the function $f(x)$ is differentiable twice and $|f''(x)| \leqslant 1$, then for $0 < h \leqslant \frac{\pi}{2r}$,

$$|f(x)-f_{h,r}(x)| \leqslant \tfrac{1}{6} rh^2. \tag{6}$$

The inequality 3.12(6) is precise.

(6) If the function $f(x)$ is such that

$$\max_x |f(x)-f_{h,r}(x)| = o(h^2) \qquad (r > 0)$$

as $h \to 0$, then $f(x) \equiv$ constant.

(7) The function $f(x)$ is convex on $[a, b]$ if and only if for $h > 0$ and $a \leqslant x-h < < x+h \leqslant b$

$$f(x) \leqslant f_{h,1}(x). \tag{7}$$

(8) If $f(x) \in L_q^* \ (1 \leqslant q < \infty)$, then

$$\left\{\int_0^{2\pi} |f(x)-f_{h,1}(x)|^q \, dx\right\}^{\frac{1}{q}} \leqslant \tfrac{1}{2} \omega_2(f; h)_{L_q^*},$$

$$\left\{\int_0^{2\pi} |f(x)-f_{h,2}(x)|^q \, dr\right\}^{\frac{1}{q}} \leqslant \tfrac{1}{2} \omega_2(f; 2h)_{L_q^*},$$

$$\left\{\int_0^{2\pi} |f_{h,r}^{(r)}(x)|^q \, dx\right\}^{\frac{1}{q}} \leqslant \frac{1}{(2h)^r} \omega_r(f; 2h)_{L_q^*}, \tag{8}$$

$$\left\{\int_0^{2\pi} |f_{h,1}(x)|^q \, dx\right\}^{\frac{1}{q}} \leqslant \left\{\int_0^{2\pi} |f(x)|^q \, dx\right\}^{\frac{1}{q}},$$

The relations 3.12.1,(1)–(3) can be verified directly; for 3.12.1(4)–(6) see A. F. Timan [10]; for 3.12.1(7) see Hardy, Littlewood and Pólya [1].

2. The following proposition is similar to theorem 3.1.2. A bounded set W of periodic functions with period $b-a$, belonging to L_q on $[a, b]$, is compact in this space if and only if as $h \to 0$

$$\sup_{f \in W} \int_a^b |f(x)-f_{h,r}(x)| \, dr \to 0. \tag{9}$$

A. N. Kolmogorov [1], $q > 1$, $r = 1$; A. N. Tulaikov [1], $q = 1$, $r = 1$. See also L. A. Lyusternik and V. I. Sobolev [1].

3. If the function $f(t)$ is defined on $(-\infty, \infty)$ and

$$\int_{-\infty}^{\infty} |f(t)|\,dt < \infty,$$

or

$$\int_{-\infty}^{\infty} |f(t)|^2\,dt < \infty,$$

then in the first case for all x, and in the second case for almost all,

$$f_{h,1}(x) = \frac{1}{\sqrt{2\pi}} \int_{-\infty}^{\infty} F(t) \frac{\sin h(t)}{ht} e^{itx}\,dt, \tag{10}$$

where $F(t)$ is the Fourier transform of the function $f(t)$.

Make use of the proposition of 3.11.22.

4. If for some $q(1 \leqslant q < \infty)$

$$\int_{-\infty}^{\infty} |f(t)|^q\,dt < \infty,$$

then

$$\left\{\int_{-\infty}^{\infty} |f(x)-f_{h,1}(x)|^q\,dx\right\}^{\frac{1}{q}} \leqslant \tfrac{1}{2}\omega_2(f;h)_{L_q}.$$

$$\left\{\int_{-\infty}^{\infty} |f(x)-f_{h,2}(x)|^q\,dx\right\}^{\frac{1}{q}} \leqslant \tfrac{1}{2}\omega_2(f;2h)_{L_q},$$

$$\left\{\int_{-\infty}^{\infty} |f_{h,r}^{(r)}(x)|^q\,dx\right\}^{\frac{1}{q}} \leqslant \frac{1}{(2h)^r}\omega_r(f;2h)_{L_q}, \tag{11}$$

$$\left\{\int_{-\infty}^{\infty} |f_{h,1}(x)|^q\,dx\right\}^{\frac{1}{q}} \leqslant \left\{\int_{-\infty}^{\infty} |f(x)|^q\,dx\right\}^{\frac{1}{q}}.$$

5. If for some $q \geqslant 1$ we have $\omega(f;t)_{L_q} \equiv 0$, then $f(x)$ is constant almost everywhere.

The inequality

$$\omega_k(f;\lambda t)_{L_q} \leqslant (\lambda+1)^k \omega_k(f;t)_{L_q} \tag{12}$$

is satisfied for any $\lambda > 0$.

If $\omega_k(f;0,\ldots,0,u_\nu,0,\ldots,0)_{L_q}$ is the partial integral modulus of smoothness on the cube of periods of the function $f(x_1,\ldots,x_m) \subset L_q(q \geqslant 1)$ which is periodic (of period 2π) in each of the variables, i.e.

$$\omega_k(f;0,\ldots,0,u_\nu,0,\ldots 0)_{L_q}$$

$$= \sup_{|h|\leqslant u_\nu} \left\{\int_0^{2\pi}\cdots\int_0^{2\pi} \left|\sum_{i=0}^{k}(-1)^i\binom{k}{i} f(x_1,\ldots,x_\nu+ih,\ldots,x_m\right|^q dx_1\ldots dx_m\right\}^{\frac{1}{q}},$$

then for any natural number n

$$\omega_k(f;0,\ldots 0,nu_\nu,0,\ldots,0)_{L_q} \leqslant n^k \omega_k(f;0,\ldots,0,u_\nu,0,\ldots,0)_{L_q}.$$

For other values of $\lambda > 0$

$$\omega_k(f; 0_1 \ldots, 0, \lambda u_\nu, 0, \ldots, 0) \leqslant (\lambda+1)^k \omega_k(f; 0, \ldots, 0, u_\nu, 0, \ldots, 0)_{L_q}.$$

The estimate 3.12(12) follows directly from 3.3(2). Make use of the identity 3.3(4).

6. If the function $f(x)$ defined on $[-1, 1]$ satisfied the conditions

$$f(1) = f(-1) = 0,$$

$$\left| f(x_1) - 2f\left(\frac{x_1+x_2}{2}\right) + f(x_2) \right| \leqslant |x_1 - x_2|, \qquad x_1, x_2 \in [-1, 1],$$

then $|f(x)| < \frac{4}{3}$. With the supplementary condition $f(-x) = f(x)$ the precise inequality $\max\limits_{-1 \leqslant x \leqslant 1} |f(x)| \leqslant \frac{5}{4}$ is satisfied.

If the function $f(x)$ is periodic and $\omega_3(f; t) \leqslant t$, then as $t \to 0$

$$\omega(f; t) \leqslant \frac{1}{\ln 3} t \ln \frac{1}{t} + O(t).$$

I. E. Gopengauz and A. F. Timan [1]. See also Yu. A. Brudnyi [1].

7. For any function $f(x, y)$ the total modulus of smoothness $\omega_2(f; u, v)$ and the mixed modulus of smoothness $\omega_{1,1}(f; u, v)$ satisfy the inequality

$$\tfrac{1}{2}\omega_{1,1}(f; u, v) \leqslant \omega_2(f; u, v). \tag{13}$$

This obvious inequality for $k = 2$ supplements the estimate 3.4(7).

8. If $\phi(x)$ is a function of bounded variation on the whole real axis $(-\infty, \infty)$ and at every point x

$$\phi(x) = \tfrac{1}{2}\{\phi(x+0) + \phi(x-0)\},$$

then

$$\sum_{k=-\infty}^{\infty} \phi(k) = \sum_{\nu=-\infty}^{\infty} \int_{-\infty}^{\infty} \phi(x) e^{2\pi i \nu x}\, dx, \tag{14}$$

subject to the condition that the series on the left converges.

Formula 3.12(4), which is known as Poisson's formula, is obtained from the expansion in a Fourier series of the function $\Phi(x) = \sum\limits_{k=-\infty}^{\infty} \phi(k+x)$ for $x = 0$.

9. If $F(x)$ is an integrable periodic function of period 2π and $D_0^{(r)}(t)$ is the sum of the series 3.5(16), then, subject to the conditions $\int_0^{2\pi} F(t)\, dt = 0$ and $0 < r < 1$, the identity

$$\frac{1}{\pi} \int_0^{2\pi} F(t) D_0^{(r)}(x-t)\, dt = \frac{1}{\Gamma(r)} \int_{-\infty}^{x} F(t)\, (x-1)^{r-1}\, dt \tag{15}$$

is satisfied.

Using formula 3.12(4), it has to be verified that for $0 < r < 1$ and $0 < x < 2\pi$ we have

$$D_0^{(r)}(x) = \frac{\pi}{\Gamma(r)} \lim_{n \to \infty} \left\{ x^{r-1} + (x+2\pi)^{r-1} + \ldots + (x+2n\pi)^{r-1} - (2\pi)^{r-1} \frac{n^r}{r} \right\}. \tag{16}$$

10. The following finite difference analogue of Rolle's theorem holds. If the continuous function $f(x)$ has $k+1$ zeros on the segment $[a, b]$, then for a sufficiently small increment h its k-th difference $\Delta_h^k f(x)$ has at least one zero on this segment.

S. N. Bernstein [3].

11. The following formula of Newton

$$f(x) = f(0) + \frac{x}{h} \Delta_h f(0) + \frac{x(x-h)}{2!h^2} \Delta_h^2 f(0) + \dots$$
$$+ \frac{x(x-h) \dots [x-(n-1)h]}{n!h^n} \Delta_h^n f(0) + R_{n+1}(x), \tag{17}$$

holds for any real continuous function, where the remainder term $R_{n+1}(x)$ is of the form

$$R_{n+1}(x) = \frac{x(x-h) \dots (x-nh)}{(n+1)!\,\delta^{n+1}} \Delta_\delta^{n+1} f(\xi) \qquad (0 < \delta < h);$$

it is the finite difference analogue of Taylor's formula.

S. N. Bernstein [3]. Make use of the proposition 3.12.10.

12. The continuous real function $f(x)$ is analytic on the segment $[a, b]$ if and only if for all integral $k > 0$

$$\sqrt[k]{\frac{\omega_k(f; t)}{t^k}} \leqslant Mk, \tag{18}$$

where M is a constant independent of k and t.

S. N. Bernstein [3]. Use formula 3.12.(17). See section 3.7.1.

13. The function $f(x)$ of the class $\overset{*}{L}_q(1 < q < \infty)$ satisfies the condition $\omega(f; t)_{\overset{*}{L}_q} = O(t)$ if and only if it is almost everywhere equal to the indefinite integral of a function of the same class.

G. H. Hardy and J. E. Littlewood [1].

14. If the function $f(x)$ satisfies the condition of the preceding theorem $(1 < q < \infty)$, then almost everywhere it is equal to a function $\phi(x)$ which satisfies the condition $\omega(\phi; t) = o\left(t^{\frac{q-1}{q}}\right)$ as $t \to 0$.

15. If the periodic function $f(z)$ of period 2π is analytic in the strip $|\mathrm{Im} z| \leqslant \delta$ and

$$c_k = \frac{1}{2\pi} \int_{-\pi}^{\pi} f(t) \mathrm{e}^{-ikt} \,\mathrm{d}t$$

are its Fourier coefficients, then for all $k = 0, \pm 1, \pm 2, \dots$

$$|c_k| \leqslant M\mathrm{e}^{-k\delta}, \tag{19}$$

where M is a constant independent of k.

Apply Cauchy's theorem on the integral round a closed contour to the function $f(z)\mathrm{e}^{ikz}$, considered in the two rectangles cut off from the strip by the axis $\mathrm{Im} z = 0$ and the straight lines $\mathrm{Re} z = 0$, $\mathrm{Re} z = 2\pi$.

16. If the sequence of numbers c_k satisfies the condition 3.12(19), then the function

$$f(z) = \sum_{k=-\infty}^{\infty} c_k \mathrm{e}^{ikx}$$

is analytic within the strip $|\mathrm{Im} z| < \delta$.

It immediately follows from theorems 3.12.15 and 3.12.16, that if the periodic function $f(x)$ is analytic or integral, then $\tilde{f}(x)$ also possesses the same property.

17. Theorem 3.8.4 can be supplemented as follows. The function $u(r, \phi)$, harmonic in the unit circle $0 \leqslant r < 1$, satisfies the condition

$$\int_0^{2\pi} |u(r, \phi)|^q \, d\phi = O(1)$$

for $q > 1$ and $r \to 1$ if and only if it is the integral 3.8(14), in which $g(t)$ is a measurable function of period 2π belonging to the class L_q^*.

See A. Zygmund [1].

18. If the function $f(x)$ is cyclically monotonic on $[a, b]$ and for any non-negative integer k the derivative $f^{(k)}(x)$ vanishes at some point $x_k \in [a, b]$, then either

$$f(x) = A \sin \frac{\pi(x-a)}{2(b-a)} \quad \text{or} \quad f(x) = A \cos \frac{\pi(x-a)}{2(b-a)} \cdot$$

Prove the periodicity of the function $f(x)$ and consider its Fourier series.

S. N. Bernstein [40].

19. For the function $f(x)$, defined on the semi-axis $[0, \infty]$, to be representable in the form

$$f(x) = \int_0^{\infty} e^{-tx} \, dF(t),$$

where $F(t)$ is a function of bounded variation on $[0, \infty]$, it is necessary and sufficient that it should be infinitely differentiable and satisfy the condition

$$\int_0^{\infty} t^n |f^{(n+1)}(t)| \, dt \leqslant Mn!$$

uniformly for all $n = 0, 1, 2, \ldots$

Widder [1]. See the proof of theorem 3.9.2.

20. Let $\phi(x)$ be a periodic function of bounded variation over the period $[0, 2\pi]$. For any integrable function $f(x)$ of period 2π the equality

$$\frac{1}{2\pi} \int_{-\pi}^{\pi} f(x)\overline{\phi(x)} \, dx = \lim_{n\to\infty} \sum_{k=-n}^{n} c_k \overline{d_k},$$

is satisfied, where the numbers c_k and d_k are respectively the complex Fourier coefficients of the functions $f(x)$ and $\varphi(x)$.

Young [1]. Make use of the fact that the partial sums of the Fourier series of functions of bounded variation are uniformly bounded, and apply the theorem on passage to the limit within a Lebesgue integral.

21. If the function $\phi(x)$ is of bounded variation on $(-\infty, \infty)$ and

$$\int_{\infty}^{\infty} |\phi(t)| \, dt < \infty$$

then for any integrable function $f(x)$ of period 2π the equality

$$\lim_{a\to-\infty,\, b\to\infty} \int_a^b f(x)\phi(x)\,dx = \lim_{n\to\infty} \sum_{k=-n}^{n} c_k \int_{-\infty}^{\infty} \phi(x)e^{ikx}\,dx,$$

holds, where the numbers c_k are the complex Fourier coefficients of $f(x)$.

G. H. Hardy [1]. Consider the function $\sum_{k=-\infty}^{\infty} \phi(x+2kx)$ and make use of theorem 3.12.20.

22. If $1 < q < \infty$ and $S_n(f; x)$ is the partial sum of order n of the Fourier series of the periodic function $f(x)$ of period 2π of the class L_q^*, then

$$\int_0^{2\pi} |S_n(f;\, x)|^q\,dx \leqslant M_q \int_0^{2\pi} |f(x)|^q\,dx \tag{20}$$

where M_q is a constant depending only on q.

A. Zygmund [1]. Use the inequality 3.11(5).

23. If the periodic function $f(x) \in W_*^{(r)} M$ (r an integer) of period 2π possesses the property that

$$\int_0^{2\pi} f(x)\,dx = 0,$$

then $\max_x |f(x)| \leqslant MK_r$, where the constant K_r is defined in section 2.11.3 (see 2.11(10)). The example of the function 2.11(9) shows that this estimate is percise.

S. N. Bernstein [14]. Use formula 3.5(15).

24. If the periodic function $f(x) \in W_*^{(r)} M$(r an integer) of period 2π possesses the property, that $f(0) = 0$, then $\max_x |f(x)| \leqslant 2MK_r$. This inequality is precise.

A. F. Timan [3].

25. If $K(t)$ is an integrable function of period 2π and

$$\int_0^{2\pi} f(t)\,dt = 0, \qquad \int_0^{2\pi} |f(t)|\,dt \leqslant 1,$$

then

$$\int_0^{2\pi} \left| \int_0^{2\pi} K(x-t)f(t)\,dt \right| dx \leqslant \tfrac{1}{2} \max_x \int_0^{2\pi} |K(x+t)-K(t)|\,dt.$$

The constant 1/2 in this inequality cannot be decreased.

S. M. Nikol'skii [3].

26. Let $1 \leqslant q < \infty$ and $f_q(x)$ be a periodic function of period 2π, which is equal to $|x|^{\frac{q-1}{q}}$ for $|x| \leqslant \pi$ if $q > 1$ and to $\ln|x|$ if $q = 1$. Show that $\omega_2(f_q; t)_{L_q}^* = O(t)$ and $\omega(f_q; t)_{L_q}^* > Ct|\ln t|^{\frac{1}{q}}$ when $1 \leqslant q \leqslant 2$, and $\omega(f_q; t)_{L_q}^* > Ct|\ln t|^{\frac{1}{2}}$ when $2 \leqslant q < \infty$ ($C > 0$.).

A. Zygmund [4].

CHAPTER IV

SOME PROPERTIES OF ALGEBRAIC POLYNOMIALS AND TRANSCENDENTAL INTEGRAL FUNCTIONS OF EXPONENTIAL TYPE

4.1. Interpolation formulae for algebraic polynomials

One of the most important properties of ordinary polynomials is that any such polynomial

$$P_n(x) = \sum_{k=0}^{n} c_k x^k \tag{1}$$

of degree n is *uniquely* determined by its values at an arbitrarily given system of $n+1$ distinct points $x_0, x_1, \ldots, x_n$. Hence

$$l_{k,n}(x) = \frac{l_{n+1}(x)}{l'_{n+1}(x_k)(x-x_k)} \quad \big(l_{n+1}(x) = (x-x_0)\ldots(x-x_n)\big) \tag{2}$$

is the only algebraic polynomial of degree n which possesses the property

$$l_{k,n}(x_i) = \begin{cases} 0, & i \neq k \\ 1, & i = k \end{cases}, \quad (i, k = 0, 1, \ldots, n), \tag{3}$$

and *the formula for the determination of $P_n(x)$ can be written in the following form*:

$$P_n(x) = \sum_{k=0}^{n} P_n(x_k) l_{k,n}(x). \tag{4}$$

The identity 4.1(4) is known as Lagrange's interpolation formula.

Among all possible distributions of the points $x_0, x_1, \ldots, x_n$ (interpolation points) the following two cases are important in various types of investigation. The first of these is the case where the interpolation points are equidistant, and the second is the case where on some segment $[a, b]$ containing them they are coincident with the zeros of the Chebyshev polynomial 2.9(7) which deviates least from zero (see section 2.9.1). We will deal with each of these two cases separately.

4.1.1. In the first of these cases, using a method which originated with Newton, formula 4.1(4) is put into a different and in some respects more convenient form. Putting

$$P_n(x) = a_0 + \sum_{k=1}^{n} a_k (x-x_0)\dots(x-x_{k-1}), \tag{5}$$

it is possible to express $a_0, a_1, \dots, a_n$ successively in terms of the values of the polynomial $P_n(x)$ at the interpolation points. If it is noted that a_n is identical with the coefficient of the highest power of the polynomial $P_n(x)$ and that

$$a_0 + \sum_{k=1}^{\nu} a_k(x-x_0)\dots(x-x_{k-1})$$

is the polynomial of degree ν which assumes at the points $x_0, x_1, \dots, x_\nu$ the same values as $P_n(x)$, we find directly from 4.1(4) that

$$a_k = \sum_{i=0}^{k} \frac{P_n(x_i)}{l'_{k+1}(x_i)} \qquad (k = 0, 1, \dots, n). \tag{6}$$

But as

$$l'_{k+1}(x_i) = (x_i-x_0)\dots(x_i-x_{i-1})(x_i-x_{i+1})\dots(x_i-x_k),$$

then, taking the interpolation points to be equidistant, we have $x_\nu = x_0+\nu h$ $(h > 0,\ \nu = 0, 1, \dots, n)$ and, consequently $l'_{k+1}(x_i) = (-1)^{k-i} h^k i!\,(k-i)!$. Thus,

$$a_k = \sum_{i=0}^{k} (-1)^{k-i} \frac{P_n(x_i)}{h^k i!\,(k-i)!}$$

and

$$P_n(x) = P_n(x_0) + \sum_{k=1}^{n} \frac{1}{h^k} \sum_{i=0}^{k} (-1)^{k-i} \frac{P_n(x_i)}{i!\,(k-i)!} (x-x_0)\dots(x-x_{k-1}).$$

Denoting by $\Delta_h^k P_n(x_0)$ the k-th difference with increment h for the polynomial $P_n(x)$ at the point x_0, i.e.

$$\Delta_h^k P_n(x_0) = \sum_{\nu=0}^{k} \binom{k}{\nu} (-1)^{k-\nu} P_n(x_0+\nu h),$$

we obtain

$$P_n(x) = \sum_{k=0}^{n} \Delta_h^k \frac{P_n(x_0)}{k!\,h^k} (x-x_0)(x-x_0-h)\dots[x-x_0(k-1)h].$$

This is Newton's interpolation formula for the determination of a polynomial in terms of its values at equidistant points, similar to Taylor's formula, which is obtained from it in the limit as $h \to 0$.

In the general case (for any set of points) formula 4.1(5), with coefficients a_k determined from 4.1(6), frequently called the finite difference interpolation formula, is more convenient for calculation than 4.1(4) if the degree of the polynomials considered is increased by the addition of new points.

4.1.2. In the second of the cases indicated in section 4.1 let us suppose for definiteness that the interpolation points are coincident with the zeros of the Chebyshev polynomial which corresponds to the segment $[-1, 1]$, i.e.

$$x_k = x_k^{(n+1)} = \cos\frac{2k+1}{2(n+1)}\pi \quad (k = 0, 1, 2, \dots, n). \tag{7}$$

Since in this case $l_{n+1}(x) = \frac{1}{2}\cos(n+1)\arccos x$ (see section 2.9.1). it follows that

$$l'_{n+1}(x_k^{(n+1)}) = \frac{n+1}{2^n\sqrt{1-[x_k^{(n+1)}]^2}}\sin\left(k+\frac{1}{2}\right)\pi = \frac{n+1}{2^n}\frac{(-1)^k}{\sin\frac{2k+1}{2(n+1)}\pi}$$

and consequently

$$P_n(x) = \frac{\cos(n+1)\arccos x}{n+1}\sum_{k=0}^{n}(-1)^k P_n(x_k^{(n+1)})\frac{\sin\frac{2k+1}{2(n+1)}\pi}{x-\cos\frac{2k+1}{2(n+1)}\pi}. \tag{8}$$

This formula can be written in a somewhat different form for values of x lying on the segment $[-1, 1]$. Let us put

$$t = \arccos x, \quad t_k^{(n)} = \frac{2k+1}{2n}\pi$$

and make use of the identity

$$(-1)^{k+1}\cos(n+1)t\sin t_k^{(n+1)} = \sin(2n+3)\frac{t+t_k^{(n+1)}}{2}\sin\frac{t-t_k^{(n+1)}}{2} + $$
$$+\sin(2n+3)\frac{t-t_k^{(n+1)}}{2}\sin\frac{t-t_k^{(n+1)}}{2}.$$

We then obtain

$$(n+1)\,l_{k,n}(x) = \frac{(-1)^k\cos(n+1)\arccos x\cdot\sin\frac{2k+1}{2(n+1)}\pi}{x-\cos\frac{2k+1}{2(n+1)}\pi}$$
$$= D_{n+1}(t+t_k^{(n+1)})+D_{n+1}(t-t_k^{(n+1)}), \tag{9}$$

where

$$D_n(t) = \frac{\sin\frac{2n+1}{2}t}{2\sin\frac{t}{2}} = \frac{1}{2}+\sum_{\nu=1}^{n}\cos\nu t \tag{10}$$

is known as Dirichlet's kernel. *Thus for the Chebyshev interpolation points on* $[-1,1]$ *we have the formula*

$$P_n(x) = \frac{1}{n+1}\sum_{k=0}^{n} P_n(x_k^{(n+1)})\,[D_n(t+t_k^{(n+1)})+D_n(t-t_k^{(n+1)})]. \tag{11}$$

In particular, putting $t = 0$ and $P_n(x) \equiv 1$ in this, we have for the Dirichlet kernel the identity

$$\frac{2}{n}\sum_{k=0}^{n-1} D_{n-1}(t_k^{(n)}) = 1. \tag{12}$$

4.1.3. In the case where some of the points are coincident, i.e. their number $m+1$ is really less than $n+1$, there are an infinite number of polynomials 4.1(1), which assume the given values at the points x_0, x_1, ..., x_m. However the given problem immediately becomes definite if in addition we are given the values of the derivatives $P_n^{(i)}(x_k)$ $(k = 1, 1, \ldots, m;\ 0 \leqslant i \leqslant \nu_k-1)$ of the polynomial $P_n(x)$ and $\nu_0+\nu_1+\ldots+\nu_m = n+1$. In this case, proceeding as in section 4.1.1 and putting

$$Q_{\nu_k-1}(x) = \sum_{i=0}^{\nu_k-1} a_i^{(k)}(x-x_k)^i,$$

and

$$P_n(x) = Q_{\nu_0-1}(x)+\sum_{k=1}^{m} Q_{x_k-1}(x)\,(x-x_0)^{\nu_0}\ldots(x-x_{k-1})^{\nu_{k-1}} \tag{13}$$

all the coefficients

$$a_i^{(k)}(i = 0, 1, \ldots, \nu_k-1;\quad k = 0, 1, \ldots, m)$$

can be successively expressed in terms of the values of the polynomial $P_n(x)$ and of its derivatives $P_n^{(i)}(x)$ $(0 \leqslant i \leqslant \nu_k-1)$ at the interpolation points. *This formula for the determination of the polynomial* $P_n(x)$ *is known as Hermite's interpolation formula.* Of the various cases which are possible here it is especially necessary to note two which in a certain sense are opposites. The first of these, which has already been considered in sections 4.1 and 4.1.1, is that where $m = n$; the second is the case where $m = 0$; it gives Taylor's formula for the polynomial $P_n(x)$.

Here let us also note the case which possesses a simple geometrical treatment, where $n = 2m+1$, $\nu_k = 2$ $(k = 0, 1, \ldots, m)$. It is left to the reader to verify the identity

$$P_{2m+1}(x) = \sum_{k=0}^{m} \frac{l_{m+1}^2(x)}{l_{m+1}'^2(x_k)\,(x-x_k)^2}\left\{P_{2m+1}(x_k)\left[1-\frac{l_{m+1}''(x_k)}{l_{m+1}'(x_k)}(x-x_k)\right]+\right.$$
$$\left. + P_{2m+1}'(x_k)\,(x-x_k)\right\}, \qquad (14)$$

to which Hermite's formula reduces in the given case.

4.1.4. In addition to the methods of specifying an algebraic polynomials considered above, it is also necessary to note an interpolation formula of a somewhat different type which permits the definition of the polynomial 4.1(1) in terms its successive derivatives at the points x_0, x_1, $\ldots$, x_n, i.e. in terms of $P_n^{(k)}(x_n)$ $(k = 0, 1, \ldots, n)$. It is obvious that whatever these points may be, on repeatedly integrating the identity

$$P_n^{(n)}(x) \equiv P_n^{(n)}(x_n),$$

we obtain:

$$P_n(x) = P_n(x_0) + \sum_{k=1}^{n} P_n^{(k)}(x_k) \int_{x_0}^{x}\int_{x_1}^{t_1}\ldots\int_{x_{k-1}}^{t_{k-1}} dt_1 \ldots dt_k,$$

or

$$P_n(x) = P_n(x_0) + \sum_{k=1}^{n} P_n^{(k)}(x_k) g_k(x), \qquad (15)$$

where $g_k(x)$ denotes the polynomial

$$g_k(x) = \int_{x_0}^{x}\int_{x_1}^{t_1}\ldots\int_{x_{k-1}}^{t_{k-1}} dt_1 \ldots dt_k. \qquad (16)$$

The identity 4.1(5) is known as *the Abel-Goncharov interpolation formula for polynomials*. It represents a generalisation of Taylor's formula

$$P_n(x) = \sum_{k=0}^{n} \frac{P_n^{(k)}(x_0)}{k!}(x-x_0)^k.$$

which corresponds to the case $x_0 = x_1 = \ldots = x_n$.

4.2. Interpolation formulae for trigonometric polynomials

On writing the trigonometric polynomial

$$T_n(x) = \frac{a_0}{2} + \sum_{k=1}^{n} (a_k \cos kx + b_k \sin kx) \not\equiv 0 \qquad (1)$$

of order n in the complex form

$$T_n(x) = \sum_{k=-n}^{n} c_k e^{ikx} = \sum_{k=-n}^{n} c_k z^k = z^{-n} \sum_{k=0}^{2n} c_{k-n} z^k$$

$$(c_k = \tfrac{1}{2}(a_k - ib_k);\ c_{-k} = \bar{c}_k)$$

it is obvious that the number of its zeros on any half-open interval of length 2π does not exceed $2n$. Hence, as in the case of algebraic polynomials, it follows that any trigonometric polynomial of order n is uniquely determined by its values at an arbitrarily given system of $2n+1$ distinct points x_0, $x_1, \ldots, x_{2n}$, lying on some half-open interval of length 2π. Hence

$$l^*_{k,n}(x) = \frac{\sin\frac{x-x_0}{2}\sin\frac{x-x_1}{2}\ldots\sin\frac{x-x_{k-1}}{2}\sin\frac{x-x_{k+1}}{2}\ldots\sin\frac{x-x_{2n}}{2}}{\sin\frac{x_k-x_0}{2}\sin\frac{x_k-x_1}{2}\ldots\sin\frac{x_k-x_{k-1}}{2}\sin\frac{x_k-x_{k+1}}{2}\ldots\sin\frac{x_k-x_{2n}}{2}} \tag{2}$$

is the only trigonometric polynomial of order n which possesses the property

$$l^*_{k,n}(x_i) = \begin{cases} 0, & i \neq k \\ 1, & i = k \end{cases} \quad (i,\ k = 0,\ 1, \ldots\ 2n); \tag{3}$$

and we have everywhere

$$T_n(x) = \sum_{k=0}^{2n} T_n(x_k) l^*_{k,n}(x). \tag{4}$$

This is Lagrange's interpolation formula for trigonometric polynomials, similar to formula 4.1(4) in the algebraic case. When the polynomial 4.2(1) is even (odd), it can be replaced by another involving the values of $T_n(x)$ at only the $n+1$ distinct points $x_0, x_1, \ldots, x_n$ of the segment $[0, \pi]$ (at the n distinct points $x_1, \ldots, x_n$ of the interval $(0, \pi)$).

In particular

$$l^*_{k,n}(x) = \frac{(\cos x - \cos x_0)\ldots(\cos x - \cos x_{k-1})(\cos x - \cos x_{k+1})\ldots(\cos x - \cos x_n)}{(\cos x_k - \cos x_0)\ldots(\cos x_k - \cos x_{k-1})(\cos x_k - \cos x_{k+1})\ldots(\cos x_k - \cos x_n)} \tag{5}$$

is the unique even trigonometric polynomial of order n possessing the property 4.2(3) for i, $k = 0, 1, \ldots, n$ and correspondingly

$$l^*_{k,n}(x) = \frac{\sin x}{\sin x_k} \times \frac{(\cos x - \cos x_1)\ldots(\cos x - \cos x_{k-1})(\cos x - \cos x_{k+1})\ldots(\cos x - \cos x_n)}{(\cos x_k - \cos x_1)\ldots(\cos x_k - \cos x_{k-1})(\cos x_k - \cos x_{k+1})\ldots(\cos x_k - \cos x_n)} \tag{6}$$

is the unique odd trigonometric polynomial of order n possessing the property 4.2(3) for $i, k = 1, 2, \ldots, n$.

The given interpolation formulae are materially simplified for certain sets of equidistant points. We note the most important cases which are possible.

4.2.1. Let

$$x_k = y_k^{(n)} = \frac{2k\pi}{2n+1} \qquad (k = 0, 1, \ldots, 2n). \tag{7}$$

The Dirichlet kernel 4.1(10) is the unique trigonometric polynomial of order n which possesses the property

$$D(y_k^{(n)}) = \begin{cases} 0, & k \neq 0, \\ n+\frac{1}{2}, & k = 0, \end{cases} \tag{8}$$

and for any trigonometric polynomial 4.2(1) *the formula*

$$T_n(x) = \frac{2}{2n+1} \sum_{k=0}^{2n} T_n(y_k^{(n)}) D_n(x-y_k^{(n)}) \tag{9}$$

is satisfied.

Hence it follows that

$$T_n(x) = \frac{2}{2n+1} \Big\{ T_n(0) D_n(x) + \\ + \sum_{k=1}^{n} T_n(y_k^{(n)}) [D_n(x-y_k^{(n)}) + D_n(x+y_k^{(n)})] \Big\} \tag{10}$$

if the polynomial $T_n(x)$ is even, and

$$T_n(x) = \frac{2}{2n+1} \sum_{k=1}^{n} T_n(y_k^{(m)}) [D_n(x-y_k^{(n)}) - D_n(x+y_k^{(n)})] \tag{11}$$

if the polynomial $T_n(x)$ is odd.

4.2.2. As in 4.1.2, let

$$t_k^{(n)} = \frac{2k+1}{2n} \pi \qquad (k = 0, 1, \ldots, 2n-1). \tag{12}$$

Since the trigonometric polynomial of order n

$$r_{k,n}(x) = \sin n(x-t_k^{(n)}) \cot \frac{x-t_k^{(n)}}{2} \tag{13}$$

possesses the property

$$r_{k,n}(t_i^{(n)}) = \begin{cases} 0, & i \neq k, \\ 2n, & i = k, \end{cases} \tag{14}$$

it follows that for any polynomial 4.2(1) the difference

$$T_n(x) - \frac{1}{2n} \sum_{k=0}^{2n-1} T_n(t_k^{(n)}) T_{k,n}(x)$$

has the same zeros as $\cos nx$. Consequently,

$$T_n(x) = C \cos nx + \frac{1}{2n} \sum_{k=0}^{2o-1} T_n(t_k^{(n)}) r_{k,n}(x), \tag{15}$$

where C is some constant which can be determined if the value of the polynomial $T_n(x)$ is known at an additional $(2n+1)$-th point. Since $r_{k,n}(x+t_k^{(n)})$ is an even polynomial, the polynomial $r_{k,n}(x)$ does not contain the term in $\sin n(x-t_k^{(n)})$, i.e. it does not contain the term in $\cos nx$. Therefore if in the identity 4.2(5) the polynomial $T_n(x)$ is of order less than n, in that case $C = 0$.

*Thus, for every polynomial of order n the following interpolation formula of M. Riesz** holds:*

$$T_n(x) = \frac{1}{2(n+1)} \sum_{k=0}^{2n+1} T_n(t_k^{(n+1)}) r_{k,n+1}(x). \tag{16}$$

We note that, as is immediately obvious from 4.1(11), *when the polynomial $T_n(x)$ is even the formula* 4.2(16) *can be written in the form*

$$T_n(t) = \frac{1}{n+1} \sum_{k=0}^{n} T_n(t_k^{(n+1)}) [D_n(t+t_k^{(n+1)}) + D_n(t-t_k^{(n+1)})], \tag{17}$$

which is similar to 4.2(10).

4.2.3. It follows from the formula of M. Riesz that for any α

$$T_n(x+\alpha) = \frac{1}{2(n+1)} \sum_{k=0}^{2n+1} T_n(t_k^{(n+1)}+\alpha) r_{k,n+1}(t).$$

Putting $x+\alpha = t$, $t_k^{(n+1)}+\alpha = \theta_k^{(n+1)}$, $r_{k,n+1}(t-\alpha) = \varrho_{k,n+1}(t)$, we have the formula

$$T_n(t) = \frac{1}{2(n+1)} \sum_{k=0}^{2n+1} T_n(\theta_k^{(n+1)}) \varrho_{k,n+1}(t) \tag{18}$$

for the determination of the polynomial $T_n(t)$ from its values at the equidistant points $\theta_k^{(n+1)}$. In particular, choosing $\alpha = \frac{\pi}{2(n+1)}$, we obtain the interpolation formula of M. Riesz for the system of equidistant points

$$\theta_k^{(n+1)} = t_k^{(n+1)} + \frac{\pi}{2(n+1)} = \frac{k+1}{n+1} \pi \quad (k = 0, 1, \ldots, 2n). \tag{19}$$

** M. Riesz [2].

In those cases where the polynomial $T_n(x)$ is odd, formula 4.2(18) with the points 4.2(19) can be written in a somewhat different form, similar to 4.2(11). For this purpose let us take the polynomial

$$\tilde{d}_{k,n+1}(t) = D_{n+1}(t-\theta_k^{(n+1)})-D_{n+1}(t+\theta_k^{(n+1)})$$

and note that it is the unique odd trigonometric polynomial of order n which possesses the property

$$d_{k,n+1}(\theta_i^{(n+1)}) = \begin{cases} 0, & i \neq k, \\ n+1, & i = k \end{cases} \quad (i, k = 0, 1, \ldots, n-1).$$

Consequently, *if* $T_n(x)$ *is any odd trigonometric polynomial of order* $\leqslant n$, *then*

$$T_n(x) = \frac{1}{n+1} \sum_{k=0}^{n-1} T_n(\theta_k^{(n+1)})[D_n(x-\theta_k^{(n+1)})-D_n(x+\theta_k^{(n+1)})]. \tag{20}$$

4.2.4. Formulae 4.2(4), 4.2(9)–(11), 4.2(16)–(18) and 4.2(20) can be used for the determination of the conjugate trigonometric polynomial $\tilde{T}_n(x)$ (see section 3.11.1) from the values of the polynomial $T_n(x)$ at the corresponding system of points. For this purpose it is only necessary to replace the fundamental polynomials $l^*_{k,n}(x)$, $r_{k,n}(x)$ and the Dirichlet kernel $D_n(x \pm x_k)$ on their right-hand sides by the conjugate trigonometric polynomials $\tilde{l}^*_{k,n}(x)$, $\tilde{r}_{k,n}(x)$ and $\tilde{D}_n(x \pm x_k)$.

4.3. Interpolation formulae for some classes of transcendental integral functions of exponential type

The integral functions

$$f(x) = \sum_{k=0}^{\infty} c_k x^k$$

with coefficients satisfying the condition

$$\overline{\lim_{k\to\infty}} \sqrt[k]{k!\,|c_k|} = \sigma < \infty,$$

possess the property, as is obvious from the inequality

$$|f(z)| \leqslant \sum_{k=0}^{\infty} \frac{1}{k!} (|z| \sqrt[k]{k!\,|c_k|})^k,$$

that for any $\varepsilon > 0$

$$|f(z)| \leqslant e^{(\sigma+\varepsilon)|z|}. \tag{1}$$

for all values of z of sufficiently great modulus.

Hence they have been given the name of integral functions of exponential type or integral functions of finite degree. By the degree of the function $f(x)$ is understood the number

$$\sigma = \overline{\lim_{k\to\infty}} \sqrt[k]{k!\,|c_k|} = \overline{\lim_{r\to\infty}} \frac{\ln M(r)}{r}, \tag{2}$$

where $M(r) = \max\limits_{|z|=r} |f(z)|$.

For certain classes of these functions interpolation formulae hold which are similar to those given for algebraic and trigonometric polynomials. Such formulae can be obtained on the model of Lagrange's formula (see sections 4.1 and 4.2), Newton's formula (see section 4.1.1) and the Abel-Goncharov formula (see section 4.1.4). However, a number of important new problems arise here. Among these a fundamental position is occupied by the question of the uniqueness of the integral function $f(x)$ which assumes the given values (or has given values of derivatives) at the points of the given sequence of points $x_k (k = 0, 1, 2, \ldots)$, and also the question of the convergence of the interpolation series

$$\begin{gathered} \sum_{k=1}^{\infty} f(x_k) \frac{l(x)}{l'(x_k)(x-x_k)}, \\ a_0 + \sum_{k=1}^{\infty} a_k(x-x_0) \ldots (x-x_{k-1}), \\ \sum_{k=0}^{\infty} f^{(k)}(x_k) g_k(x) \end{gathered} \tag{3}$$

of Lagrange, Newton and Abel-Goncharov.

Without stopping here for the consideration of these questions in all their generality**, we will only touch upon the single case of equidistant points, which is important in the theory of approximation of functions.

4.3.1. A natural generalisation of trigonometric functions is given by the integral functions $G(x)$ considered in section 1.7, which satisfy the condition 1.7(1) and are integral functions of exponential type, bounded on the real axis. The class of all integral functions $G_\sigma(x)$ of finite degree $\leqslant \sigma$ and bounded on the real axis is denoted by B_σ. We note that every function $G_\sigma(x) \in B_\sigma$ satisfies the inequality 1.7(1). In fact, since for any $\varepsilon > 0$ we have $|G_\delta(z)| \leqslant A(\varepsilon) e^{(\sigma+\varepsilon)|z|}$, where $A(\varepsilon)$ is a constant depending

** For information on these questions the reader can refer to the works of A. O. Gel'fond [1, 2], I. I. Ibramov and M. V. Keldysh [1], I. I. Ibramov [1] and the monographs of M. A. Evgrafov and B. Ya. Levin [1].

only on ε, the function $\Phi(z) = G_\sigma(z)e^{i(\sigma+\varepsilon)z}$ satisfies for $y \geqslant 0$ the condition $|\Phi(iy)| \leqslant A(\varepsilon)$. In addition, $|\Phi(x)| \leqslant \sup_{-\infty<x<\infty} |G_\sigma(x)|$ on the real axis. Considering the two angles into which the imaginary axis divides the upper half-plane, and applying the Phragmén-Lindelöf theorem to the function $\Phi(z)$, we find that for any $y \geqslant 0$

$$|\Phi(x+iy)| \leqslant M(\varepsilon) = \max \{A(\varepsilon), \sup_{-\infty<x<\infty} |G_\sigma(x)|\}.$$

The application of the Phragmén-Lindelöf theorem to the function $\Phi(x)$, considered in the whole of the upper half-plane, shows that $M(\varepsilon) = \sup_{-\infty<x<\infty} |G_\sigma(x)|$. Hence $|G_\sigma(x+iy)| \leqslant \sup_{-\infty<x<\infty} |G_\sigma(x)| e^{(\sigma+\varepsilon)y} (y \geqslant 0)$ for any $\varepsilon > 0$. From the same considerations it follows that $|G_\sigma(x-iy)| \leqslant \sup_{-\infty<x<\infty} |G_\sigma(x)| e^{(\sigma+\varepsilon)y} (y \geqslant 0)$. On passing to the limit as $\varepsilon \to 0$, we find that $G_\sigma(x)$ satisfies the inequality 1.7(1).

It follows from Taylor's theorem that if the function $G_\sigma(x)$ of B_σ has zeros at the set of equidistant points $\frac{k\pi}{\sigma} + \alpha$ $(k = 0, \pm 1, \pm 2, \ldots)$ then $G_\sigma(x) = C \sin \sigma(x-\alpha)$. It is sufficient merely to note that the quotient

$$\phi(z) = \frac{G_\sigma(z)}{\sin \sigma(z-\alpha)} \tag{4}$$

is an integral function, which in virtue of the inequality 1.7(1) and the formula

$$\phi(z) = \frac{1}{2\pi i} \int_{\left|\zeta - \frac{k\pi}{\sigma} - \alpha\right| = \frac{\pi}{2\sigma}} \frac{\phi(\zeta)}{\zeta - z} d\zeta \quad \left(\left|z - \frac{k\pi}{\sigma} - \alpha\right| \leqslant \frac{\pi}{4\sigma}\right)$$

is bounded throughout the whole plane of the complex variable and is consequently identically equal to some constant. It follows from this remark that *any integral function* $G_\sigma(x)$ *which satisfies condition* 1.7(1) *is uniquely determined by its values at the equidistant points* $x_k = \frac{k\pi}{\sigma} + \alpha$ $(k = 0, \pm 1, \ldots)$ *and by the value of the derivative* $G'_\sigma(\alpha)$ *at the point* α. The last assertion remains valid if in place of B_σ we consider the wider class of integral functions $G_\sigma(x)$ which for some natural number m satisfy the condition

$$G_\sigma(z) = 0[(1+|z|^m)e^{\sigma|y|}] \quad (z = x+iy) \tag{5}$$

and for which on the real axis

$$\lim_{x \to \pm\infty} \frac{G_\sigma(x)}{x} = 0. \tag{6}$$

In fact, if such a function has zeros at the equidistant points $x_k = \frac{k\pi}{\sigma} + \alpha$ $(k = 0, \pm 1, \pm 2, \ldots)$, the ratio 4.3(4) is an integral function which is $O(|z|^m)$ as $|z| \to \infty$. In virtue of the generalisation of Liouville's theorem this ratio is identically equal to some polynomial $P_m(z) = a_0 + a_1 z + \ldots + a_m z^m$, i.e. $G_\sigma(z) = P_m(z) \sin \sigma(z-\alpha)$.

It follows from 4.3(6) that $a_k = 0$ for $k \geqslant 1$ and hence, as in the case of the class B_σ, we have $G_\sigma(z) = a_0 \sin \sigma(z-\alpha)$.

4.3.11. Let us denote by $B_\sigma^{(1)}$ the class of all integral functions which besides conditions 4.3(5) and 4.3(6) also satisfy the condition

$$G_0\left(\frac{k\pi}{\sigma} + \alpha\right) = 0(1) \quad (k = 0, \pm 1, \pm 2, \ldots). \tag{7}$$

This class of functions contains B_σ as a proper part. In order to verify this it is sufficient to put $\alpha = 0$ and consider the function

$$f(x) = \sum_{k=1}^{\infty} \frac{ax \sin \sigma x}{k\pi(\sigma x - k\pi)}. \tag{8}$$

For this function $f\left(\frac{k\pi}{\sigma}\right) = (-1)^k$ when $k > 0$, $f\left(\frac{k\pi}{\sigma}\right) = 0$ when $k \leqslant 0$, and

$$|f(z)| \leqslant |\sigma z \sin \sigma z| \sum_{k=1}^{\infty} \frac{1}{|\sigma z - k\pi| k\pi}$$

$$= O\left\{|z \sin \sigma z| \sum_{k=1}^{\infty} \frac{1}{k(|\sigma x - k\pi| + \sigma|y|)}\right\}.$$

If $k_x = k(x)$ is an integer, for which $|\sigma x - k_x \pi| < \frac{\pi}{2}$, then

$$|f(z)| = |z| e^{\sigma|y|} \cdot O\left(\sum_{k=1}^{\infty}{}' \frac{1}{k|\sigma x - k\pi|}\right) + O(e^{\sigma|y|}),$$

where the prime following the summation sign indicates that the term with index $k(x)$ is omitted. Now from

$$\sum_{k=1}^{\infty}{}' \frac{1}{k|\sigma x - k\pi|} = O\left[\frac{\ln(2+|x|)}{1+|x|}\right]$$

we obtain

$$|f(z)| = O\left\{\frac{|z|}{1+|x|}\ln(2+|x|)e^{\sigma|y|}\right\}. \tag{9}$$

Thus the function $f(x)$ satisfies the conditions 4.3(5)–(7) and hence belongs to $B_\sigma^{(1)}$. In addition it is immediately obvious that it is bounded on the real axis, as for $n > 1$

$$\left|f\left(\frac{2n+1}{2\sigma}\pi\right)\right| = \frac{1}{\pi}\sum_{k=1}^{\infty}\left|\frac{1}{n-k+\frac{1}{2}}+\frac{1}{k}\right| = \frac{3}{\pi}\ln n+O(1).$$

Passing to the construction of an interpolation formula for the functions of the class $B_\sigma^{(1)}$, it has first of all to be noted that in virtue of section 4.3.1

$$g_{\sigma,k}(z)\begin{cases}\dfrac{\sigma(z-a)\sin[\sigma(z-a)-k\pi]}{k\pi[\sigma(z-a)-k\pi]}, & k\neq 0,\\[2ex] \dfrac{\sin\sigma(z-a)}{\sigma(z-a)}, & k=0\end{cases} \tag{10}$$

sis the only integral function of this class which satisfies the condition

$$g_{\sigma,k}(x_i) = \begin{cases}0, & i\neq k,\\ 1, & i=k,\end{cases} \qquad g'_{\sigma,k}(a) = 0 \ (k = 0, \pm 1, \pm 2, \ldots). \tag{11}$$

If $G_\sigma(x)\in B_\sigma^{(1)}$, then by the condition 4.3(7) the series

$$\sum_{k=-\infty}^{\infty} G_\sigma(x_k)g_{\sigma,k}(z) \tag{12}$$

converges absolutely for every z and its partial sums are uniformly bounded in any finite part of the plane. Hence it follows from Vitali's theorem that its sum represents an integral function $F(x)$. Moreover, since in virtue of 4.3(7)

$$|F(z)| = O\left\{|\sin\sigma(z-a)|\sum_{\substack{k=-\infty\\k\neq 0}}^{\infty}\left|\frac{1}{\sigma(z-a)-k\pi}+\frac{1}{k\pi}\right|+1\right\},$$

we find on using the estimate 4.3(9) that $F(z)$ satisfies conditions 4.3(5) and 4.3(6). Since the difference $G_\sigma(z)-F(z)$ becomes zero at all the points x_k, we conclude from the discussion in section 4.3.1 that $G_\sigma(z)-F(z) \equiv C\sin\sigma(z-a)$.

Taking account of the fact that $C = \frac{1}{\sigma} G'_\sigma(\alpha)$, we obtain the formula**

$$G_\sigma(x) = \frac{\sin\sigma(x-\alpha)}{\sigma} G'_\sigma(\alpha) + \sum_{k=-\infty}^{\infty} G_0(x_k) g_{\sigma,k}(x) \qquad (13)$$

for the function $G_\sigma(x) \in B_\sigma^{(1)}$ in terms of $G'_\sigma(\alpha)$ and its values at the points $x_k = \frac{k\pi}{\sigma} + \alpha$.

4.3.12. *For* $\alpha = 0$ *this interpolation formula assumes the form*

$$G_\sigma(x) = \frac{\sin\sigma x}{\sigma x} G_\sigma(0) + \sin\sigma x \sum_{\substack{k=-\infty \\ k\neq 0}}^{\infty} (-1)^k G_\sigma\left(\frac{k\pi}{\sigma}\right)\left\{\frac{1}{\sigma x - k\pi} + \frac{1}{k\pi}\right\} + \frac{\sin\sigma x}{\sigma} G'_\sigma(0), \qquad (14)$$

which in the case *where the function* $G_\sigma(x)$ *is even* can be written

$$G_\sigma(x) = 2\sigma x \sin\sigma x \sum_{k=1}^{\infty} \frac{(-1)^k}{\sigma^2 x^2 - k^2\pi^2} G_\sigma\left(\frac{k\pi}{\sigma}\right) + \frac{\sin\sigma x}{\sigma x} G_\sigma(0). \qquad (15)$$

4.3.13. For $\alpha = \frac{\pi}{2\sigma}$ formula 4.3(13) assumes the form

$$G_\sigma(x) = -\frac{\cos\sigma x}{\sigma x - \frac{\pi}{2}} G_\sigma\left(\frac{\pi}{2\sigma}\right) - \cos\sigma x \sum_{\substack{k=-\infty \\ k\neq 0}}^{\infty} (-1)^k G_\sigma\left(\frac{2k+1}{2\sigma}\pi\right) \times$$

$$\times \left\{\frac{1}{\sigma x - \left(k+\frac{1}{2}\right)\pi} + \frac{1}{k\pi}\right\} - \frac{\cos\sigma x}{\sigma} G'_\sigma\left(\frac{\pi}{2\sigma}\right). \qquad (16)$$

If the function $G_\sigma(x)$ *is odd then, putting* $x = 0$, *we obtain the identity*

$$\frac{\pi}{\sigma} G'_\sigma\left(\frac{\pi}{2\sigma}\right) = G_\sigma\left(\frac{\pi}{2\sigma}\right) - \sum_{k=1}^{\infty} \frac{(-1)^k}{k(k+1)} G_\sigma\left(\frac{2k+1}{2\sigma}\pi\right). \qquad (17)$$

** S. N. Bernstein [15]. As was shown by S. N. Bernstein [25], formula 4.3(13) remains valid if instead of 4.3(7) we assume the weaker condition

$$\sum_{\substack{k=-\infty \\ k\neq 0}}^{\infty} \frac{|G_\sigma(x_k)|}{k^2} < \infty .$$

Using this identity, and observing that in virtue of 4.3(16)

$$G_\sigma(x) = -\frac{\cos \sigma x}{\sigma x - \frac{\pi}{2}} G_\sigma\left(\frac{\pi}{2\sigma}\right) -$$

$$- \cos \sigma x \sum_{k=1}^{\infty} (-1)^k G_\sigma\left(\frac{2k+1}{2\sigma}\pi\right)\left\{\frac{1}{\sigma x - \left(k+\frac{1}{2}\right)\pi} + \frac{1}{k\pi}\right\} -$$

$$- \cos \sigma x \sum_{k=0}^{\infty} (-1)^k G_\sigma\left(\frac{2k+1}{2\sigma}\pi\right)\left\{\frac{1}{\sigma x + \left(k+\frac{1}{2}\right)\pi} - \frac{1}{(k+1)\pi}\right\} -$$

$$- \frac{\cos \sigma x}{\sigma} G'_\sigma\left(\frac{\pi}{2\sigma}\right),$$

we obtain *an interpolation formula for the determination of an odd integral function* $G_\sigma(x) \in B_\sigma^{(1)}$ *from its values at the points* $x_k = \frac{2k+1}{2\sigma}\pi \, (k = 0, \pm 1, \ldots)$:

$$G_\sigma(x) = 2\sigma x \cos \sigma x \sum_{k=0}^{\infty} \frac{(-1)^{k+1}}{\sigma^2 x^2 - \left(k+\frac{1}{2}\right)^2 \pi^2} G_\sigma\left(\frac{2k+1}{2\sigma}\right). \tag{18}$$

4.3.2. Formula 4.3(13) can be simplified by imposing other limitations on the function $G_\sigma(x)$. Thus, for example, *if the integral function* $G(x)$ *is of degree* $p < \sigma$ *and*

$$\sum_{\substack{k=-\infty \\ k \neq 0}}^{\infty} \frac{|G(x_k)|}{k} < \infty \qquad \left(x_k = \frac{k\pi}{\sigma} + a\right), \tag{19}$$

then

$$G(x) = \sin \sigma(x-a) \sum_{k=-\infty}^{\infty} \frac{(-1)^k}{\sigma(x-a) - k\pi} G(x_k). \tag{20}$$

Without loss of generality it is possible to put $a = 0$ and $\sigma = 1$. In virtue of condition 4.3(19) the series

$$\sin z \sum_{k=-\infty}^{\infty} \frac{(-1)^k}{z - k\pi} G(x_k)$$

converges absolutely and its partial sums are uniformly bounded in any finite part of the plane. Consequently the sum of the given series is some integral function $f(z)$. If $z = re^{i\theta}$, then for any θ

$$|f(re^{i\theta})| \leqslant |\sin z| \sum_{k=-\infty}^{\infty} \frac{|G(k\pi)|}{\sqrt{(r\cos\theta-k\pi)^2+r^2\sin^2\theta}}$$

$$= O\left\{|\sin z| \sum_{k=-\infty}^{\infty} \frac{|G(k\pi)|}{|r\cos\theta-k\pi|+r|\sin\theta|}\right.$$

$$= O\left\{|\sin z| \sum_{|k|\leqslant 2\left[\frac{r|\cos\theta|}{\pi}\right]} \frac{|G(k\pi)|}{|r\cos\theta-k\pi|+r|\sin\theta|} + \right.$$

$$\left. +|\sin z| \sum_{|k|>2\left[\frac{r|\cos\theta|}{\pi}\right]} \frac{|G(k\pi)|}{r\cos\theta-k\pi|+r|\sin\theta}\right\}$$

$$= O\{\min([|\cot\theta|+1]e^{r|\sin\theta|}, [r+1]e^{r}\}. \tag{21}$$

By hypothesis, for sufficiently small $\delta > 0$

$$G(re^{i\theta}) = O[e^{p(1+\varepsilon)r}] = O[e^{r(1-\delta)}].$$

If a value $\theta \neq \pm\frac{\pi}{2}$ is chosen such that $e^{r(1-\delta)} = O(e^{r|\sin\theta|})$, then for the difference $\phi(z) = G(z)-f(z)$ we have the estimate

$$\phi(re^{i\theta}) = O(e^{r|\sin\theta|}). \tag{22}$$

Since $\phi(z)$ vanishes at all the points $x_k = k\pi$, the ratio

$$\psi(z) = \frac{\phi(z)}{\sin z}$$

is an integral function. Moreover, as for any $\varepsilon > 0$

$$f(z) = O(re^{r}) = O(e^{r^{1+\varepsilon}}) \qquad (r \to \infty),$$

it follows that $\phi(z) = O(e^{r^{1+\varepsilon}})$, and consequently, also

$$\psi(z) = O(e^{r^{1+\varepsilon}}). \tag{23}$$

Let us choose $\varepsilon < \dfrac{\pi}{2|\theta|} - 1$ if $|\theta| < \dfrac{\pi}{2}$, and $\varepsilon < \dfrac{\pi}{2(\pi-|\theta|)}$ if $\dfrac{\pi}{2} < |\theta| < \pi$.

The function $\psi(z)$ satisfies condition 4.3(23) and in virtue of 4.3(22) is bounded on the rays $\pm\theta$, $\pi\pm\theta$. Hence by the well-known theorem of Phragmén and Lindelöf, and Liouville's theorem, we conclude that $\psi(z)$ is constant. In consequence of the fact that the degree p of the function $G(z)$ is less than $\sigma = 1$ and as $f(iy) = o(e^{y})$ as $y \to \infty$, we have $\psi(iy) \to 0$ as $y \to \infty$. Hence $\psi(z) \equiv 0$, i.e. $G(z) = f(z)$. This proves the interpolation formula 4.3(20).

4.3.3. The proposition of section 4.3.2 enables us easily to obtain remarkable theorem due to M. L. Cartwright [1]. *If the integral function $F(x)$ of finite degree $p < \sigma$ satisfies the condition*

$$\left|F\left(\frac{k\pi}{\sigma}\right)\right| \leqslant M \qquad (k = 0, \pm 1, \pm 2, \ldots),$$

then it is bounded on the whole real axis. To prove this it is sufficient to select any $\varrho < \sigma - p$ and for an arbitrary value of w to consider the function

$$G(x) = F(x)\frac{\sin\varrho(x-w)}{\varrho(x-w)},$$

which is of finite degree $p+\varrho < \sigma$ and satisfies the condition 4.3(19) for $a = 0$. Consequently the theorem of section 4.3.2 is applicable to $G(x)$, i.e.

$$F(x)\frac{\sin\varrho(x-w)}{\varrho(x-w)} = \sin\sigma x \sum_{k=-\infty}^{\infty} (-1)^k \frac{\sin\varrho\left(\frac{k\pi}{\sigma}-w\right)}{\varrho\left(\frac{k\pi}{\sigma}-w\right)(\sigma x - k\pi)} F\left(\frac{k\pi}{\sigma}\right).$$

Putting $w = x$, we see that *for any integral function $F(x)$ of degree $p < \sigma$ the formula*

$$F(x) = \frac{\sin\sigma x}{\sigma} \sum_{k=-\infty}^{\infty} (-1)^k \frac{\sin\varrho\left(x-\frac{k\pi}{\sigma}\right)}{\varrho\left(x-\frac{k\pi}{\sigma}\right)^2} F\left(\frac{k\pi}{\sigma}\right) \tag{24}$$

holds for any positive number $\varrho < \sigma - p$.

We conclude from 4.3(24) that

$$|F(x)| \leqslant M\frac{|\sin\sigma x|}{\sigma\varrho} \sum_{k=-\infty}^{\infty} \frac{\left|\sin\varrho\left(x-\frac{k\pi}{\sigma}\right)\right|}{\left(x-\frac{k\pi}{\sigma}\right)}$$

$$= M\frac{|\sin\sigma x|}{\sigma\varrho} \sum_{|k|\leqslant 2\left[\frac{\sigma|x|}{\pi}\right]} \frac{\left|\sin\varrho\left(x-\frac{k\pi}{\sigma}\right)\right|}{\left(x-\frac{k\pi}{\sigma}\right)^2} +$$

$$+M\frac{|\sin \sigma x|}{\sigma \varrho} \sum_{|k|>2\left[\frac{\sigma |x|}{\pi}\right]} \frac{\left|\sin \varrho\left(x-\frac{k\pi}{\sigma}\right)\right|}{\left(x-\frac{k\pi}{\sigma}\right)^2},$$

$$|F(x)| \leqslant B(\sigma, p) M. \tag{25}$$

The example of 4.3(8) shows that M. L. Cartwright's theorem no longer applies for $p = \sigma$. This is connected with the fact that the constant $B(\sigma, p)$ in the inequality 4.3(25) increases without limit when $p \to \sigma$. S. N. Bernstein [39] proved, that as $p \to \sigma$ the asymptotic equality

$$B(\sigma, p) = \frac{2}{\pi} \ln \frac{\sigma}{\sigma - p} + O(1) \tag{26}$$

holds for $B(\sigma, p)$.

4.4. On the phenomenon of interference in the behaviour of integral functions of finite degree.

As is shown by the example of 4.3(8), in the general case, from the boundedness of the sequence of values of the integral function $F(x)$ of degree σ at the points $\frac{\nu\pi}{\sigma}$ it does not follow that it is uniformly bounded on the whole real axis. In this connection it appears that for certain classes of integral functions of degree σ, bounded at the sequence of points $\frac{\nu\pi}{\sigma}$ $(\nu = 0, \pm 1, \pm 2, \ldots)$, the sum

$$\sum_{k=1}^{m} a_k F(x+r_k)$$

is uniformly bounded on the real axis for special values of $a_k \neq 0$ and $\tau_k (k = 1, \ldots m)$.

This property, which it is natural to call the phenomenon of interference, is, as will be seen later, possessed by all functions of the class $B_\sigma^{(1)}$.

4.4.1. Let $\varrho(x)$ be a real function defined on the whole real axis and satisfying for some $\sigma_0 > \sigma$ the condition

$$\int_{-\infty}^{\infty} e^{\sigma_0 |x|} |d\varrho(x)| < \infty, \tag{1}$$

which characterises its behaviour at infinity and the finiteness of its total variation in the whole of its interval of definition. The condition 4.4(1) ensures the existence of the operator

$$R(f;z) = \int_{-\infty}^{\infty} f(z+t)\, d\varrho(t) \tag{2}$$

on the class of all integral functions of degree $\leqslant \sigma$, which puts into correspondence with any such function $f(z)$ a certain other function $R(z) = R(f;z)$ of degree not higher than σ. We shall say that $\varrho(x)$ gives rise to interference on the class $B_\sigma^{(1)}$, if for any function $F(z) \in B_\sigma^{(1)}$ the function $R(z) = R(F;z)$ is bounded on the real axis.

*In order that the function $\varrho(x)$ should give rise to interference on the class $B_\sigma^{(1)}$, it is necessary and sufficient that***

$$\int_{-\infty}^{\infty} e^{i\sigma t}\, d\varrho(t) = 0. \tag{3}$$

In particular, considering, for example, the functions

$$\varrho(x) = \begin{cases} 0, & x < -\dfrac{\pi}{2\sigma}, \\ \dfrac{1}{2}, & -\dfrac{\pi}{2\sigma} \leqslant x < \dfrac{\pi}{2\sigma}, \\ 1, & x \geqslant \dfrac{\pi}{2\sigma}, \end{cases} \qquad \varrho(x) = \begin{cases} 1, & x < 0, \\ 0, & 0 \leqslant x < \dfrac{2\pi}{\sigma}, \\ 1, & x \geqslant \dfrac{2\pi}{\sigma}, \end{cases} \tag{4}$$

which satisfy the condition 4.4(3), we observe interference of the type $F\left(x - \dfrac{\pi}{2\sigma}\right) + F\left(x + \dfrac{\pi}{2\sigma}\right)$ and $F(x) - F\left(x + \dfrac{2\pi}{\sigma}\right)$, first observed by S. N. Bernstein [39].

Using the condition 4.4(3) it is possible to indicate a series of other interference phenomena, for example, for

$$\varrho(x) = \begin{cases} 0, & x < \dfrac{2(3r-1)}{3\sigma}\pi, \\ \dfrac{1}{3}, & \dfrac{2(3r-1)}{3\sigma}\pi \leqslant x < 0, \\ \dfrac{2}{3}, & 0 \leqslant x < \dfrac{2(3s+1)}{3\sigma}\pi, \\ 1, & x \geqslant \dfrac{2(3s+1)}{3\sigma}\pi, \end{cases} \qquad \varrho(x) = \begin{cases} 0, & x < \dfrac{(6r-1)\pi}{3\sigma}, \\ \dfrac{1}{3}, & \dfrac{(6r-1)\pi}{3\sigma} \leqslant x < 0, \\ 0, & 0 \leqslant x < \dfrac{(6s+1)\pi}{3\sigma}, \\ \dfrac{1}{3}, & x \geqslant \dfrac{(6s+1)\pi}{3\sigma}, \end{cases}$$

** A. F. Timan [11]. This theorem also admits of generalisation to certain other classes of unbounded functions. It has received further development in the work of Boas [1, 3].

(r, s integers, $r \leqslant s$) or in the more general case for

$$\varrho(x) = \sum_{\nu=0}^{k} \lambda_\nu^{(n)},$$

when $x_k \leqslant x < x_{k+1}$ $(k = 0, 1, \ldots, n)$, $x_0 = -\infty$, $x_{n+1} = \infty$ and

$$\sum_{k,j=1}^{n} \lambda_k^{(n)} \lambda_j^{(n)} \cos\sigma(x_k - x_j) = 0.$$

In order to prove the theorem formulated, let us consider the functions

$$\gamma_k(x) = \begin{cases} \dfrac{1}{\sigma x - k\pi}, & |\sigma x - k\pi| \geqslant \dfrac{\pi}{2}, \\ 0, & |\sigma x - k\pi| < \dfrac{\pi}{2}, \end{cases}$$

$$\gamma(F; x) = \sum_{|k|\geqslant 1} \left[\gamma_k(x) + \frac{1}{k\pi}\right](-1)^k F\left(\frac{k\pi}{\sigma}\right).$$

If $\varrho(x)$ satisfies condition 4.4(3), then

$$\int_{\infty-}^{\infty} \gamma(F; x) \sin\sigma(x+t)\, d\varrho(t) \equiv 0.$$

Hence, using formula 4.3(14), we find

$$R(F; x) = \int_{-\infty}^{\infty} \left\{ \sum_{k=-\infty}^{\infty} (-1)^k F\left(\frac{k\pi}{\sigma}\right) \sin\sigma(x+t) \times \right.$$
$$\left. \times \left[\frac{1}{\sigma(x+t) - k\pi} - \gamma_k(x)\right]\right\} d\varrho(t).$$

Consequently, if $\left|F\left(\dfrac{k\pi}{\sigma}\right)\right| \leqslant M$, then

$$|R(F; x)| \leqslant M \int_{-\infty}^{\infty} \sum_{k=-\infty}^{\infty} |\sin\sigma(x+t)| \left|\frac{1}{\sigma(x+t) - k\pi} - \gamma_k(x)\right| \cdot |d\varrho(t)|.$$

Now denoting by $k_x = k(x)$ the integer for which $|\sigma x - k_x\pi| < \dfrac{\pi}{2}$, we have

$$|R(F; x)| \leqslant M\sigma \int_{-\infty}^{\infty} \sum_{k=-\infty}^{\infty}{}' \frac{1}{|\sigma(x+t) - k\pi| \cdot |\sigma x - k\pi|} |d\varrho(t)| + C(x),$$

where the prime indicates that the indices $k(x)$ and $k(x+t)$ are omitted from the summation inside the integral, and $C(x)$ is a function which is bounded on the whole real axis. But as the function

$$\Phi(x, t) = \sum_{k=-\infty}^{\infty}{}' \frac{1}{|\sigma(x+t)-k\pi|\cdot|\sigma x-k\pi|}$$

is uniformly bounded for all possible values of x and t, it follows that in virtue of the condition at infinity which $\varrho(x)$ satisfies, the function $R(z) = R(F;z)$ is bounded on the real axis.

If condition 4.4(3) is not satisfied, there is a sequence of real values $\{x_n\}$ $(n = 1, 2, \ldots)$ increasing to infinity such that

$$\int_{-\infty}^{\infty} \sin\sigma(x_n+t)\,d\varrho(t) = a \neq 0.$$

In this case we consider the function 4.3(8). For this

$$R(F; x) = \int_{-\infty}^{\infty} \sin\sigma(x+t)\sum_{k=1}^{\infty}\left\{\frac{1}{k\pi}+\frac{1}{\sigma(x+t)-k\pi}\right\}d\varrho(t)$$

and hence

$$R(F; x_n) = \int_{-\infty}^{\infty} \sin\sigma(x_n+t)\ln(1+|x_n+t|)\,d\varrho(t)+O(1).$$

It follows that

$$|R(f; x_n)| = \left|\int_{-\frac{1}{2}x_n}^{\frac{1}{2}x_n} \sin\sigma(x_n+t)\ln(1+|x_n|)\,d\varrho(t)\right|+O(1).$$

Thus as $n \to \infty$

$$|R(f; x_n)| > \tfrac{1}{2}|a|\ln(1+|x_n|)+O(1),$$

and consequently in this case there is no interference.

4.4.2. We have the simplest type of interference when for some τ the sum $f(x)+f(x+\tau)$ or the difference $f(x)-f(x+\tau)$ is bounded.

It follows immediately from theorem 4.4.1 that the sum $f(x)+f(x+\tau)$ (the difference $f(x)-f(x+\tau)$) is bounded on the whole real axis for any function $f(x) \in B_\sigma^{(1)}$ if and only if $\tau = \frac{2p+1}{\sigma}\pi\left(\tau = \frac{2p\pi}{\sigma}\right)$, where p is some integer.

Thus for any fixed odd m the sum $f\left(x-\frac{m\pi}{2\sigma}\right)+f\left(x+\frac{m\pi}{2\sigma}\right)$ is bounded on the interval $-\infty < x < \infty$ whatever the function $f(x) \in B_\sigma^{(1)}$ may be.

The following more precise assertion is valid. For every function $f(x) \in B_\sigma^{(1)}$ and any odd m the inequality

$$\frac{1}{2}\left|f\left(x-\frac{m\pi}{2\sigma}\right)+f\left(x+\frac{m\pi}{2\sigma}\right)\right| \leqslant \frac{4M}{\pi m}+\frac{8mM}{\pi}\sum_{\nu=1}^{\frac{m-1}{2}}\frac{1}{m^2-4\nu^2} \tag{5}$$

holds, where $M=\sup_k\left|f\left(\frac{k\pi}{\sigma}\right)\right|$ $(k=0, \pm 1, \pm 2, \ldots)$. For any such m there exists a function $f(x) \in B_\sigma$ for which the left-hand side of 4.4(5) is exactly equal to the right for $x=0$.

It follows from 4.3(14) that

$$\left|f\left(x-\frac{m\pi}{2\sigma}\right)+f\left(x+\frac{m\pi}{2\sigma}\right)\right|$$
$$=\left|\cos\sigma x\sum_{k=-\infty}^{\infty}(-1)^k f\left(\frac{k\pi}{\sigma}\right)\left[\frac{1}{\sigma x-\left(k-\frac{m}{2}\right)\pi}-\frac{1}{\sigma x-\left(k+\frac{m}{2}\right)\pi}\right]\right|.$$

Let us suppose that $\cos\sigma x \neq 0$ and that k_x is an integer for which $|\sigma x-k_x\pi|<\frac{\pi}{2}$, i.e. $\sigma x=k_x\pi+\delta\pi$, $|\delta|<\frac{1}{2}$.

Using the inequality

$$\left|f\left(x-\frac{m\pi}{2\sigma}\right)+f\left(x+\frac{m\pi}{2\sigma}\right)\right| \leqslant$$
$$\leqslant |\cos\sigma x|\, M\sum_{k=-\infty}^{\infty}\left|\frac{1}{\sigma x-\left(k-\frac{m}{2}\right)\pi}-\frac{1}{\sigma x-\left(k+\frac{m}{2}\right)\pi}\right|$$

and observing that the difference

$$\frac{1}{\sigma x-\left(k+p+\frac{1}{2}\right)\pi}-\frac{1}{\sigma x-\left(k-p-\frac{1}{2}\right)\pi} \qquad (m=2p+1)$$

is positive for $-\infty<k\leqslant k_x-p-1$, $k_x+p+1\leqslant k<\infty$ and negative for $k_x-p\leqslant k\leqslant k_x+p$, we obtain

$$\left|f\left(x-\frac{m\pi}{2\sigma}\right)+f\left(x+\frac{m\pi}{2\sigma}\right)\right|$$
$$\leqslant M\cos\delta\pi\left\{\sum_{k=-\infty}^{k_x-p-1}\left[\frac{1}{\sigma x-\left(k+p+\frac{1}{2}\right)\pi}-\frac{1}{\sigma x-\left(k-p-\frac{1}{2}\right)\pi}\right]-\right.$$

$$-\sum_{k_x=k-p}^{k_x+p}\left[\frac{1}{\sigma x-\left(k+p+\frac{1}{2}\right)\pi}-\frac{1}{\sigma x-\left(k-p-\frac{1}{2}\right)\pi}\right]+$$

$$+\sum_{k=k_x+p+1}^{\infty}\left[\frac{1}{\sigma x-\left(k+p+\frac{1}{2}\right)\pi}-\frac{1}{\sigma x-\left(k-p-\frac{1}{2}\right)\pi}\right]\Bigg\}.$$

But as

$$\frac{1}{\sigma x-\left(k+p+\frac{1}{2}\right)\pi}-\frac{1}{\sigma x-\left(k-p-\frac{1}{2}\right)\pi}$$

$$=\sum_{\nu=-p}^{p}\left[\frac{1}{\sigma x-\left(k+\nu+\frac{1}{2}\right)\pi}-\frac{1}{\sigma x-\left(k+\nu-\frac{1}{2}\right)\pi}\right],$$

it follows that

$$\left|f\left(x-\frac{m\pi}{2\sigma}\right)+f\left(x+\frac{m\pi}{2\sigma}\right)\right|\leqslant$$

$$\leqslant M\cos\delta\pi\left\{\sum_{\nu=-p}^{p}\sum_{k=-\infty}^{k_x-p-1}\left[\frac{1}{\sigma x-\left(k+\nu+\frac{1}{2}\right)\pi}-\frac{1}{\sigma x-\left(k+\nu-\frac{1}{2}\right)\pi}\right]-\right.$$

$$-\sum_{\nu=-p}^{p}\sum_{k=k_x-p}^{k_x+p}\left[\frac{1}{\sigma x-\left(k+\nu+\frac{1}{2}\right)\pi}-\frac{1}{\sigma x-\left(k+\nu-\frac{1}{2}\right)\pi}\right]+$$

$$\left.+\sum_{\nu=-p}^{p}\sum_{k=k_x+p+1}^{\infty}\left[\frac{1}{\sigma x-\left(k+\nu+\frac{1}{2}\right)\pi}-\frac{1}{\sigma x-\left(k+\nu-\frac{1}{2}\right)\pi}\right]\right\},$$

whence

$$\left|f\left(x-\frac{m\pi}{2\sigma}\right)+f\left(x+\frac{m\pi}{2\sigma}\right)\right|$$

$$\leqslant M\cos\delta\pi\left\{\sum_{\nu=-p}^{p}\frac{1}{\sigma x-(k_x-p+\nu-\frac{1}{2})\pi}+\right.$$

$$+\sum_{\nu=-p}^{p}\left[\frac{1}{\sigma x-(k_x-p+\nu-\frac{1}{2})\pi}-\frac{1}{\sigma x-(k_x+p+\frac{1}{2})\pi}\right]-$$

$$\left.-\sum_{\nu=-p}^{p}\frac{1}{\sigma x-(k_x+p+\nu+\frac{1}{2})\pi}\right\},$$

or

$$\left|f\left(x-\frac{m\pi}{2\sigma}\right)+f\left(x+\frac{m\pi}{2\sigma}\right)\right| \leqslant$$

$$\leqslant 2M\cos\delta\pi \sum_{\nu=-p}^{p}\left[\frac{1}{(k_x-\nu+p+\frac{1}{2})\pi-\sigma x}+\frac{1}{\sigma x-(k_x-\nu-p-\frac{1}{2})\pi}\right]$$

$$=\frac{2M(2p+1)}{\pi}\cos\delta\pi\sum_{\nu=-p}^{p}\frac{1}{(p+\frac{1}{2})^2-(\nu+\delta)^2}. \tag{6}$$

Using the inequality

$$\left|\frac{\sin t}{t}\right| \geqslant \frac{2\sqrt{2}}{\pi} \quad \left(|t|\leqslant\frac{\pi}{4}\right),$$

it is easily verified that for any $a \geqslant 1$ and $|\delta| < \frac{1}{2}$

$$\frac{a\cos\delta\pi}{a^2-4\delta^2} \leqslant \frac{1}{a}.$$

Hence since

$$m\left\{\frac{1}{m^2-4(\nu+\delta)^2}+\frac{1}{m^2-4(\nu-\delta)^2}\right\}$$

$$=\frac{m-2\nu}{(m-2\nu)^2-4\delta^2}+\frac{m+2\nu}{(m+2\nu)^2-4\delta^2},$$

we obtain from 4.4(6) the inequality

$$\left|f\left(x-\frac{m\pi}{2\sigma}\right)+f\left(x+\frac{m\pi}{2\sigma}\right)\right| \leqslant \frac{8M}{\pi m}+\frac{16M}{\pi}m\sum_{\nu=1}^{p}\frac{1}{m^2-4\nu^2},$$

i.e. 4.4(5) is satisfied.

In order to verify that the inequality 4.4(5) is precise, let us consider the function

$$f(x) = 2\sum_{\nu=-\frac{m-1}{2}}^{\frac{m-1}{2}}\frac{\sin\sigma x}{\sigma x-\nu\pi}-\cos\sigma x.$$

which obviously belongs to the class B_σ; here $M = 1$. For this function

$$\left|f\left(-\frac{m\pi}{2\sigma}\right)+f\left(\frac{m\pi}{2\sigma}\right)\right|$$

$$=\frac{1}{\pi}\left|\frac{1}{2}\sum_{\nu=-\frac{m-1}{2}}^{\frac{m-1}{2}}\left\{\frac{1}{\nu-\frac{m}{2}}-\frac{1}{\nu+\frac{m}{2}}\right\}\right| = \frac{8}{\pi m}+\frac{16m}{\pi}\sum_{\nu=1}^{\frac{m-1}{2}}\frac{1}{m^2-4\nu^2}.$$

4.4.3. Putting $m = 1$ in the theorem which has just been proved, we obtain the inequality

$$\frac{1}{2}\left|f\left(x-\frac{\pi}{2\sigma}\right)+f\left(x+\frac{\pi}{2\sigma}\right)\right| \leqslant \frac{4}{\pi}M, \tag{7}$$

first discovered by S. N. Bernstein [39] for functions of the class $B_\sigma^{(1)}$.

From the inequality 4.4(7) there immediately follows the following proposition, which supplements the theorem of M. L. Cartwright (see section 4.3.4). *If the function $f(x)$ belongs to the class $B_\sigma^{(1)}$ and is non-negative, then it is bounded on the whole real axis and*

$$f(x) \leqslant \left(1+\frac{4}{\pi}\right)M, \tag{8}$$

where $M = \sup_k f\left(\frac{k\pi}{\sigma}\right)$ $(k = 0, \pm 1, \pm 2, \ldots)$.

In order to verify this it is sufficient to apply the inequality 4.4(7) to the function $2f(x)-M$. We note that the application of the more general inequality 4.4(5) would lead to a rougher estimate than 4.4(8). It follows from theorem 4.4.2 that *for a given value of M and for $m = 2p+1$ increasing without limit the asymptotic equality***

$$\frac{1}{2}\sup_{f\in B_\sigma^{(1)}}\left|f\left(x-\frac{m\pi}{2\sigma}\right)+f\left(x+\frac{m\pi}{2\sigma}\right)\right| = \frac{2M}{\pi}\ln m+O(1) \tag{9}$$

is satisfied, showing to what extent the displacement of the argument affects the qualitative deterioration of the interference in the class $B_\sigma^{(1)}$.

4.5. Some integral representations for algebraic polynomials and trigonometric polynomials

There exist simple integral analogues of some of the interpolation formulae considered in sections 4.1 and 4.2.

4.5.1. For any function $f(x)$ integrable over $[0, 2\pi]$, the integral

$$\frac{1}{\pi}\int_0^{2\pi} f(t)\,D_n(x-t)\,dt \tag{1}$$

represents the trigonometric polynomial 4.2(1), which is identical with the partial sum $S_n(f;x)$ of the Fourier series of $f(x)$. The coefficients a_k and

** A. F. Timan [1].

$b_k (k = 0, 1, \ldots, n)$ of this polynomial are connected with the function $f(x)$ by the formula

$$a_k - ib_k = \frac{1}{\pi} \int_0^{2\pi} f(t) e^{-ikt} dt. \tag{2}$$

The operator 4.5(1), which maps the space of all the functions integrable in $[0, 2\pi]$ onto its $(2n+1)$-dimensional space of all trigonometric polynomials of order $\leqslant n$, possesses the property that for every such polynomial $T_n(x)$ the identity

$$T_n(x) = \frac{1}{\pi} \int_0^{2\pi} T_n(t) D_n(x-t) dt, \tag{3}$$

similar to 4.2(9), is satisfied. The operator

$$\frac{2}{2n+1} \sum_{k=0}^{2n} f(y_k^{(n)}) D_n(x - y_k^{(n)}), \tag{4}$$

where $y_k^{(n)}$ is the set of points 4.2(7), possesses the same property. The integral 4.5(1) and the sum 4.5(4) are the two most important cases of the Dirichlet—Stieltjes integral

$$\frac{1}{\pi} \int_0^{2\pi} f(t) D_n(x-t) d\varrho_n(t).$$

in the theory of approximation of functions. In the study of the behaviour of the sequence of integrals 4.5(1) and the sum 4.5(4), when n increases without limit, an essential part is played by the upper bounds

$$L_n^* = \sup_{|f(t)| \leqslant 1} \frac{1}{\pi} \int_0^{2\pi} f(t) D_n(x-t) dt = \frac{1}{\pi} \int_0^{2\pi} |D_n(t)| dt, \tag{5}$$

$$L_n^*(x) = \sup_{|f(t)| \leqslant 1} \frac{2}{2n+1} \sum_{k=0}^{2n} f(y_k^{(n)}) D_n(x - y_k^{(n)})$$
$$= \frac{2}{2n+1} \sum_{k=0}^{2n} |D_n(x - y_k^{(n)})|, \tag{6}$$

called respectively Lebesgue's constant and the interpolation function of Lebesgue for the system of points 4.2(7). It will be seen in what follows that there exists a close connection between these quantities (see 4.12(5), 4.12(8)).

4.5.2. For any function $\phi(t)$ integrable over $[0, 2\pi]$, the integral

$$\frac{1}{\pi}\int_0^\pi \phi(u)[D_n(t+u)+D_n(t-u)]\,du \tag{7}$$

represents the even trigonometric polynomial

$$T_n(t) = \frac{c_0}{2} + \sum_{k=1}^{n} c_k \cos kt,$$

the coefficients of which are connected with $\phi(t)$ by the formula

$$c_k = \frac{2}{\pi}\int_0^\pi \phi(u) \cos ku\, du.$$

If the function $f(x)$ is defined on the segment $[-1, 1]$, then, considering $\phi(t) = f(\cos t)$, we note that

$$\frac{1}{\pi}\int_0^\pi f(\cos u)[D_n(t+u)+D_n(t-u)]\,du$$

is the algebraic polynomial

$$P_n(x) = T_n(\arccos x) = \frac{c_0}{2} + \sum_{k=1}^{n} c_k \cos k \arccos x,$$

where

$$c_k = \frac{2}{\pi}\int_{-1}^{1} \frac{f(t) \cos k \arccos t}{\sqrt{1-t^2}}\,dt.$$

The operator 4.5(7), which maps the space of all functions integrable on $[0, \pi]$ onto the $(n+1)$-dimensional space of all even trigonometric polynomials of the n-th order (algebraic polynomials of degree $\leqslant n$), possesses the property that for every such polynomial $P_n(x)$ the identity

$$P_n(\cos t) = \frac{1}{\pi}\int_0^\pi P_n(\cos u)[D_n(t+u)+D_n(t-u)]\,du, \tag{8}$$

similar to 4.1(11), holds.

The operator

$$\frac{1}{n+1}\sum_{k=0}^{n} \phi(t_k^{(n+1)})[D_n(t+t_k^{(n+1)})+D_n(t-t_k^{(n+1)})] \tag{9}$$

possesses the same property, where $t_k^{(n+1)}$ is the system of Chebyshev points (see section 4.1.2).

As in section 4.5.1, the integral 4.5(7) and the sum 4.5(9) are connected with their Lebesgue functions

$$L_n(t) = \frac{1}{n}\int_0^\pi |D_n(t+u)+D_n(t-u)|\,\mathrm{d}u \tag{10}$$

and

$$\bar{L}_n(t) = \frac{1}{n+1}\sum_{k=0}^{n} |D_n(t+t_k^{(n+1)})+D_n(t-t_k^{(n+1)})|, \tag{11}$$

which here play the same part as the quantities 4.5(5) and 4.5(6) do in the periodic case.

4.6. An integral representation for some classes of transcendental integral functions of exponential type. The Wiener–Paley theorem

In a number of important cases transcendental integral functions of exponential type admit of a representation similar to the expression

$$T_n(x) = \sum_{k=-n}^{n} c_k \mathrm{e}^{ikx} \tag{1}$$

for the simplest of them, the trigonometric polynomials.

4.6.1. Let us consider the class of all integral functions of degree $\leqslant \sigma$ for which

$$\int_{-\infty}^{\infty} |G(x)|^2\,\mathrm{d}x < \infty. \tag{2}$$

The following fundamental theorem, established by Wiener and Paley**, holds. *The integral function $G(x)$ is of finite degree $\leqslant \sigma$ and satisfies condition* 4.6(2) *if and only if it can be represented in the form*

$$G(x) = \frac{1}{\sqrt{2\pi}}\int_{-\sigma}^{\sigma} g(t)\mathrm{e}^{itx}\,\mathrm{d}t, \tag{3}$$

where $g(t)$ is some function of integrable square on the segment $-\sigma \leqslant t \leqslant \sigma$.

If the function $G(x)$ admits the representation 4.6(3), then it is an integral function since the integral

$$\frac{1}{\sqrt{2\pi}}\int_{-\sigma}^{\sigma} g(t)\mathrm{e}^{itz}\,\mathrm{d}t$$

** Paley and Wiener [1].

where $z = x+iy$ is an integral function identical with $G(x)$ on the real axis. Moreover,

$$|G(x+iy)| \leqslant \frac{1}{\sqrt{2\pi}} \left\{ \int_{-\sigma}^{\sigma} |g(t)|^2 \, dt \right\}^{\frac{1}{2}} \left\{ \int_{-\sigma}^{\sigma} e^{-2ty} \, dt \right\}^{\frac{1}{2}} \leqslant Ce^{\sigma|y|}.$$

Consequently $G(x)$ is of finite degree not greater than σ. Since $G(x)$ is the Fourier transform of a function which is equal to $g(t)$ on the segment $[-\sigma, \sigma]$ and to zero for $|t| > \sigma$, then in virtue of theorem 3.11.21

$$\int_{-\infty}^{\infty} |G(x)|^2 \, dx = \int_{-\sigma}^{\sigma} |g(t)|^2 \, dt \tag{4}$$

and $G(x)$ satisfies condition 4.6(2).

In order to prove the second part of the theorem it is necessary to establish that *if the function $G(x)$ which satisfies the condition 4.6(2) is an integral function of finite degree $\leqslant \sigma$, then its Fourier transform*

$$g(x) = \frac{1}{\sqrt{2\pi}} \frac{d}{dx} \int_{-\infty}^{\infty} G(t) \frac{e^{-itx} - 1}{-it} \, dt$$

*is equal to zero for almost all x for which $|x| > \sigma$***.

4.6.11. Since

$$G(x) = \sum_{k=0}^{\infty} \frac{a_k}{k!} x^k$$

is an integral function of degree $\leqslant \sigma$, it follows that $\overline{\lim_{k\to\infty}} \sqrt[k]{|a_k|} \leqslant \sigma$. Therefore the function***

$$F(z) = \sum_{k=0}^{\infty} \frac{a_k}{z^{k+1}} \tag{5}$$

is analytic everywhere outside the circle $|z| \leqslant \sigma$. We shall show, that in *any open half-plane D which does not contain this circle the function $F(z)$ is expressed in terms of $G(z)$ by the formula*

$$F(z) = \int_{0}^{\infty} G(\zeta) e^{-\zeta z} \, d\zeta, \tag{6}$$

in which the integration is carried out along the ray which emerges from the point $\zeta = 0$ in a direction perpendicular to the boundary of D.

** The proof given is due to Pólya and Plancherel [1].

*** The function 4.6(5) is called the Borel transform of $G(z)$.

If $-\theta$ is the angle which the given ray makes with the positive direction of the real axis, and if $z = x+iy$ belongs to some closed region $Q \subset D$, then for sufficiently small $\delta > 0$ we have for the integrand in 4.6(6) the estimate

$$|G(\zeta)\,e^{-\zeta z}| = O\{e^{(\sigma+\delta)|\zeta|}\,e^{-(x\cos\theta+y\sin\theta)|\zeta|}\}$$
$$= O\{e^{(\sigma+\delta)|\zeta|}\,e^{-(\sigma+2\delta)|\zeta|}\} = O(e^{-\delta|\zeta|}).$$

Hence we conclude that the integral 4.6(6) converges absolutely and uniformly in any closed region Q which lies within D, and consequently represents a function analytic in D. Let us expand the integrand in the series

$$G(\zeta)\,e^{-\zeta z} = \sum_{k=0}^{\infty} \frac{a_k}{k!}\,\zeta^k e^{-\zeta z}. \tag{7}$$

If $z = x+iy$ lies within the half-plane D, then for some $\delta > 0$ we have $x\cos\theta+y\sin\theta > \sigma+2\delta$. Moreover, a $C > 0$ can be found such that

$$|a_k| \leqslant C(\sigma+\delta)^k.$$

Hence

$$\left|\frac{a_k}{k!}\,\zeta^k e^{-\zeta z}\right| \leqslant C\frac{(\sigma+\delta)^k}{k!}|\zeta|^k e^{-(\sigma+2\delta)|\zeta|}$$

and consequently the series 4.6(7) can be integrated term by term along the ray $\zeta = |\zeta|\,e^{-i\theta}$, whereupon we obtain 4.6(6).

Thus the integral on the right-hand side of 4.6(6) can be continued analytically throughout the whole of the domain $|z| > \sigma$. By the uniqueness theorem this analytic continuation is the function $F(z)$.

4.6.12. Using formula 4.6(6), we will now show that if $G(x)$ is an integral function of finite degree $\leqslant \sigma$ satisfying condition 4.6(2), then the actual domain of analyticity of the function $F(z)$ is the whole of the z-plane with the exclusion of the segment of the imaginary axis from $-i\sigma$ to $+i\sigma$.

By what was proved in section 4.6.11, in the half-planes $x > \sigma$ and $x < -\sigma$ the function $F(x)$ is representable by formula 4.6(6), i.e. for $x > \sigma$

$$F(x+iy) = \int_0^{\infty} G(t)\,e^{-t(x+iy)}\,dt \tag{8}$$

and for $x < -\sigma$

$$F(x+iy) = -\int_{-\infty}^{0} G(t)e^{-t(x+iy)}\,dt. \tag{9}$$

But since for any $\varepsilon > 0$ the integrals 4.6(8) and 4.6(9), in virtue of condition 4.6(2), converge absolutely and uniformly for $x \geqslant \varepsilon$ and $x \leqslant -\varepsilon$ respectively, it follows that as functions of $z = x+iy$ they are analytic throughout the whole of the plane, with the exception of the imaginary axis $x = 0$. Thus they serve as the analytic continuations of the function $F(z)$ within the circle $|z| \leqslant \sigma$ with a cut along the diameter from the point $-i\sigma$ to the point $+i\sigma$.

4.6.13. For any $x > 0$, as is obvious from 4.6(8), $F(x+iy)$ is the Fourier transform of a function equal to zero for $t < 0$ and to $\sqrt{2\pi}\,G(t)e^{-ix}$ for $t \geqslant 0$, and correspondingly from 4.6(9) $F(-x+iy)$ is the Fourier transform of a function equal to zero for $t > 0$ and to $-\sqrt{2\pi}\,G(t)e^{tx}$ for $t \leqslant 0$.

Considering the Fourier transforms

$$g_+(y) = \frac{d}{dy}\int_0^\infty G(t)\frac{e^{-ity}-1}{-it}\,dt, \qquad g_-(y) = \frac{d}{dy}\int_{-\infty}^0 G(t)\frac{e^{-ity}-1}{it}\,dt,$$

we see by Plancherel's theorem (see 3.11.21) that for any $x > 0$

$$\frac{1}{2\pi}\int_{-\infty}^\infty |F(x+iy)-g_+(y)|^2\,dy = \int_0^\infty |1-e^{-tx}|^2\,|G(t)|^2\,dt,$$

$$\frac{1}{2\pi}\int_{-\infty}^\infty |F(-x+iy)-g_-(y)|^2\,dy = \int_{-\infty}^0 |1-e^{tx}|^2|\,|G(t)|^2\,dt.$$

Hence it follows that these integrals tend to zero with x. But for $|y| > \sigma$

$$\lim_{x\to 0} F(x+iy) = \lim_{x\to 0} F(-x+iy) = F(iy) \qquad (x > 0).$$

Hence for almost all $|y| > \sigma$

$$F(iy) = g_+(y) = g_-(y)$$

and for the Fourier transform of the function $G(t)$ we have

$$g(y) = \frac{1}{\sqrt{2\pi}}\frac{d}{dy}\int_{-\infty}^\infty G(t)\frac{e^{-ity}-1}{-it}\,dt$$

$$= \frac{1}{\sqrt{2\pi}}\frac{d}{dy}\int_0^\infty G(t)\frac{e^{-ity}-1}{-it}\,dt + \frac{1}{\sqrt{2\pi}}\frac{d}{dy}\int_{-\infty}^0 G(t)\frac{e^{-ity}-1}{-it}\,dt$$

$$= g_+(y) - g_-(y) = 0.$$

This proves the last proposition of section 4.6.1, i.e. the second part of the Wiener–Paley theorem.

4.6.2. Let us now consider the set of all integral functions $G(x)$ of degree $\leqslant \sigma$ which belong to Wiener's class (see section 3.11.24), i.e. those for which

$$\int_{-\infty}^{\infty} \frac{|G(x)|^2}{1+x^2} \, dx < \infty.$$

It follows immediately from theorem 4.6.1 that for such functions

$$\frac{G(x)-G(0)}{x} = \frac{1}{\sqrt{2\pi}} \int_{-\sigma}^{\sigma} g(t) e^{itx} \, dt.$$

In particular, *any integral function $G(x)$ of degree $\leqslant \sigma$ which is bounded on the whole real axis is representable in the form*

$$G(x) = G(0) + \frac{x}{\sqrt{2\pi}} \int_{-\sigma}^{\sigma} g(t) e^{itx} \, dt, \tag{10}$$

where $g(t)$ is some function of integrable square on $[-\sigma, \sigma]$. In virtue of the definition of the conjugate function $\tilde{G}(x)$ given in 3.11.2, it follows from this that *for any integral function $G(x)$ of Wiener's class of degree* $\leqslant \sigma$

$$\tilde{G}(x) = \frac{x}{i\sqrt{2\pi}} \int_{-\sigma}^{\sigma} g(t) \operatorname{sign} t \cdot e^{itx} \, dt. \tag{11}$$

It follows from 4.6(11) that if the integral function $G(x)$ of degree $\leqslant \sigma$ belongs to Wiener's class, then $\tilde{G}$ also is an integral function of degree $\leqslant \sigma$ belonging to the same class, and $\dfrac{\tilde{G}(x)}{x} \to 0$ as $x \to \neq \infty$.

4.7. Some interpolation and integral identities for the derivatives of algebraic polynomials and transcendental integral functions of exponential type

Commencing with the known integral and interpolation representations of the integral functions studied here, it is possible to obtain a series of similar formulae for their derivatives which are useful in the theory of approximation of functions.

4.7.1. Thus, for example, on differentiating both sides of 4.1(4) we obtain the formula

$$P_n'(x) = \sum_{k=0}^{n} P_n(x_k) l_{n,k}'(x), \tag{1}$$

which expresses the derivative of any algebraic polynomial $P_n(x)$ of order $\leqslant n$ in terms of its values at a system of $n+1$ distinct points x_k.

Of the similar interpolation identities for an arbitrary trigonometric polynomial it is necessary first of all to note those which correspond to the systems of equidistant points 4.2(7) and 4.2(12). The first follows from 4.2(9) and is of the form

$$T_n'(x) = \frac{2}{2n+1} \sum_{k=0}^{2n} T_n(y_k^{(n)})\, D_n'(x-y_k^{(n)}); \tag{2}$$

the second (M. Riesz's identity)

$$T_n'(x) = \frac{1}{4n} \sum_{k=0}^{2n-1} T_n(x+t_k^{(n)}) \frac{(-1)^k}{\sin^2 \dfrac{t_k^{(n)}}{2}} \tag{3}$$

is obtained if in 4.2(15) instead of the polynomial $T_n(x)$ we considei the polynomial $T_n(n+t)$, differentiate both sides with respect to x, and then put $x=0$.

4.7.2. The identity similar to 4.7(3) for any integral function $G_\sigma(x)$ of the class B_σ is of the form

$$G_\sigma'(t) = \frac{\sigma}{\pi^2} \sum_{k=-\infty}^{\infty} \frac{(-1)^k}{(k+\frac{1}{2})^2} G_0\left(t+\frac{2k+1}{2\sigma}\pi\right). \tag{4}$$

This follows from 4.3(14) if instead of the function $G_\sigma(x)$ we consider the function $G_\sigma(x+t)$, differentiate both sides with respect to** x and then put $x = -\dfrac{\pi}{2\sigma}$.

4.7.3. In order to obtain a formula similar to 4.7(4) for the derivative of the conjugate function $\tilde{G}_\sigma(x)$, it is convenient to use the integral representation 4.6(11), from which it follows that

$$\tilde{G}_\sigma'(x) = \frac{1}{i\sqrt{2\pi}} \int_{-\sigma}^{\sigma} g(t) \operatorname{sign} t \cdot e^{itx}\, dt + \frac{x}{\sqrt{2\pi}} \int_{-\sigma}^{\sigma} g(t)|t|\, e^{itx}\, dt$$

$$= \frac{1}{i\sqrt{2\pi}} \int_{-\sigma}^{\sigma} g(t) \frac{\partial}{\partial t} (|t|\, e^{itx})\, dt.$$

** Term-by-term differentiation is possible here since the series 4.3(12) converges absolutely and uniformly in any finite part of the complex plane.

But as for $|t| \leqslant \sigma$

$$|t| = \frac{\sigma}{2} - \frac{2\sigma}{\pi^2} \sum_{k=-\infty}^{\infty} \frac{1}{(2k+1)^2} e^{i\frac{2k+1}{\sigma}\pi t}$$

and consequently

$$|t| e^{itx} = \frac{\sigma}{2} e^{itx} - \frac{2\sigma}{\pi^2} \sum_{k=-\infty}^{\infty} \frac{1}{(2k+1)^2} e^{\left(\frac{2k+1}{\sigma}\pi + x\right) it},$$

it follows that

$$\tilde{G}'_\sigma(x) = \frac{\sigma}{2i\sqrt{2\pi}} \int_{-\sigma}^{\sigma} g(t)\, de^{itx} -$$

$$- \frac{2\sigma}{i\pi^2\sqrt{2\pi}} \int_{-\sigma}^{\sigma} g(t) \frac{\partial}{\partial t} \sum_{k=-\infty}^{\infty} \frac{1}{(2k+1)^2} e^{\left(\frac{2k+1}{\sigma}\pi + x\right) it} dt$$

$$= \frac{\sigma}{2i\sqrt{2\pi}} \int_{-\sigma}^{\sigma} g(t)\, de^{itx} -$$

$$- \frac{2\sigma}{i\pi^2\sqrt{2\pi}} \sum_{k=-\infty}^{\infty} \frac{1}{(2k+1)^2} \int_{-\sigma}^{\sigma} g(t)\, de^{\left(\frac{2k+1}{\sigma}\pi + x\right) it}.$$

Taking into account the fact that in virtue of 4.6(10)

$$G_\sigma(x) = G_\sigma(0) + \frac{1}{i\sqrt{2\pi}} \int_{-\sigma}^{\sigma} g(t)\, de^{itx},$$

we thus obtain

$$\tilde{G}'_\sigma(x) = \frac{\sigma}{2}[G_\sigma(x) - G_\sigma(0)] -$$

$$- \frac{2\sigma}{\pi^2} \sum_{k=-\infty}^{\infty} \frac{1}{(2k+1)^2} \left[G_\sigma\left(\frac{2k+1}{\sigma}\pi + x\right) - G_\sigma(0)\right].$$

Since

$$\sum_{k=0}^{\infty} \frac{1}{(2k+1)^2} = \frac{\pi^2}{8} \tag{5}$$

we arrive at the result that *for any integral function $G_\sigma(x)$ of degree $\leqslant \sigma$ of Wiener's class the identity*

$$\tilde{G}'_\sigma(x) = \frac{\sigma}{2} G_\sigma(x) - \frac{2\sigma}{\pi^2} \sum_{k=-\infty}^{\infty} \frac{1}{(2k+1)^2} G_\sigma\left(\frac{2k+1}{\sigma}\pi + x\right) \tag{6}$$

*is satisfied***.

In a similar way it would be possible to obtain also the identity 4.7(4)**.

** N. I. Akhiezer [3].

4.7.4. Let us note two integral analogues of the interpolation identities for trigonometric polynomials derived above. If we differentiate 4.5(3) we obtain the integral analogue of 4.7(2):

$$T_n'(x) = \frac{1}{\pi}\int_0^{2\pi} T_n(t) D_n'(x-t)\,dt. \tag{7}$$

Writing this identity in the form (see 4.1(10))

$$T_n'(x) = \frac{1}{\pi}\int_0^{2\pi} T_n(t) \sum_{k=1}^{n} k \sin k(t-x)\,dt$$

and noting that

$$\int_0^{2\pi} T_n(x+t) \sum_{k=1}^{n-1} k \sin (2n-k) t\,dt = 0,$$

we have

$$T_n'(x) = \frac{1}{\pi}\int_0^{2\pi} T_n(x+t)\left\{\sum_{k=1}^{n} k \sin kt + \sum_{k=1}^{n-1} k \sin (2n-k)t\right\} dt,$$

i.e. we obtain the integral analogue of 4.7(3)**:

$$T_n'(x) = \frac{1}{\pi}\int_0^{2\pi} T_n(x+t) \sin nt \left(\frac{\sin n\frac{t}{2}}{\sin\frac{t}{2}}\right)^2 dt. \tag{8}$$

4.7.5. The identities 4.7(7) and 4.7(8) are particular cases of a more general integral representation of the derivative of a trigonometric polynomial. For any trigonometric polynomial $T_n(x)$ of order $\leqslant n$,

$$T_n'(x) = \frac{2^{2r}\cdot n^2}{\pi}\int_0^{2\pi} T_n(x+t)\cdot H_{r,n}(t)\left(\frac{\sin\frac{nt}{2}}{\sin\frac{t}{2}}\right)^{2r} dt \tag{9}$$

for any integer $r \geqslant o$, where $H_{r,n}(t)$ is a trigonometric polynomial which possesses the property*** $|H_{r,n}(t)| \leqslant 1$.

Since in the case $r = 0$, for $H_{0,n} = -\frac{1}{n^2}D_n'(t)$, 4.7(9) is identical with 4.7(7), it is sufficient to prove that if this identity is true for $r = k$ then it is true also for $r = k+1$. Suppose 4.7(9) true for a given r. Let us

** A. Zygmund [1].
*** Yu. A. Brudnyi [2].

consider the arbitrary trigonometric polynomial $T_{n+1}(x)$ of order $\leqslant n+1$ and the trigonometric polynomial

$$\Phi_{2n}(x) = T_{n+1}(x)\left(\frac{\sin\frac{nx}{2}}{n\sin\frac{x}{2}}\right)^2 = \frac{2}{n^2}T_{n+1}(x)\sum_{k=0}^{n-1}D_k(x) \tag{10}$$

(see 4.1(10)), the order of which does not exceed $2n$. By supposition,

$$\Phi'_{2n}(0) = \frac{2^{2r}\cdot(2n)^2}{\pi}\int_0^{2\pi}\Phi_{2n}(t)H_{r,2n}(t)\left(\frac{\sin nt}{2n\sin\frac{t}{2}}\right)^{2r}dt,$$

and consequently

$$\Phi'_{2n}(0) = \frac{2^{2r+2}\cdot n^2}{\pi}\int_0^{2\pi}T_{n+1}(t)H_{r,2n}(t)\cos^{2r}\frac{nt}{2}\left(\frac{\sin\frac{nt}{2}}{n\sin\frac{t}{2}}\right)^{2r+2}dt.$$

But as $\Phi'_{2n}(0) = T'_{n+1}(0)$, we obtain, on putting $H_{r+1,n}(t) = H_{r,2n}(t)$ $\cos^{2r}\frac{nt}{2}$,

$$T'_{n+1}(0) = \frac{2^{2r+2}n^2}{\pi}\int_0^{2\pi}T_{n+1}(t)H_{r+1,n}(t)\left(\frac{\sin\frac{nt}{2}}{n\sin\frac{t}{2}}\right)^{2r+2}dt.$$

Applying the last identity to the polynomial $T_{n+1}(x+t)$, we find that for any trigonometric polynomial of order $\leqslant n+1$

$$T'_{n+1}(x) = \frac{2^{2r+2}n^2}{\pi}\int_0^{2\pi}T_{n+1}(x+t)H_{r+1,n}(t)\left(\frac{\sin\frac{nt}{2}}{n\sin\frac{t}{2}}\right)^{2r+2}dt.$$

This proves 4.7(9) for any integer $r \geqslant 0$.

4.7.6. An identity similar to 4.7(9) can be obtained for any integral function of exponential type which is bounded on the whole real axis**. *For any integral function $G_\sigma(x)$ of degree $\leqslant \sigma$, bounded on the whole real axis, we have for any integer $r \geqslant 0$*

$$G'_\sigma(x) = C_r\sigma^2\int_{-\infty}^{\infty}G_\sigma(x+t)h_{r,\sigma}(t)\left(\frac{\sin\sigma t}{\sigma t}\right)^{2r+2}dt, \tag{11}$$

where $|h_{r,\sigma}(t)| \leqslant 1$, and the constant C_r depends only on r.

** Yu. A. Brudnyi [2].

Since the function

$$G_{\sigma,r}(t) = G_\sigma(t)\left(\frac{\sin\sigma t}{\sigma t}\right)^{2r+2}$$

is of degree $\leqslant (2r+3)\sigma$ and satisfies condition 4.6(2), it follows that in virtue of theorem 4.6.1

$$G_{\sigma,r}(x) = \frac{1}{\sqrt{2\pi}} \int\limits_{-(2r+3)\sigma}^{(2r+3)\sigma} g(t)\mathrm{e}^{itx}\,\mathrm{d}t,$$

where $g(t)$ is some function of integrable square on $[-(2r+3)\sigma, (2r+3)\sigma]$. Moreover, since the Fourier transform of the function $\frac{\sin\lambda t}{t}$ is equal to $\frac{1}{2}\sqrt{2\pi}$ for $-\lambda < t < \lambda$ and to zero for $|t| > \lambda$, we have by formula 3.11(16)

$$\frac{1}{\pi}\int\limits_{-\infty}^{\infty} G_{\sigma,r}(-x-t)\frac{\sin(2r+3)\sigma t}{t}\,\mathrm{d}t = \frac{1}{\sqrt{2\pi}}\int\limits_{-(2r+3)\sigma}^{(2r+3)\sigma} g(t)\mathrm{e}^{itx}\,\mathrm{d}t = G_{\sigma,r}(x).$$

If follows from this that

$$G'_{\sigma,r}(x) = \frac{1}{\pi}\int\limits_{-\infty}^{\infty} G_{\sigma,r}(u-x)\frac{\mathrm{d}}{\mathrm{d}u}\left\{\frac{\sin(2r+3)\sigma u}{u}\right\}\mathrm{d}u.$$

Consequently, if we take account of the fact that $G'_{\sigma,r}(0) = G'_{\sigma,r}(0)$, and $\left|\frac{\mathrm{d}}{\mathrm{d}u}\left\{\frac{\sin\lambda u}{u}\right\}\right| \leqslant \lambda^2$ (see, for example, section 4.8.2), we obtain the equality

$$G'_\sigma(0) = G'_{\sigma,r}(0) = \frac{(2r+3)^2\sigma^2}{\pi}\int\limits_{-\infty}^{\infty} G_{\sigma,r}(u)h_{r,\sigma}(u)\,\mathrm{d}u$$

$$= \frac{(2r+3)^2\sigma^2}{\pi}\int\limits_{-\infty}^{\infty} G_\sigma(u)h_{r,\sigma}(u)\left(\frac{\sin\sigma u}{\sigma u}\right)^{2r+2}\mathrm{d}u,$$

where $|h_{r,\sigma}(u)| \leqslant 1$. Replacing here the function $G_\sigma(t)$ by $G_\sigma(t+x)$, we obtain the identity 4.7(11)**.

4.8. Some extremal properties of algebraic polynomials and transcendental integral functions of finite degree

Rational integral functions (algebraic polynomials) and transcendental integral functions of exponential type possess important extremal properties, which can be discovered by considering the interpolation and integral representations obtained for them above.

** The identity 4.7(9) can be obtained in a similar manner directly from 4.7(8).

4.8.1. Let us first of all consider a problem for transcendental integral functions of finite degree, which is similar to the question of the algebraic polynomial which deviates least from zero (see section 2.19.12), and let us prove**, that *if*

$$G(x) = c_0 + c_1 x + c_2 x^2 + c_3 x^3 + \dots \tag{1}$$

is an integral function of degree $\leqslant \sigma$ *then*

$$\sup_{-\infty<x<\infty} |G(x)| \geqslant \frac{|c_1|}{\sigma}. \tag{2}$$

In the class of all such functions $G(x)$ the inequality 4.8(2) cannot be improved, as for the function $\frac{|c_1|}{\sigma}\sin\sigma x$ it becomes an equality.

Thus *among all integral functions* $G(x)$ *of degree* $\leqslant \sigma$ *for which* $G'(0) = 1$, *the function* $\frac{\sin\sigma x}{\delta}$ *deviates least from zero on the whole real axis***.

Let us consider the class of all integral functions of degree $\leqslant \sigma$ for which $G'(0) = 1$. Supposing that the function $G(x)$ is bounded, let us apply to it the identity 4.7(4), in virtue of which

$$G'(0) = \sigma \sum_{k=-\infty}^{\infty} \frac{(-1)^k}{(k+\frac{1}{2})^2\pi^2} G\left(\frac{2k+1}{2\sigma}\pi\right). \tag{3}$$

If we make use of the condition $G'(0) = 1$ and take account of 4.7(5), we obtain

$$\sigma \sum_{k=-\infty}^{\infty} \frac{(-1)^k}{\left(k+\frac{1}{2}\right)^2 \pi^2}\left[G\left(\frac{2k+1}{2\sigma}\pi\right) - \frac{(-1)^k}{\sigma}\right] = 0.$$

From the last relation we conclude that either $\sup\limits_{-\infty<x<\infty} |G(x) > \frac{1}{\sigma}$, or for all $k = 0, \pm 1, \pm 2, \dots$

$$G\left(\frac{2k+1}{2\sigma}\pi\right) = \frac{(-1)^k}{\sigma}. \tag{4}$$

But subject to the condition $G'(0) = 1$ the system of equations 4.8(4) is possible only when $G(x) = \frac{\sin\sigma x}{\sigma}$ (see 4.3.1). Hence we arrive at the result that

$$\sup_{-\infty<x<\infty} |G(x)| \geqslant \frac{1}{\sigma}.$$

This shows that the function $G(x) = \frac{c_1 \sin\sigma x}{\sigma}$ is the unique integral func-

** S. N. Bernstein [15].

tion 4.8(1) of degree $\leqslant \sigma$ for which the inequality 4.8(2) becomes an equality.

4.8.2. From theorem 4.8.1 there immediately follows a remarkable inequality for integral functions of finite degree, established by S. N. Bernstein [15]. *If the integral function $G(x)$ of degree $\leqslant \sigma$ does not exceed the constant M in modulus on the whole real axis, then its derivative of any order satisfies the inequality*

$$\sup_{-\infty<x<\infty} |G^{(k)}](x)| \leqslant M\sigma^k \quad (k=0,1,2,\ldots). \tag{5}$$

It is sufficient to verify the validity of the inequality 4.8(5) only for $k=1$. If for some function $G(x)$ of degree $\leqslant \sigma$ it should happen that at some point x_0

$$|G'(x_0)| > \sigma M,$$

the considering the function $\Phi(x) = \dfrac{1}{\sigma M}\, G(x+x_0)$ we should have $|\Phi'(0)| > 1$. But then by theorem 4.8.1

$$\sup_{-\infty<x<\infty} |\Phi(x)| > \frac{1}{\sigma}$$

i.e. $\sup\limits_{-\infty<x<\infty} |G(x)| > M$. Consequently the inequality 4.8(5) follows from the condition $\sup\limits_{-\infty<x<\infty} |G(x)| \leqslant M$.

4.8.21. Let us note that Bernstein's inequality 4.8(5) could have been obtained directly from the identity 4.7(4).

From the similar identity 4.7(6) for the conjugate function it follows that

$$\sup_{-\infty<x<\infty} |\tilde{G}'_\sigma(x)| \leqslant \left[\frac{\sigma}{2} + \frac{2\sigma}{\pi^2} \sum_{k=-\infty}^{\infty} \frac{1}{(2k+1)^2}\right] \sup_{-\infty<x<\infty} |G(x)|,$$

i.e. for any integral function $G_\sigma(x)$ of the class B_σ the inequality

$$\sup_{-\infty<x<\infty} |\tilde{G}'(x)| \leqslant \sigma \sup_{-\infty<x<\infty} |G_\sigma(x)|. \tag{6}$$

holds**.

4.8.22. Special attention should be paid to the particular case where the function $G(x)$ is the trigonometric polynomial 4.2(1). *The inequalities****

$$(a)\ \max_x |T'_n(x)| \leqslant n \cdot \max_x |T_n(x)|,$$
$$(b)\ \max_x |\tilde{T}'_n(x)| \leqslant n \cdot \max_x |T_n(x)|, \tag{7}$$

are satisfied for any trigonometric polynomial $T_n(x)$ of order $\leqslant n$.

** N. I. Akhiezer [3].

*** The inequality (7a) is due to S. N. Bernstein [1], the inequality (7b) to Szegö [1]. For later generalisations see S. N. Bernstein [34], N. I. Akhiezer [3] and Rogosinski [2].

4.8.3. Let us note that the inequalities 4.8(5) characterise the class of integral functions of degree $\leqslant \sigma$ which do not exceed the number M in modulus on the real axis. *If the function* $G(x)$, *defined on* $(-\infty, \infty)$, *is infinitely diffenentiable and satisfies condition* 4.8(5), *then it is an integral function of degree* $\leqslant \sigma$ *and*

$$|G(x+iy)| \leqslant M e^{\sigma|y|}. \tag{8}$$

It is not difficult to confirm the validity of this assertion by considering the Maclaurin series

$$\sum_{k=0}^{\infty} \frac{1}{k!} G^{(k)}(0) x^k$$

and the Taylor series

$$\sum_{k=0}^{\infty} \frac{G^{(k)}(x)}{k!} (iy)^k,$$

the radius of convergence of which is infinite and which for $y = 0$ has $G(x)$ as its sum.

4.8.31. The assertion that a function $G(x)$ infinitely differentiable on $(-\infty, \infty)$ which satisfies the inequality 4.8(5) is an integral function of finite degree can be strengthened by assuming that this inequality is satisfied *only for some unbounded subsequence of values of*** k. In fact, let $G(x)$ be a function which is infinitely differentiable on the real axis and for some sequence of natural numbers n_k

$$\sup_{-\infty<x<\infty} |G^{(n_k)}(x)| \leqslant M\sigma^{n_k}. \tag{9}$$

If $n_k \leqslant n \leqslant n_{k+1}$, then by the inequalities 3.10(8)

$$\sup_{-\infty<x<\infty} |G^{(n)}(x)| \leqslant 2e^{n-m_k} \{ \sup_{-\infty<x<\infty} |G^{(n_k)}(x)| \}^{1-\frac{n-n_k}{n_{k+1}-n_k}} \times$$

$$\times \{ \sup_{-\infty<x<\infty} |G^{(n_{k+1})}(x)| \}^{\frac{n-n_k}{n_{k+1}-n_k}}$$

$$\leqslant 2e^{n-n_k} M \sigma^{n_k\left(1-\frac{n-n_k}{n_{k+1}-n_k}\right)} \sigma^{n_{k+1}\frac{n-n_k}{n_{k+1}-n_k}} \leqslant 2M(e\sigma)^n.$$

Hence in virtue of theorem 4.8.3 it follows that $G(x)$ is an integral function of degree $\leqslant e\sigma$. If instead of the inequality 3.10(8) use is made of the more precise inequality of A. N. Kolmogorov***, it is possible to

** S. N. Bernstein [44].
*** See the footnote ** on page 152.

show that the degree of the function $G(x)$ which satisfies condition 4.8(9) in fact never exceeds σ.

4.8.4. It is obvious from the identities 4.7(4) and 4.7(6), that if for some trigonometric polynomial $T_n(x)$ the inequality 4.8(7a) or the inequality 4.8(7b) becomes an equality and if t_0 is the value for which $|T_n'(t_0)| = \sup\limits_{-\infty<t<\infty} |T_n'(t)|$ or $|\tilde{T}_n'(t_0)| = \sup\limits_{-\infty<t<\infty} |\tilde{T}_n'(t)|$ respectively, then $T_n(x) = M \sin n(x-t_0)$. In the first case we consider the function $\Phi(x) = \dfrac{T_n(x+t_0)}{T_n'(t_0)}$. Since $\Phi'(0) = 1$ and $\sup\limits_{-\infty<x<\infty} |\Phi(x)| = \dfrac{1}{n}$, then by what was proved in section 4.8.1 $\Phi(x) = \dfrac{1}{n} \sin nx$, i.e. $T_n(x+t_0) = M \sin nx$; similarly in the second case.

4.8.5. In connection with the inequality 4.8(7a) let us note another useful estimate for the derivative of a trigonometric polynomial, which although less precise as regards the constant is nevertheless of a somewhat different character and shows that the order of increase of the modulus of the derivative of a trigonometric polynomial $T_n(x)$ at a given point as $n \to \infty$ is majorised by a quantity which in a certain sense depends only on its value in the neighbourhood of this point. Let $\phi(x) > 0$ be an arbitrary continuous periodic function of period 2π, and**

$$\sigma_n^{(r)}(\phi; x) = \left| \phi(x) - \frac{1}{\pi \gamma_n^{(r)}} \int_0^{2\pi} \phi(x+t) \left(\frac{\sin \frac{nt}{2}}{n \sin \frac{t}{2}} \right)^{2r+4} dt \right|, \tag{10}$$

where

$$\gamma_n^{(r)} = \frac{1}{\pi} \int_{\pi}^{\pi} \left| \frac{\sin \frac{nt}{2}}{n \sin \frac{t}{2}} \right|^{2r+4} dt. \tag{11}$$

** For any integral $r \geqslant 0$,

$$\delta_n^{(r)}(\phi; x) \leqslant \frac{1}{\pi \gamma_n^{(r)}} \int_0^{\pi} \omega_2(\phi; t) \left(\frac{\sin \frac{nt}{2}}{n \sin \frac{t}{2}} \right)^{2r+4} dt \leqslant$$

$$\leqslant \frac{n^2 \omega_2 \left(\phi; \frac{1}{n} \right)}{\pi \gamma_n^{(r)}} \int_0^{\pi} \left(t + \frac{1}{n} \right)^2 \left(\frac{\sin \frac{nt}{2}}{n \sin \frac{t}{2}} \right)^{2r+4} dt = O\left[\omega_2 \left(\phi; \frac{1}{n} \right) \right],$$

i.e. $\delta_n^{(r)}(\phi; x)$ tends to zero uniformly as $n \to \infty$.

If the trigonometric polynomial $T_n(x)$ *of order* $\leqslant n$ *satisfies the inequality* $|T_n(x)| \leqslant \phi(x)$, *then for any integral* $r \geqslant 0$

$$|T_n'(x)| < C_r n\left[1+\frac{\delta_n^{(r)}[\phi;x]}{\phi(x)}\right]\phi(x), \tag{12}$$

where

$$C_r = \frac{2}{\pi}\int_{-\infty}^{\infty}\left(\frac{\pi \sin t}{t}\right)^{2r+4} dt.$$

The truth of this assertion follows from the identity 4.7(9), in virtue of which

$$|T_n'(x)| \leqslant \frac{2^{2r+4}n^2}{\pi}\int_0^{2\pi}\phi(x+t)\left(\frac{\sin\frac{nt}{2}}{n\sin\frac{t}{2}}\right)^{2r+4} dt.$$

From this there immediately follows the inequality

$$|T_n'(x)| \leqslant 2^{2r+4}n^2\gamma_n^{(r)}[\phi(x)+\delta_n^{(r)}(\phi;x)].$$

It only remains to take into account the fact that

$$\gamma_n^{(r)} = \frac{1}{\pi}\int_{-\pi}^{\pi}\left|\frac{\sin\frac{nt}{2}}{n\sin\frac{t}{2}}\right|^{2r+4} dt < \frac{2}{2^{2r+4}n\pi}\int_{-\infty}^{\infty}\left(\frac{\pi\sin u}{u}\right)^{2r+4} du = \frac{Cr}{2^{2r+4}n}.$$

Let us note that an inequality of the type 4.8(12) holds for any integral function of the class B_σ. It follows directly from the identity 4.7(11) that *if for any real values of x the integral function* $G_\sigma(x)$ *of degree* $\leqslant \sigma$ *satisfies the inequality* $|G_\sigma(x)| \leqslant \phi(x)$, *where* $\phi(x)$ *is some function which is non-negative, measurable and bounded on the whole real axis* $-\infty < x < \infty$, *then for any integral* $r > 0$

$$|G_\sigma'(x)| \leqslant C_r^*\sigma\{\phi(x)+\eta_\sigma^{(r)}(\phi;r)\},$$

where C_r^* *is a constant depending only on r, and***

$$\eta_\sigma^{(r)}(\phi;x) = \left|\phi(x)-\frac{\sigma}{\int_{-\infty}^{\infty}\left(\frac{\sin t}{t}\right)^{2r+2}dt}\int_{-\infty}^{\infty}\phi(x+t)\left(\frac{\sin\sigma t}{\sigma t}\right)^{2r+2}dt\right|.$$

** The relation $\eta_\sigma^{(r)}(\phi;x) = 0\left[\omega_2\left(\phi;\frac{1}{\sigma}\right)\right]$ holds for any integer $r > 0$, and consequently, if the function $\phi(x)$ is uniformly continuous on $(-\infty, \infty)$, then $\eta_\sigma^{(r)}(\phi;x)$ tends to zero uniformly as $\sigma \to \infty$.

4.8.6. From the inequalities 4.8(7) for an arbitrary trigonometric polynomial there follows, with the help of elementary considerations, a series of similar inequalities. Let us consider the set of all functions $f(x)$ which satisfy the conditions:

(a) $f(x)$ is real and analytic on the real axis;

(b) A point x_0 and a number $o > 0$ exist such that for any integer $k > 0$

$$|f^{(2k+1)}(x_0)| \leqslant \sigma^{2k} |f'(x_0)| ; \tag{13}$$

(c) The inequality 4.8(13) can become an equality only for all $k > 0$ simultaneously, and in this case $f^{(2k+1)}(x_0) = (-1)^k \sigma^{2k} f'(x_0)$.

As follows from 4.8(7) and section 4.8.1, any trigonometric polynomial of order $\leqslant \sigma$ satisfies these three conditions. Moreover, it is obvious that such properties may also be possessed by certain integral functions which are not of finite degree.

If we make use of the expansion

$$\frac{1}{2}\{f(x_0+h)-f(x_0-h)\} = hf'(x_0)+\frac{h^3}{3!}f'''(x_0)+\dots,$$

which holds for sufficiently small h, and consider also the expansion of $\sin \sigma h$ in powers of h, we obtain:

$$\frac{\sigma}{2}\{f(x_0+h)-f(x_0-h)\} = f'(x_0)\sin \sigma h +$$

$$+\frac{\sigma h^3}{3!}[\sigma^2 f'(x_0)+f''(x_0)]-\frac{\sigma h^5}{5!}[\sigma^4 f'(x_0)-f^{(3)}(x_0)]+\dots$$

In virtue of conditions (b) and (c) it follows from this that there exists a certain $\delta > 0$, such that for $0 < h < \delta$

$$|f'(x_0)| \leqslant \frac{\sigma}{2\sin \sigma h}|f(x_0+h)-f(x_0-h)|. \tag{14}$$

Taking σ to be an integer, it is easily seen that in the case where $f(x)$ is a trigonometric polynomial of order $\leqslant \sigma$, and x_0 is the point at which its derivative attains the maximum of its modulus, the inequality 4.8(14) holds for all $h < \dfrac{\pi}{\sigma}$ and becomes an equality only for $f(x) = M \cos \sigma(x-\tau_0)$.

In fact, if for some $0 < h < \dfrac{\pi}{\sigma}$ it happened that

$$|f'(x_0)| = \frac{\sigma}{2\sin \sigma h}|f(x_0+h)-f(x_0-h)|,$$

i.e.

$$\left|\int_{-h}^{h} f'(x_0)\cos\sigma t\,\mathrm{d}t\right| = \left|\int_{-h}^{h} f'(x_0+t)\,\mathrm{d}t\right|,$$

then a value t_0 $\left(0<|t_0|\leqslant\frac{\pi}{\sigma}\right)$ could be found for which

$$f'(x_0+t_0) = f'(x_0)\cos\sigma t_0.$$

Hence, since $|f'(x_0+t)|\leqslant|f'(x_0)|$, it would follow that the polynomial $f'(x_0+t)-f'(x_0)\cos\sigma t$ of order $\leqslant\sigma$ had, besides a zero of multiplicity two at the point $t=0$, also a zero on each of the segments $\left[\frac{\nu\pi}{\sigma},\frac{(\nu+1)\pi}{\sigma}\right]$ ($\nu=1,\ldots,2\sigma-1$), i.e. $f'(x_0+t)\equiv f'(x_0)\cos\sigma t$. Similarly it is possible to obtain the inequality

$$|f'(x_0)|\geqslant\frac{\sigma}{2\operatorname{sh}\sigma h}|f(x_0+h)-f(x_0-h)|, \tag{15}$$

contrary to 4.8(14) which holds for any integral function satisfying condition (b) and reduces to an equality for $f(x)=\mathrm{e}^{\sigma x}$.

If instead of condition (b) it is required that at the point x_0

$$|f^{(2k+1)}(x_0)|\leqslant\sigma^{2k+1}|f(x_0)| \qquad (k=0,1,2,\ldots),$$

then from the same considerations we obtain the inequality

$$|f(x_0+h)-f(x_0-h)|\leqslant 2\sin\sigma h|f(x_0)| \qquad (h>0). \tag{16}$$

For integral σ in the case where $f(x)$ is a trigonometric polynomial of order $\leqslant\sigma$, and x_0 is the point at which it attains its maximum modulus, it is easily seen as in the previous case that this inequality holds for all x and $h\leqslant\frac{\pi}{2\sigma}$ if on the left-hand side x_0 is replaced by x.

Thus, for any trigonometric polynomial $T_n(x)$ of order $\leqslant n$ the inequalities

$$\max_x|T_n'(x)|\leqslant\frac{n}{2\sin nh}\max_x|T_n(x+h)-T_n(x-h)|\leqslant n\max_x|T_n(x)| \tag{17}$$

hold respectively for all $h<\frac{\pi}{n}$ and for all $h\leqslant\frac{\pi}{2n}$. If we consider differences of higher orders

$$\Delta_h^{(r)}T_n(x)=\sum_{k=}^{r}(-1)^k\binom{r}{k}T_n[x+(r-2k)h],$$

then by induction with respect to r there follow directly from 4.8(17) the more general inequalities**

$$\max_x |T_n^{(r)}(x)| \leqslant \left(\frac{n}{2\sin nh}\right)^r \max_x |\Delta_h^{(r)} T_n(x)| \leqslant n^r \max_x |T_n(x)|, \tag{18}$$

which hold for the same ranges of variation of h.

4.8.61. Similar considerations lead to similar inequalities for any integral functions of the class β_σ. However, these inequalities can also be obtained by starting from other considerations***.

Let $\varrho(t)$ be an arbitrary function of bounded variation on $[-\sigma, \sigma]$, on the same segment let $\mu(t)$ be the sum of the uniformly convergent series $\sum_{n=-\infty}^{\infty} c_n e^{\frac{in\pi t}{\sigma}}$ where $(-1)^n c_n \geqslant 0$, let $\phi(t)$ be continuous on $[-\sigma, \sigma]$, and let

$$f(x) = \int_{-\sigma}^{\sigma} \phi(t) e^{ixt}\, d\varrho(t), \tag{19}$$

$$F(x) = \int_{-\sigma}^{\sigma} \mu(t)\phi(t) e^{ixt}\, d\varrho(t). \tag{20}$$

On replacing the function $\mu(t)$ by its Fourier series and integrating term by term, we obtain for $F(x)$ the representation

$$F(x) = \sum_{n=-\infty}^{\infty} c_n f\left(\frac{n\pi}{\sigma} + x\right), \tag{21}$$

from which follows****

$$\sup_x |F(x)| \leqslant \sum_{n=-\infty}^{\infty} |c_n| \sup_x |f(x)| = \mu(\sigma) \sup_x |f(x)|. \tag{22}$$

From 4.8(21) with the help of Minkowski's inequality it follows also that if for some $q \geqslant 1$

$$\int_{-\infty}^{\infty} |f(x)|^q\, dx < \infty, \tag{23}$$

then

$$\left\{\int_{-\infty}^{\infty} |F(x)|^q\, dx\right\}^{\frac{1}{q}} \leqslant \mu(\sigma) \left\{\int_{-\infty}^{\infty} |f(x)|^q\, dx\right\}^{\frac{1}{q}}. \tag{24}$$

** For the first of these see S. V. Stechkin [1], and also S. M. Nikol'skii [10] in the case $h = \pi/2n$. For the second when $r = 1$ see S. N. Bernstein [38].

*** Boas [2].

**** Civin [1]. See also M. G. Krein [3].

Similarly, if the function $f(x)$ is periodic (for example, of period 2π) and

$$\int_0^{2\pi} |f(x)|^q \, dx < \infty .$$

then

$$\left\{\int_0^{2\pi} |F(x)|^q \, dx\right\}^{\frac{1}{q}} \leqslant \mu(\sigma)\left\{\int_0^{2\pi} |f(x)|^q \, dx\right\}^{\frac{1}{q}}. \tag{25}$$

In order to obtain from this inequalities similar to those which were given in section 4.8.6 for trigonometric polynomials, it must be noted that on account of the identity

$$\int_0^{\sigma} \mu(t) \cos\frac{n\pi t}{\sigma} \, dt = -\frac{\sigma}{n\pi}\int_0^{\frac{\sigma}{n}} \sin\frac{n\pi t}{\sigma}\sum_{k=0}^{n-1}(-1)^k \mu'\left(t+\frac{k\sigma}{n}\right) dt$$

it is poossible to take as $\mu(t)$ any non-negative even function which is convex on $[-\sigma, \sigma]$.

Let $0 < h < \dfrac{\pi}{\sigma}$, $\phi(t) = (2i \sin ht)^r (r = 1, 2, \ldots)$ and

$$\Phi(x) = \int_{-\sigma}^{\sigma} e^{ixt} \, d\varrho(t).$$

Then by 4.8(19) and 4.8(20) we have $f(x) = \Delta_h^r \Phi(x)$, and for $\mu(t) = \left(\dfrac{t}{2 \sin ht}\right)^r$ we have $F(x) = \Phi^{(r)}(x)$. In this particular case, for $\Phi(x) = T_n(x)$ the inequality 4.8(22) includes the first of the inequalities 4.8(18). The second can be obtained in a similar way.

Considering the same particular case and using the estimate 4.8(25), we also find** that *for any trigonometric polynomial $T_n(x)$ of order $\leqslant n$, for every $q (1 \leqslant q \leqslant \infty)$ and $0 < h < \dfrac{\pi}{n}$.*

$$\left\{\int_0^{2\pi} |T_n^{(r)}(x)|^q \, dx\right\}^{\frac{1}{q}} \leqslant \left(\frac{n}{2 \sin nh}\right)^r \left\{\int_0^{2\pi} |\Delta_h^{(r)} T_n(x)|^q \, dx\right\}^{\frac{1}{q}}.$$

In particular,

$$\left\{\int_0^{2\pi} |T_n^{(r)}(x)|^q \, dx\right\}^{\frac{1}{q}} \leqslant n^r \left\{\int_0^{2\pi} |T_n(x)|^q \, dx\right\}^{\frac{1}{q}}. \tag{26}$$

** S. M. Nikol'skii [10], the case $h = \dfrac{\pi}{2n}$.

If now $G(x)$ is a certain integral function of the class B_σ, then for any $\varepsilon > 0$ the function $G(\varepsilon; x) = G(x)\dfrac{\sin \varepsilon x}{\varepsilon x}$ is an integral function of degree $\leqslant \sigma+\varepsilon$ which satisfies the condition

$$\int_{-\infty}^{\infty} |G(\varepsilon; x)|^2 \, dx < \infty .$$

Consequently, by the Wiener–Paley theorem (see section 4.6.1) it can be represented in the form

$$G(\varepsilon; x) = \int_{-\sigma-\varepsilon}^{\sigma+\varepsilon} g(\varepsilon; t) e^{ixt} \, dt .$$

From this

$$\Phi(\varepsilon; x) = \Delta_h^{(r)} G(\varepsilon; x) = \int_{-\sigma-\varepsilon}^{\sigma+\varepsilon} (2i \sin ht)^r g(\varepsilon, t) e^{ixt} \, dt .$$

If, in addition, for sufficiently small $\varepsilon > 0$ the function $\mu(t)$ is non-negative, even and convex on $[-\sigma-\varepsilon, \sigma+\varepsilon]$, then by what was proved above the inequality

$$\sup_x |F(\varepsilon; x)| \leqslant \mu(\sigma+\varepsilon) \sup_x |\Phi(\varepsilon; x)| ,$$

is satisfied for

$$F(\varepsilon; x) = \int_{-\sigma-\varepsilon}^{\sigma+\varepsilon} \mu(t)(2i \sin ht)^r g(\varepsilon; t) e^{ixt} \, dt .$$

For $0 < h < \dfrac{\pi}{\sigma}$ let us put $\mu(t) = \left(\dfrac{t}{2 \sin ht}\right)^r$ $(r = 1, 2, \ldots)$. Then we obtain $F(\varepsilon; x) = G^{(r)}(\varepsilon; x)$, and thus

$$\sup_x |G^{(r)}(\varepsilon; x)| \leqslant \left[\frac{\sigma+\varepsilon}{2 \sin h(\sigma+\varepsilon)}\right]^r \sup_x |\Delta_h^{(r)} G(\varepsilon; x)| .$$

Passing to the limit as $\varepsilon \to 0$, and taking into account that $\sup\limits_{-\infty < x < \infty} \left| \dfrac{d}{dx}\left(\dfrac{\sin \varepsilon x}{\varepsilon x}\right)\right| = O(\varepsilon)$, we finally have that *for any integral function* $G(x) \in B_\sigma$ *the inequality***

$$\sup_x |G^{(r)}(x)| \leqslant \left(\frac{\sigma}{2 \sin \sigma h}\right)^r \sup_x |\Delta_k^{(r)} G(x)| \tag{27}$$

holds for $0 < h < \dfrac{\pi}{2}$.

** S. N. Bernstein [38] for the case $r = 1$. S. M. Nikol'skii [10] for the case $h = \dfrac{\pi}{2\sigma}$.

With the condition 4.8(23) for $G(x)$ (if the estimate 4.8(24) is used instead of 4.8(22)) we arrive at a similar inequality in the metric L_q $(1 \leqslant q < \infty)$.

*If $G(x)$ is an integral function of degree $\leqslant \sigma$ which belongs to the class L_q $(q \geqslant 1)$ on $(-\infty, \infty)$, then***

$$\left\{\int_{-\infty}^{\infty} |G^{(r)}(x)|^q \, dx\right\}^{\frac{1}{q}} \leqslant \left(\frac{\sigma}{2 \sin \sigma h}\right)^r \left\{\int_{-\infty}^{\infty} |\Delta_h^{(r)} G(x)|^q \, dx\right\}^{\frac{1}{q}}.$$

In particular,

$$\left\{\int_{-\infty}^{\infty} |G^{(r)}(x)|^q \, dx\right\}^{\frac{1}{q}} \leqslant \sigma^r \left\{\int_{-\infty}^{\infty} |G(x)|^q \, dx\right\}^{\frac{1}{q}}. \tag{28}$$

4.8.62. The inequalities derived in section 4.8.61 also extend to the case of trigonometric polynomials and other integral functions of finite degree of several variables. Thus, for example, if $T_{mn}(x, y)$ is a trigonometric polynomial of order $\leqslant m$ with respect to x and of order $\leqslant n$ with respect to y, then for any $q (1 \leqslant q \leqslant \infty)$ and $0 < h < \frac{\pi}{m}$, $0 < \delta < \frac{\pi}{n}$ we have

$$\left\{\int_0^{2\pi}\int_0^{2\pi} \left|\frac{\partial^{k+l} T_{mn}(x, y)}{\partial x^k \, \partial y^l}\right|^q dx\, dy\right\}^{\frac{1}{q}}$$

$$\leqslant \left(\frac{m}{2 \sin mh}\right)^k \left\{\int_0^{2\pi} dy \int_0^{2\pi} \left|\frac{\partial^l \Delta_h^{(k)} T_{mn}(x, y)}{\partial y^l}\right|^q dx\right\}^{\frac{1}{q}} \leqslant$$

$$\leqslant \left(\frac{m}{2 \sin mh}\right)^k \left(\frac{n}{2 \sin n\delta}\right)^l \left\{\int_0^{2\pi}\int_0^{2\pi} |\Delta_h^{(k)} \Delta_\delta^{(l)} T_{mn}(x, y)|^q \, dx\, dy\right)^{\frac{1}{q}} \tag{29}$$

In particular,

$$\left\{\int_0^{2\pi}\int_0^{2\pi} \left|\frac{\partial^{k+l} T_{mn}(x, y)}{\partial x^k \, \partial y^l}\right|^q dx\, dy\right\}^{\frac{1}{q}} \leqslant m^k n^l \left\{\int_0^{2\pi}\int_0^{2\pi} |T_{mn}(x, y)|^q \, dx\, dy\right\}^{\frac{1}{q}}. \tag{30}$$

4.8.7. It follows from theorem 4.8.22 (inequality 4.8(7a)), that *if $P_n(x)$ is an algebraic polynomial of order $\leqslant n$, then*

$$|P_n'(\cos t) \sin t| \leqslant \max_u |P_n(\cos u)|,$$

** S. M. Nikol'skii [10] for the case $h = \frac{\pi}{2\sigma}$. The inequality 4.8(22) can also be obtained from the identity 4.7(4) if account is taken of the fact that in the case considered $G(x) \in B_\sigma$ (see 4.9.53).

i.e. for all** $x \in [-1, 1]$

$$|P_n'(x)| \leqslant \frac{n}{\sqrt{1-x^2}} \max_{-1 \leqslant y \leqslant 1} |P_n(y)|. \tag{31}$$

The inequality 4.8(31) becomes weak near the ends of the segment $[-1, 1]$. Hence another inequality due to A. A. Markov [1] is also of interest. For all $x \in [-1, 1]$

$$|P_n'(x)| \leqslant n^2 \max_{-1 \leqslant y \leqslant 1} |P_n(y)|. \tag{32}$$

In order to prove 4.8(32), we consider*** the interpolation formula 4.1(4) for $P_n'(x)$ with the set of points $x_k^{(n)}$ at the zeros of a Chebyshev polynomial (see 4.1(7)):

$$P_n'(x) = \frac{\cos n \arccos x}{n} \sum_{k=0}^{n-1} (-1)^k P_n'(x_k^{(n)}) \sqrt{1-x_k^{(n)2}} \frac{1}{x-x_k^{(n)}}.$$

Hence in virtue of 4.8(31) we obtain

$$|P_n'(x)| \leqslant |\cos n \arccos x| \max_{-1 \leqslant y \leqslant 1} |P_n(y)| \sum_{k=0}^{n-1} \frac{1}{|x-x_k^{(n)}|}.$$

If $x_0^{(n)} < |x| \leqslant 1$, then all the differences $x-x_k^{(n)}$ $(k = 0, 1, \ldots, n-1)$ have the same sign. Consequently, in this case

$$|P_n'(x)| \leqslant \max_{-1 \leqslant y \leqslant 1} |P_n(y)| \left| \sum_{k=0}^{n-1} \frac{\cos n \arccos x}{x-x_k^{(n)}} \right|$$
$$= \left| \frac{n \sin n \arccos x}{\sqrt{1-x^2}} \right| \max_{-1 \leqslant y \leqslant 1} |P_n(y)| \leqslant n^2 \max_{-1 \leqslant y \leqslant 1} |P_n(y)|.$$

If, however, $x \leqslant x_0^{(n)} = \cos \frac{\pi}{2n}$, then 4.8(32) follows directly from 4.8(31), as $n \sin \frac{\pi}{2n} \geqslant 1$. Thus *the inequality*

$$|P_n'(x)| \leqslant n \min \left\{ \frac{1}{\sqrt{1-x^2}}, n \right\} \max_{-1 \leqslant y \leqslant 1} |P_n(y)| \tag{33}$$

is always satisfied on the segment $-1 \leqslant x \leqslant 1$.

4.8.71. Let us note an additional estimate for the derivative of an algebraic polynomial which follows directly from theorem 4.8.5. *If the*

** S. N. Bernstein [1].

*** Following I. Schur [1].

function $\phi(x) \mid > 0$ *is continuous on the segment* $[-1, 1]$ *and* $|P_n(x)| \leqslant \phi(x)$, *everywhere on this segment,* then for any** $r \geqslant 0$

$$|P_n'(x)| \leqslant C_r n\left[1+\frac{\delta_n^{(r)}(f;t)}{\phi(x)}\right]\frac{\phi(x)}{\sqrt{1-x^2}}, \tag{34}$$

where $x = \cos t$, $f(t) = \phi(\cos t)$, and $\delta_n^{(r)}(f;t)$ and C_r are defined as in section 4.8.5.

4.8.72. Later on (see section 6.2) an essential part is played by the particular case for which

$$\phi(x) = \phi_n(x) = \frac{1}{n^r}\left(\sqrt{1-x^2}+\frac{1}{n}\right)^r \omega\left[\frac{1}{n}\left(\sqrt{1-x^2}+\frac{1}{n}\right)\right],$$

where $\omega(u)$ is a certain modulus of continuity (see section 3.2). The following statement holds for this case. *If everywhere on the segment* $-1 \leqslant x \leqslant 1$ *the algebraic polynomial* $P_n(x)$ *satisfies the inequality*

$$|P_n(x)| \leqslant \frac{1}{n^r}\left(\sqrt{1-x^2}+\frac{1}{n}\right)^r \omega\left[\frac{1}{n}\left(\sqrt{1-x^2}+\frac{1}{n}\right)\right] \tag{35}$$

for integral $r \geqslant 0$, *then*

$$|P_n'(x)| \leqslant \frac{C}{n^{r-1}}\left(\sqrt{1-x^2}+\frac{1}{n}\right)^{r-1} \omega\left[\frac{1}{n}\left(\sqrt{1-x^2}+\frac{1}{n}\right)\right], \tag{36}$$

*where C is a constant independent of x and n****. To prove this let us evaluate the quantity $\delta_n^{(r)}(f;t)$ which occurs in 4.8(34). Since

$$\begin{aligned} f(t) = f_n(t) &= \frac{1}{n^r}\left(|\sin t|+\frac{1}{n}\right)^r \omega\left[\frac{1}{n}\left(|\sin t|+\frac{1}{n}\right)\right] \leqslant \\ &\leqslant \frac{2^{r-1}}{n^r}\left(|\sin t|^r+\frac{1}{n^r}\right)\left[\omega\left(\frac{|\sin t|}{n}\right)+\omega\left(\frac{1}{n^2}\right)\right] \leqslant \\ &\leqslant 3\cdot 2^{r-1}\left[\left(\frac{|\sin t|}{n}\right)^r \omega\left(\frac{|\sin t|}{n}\right)+\frac{1}{n^{2r}}\omega\left(\frac{1}{n^2}\right)\right], \end{aligned}$$

then, by 4.8(10),

$$\delta_n^{(r)}(f;t) \leqslant \frac{3\cdot 2^{r-1}}{n^r}\left\{|\sin t|^r \omega\left(\frac{|\sin t|}{n}\right)+\right.$$

** Yu. A. Brudnyi [12]. An inequality of the type 4.8(34) was first established by S. N. Bernstein [11]. The fundamental result of this note of S. N. Bernstein was later developed in a series of papers by V. C. Videnskii [1–5].

*** It will be seen from the proof that for the validity of the inequality 4.8(36) it is sufficient that the function $\omega(t)$ should be non-decreasing and satisfy the condition $\omega(t_1+t_2) \leqslant M\{\omega(t_1)+\omega(t_2)\}$. For the particular case $\omega(t) = t^\alpha (0\ 2\alpha \leqslant 1)$ the inequality 4.8(36) was proved in a different way by V. K. Dzyadyk [3].

$$+\frac{1}{\pi\gamma_n^{(r)}}\int_{-\pi}^{\pi}|\sin(t+u)|^r\omega\left(\frac{|\sin(t+u)|}{n}\right)\left(\frac{\sin\frac{nu}{2}}{n\sin\frac{u}{2}}\right)^{2r+4}du\Bigg\}+$$

$$+O\left[\frac{1}{n^{2r}}\omega\left(\frac{1}{n^2}\right)\right].$$

Moreover,

$$\frac{1}{\pi\gamma_n^{(r)}}\int_{-\pi}^{\pi}|\sin(t+u)|^r\omega\left(\frac{|\sin(t+u)|}{n}\right)\left(\frac{\sin\frac{nu}{2}}{n\sin\frac{u}{2}}\right)^{2r+4}du\leqslant$$

$$\leqslant 3\cdot 2^{r-1}|\sin t|^r\omega\left(\frac{|\sin t|}{n}\right)+\frac{3\cdot 2^{r-1}}{\pi\gamma_n^{(r)}}\int_{-\pi}^{\pi}|u|^r\omega\left(\frac{|u|}{n}\right)\left(\frac{\sin\frac{nu}{2}}{n\sin\frac{u}{2}}\right)^{2r+4}du.$$

Let us note also that

$$\frac{1}{\pi\gamma_n^{(r)}}\int_{0}^{\frac{1}{n}}u^r\omega\left(\frac{u}{n}\right)\left(\frac{\sin\frac{nu}{2}}{n\sin\frac{u}{2}}\right)^{2r+4}du\leqslant\frac{1}{n^r}\omega\left(\frac{1}{n^2}\right),$$

and since for $0<u\leqslant\pi$ we have $\sin\frac{u}{2}\geqslant\frac{u}{\pi}$, it follows from 3.2(6) that

$$\frac{1}{\pi\gamma_n^{(r)}}\int_{\frac{1}{n}}^{\pi}u^r\omega\frac{u}{n}\left(\frac{\sin\frac{nu}{2}}{n\sin\frac{u}{2}}\right)^{2r+4}du\leqslant\frac{\pi^{2r+3}}{n^{2r+4}\gamma_n^{(r)}}\int_{\frac{1}{n}}^{\pi}\omega\left(\frac{u}{n}\right)\frac{du}{u^{r+4}}$$

$$\leqslant\frac{2\pi^{2r+3}}{n^{2r+3}\gamma_n^{(r)}}\omega\left(\frac{1}{n^2}\right)\int_{\frac{1}{n}}^{\pi}u^{-r-3}\,du.$$

Since

$$\gamma_n^{(r)}=\frac{1}{\pi}\int_{-\pi}^{\pi}\left(\frac{\sin\frac{nu}{2}}{n\sin\frac{u}{2}}\right)^{2r+4}du=\frac{4}{\pi}\int_{0}^{\frac{\pi}{2}}\left(\frac{\sin nu}{n\sin u}\right)^{2r+4}du>$$

$$> \frac{4}{\pi} \int_0^{\frac{\pi}{2}} \left(\frac{\sin nu}{nu} \right)^{2r+4} du > \int_0^{\frac{\pi}{2n}} \left(\frac{\sin nu}{nu} \right)^{2r+4} du$$

$$= \frac{4}{n\pi} \int_0^{\frac{\pi}{2}} \left(\frac{\sin u}{u} \right)^{2r+4} du,$$

the last inequality gives the estimate

$$\frac{1}{\pi \gamma_n^{(r)}} \int_{\frac{1}{n}}^{\pi} u^r \omega\left(\frac{u}{n}\right) \left(\frac{\sin \frac{nu}{2}}{n \sin \frac{u}{2}} \right)^{2r+4} du = O\left[\frac{1}{n^r} \omega\left(\frac{1}{n^2}\right)\right].$$

Thus

$$\delta_n^{(r)}(f;\, t) = O\left[\left(\frac{|\sin t|}{n}\right)^r \omega\left(\frac{|\sin t|}{n}\right) + \frac{1}{n^{2r}} \omega\left(\frac{1}{n^2}\right)\right].$$

On substituting in 4.8(34) we obtain the inequality

$$|P_n'(x)| \leqslant \frac{M}{n^{r-1}\sqrt{1-x^2}} \left(\sqrt{1-x^2} + \frac{1}{n}\right)^r \omega\left[\frac{1}{n}\left(\sqrt{1-x^2} + \frac{1}{n}\right)\right], \quad (37)$$

where M is a constant independent of x and n.

The validity of 4.8(36) on the segment $|x| \leqslant 1 - \frac{1}{n^2}$ follows directly from 4.8(37). In order to extend this inequality to the remaining part of the segment $[-1, 1]$ it is sufficient to consider the function

$$W_n(z) = \frac{n^2}{n^2-1}\left[z + \sqrt{z^2 - \left(1 - \frac{1}{n^2}\right)^2}\right],$$

which conformally maps the z-plane with a slit along the segment $\left[-1 + \frac{1}{n^2}, 1 - \frac{1}{n^2}\right]$ of the real axis onto the exterior of the unit circle. The quotient

$$\frac{P_n'(z)}{[W_n(z)]^n \left(\sqrt{1 - \frac{1}{n^2} - z^2} + \frac{1}{n}\right)^r},$$

analytic throughout the whole of the z-plane outside this segment, attains on it the maximum of its modulus, which in virtue of 4.8(37) and 3.2(6) is $O\left[\frac{1}{n^{r-2}}\omega\left(\frac{1}{n^2}\right)\right]$. In addition, for $1-\frac{1}{n^2}\leqslant |x|\leqslant 1$ the relation $|W_n(x)|^n = O(1)$ is satisfied. Hence

$$P_n'(x) = O\left[\frac{1}{n^{2r-2}}\omega\left(\frac{1}{n^2}\right)\right]$$

uniformly for all x for which $1-\frac{1}{n^2}\leqslant |x|\leqslant 1$.

Thus the inequality 4.8(36) holds on the whole segment $[-1, 1]$.

The last part of the proof can be carried through without the assistance of the methods of the theory of functions of a complex variable. If $r = 0$, then on account of 3.2(6) we conclude from 4.8(37) that on the segment $|x|\leqslant 1-\frac{1}{n^2}$

$$|P_n'(x)|\leqslant \frac{M_n}{\sqrt{1-x^2}}\,\omega\left(\frac{\sqrt{1-x^2}}{n}\right)+\frac{M_n}{\sqrt{1-x^2}}\,\omega\left(\frac{1}{n^2}\right)\leqslant M_1 n^2\,\omega\left(\frac{1}{n^2}\right). \tag{38}$$

Let us now apply theorem 2.9.11 (see the inequality 2.9(9)). Since (see 2.13.8)

$$\cos n \arccos\frac{2x-a-b}{b-a} = \frac{1}{2}\left(R^n+\frac{1}{R^n}\right),$$

where for $h = \frac{1}{2}(b-a)$ and $y = x-\frac{a+b}{2}$ we have

$$R = \frac{y+\sqrt{y^2-h^2}}{h},$$

it follows that outside the given segment $[a, b]$ the inequality 2.9(9) assumes the form

$$|P_n(x)|\leqslant \frac{1}{2}\max_{a\leqslant u\leqslant b}|P_n(u)|\left(R^n+\frac{1}{R^n}\right). \tag{39}$$

In our case $a = -1+\frac{1}{n^2}$, $b = 1-\frac{1}{n^2}$, $1-\frac{1}{n^2}\leqslant |x|\leqslant 1$, and consequently, $R = 1+O\left(\frac{1}{n}\right)$. Hence there exists a constant M_2 independent of n and such that

$$\max_{a-\frac{1}{n^2}\leqslant x\leqslant b+\frac{1}{n^2}}|P_n(x)|\leqslant M_2\max_{a\leqslant x\leqslant b}|P_n(x)|. \tag{40}$$

Since the estimate 4.8(38) has already been established for the segment

$\left[-1+\frac{1}{n^2},\ 1-\frac{1}{n^2}\right]$, it is true for all $x \in [-1, 1]$ apart from the value of the constant M_1. From it and from 4.8(37) the inequality 4.8(36) for $r = 0$ follows directly. Now suppose the theorem formulated to be true for all $r \leqslant k$ and put $r = k+1$. From 4.8(37) the validity of the inequality 4.8(36) for $r = k+1$ and $|x| \leqslant 1-\frac{1}{2n^2}$ follows. In 4.8(36) for $r = k+1$ let us put $x = \left(1-\frac{1}{2n^2}\right)u$. Then for all $|u| \leqslant 1$

$$\left(1-\frac{1}{2n^2}\right)\left|P_n'\left[\left(1-\frac{1}{2n^2}\right)u\right]\right| \leqslant$$
$$\leqslant \frac{C}{n^k}\left[\sqrt{1-\left(1-\frac{1}{2n^2}\right)^2 u^2}+\frac{1}{n}\right]^k \omega\left[\frac{1}{n}\left\{\sqrt{1-\left(1-\frac{1}{2n^2}\right)^2 u^2}+\frac{1}{n}\right\}\right],$$

or

$$\left|P_n'\left[\left(1-\frac{1}{2n^2}\right)u\right]\right| \leqslant$$
$$\leqslant \frac{C^*}{n^k}\left(\sqrt{1+\frac{1}{n^2}-u^2}+\frac{1}{n}\right)^k \omega\left[\frac{1}{n}\left(\sqrt{1+\frac{1}{n^2}-u^2}+\frac{1}{n}\right)\right]. \tag{41}$$

By supposition, for any ν $(0 \leqslant \nu \leqslant k)$ and $|u| \leqslant 1$

$$\left(1-\frac{1}{2n^2}\right)^\nu \left|P_n^{(\nu+1)}\left[\left(1-\frac{1}{2n^2}\right)u\right]\right| \leqslant$$
$$\leqslant \frac{C_\nu^*}{n^{k-\nu}}\left(\sqrt{1-u^2}+\frac{1}{n}\right)^{k-\nu} \omega\left[\frac{1}{n}\left(\sqrt{1-u^2}+\frac{1}{n}\right)\right].$$

Putting $\nu = k$ in this and applying the inequality 4.8(36) in the case $r = 0$, we obtain

$$\left|P_n^{(k+2)}\left[\left(1-\frac{1}{2n^2}\right)u\right]\right| = O\left[n^2\omega\left(\frac{1}{n^2}\right)\right] \tag{42}$$

uniformly with respect to all $u \in [-1, 1]$. In consequence of the inequality 4.8(39) we see as before that this estimate remains valid over the wider segment $-1-\frac{1}{n^2} \leqslant u \leqslant 1+\frac{1}{n^2}$. Consequently, integrating the relation 4.8(42) $k+1$ times with respect to u between the limits ± 1, we find that, subject to the condition $1 \leqslant |u| \leqslant 1+\frac{1}{n^2}$,

$$\left|P_n'\left[\left(1-\frac{1}{2n^2}\right)u\right]\right| = O\left[\frac{1}{n^{2k}}\omega\left(\frac{1}{n^2}\right)\right].$$

Thus 4.8(41) with appropriate choice of the constant C^* is valid for all $|u| \leqslant 1+\frac{1}{n^2}$, i.e. for all $x \in [-1, 1]$

$$|P_n'(x)| \leqslant \frac{C}{n^k}\left(\sqrt{1-x^2}+\frac{1}{n}\right)^k \omega\left[\frac{1}{n}\left(\sqrt{1-x^2}+\frac{1}{n}\right)\right].$$

This proves the validity of the inequality 4.8(36) for the whole segment $[-1, 1]$ and for $r = k+1$.

4.8.73. Theorem 4.8.72 also remains valid for all negative integral values of r. If r is negative, then it follows from 4.8(35) that

$$n^r |P_n(x)| \leqslant \left(\sqrt{1-x^2}\right) \omega\left[\frac{1}{n}\left(\sqrt{1-x^2}+\frac{1}{n}\right)\right].$$

Putting $m = \left[\frac{-r+1}{2}\right]$, we have

$$n^r |P_n(x)(1-x^2)^m| \leqslant \left(\sqrt{1-x^2}\right)^{r+2m} \omega\left[\frac{1}{n}\left(\sqrt{1-x^2}+\frac{1}{n}\right)\right].$$

Here we already have $r+2m \geqslant 0$, and in virtue of 4.8(36)

$$n^r|-2mx(1-x^2)^{m-1}P_n(x)+(1-x^2)^m P_n'(x)| \leqslant Cn\left(\sqrt{1-x^2}+\frac{1}{n}\right)^{r+2m-1} \omega\left[\frac{1}{n}\left(\sqrt{1-x^2}+\frac{1}{n}\right)\right].$$

Hence

$$n^r |P_n'(x)| \leqslant 2m\left(\sqrt{1-x^2}\right)^{r-2} \omega\left[\frac{1}{n}\left(\sqrt{1-x^2}+\frac{1}{n}\right)\right] + Cn\frac{\left(\sqrt{1-x^2}+\frac{1}{n}\right)^{r+2m-1} \omega\left[\frac{1}{n}\left(\sqrt{1-x^2}+\frac{1}{n}\right)\right]}{(1-x^2)^m}.$$

It follows from this that for values of x on the segment $\left[-1+\frac{1}{n^2}, 1-\frac{1}{n^2}\right]$ we have

$$n^{r-1} |P_n'(x)| \leqslant M^*\left(\sqrt{1-x^2}+\frac{1}{n}\right)^{r-1} \omega\left[\frac{1}{n}\left(\sqrt{1-x^2}+\frac{1}{n}\right)\right]$$

and

$$|P_n'(x)| = O\left[\frac{1}{n^{2r-2}} \omega\left(\frac{1}{n^2}\right)\right].$$

Thanks to the inequality 4.8(39), the second of these estimates is true on the whole of the segment $-1 \leqslant n \leqslant 1$. Consequently the first also holds for any $x \in [-1, 1]$, i.e. 4.8(36) is satisfied.

4.8.8. From A. A. Markov's inequality (see 4.8(32)) for the derivatives of any order of the polynomial $P_n(x)$ it is possible to obtain the following estimate:

$$|P_n^{(r)}(x)| \leqslant n^{2r} \max_{-1 \leqslant u \leqslant 1} |P_n(u)|. \tag{43}$$

However, this estimate, precise as regards order (when $n \to \infty$) for every fixed value of r, loses significance if r increases without limit. Using the interpolation formula 4.1(4), it is not difficult to show that *for any integral* $r \geqslant 0$

$$|P^{(r)}(x)| \leqslant 2^r T_n^{(r)}(1) \max_{-1 \leqslant u \leqslant 1} |P_n(u)|, \tag{44}$$

*where*** $T_n(x) = \cos n \arccos x$. In fact, from formula 4.1(4) it follows that

$$P_n^{(r)}(x) = \sum_{k=0}^{n} P_n(x_k) l_{k,n}^{(r)}(x). \tag{45}$$

Let us put $x_k = \cos\dfrac{k\pi}{n}$. Since $x_k \leqslant 1$ for any k, it follows from the definition of the polynomials $l_{k,n}(x)$ (see 4.1(2)) that $(-1)^k l_{k,n}^{(r)}(1) \geqslant 0$. Hence

$$|P_n^{(r)}(1)| \leqslant \sum_{k=0}^{n} (-1)^k |P_n(x_k)| l_{k,n}^{(r)}(1) \leqslant \max_{-1 \leqslant u \leqslant 1} |P_n(u)| \sum_{k=0}^{n} (-1)^k l_{k,n}^{(r)}(1).$$

But since $T_n(x_k) = \cos n \arccos x_k = (1)^k$, then in virtue of the identity 4.8(45) we have for the system of points considered

$$\sum_{k=0}^{n} (-1)^n l_{k,n}^{(r)}(1) = T_n^{(r)}(1).$$

Consequently, for any polynomial $P_n(x)$ of degree $\leqslant n$

$$|P_r^{(r)}(1)| \leqslant T_n^{(r)}(1) \max_{-1 \leqslant u \leqslant 1} |P_n(u)|. \tag{46}$$

** The idea of the given proof of the inequality 4.8(44) is due to S. N. Bernstein [31]. Later it was again proposed by Wang (see Mandelbrojt [2], page 212). In fact, as was shown by V. A. Markov [1], the inequality 4.8(44) also remains valid without the factor 2^r on the right-hand side. The original proof was extremely complicated, only 45 years later were two simple proofs of V. A. Markov's inequality found by S. N. Bernstein [19] and Schaeffer and Duffin [1]. This inequality was generalised in two different directions by Schaeffer and Duffin [2] and by V. S. Videnskii [4].

For the polynomial

$$Q_n(u) = P_n\left[\frac{1\pm x}{2}(1+u)-1\right]$$

we obtain from this

$$|Q_n^{(r)}(1)| = \left(\frac{1\pm x}{2}\right)^r |P_n^{(r)}(\pm x)| \leqslant T_n^{(r)}(1) \max_{-1\leqslant t\leqslant 1} |P_n(t)|.$$

This proves the inequality 4.8(44) for all $x \in [-1, 1]$.

From the identity

$$(x^2-1)\,T_n^{(r+1)}(x)+(2r-1)\,xT_n^{(r)}(x) = [n^2-(r-1)^2]\,T_n^{(r-1)}(x) \tag{47}$$

satisfied by the Chebyshev polynomials**, there follows the recurrence relation

$$(2r-1)T_n^{(r)}(1) = [n^2-(r-1)^2]\,T_n^{(r-1)}(1).$$

This shows that for $1 \leqslant r \leqslant n$

$$T_n^{(r)}(1) = \frac{n^2(n^2-1)\dots[n^2-(r-1)^2]}{1\cdot 3\dots(2r-1)}.$$

Thus the inequality obtained indicates that

$$|P_n^{(r)}(x)| < 4^r \frac{r!}{(2r)!}\, n^{2r} \max_{-1\leqslant u\leqslant 1} |P_n(u)|,$$

i.e. the estimate

$$|P_n^{(r)}(x)| < \frac{1}{2}\left(\frac{4n^2}{r}\right)^r \max_{-1\leqslant u\leqslant 1} |P_n(u)|, \tag{48}$$

which as regards order is better than 4.8(43), is satisfied.

4.8.81. In the general case the inequality 4.8(46) cannot be improved, since for the polynomials $P_n(x) = T_n(x)$ it becomes an equality. It is useful to compare this estimate for the derivatives of the algebraic polynomial $P_n(x)$ at an end of the segment with the inequality obtained in section 2.9.12 for its coefficients, i.e. for the derivatives at the centre of the given segment, and this shows that *for any polynomial $P_n(x)$ of degree $\leqslant n$*

$$|P_n^{(r)}(0)| \leqslant n^r \cdot \max_{-1\leqslant x\leqslant 1} |P_n(x)| \tag{49}$$

for any $r \geqslant 0$. (For $r = 1$ this estimate is a consequence of the inequality 4.8(31).) By the substitution $x = \dfrac{2t-a-b}{b-a}$ the segment $a \leqslant t \leqslant b$ is trans-

** For $r = 1$ it can be verified directly (see 2.13.16), and for the remaining values of r it can be obtained by successive differentiation.

formed into $-1 \leqslant x \leqslant 1$, and its centre $t_0 = \frac{1}{2}(a+b)$ into the point $x = 0$. Hence when applied to any finite segment $[a, b]$ the inequality 4.8(49) assumes the form

$$\left| P_n^{(r)}\left(\frac{a+b}{2}\right) \right| \leqslant \left(\frac{2n}{b-a}\right)^r \max_{a \leqslant t \leqslant b} |P_n(t)|. \tag{50}$$

Since every interior point x of the interval $[-1,1]$ can be considered as the centre of some part of it, the inequality 4.8(50) enables us to assert that when the degree n of the polynomial $P_n(x)$ increases without limit the order of increase of the modulus of its r-th derivative at any fixed interior point is not higher than n^r. In addition to this, as has already been mentioned, at the ends of the segment this upper bound is a quantity of the order $O(n^{2r})$. The two given estimates of the order of increase of the modulus of the derivative of a polynomial within and at the ends of a segment are contained in the inequality

$$|P_n^{(k)}(x)| \leqslant \frac{C_k n^k}{\left(\sqrt{1-x^2} + \frac{1}{n}\right)^k} \max_{-1 \leqslant u \leqslant 1} |P_n(u)| \tag{51}$$

(C_k constant), which follows from the proposition of section 4.8.73 if we consider there the sequence $r = -1, -2, -3, \ldots, -k$; $\omega(t) = t \max\limits_{-1 \leqslant u \leqslant 1} |P_n(u)|$.

4.9. Integral and interpolation norms of integral functions of finite degree in various metrics

The upper bounds L_n^* and $\sup\limits_x L_n^*(x)$ (Lebesgue's constants), defined in section 4.5.1 (see 4.5(5) and 4.5(6)) are, apart from the factor $\frac{1}{\pi}$, the integral and interpolation norms

$$\|D_n\|_q = \left\{ \int_0^{2\pi} |D_n(t)|^q \, dt \right\}^{\frac{1}{q}},$$

$$((D_n))_q = \sup_x \left\{ \frac{2\pi}{2n+1} \sum_{k=0}^{2n} |D_n(x - y_k^{(n)})|^q \right\}^{\frac{1}{q}}$$

of the trigonometric polynomial $D_n(t)$ (Dirichler's kernel) in the metric L_q for $q = 1$. In this section we will consider some relations between such norms for any trigonometric polynomials and other integral functions of exponential type.

4.9.1. Let $T_n(x)$ be an arbitrary trigonometric polynomial of order $\leqslant n$, N some positive integer, $x_k = \frac{2k\pi}{N}$ and

$$\| T_n \|_q = \left\{ \int_0^{2\pi} | T_n(t) |^q \, dt \right\}^{\frac{1}{q}}, \tag{1}$$

$$((T_n))_q = \sup_x \left\{ \frac{2\pi}{N} \sum_{k=0}^{N-1} | T_n(x-x_k) |^q \right\}^{\frac{1}{q}}. \tag{2}$$

For any $q(1 \leqslant q \leqslant \infty)$ *the relation***

$$\| T_q \| \leqslant ((T_n))_q \leqslant \left(1 + \frac{2n\pi}{N} \right) \| T_n \|_q \tag{3}$$

holds between the norms 4.9(1) *and* 4.9(2).

The first inequality follows from the relations

$$\| T_n \|_q = \left\{ \sum_{k=0}^{N-1} \int_{x_k}^{x_{k+1}} | T_n(t) |^q \, dt \right\}^{\frac{1}{q}} = \left\{ \int_0^{x_1} \sum_{k=0}^{N-1} | T_n(t+x_k) |^q \, dt \right\}^{\frac{1}{q}} \leqslant ((T_n))_q .$$

Also, we have

$$\| T_n \|_q^q = \sum_{k=0}^{N-1} \int_{x_k}^{x_{k+1}} | T_n(t) |^q \, dt = \frac{2\pi}{N} \sum_{k=0}^{N-1} | T_n(\xi_k) |^q , \tag{4}$$

where $x_k \leqslant \xi_k \leqslant x_{k+1} (k = 0, 1, \dots, N-1)$. In virtue of the inequalities of Minkowski and Hölder,

$$\left\{ \sum_{k=0}^{N-1} | T_n(x_k) |^q \right\}^{\frac{1}{q}} - \left\{ \sum_{k=0}^{N-1} | T_n(\xi_k) |^q \right\}^{\frac{1}{q}} \leqslant$$

$$\leqslant \left\{ \sum_{k=0}^{N-1} | T_n(\xi_k) - T_n(x_k) |^q \right\}^{\frac{1}{q}} \leqslant \left\{ \sum_{k=0}^{N-1} \left[\int_{x_k}^{x_{k+1}} | T_n'(t) | \, dt \right]^q \right\}^{\frac{1}{q}} \leqslant$$

$$\leqslant \left(\frac{2\pi}{N} \right)^{\frac{q-1}{q}} \left\{ \int_0^{2\pi} | T_n'(t) |^q \, dt \right\}^{\frac{1}{q}}. \tag{5}$$

Then applying the inequality 4.8(26), we obtain from 4.9(4) and 4.9(5)

$$\left\{ \frac{2\pi}{N} \sum_{k=0}^{N-1} | T_n(x_k) |^q \right\}^{\frac{1}{q}} \leqslant \left(1 + \frac{2n\pi}{N} \right) \| T_n \|_q .$$

** S. M. Nikol'skii [13]. For the case $N = 2n+1$, $q = 1$, see S. M. Lozinskii [1]

Replacing here the polynomial $T_n(t)$ by $T_n(x-t)$, we obtain, since $\| T_n(x-t) \|_q = \| T_n(t) \|_q$, the second of the relations 4.9(3).

4.9.2. The norms of trigonometric polynomials in different metrics L_q $(1 \leqslant q \leqslant \infty)$ are connected with one another by simple inequalities. If $1 \leqslant q \leqslant q' \leqslant \infty$, then by Hölder's inequality

$$\| T_n \|_q \leqslant (2\pi)^{\frac{1}{q}-\frac{1}{q'}} \| T_n \|_{q'}. \tag{6}$$

On the other hand, there exists an inequality in a certain sense converse to 4.9(6). *For any positive q and $q' \leqslant q_1$ we have for every trigonometric polynomial $T_n(x)$ of order $\leqslant n$*

$$\| T_n \|_{q'} \leqslant \left(\frac{nq_0+1}{2\pi}\right)^{\frac{1}{q}-\frac{1}{q'}} \| T_n \|_q, \tag{7}$$

*where q_0 is the least even number greater than or equal to*** q.

In order to obtain the inequality 4.9(7) let us first verify that *for any trigonometric polynomial $T_n(x)$ of order $\leqslant n$*

$$\sqrt{\frac{2\pi}{2n+1}} \max_x | T_n(x) | \leqslant \| T_n \|_2, \tag{8}$$

this inequality being precise for every fixed value of n. In fact, in virtue of the identity 4.5(3) and the Bunyakovskii–Schwarz inequality

$$| T_n(x) | \leqslant \frac{1}{\pi} \int_0^{2\pi} | T_n(x+t) | \, | D_n(t) | \, dt \leqslant \frac{1}{\pi} \| T_n \|_2 \| D_n \|_2 \tag{9}$$

If we take into account that

$$\| D_n \|_2 = \sqrt{(n+\tfrac{1}{2})\pi}, \tag{10}$$

we obtain 4.9(8). For the polynomial $T_n(x) = D_n(x)$ (Dirichlet's kernel) the left-hand side of 4.9(8) is exactly equal to the right.

If we apply the inequality 4.9(8) to the polynomial $\{T_n(x)\}^{\frac{q_0}{2}}$, we obtain

$$\max_x | T_n(x) |^{\frac{q_0}{2}} \leqslant \left(\frac{nq_0+1}{2\pi}\right)^{\frac{1}{2}} \left\{ \int_0^{2\pi} | T_n(x) |^{q_0} dx \right\}^{\frac{1}{2}} \leqslant$$

$$\leqslant \left(\frac{nq_0+1}{2\pi}\right)^{\frac{1}{2}} \max_x | T_n(x) |^{\frac{q_0-q}{2}} \| T_n \|_q^{\frac{1}{2}},$$

** From 4.9(7) there follows the estimate $\| T_n \|_{q'} < 2n^{\frac{1}{q}-\frac{1}{q'}} \| T_n \|_q$ $(n \geqslant 1)$, which for $q' = \infty$, $q = 1$ was first obtained by Jackson [4], and for all other values $q' \geqslant q > 1$ by S. M. Nikol'skii [13].

i.e.

$$\max_x |T_n(x)| \leqslant \left(\frac{nq_0+1}{2\pi}\right)^{\frac{1}{q}} \| T_n \|_q. \tag{11}$$

Hence subject to the condition $q' \geqslant q$

$$\int_0^{2\pi} |T_n(x)|^{q'} \mathrm{d}x = \int_0^{2\pi} |T_n(x)|^{q'-q} |T_n(x)|^q \mathrm{d}x \leqslant \left(\frac{nq_0+1}{2\pi}\right)^{\frac{q'-q}{q}} \| T_n \|_q^{q'}. \tag{12}$$

This proves the inequality 4.9(7).

Considering the sequence of trigonometric Fejér kernels

$$T_{n-1}(x) = \left(\frac{\sin\frac{nx}{2}}{n\sin\frac{x}{2}}\right)^2, \tag{13}$$

for which $\| T_{n-1} \|_q^q = \pi\gamma_n^{(q-2)}$, where $\gamma_n^{(r)}$ is defined by 4.8(11), and also considering the upper and lower estimates given for $\gamma_n^{(r)}$ in sections 4.8.5 and 4.8.72, we see that for the polynomials 4.9(13) there exists a constant $C > 0$ independent of n with the property

$$\| T_{n-1} \|_{q'} > Cn^{\frac{1}{q}-\frac{1}{q'}} \| T_{n-1} \|_q. \tag{14}$$

Thus this example shows that in the class of all trigonometric polynomials the estimate 4.9(7) is precise as regards order (as $n \to \infty$).

4.9.3. From the estimates obtained in section 4.9.2 for integral norms and the relations 4.9(3) there immediately follow similar inequalities for the interpolation norms of trigonometric polynomials in various metrics L_q. Thus, for example, the estimate

$$((T_n))_{q'} \leqslant \left(1+\frac{2n\pi}{N}\right)\left(\frac{nq_0+1}{2\pi}\right)^{\frac{1}{q}-\frac{1}{q'}} ((T_n))_q \quad (1 \leqslant q \leqslant q' \leqslant \infty) \tag{15}$$

serves as the interpolation analogue of the inequality 4.9(9).

4.9.4. Inequalities similar to those given in sections 4.9.1–4.9.3 hold also for trigonometric polynomials $T_{n_1}, \ldots, {}_{n_m}(x_1, \ldots, x_m)$ of many variables if their integral and interpolational norms in L_q $(1 \leqslant q \leqslant \infty)$ are defined as follows:

$$\|T_{n_1, \ldots, n_m}\|_q = \left\{\int_0^{2\pi} \cdots \int_0^{2\pi} |T_{n_1, \ldots, n_m}(x_1, \ldots, x_m)|^q \mathrm{d}x_1, \ldots, \mathrm{d}x_m\right\}^{\frac{1}{q}},$$

$$(T_{n_1, \dots, n_m}))_q = \sup_{u_1, \dots, u_m} \left\{ \prod_{j=1}^{m} \frac{2\pi}{N_j} \sum_{k_1=0}^{N_1-1} \dots \sum_{k_m=0}^{N_m-1} |T_{n_1, \dots, n_m}(u_1 - x_{k_1}^{(1)}, \dots, u_m - x_{k_m}^{(m)}|^q \right\}^{\frac{1}{q}}, \quad (16)$$

where N_j is an integer, and $x_k^{(j)} = \frac{2k\pi}{N_j}$. They can be obtained as in the case $m = 1$. Let us derive some of them**.

For any $q(1 \leqslant q \leqslant \infty)$

$$|T_{n_1, \dots, n_m}\|_q \leqslant ((T_{n_1, \dots, n_m}))_q \leqslant \prod_{k=1}^{m} \left(1 + \frac{2n_k \pi}{N_k}\right) \|T_{n_1, \dots, n_m}\|_q. \quad (17)$$

If $1 \leqslant q \leqslant q' \leqslant \infty$, *then*

$$\|T_{n_1, \dots, n_m}\|_{q'} \leqslant C^m \left(\prod_{k=1}^{m} \frac{n_k q_0 + 1}{2\pi} \right)^{\frac{1}{q} - \frac{1}{q'}} \|T_{n_1, \dots, n_m}\|_q. \quad (18)$$

Besides these relations, which are quite similar to those derived above for the case $m = 1$, it is necessary to note a new estimate which follows from them, restricted to polynomials of several variables and establishing a connection between their "total" norm 4.9(16) and their "partial" norms of the form

$$\|T_{(n_1, \dots, n_r) n_{r+1}, \dots, n_m}\|_{q, q'} = \left\{ \int_0^{2\pi} \dots \int_0^{2\pi} dx_{r+1} \dots, dx_m \times \right.$$
$$\left. \times \left[\int_0^{2\pi} \dots \int_0^{2\pi} |T_{n_1, \dots, n_m}(x_1, \dots, x_m)|^q \, dx_1, \dots, dx_r \right]^{\frac{q'}{q}} \right\}^{\frac{1}{q'}}. \quad (19)$$

If $1 \leqslant q \leqslant q' \leqslant \infty$ *and* $1 \leqslant r \leqslant m$, *then****

$$\|T_{(n_1, \dots, n_r) n_{r+1}, \dots, n_m}\|_{q, q'} \leqslant C^m \left(\prod_{k=r+1}^{m} n_k \right)^{\frac{1}{q} - \frac{1}{q'}} \|T_{n_1, \dots, n_m}\|_q, \quad (20)$$

where C is an absolute constant. The estimate 4.9(20) is a direct consequence of 4.9(17) since

$$\prod_{k=1}^{m} \left(1 + \frac{2n_k T}{N_k}\right) \| T_{n_1, \dots, n_m} \|_q \geqslant ((T_{n_1, \dots, n_m}))$$
$$= \left(\prod_{j=1}^{m} \frac{2\pi}{N_j} \right)^{\frac{1}{q}} \sup_{u_1, \dots, u_m} \left\{ \sum_{k_1=0}^{N_1-1} \dots \sum_{k_m=0}^{N_m-1} |T_{n_1, \dots, n_m}(u_1 - x_{k_1}^{(1)}, \dots, u_m - x_{k_m}^{(m)}|^q \right\}^{\frac{1}{q}} \geqslant$$

** S. M. Nikol'skii [13].
*** S. M. Nikol'skii [13], in the case $q' = \infty$.

$$\geqslant \left(\prod_{j=1}^{m} \frac{2\pi}{N_j}\right)^{\frac{1}{q}} \sup_{u_1, \dots, u_m} \sup_{x_{k_{r+1}}, \dots x_{k_m}} \Bigg\{\sum_{k_1=0}^{N_1-1} \cdots \sum_{k_r=0}^{N_r-1} \times$$

$$\times |T_{n_1, \dots, n_m}(u_1 - x_{k_1}^{(1)}, \dots u_m - x_{k_m}^{(m)})|^q\Bigg\}^{\frac{1}{q}}$$

$$\geqslant \left(\prod_{j=r+1}^{m} \frac{2\pi}{N_j}\right)^{\frac{1}{q}} \|T_{(n_1, \dots, n_r), n_{r+1}, \dots, n_m}\|_{l,\infty}.$$

If in the last inequality we put, for example, $N_j = n_j$, then we obtain 4.9(20) for the case $q' = \infty$:

$$\|T_{(n_1, \dots, n_r)n_{r+1}, \dots, n_m}\|_{q,\infty} \leqslant C^m\Big(\prod_{k=r+1}^{m} n_k\Big)^{\frac{1}{q}} \|T_{n_1, \dots, n_m}\|_q. \tag{21}$$

Hence for any $q' \geqslant q$ we find

$$\int_0^{2\pi} \cdots \int_0^{2\pi} \left[\int_0^{2\pi} \cdots \int_0^{2\pi} |T_{n_1, \dots, n_m}(x_1, \dots, x_m)|^q \, dx_1, \dots, dx_r\right]^{\frac{q'}{q}} dx_{r+1} \dots dx_m$$

$$= \int_0^{2\pi} \cdots \int_0^{2\pi} \left[\int_0^{2\pi} \cdots \int_0^{2\pi} |T_{n_1, \dots, n_m}(x_1, \dots, x_m)|^q \, dx_1 \dots dx_r\right]^{\frac{q'-q}{q}} \times$$

$$\times \int_0^{2\pi} \cdots \int_0^{2\pi} |T_{n_1, \dots, n_m}(x_1, \dots, x_m)|^q \, dx_1 \dots dx_r \, dx_{r+1} \dots dx_m \leqslant$$

$$\leqslant \|T_{(n_1, \dots, n_r)n_{r+1}, \dots, n_m}\|_{q,\infty}^{q'-q} \|T_{n_1, \dots, n_m}\|_q^q \leqslant$$

$$\leqslant C^{m(q'-q)}\Big(\prod_{k=r+1}^{m} n_k\Big)^{\frac{q'-q}{q}} \|T_{n_1, \dots, n_m}\|_q^{q'},$$

i.e. 4.9(20) holds.

Considering the polynomials $T_{n_1, \dots, n_m}(x_1, \dots, x_m) = \prod_{k=1}^{m} T_{n_k}(x_k)$, where $T_n(x)$ is a Fejér kernel (see 4.9.13), we see that the estimate 4.9(20) is precise as regards order (as $n_k \to \infty$).

In conclusion we note another inequality

$$\|T_{(n_1, \dots, n_r)n_{r+1}, \dots, n_m}\|_{q,q'} \leqslant \|T_{n_1, \dots, n_r(n_{r+1}, \dots, n_m)}\|_{q',q} \qquad (q' \geqslant q) \tag{22}$$

which immediately follows from the definition of the norm 4.9(19) if use is made of Minkowski's inequality for multiple integrals.

4.9.5. Relations similar to those derived above for trigonometric polynomials can also be obtained from similar considerations for other integral functions of exponential type. Let $G_\sigma(x)$ be an integral function of finite degree $\leqslant \sigma$ for which

$$\int_{-\infty}^{\infty} |G_\sigma(x)|^q \, dx < \infty . \tag{23}$$

Let us put

$$||G_\sigma||_q = \{|G_\sigma(x)|^q \, dx\}^{\frac{1}{q}} < \infty \tag{24}$$

and for some system of equidistant points $x_k = kh$ ($h > 0$, $k = 0, \pm 1, \pm 2, \ldots$)

$$((G_\sigma))_q = \sup_x \Big\{h \sum_{k=-\infty}^{\infty} |G_\sigma(x - x_k)|^q\Big\}^{\frac{1}{q}} . \tag{25}$$

4.9.51. *For any $h > 0$ and $q \geqslant 1$***

$$||G_\sigma||_q \leqslant ((G_\sigma))_q \leqslant (1+\sigma h)||G_\sigma||_q . \tag{26}$$

The proof of these inequalities is the same as in section 4.9.1, except that in this case instead of finite sums we have sums of series.

4.9.52. If $G_\sigma(x)$ satisfies the condition 4.9(23) for $q = 2$, then by the Wiener-Paley theorem (see 4.6.1)

$$G_\sigma(x) = \frac{1}{\sqrt{2\pi}} \int_{-\sigma}^{\sigma} g(t) e^{itx} \, dt . \tag{27}$$

Hence in virtue of the Bunyakovskii–Schwarz inequality and 4.6(4),

$$\sup_x |G_\sigma(x)| \leqslant \frac{1}{\sqrt{\pi}} \sigma^{\frac{1}{2}} ||G_\sigma||_2 . \tag{28}$$

This estimate*** cannot be improved. For the function $G_\sigma(x) = \dfrac{\sin \sigma x}{x}$ the two sides of 4.9(28) are equal.

4.9.53. The inequality 4.9(28) can be written in a more general form. From it, exactly as in section 4.9.2, it follows directly that**** *for any positive q and $q' \geqslant q$ we have, for every integral function of degree $\leqslant \sigma$ which satisfies condition* 4.9(23),

$$||G_\sigma||_{q'} \leqslant \left(\frac{q_0 \sigma}{2\pi}\right)^{\frac{1}{q} - \frac{1}{q'}} ||G_\sigma||_q . \tag{29}$$

** S. M. Nikol'skii [13].

*** I. I. Ibragimov [3].

**** For the particular case $0 < q < 2$ see I.I. Ibragimov [5]. From 4.9(29) there follows the estimate $||G_\sigma||_{q'} < 2\sigma^{\frac{1}{q} - \frac{1}{q'}} ||G_\sigma||_q$, which for $q' \geqslant q \geqslant 1$ was first obtained by S. M. Nikol'skii [13].

The example of the function

$$G_\sigma(x) = \left(\frac{\sin\dfrac{\sigma x}{2}}{x}\right)^2 \tag{30}$$

for $q > \frac{1}{2}$ shows that this estimate is precise as regards order (as $\sigma \to \infty$).

4.9.54. If $1 \leqslant q \leqslant q' \leqslant \infty$, then

$$((G_\sigma))_{q'} \leqslant \left(\frac{q_0 \sigma}{2\pi}\right)^{\frac{1}{q}-\frac{1}{q'}} (1+\sigma h)\,((G_\sigma))_q. \tag{31}$$

This inequality follows directly from 4.9(26) and 4.9(29).

4.9.55. In a number of problems an important part is played by integral functions $G_{\sigma_1,\dots,\sigma_m}(x_1,\dots,x_m)$ of many variables, which are of finite degree in these variables. Such functions, as in section 4.3, are defined as integral functions which possess the property, that for any $\varepsilon > 0$

$$|G_{\sigma_1,\dots,\sigma_m}(z_1,\dots,z_m)| \leqslant M(\varepsilon)\,e^{\sum\limits_{k=1}^{m}(\sigma_k+\varepsilon)|z_k|} \tag{32}$$

for all values of $z_k = x_k + iy_k$. If by analogy with 4.9(24) and 4.9(25) we put

$$\|G_{\sigma_1,\dots,\sigma_m}\|_q = \left\{\int_{-\infty}^{\infty}\cdots\int_{-\infty}^{\infty} |G_{\sigma_1,\dots,\sigma_m}(x_1,\dots,x_m)|^q\,dx_1,\dots,dx_m\right\}^{\frac{1}{q}},$$

$$\|G_{(\sigma_1,\dots,\sigma_r)\sigma_{r+1},\dots,\sigma_m}\|_{q,q'}$$

$$= \left\{\int_{-\infty}^{\infty}\cdots\int_{-\infty}^{\infty}\left[\int_{-\infty}^{\infty}\cdots\int_{-\infty}^{\infty} |G_{\sigma_1,\dots,\sigma_m}(x_1,\dots,x_m)|^q\,dx_1,\dots,dx_r\right]^{\frac{q'}{q}} dx_{r+1},\dots,dx_m\right\}^{\frac{1}{q'}}$$

$$(q' \geqslant q)$$

and for some systems of equidistant points $x_k^\nu = \nu h_k$ ($h_k > 0$; $k = 1,\dots, m$; $\nu = 0, \pm 1, \pm 2, \pm 3, \dots$)

$$((G_{\sigma_1,\dots,\sigma_m}))_q$$

$$= \sup_{u_1,\dots,u_m}\left\{\prod_{k=1}^{m} h_\kappa \sum_{k_1=-\infty}^{\infty}\cdots\sum_{k_m=-\infty}^{\infty} |G_{\sigma_1,\dots,\sigma_m}(u_1 - x_{k_1}^{(1)},\dots, u_m - x_{k_m}^{(m)}|^q\right\}^{\frac{1}{q}},$$

then we have for these norms relations similar to those derived in section 4.9.4.

For any q $(1 \leqslant q \leqslant \infty)$,

$$\|G_{\sigma_1,\dots,\sigma_m}\|_q \leqslant ((G_{\sigma_1,\dots,\sigma_m}))_q \leqslant \prod_{k=1}^{m}(1+\sigma_k h_k)\|G_{\sigma_1,\dots,\sigma_m}\|_q. \tag{33}$$

With the condition $1 \leqslant q \leqslant q' \leqslant \infty$**,

$$\|G_{\sigma_1,\ldots,\sigma_m}\|_{q'} \leqslant C^m\Big(\prod_{k=1}^{m} \sigma_k\Big)^{\frac{1}{q}-\frac{1}{q'}} \|G_{\sigma_1,\ldots,\sigma_m}\|_q, \tag{34}$$

where C is an absolute constant. Observing that for any non-negative numbers $a_1, \ldots, a_n$ the quantity $\Big(\sum_{k=1}^{n} a_k^q\Big)^{\frac{1}{q}}$ is a non-increasing function of q***
we obtain from 4.9(33)

$$\|G_{\sigma_1,\ldots,\sigma_m}\|_{q'} \leqslant ((G_{\sigma_1,\ldots,\sigma_m}))_{q'}$$

$$= \Big(\prod_{k=1}^{m} h_k\Big)^{\frac{1}{q'}} \sup_{u_1,\ldots,u_m} \Big\{\sum_{k_1} \cdots \sum_{k_m} |G_{\sigma_1,\ldots,\sigma_m}(u_1 - x_{k_1}^{(1)}, \ldots, u_m - x_{k_m}^{(m)}|^{q'}\Big\}^{\frac{1}{q'}} <$$

$$\leqslant \Big(\prod_{k=1}^{m} h_k\Big)^{\frac{1}{q'}-\frac{1}{q}} ((G_{\sigma_1,\ldots,\sigma_m}))_q \leqslant$$

$$\leqslant \Big(\prod_{k=1}^{m} h_k\Big)^{\frac{1}{q'}-\frac{1}{q}} \prod_{k=1}^{m} (1+\sigma_k h_k) \|G_{\sigma_1,\ldots,\sigma m}\|_q.$$

Putting $\sigma_k h_k = 1$ and $q' = \infty$, we obtain the inequality

$$|G_{\sigma_1,\ldots,\sigma_m}(x_1, \ldots, x_m)| \leqslant 2^m\Big(\prod_{k=1}^{m} \sigma_k\Big)^{\frac{1}{q}} \|G_{\sigma_1,\ldots,\sigma_m}\|_q,$$

from which it immediately follows that *the constant C in* 4.9(34) *is not greater than* $2^{1-\frac{q}{q'}}$.

If $1 \leqslant q \leqslant q' \leqslant \infty$, then

$$\|G_{(\sigma_1,\ldots,\sigma_r),\sigma_{r+1},\ldots,\sigma_m}\|_{q,q'} \leqslant C^m\Big(\prod_{k=r+1}^{m} \sigma_k\Big)^{\frac{1}{q}-\frac{1}{q'}} \|G_{\sigma_1,\ldots,\sigma_m}\|_q. \tag{35}$$

4.9.6. Considering the algebraic polynomials $Pn(x)$ of degree n and their integral norms with respect to a certain given segment $[a, b]$:

$$\|P_n\|_q = \Big\{\int_a^b |P_n(x)|^q \, dx\Big\}^{\frac{1}{q}} \quad (1 \leqslant q \leqslant \infty),$$

it is possible, as in section 4.9.2, to arrive at similar estimates.

** See S. M. Nikol'skii [13], where for the constant C in the inequalities 4.9(34) and 4.9(35) there is given the estimate $C \leqslant 2$, and the relation 4.9(35) is derived for the case $q' = \infty$.

*** This is obvious in the case $n = 2$, and for other values of n it is proved by induction.

*If $P_n(x)$ is an algebraic polynomial of degree $\leqslant n$, then subject to the condition $1 \leqslant q \leqslant q' \leqslant \infty$ we have***

$$\|P_n\|_{q'} \leqslant \left[\frac{2(q+1)}{b-a}\right]^{\frac{1}{q}-\frac{1}{q'}} n^{2\left(\frac{1}{q}-\frac{1}{q'}\right)} \|P_n\|_q. \tag{36}$$

Let us assume that x_0 is the point of the segment $[a, b]$ at which $|P_n(x_0)| = \max\limits_{a\leqslant x\leqslant b} |P_n(x)|$. For all $x \in [a, b]$

$$|P_n(x)-P_n(x_0)| \leqslant |x-x_0| \max_{a\leqslant u\leqslant b} |P_n'(u)|.$$

In virtue of the inequality 4.8(32), which when adapted to the case of the segment $[a, b]$ assumes the form

$$\max_{a\leqslant u\leqslant b} |P_n'(u)| \leqslant \frac{2n^2}{b-a} \max_{a\leqslant u\leqslant b} |P_n(u)|, \tag{37}$$

we have

$$|P_n(x_0)| - |P_n(x) \leqslant \frac{2n^2}{b-a} |x-x_0| \, |P_n(x_0)|.$$

From this, if we assume for definiteness that $x_0 \leqslant \frac{1}{2}(a+b)$, there follows the estimate

$$|P_n(x_0)|^q \int\limits_{x_0}^{x_0+\frac{b-a}{2n^2}} \left(1-\frac{2n^2}{b-a}|x-x_0|\right)^q dx \leqslant \int\limits_a^b |P_n(x)|^q \, dx$$

i.e.

$$\max_{a\leqslant u\leqslant b} |P_n(u)| \leqslant \left[\frac{2(q+1)}{b-a}\right]^{\frac{1}{q}} n^{\frac{2}{q}} \|P_n\|_q.$$

It only remains to observe that

$$\int\limits_a^b |P_n(x)|^{q'} \, dx = \int\limits_a^b |P_n(x)|^{q-q'} |P_n(x)|^q \, dx \leqslant \left[\frac{2(q+1)}{b-a}\right]^{\frac{q'-q}{q}} n^{\frac{2}{q}(q'-q)} \|P_n\|^{q'}.$$

It is not difficult to see that the inequality 4.9(36) cannot be improved as regard order (as $n \to \infty$).

Thus in the case $a = -1$, $b = 1$ it is sufficient to consider, for example, the sequence of polynomials

$$P_{4n-4}(x) = \left(\frac{\sin n \arccos x}{\sqrt{1-x^2}}\right)^4. \tag{38}$$

** Inequalities of this type for the norms of trigonometric polynomials on segments of length less than 2π were established by N. K. Bari [2].

4.10. The connection between trigonometric polynomials and other integral functions of the classes B_σ

In the study of integral functions of the classes B_σ it may be noticed that all such functions possess many properties which are possessed by trigonometric polynomials (see, for example, sections 4.8.2, 4.8.4, 4.8.6, 4.9.5)**. The reason for this consists in the fact that there exists a close constructive bond between trigonometric polynomials and all other integral functions which are bounded on the real axis. This connection is expressed in the following remarkable theorem***.

4.10.1. *For any integral function of exponential type bounded on the real axis, it is possible to give a sequence of trigonometric polynomials*

$$\sum_{\nu=0}^{n} (a_\nu^{(n)} \cos \nu\lambda_n x + b_\nu^{(n)} \sin \nu\lambda_n x) \qquad (\lambda_n > 0) \tag{1}$$

which converges to it uniformly on any finite segment of the real axis.

Let $G_\sigma(x)$ be an arbitrary integral function of degree $\leqslant \sigma$ which is bounded on $(-\infty, \infty)$. For any natural number n let us consider the series

$$\frac{\sin \sigma x}{n \sin \dfrac{\sigma x}{n}} G_\sigma(0) + n \sin \frac{\sigma x}{n} \sin \sigma x \cdot \sum_{\substack{k=-\infty \\ k\neq 0}}^{\infty} \frac{(-1)^k \sin \dfrac{k\pi}{n}}{k^2\pi^2 \sin \dfrac{\sigma x - k\pi}{n}} G_\sigma\left(\frac{k\pi}{\sigma}\right) + \frac{\sin \sigma x}{\sigma} G'_\sigma(0). \tag{2}$$

It is not difficult to see that this series converges uniformly with respect to all $n > 0$ and all values of x which belong to some given finite segment; it is sufficient merely to note that

$$\left| \frac{n \sin \dfrac{\sigma x}{n} \sin(\sigma x - k\pi) \sin \dfrac{k\pi}{n}}{\sin \dfrac{\sigma x - k\pi}{n}} \right| \leqslant |\sigma x \sin \sigma x| + \sigma^2 x^2 \left| \frac{\sin(\sigma x - kx)}{n \sin \dfrac{\sigma x - k\pi}{n}} \right| \leqslant |(\sigma x \sin \sigma x| + \sigma^2 x^2.$$

** See S. N. Bernstein [39].

*** B. M. Levitan [1]. The idea of the given proof is based on the use of the interpolation formula 4.3(14), due to N. I. Akhiezer and V. A. Marchenko [1].

Let us put $t = \dfrac{\sigma x}{n}$. Since the quotient $\dfrac{\sin nt}{\sin t}$ is a trigonometric polynomial of order $n-1$, every term in the sum 4.10(2) represents a trigonometric polynomial of order $\leqslant n$ with respect to t:

$$T_{k,n}(t) = \sum_{\nu=0}^{n} (a_{\nu,k}^{(n)} \cos \nu t + b_{\nu,k}^{(n)} \sin \nu t).$$

As the series 4.10(2) converges uniformly on the segment $0 \leqslant t \leqslant 2\pi$, then, as follows from considerations similar to those of section 2.1.1, its sum is a certain trigonometric polynomial

$$T_n(t) = \sum_{\nu=0}^{n} (a_\nu^{(n)} \cos \nu t + b_\nu^{(n)} \sin \nu t)$$
$$= \sum_{\nu=0}^{n} \left(a_\nu^{(n)} \cos \nu \frac{\sigma x}{n} + b_\nu^{(n)} \sin \nu \frac{\sigma x}{n} \right) \tag{3}$$

of order $\leqslant n$. Observing that on any finite segment of variation of x the series 4.10(2) converges uniformly with respect to all $n > 0$, and taking the limit of each term as $n \to \infty$, we obtain

$$\lim_{n\to\infty} T_n\left(\frac{\sigma x}{n}\right) = \frac{\sin \sigma x}{\sigma x} G_\sigma(0) +$$
$$+ \sin \sigma x \sum_{\substack{k=-\infty \\ k\neq 0}}^{\infty} (-1)^k \left\{ \frac{1}{\sigma x - k\pi} + \frac{1}{k\pi} \right\} G_\sigma\left(\frac{k\pi}{\sigma}\right) + \frac{\sin \sigma x}{\sigma} G'_\sigma(0),$$

or by 4.3(14)

$$T_n\left(\frac{\sigma x}{n}\right) \to G_\sigma(x) \tag{4}$$

as $n \to \infty$, uniformly on any finite segment of the real axis.

4.10.2. In the class of all integral functions of exponential type the trigonometric polynomials are characterised by the fact that they are the only functions which have a finite period**. *If the integral function $G_\sigma(x)$ of degree $\leqslant \sigma$ has the period $\dfrac{2\pi}{\lambda}$, then it is the trigonometric polynomial*

$$\sum_{k=0}^{\left[\frac{\sigma}{\lambda}\right]} (a_k \cos k\lambda x + b_k \sin k\lambda x) \tag{5}$$

** S. N. Bernstein [23].

of order $\leqslant \left[\frac{\sigma}{\lambda}\right]$. In fact, let

$$G_\sigma(x) = \sum_{k=0}^{\infty} (a_k \cos k\lambda x + b_k \sin k\lambda x)$$

be the expansion of the function $G_\sigma(x)$ in a Fourier series. Since $G_\sigma(x)$ is an integral function its Fourier series can be differentiated term by term any number of times. Applying Bessel's inequality after r successive differentiations, we obtain

$$(k\lambda)^{2r} (a_k^2 + b_k^2) \leqslant \frac{\lambda}{\pi} \int_{-\frac{\pi}{\lambda}}^{\frac{\pi}{\lambda}} |G_\sigma^{(r)}(x)|^2 \, dx. \tag{6}$$

Moreover, as the function $G_\sigma(x)$ is periodic, it is bounded on the real axis. Therefore in virtue of the inequality 4.8(5) we have:

$$\frac{\lambda}{\pi} \int_{-\frac{\pi}{\lambda}}^{\frac{\pi}{\lambda}} |G_\sigma^{(r)}(x)|^2 \, dx \leqslant 2\sigma^{2r} \sup_x |G_\sigma(x)|. \tag{7}$$

Comparing 4.10(6) and 4.10(7), we obtain the relation

$$a_k^2 + b_k^2 \leqslant 2 \left(\frac{\sigma}{k\lambda}\right)^{2r} \sup_x |G_\sigma(x)|,$$

which must hold for any natural number r. But this is possible only when $a_k^2 + b_k^2 = 0$ for all $k > \left[\frac{\sigma}{\lambda}\right]$. Consequently

$$G_\sigma(x) = \sum_{k=0}^{\left[\frac{\sigma}{\lambda}\right]} (a_k \cos k\lambda x + b_k \sin k\lambda x).$$

4.10.3. Let us apply theorem 4.10.2 for the construction of some processes of approximation to integral functions of the class B_σ by trigonometric polynomials. Let $G_\sigma(x) \in B_\sigma$ and r be some non-negative whole number. For an arbitrary natural number n let us consider the series**

$$\sum_{k=-\infty}^{\infty} G_\sigma\left(x + k\frac{n\pi}{\sigma}\right) \left(\frac{\sin \frac{\sigma x}{n}}{\frac{\sigma x}{n} + k\pi}\right)^{2r+2} \tag{8}$$

** N. I. Akhiezer and V. A. Marchenko [1].

Every term of the series 4.10(8) is an integral function of degree $\leqslant \sigma + \frac{\sigma}{n}(2r+2)$. In addition, this series converges uniformly in any finite part of the complex plane and its symmetric partial sums are bounded in modulus on the whole real axis by one and the same number. Hence by 4.8(8) we conclude that the sum of the series 4.10(8) is some integral function of order $\leqslant \sigma + \frac{\sigma}{n}(2r+2)$. It is easily verified that the function obtained has period $\frac{n\pi}{\sigma}$. Consequently (see section 4.10.2) it is the trigonometric polynomial

$$T_{m,r}(G;x) = \sum_{k=0}^{m} \left(a_k \cos 2k\frac{\sigma}{n}x + b_k \sin 2k\frac{\sigma}{n}x \right) \tag{9}$$

of order $\leqslant m = \left[\frac{n}{2}\right] + r + 1$.

The series 4.10(8) converges uniformly for all integral $n > 0$ and for all values of x which belong to a finite segment. Passing to the limit in each term as $n \to \infty$, we obtain

$$T_{m,r}(G_\sigma; x) \to G_\sigma(x) \quad \text{as} \quad n \to \infty,$$

uniformly on any finite segment of the real x-axis.

4.10.31. It immediately follows from the representation of the polynomials $T_{m,r}(G_\sigma; x)$ in the form of the series 4.10(8) that *for any integral* $r \geqslant 0$ *these polynomials are always non-negative for non-negative* $G_\sigma(x)$. In addition, for $r = 0$**

$$|T_{m,0}(G;x)| \leqslant \sup_{-\infty < t < \infty} |G_\sigma(t)|. \tag{10}$$

4.11. Non-negative polynomials and transcendental integral functions of exponential type

In conclusion we will consider some specific properties of algebraic polynomials and transcendental integral functions of finite degree, which are of constant sign.

4.11.1. If the polynomial

$$P_n(x) = \sum_{k=0}^{n} c_k x^k \tag{1}$$

** In the case $r = 0$ the polynomials considered are described as Levitan polynomials.

assumes non-negative values everywhere on the interval $-\infty < x < \infty$, then its degree n is an even number $2m$ and to each zero $x_k + iy_k$ $(k = 1, \ldots, m)$ there corresponds the conjugate zero $x_k - iy_k$. Therefore

$$P_n(x) = c_n \prod_{k=1}^{m} [(x-x_k)^2 + y_k^2].$$

Hence, using the identity

$$(a^2+b^2)(c^2+d^2) = (ac+bd)^2 + (ad-bc)^2, \tag{2}$$

it follows that *any polynomial $P_n(x)$ which is non-negative on the whole real axis can be written in the form*

$$P_n(x) = A_m^2(x) + B_m^2(x) \tag{3}$$

where $A_m(x)$ and $B_m(x)$ are real algebraic polynomials of degree $\leqslant m$.

4.11.2. The representation 4.11(3) no longer applies in the wider class of polynomials which satisfy the condition $P_n(x) \geqslant 0$ on the semi-axis $(0, \infty)$. However, since in this case also $P_n(x)$ assumes only real values everywhere on the real axis and does not possess zeros of odd multiplicity, it follows that

$$P_n(x) = \prod_{k=1}^{r} (a_k + xb_k) \prod_{\nu=1}^{l} (x^2 + h_\nu x + q_\nu),$$

where $a_k \geqslant 0$, $b_k \geqslant 0$ $(k = 1, 2, \ldots, r)$, $p_\nu^2 - 4q_\nu \leqslant 0$ $(\nu = 1, 2, \ldots, l)$, $2l + r \leqslant n$.

Hence we see that** *any polynomial $P_n(x)$ non-negative on the semi-axis $(0, \infty)$ can be written in the form*

$$P_n(x) = A_m^2(x) + xB_m^2(x), \tag{4}$$

where $A_m(x)$ and $B_m(x)$ are algebraic polynomials with real coefficients determined from the relation

$$A_m(x) + i\sqrt{x}\, B_m(x)$$
$$= \prod_{k=1}^{r} \left(\sqrt{a_k} - i\sqrt{b_k}\sqrt{x}\right) \prod_{\nu=1}^{l} \left\{(x - \sqrt{q\nu}) + i\sqrt{x}\sqrt{2\sqrt{q\nu} + h\nu}\right\}.$$

4.11.3. In order to characterise in a similar manner polynomials which preserve their sign on a finite interval, let us first prove the following proposition. *If the function*

$$R_{2n}(z) = \sum_{k=-n}^{n} c_k z^k \quad (c_r \neq 0) \tag{5}$$

** V. S. Videnskii [3]. See also Shohat and Tamarkin [1].

assumes real non-negative values on the circle $z = e^{it}$, *then it can be written in the form*

$$R_{2n}(z) = C\prod_{k=1}^{r}(z-z_k)\left(\frac{1}{z}-\bar{z}_k\right)\prod_{k=1}^{s}(z-w_k)\left(\frac{1}{z}-\bar{w}_k\right), \tag{6}$$

where $C > 0$, $|z_k| < 1$ *and* $|w_k| = 1$. As the function $R_{2n}(z)$ is real on the circle $|z| = 1$, then in virtue of the Riemann-Schwarz principle of symmetry

$$R_{2n}\left(\frac{1}{\bar{z}}\right) = \overline{R_{2n}}(z),$$

and consequently its zeros are arranged symmetrically with respect to this circumference. Denoting by z_k $(k = 1, 2, \ldots, r)$ those of them which lie within the circle $|z| < 1$, and by w_k $(k = 1, \ldots, 2(n-r))$ those which lie on its boundary, we have

$$R_{2n}(z) = \frac{C_n}{z^n}\prod_{k=1}^{r}(z-z_k)\left(z-\frac{1}{\bar{z}_k}\right)\prod_{k=1}^{2(n-r)}(z-w_k)$$

$$= \prod_{k=1}^{r}(z-z_k)\left(\frac{1}{z}-\bar{z}_k\right)\frac{C_n^*}{z^{n-r}}\prod_{k=1}^{2(n-r)}(z-w_k).$$

It is easily seen that because of the positiveness of the expression $R_{2n}(e^{it})$ all the zeros w_k are zeros of even multiplicity. Hence

$$R_{2n}(z) = \prod_{k=1}^{r}(z-z_k)\left(\frac{1}{z}-\bar{z}_k\right)\cdot\frac{C_n^*}{z^{n-r}}\prod_{k=1}^{n-r}(z-w_k)^2$$

$$= C\prod_{k=1}^{r}(z-z_k)\left(\frac{1}{z}-\bar{z}_k\right)\prod_{k=1}^{s}(z-w_k)\left(\frac{1}{z}-\bar{w}_k\right).$$

Putting $z = e^{it}$, we obtain $C > 0$.

4.11.31. Now let $P_n(x)$ be the algebraic polynomial 4.11(1) non-negative on some finite segment $[a, b]$. Let us consider the case where $a = -1$, $b = 1$. In this case, putting $x = \cos t = \frac{1}{2}(e^{it}+e^{-it})$, we obtain the non-negative function

$$R_{2n}(e^{it}) = P_n(\cos t) = \sum_{k=-n}^{n} d_k e^{ikt}.$$

By formula 4.11(6)

$$R_{2n}(e^{it}) = C\prod_{k=1}^{r}|e^{it}-z_k|^2\prod_{k=1}^{s}|e^{it}-w_k|^2,$$

or

$$P_n(\cos t) = |Q_{r+s}(e^{it})|^2, \tag{7}$$

where the $Q_{r+s}(z)$ are algebraic polynomials. Since $P_n(\cos t)$ is a non-negative even trigonometric polynomial, all the coefficients are real. Therefore to all the zeros w_k, just as to all the zeros z_k, there correspond conjugate zeros $\overline{w}_k$ and $\overline{z}_k$. It follows from this that all the coefficients of the polynomial are real. If we also take account of the fact that

$$Q_{r+s}(\cos t+i\sin t) = A(\cos t)+i\sin t B(\cos t),$$

where $A(x)$ and $B(x)$ are *polynomials with real coefficients*, then on returning to the variable x we obtain from 4.11(7) that *any algebraic polynomial $P_n(x)$ which is non-negative on the segment $-1 \leqslant x \leqslant 1$ can be written in the form***

$$P_n(x) = A^2(x)+(1-x^2)B^2(x). \tag{8}$$

In the general case, when the polynomial $P_n(x)$ is non-negative on the segment $[a, b]$, the given representation assumes the form

$$P_n(x) = A^2(x)+(x-a)(b-x)B^2(x). \tag{9}$$

4.11.4. By using formula 4.11(6), it is easy to obtain the following theorem, due to Fejér and Riesz***, concerning trigonometric polynomials with constant sign.

If $T_n(t)$ is a trigonometric polynomial of order $\leqslant n$ which assumes on the real axis only negative real values, then

$$|T_n(t)| = \Big|\sum_{k=0}^{n} \gamma_k e^{ikt}\Big|^2, \tag{10}$$

where $g_n(z) = \sum_{k=0}^{n} \gamma_k e^{ikz}$ is a function with no zeros in the lower half-plane. Writing the polynomial $T_n(t)$ in the complex form (see section 4.2), we have $T_n(t) = R_{2n}(e^{it})$ (see 4.11(5)). Hence in virtue of proposition 4.11.3 (formula 4.11(6))

$$T_n(t) = R_{2n}(e^{it}) = C\prod_{k=1}^{r} |e^{it}-z_k|^2 \prod_{k=1}^{s} |e^{it}-w_k|^2,$$

i.e.

$$T_n(t) = \Big|\sum_{k=0}^{r+s} \gamma_k e^{ikt}\Big|^2.$$

It only remains to note that $|z_k| < 1$ and $|w_k| = 1$.

** See G. Pólya and G. Szegö [1], Part 2.
*** See Fejèr [4].

4.11.5. A theorem similar to 4.11.4 also holds for any integral function of the class B_σ**. *If the function* $G_\sigma(x) \in B_\sigma$ *is non-negative on the real axis, then there exists a function* $g(x) \in B_{\sigma/2}$ *such that*

$$G_\sigma(x) = |g(x)|^2 \tag{11}$$

The validity of this assertion can be verified by using the results of sections 4.10 and 4.11.4. Let us consider the sequence of trigonometric polynomials $T_{m,0}(G_\sigma; x)$ defined in section 4.10.3. If $G_\sigma(x) \geqslant 0$ then in virtue of the remark of section 4.10.31 we have $T_{m,0}(G_\sigma; x) \geqslant 0$. By theorem 4.11.4

$$T_{m,0}(G_\sigma; x) = |g_n(x)|^2, \tag{12}$$

where

$$g_n(z) = e^{-\frac{im\sigma z}{n}} \sum_{k=0}^{m} \beta_k^{(n)} e^{\frac{2ik\sigma z}{n}}$$

Since $m \leqslant \left[\frac{n}{2}\right]+1$ (see 4.10(9)), $g_n(x)$ is an integral function of degree $\leqslant \frac{1}{n}\left(\left[\frac{n}{2}\right]+1\right)\sigma$. Moreover, if $\sup_{-\infty<x<\infty} |G(x)| = M$ then as a result of the inequality 4.10(10)

$$\sup_{-\infty<x<\infty} |g_n(x)|^2 \leqslant M. \tag{13}$$

The estimate 4.11(13) makes it possible to use Vitali's theorem and choose a subsequence $g_{n_k}(z)$ which converges for all $z = x+iy$. It follows from sections 4.8.3 and 4.11(13) that the limit function

$$g(z) = \lim_{n_k\to\infty} g_{n_k}(z)$$

is an integral function of degree $\leqslant \frac{\sigma}{2}$. On the other hand, by what was proved in section 4.10.3,

$$G(x) = \lim_{n_k\to\infty} T_{m_k,0}(G_\sigma; x) = \lim_{n_k\to\infty} |g_{n_k}(x)|^2 = |g(x)|^2.$$

Since it can be assumed that the function $g_n(z)$ has no zeros in the lower half-plane (see section 4.11.4), we conclude that in this case the function $g(z)$ possesses the same property.

** M. G. Krein [3]. This investigation was later continued by N. I. Akhiezer [4], who obtained necessary and sufficient conditions for an integral function of finite degree σ which is non-negative on the real axis to be representable for all real values of the variable as the square of the modulus of a certain integral function of degree $\sigma/2$.

4.11.6. Here we will indicate an application of theorem 4.11.5 to the extremal problems considered in section 4.8, and give the following theorem**. If

$$G_\sigma(x) = c_0+c_1x+c_2x^2+\ldots \tag{14}$$

is an integral function of degree $\leqslant \sigma$ *which is monotonic on the whole real axis, then*

$$\sup_{-\infty<x<\infty} |G_\sigma(x)| \geqslant \frac{|c_1|\pi}{\sigma}. \tag{15}$$

Supposing that $G_\sigma(x)$ is non-decreasing and $\sup\limits_{-\infty<x<\infty} |G_\sigma(x)| < \infty$, let us examine the function $G'_\sigma(x)$, which is non-negative and in virtue of 4.8(5) belongs to B_σ. Without loss of generality it is also possible to assume that $G_\sigma(-x) = -G_\sigma(x)$ and $G'_\sigma(0) = 1$. Theorem 4.11.5, which is applicable to the function $G'_\sigma(x)$, gives

$$G'_\sigma(x) = |g_{\frac{\sigma}{2}}(x)|^2.$$

Since

$$G_\sigma(x) = \int_0^x |g_{\frac{\sigma}{2}}(t)|^2 \mathrm{d}t$$

and $G_\sigma(x) \in B_\sigma$, it follows that

$$\int_{-\infty}^{\infty} |g_{\frac{\sigma}{2}}(t)|^2 \,\mathrm{d}t < \infty.$$

Thus the function $g_{\frac{\sigma}{2}}(x)$ can be represented in the form (see theorem 4.6.1)

$$g_{\frac{\sigma}{2}}(x) = \frac{1}{\sqrt{2\pi}} \int_{-\frac{\sigma}{2}}^{\frac{\sigma}{2}} \phi(t) \mathrm{e}^{itx} \mathrm{d}t. \tag{16}$$

Then (see 4.6(4))

$$\int_{-\infty}^{\infty} |g_{\frac{\sigma}{2}}(t)|^2 \mathrm{d}t = \int_{-\frac{\sigma}{2}}^{\frac{\sigma}{2}} |\phi(t)|^2 \mathrm{d}t.$$

** S. N. Bernstein [28].

Since by supposition $G_\sigma(x)$ is odd, the derivative $G'_\sigma(x)$ is even, i.e. $|g_{\frac{\sigma}{2}}(-x)|^2 = |g_{\frac{\sigma}{2}}(x)|^2$. Consequently

$$\sup_{-\infty<x<\infty} |G_\sigma(x)| = \int_0^\infty |g_{\frac{\sigma}{2}}(t)|^2\,dt = \frac{1}{2}\int_{-\frac{\sigma}{2}}^{\frac{\sigma}{2}} |\phi(t)|^2\,dt.$$

But by the Bunyakovskii–Schwarz inequality it follows from 4.11(16) that

$$\int_{-\frac{\sigma}{2}}^{\frac{\sigma}{2}} |\phi(t)|^2\,dt \geqslant \frac{1}{\sigma}\left|\int_{-\frac{\sigma}{2}}^{\frac{\sigma}{2}} \phi(t)\,dt\right|^2 = \frac{2\pi}{\sigma}\,|g_{\frac{\sigma}{2}}(0)|^2 = \frac{2\pi}{\sigma}\,G'_\sigma(0).$$

Hence by the condition $G'_\sigma(0) = 1$ we obtain the estimate

$$\sup_{-\infty<x<\infty} |G_\sigma(x)| \geqslant \frac{\pi}{\sigma}. \tag{17}$$

It should be noted that the result obtained is definitive, as in the case where $\phi(t) \equiv \frac{\sqrt{2\pi}}{\sigma}$, i.e.

$$g_{\frac{\sigma}{2}}(x) = \frac{1}{\sigma}\int_{-\frac{\sigma}{2}}^{\frac{\sigma}{2}} e^{itx}\,dt = \frac{2}{\sigma x}\sin\frac{\sigma x}{2},$$

the two sides of 4.11(17) are equal.

4.11.61. As in section 4.8.2, the statement just proved leads directly to a new inequality** for the derivatives of integral functions of the class B_σ on the assumption that they are monotonic. *If the integral function $G_\sigma(x)$ of degree $\leqslant \sigma$ is monotonic on the whole real axis, then*

$$\sup_{-\infty<x<\infty} |G'_\sigma(x)| \leqslant \frac{\sigma}{\pi} \sup_{-\infty<x<\infty} |G_\sigma(x)|. \tag{18}$$

This inequality forms an essential supplement to 4.8(5).

4.12. Various problems and theorems

1. Let $P_n(x)$ be an algebraic polynomial of degree $n \geqslant 2$ which has only real zeros and is such that $P_n(x) \neq 0$ for $-1 < x < 1$, $P_n(+1) = P_n(-1) = 0$, $\max\limits_{-1\leqslant x\leqslant 1} |P_n(x)| = 1$.

** S. N. Bernstein [28].

If $-1 < a < b < 1$ and $P_n(a) = P_n(b) = d$, then $b-a \leqslant 2\sqrt{1-|d|}$. This inequality is precise and becomes an equality only for the polynomial $P_2(x) = 1-x^2$.

P. Erdös [1].

2. If $P_n(x)$ is a polynomial which satisfies the conditions of the preceding proposition, then

$$\int_{-1}^{1} P_n(x)\,\mathrm{d}x \leqslant \tfrac{4}{3}. \tag{1}$$

This inequality is precise and becomes an equality only for the polynomial $P_2(x) = 1-x^2$.

P. Erdös.

3. Let $x_0 < x_1 < \ldots < x_n$ be arbitrary integral-valued points of the real axis. Among the values of any polynomial 2.9(1) at these points there is at least one which in absolute value is greater than or equal to $\dfrac{n!}{2^n}$.

G. Pólya and G. Szegö [1]. Make use of formula 4.1(4).

4. Let $m = 2\left[\dfrac{n+2}{2}\right]$ and

$$K_n(t) = \frac{a_0}{2} + \sum_{k=1}^{n} a_k \cos kt \tag{2}$$

be an arbitrary even trigonometric polynomial. It satisfies the inequality

$$\inf_{\substack{a_{2j+1} \\ 1|\leqslant 2j+1|\leqslant n}} \int_0^\pi |K_n(t)|\,\mathrm{d}t \geqslant \frac{2}{\pi} \left| \sum_{\nu=0}^{\infty} \frac{1}{2\nu+1} \sum_{k=\nu m+1}^{(\nu+1)m-1} \frac{a_{|(2\nu+1)m-2k|}}{k} \right|, \tag{3}$$

in which the constant $\dfrac{2}{\pi}$ cannot be replaced by a greater one for the whole set of values $n = 0, 1, 2, \ldots$

A. F. Timan [17].

5. Let $m = 2\left[\dfrac{n+1}{2}\right]+1$ and $K_n(t)$ be the arbitrary trigonometric polynomial 4.12(2). The inequality

$$\inf_{\substack{a_{2j} \\ 0\leqslant 2j\leqslant n}} \int_0^\pi |K_n(t)|\,\mathrm{d}t \geqslant \frac{2}{\pi} \left| \sum_{\nu=0}^{\infty} \frac{1}{2\nu+1} \sum_{k=\nu m+1}^{(\nu+1)m-1} \frac{a_{|(2\nu+1)m-2k|}}{k} \right|, \tag{4}$$

is satisfied, where the constant $\dfrac{2}{\pi}$ cannot be replaced by a greater one for the whole set of values of n.

A. F. Timan [17].

6. If $D_n(t)$ is a Dirichlet kernel (see 4.1(10)), then as $n \to \infty$

$$\int_0^\pi |D_n(t)|\,\mathrm{d}t = \frac{2}{\pi} \ln n + O(1). \tag{5}$$

Fejér [2]. Taking into account that for $0 < t \leqslant \frac{\pi}{2}$ the difference $\frac{1}{\sin t} - \frac{1}{t}$ is bounded, we have

$$\int_0^{\pi} |D_n(t)|\,dt = \frac{1}{2}\int_0^{\pi} \frac{\left|\sin\left(n+\frac{1}{2}\right)t\right|}{\sin\frac{t}{2}}\,dt = \int_0^{\pi} \frac{\left|\sin\left(n+\frac{1}{2}\right)t\right|}{t}\,dt + O(1)$$

$$= \sum_{k=0}^{n-1} (-1)^k \int_{\frac{2k\pi}{2n+1}}^{\frac{(2k+1)\pi}{2n+1}} \frac{\sin\left(n+\frac{1}{2}\right)t}{t}\,dt + O(1)$$

$$= \int_0^{\frac{2\pi}{2n+1}} \sum_{k=0}^{n-1} \frac{\sin\left(n+\frac{1}{2}\right)t}{t+\frac{2k\pi}{2n+1}}\,dt + O(1)$$

$$= \frac{2n+1}{2\pi} \int_0^{\frac{2\pi}{2n+1}} \sin\left(n+\frac{1}{2}\right)t \sum_{k=1}^{n-1} \frac{1}{k}\,dt + O(1) = \frac{2}{\pi}\ln n + O(1). \tag{6}$$

7. If $\sigma \geqslant 1$, then

$$\int_{-\infty}^{\infty} \frac{|\cos(\sigma-1)t - \cos\sigma t|}{t^2}\,dt = \frac{4}{\pi}\ln\sigma + O(1). \tag{7}$$

8. If $y_k^{(n)}$ is the system of points 4.2(7), then

$$\frac{2}{2n+1}\sum_{k=0}^{2n} \left|D_n(x-y_k^{(n)})\right| = \frac{2}{\pi}\left|\sin\frac{2n+1}{2}x\right|\ln n + O(1). \tag{8}$$

S. M. Nikol'skii [1].

9. For $n \to \infty$ and $|y| \leqslant \pi$ the asymptotic equation

$$\int_0^{\pi} |D_n(t+y) + D_n(t-y)|\,dt = \frac{2}{\pi}\ln\frac{1+n|y|}{\left(|y|+\frac{1}{n}\right)|\cos ny|} + O(1) \tag{9}$$

is satisfied uniformly with respect to y.

A. F. Timan [2].

10. If $t_k^{(n)}$ are the points 4.2(12), then for $|y| \leqslant \pi$

$$\frac{1}{n}\sum_{k=0}^{n-1} \left|D_n(t_k^{(n)}+y) + D_n(t_k^{(n)}-y)\right| = \frac{2}{\pi}|\cos ny|\ln n + O(1) \tag{10}$$

uniformly with respect to y.

11. If for the function $f(x)$, defined on the whole real axis $-\infty < x < \infty$, we have for some $q(1 \leqslant q \leqslant \infty)$

$$\left\{ \int_{-\infty}^{\infty} |f(x)|^q \, dx \right\}^{\frac{1}{q}} < \infty,$$

then for every natural number r, $\sigma > 0$ and $0 < \theta \leqslant 1$, the integrals

$$\int_{-\infty}^{\infty} f(t) \left[\frac{\sin \frac{\sigma}{2r}(n-t)}{x-t} \right]^{2r} dt,$$

$$\frac{1}{\sigma\theta\pi} \int_{-\infty}^{\infty} f(t) \frac{\cos \sigma(1-\theta)(x-t) - \cos \sigma(x-t)}{(x-t)^2} dt \tag{11}$$

are integral functions of finite degree $\leqslant \sigma$. If $f(x)$ belongs to the class $B_{\sigma(1-\theta)}$, then the second of these integrals is identically equal to $f(x)$.

Expand the integrand in a series of powers of x and make use of the inequality 4.8(28). For the proof of the second assertion use theorem 4.6.2 (see 4.6(10)) with the identity

$$\frac{\cos(1-\theta)x - \cos x}{x^2} = \frac{1}{2\theta} \int_{-\infty}^{\infty} \phi(t) \cos xt \, dt$$

where

$$\phi(t) = \begin{cases} 1, & |t| \leqslant 1-\theta, \\ \frac{1}{\theta}(1-|t|), & 1-\theta \leqslant |t| \leqslant 1, \\ 0, & |t| \geqslant 1, \end{cases}$$

and apply formula 3.11(16).

12. The estimate

$$\sum_{k=1}^{n} \frac{\sin kx}{k} = O(1) \tag{12}$$

holds uniformly with respect to x and n.

If $0 < \frac{2p\pi}{2n+1} \leqslant x < \frac{2(p+1)\pi}{2n+1} < \pi$, then

$$\sum_{k=1}^{n} \frac{\sin kx}{k} = \int_0^x \left(\frac{1}{2} + \sum_{k=1}^{n} \cos kt \right) dt + O(1)$$

$$= \frac{1}{2} \int_0^x \frac{\sin \frac{2n+1}{2} t}{\sin \frac{t}{2}} dt + O(1) = \int_0^x \frac{\sin \left(n+\frac{1}{2}\right) t}{t} dt + O(1) =$$

$$= \sum_{k=0}^{p} \int_{\frac{2k\pi}{2n+1}}^{\frac{2(k+1)\pi}{2n+1}} \frac{\sin\left(n+\frac{1}{2}\right)t}{t}\,dt + O(1)$$

$$= \int_{0}^{\frac{2\pi}{2n+1}} \sum_{k=0}^{p} \frac{(-1)^k \sin\left(n+\frac{1}{2}\right)t}{t+\frac{2k\pi}{2n+1}}\,dt + O(1)$$

$$= \frac{2n+1}{2\pi} \int_{0}^{\frac{2\pi}{2n+1}} \sin\frac{2n+1}{2}t\,dt \sum_{k=1}^{p} \frac{(-1)^k}{k} + O(1) = O(1).$$

13. If $|x| \leqslant \pi$, then

$$\sum_{k=1}^{n} \frac{\cos kx}{k} = \min\left\{\ln\frac{1}{|x|}, \ln n\right\} + O(1) \tag{13}$$

uniformly with respect to x and n.

We have

$$\sum_{k=1}^{n} \frac{\cos kx}{k} = -\int_{0}^{x} \sum_{k=1}^{n} \sin kt\,dt + \ln n + O(1).$$

$$= \ln n - \int_{0}^{x} \frac{\cos\frac{t}{2} - \cos\frac{2n+1}{2}t}{2\sin\frac{t}{2}}\,dt + O(1),$$

and if $0 \leqslant 2p\pi < nx \leqslant 2(p+1)\pi$

$$\int_{0}^{x} \frac{\cos\frac{t}{2} - \cos\frac{2n+1}{2}t}{2\sin\frac{t}{2}}\,dt = 2\int_{0}^{x} \frac{\sin^2\frac{nt}{2}}{t}\,dt + O(1)$$

$$= 2\int_{0}^{nx} \frac{\sin^2\frac{t}{2}}{t}\,dt + O(1) = 2\sum_{k=1}^{p+1} \int_{2k\pi}^{2(k+1)\pi} \frac{\sin^2\frac{t}{2}}{t}\,dt + O(1)$$

$$= 2\int_{0}^{2\pi} \sin^2\frac{t}{2} \sum_{k=1}^{p+1} \frac{1}{t+2k\pi}\,dt + O(1) = \frac{1}{\pi}\int_{0}^{2\pi} \sin^2\frac{t}{2}\,dt \sum_{k=1}^{p+1} \frac{1}{k} + O(1)$$

$$= \ln(p+1) + O(1) = \ln(1+nx) + O(1).$$

14. For any algebraic polynomial $P_n(x)$ of degree $n \geqslant 2$ for which $P_n(a) = P_n(b) = 0$ $(-\infty < a < b < \infty)$, the inequality

$$\max_{a \leqslant x \leqslant b} |P_n'(x)| \leqslant \frac{2n\cot\frac{\pi}{2n}}{b-a} \max_{a \leqslant x \leqslant b} |P_n(x)| \tag{14}$$

is satisfied.

I. Schur [1].

15. If the algebraic polynomial $P_n(x)$ of degree n has only real zeros, situated outside the interval (a, b), then

$$\max_{a\leqslant x\leqslant b} |P_n'(x)| < \frac{en}{b-a} \max_{a\leqslant x\leqslant b} |P_n(x)| \tag{15}$$

This estimate cannot be improved. P. Erdös [2].

16. If all the coefficients of the algebraic polynomial

$$P_n(x) = \sum_{k=0}^{n} c_k x^k$$

of degree $\leqslant n$ are real and non-negative, then for any $x > 0$

$$P_n'(x) \leqslant \frac{n}{x} P_n(x). \tag{16}$$

With the supplementary condition $P_n(0) = 0$, the inequality

$$P_n'(x) \geqslant \frac{1}{x} P_n(x) \tag{17}$$

is satisfied for $x > 0$. If $c_k > 0 (k = 0, 1, \ldots, n)$ and $\frac{c_k}{c_k+1} \geqslant \delta > 0$ $(k = 0, 1, \ldots, n)$, then for values of x such that $0 < x \leqslant \delta$,

$$P_n'(x) \leqslant \frac{n}{2x} P_n(x). \tag{18}$$

The inequalities 4.12(16) and 4.12(17) become equalities for the polynomials $P_n(x) = x^n$ and $P_n(x) = x$ respectively. For $x = 1$ the polynomial $P_n(x) = \sum_{k=0}^{n} x^k$ the inequality 4.12(18) becomes an equality.

A. Soble [1].

17. S. N. Bernstein's inequality for the derivative of a trigonometric polynomial can be extended to the case of fractional derivatives. For any non-negative r there exists a constant $C(r)$ such that, for every trigonometric polynomial $T_n(t)$ of order $\leqslant n$ the inequality

$$\max_{n} |T_n^{(r)}(x)| \leqslant C(r) n^r \max_{x} |T_n(x)| \tag{19}$$

is satisfied.

Civin [1].

18. The first of the inequalities of 4.8(18) can be obtained by passing to the limit as $\delta \to 0$ in the following assertion. If $0 < \delta < \mathrm{h} < \frac{\pi}{n}$, then for every trigonometric polynomial $T_n(x)$ of order $\leqslant n$

$$\left(\frac{n}{2\sin n\delta}\right)^r \max_{x} |\Delta_\delta^r T_n(x)| \leqslant \left(\frac{n}{2\sin nh}\right)^r \max_{x} |\Delta_h^r T_n(x)| \tag{20}$$

for any natural number r.

Apply the inequality 4.8(29) to the function $\mu(t) = \left(\frac{\sin \delta t}{\sin ht}\right)^r$.

Boas [2].

19. If $f(x)$ is an integral function of degree $\leqslant \pi$, which possesses the property that $f(x) = o|x|$ as $|x| \to \infty$, and $R(f; z)$ is the transform 4.4(2), where $\varrho(t)$ satisfies the conditions 4.4(1) and 4.4(3) for $\sigma = \pi$, then

$$\sup_{-\infty<x<\infty} |R(f; x)| \leqslant \sup_{u} \sum_{k=-\infty}^{\infty} |c_k(u)| \sup_{n} |f(n)|,$$

where $c_k(u)$ are the Fourier coefficients of the function $\int_{-\infty}^{\infty} e^{i(u+t)x} \, d\varrho(t)$.

Boas [1]. This estimate cannot be improved.

20. If $P_n(x)$ is an algebraic polynomial of degree $\leqslant n$, then for $|x| \leqslant 1$

$$|P_n'(x)| \leqslant \frac{n^2}{2} \omega\left(P_n; 2\sin\frac{\pi}{2n}\right) \tag{21}$$

and for $|x| < 1$

$$|P_n'(x)| \leqslant \frac{n}{2\sqrt{1-x^2}} \omega\left(P_n; 2\sin\frac{\pi}{2n}\right). \tag{22}$$

The inequality 4.12(22) follows directly from 4.8(18) for $r=1$. In order to obtain 4.12(21), see section 4.8.7. S. B. Stechkin [1].

21. For $k=1$ the inequalities 4.8(6) and 4.8(5) are included in the following assertion. If $G_\sigma(x)$ is an integral function of finite degree $\leqslant \sigma$, then for any real α and any q ($1 \leqslant \leqslant q \leqslant \infty$)

$$\left\{ \int_{-\infty}^{\infty} |\sin \alpha G_\sigma'(x) + \cos \alpha \tilde{G}_\sigma'(x)|^q dx \right\}^{\frac{1}{q}} \leqslant \sigma \left\{ \int_{-\infty}^{\infty} |G_\sigma(x)|^q \, dx \right\}^{\frac{1}{q}}, \tag{23}$$

$$\left\{ \int_{-\infty}^{\infty} |\sin \alpha G_\sigma'(x) - \sigma \cos \alpha G_\sigma(x)|^q dx \right\}^{\frac{1}{q}} \leqslant \sigma \left\{ \int_{-\infty}^{\infty} |G_\sigma(x)|^q \, dx \right\}^{\frac{1}{q}}. \tag{24}$$

N. I. Akhiezer [3].

22. If $P_n(x)$ is an algebraic polynomial of degree n, all the zeros of which lie on the segment $[-1, 1]$, then

$$\max_{-1\leqslant x\leqslant 1} |P_n'(x)| > \tfrac{1}{6} \sqrt{n} \max_{-1\leqslant x\leqslant 1} |P_n(x)|. \tag{25}$$

P. Turan [1].

23. If $G_\sigma(x)$ is an integral function of degree $\leqslant \sigma$ and $\int_{-\infty}^{\infty} |G_\sigma(x)|^2 \, dx < \infty$, then for $q \geqslant 0, n \geqslant 0$,

$$\left\{ \int_{-\infty}^{\infty} |G_\sigma^{(n)}(x)|^q \, dx \right\}^{\frac{1}{q}} \leqslant \left[\frac{\sigma}{(2n+1)\pi} \right]^{\frac{q-2}{2q}} \sigma^n \left\{ \int_{-\infty}^{\infty} |G_\sigma(x)|^2 \, dx \right\}^{\frac{1}{2}}. \tag{26}$$

I. I. Ibragimov [3], in the case $q=\infty$. Use theorem 4.6.1.

24. Let $G_{\sigma_1,\dots,\sigma_m}(x_1,\dots,x_m)$ be an integral function which is of finite degree with respect to each of the variables (see section 4.9.55). For any q $(1 \leqslant q \leqslant \infty)$

$$\left\{\int_{-\infty}^{\infty}\cdots\int_{-\infty}^{\infty}\left|\frac{\partial^{k_1+\dots k_m}G_{\sigma_1,\dots,\sigma_m}(x_1,\dots,x_m)}{\partial x_1^{k_1}\dots\partial x_m^{k_m}}\right|^q dx_1\dots, dx_m\right\}^{\frac{1}{q}}$$

$$\leqslant \prod_{\nu=1}^{m}\sigma_\nu^{k_\nu}\left\{\int_{-\infty}^{\infty}\cdots\int_{-\infty}^{\infty}G_{\sigma_1,\dots,\sigma_m}(x_1,\dots,x_m)|^q\,dx_1,\dots,dx_m\right\}^{\frac{1}{q}}. \qquad (27)$$

S. N. Bernstein [37], S. M. Nikol'skii [13].

25. If the integral function $G_\sigma(x)$ is such that as $|x|\to\infty$

$$G_\sigma(x) = O(|x|^{2r}), \qquad (28)$$

then for its deviation from the trigonometric polynomial 4.10(8) we have the estimate

$$G_\sigma(x)-T_{m,r}(G_\sigma;x) = O\left\{x^{2r}\left[-1\left(n\frac{\sin\frac{\sigma x}{n}}{\sigma x}\right)^2\right]\right\} \qquad (29)$$

which holds uniformly for all x on the real axis.

N. I. Akhiezer and V. A. Marchenko [1].

CHAPTER V

DIRECT THEOREMS OF THE CONSTRUCTIVE THEORY OF FUNCTIONS

5.1. The effect of the differential properties of functions on the rapidity of decrease to zero of their best approximations. Jackson's theorem

In chapter II it was noted (sections 2.1, 2.2.1, 2.2.2, 2.2.4, 2.6.1, 2.6.2) that, thanks to the concept of the best approximation, the fundamental theorem of Weierstrass (theorem 1.1), and other propositions of the same type, obtained in chapter I, could be formulated as theorems on the approach to zero of best approximations. Of fundamental significance for the constructive theory of functions was the discovery by Jackson** of the possibility of materially improving the classical theorem of Weierstrass and the new formulation of it in the form of the inequality

$$E_n(f) = E_n(f; a, b) \leqslant C\omega\left(f; \frac{b-a}{n}\right), \tag{1}$$

where C is an absolute constant, $f(x)$ is an arbitrary function continuous on the given finite segment $[a, b]$, and $E_n(f)$ and $\omega(f; t)$ are respectively the best approximation by the algebraic polynomials $P_n(x)$ of degree $\leqslant n$ (see 2.1(3)) and the modulus of continuity of $f(x)$ on $[a,b]$ (see section 3.1.1).

5.1.1. Let us give a simple proof of the inequality 5.1(1). Without loss of generality we can take $a = -1$, $b = 1$. Let us consider the sequence of trigonometric polynomials

$$T_{2n-2}(u) = \left(\frac{\sin \dfrac{nu}{2}}{n \sin \dfrac{u}{2}}\right)^4 \tag{2}$$

** D. Jackson [1].

of order $2n-2$ (see 4.7(10)). The integral

$$\frac{1}{\pi\gamma_n^{(0)}}\int_0^{\pi} f(\cos t)\,[T_{2n-2}(t+y)+T_{2n-2}(t-y)]\,dt, \tag{3}$$

where $\gamma_n^{(0)}$ is determined from 4.8(11), represents a certain even trigonometric polynomial of order $\leqslant 2n-2$. Hence if we put $x = \cos y$, then for $x\in[-1, 1]$ the given integral is an algebraic polynomial $Q_{2n-2}(f; x)$ of degree $\leqslant 2n-2$. Since

$$\frac{1}{\pi\gamma_n^{(0)}}\int_0^{\pi}\{T_{2n-2}(t+y)+T_{2n-2}(t-y)\}\,dt = 1$$

and consequently

$$f(x) = \frac{1}{\pi\gamma_n^{(0)}}\int_0^{\pi} f(\cos y)\{T_{2n-2}(t+y)+T_{2n-2}(t-y)\}\,dt,$$

it follows that for the difference between the integral 5.1(3) and $f(x)$ we obtain the estimate

$$|f(x)-Q_{2n-2}(f; x)|$$
$$\leqslant \frac{1}{\pi\gamma_n^{(0)}}\int_0^{\pi}|f(\cos t)-f(\cos y)|\,\{T_{2n-2}(t+y)+T_{2n-2}(t-y)\}\,dt.$$

It follows from this that for any $x\in[-1, 1]$

$$|f(x)-Q_{2n-2}(f; x)|$$
$$\leqslant \frac{1}{2\pi\gamma_n^{(0)}}\int_{-\pi}^{\pi}\omega(|\cos t-\cos y|)\,\{T_{2n-2}(t+y)+T_{2n-2}(t-y)\}\,dt, \tag{4}$$

where $\omega(h) = \omega(f; h)$. Let us evaluate the integral on the right-hand side of the last inequality. We have

$$I_n(\omega; \pm y) = \frac{1}{2\pi n^4\gamma_n^{(0)}}\int_{-\pi}^{\pi}\omega\left(2\left|\sin\frac{t+y}{2}\sin\frac{t-y}{2}\right|\right)\left(\frac{\sin n\dfrac{t\pm y}{2}}{\sin\dfrac{t\pm y}{2}}\right)^4 dt$$

$$= \frac{1}{\pi n^4\gamma_n^{(0)}}\int_{-\frac{\pi}{2}}^{\frac{\pi}{2}}\omega(2)\sin t\sin(t\pm y)\left(\frac{\sin nt}{\sin t}\right)^4 dt.$$

In virtue of the properties of the modulus of continuity (see section 3.2)

$$I_n(\omega;\pm y)\leqslant\frac{4}{\pi n^4\gamma_n^{(0)}}\left\{\int_0^{\frac{\pi}{2}}\omega(t^2)\left(\frac{\sin nt}{\sin t}\right)^4 dt+\right.$$

$$\left.+\int_0^{\frac{\pi}{2}}\omega(t|\sin y|)\left(\frac{\sin nt}{\sin t}\right)^4 dt\right\}.$$

Moreover,

$$\int_0^{\frac{\pi}{2}}\omega(t^2)\left(\frac{\sin nt}{\sin t}\right)^4 dt<\left(\frac{\pi}{2}\right)^4\int_0^{\frac{\pi}{2}}\omega(t^2)\left(\frac{\sin nt}{\sin t}\right)^4 dt$$

$$<\left(\frac{\pi}{2}\right)^4 n^3\omega\left(\frac{1}{n^2}\right)\int_0^{\infty}(t^2+1)\left(\frac{\sin t}{t}\right)^4 dt$$

and similarly

$$\int_0^{\frac{\pi}{2}}\omega(t|\sin y|)\left(\frac{\sin nt}{\sin t}\right)^4 dt<\left(\frac{\pi}{2}\right)^4 n^3\omega\left(\frac{|\sin y|}{n}\right)\int_0^{\infty}(t+1)\left(\frac{\sin t}{t}\right)^4 dt.$$

Consequently, if we put

$$M_0=\int_0^{\infty}\left\{\left|\frac{\sin t}{t}\right|^2+\left|\frac{\sin t}{t}\right|^3+\left|\frac{\sin t}{t}\right|^4\right\}dt,$$

then

$$I_n(\omega;\pm y)<\frac{\pi^3 M_0}{4n\gamma_n^{(0)}}\left[\omega\left(\frac{|\sin y|}{n}\right)+\omega\left(\frac{1}{n^2}\right)\right].$$

Using the estimate from below for $\gamma_n^{(0)}$ derived in section 4.8.72, we obtain

$$I_n(\omega;\pm y)\leqslant M\left[\omega\left(\frac{|\sin y|}{n}\right)+\omega\left(\frac{1}{n^2}\right)\right],\tag{5}$$

where M is an absolute constant. Substituting in the inequality 5.1(4) we find

$$|f(x)-Q_{2n-2}(f;x)|\leqslant 2M\left[\omega\left(\frac{\sqrt{1-x^2}}{n}\right)+\omega\left(\frac{1}{n^2}\right)\right].\tag{6}$$

i.e. for all $n = 1, 2, \ldots$

$$E_{2n-2}(f) \leqslant \max_{-1 \leqslant x \leqslant 1} |f(x) - Q_{2n-2}(f; x)| \leqslant 4M\omega\left(\frac{1}{n}\right). \tag{7}$$

Since $E_{2n-1}(f) \leqslant E_{2n-2}(f)$, this proves Jackson's inequality 5.1(1) for the segment $[-1,1]$.

5.1.2. An inequality similar to 5.1(1) holds also for the best approximation $A_\sigma(f)$ of a function $f(x)$, bounded and uniformly continuous on the whole real axis (see section 2.6.2)**.

In fact, if the bounded function $f(x)$ is uniformly continuous on $(-\infty, \infty)$ and $\omega(f;t)$ is its modulus of continuity, then by the inequality 5.1(1), for any $\sigma > 0$,

$$E_n\left(f; -\frac{n}{\sigma}, \frac{n}{\sigma}\right) \leqslant C\omega\left(f; \frac{2}{\sigma}\right). \tag{8}$$

On making use of the inequality 2.6(12), we obtain from this

$$A_\sigma(f) \leqslant 2C\omega\left(f; \frac{1}{\sigma}\right). \tag{9}$$

In particular, if $f(x)$ is a continuous periodic function of period 2π, then for its best approximation by trigonometric polynomials of order $\leqslant n$ (see sections 2.2.1 and 2.6.21) the estimate

$$E_n^*(f) \leqslant 2C\omega\left(f; \frac{1}{n+1}\right) \tag{10}$$

applies.

5.1.3. The inequality 5.1(9), which is an improvement on theorem 1.7.1, is a particular case of the following more general assertion. *If* $\sup_{-\infty<x<\infty} [f(x)] < \infty$, *then for any integral value of* k *and* $\sigma > 0$ *the inequality****

$$A_\sigma(f) \leqslant C_k\omega_k\left(f; \frac{1}{\sigma}\right), \tag{11}$$

is satisfied, where C_k *is a constant depending only on* k, *and* $\omega_k(f; t)$ *is the corresponding modulus of smoothness.*

** S. N. Bernstein [23].

*** N. I. Akhiezer [3], in the case $k=2$.

In order to prove this let us consider for some integral $r \geqslant \frac{1}{2}(k+2)$ the integral function

$$g_\sigma(x) = \left(\frac{\sin \dfrac{\sigma x}{2r}}{x}\right)^{2r} \tag{12}$$

and put

$$\gamma_{\sigma,r} = \int_{-\infty}^{\infty} \left(\frac{\sin \dfrac{\sigma t}{2r}}{t}\right)^{2r} \mathrm{d}t. \tag{13}$$

The integral

$$\frac{1}{\gamma_{\sigma,r}} \int_{-\infty}^{\infty} g_\sigma(t) \sum_{\nu=1}^{k} (-1)^{\nu-1} \binom{k}{\nu} f(x+\nu t)\,\mathrm{d}t = \frac{1}{\gamma_{\sigma,r}} \int_{-\infty}^{\infty} G_{\sigma,k}(x-t) f(t)\,\mathrm{d}t,$$

where

$$G_{\sigma,k}(u) = \sum_{\nu=1}^{k} (-1)^{\nu-1} \frac{1}{\nu} \binom{k}{\nu} g_\sigma\left(\frac{u}{\nu}\right),$$

is an integral function $Q_\sigma(f;x)$ of finite degree $\leqslant \sigma$**. For the difference between $f(x)$ and $Q_\sigma(f;x)$ we have the estimate

$$|f(x)-Q_\sigma(f;x)| \leqslant \frac{1}{\gamma_{\sigma,r}} \int_{-\infty}^{\infty} g_\sigma(t) \left| \sum_{\nu=0}^{k} (-1)^{k-\nu} \binom{k}{\nu} f(x+\nu t) \right| dt$$

$$\leqslant \frac{1}{\gamma_{\sigma,r}} \int_{-\infty}^{\infty} g_\sigma(t)\, \omega_k(f;t)\,dt.$$

** We note that $Q_\sigma(f;x)$ is an integral function of degree $\leqslant \sigma$ whenever $f(x) = O(|x|^{2r-2}+1)$. To verify this it is sufficient to consider the integral $F(z) = \frac{1}{\gamma_{\sigma,r}} \int_{-\infty}^{\infty} g_\sigma(z-t) f(t)\,\mathrm{d}t$. The function $F(z)$ is the limit of the sequence of integrals

$$\frac{1}{\gamma_{\sigma,r}} \int_{-n}^{n} g_\sigma(z-t) f(t)\,\mathrm{d}t \qquad (n = 1, 2, \ldots)$$

which converges uniformly in any finite part of the complex plane of $z=x+iy$. Since

$$\frac{1}{\gamma_{\sigma,r}} \int_{-n}^{n} g_\sigma(z-t) f(t)\,\mathrm{d}t = \sum_{k=0}^{\infty} \frac{c_k^{(n)}}{k!} z^k,$$

Using the properties of the modulus of smoothness (see 3.3,(5)) and 3.12(12),
we obtain from this

$$|f(x)-Q_\sigma(f,x)| \leqslant \frac{\sigma^k}{\gamma_{\sigma,r}}\omega_k\left(f;\frac{1}{\sigma}\right)\int_{-\infty}^{\infty} g_\sigma(t)\left(t+\frac{1}{\sigma}\right)^k dt$$

$$\leqslant \omega_k\left(f;\frac{1}{\sigma}\right)\left\{\frac{2^k}{\gamma_{\sigma,r}}\int_{-\frac{1}{\sigma}}^{\frac{1}{\sigma}} g_\sigma(t)\,dt+\frac{2^k\sigma^k}{\gamma_{\sigma,r}}\int_{|t|\geqslant\frac{1}{\sigma}} t^k g_\sigma(t)\,dt\right\}.$$

Since $2r \geqslant k+2$, we have

$$\frac{\sigma^k}{\gamma_{\sigma,r}}\int_{-\infty}^{\infty} t^k\left(\frac{\sin\frac{\sigma t}{2r}}{t}\right)^{2r} dt = O(1),$$

uniformly with respect to σ. Consequently

$$A_\sigma(f) \leqslant \sup_{-\infty<x<\infty}|f(x)-Q_\sigma(f,x)| \leqslant C_k\omega_k\left(f;\frac{1}{\sigma}\right).$$

5.1.31. If for some $q \geqslant 1$

$$\left\{\int_{-\infty}^{\infty}|f(x)|^q dx\right\}^{\frac{1}{q}} < \infty, \tag{14}$$

then, applying Minkowski's inequality for integrals we obtain

$$\left\{\int_{-\infty}^{\infty}|f(x)-Q_\sigma(f;x)|^q dx\right\}^{\frac{1}{q}} \leqslant \frac{1}{\gamma_{\sigma,r}}\int_{-\infty}^{\infty} g_\sigma(t)\omega_k(f;t)_{L_q}dt.$$

---

where

$$|c_k^{(n)}| = \frac{1}{\gamma_{\sigma,r}}\left|\int_{-n}^{n} g_\sigma^{(k)}(t)f(t)\,dt\right| = O(n^{2r-2}\sigma^k),$$

each of these integrals is an integral function. Hence $F(z)$ is itself an integral function. Thus (see section 4.8.3)

$$|F(z)| \leqslant \frac{1}{\gamma_{\sigma,r}}\left\{\int_{|x-t|\leqslant 1}|g_\sigma(z-t)f(t)|\,dt+\int_{|x-t|>1}|g_\sigma(z-t)f(t)|\,dt\right\}$$

$$\leqslant \frac{e^{\sigma|y|}}{\gamma_{\sigma,r}}\left\{\int_{|x-t|\leqslant 1}|f\ t)|\,dt+\int_{|x-t|>1}\frac{|f(t)|}{|x-t|^{2r}}dt\right\} \leqslant C_{\sigma,r}|x|^{2r-2}e^{\sigma|y|}.$$

Consequently, whatever $\varepsilon > 0$ may be, for all sufficiently great values of $|z|$ the inequality $|F(z)| \leqslant e^{(\sigma+\varepsilon)|z|}$ is satisfied, that is, $F(z)$ is an integral function of degree $\leqslant .\sigma$

Hence as in section 5.1.3 we conclude that** for any $q(1 \leqslant q \leqslant \infty)$

$$A_\sigma(f)_{Lq} \leqslant \left\{ \int_{-\infty}^{\infty} |f(x) - Q_\sigma(f; x)|^q \mathrm{d}x \right\}^{\frac{1}{q}} \leqslant C_k \omega_k \left(f; \frac{1}{\sigma} \right)_{Lq} \tag{15}$$

for any integer k.

5.1.32. Let us note that in the case when the function $f(x)$ is periodic the integral function $Q_\sigma(f; x)$ has the same period and, consequently, (see 4.10.2), is a trigonometric polynomial. Hence it follows from theorem 5.1.3 that *for any bounded periodic function $f(x)$ of period 2π the best approximation by trigonometric polynomials* (see 2.2.1) *satisfies the inequality****

$$E_n^*(f) \leqslant C_k \omega_k \left(f; \frac{1}{n+1} \right). \tag{16}$$

A similar estimate also holds for the best integral approximation (see section 5.11.1).

5.1.4. The relation 5.1(11) (or the more general 5.1(15)) indicates a deep connection between the differential properties of functions and their best approximations. From this it follows with the help of the inequality 3.3(1) that *if the bounded function $f(x)$, defined on $(-\infty, \infty)$, has a bounded derivative $f^{(r)}(x)$ of order $r (r > 0$ an integer) then for any integer k and $\sigma > 0$*

$$A_\sigma(f) \leqslant \frac{C_{k+r}}{\sigma^r} \omega_k \left(f^{(r)}; \frac{1}{\sigma} \right), \tag{17}$$

and in the periodic case

$$E_n^*(f) \leqslant \frac{C_{k+r}}{(n+1)^r} \omega_k \left(f^{(r)}; \frac{1}{n+1} \right). \tag{18}$$

Similarly *if for some $q \geqslant 1$ the function $f(x)$ satisfies the condition* 5.1(14) *and has a derivative $f^{(r)}(x)$ of order r which satisfies the same condition then for any integral k and $\sigma > 0$*

$$A_\sigma(f)_{Lq} \leqslant \frac{C_{k+r}}{\sigma^r} \omega_k \left(f^{(r)}; \frac{1}{\sigma} \right)_{Lq}. \tag{19}$$

** N. I. Akhiezer [3] for $k = 1, q < \infty$; $k = 2$. A. F. Timan and M. F. Timan [2] for $k \geqslant 3$.

*** N. I. Akhiezer [3] for $k = 2$. D. Jackson [1] for $k = 1$. S. B. Stechkin [2] for $k \geqslant 3$.

5.1.41. It follows from the inequality 5.1(17) that if the bounded function $f(x)$, defined on $(-\infty, \infty)$, has a bounded derivative $f^{(r)}(x)$ of order r, then for any $\sigma > 0$

$$A_\sigma(f) \leqslant \frac{B_r}{\sigma^r} \sup_{-\infty<x<\infty} |f^{(r)}(x)|, \tag{20}$$

and in the periodic case

$$E_n^*(f) \leqslant \frac{B_r}{(n+1)^r} \sup_x |f^{(r)}(x)|, \tag{21}$$

where B_r is a constant depending only on r.

With the help of theorem 3.5.7 these inequalities are easily extended to the case of derivatives of any fractional order (see section 3.5.6). For this it is only necessary to apply the estimates 5.1(17) and 5.1(18) (if $k = 1$) for a derivative of order $[r]$, and then to use theorem 3.5.7.

5.1.5. The relation 5.1(1) is a particular case of the following general theorem of Jackson**. *If the function $f(x)$, defined on the finite segment $[a, b]$, has an r-th continuous derivative, then for $n > r$*

$$E_n(f) = E_n(f; a, b) \leqslant M_r \left(\frac{b-a}{n}\right)^r \omega\left(f^{(r)}; \frac{b-a}{n}\right), \tag{22}$$

where M_r is a constant depending only on r.

In particular, if $\omega(f^{(r)}; t) \leqslant t$, then

$$E_n(f) \leqslant M_r \left(\frac{b-a}{n}\right)^{r+1}. \tag{23}$$

It must be noted that the last estimate remains valid for a wider class of functions. It is true whenever $\omega_2\ (f^{(r)}; t) \leqslant t$ (the r-th derivative is quasi-smooth)***.

We omit proofs of these propositions, since they follow from more precise inequalities, established in the following section, which more fully reflect the differential properties of functions defined on a finite segment.

5.2. A strong form of Jackson's theorem on the best approximation of continuous functions by algebraic polynomials on a finite segment of the real axis

In this section we shall show that in the case of a finite segment Jackson's inequality 5.1(22) can be materially improved if in place of uniform

** D. Jackson [1].
*** A. F. Timan and V. K. Dzyadyk [1].

approximation by polynomials we consider approximation in which account is taken of the position of the point x on the given segment. For definiteness we consider the segment $[-1, 1]$

5.2.1. *If the function $f(x)$, defined on $[-1, 1]$, has an r-th ($r \geqslant 0$ an integer) continuous derivative, then there exists a constant M_r, not depending on f, x and n, such that for any $n > r$ there is an ordinary polynomial $P_n(x) = \sum_{k=0}^{n} a_k x^k$ of degree not higher than n which satisfies for every $x \in [-1, 1]$ the inequality*

$$|f(x)-P_n(x)| \leqslant M_r \left[\frac{1}{n}\left(\sqrt{1-x^2}+\frac{|x|}{n}\right)\right]^r \omega\left[\frac{1}{n}\left(\sqrt{1-x^2}+\frac{|x|}{n}\right)\right], \qquad (1)$$

*where $\omega(t) = \omega(f^{(r)}; t)$ is the modulus of continuity of the r-th derivative***.

The inequality 5.2(1) shows that, retaining the best order of approximation established by theorem 5.1.5 throughout the interval as a whole, it is possible to substantially raise this order near its ends. Thus the character of the best approximation near the ends of the segment is here the same as the character of the increase of the order of the maximal growth of the derivative of an algebraic polynomial as $|x| \to 1$ in the inequalities of S. N. Bernstein and A. A. Markov (see 4.8(31) and 4.8(32), and also 4.8(51).) The last fact is of extreme importance in the study of the constructive characteristics of the differential properties of functions defined on a finite segment (see section 6.2).

We note that in the process of proving the inequality 5.1(1) we have in fact already established theorem 5.2.1 for the case $r = 0$ (see 5.1(6).

5.2.11. In the general case, the proof of theorem 5.2.1 is based on the following lemma. *For every integer p such that $0 \leqslant p \leqslant r$, and for any function $\omega(t)$ which is a modulus of continuity the relation*

$$\int_{-\pi}^{\pi} |\cos t-\cos y|^p \omega(|\cos t-\cos y|) \left(\frac{\sin n\dfrac{t\pm y}{2}}{n\sin\dfrac{t\pm y}{2}}\right)^{2r+4} dt$$

$$= \frac{1}{n^{p+1}} O\left\{\left(|\sin y|+\frac{|\cos y|}{n}\right)^p \omega\left[\frac{1}{n}\left(|\sin y|+\frac{|\cos y|}{n}\right)\right]\right\} \qquad (2)$$

is satisfied, uniformly with respect to y, as $n \to \infty$.

** A. F. Timan [7].

Let us denote the left-hand side of 5.2(2) by $I_{n,r}^{(p)}(\omega;\pm y)$. Then

$$I_{n,r}^{(p)}(\omega;\pm y)$$

$$=\frac{2^p}{n^{2r+4}}\int_{-\pi}^{\pi}\left|\sin\frac{t\mp y}{2}\right|^p\omega\left(2\left|\sin\frac{t+y}{2}\sin\frac{t-y}{2}\right|\right)\frac{\sin^{2r+4}n\frac{t\pm y}{2}}{\left|\sin\frac{t\pm y}{2}\right|^{2r-p+4}}\,\mathrm{d}t$$

$$=\frac{2^{p+1}}{n^{2r+4}}\int_{-\frac{\pi}{2}}^{\frac{\pi}{2}}|\sin t\cos y\mp\cos t\sin y|^p\times$$

$$\times\,\omega\big(2|\sin t\sin(t\mp y)|\big)\frac{\sin^{2r+4}nt}{|\sin t|^{2r-p+4}}\,\mathrm{d}t.$$

In virtue of the properties of the modulus of continuity already applied in section 5.1.1, we have

$$I_{n,r}^{(p)}(\omega;\pm y)$$

$$\leqslant\frac{2^{p+3}}{n^{2r+4}}\sum_{k=0}^{p}\binom{p}{k}|\cos y|^k|\sin y|^{p-k}\int_0^{\frac{\pi}{2}}\omega(t^2|\cos y|)\frac{\sin^{2r+4}nt}{(\sin t)^{2r-p-k+4}}\,\mathrm{d}t+$$

$$+\frac{2^{p+3}}{n^{2r+4}}\sum_{k=0}^{p}\binom{p}{k}|\cos y|^k|\sin y|^{p-k}\int_0^{\frac{\pi}{2}}\omega(t|\sin y|)\frac{\sin^{2r+4}nt}{(\sin t)^{2r-p-k+4}}\,\mathrm{d}t.$$

Moreover, since

$$\int_0^{\frac{\pi}{2}}\omega(t^2|\cos y|)\frac{\sin^{2r+4}nt}{(\sin t)^{2r-p-k+4}}\,\mathrm{d}t$$

$$\leqslant\left(\frac{\pi}{2}\right)^{2r-p-k+4}\int_0^{\frac{\pi}{2}}\omega(t^2|\cos y|)\frac{\sin^{2r+4}nt}{t^{2r-p-k+4}}\,\mathrm{d}t$$

$$\leqslant\left(\frac{\pi}{2}\right)^{2r-p-k+4}n^{2r-p-k+3}\omega\left(\frac{|\cos y|}{n^2}\right)\int_0^{\infty}(1+t^2)\frac{\sin^{2r+4}}{t^{2r-p-k+4}t}\,\mathrm{d}t,$$

and similarly

$$\int_0^{\frac{\pi}{2}}\omega(t|\sin y|)\frac{\sin^{2r+4}nt}{(\sin t)^{2r-p-k+4}}\,\mathrm{d}t$$

$$\leqslant\left(\frac{\pi}{2}\right)^{2r-p-k+4}n^{2r-p-k+3}\omega\left(\frac{\sin y}{n}\right)\int_0^{\infty}(1+t)\frac{\sin^{2r+4}t}{t^{2r-p-k+4}}\,\mathrm{d}t,$$

it follows that

$$I_{n,r}^{(p)}(\omega; \pm y) = O\left\{\sum_{k=0}^{p}\binom{p}{k}\left[\frac{|\cos y|^k|\sin y|^{p-k}}{n^{p+k+1}}\,\omega\left(\frac{|\cos y|}{n^2}\right)+\right.\right.$$

$$\left.\left.+\frac{|\cos y|^k|\sin y|^{p-k}}{n^{p+k+1}}\,\omega\left(\frac{|\sin y|}{n}\right)\right]\right\},$$

or

$$I_{n,r}^{(p)}(\omega; \pm y) = \frac{1}{n^{p+1}}\,O\left\{\left(|\sin y| + \frac{|\cos y|}{n}\right)^p\left[\omega\left(\frac{|\sin y|}{n}\right)+\omega\left(\frac{|\cos y|}{n^2}\right)\right]\right\}.$$

Since all the estimates given are uniform with respect to y, this proves the relation 5.2(2).

5.2.12. Let us pass to the proof of theorem 5.2.1. For this purpose we assume the result for all k which satisfy the inequality $0 \leqslant k \leqslant r-1$ and proceed to prove that it follows for $k = r$. In virtue of the hypotheses of the theorem, the ν-th $(1 \leqslant \nu \leqslant r)$ derivative of the function $f(x)$ has an $(r-\nu)$-th derivative with modulus of continuity $\omega(t)$. Hence on the assumption that our assertion has already been proved for the values of $k\,(0 \leqslant k \leqslant r-1)$, there exists a certain constant C_r such that for every $n \geqslant r+1$ there can be found r polynomials $Q_n(f^{(\nu)}; x)$ $(1 \leqslant \nu \leqslant r)$ of degree $\leqslant n$ which satisfy the inequalities

$$|f^{(\nu)}(x)-Q_n(f^{(\nu)}; x)|$$
$$\leqslant \frac{C_r}{n^{r-\nu}}\left(\sqrt{1-x^2}+\frac{|x|}{n}\right)^{r-\nu}\omega\left[\frac{1}{n}\left(\sqrt{1-x^2}+\frac{|x|}{n}\right)\right]. \tag{3}$$

For every k $(1 \leqslant k \leqslant r)$ let us consider the integral

$$V_k(x) = \frac{1}{\pi\gamma_n^{(r)}}\int_0^{\pi}(\cos t-\cos y)^k\{T_{(r+2)(n-1)}(t+y)+T_{(r+2)(n-1)}(t-y)\}\,dt,$$

where

$$y = \arccos x, \qquad T_{(r+2)(n-1)}(u) = \left(\frac{\sin\dfrac{nu}{2}}{n\sin\dfrac{n}{2}}\right)^{2r+4},$$

$$\gamma_n^{(r)} = \frac{1}{\pi}\int_{-\pi}^{\pi}T_{(r+2)(n-1)}(u)\,du. \tag{4}$$

Since $T_{(r+2)(n-1)}(u)$ is an even trigonometric polynomial of order

$(r+2)(n-1)$ (see 4.7(10)), it follows that $V_k(x)$ is an algebraic polynomial of degree $\leqslant k$. Moreover, for any function $f(x)$ considered let us put

$$W_{(r+2)(n-1)}(f; x) = \frac{1}{\pi\gamma_n^{(r)}}\int_0^\pi f(\cos t)\,\{T_{(r+2)(n-1)}(t+y)+T_{(r+2)(n-1)}(t-y)\}\,dt. \tag{5}$$

It is obvious also that $W_{(r+2)(n-1)}(f; x)$ is an algebraic polynomial of degree $\leqslant (r+2)(n-1)$. From $W_{(r+2)(n-1)}(f; x)$ and $V_k(x)$ $(1 \leqslant k \leqslant r)$ let us form the polymial

$$P_{(r+2)(n-1)}(x) = W_{(r+2)(n-1)}(f; x) - \sum_{k=1}^{r}\frac{1}{k!}Q_n(f^{(k)}; x)V_k(x), \tag{6}$$

the degree of which is not higher than $(r+2)$ $(r-1)$ when $n > 1$, and not higher than $r+1$ when $n = 1$. Let us estimate the deviation of $P_{(r+2)(n-1)}(x)$ from the function $f(x)$. On account of the formula

$$f(n)-f(x) = \sum_{\nu=1}^{r}\frac{1}{\nu!}f^{(\nu)}(x)\,(u-x)^\nu + O\,\{|u-x|^r\omega(|u-x|)\} \tag{7}$$

we find from the identity

$$\begin{aligned} f(x)-P_{(r+2)(n-1)}(x) = \frac{1}{\pi\gamma_n^{(r)}}\Bigg\{&\int_0^\pi [f(\cos t)-f(\cos y)]\times \\ &\times [T_{(r+2)(n-1)}(t+y)+T_{(r+2)(n-1)}(t-y)]\,dt- \\ &-\sum_{k=1}^{r}\frac{1}{k!}Q_n(f^{(k)}; x)\int_0^\pi(\cos t-\cos y)^k[T_{(r+2)(n-1)}(t+y)+ \\ &+T_{(r+2)(n-1)}(t-y)]\,dt\Bigg\} \end{aligned} \tag{8}$$

that

$$\begin{aligned} |f(x)-P_{(r+2)(n-1)}(x)| = \frac{1}{\pi\gamma_n^{(r)}}&\Bigg|\sum_{k=1}^{r}\frac{1}{k!}[f^{(k)}(x)-Q_n(f^{(k)}; x)]\times \\ &\times\int_0^\pi(\cos t-\cos y)^k\{T_{(r+2)(n-1)}(t+y)+T_{(r+2)(n-1)}(t-y)\}\,dt\Bigg| + \\ &+O\left\{\frac{1}{\gamma_n^{(r)}}\int_0^\pi|\cos t-\cos y|^r\omega(|\cos t-\cos y|)\times\right. \\ &\left.\times[T_{(r+2)(n-1)}(t+y)+T_{(r+2)(n-1)}(t-y)]\,dt\right\}. \end{aligned}$$

Hence by 5.2(3) and the lemma of 5.2.11 we have the relation

$$f(x)-P_{(r+2)(n-1)}(x)=\frac{1}{n^{r+1}\gamma_n^{(r)}}O\left\{\left(\sqrt{1-x^2}+\frac{|x|}{n}\right)^r\cdot\omega\left[\frac{1}{n}\left(\sqrt{1-x^2}+\frac{|x|}{n}\right)\right]\right\}$$

uniformly for all $x\in[-1,1]$. Taking into account the estimate from below given for $\gamma_n^{(r)}$ in section 4.8.72, we arrive at the inequality

$$f(x)-P_{(r+2)(n-1)}(x)\leqslant\frac{M_r}{n^r}\left(\sqrt{1-x^2}+\frac{|x|}{n}\right)^r\omega\left[\frac{1}{n}\left(\sqrt{1-x^2}+\frac{|x|}{n}\right)\right],$$

in which M_r is a constant depending only on r. Thus the correctness of theorem 5.2.1 in the case $k=r$ follows. Since the inequality 5.2(1) has already been proved for the case $r=0$ (see 5.1(6)), it holds in consequence for any integral r.

5.2.2. It follows from theorem 5.2.1, that *if the function $f(x)$ defined on $[-1,1]$ has a bounded derivative of order r, then there exists a sequence of algebraic polynomials $P_n(x)$ $(n=1, 2, \ldots)$ which satisfy the relation*

$$f(x)-P_n(x)=O\left\{\frac{1}{n^r}\left(\sqrt{1-x^2}+\frac{|x|}{n}\right)^r\right\}\tag{9}$$

uniformly for all $x\in[-1,1]$.

We note that with the help of theorem 3.5.7 this assertion is immediately extended also to the case of derivatives of any fractional order $r>0$.

5.2.3. Propositions of the type 5.2.1 can also be obtained for other classes of functions. As an example let us consider the class of functions $f(x)$ with a given modulus of smoothness $\omega_2(f;t)$ on $[-1,1]$. We shall show that *if $W_2(f;t)=\omega_2(t)$, then there exists a sequence of algebraic polynomials $P_n(x)$ $(n=1,2,\ldots)$ of degree $\leqslant n$ which satisfy for every $x\in[-1,1]$ the inequality***

$$|f(x)-P_n(x)|\leqslant c\omega_2\left(\frac{\sqrt{1-x^2}}{n}+\frac{1}{n^2}\right),\tag{10}$$

where C does not depend on x and n.

In virtue of the remark 3.5.71 it is possible without loss of generality to suppose in the proof that $f(+1)=f(-1)=0$ and that $f(x)$ is defined on the whole real axis as an odd function with respect to the points ±1 with period equal to four.

** V. K. Dzyadyk [4]. The proof given closely resembles that of theorem 5.1.3 for $k=2$.

Retaining the notation used in section 5.1.3, let us consider for $r \geqslant 4$ the integral

$$Q_\sigma^*(f;y) = \frac{1}{\gamma_{\sigma,r}} \int_{-\infty}^{\infty} \left\{2f\left[\cos\left(y+\frac{t}{\sqrt{2}}\right)\right] - f[\cos(y+t)]\right\} g_\sigma(t)\,dt, \tag{11}$$

which is an even integral function of degree $\leqslant \sigma\sqrt{2}$ of period 2π.

By theorem 4.10.2, $Q_\sigma^*(f;y)$ is an even trigonometric polynomial of order $\leqslant \sigma\sqrt{2}$.

It is obvious that

$$|f(\cos y) - Q_\sigma^*(f;y)|$$

$$\leqslant \frac{1}{\gamma_{\sigma,r}} \int_{-\infty}^{\infty} \left| f(\cos y) - f\left[\cos\left(y+\frac{t}{\sqrt{2}}\right)\right] - f\left[\cos\left(y-\frac{t}{\sqrt{2}}\right)\right] + \right.$$

$$\left. + \frac{1}{2} f[\cos(y-t)] + \frac{1}{2} f[\cos(y-t)] \right| g_\sigma(t)\,dt.$$

Since

$$\cos(y+t) = \alpha(t,y) + \varepsilon(t,y),$$

where

$$\alpha(t,y) = \cos y - t\sin y - t^2\frac{\cos y}{2} + t^3\frac{\sin y}{6}, \qquad |\varepsilon(t,y)| \leqslant \frac{t^4}{4},$$

then

$$|f(\cos y) - Q_\sigma^*(f,y)|$$

$$\leqslant \frac{1}{\gamma_{\sigma,r}} \int_{-\infty}^{\infty} \left| f(\cos y) - f\left[\alpha\left(\frac{t}{\sqrt{2}}, y\right)\right] - f\left[\alpha\left(-\frac{t}{\sqrt{2}}, y\right)\right] \right.$$

$$\left. + \frac{1}{2} f[\alpha(t,y)] + \frac{1}{2} f[\alpha(-t,y)] \right| g_\sigma(t)\,dt +$$

$$+ \frac{6}{\gamma_{\sigma,r}} \int_{-\infty}^{\infty} \omega(f;t^4)\, g_\sigma(t)\,dt. \tag{12}$$

If now we note that

$$\tfrac{1}{2}\{\alpha(t,y) + \alpha(-t,y)\} = \cos y - \tfrac{1}{2}t^2\cos y,$$

we obtain the inequality (see section 3.5.71)

$$|f(\cos y)-Q_\sigma^*(f;y)|$$

$$\leqslant \frac{5}{\gamma_{\sigma,r}}\cdot\int_{-\infty}^{\infty}\omega_2\left(\frac{|t\sin y|}{\sqrt{2}}+\frac{|t^3\sin y|}{12\sqrt{2}}\right)g_\sigma(t)\,dt+$$

$$+\frac{1}{2}\int_{-\infty}^{\infty}\omega_2\left(|t\sin y|+\frac{|t^3\sin y|}{6}\right)g_\sigma(t)\,dt+$$

$$+\int_{-\infty}^{\infty}\omega_2\left(\frac{t^2|\cos y|}{4}\right)g_\sigma(t)\,dt\Bigg\}+\frac{6}{\gamma_{\sigma,r}}\int_{-\infty}^{\infty}\omega(f;t^4)g_\sigma(t)\,dt,$$

from which in virtue of the properties of moduli of smoothness (see sections 3.3 and 3.3.1) it follows that:

$$|f(\cos y)-Q_\sigma^*(f;y)|$$

$$\leqslant\frac{6}{\gamma_{\sigma,r}}\left\{\omega_2\left(\frac{|\sin y|}{\sigma}\right)\int_{-\infty}^{\infty}[(|\sigma t|+1)^2+(|\sigma t|^3+1)^2]g_\sigma(t)\,dt+\right.$$

$$\left.+\omega_2\left(\frac{1}{\sigma^2}\right)\int_{-\infty}^{\infty}(\sigma^2t^2+1)^2g_\sigma(t)\,dt+\omega\left(f;\frac{1}{\sigma^4}\right)\int_{-\infty}^{\infty}(\sigma^4t^4+1)g_\sigma(t)\,dt\right\}.$$

Putting here $\cos y = x$, $\sigma=\dfrac{n}{\sqrt{2}}$, $Q_\sigma^*(f;y)=P_n(x)$ and using the relation $\omega\left(f;\dfrac{1}{\sigma^4}\right)=O\left[\omega_z\left(f;\dfrac{1}{\sigma^2}\right)\right]$ (see section 3.3.2), we arrive at the inequality 5.2(10).

5.2.4. It follows from the inequality 5.2(10) that for the best approximation $E_n(f;a,b)$ of the function $f(x)$ which satisfies on $[a,b)]$ the condition $\omega_2(f;t)\leqslant Mt$, the estimate

$$E_n(f;a,b)\leqslant C_0M\frac{b-a}{n},\tag{13}$$

obtains, where C_0 is a certain absolute constant. This inequality can be generalised. If the function $f(x)$, defined on $[a,b]$, has there a derivative $f^{(r)}(x)$ of order $r(r>0$ an integer) which satisfies the condition $\omega_2(f^{(r)};t)\leqslant$ $\leqslant M\cdot t$, then for all $n>r$

$$E_n(f;a,b)\leqslant C_r\cdot M\left(\frac{b-a}{n}\right)^{r+1}.\tag{14}$$

Let $P_{n-1}(f'; x)$ be the algebraic polynomial of order $\leqslant n-1$ which deviates least from the derivative $f'(x)$ on $[a, b]$. The difference

$$F_n(x) = f(x) - \int_a^x P_{n-1}(f'; t)\,dt$$

possesses the property that $|F_n'(x)| \leqslant E_{n-1}(f'; a, b)$, i.e.

$$\omega(F_n; t) \leqslant E_{n-1}(f'; a, b)\, t.$$

Hence by the inequality 5.1(1)

$$E_n(F_n; a, b) \leqslant C\frac{b-a}{n} E_{n-1}(f'; a, b).$$

Since

$$E(f; a, b) \leqslant \max_{a\leqslant x\leqslant b}\left| f(x) - P_n(F_n; x) - \int_a^x P_{n-1}(f'; t)\,dt \right|,$$

we obtain the relation

$$E_n(f; a, b) \leqslant C\frac{b-a}{n} E_{n-1}(f'; a, b), \tag{15}$$

after r-fold application of which, on account of the estimate

$$E_{n-r}(f^{(r)}; a, b) \leqslant C_0 M \frac{b-a}{n-r},$$

we arrive at the inequality 5.2(14).

5.2.5. Let us note that the situation revealed by theorem 5.2.1 holds also in a number of other cases. In particular, there is a similar theorem if, instead of functions defined on a finite segment of the real axis, functions are considered which are defined outside its limits, and instead of approximation by algebraic polynomials there is considered approximation by arbitrary integral functions of exponential type. *If the function $f(x)$, defined for all real values of x outside the interval $(-1, 1)$, is bounded and has r derivatives, where the r-th derivative $f^{(r)}(x)$ is bounded and uniformly continuous, then there exists a constant M_r, not depending on x and σ, such that for any $\sigma \geqslant 1$ there can be found an integral function $G_\sigma(x)$ of degree $\leqslant \sigma$ which satisfies for every x ($|x| \geqslant 1$) the inequality*

$$|f(x) - G_\sigma(x)| \leqslant M_r \left[\frac{1}{\sigma}\left(\frac{\sqrt{x^2-1}}{|x|} + \frac{1}{\sigma}\right)\right]^r \omega\left[\frac{1}{\sigma}\left(\frac{\sqrt{x^2-1}}{|x|} + \frac{1}{\sigma}\right)\right], \tag{16}$$

*where $\omega(t) = \omega(f^{(r)}; t)$ is the modulus of continuity of the r-th derivative***.

** Yu. A. Brudnyi [2]. The idea of the proof given here is due to N. I. Akhiezer. A similar theorem on the class of functions whose r-th derivative has a given modulus of smoothness of the second order was proved by R. M. Trigub.

Let

$$K(u) = \int_{-1}^{1} \phi(v)\,\mathrm{e}^{iuv}\,\mathrm{d}v, \tag{17}$$

where $\phi(v)$ has on $[-1,1]$ an absolutely continuous derivative of order $p-1$ and

$$\begin{aligned}
&(1)\ \phi(-v) = \phi(v), \quad 0 \leqslant v \leqslant 1;\\
&(2)\ \phi(0) = \tfrac{1}{2}\pi;\\
&(3)\ \phi(1) = \phi'(1) = \ldots = \phi^{(p-1)}(1) = 0 \quad (p \geqslant 2);\\
&(4)\ \phi''(0) = \phi^{(\mathrm{IV})}(0) = \ldots = \phi^{(2q)}(0) = 0 \quad (2q < p-2).
\end{aligned} \tag{18}$$

Integrating 5.2(17) by parts, we see that in virtue of the properties of the function $\phi(v)$

$$K(u) = \left(\frac{-1}{iu}\right)^p \int_{-1}^{1} \phi^{(p)}(v)\,\mathrm{e}^{iuv}\,\mathrm{d}v = O\left(\frac{1}{1+|u|^p}\right).$$

Hence, as $2q+1 \leqslant p-2$, we have for $m = 1, 2, \ldots, 2q+1$ a function $|K(u)u^m|$ integrable on $(-\infty, \infty)$ and

$$\int_{-\infty}^{\infty} K(u)u^m\,\mathrm{d}u = \frac{1}{(-i)^m}\frac{\mathrm{d}^m}{\mathrm{d}t^m}\left\{\int_{-\infty}^{\infty} K(u)\mathrm{e}^{iut}\,\mathrm{d}u\right\}_{t=0} = \frac{2\pi}{(-i)^m}\phi^{(m)}(0) = 0.$$

Let us assume to begin with that the function $f(x)$ is even, and let us consider the function $F(u) = f(\sqrt{u^2+1})$ defined on the whole real axis $-\infty < u < \infty$. The integral

$$g_\sigma(F, u) = \int_{-\infty}^{\infty} F\left(u+\frac{t}{\sigma}\right)K(t)\,\mathrm{d}t = \sigma\int_{-\infty}^{\infty} F(t)K[\sigma(t-u)]\,\mathrm{d}t \tag{19}$$

is an even integral function of degree $\leqslant \sigma$, since for all u

$$g_\sigma(F; u) = \sum_{\nu=0}^{\infty}\frac{(\sigma u)^{2\nu}}{(2\nu)!}\int_{-\infty}^{\infty} F\left(\frac{t}{\sigma}\right)K^{(2\nu)}(t)\,\mathrm{d}t = \sum_{\nu=0}^{\infty} c_\nu u^{2\nu}.$$

Let us consider the integral function

$$g(F; \sqrt{x^2-1}) = G_\sigma(f; x) = \sum_{\nu=0}^{\infty} c_\nu(x^2-1)^\nu,$$

the degree of which $\leqslant \sigma$, and estimate the difference between $f(x)$ and $G_\sigma(f; x)$. It follows from 5.2(18) that for $u = \sqrt{x^2-1}$

$$|f(x)-G_\sigma(f; x)| = |F(u)-g_\sigma(F; u)| = \left|\int_{-\infty}^{\infty}\left\{F(u)-F\left(u+\frac{t}{\sigma}\right)\right\}K(t)\,\mathrm{d}t\right|.$$

Putting $\sqrt{\left(u+\frac{t}{\sigma}\right)^2+1} = y$ and $\sqrt{u^2+1} = x$, we have

$$F\left(u+\frac{t}{\sigma}\right)-F(u) = f(y)-f(x) = \sum_{k=1}^{r} \frac{(y-x)^k}{k!} f^{(k)}(x)+D_r(f;x,y),$$

where

$$|D_r(f;x,y)| \leqslant \frac{|y-x|^r}{r!}\,\omega(|y-x|).$$

Hence

$$|f(x)-G_\sigma(f;x)| \leqslant \left| \sum_{\nu=1}^{r} \frac{f^{(\nu)}(x)}{\nu!} \int_{-\infty}^{\infty} (y-x)^\nu K(t)\,dt \right| +$$

$$+\frac{1}{r!}\int_{-\infty}^{\infty} |y-x|^r \omega(|y-x|) K(t)|\,dt.$$

But

$$y-x = \sqrt{\left(u+\frac{t}{\sigma}\right)^2+1} - \sqrt{u^2+1} = \sum_{\nu=1}^{m} \frac{1}{\nu!}\left(\frac{t}{\sigma}\right)^\nu \frac{d^\nu}{du^\nu}\left(\sqrt{u^2+1}\right)+R_m,$$

where

$$R_m = \frac{t^{m+1}}{(m+1)!\,\sigma^{m+1}} \frac{d^{m+1}}{dx^{m+1}}\left(\sqrt{x^2+1}\right)\Big|_{x=u+\theta\frac{t}{\sigma}} \qquad (0<\theta<1).$$

Since

$$\frac{d^\nu}{du^\nu}\left(\sqrt{1+u^2}\right) = \frac{P_\nu(u)}{(1+u^2)^{\nu-\frac{1}{2}}},$$

where the $P_\nu(u)$ are algebraic polynomials of degree $\leqslant \nu$, then for $\nu \leqslant 1$ these derivatives are bounded on the whole axis $-\infty < u < \infty$.

Thus

$$y-x = \sum_{\nu=1}^{m} \frac{t^\nu}{\nu!\,\sigma^n} \gamma_\nu(u) + \frac{t^{m+1}}{(m+1)!\,\sigma^{m+1}} \gamma_{m+1}(u,t),$$

where $\gamma_\nu(u)$ and $\gamma_{m+1}(u,t)$ are bounded functions of their arguments Consequently, for any natural numbers $k \leqslant r$ and $m = 2r+3-k$,

$$(y-x)^k = \sum_{\nu=1}^{2r+2} \frac{t^\nu}{\sigma^\nu} \gamma_{\nu,k}(u) + \frac{|t|^{(2r+4-k)k}+1}{\sigma^{2r+2}} \gamma_k^*(u,t),$$

where $\gamma_{\nu,k}(u)$ and $\gamma_k^*(u, t)$ are bounded functions of their arguments. For the determination of the kernel $K(t)$ let us put in the conditions 5.2(18) $p = (r+2)^2+2$, $q = r+1$. Then for $k \leqslant r$ we have

$$\left|\int_{-\infty}^{\infty} (y-x)^k K(t)\,dt\right| = \left|\int_{-\infty}^{\infty} \frac{|t|^{(2r+4-k)k}+1}{\sigma^{2r+2}} \gamma_k^*(u, t)\,K(t)\,dt\right|$$

$$\leqslant \frac{1}{\sigma^{2r+2}} \int_{-\infty}^{\infty} (|t|^{(r+2)^2}+2)|\gamma_k^*(u, t)|\,|K(t)|\,dt \leqslant \frac{C_{k,r}}{\sigma^{2r+2}}.$$

Hence in virtue of the boundedness of $f(x)$ and $f^{(r)}(x)$ for $|x| \geqslant 1$ we have $\sup\limits_{|x|\geqslant 1} |f^{(k)}(x)| < \infty$ for any $k (0 \leqslant k \leqslant r)$,

$$\left|\sum_{\nu=1}^{r} \frac{f^{(\nu)}(x)}{\nu!} \int_{-\infty}^{\infty} (y-x)^\nu K(t)\,dt\right| \leqslant \sum_{\nu=1}^{r} \frac{C_{\nu,r}}{\nu!\,\sigma^{2r+2}} |f^{(\nu)}(x)| \leqslant \frac{C_r(f)}{\sigma^{2r+2}}$$

and

$$|f(x)-G_\sigma(f; x)| \leqslant \frac{C_r(f)}{\sigma^{2r+2}} + \frac{1}{r!} \int_{-\infty}^{\infty} |y-x|^r \omega(|y-x|)\,|K(t)|\,dt.$$

It remains to take account of 3.2(8) and note that, since

$$y-x = \frac{t}{\sigma} \frac{u}{\sqrt{1+u^2}} + \frac{t^2}{\sigma^2} \gamma_2(u, t),$$

it follows that

$$|y-x|^r \omega(|y-x|) \leqslant C_r(|t|+1)^{2r+2} \left(\frac{|u|}{\sigma\sqrt{1+u^2}} + \frac{1}{\sigma^2}\right)^r \omega\left(\frac{|u|}{\sigma\sqrt{1+u^2}} + \frac{1}{\sigma^2}\right)$$

$$= C_r(|t|+1)^{2r+2} \left(\frac{\sqrt{x^2-1}}{\sigma|x|} + \frac{1}{\sigma^2}\right)^2 \omega\left(\frac{\sqrt{x^2-1}}{\sigma|x|} + \frac{1}{\sigma^2}\right)$$

and

$$\int_{-\infty}^{\infty} (|t|+1)^{2r+2} |K(t)|\,dt < \infty.$$

Now let the function $f(x)$ be odd. Let us consider the quotient $\Psi(x) = \dfrac{f(x)}{g(x)}$, where $g(x)$ is the function 1.7(3).

Since for $\nu = 0, 1, \ldots, r$

$$\sup_{|x|\geqslant 1} \left|\frac{d^\nu}{dx^\nu} \frac{1}{g(x)}\right| < \infty,$$

the even function $\Psi(x)$ has r derivatives, bounded outside the interval (—1,1), and as

$$\Psi^{(r)}(x) = \frac{f^{(r)}(x)}{g(x)} + \sum_{\nu=0}^{r-1} \binom{r}{\nu} f^{(\nu)}(x) \frac{d^{r-\nu}}{dx^{r-\nu}} \left\{\frac{1}{g(x)}\right\},$$

where the second term on the right-hand side has a bounded derivative, it follows that for $|x_1| \geqslant 1$, $|x_2| \geqslant 1$

$$\begin{aligned} &|\Psi^{(r)}(x_1) - \Psi^{(r)}(x_2)| \\ &\quad \leqslant \frac{1}{|g(x_1)|} |f^{(r)}(x_1) - f^{(r)}(x_2)| + |f^{(r)}(x_2)| \left|\frac{1}{g(x_1)} - \frac{1}{g(x_2)}\right| + C_r |x - x_1| \\ &\quad \leqslant C_r^* \{\omega(|x_1 - x_2|) + (|x_1 - x_2|)\}. \end{aligned}$$

It now only remains to use the relation 3.2(8), in view of the fact that the theorem has already been proved for even functions and the fact that $g(x)$ is an integral function of the first degree satisfying the inequality $|g(x)| < 1$.

Since every function defined outside the interval $(-1,1)$ is the sum of an even and an odd function, the theorem is automatically proved in the general case.

5.3. Some direct theorems for functions of many variables

More precise forms of Weierstrass's theorem, similar to those given in section 5.1, are also possible for functions of many variables.

5.3.1. Let us begin by considering the function $f(x_1, \ldots, x_m)$, defined throughout the whole space of the variables $x_1, \ldots, x_m$ and of period 2π in each of these variables.

If $1 \leqslant q \leqslant \infty$ *and*

$$\|f\|_{L_q} = \left\{\int_0^{2\pi} \cdots \int_0^{2\pi} |f(t_1, \ldots, t_m)|^q \, dt_1 \ldots dt_m\right\}^{\frac{1}{q}} < \infty$$

then for any natural numbers $k_1, \ldots, k_m, n_1 \geqslant 0, \ldots, n_m \geqslant 0$

$$E^*_{n_1, \ldots, n_m}(f)_{L_q} \leqslant C_m^{(k_1, \ldots, k_m)} \sum_{\nu=1}^{m} \omega_{k_\nu}\left(f; 0, \ldots, \frac{1}{n_\nu + 1}, \ldots, 0\right)_{L_q}, \tag{1}$$

where $C_m^{(k_1, \ldots, k_m)}$ *is a constant depending only on* m *and* $k_1, \ldots, k_m$, *and* $\omega_{k_\nu}(f; 0, \ldots, h_\nu, \ldots, 0)_{L_q}$

$$= \sup_{|t| \leqslant h_\nu} \left\{\int_0^{2\pi} \cdots \int_0^{2\pi} \left|\sum_{j=0}^{k_\nu} (-1)^j \binom{k_\nu}{j} f(x_1, \ldots, x_\nu + jt, \ldots, x_m)\right|^q dx_1 \ldots, dx_m\right\}^{\frac{1}{q}} \tag{2}$$

is the corresponding partial modulus of smoothness of the function f (see sections 2.2.4, 3.4.32, 3.4.34 and 3.12.5).

Since in the proof of the inequality 5.3(1) the number of variables is not of great importance, we will give it for $m = 2$. For this purpose let us consider the trigonometric polynomials

$$\sigma_{(k_1+2)(n_1-1),\,(k_2+2)(n_2-1)}(f;\, x_1, x_2)$$

$$= \frac{1}{\pi^2 \gamma_{n_1}^{(k_1)} \gamma_{n_2}^{(k_2)}} \int_0^{2\pi}\int_0^{2\pi} T_{(k_1+2)(n_1-1)}(t_1)\, T_{(k_2+2)(n_2-1)}(t_2) \times$$

$$\times \sum_{i=1}^{k_1}\sum_{j=1}^{k_2} (-1)^{i+j} \binom{k_1}{i}\binom{k_2}{j} f(x_1+it_1,\, x_2+it_2)\, dt_1\, dt_2 \tag{3}$$

(where $T_{(r+2)(n-1)}(t)$ is the polynomial 5.2(4)) of order $(k_1+2)\,(n_1-1)$ with respect to x_1, and of order $(k_2+2)\,(n_2-1)$ with respect to x_2.

We have

$$\sigma_{(k_1+2)(n_1-1),\,(k_2+2)(n_2-1)}(f;\, x_1, x_2) - f(x_1, x_2)$$

$$= \frac{1}{\pi^2 \gamma_{n_1}^{(k_1)} \gamma_{n_2}^{(k_2)}} \int_0^{2\pi}\int_0^{2\pi} T_{(k_1+2)(n_1-1)}(t_1)\, T_{(k_2+2)(n_2-1)}(t_2) \times$$

$$\times \left\{ \sum_{i=1}^{k_1}\sum_{j=1}^{k_2} (-1)^{i+j} \binom{k_1}{i}\binom{k_2}{j} f(x_1+it_1,\, x_2+jt_2) - \right.$$

$$\left. - \sum_{j=1}^{k_2} (-1)^{j-1} \binom{k_2}{j} f(x_1,\, x_2+jt_2) \right\} dt_1\, dt_2 +$$

$$+ \frac{1}{\pi^2 \gamma_{n_1}^{(k_1)} \gamma_{n_2}^{(k_2)}} \int_0^{2\pi}\int_0^{2\pi} T_{(k_1+2)(n_1-1)}(t_1)\, T_{(k_2+2)(n_2-1)}(t_2) \times$$

$$\times \left\{ \sum_{j=1}^{k_2} (-1)^{j-1} \binom{k_2}{j} f(x_1,\, x_2+jt_2) - f(x_1, x_2) \right\} dt_1\, dt_2. \tag{4}$$

Let

$$\phi_1(x_1, x_2) = \frac{1}{\pi^2 \gamma_{n_1}^{(k_1)} \gamma_{n_2}^{(k_2)}} \int_0^{2\pi}\int_0^{2\pi} T_{(k_1+2)(n_1-1)}(t_1)\, T_{(k_2+2)(n_2-1)}(t_2) \times$$

$$\times \left\{ \sum_{j=1}^{k_2} (-1)^{j-1} \binom{k_2}{j} f(x_1,\, x_2+jt_2) - f(x_1, x_2) \right\} dt_1,\, dt_2$$

$$=\frac{1}{\pi\gamma_{n_2}^{(k_2)}}\int_0^{2\pi}T_{(k_2+2)(n_2-1)}(t_2)\left\{\sum_{j=1}^{k_2}(-1)^{j-1}\binom{k_2}{j}f(x_1,x_2+jt_2)-\right.$$

$$\left.-f(x_1,x_2)\right\}dt_2,$$

$$\phi_2(x_1,x_2)=\frac{1}{\pi^2\gamma_{n_1}^{(k_1)}\gamma_{n_2}^{(k_2)}}\int_0^{2\pi}\int_0^{2\pi}T_{(k_1+2)(n_1-1)}(t_1)T_{k_2+2)(n_2-1)}(t_2)\times$$

$$\times\left\{\sum_{i=1}^{k_1}\sum_{j=1}^{k_2}(-1)^{i+j}\binom{k_1}{i}\binom{k_2}{j}f(x_1+it_1,x_2+jt_2)-\right.$$

$$\left.-\sum_{j=1}^{k_2}(-1)^{j-1}\binom{k_2}{j}f(x_1,x_2+jt_2)\right\}dt_1,dt_2$$

$$=\frac{1}{\pi\gamma_{n_2}^{(k_2)}}\int_0^{2\pi}T_{(k_2+2)(n_2-1)}(t_2)\Phi_1(x_1,x_2,t_2)\,dt_2,$$

where

$$\Phi_1(x_1,x_2,t_2)=\frac{1}{\pi\gamma_{n_1}^{(k_1)}}\int_0^{2\pi}T_{(k_1+2)(n_1-1)}(t_1)\times$$

$$\times\left\{\sum_{i=1}^{k_1}\sum_{j=1}^{k_2}(-1)^{i+j}\binom{k_1}{i}\binom{k_2}{j}f(x_1+it_1,x_2+jt_2)-\right.$$

$$\left.-\sum_{j=1}^{k_2}(-1)^{j-1}\binom{k_2}{j}f(x_1,x_2+jt_2)\right\}dt_1.$$

Applying Minkowski's inequality for integrals, we obtain

$$\|\phi_1\|_{L_q}=\frac{1}{\pi\gamma_{n_2}^{(k_2)}}\left\{\int_0^{2\pi}\int_0^{2\pi}\left|\int_0^{2\pi}T_{(k_2+2)(n_2-1)}(t_2)\times\right.\right.$$

$$\left.\left.\times\left[\sum_{j=1}^{k_2}(-1)^{j-1}\binom{k_2}{j}f(x_1,x_2+jt_2)-f(x_1,x_2)\right]dt_2\right|^q dx_1\,dx_2\right\}^{\frac{1}{q}}$$

$$\leqslant\frac{1}{\pi\gamma_{n_2}^{(k_2)}}\int_{-\pi}^{\pi}T_{(k_2+2)(n_2-1)}(t_2)\times$$

$$\times\left\{\int_0^{2\pi}\int_0^{2\pi}\left|\sum_{j=0}^{k_2}(-1)^j\binom{k_2}{j}f(x_1, x_2+jt_2)\right|^q dx_1 dx_2\right\}^{\frac{1}{q}} dt_2$$

$$\leqslant \frac{2}{\pi\gamma_{n_2}^{(k_2)}}\int_0^{\pi} T_{(k_2+2)(n_2-1)}(t_2)\omega_{k_2}(f; 0, t_2)_{L_q} dt_2,$$

$$\|\phi_2\|_{L_q} \leqslant \frac{1}{\pi\gamma_{n_2}^{(k_2)}}\int_0^{2\pi} T_{(k_2+2)(n_2-1)}(t_2)\left\{\int_0^{2\pi}\int_0^{2\pi}|\Phi_1(x_1, x_2, t_2)|^q dx_1 dx_2\right\}^{\frac{1}{q}} dt_2,$$

$$\left\{\int_0^{2\pi}\int_0^{2\pi}|\Phi_1(x_1, x_2, t_2)|^q dx_1, dx^2\right\}^{\frac{1}{q}}$$

$$= \frac{1}{\pi\gamma_{n_1}^{(k_1)}}\left\{\int_0^{2\pi}\int_0^{2\pi}\left|\int_0^{2\pi} T_{(k_1+2)(n_1-1)}(t_1)\sum_{i=0}^{k_1}\sum_{j=1}^{k_2}(-1)^{i+j-1}\times\right.\right.$$

$$\left.\left.\times\binom{k_1}{i}\binom{k_2}{j}f(x_1+it_1, x_2+it_2)\,dt_1\right|^q dx_1 dx_2\right\}^{\frac{1}{q}}$$

$$\leqslant \frac{1}{\pi\gamma_{n_1}^{(k_1)}}\int_{-\pi}^{\pi} T_{(k_1+2)(n_1-1)}(t_1)\left\{\int_0^{2\pi}\int_0^{2\pi}\left|\sum_{j=1}^{k_2}\sum_{i=0}^{k_1}(-1)^{i+j-1}\times\right.\right.$$

$$\left.\left.\times\binom{k_1}{i}\binom{k_2}{j}f(x_1+it_1, x_2+it_2)\right|^q dx_1 dx_2\right\}^{\frac{1}{q}} dt_1$$

$$\leqslant \frac{2^{k_2+1}}{\pi\gamma_{n_1}^{(k_1)}}\int_0^{\pi} T_{(k_1+2)(n_1-1)}(t_1)\omega_{k_1}(f; t_1, 0)_{L_q} dt_1.$$

Hence for the difference 5.3(4) we have the estimate

$$\|f(x_1, x_2)-\sigma_{(k_1+2)(n_1-1),(k_2+2)(n_2-1)}(f; x_1, x_2)\|_{L_q}$$

$$\leqslant \frac{2^{k_2+1}}{\pi\gamma_{n_1}^{(k_1)}}\int_0^{\pi} T_{(k_1+2)(n_1-1)}(t)\omega_{k_1}(f; t, 0)_{L_q} dt +$$

$$+\frac{2}{\pi\gamma_{n_1}^{(k_1)}}\int_0^{\pi} T_{(k_2+2)(n_2-1)}(t)\omega_{k_2}(f; 0, t)_{L_q} dt. \qquad (5)$$

In virtue of the properties of partial moduli of smoothness (see 3.12.5)

$$\frac{1}{\pi\gamma_{n_1}^{(k_1)}}\int_0^{\pi} T_{(k_1+2)(n_1-1)}(t)\,\omega_{k_1}(f;\,t,\,0)_{L_q}\,\mathrm{d}t$$

$$\leqslant \omega_{k_1}\left(f;\frac{1}{n_1+1},0\right)_{L_q}\frac{(n_1+1)^{k_1}}{\pi\gamma_{n_1}^{(k_1)}}\int_0^{\pi} T_{(k_1+2)(n_1-1)}(t)\left(t+\frac{1}{n_1+1}\right)^{k_1}\mathrm{d}t$$

$$\leqslant \omega_{k_1}\left(f;\frac{1}{n_1+1},0\right)_{L_q}\left\{\frac{2^{k_1}}{\pi\gamma_{n_1}^{(k_1)}}\int_0^{\frac{1}{n_1-1}} T_{(k_1+2)(n_1-1)}(t)\,\mathrm{d}t+\right.$$

$$\left.+\frac{2^{k_1}(n_1+1)^{k_1}}{\pi\gamma_{n_1}^{(k_1)}}\int_{\frac{1}{n_1+1}}^{\pi} t^{k_1}T_{(k_1+2)(n_1-1)}(t)\,\mathrm{d}t\right\}.$$

Now

$$\frac{2^{k_1}(n_1+1)^{k_1}}{\gamma_{n_1}^{(k_1)}}\int_{\frac{1}{n_1+1}}^{\pi} t^{k_1}T_{(k_1+2)(n_1-1)}(t)\,\mathrm{d}t$$

$$=\frac{2^{k_1}(n_1+1)^{k_1}}{\gamma_{n_1}^{k_1}}\int_{\frac{1}{n_1+1}}^{\pi} t^{k_1}\left(\frac{\sin\dfrac{n_1 t}{2}}{n_1\sin\dfrac{t}{2}}\right)^{2k_1+4}\mathrm{d}t\leqslant M_{k_1},$$

where M_{k_1} is a constant, depending only on k_1.

Consequently

$$\frac{1}{\pi\gamma_{n_1}^{(k_1)}}\int_0^{\pi} T_{(k_1+2)(n_1-1)}(t)\,\omega_{k_1}(f;\,t,\,0)_{L_q}\,\mathrm{d}t\leqslant(2^{k_1}+M_{k_1})\,\omega_{k_1}\left(f;\frac{1}{n_1+1},\,0\right)_{L_q},$$

and similarly

$$\frac{1}{\pi\gamma_{n_2}^{(k_2)}}\int_0^{\pi} T_{(k_2+2)(n_2-1)}(t)\,\omega_{k_2}(f;\,0,\,t)_{L_q}\,\mathrm{d}t\leqslant(2^{k_2}+M_{k_2})\,\omega_{k_2}\left(f;\,0,\frac{1}{n_2+1}\right)_{L_q}.$$

If we substitute these estimates in 5.3(5), we arrive at the inequality 5.3(1) for the case of functions of two variables which we are considering.

5.3.11. If the function $f(x_1, \ldots, x_m)$, which satisfies the conditions of theorem 5.3.1, has for almost all $x_1, \ldots, x_{\nu-1}, x_{\nu+1}, \ldots, x_m$ an absolutely continuous (with respect to x_ν) partial derivative $\frac{\partial^{k_\nu-1}f}{\partial x_\nu^{k_\nu-1}}$ and a partial derivative $\frac{\partial^{k_\nu}f(x_1, \ldots, x_m)}{\partial x_\nu^{k_\nu}}$ of order k_ν for which

$$\left\| \frac{\partial^{k_\nu}f(x_1, \ldots, x_m)}{\partial x_\nu^{k_\nu}} \right\|_{L_q} < \infty, \tag{6}$$

for $\nu = 1, \ldots, m$, then from the identity 3.3(4) with the help of Minkowski's inequality for integrals it follows that

$$\omega_{k_\nu+j_\nu}(f; 0; \ldots, t_\nu, \ldots, 0)_{L_q} \leqslant t_\nu^{k_\nu} \omega_{j_\nu}\left(\frac{\partial^{k_\nu}f}{\partial x_\nu^{k_\nu}}; 0, \ldots, t_\nu, \ldots, 0\right)_{L_q}$$

Hence from the inequality 5.3(1) there follows for such functions the estimate**

$$E^*_{n_1, \ldots, n_m}(f)_{L_q} \leqslant C_m^{(k_1, \ldots, k_m)} \sum_{\nu=1}^{m} \frac{1}{(n_\nu+1)k_\nu} \omega\left(\frac{\partial^{k_\nu}f}{\partial x_\nu^{k_\nu}} 0, \ldots, \frac{1}{n_\nu+1}, \ldots, 0\right)_{L_q}. \tag{7}$$

In particular, if $f(x_1, \ldots, x_m) \in W^{(k_1, \ldots, k_m)}H^{(\alpha_1, \ldots, \alpha_m)}(L_q)$ $(1 \leqslant q \leqslant \infty;$ $0 < \alpha_1, \ldots, \alpha_m \leqslant 1)$ (see 3.5.8), then***

$$E^*_{n_1, \ldots, n_m}(f)_{L_q} = O\left(\sum_{\nu=1}^{m} \frac{1}{n_\nu^{k_\nu+\alpha_\nu}}\right). \tag{8}$$

5.3.2. Let the function $f(x_1, \ldots, x_m)$ be defined in the closed parallelepiped $-\infty < a_\nu \leqslant x_\nu \leqslant b_\nu < \infty$ $(\nu = 1, \ldots, m)$ and have there a continuous partial derivative $\frac{\partial^{k_\nu}f(x_1, \ldots, x_m)}{\partial x_\nu^{k_\nu}}$ of order k_ν with respect to each variable x_ν. If for definiteness we suppose that $a_\nu = -1 (b_\nu = 1, \ldots, m)$, then the function $\phi(t_1, \ldots, t_m) = f(\cos t_1, \ldots, \cos t_m)$ has, as is immediately obvious, a continuous partial derivative $\frac{\partial^{k_\nu}\phi(t_1, \ldots, t_m)}{\partial t_\nu^{k_\nu}}$ of order k_ν with respect to each variable t_ν and for a certain value of the constant M_{k_ν}

$$\omega\left(\frac{\partial^{k_\nu}\phi}{\partial t_\nu^{k_\nu}}; 0, \ldots, h_\nu, \ldots, 0\right) \leqslant M_{k_\nu} \omega\left(\frac{\partial^{k_\nu}f}{\partial x_\nu^{k_\nu}}; 0, \ldots, h_\nu, \ldots, 0\right). \tag{9}$$

Applying to the function $\phi(t_1, \ldots, t_m)$ the inequality 5.3(7) and taking 5.3(9) into consideration, we obtain

$$E^*_{n_1, \ldots, n_m}(\phi)_{L_\infty} \leqslant C_m^{(k_1, \ldots, k_m)} \sum_{\nu=1}^{m} \frac{M_{k_\nu}}{(n_\nu+1)^{k_\nu}} \omega\left(\frac{\partial^{k_\nu}f}{\partial x_\nu^{k_\nu}}; 0, \ldots, \frac{1}{n_\nu+1}, \ldots, 0.\right.$$

** S. N. Bernstein [42], in the case $q = \infty$, $m = 2$.
*** S. N. Bernstein [36], in the case $q = \infty$, $m = 2$.

Since the function $\phi(t_1, \ldots, t_m)$ is even, it can be supposed that the trigonometric polynomial which deviates least from it contains only cosines. Hence it follows that the best uniform approximation $E_{n_1, \ldots, n_m}(f)$ of the function $f(x_1, \ldots, x_m)$ by the algebraic polynomials $P_{n_1, \ldots, n_m}(x_1, \ldots, x_m)$ (see section 2.2.4) on the parallelepiped considered satisfies the inequality

$$E_{n_1, \ldots, n_m}(f) \leqslant B_m^{(k_1, \ldots, k_m)} \times$$

$$\times \sum_{\nu=1}^{m} \frac{1}{(n_\nu+1)^{k_\nu}} \omega\left(\frac{\partial^{k_\nu} f}{\partial x_\nu^{k_\nu}}; 0, \ldots, 0, \frac{1}{n_\nu+1}, 0, \ldots, 0\right) \tag{10}$$

5.3.3. A theorem similar to 5.3.1 holds also for the best approximations by integral functions of many variables, which are of finite degree in each of these variables (see section 4.9.55). Let the function $f(x_1, \ldots, x_m)$ be defined throughout the whole space of the variables $x_1, \ldots, x_m$ ($-\infty < x_1, \ldots, x_m < \infty$), and for some q ($1 \leqslant q \leqslant \infty$) let

$$\|f\|_{L_q} = \left\{ \int_{-\infty}^{\infty} \cdots \int_{-\infty}^{\infty} |f(x_1, \ldots, x_m)|^q \, \mathrm{d}x, \ldots, \mathrm{d}x_m \right\}^{\frac{1}{q}} < \infty .$$

For the lower bound

$$A_{\sigma, \ldots, \sigma_m}(f)_{L_q} = \inf_{G_{\sigma_1, \ldots, \sigma_m}} \|f(x_1, \ldots, x_m) - G_{\sigma_1, \ldots, \sigma_m}(x_1, \ldots, x_m)\|_{L_q} \tag{11}$$

with respect to all possible integral functions $G_{\sigma_1, \ldots, \sigma_m}(x_1, \ldots, x_m)$ *which satisfy the condition* 4.9(32), *this being the best approximation of* $f(x_1, \ldots, x_m)$ *by such functions in the metric* L_q, *there holds the inequality*

$$A_{\sigma_1, \ldots, \sigma_m}(f)_{L_q} \leqslant C_m^{(k_1, \ldots, k_m)} \sum_{\nu=1}^{m} \omega_{k_\nu}\left(f; 0, \ldots, 0, \frac{1}{\sigma_\nu}, \ldots, 0\right)_{L_q}, \tag{12}$$

where $G_m^{(k_1, \ldots, k_m)}$ *is a constant depending only on* m *and* $k_1, \ldots, k_m$, *and*

$$\omega_{k_\nu}(f; 0, \ldots, h_\nu, \ldots, 0)_{L_q}$$

$$= \sup_{|t| \leqslant h_\nu} \left\{ \int_{-\infty}^{\infty} \cdots \int_{-\infty}^{\infty} \left| \sum_{j=0}^{k_\nu} (-1)^j \binom{k_\nu}{j} f(x_1, \ldots, x_\nu + jt, \ldots, x_m) \right|^q \mathrm{d}x \ldots \mathrm{d}x_m \right\}^{\frac{1}{q}}$$

is the corresponding partial modulus of smoothness of the function f.

The proof of the inequality 5.3(12) is carried out as in the periodic case. Thus it is necessary to replace the polynomials 5.2(4) by the integral functions $g\sigma_\nu(x_\nu)$ defined in section 5.1.3, and the trigonometric polynomials 5.3(3) for $m = 2$ by the integral functions

$$\frac{1}{\gamma_{\sigma_1,k_1}\cdot\gamma_{\sigma_2,k_2}}\int_{-\infty}^{\infty}\int_{-\infty}^{\infty} f(x_1+t_1,\, x_2+t_2)\, G_{\sigma_1,k_1}(t_1)\, G_{\sigma_2,k_2}(t_2)\, dt\, dt_2$$

(see 5.1.3).

We note also that the remark 5.3.11 (see 5.3(7) and 5.3(8)) carries over as a whole to the best approximation $A\sigma_1, \ldots, \sigma_m(f)_{Lq}$.

5.4. The rapidity of the decrease to zero of the best approximation of analytic functions

It follows from theorem 5.1.5 that if the function $f(x)$, defined on $[a, b]$, is differentiable an infinite number of times, then for any r

$$\lim_{n\to\infty} n^r E_n(f; a, b) = 0. \tag{1}$$

In particular, the relation 5.4(1) holds for any function analytic on $[a, b]$. However, in the last case it is possible to say a great deal more about the rapidity of the decrease to zero of the best approximation of the function $f(x)$.

5.4.1. It turns out that if the function $f(x)$ is analytic on $[a, b]$ then the sequence of numbers $E_n(f; a, b)$, as n increases, decreases to zero not more slowly than the terms of a certain decreasing geometrical progression**.

If for $R > \frac{b-a}{2}$ *the function* $f(x)$ *belongs to the class* $A_R(a, b)$ (see section 3.7.21), *then*

$$\overline{\lim_{n\to\infty}} \sqrt[n]{E_n(f; a, b)} \leqslant \frac{b-a}{2R}. \tag{2}$$

For definiteness let us consider the segment $[-1,1]$. The transition to any other segment is accomplished by a well-known linear transformation. Putting $x = \cos y$, let us consider the sequence of algebraic polynomials

$$P_n(x) = \frac{1}{\pi}\int_0^{\pi} f(\cos t)[D_n(t+y)+D_n(t-y)]\, dt$$

$$= \frac{c_0}{2} + \sum_{k=1}^{n} c_k \cos k \arccos x, \tag{3}$$

** S. N. Bernstein [1], [15].

where $D_n(t)$ is a Dirichlet kernel (see section 4.5.2). Since the function $f(x) = f(\cos y)$ is analytic, it follows that on $[-1,1]$ it may be expanded in a Fourier-Chebyshev series, the partial sum of which is $P_n(x)$. Hence

$$|f(x)-P_n(x)| = \left| \sum_{k=n+1}^{\infty} c_k \cos k \arccos x \right|.$$

Consequently

$$E_n(f) \leqslant \max_{|x| \leqslant 1} |f(x)-P_n(x)| \leqslant \sum_{k=n+1}^{\infty} |c_k|.$$

If we take into consideration the inequality 3.7(9) for the coefficients c_k, we find that for any $r < R$ $(r > 1)$

$$E_n(f) \leqslant \frac{2M(r)}{r-1} \frac{1}{r^n} \tag{4}$$

i.e. 5.4(2) is satisfied.

5.4.11. It should be noted that the property of the best approximation of a function analytic on a given segment, obtained in section 5.4.1, which indicates that

$$\varlimsup_{n\to\infty} \sqrt[n]{E_n(f; a, b)} < 1,$$

can no longer hold when we pass to best approximations by polynomials with only integral coefficients. Moreover if (a, b) contains the integer m and if the radius of convergence of the Taylor series in powers of $x-m$ of a function which is analytic on $[a, b]$ is greater than unity, then the upper limit of the n-th root of the best approximation of $f(x)$ on this segment by polynomials of the n-th order with integral coefficients will always equal unity as $n \to \infty$. In fact, if the above-mentioned upper limit were less than unity, then in the neighbourhood of the point m the function $f(x)$ would be the sum of a series of polynomials with integral coefficients, which could be differentiated term by term any number of times. Hence it would follow that all coefficients of the given Taylor series for $f(x)$ are integers, i.e. the radius of convergence of this series would not be greater than unity.

Moreover, simple considerations similar to those of section 1.5.2 show that in those cases where the segment $[a, b]$ does not contain integers, the best approximation of any analytic function by polynomials of the n-th order with integral coefficients, as also the best approximation by any algebraic polynomials of the same order, is $O(\varrho^n)$ for some $\varrho < 1$ as $n \to \infty$.

5.4.12. In order to verify this, it is possible without loss of generality to take $[a, b] = [\delta, 1-\delta]$ $(0 < \delta < \frac{1}{2})$. It is sufficient to establish that** *for any polynomial*

$$P_n(x) = \sum_{k=0}^{n} a_k x^k = \sum_{k=0}^{n} d_k x^k (1-x)^{n-k}$$

there can be found a polynomial $Q_n(x) = \sum_{k=0}^{n} c_k x^k$ *with integral coefficients such that, for some* ϱ $(0 < \varrho < 1)$ *which does not depend on n,*

$$\max_{x \in [\delta, 1-\delta]} |P_n(x) - Q_n(x)| = O(\varrho^n) \quad as \quad n \to \infty. \tag{5}$$

The existence of the polynomial $Q_n(x)$ follows from the inequality

$$\max_{\delta \leqslant x \leqslant 1-\delta} |P_n(x) - \sum_{k=0}^{n} [\mathrm{d}_k] x^k (1-x)^{n-k}| \leqslant (n+1)(1-\delta)^n.$$

5.4.13. We note here by the way that, using the proposition of section 5.4.12, we can also verify that for the best approximations by algebraic polynomials with integral coefficients of functions differentiable a given finite number of times on segments not containing integers, the same estimates of order apply as were obtained previously for the uniform best approximations by any algebraic polynomials*** (see sections 5.1 and 5.1.5). This remark also applies to the estimates 5.2(1) and 5.2(10).

5.4.2. Considering instead of the polynomials 5.4(3) the partial sums of a trigonometric Fourier series and using the estimate 3.12(19) for its coefficients, it can be verified that *if the periodic function* $f(x)$ *is analytic, then for some* $\delta > 0$

$$\overline{\lim_{n \to \infty}} \sqrt[n]{E_n^*(f)} \leqslant \mathrm{e}^{-\delta}. \tag{6}$$

Here in the estimate obtained 2δ is the width of the strip $|\mathrm{Im}\, z| \leqslant \delta$ within which the function $f(x+iy)$ is analytic (see 3.12.15).

** L. V. Kantorovich [1].

*** A. O. Gel'fond [3] showed that, with certain limitations on $f(x)$, a similar proposition holds also in the case where the interval (a, b) does not contain integers. For the generalisation to the case of functions of many variables, see the work of G. A. Zhirnova [1].

5.4.3. Thus *if the function $f(x)$ is an integral function, then whatever may be the length of the segment $[a, b]$, for its best approximation by algebraic polynomials, we have the equality***.

$$\lim_{n\to\infty} \sqrt[n]{E(f; a, b)} = 0. \tag{7}$$

5.4.4. The relation 5.4(6), or even the significantly weaker assertion of Weierstrass's theorem that

$$\lim_{n\to\infty} E_n(f; a, b) = 0,$$

valid for the fixed segment $[a, b]$, cannot in general be satisfied if as n increases the difference $b-a$ increases without limit. Moreover, in a number of cases when $b-a \to \infty$, if there is a definite connection between the properties of the integral function $f(x)$ and the nature of the increase of the length of the segment, the best approximation $E_n(f; a, b)$ correspondingly tends to zero.

Let us consider one of these cases***. *Let the integral function $f(x)$ belong to the class B_σ for some $\sigma > 0$* (see section 4.3.1). *The inequality*

$$E_n\left(f; -\frac{n}{p}, \frac{n}{p}\right) < 2\sqrt{\frac{p}{p-\sigma}}\, e^{-\frac{2n}{3}\left(\frac{p-\sigma}{p}\right)^{3/2}} \sup_{-\infty<x<\infty} |f(x)|. \tag{8}$$

is satisfied for any $p > \sigma$ for all natural numbers n.

In order to verify the correctness of this estimate, let us put $\lambda = \dfrac{n}{p}$

** S. N. Bernstein [15]. This assertion no longer applies on passing to best approximations by algebraic polynomials $P_n(x)$ with integral coefficients, even when the segment $[a, b]$ does not contain integers. It is sufficient to consider the function $f(x) = C$, where C is an arbitrary irrational algebraic number. For any polynomial $P_n(x)$ of degree $\leqslant n$ with integral coefficients, and any rational number $\dfrac{p}{q} \in [a, b]$, we have, in virtue of Liouville's theorem on approximation of algebraic numbers by rationals,

$$\max_{a\leqslant x\leqslant b} |C-P_n(x)| \geqslant \left|C-P_n\left(\frac{p}{q}\right)\right| = \left|C-\frac{r}{q^n}\right| \geqslant \frac{A}{q^{mn}},$$

where r is an integer and m and A are constants which do not depend on n. In the case where a is an integer and $b = a+1$ it is not difficult to make this remark more precise by using the asymptotic estimate of the growth of the function $\pi(x)$ from the theory of the distribution of prime numbers. It can be shown that if the function $f(x)$ defined on $[0, 1]$ is not an algebraic polynomial with integral coefficients, then the upper limit of the n-th root of its best uniform approximation by polynomials on $[0, 1]$ of degree $\leqslant n$ with integral coefficients is never less than $\dfrac{1}{e}$.

*** S. N. Bernstein [26].

and consider the polynomials $P_n(x)$ of the type 5.4(3), which correspond to the segment $[-\lambda, \lambda]$, i.e. for $x = \lambda \cos y$ let us put

$$P_n(x) = \frac{1}{\pi} \int_0^{\pi} f(\lambda \cos t)[D_n(t+y)+D_n(t-y)]\,dt. \tag{9}$$

As in section 5.4.1, we have

$$|f(x)-P_n(x)| \leqslant \sum_{k=n+1}^{\infty} |c_k|, \tag{10}$$

where

$$c_k = \frac{2}{\pi} \int_0^{\pi} f(\lambda \cos t) \cos kt\,dt.$$

Since $f(x) \in B_\sigma$, it follows by theorem 4.8.3 that

$$|f(x+iy)| \leqslant e^{\sigma|y|} \sup_{-\infty<t<\infty} |f(t)|.$$

Consequently, for any $\delta > 0$ we have

$$|f(x+iy)| \leqslant e^{\sigma\delta\lambda} \sup_{-\infty<t<\infty} |f(t)|.$$

on the ellipse with foci at the points $\pm\lambda$ and with minor axis $2\delta\lambda$. As a result of this, we obtain as in section 3.7.3 the estimate

$$|c_k| \leqslant \frac{2e^{\sigma\delta\lambda}}{\left(\delta+\sqrt{1+\delta^2}\right)^k} \sup_{-\infty<t<\infty} |f(t)|,$$

from which in virtue of 5.4(10) it follows that

$$E_n\left(f; -\frac{n}{p}, \frac{n}{p}\right) \leqslant \max_{|x|\leqslant\lambda} |f(x)-P_n(x)| < \frac{2e^{\sigma\delta\lambda}}{\delta\left(\delta+\sqrt{1+\delta^2}\right)^n} \sup_{-\infty<t<\infty} |f(t)|.$$

We now note that since

$$\delta = \operatorname{sh} \ln\left(\delta+\sqrt{1+\delta^2}\right)$$

and

$$\alpha > \operatorname{sh}\alpha - \tfrac{1}{3}(\operatorname{sh}\alpha)^3,$$

for any $\alpha > 0$, it follows that

$$\delta+\sqrt{1+\delta^2} > e^{\delta-\frac{1}{3}\delta^3}.$$

Thus

$$E_n\left(f; -\frac{n}{p}, \frac{n}{p}\right) < \frac{2}{\delta} e^{\sigma\delta\lambda - n\delta + \frac{n}{3}\delta^3} \sup_{-\infty<t<\infty} |f(t)|.$$

Putting here $\delta = \sqrt{\dfrac{p-\sigma}{p}}$, we obtain the inequality 5.4(8).

5.4.5. We see that all integral functions $f(x)$ of the classes B_σ possess the property that for any $p > \delta$

$$\lim_{n\to\infty} E_n\left(f; -\frac{n}{p}, \frac{n}{p}\right) = 0. \tag{11}$$

From considerations similar to those of section 5.5.4, there follows a similar proposition for any integral functions of exponential type**. *If $f(x)$ is an integral function of finite degree $\leqslant \sigma$ and the constant $\gamma > 0$ satisfies the equation*

$$\frac{1}{\gamma}\sqrt{1+\gamma^2} = \ln\left(\gamma+\sqrt{1+\gamma^2}\right), \tag{12}$$

then 5.4(11) *holds for all $p > \gamma\sigma$.*

Since the function $f(x)$ is of degree $\leqslant \sigma$, then (see 4.3(1)) for any $\tau > \sigma$ an r can be found such that for all $|z| > r$ we have

$$|f(z)| < e^{\tau|z|}.$$

Consequently, for any $\lambda > r$ and $\delta > 0$

$$|f(x+iy)| \leqslant e^{\tau\lambda\sqrt{1+\gamma^2}}$$

on the ellipse with foci at the points $\pm\lambda$ and with minor axis $2\gamma\lambda$. Hence, as already mentioned in section 5.4.4, the coefficients c_k of the polynomial 5.4(9) satisfy the inequality

$$|c_k| \leqslant \frac{2e^{\tau\lambda\sqrt{1+\gamma^2}}}{\left(\gamma+\sqrt{1+\gamma^2}\right)^k},$$

or if $\beta > 0$ is such that

$$\frac{1}{\beta}\sqrt{1+\gamma^2} - \ln\left(\gamma+\sqrt{1+\gamma^2}\right) = -\varrho < 0,$$

then for $k > \tau\lambda\beta$

$$|c_k| < 2\left(\frac{e^{\frac{\sqrt{1+\gamma^2}}{\beta}}}{\gamma+\sqrt{1+\gamma^2}}\right)^k = 2e^{-k\rho}.$$

Thus if $n > \tau\lambda\beta$, then in virtue of the inequality 5.4(10) we have for all $x \in [-\lambda_1 \lambda]$

$$|f(x)-P_n(x)| < \frac{2e^{-n\rho}}{1-e^{-\rho}},$$

** S. N. Bernstein [27].

i.e.

$$E(f;\,-\lambda,\lambda) < \frac{2e^{-n\rho}}{1-e^{-\rho}}.$$

Putting $\beta = \dfrac{\gamma}{1-\alpha}$ $(0<\alpha<1)$ and noting that the constant γ satisfies the equation 5.4(12), we see that

$$\varrho = \frac{\alpha}{\gamma}\sqrt{1+\gamma^2} > 0,$$

and hence it is possible to take as β any number as close as desired to γ. As a result of this we may assert that for any $\tau>\sigma$ and $\beta>\gamma$ we have for $\lambda < \dfrac{n}{\beta\tau}$

$$\lim_{n\to\infty} E_n(f;\,-\lambda,\lambda) = 0,$$

i.e. 5.4(11) holds for any $p>\gamma\sigma$.

We note also that from the point of view considered here the given constant γ is the best and cannot be decreased in general. For every $p<\gamma\sigma$ there can be found an integral function $f(x)$ of degree $\leqslant\sigma$ for which $E_n\left(f;\,-\dfrac{n}{p},\dfrac{n}{p}\right)$ increases without limit with n (see, for example, 5.11.21).

5.4.6. By the use of the proposition of section 5.4.5, it is easy to extend the relations 2.6(8), proved in section 2.6.22 for bounded functions, to arbitrary functions $f(x)$ defined on the whole real axis, provided that σ is sufficiently great. If the function $f(x)$, defined on $(-\infty,\infty)$, is such that for some finite $\sigma_0>0$ the lower bound

$$A_{\sigma_0}(f) = \inf_{g_{\sigma_0}(x)}\ \sup_{-\infty<x<\infty} |f(x)-g_{\sigma_0}(x)|$$

is finite for all possible integral functions $g_{\sigma_0}(x)$ of degree $\leqslant\sigma_0$, then for any $\sigma>\gamma\sigma_0$, where $\gamma>0$ is a root of equation 5.4(12),

$$\lim_{n\to\infty} E_n\left(f;\,\frac{n}{\sigma\pm 0}\right) = \lim_{\varepsilon\to 0}\ \overline{\lim_{n\to\infty}}\, E_n\left(f;\,-\frac{n}{\sigma\pm\varepsilon},\frac{n}{\sigma\pm\varepsilon}\right)$$
$$= A_{\sigma\pm 0}(f) \quad (\varepsilon>0).$$

Since, by supposition, there exists some integral function $g_{\sigma_0}(x)$ of degree $\leqslant\sigma_0$ for which $\sup\limits_{-\infty<x<\infty}|f(x)-g_{\sigma_0}(x)|<\infty$, it follows in virtue of theorem 2.6.22 that

$$A_{\sigma\pm 0}(f) = A_{\sigma\pm 0}(f-g_{\sigma_0}) = \lim_{n\to\infty} E_n\left(f-g_{\sigma_0};\,\frac{n}{\sigma\pm 0}\right).$$

If in addition $\sigma > \gamma\sigma_0$, then, by what was proved in section 5.4.5,

$$\lim_{n\to\infty} E_n\left(g_{\sigma_0}; \frac{n}{\sigma \pm 0}\right) = 0.$$

Consequently

$$A_{\sigma+0}(f) = \lim_{n\to\infty} E_n\left(f; \frac{n}{\sigma \pm 0}\right)$$

for all $\sigma > \gamma\sigma_0$.

5.5. More precise forms of Jackson's theorem. The estimates of N. I. Akhiezer, M. G. Krein and J. Favard for differentiable periodic functions

If use is made of the notation introduced in section 3.2.7, then Jackson's inequality 5.1(1) is equivalent to the assertion that for modulus of continuity $\omega(t) \not\equiv 0$ and for $n > 0$

$$C_n^{(\omega)}(a, b) = \sup_{f \in H_\omega} \frac{E_n(f; a, b)}{\omega\left(\frac{b-a}{n}\right)} \leqslant C < \infty, \tag{1}$$

where C is an absolute constant. Similarly, for the upper bound

$$C_\sigma^{(\omega)} = \sup_{f \in H_\omega} \frac{A_\sigma(f)}{\omega\left(\frac{1}{\sigma}\right)} \tag{2}$$

the inequality 5.1(9) holds for the best approximation on the whole real axis considered in section 5.1.2. A similar formulation of the inequality 5.1(10) corresponding to the periodic case is connected with the upper bound

$$C_n^{(\omega)*} = \sup_{f \in H_\omega^*} \frac{E_n^*(f)}{\omega\left(\frac{1}{n+1}\right)}, \tag{3}$$

where H_ω^* is the class of all periodic functions $f(x)$ with period 2π of H_ω. The other theorems of section 5.1 can also be formulated in this form. Thus, for example, for the class $W_*^{(r)}(M)$ of all the periodic functions $f(x)$ of period 2π for which there exists an $(r—1)$-th absolutely continuous derivative $f^{(r-1)}(x)$ $(r \geqslant 1)$, and $|f^{(r)}(x)| \leqslant M$ almost everywhere, we obtain from 5.1(21) (for $M = 1$) the relation

$$C_{r,n}^* = \sup_{f \in W_*^{(r)}(1)} (n+1)^r E_n^*(f) \leqslant B_r < \infty, \tag{4}$$

where B_r is a constant depending only on r.

Concrete applications of the estimates of the best approximation of the functions which belong to the given classes requires a knowledge of the numerical values of the constants in the corresponding inequalities of this type. It is not difficult to obtain these constants directly from the proofs given in section 5.1** if best possible values are not important. The more delicate problem of the exact or asymptotically exact (as $n \to \infty$) value of the magnitude of the upper bounds of the type 5.5(1)–(4) requires special consideration.

5.5.1. The first such more precise form of Jackson's theorem was obtained by J. Favard [1], and also independently of him by N. I. Akhiezer and M. G. Krein [1] for the class $W_*^{(r)}(M)$. This result gives the exact value of the upper bound 5.5(4). *For any integral non-negative n and r the equality*

$$C_{r,n}^* = \sup_{f \in W_*^{(r)} 1} (n+1) E_n^*(f) = K_r, \tag{5}$$

is satisfied, where K_r *is the constant of* 2.11(10)***. In fact, if $f(x) \in W_*^{(r)} M$ and

$$\frac{a}{2} + \sum_{k=1}^{\infty} (a_k \cos kx + b_k \sin kx) \tag{6}$$

is the Fourier series of the function $f(x)$, then for $r \geqslant 1$, by the formula 3.5(15),

$$f(x) = \frac{a}{2} + \frac{1}{\pi} \int_0^{2\pi} f^{(r)}(t) D_0^{(r)}(x-t) \, dt.$$

Putting

$$T_n(x) = U_n(f; x; \lambda) = \frac{a}{2} + \sum_{k=1}^{n} \lambda_k^{(n)} (a_k \cos kx + b_n \sin kx), \tag{7}$$

where $\lambda_k^{(n)}$ $(n = 1, 2, \ldots; k = 1, 2, \ldots, n)$ is a certain arbitrary system of numbers, let us consider the difference $f(x) - T_n(x)$. Since

$$T_n(x) = U_n(f; x; \lambda) = \frac{a_0}{2} + \frac{1}{\pi} \int_0^{2\pi} f^{(r)}(x-t) \sum_{k=1}^{n} \lambda_k^{(n)} \frac{\cos\left(kt - \frac{r\pi}{2}\right)}{k^r} \, dt,$$

** Simple calculations show that for the constant C in the inequality 5.5(1) and for the constant B_r in the inequality 5.5(4) (for integral r) we have the estimates $C \leqslant 12$ and $B_r \leqslant 12^r$ (see I. P. Natanson [2]).

*** $W_*^{(1)} M$ is the class of periodic functions for which $|f(x)| \leqslant M$ almost everywhere. The case of fractional $r(0 < r < 1))$ was investigated by V. K. Dzyadyk [2] and Sun Yun-shen [1]($r > 6$).

then for any value of $\lambda_0^{(n)}$

$$|f(x)-T_n(x)|$$

$$=\frac{1}{\pi}\left|\int_0^{2\pi}\left\{D_0^{(r)}(t)-\lambda_0^{(n)}-\sum_{k=1}^{n}\frac{\lambda_k^{(n)}}{k^r}\cos\left(kt-\frac{r\pi}{2}\right)\right\}f^{(r)}(x-t)\,dt\right|$$

$$\leqslant\frac{M}{\pi}\int_0^{2\pi}\left|D_0^{(r)}(t)-\lambda_0^{(n)}-\sum_{k=1}^{n}\frac{\lambda_k^{(n)}}{k^r}\cos\left(kt-\frac{r\pi}{2}\right)\right|dt.$$

We can choose the numbers $\lambda_0^{(n)}$ and $\lambda_k^{(n)}$ $(k=1,2,\ldots,n)$ in such a way that the integral on the right-hand side of the last inequality will assume the least possible value. By theorem 2.11.5 (with respect to the function $F(t)=D_0^{(r)}(t)$) in the case when r is even, it is necessary for this purpose to put

$$(-1)^{\frac{r}{2}}\lambda_k^{(n)}=\lambda_r\left(\frac{k}{n+1}\right)$$

$$=1-k^r\sum_{\nu=1}^{\infty}(-1)^{\nu+1}\left\{\frac{1}{[2\nu(n+1)-k]^r}+\frac{1}{[2\nu(n+1)+k]^r}\right\},\tag{8}$$

and in the case when r is odd (see 2.13.132)

$$(-1)^{\frac{r-1}{2}}\lambda_k^{(n)}=\lambda_r\left(\frac{k}{n+1}\right)$$

$$=1-k^r\sum_{\nu=1}^{\infty}\left\{\frac{1}{[2\nu(n+1)-k]^r}-\frac{1}{[2\nu(n+1)+k]^r}\right\}.\tag{9}$$

By theorems 2.11.5 and 2.13.32, for a system of factors chosen in this way the deviation of every function $f(x)\in W_*^{(r)}M$ from the corresponding polynomial 5.5(7) will for any integral $r\geqslant 1$ satisfy the inequality

$$|f(x)-U_n(f;x;\lambda)|\leqslant\frac{MK_r}{(n+1)^r}.\tag{10}$$

It follows that for every integer $n\geqslant 0$

$$\sup_{f\in W_*^{(r)}M}E_n^*(f)\leqslant\frac{MK_r}{(n+1)^r}.\tag{11}$$

It remains to note that if $f(x)$ is the function 2.11(9), then $f^{(r)}(x)=(n+1)^r$ sign sin $(n+1)x$. Hence $\frac{M}{(n+1)^r}f(x)\in W_*^{(r)}M$, and since for this function $E_n^*(f)=K_r$ (see section 2.11.3), the sign of equality occurs in the relation 5.5(11).

5.5.2. Whatever the system of numbers $\lambda_n^{(n)}$ may be, the equation 5.5(7) establishes a correspondence between every integrable function $f(x)$ of period 2π and a trigonometric polynomial $U_n(f; x; \lambda)$. Since

$$U_n(f; x; \lambda) = \frac{1}{\pi}\int_0^{\pi 2} f(t)\left\{\frac{1}{2}+\sum_{k=1}^{n} \lambda_k^{(n)} \cos k\, x-t)\right\} dt, \tag{12}$$

this correspondence is always additive and homogeneous. It thus defines a certain linear method of approximation to the functions considered. It follows from the proof of theorem 5.5.1 that if $\lambda_k^{(n)}$ $(k = 0, 1, 2, \ldots, n)$ is the system of numbers 5.5(8) (when r is even) or the system of numbers 5.5(9) (when r is odd), then for every integer $r \geqslant 1$

$$\sup_{f\in W_*^{(r)}M} \max_x |f(x)-U_n(f; x; \lambda)| = \sup_{f\in W_*^{(r)}M} E_n^*(f).$$

This indicates that, if considered for the whole class $W_*^{(r)}M$, the method of approximation which corresponds to such a system of numbers gives precisely the same uniform deviation as it would be possible to obtain by taking for each function $f(x) \in W_*^{(r)}M$ its best Chebyshev polynomial approximation. In other words, *this triangular matrix of numbers* $\lambda_k^{(n)}$ $(k = 0, 1, 2, \ldots, n)$ *defines the best linear method of approximation to the considered class of functions.*

5.5.3. In the case when for some $n \geqslant 0$ the first $2n+1$ Fourier coefficients a_k $(k = 0, 1, \ldots, n)$ and b_k $(k = 1, 2, \ldots, n)$ of the function $f(x)$ are equal to zero, we have $U_n(f; x; \lambda) \equiv 0$. Hence it follows from the inequality 5.5(10) that *if for some integral* $r \geqslant 1$ *such a function belongs to the class* $W_*^{(r)}1$, *then***

$$|f(x)| \leqslant \frac{K_r}{(n+1)^r}. \tag{13}$$

5.5.4. It is immediately obvious from the definition of the numbers K_r (see 2.11(10)) that $K_1 > K_3 > K_5 > \ldots > \frac{4}{\pi}$ and $K_0 < K_2 < K_4 < < \ldots < \frac{4}{\pi}$. Here $K_0 = 1$ and $K_1 = \frac{\pi}{2}$. Consequently, for any integer r

$$1 \leqslant K_r \leqslant \frac{\pi}{2}. \tag{14}$$

This remark shows that the constant B_r in the inequality 5.5(4) or 5.1(21) is in reality bounded with respect to r, and $B_r \leqslant \frac{\pi}{2}$.

** J. Favard [1]. S. N. Bernstein [14], in the case $n = 0$, $r \geqslant 1$; (see 3.12.23); H. Bohr [1], in the case $n \geqslant 1$, $r = 1$.

It is also possible to give similar estimates for the constant in the other inequalities of section 5.1. Thus, for example, *if the periodic function* $f(x)$ *of period* 2π *has the continuous r-th* (r an integer) *derivative* $f^{(r)}(x)$, *then*

$$E^*_{n-1}(f) < \frac{5}{2n^r}\omega_2\left(f^{(r)}; \frac{1}{n}\right) \quad (n \geqslant 1). \tag{15}$$

It is not difficult to obtain a similar estimate which improves on the precision of 5.1(18) for $k = 2$, if use is made of the last two inequalities of 3.12(4). Putting $h = \frac{1}{2n}$, let us write the function $f(x)$ in the form of the sum

$$f(x) = \{f(x) - f_{h,2}(x)\} + f_{h,2}(x),$$

where $f_{h,2}(x)$ is the second function of V. A. Steklov for $f(x)$ (see section 3.12.1). Since by 3.12(4)

$$|f^{(r)}(x) - f^{(r)}_{h,2}(x)| \leqslant \frac{1}{2}\omega_2(f^{(r)}; 2h) \quad \text{and} \quad |f^{(r+2)}_{h,2}(x)| \leqslant \frac{1}{4h^2}\omega_2(f^{(r)}; 2h),$$

it follows that

$$E^*_{n-1}(f - f_{h,2}) \leqslant \frac{B_r}{2n^r}\omega_2(f^{(r)}; 2h) \leqslant \frac{\pi}{4n^r}\omega_2\left(f^{(r)}; \frac{1}{n}\right)$$

and

$$E^*_{n-1}(f_{h,2}) \leqslant \frac{B}{4h^2 n^{r+2}}\omega_2(f^{(r)}; 2h) \leqslant \frac{\pi}{2n^r}\omega_2\left(f^{(r)}; \frac{1}{n}\right).$$

Consequently

$$E^*_{n-1}(f) \leqslant E^*_{n-1}(f - f_{h,2}) + E^*_{n-1}(f_{h,2}) < \frac{5}{2n^r}\omega_2\left(f^{(r)}; \frac{1}{n}\right).$$

Similarly, by considering the first function of V. A. Steklov $f_{h,1}(x)$, it is possible to obtain the estimate**

$$E^*_{n-1}(f) < \frac{3}{n^r}\omega\left(f^{(r)}; \frac{1}{n}\right), \tag{16}$$

which gives one possible value of the constant on the right-hand side of the inequality 5.1(18) for $k = 1$.

** N. I. Akhiezer [1]. In this monograph there was first given a systematic application of the functions of V. A. Steklov to problems of harmonic approximation. Even earlier A. N. Kolmogorov [1] applied these means to the investigation of the conditions of compactness of families of functions (see 3.12.2).

5.5.5. If the r-th derivative of the function $f(x)$ belongs to the class $L_q^*(q \geqslant 1)$ and

$$\left\{\int_0^{2\pi} |f^{(r)}(x)|^q \, dx\right\}^{\frac{1}{q}} \leqslant M,$$

then, using Minkowski's inequality for integrals and reasoning exactly as in section 5.5.1, we obtain the result that for the system of numbers 5.5(8) when r is even, or 5.5(9) when r is odd,

$$\left\{\int_0^{2\pi} |f(x) - U_n(f; x; \lambda)|^q \, dx\right\}^{\frac{1}{q}} \leqslant \frac{MK_r}{(n+1)^r},$$

i.e.

$$E_n^*(f)_{Lq} \leqslant \frac{MK_r}{(n+1)^r}. \tag{17}$$

Hence with the help of the inequalities 3.12(8), just as in section 5.5.4, it is possible to obtain the similar estimates

$$\begin{aligned} E_n^*(f)_{Lq} &< \frac{3}{n^r}\,\omega\left(f^{(r)}; \frac{1}{n}\right)_{Lq}, \\ E_n^*(f)_{Lq} &< \frac{5}{2n^r}\,\omega_2\left(f^{(r)}; \frac{1}{n}\right)_{Lq} \end{aligned} \tag{18}$$

for differentiable functions.

5.6. On the asymptotic behaviour of the upper bounds of best approximations of classes of functions differentiable a given finite number of times. S. N. Bernstein's theorem

Formula 5.5(5), obtained for the best approximations of periodic functions by trigonometric polynomials considered in section 5.5.1, shows that as n increases without limit the upper bound $\sup\limits_{f \in W_*^{(r)} 1} n^r E_n^*(f)$ possesses a limit. It appears that this is true in certain other cases also. In this section the classes $W^{(r)}H^{(\alpha)}$ (see section 3.5.3) are investigated from this point of view and an important asymptotic relation is established for the corresponding upper bounds for the best approximations by algebraic polynomials on a finite segment and for the best approximations by transcendental integral functions on the whole real axis.

5.6.1. In addition to the class $W^{(r)}H^{(\alpha)}M$ of all continuous functions $f(x)$ defined on the real axis $-\infty < x < \infty$ and having an r-th (r an integer) derivative $f^{(r)}(x)$ which satisfies the Lipschitz condition $\omega(f^{(r)}; t) \leqslant Mt^{\alpha}$ $(0 < \alpha \leqslant 1)$, let us consider the part of it, $W_{*}^{(r)}H^{(\alpha)}M$, which consists of all periodic functions of period 2π occurring in $W^{(r)}H^{\alpha}M$, and also the class $W_{a}^{(r)}H^{(\alpha)}M$ of functions considered on $[-a, a]$ and possessing an r-th derivative which satisfies on this segment a Lipschitz condition of order α with constant M. Let us note here that *for a given value of a, whatever the function $f(x) \in W_{a}^{(r)}H^{(\alpha)}M$ may be, there exists a periodic function $F_r(x) \in W^{(r)}H^{(\alpha)}M$ which differs from $f(x)$ on the segment $-a \leqslant x \leqslant a$ by an algebraic polynomial of degree $\leqslant r$.*

In fact, let us continue the function $f(x) = f^{(r)}(x) - \frac{1}{2a}\int_{-a}^{a} f^{(r)}(t)\,dt$ periodically on to the whole real axis, preserving the modulus of continuity, as indicated in the remark of section 3.5.71. Then the mean value over a period of the function $F(x)$ after continuation is equal to zero, the period is equal to $4a$ and

$$\max_{-\infty<x<\infty} |F(x)| = \max_{|x|\leqslant a}\left| f^{(r)}(x) - \frac{1}{2a}\int_{-a}^{a} f^{(r)}(t)\,dt \right| \leqslant 2^{\alpha}Ma^{\alpha}. \tag{1}$$

Let

$$F(x) = \sum_{\substack{k=-\infty \\ k\neq 0}}^{\infty} c_k e^{i\frac{k\pi}{2a}x}$$

be the Fourier series of $F(x)$. The function

$$F_r(x) = \sum_{\substack{k=-\infty \\ k\neq 0}}^{\infty} \left(\frac{2a}{ik\pi}\right)^{r} c_k e^{i\frac{k\pi}{2a}x} \tag{2}$$

belongs to the class $W^{(r)}H^{(\alpha)}M$ and possesses the property that

$$F_r^{(r)}(x) = f^{(r)}(x) - \frac{1}{2a}\int_{-a}^{a} f^{(r)}(t)\,dt \quad \text{for} \quad x \in [-a, a].$$

Hence for $-a \leqslant x \leqslant a$

$$F_r(x) = f(x) + \sum_{\nu=0}^{r} b_\nu x^\nu .$$

Thus by the remark quoted above

$$C_{r,n}^{(a)}(-1, 1) = \sup_{f \in W^{(r)}H^{(a)}1} n^{r+a}E_n(f;-1, 1)$$
$$= \sup_{f \in W^{(r)}H^{(a)}1} n^{r+a}E_n(f;-1, 1) \tag{3}$$

for $n \geqslant r$.

5.6.2. Let us prove that *as n increases without limit the quantity* $C_{r,n}^{(a)}(-1, 1)$ *has a limit and*

$$\lim_{n\to\infty} C_{r,n}^{(a)}(-1, 1) = C_r^{(a)} = \sup_{f \in W^{(r)}H^{(a)}1} A_1(f), \tag{4}$$

where $A_1(f)$ *is the best approximation of the function* $f(x)$ *by integral functions of finite degree* $\leqslant 1$ *on* $(-\infty, \infty)$** (see section 2.6.2).

Let $f(x)$ be an arbitrary function of the class $W^{(r)}H^{(\alpha)}M$, $f_n(t) = f\left(\frac{n}{\sigma}, t\right)$ and $P_n(f; x)$ an algebraic polynomial of degree $\leqslant n$ which deviates least from the function $f(x)$ on $\left[-\frac{n}{\sigma}, \frac{n}{\sigma}\right]$. From the definition of the constant $C_{r,n}^{(\alpha)}$ it immediately follows that

$$E_n\left(f;-\frac{n}{\sigma}, \frac{n}{\sigma}\right) = \max_{|x|\leqslant \frac{n}{\sigma}} |f(x)-P_n(f;x)|$$
$$= \max_{|t|\leqslant 1} \left|f_n(t)-P_n\left(f;\frac{n}{\sigma}t\right)\right| = E_n(f_n;-1,1)$$
$$\leqslant \left(\frac{n}{\sigma}\right)^{r+\alpha} \frac{MC_{r,n}^{(\alpha)}(-1,1)}{n^{r+\alpha}} = \frac{MC_{r,n}^{(\alpha)}(-1,1)}{\sigma^{r+\alpha}} \tag{5}$$

for any $\sigma > 0$ and $f(x) \in W^{(r)}H^{(\alpha)}M$. But since (see section 5.4.6***)

$$A_1(f) = \lim_{\varepsilon\to 0} \varlimsup_{n\to\infty} E_n\left(f;-\frac{n}{1+\varepsilon}, \frac{n}{1+\varepsilon}\right) \qquad (\varepsilon > 0)$$

we obtain, from 5.6(5), after passing to the limit, the inequality

$$A_1(f) \leqslant M \varliminf_{n\to\infty} C_{r,n}^{(\alpha)}(-1,1),$$

** S. N. Bernstein [35].

*** Since for any function $f(x) \in W^{(r)}H^{(\alpha)}M$ the relation $f(x) = O(|x|^{r+\alpha}+1)$ holds, then in virtue of the footnote to page 258, the estimates of section 5.1.3 and the inequality 3.3(1), the limit equality of section 5.4.6 is applicable to it for all $\sigma > 0$.

from which it follows for $M=1$ that

$$C_r^{(\alpha)} \leqslant \varliminf_{n\to\infty} C_{r,n}^{(\alpha)}(-1,1) \tag{6}$$

Now let $f(x)$ be an arbitrary function of the class $W_1^{(r)} H^{(\alpha)} M$ and $F_r(x)$ the function corresponding to it in the class $W^{(r)} H^{(\alpha)} M$, defined in section 5.6.1. If $n \geqslant r$ $(\varepsilon > 0)$, $\Phi_r(x) = F_r\left(\frac{1+\varepsilon}{n}x\right)$ and $P_n(F_r; x)$ is the algebraic polynomial of degree $\leqslant n$ which deviates least from $F_r(x)$ on $[-1,1]$, then

$$\begin{aligned} E_n(f;-1,1) = E_n(F_r;-1,1) &= \max_{|t|\leqslant 1} |F_r(t)-P_n(F_r;t)| \\ &= \max_{|x|\leqslant \frac{n}{1+\varepsilon}} \left| F_r\left(\frac{1+\varepsilon}{n}x\right) - P_n\left(F_r; \frac{1+\varepsilon}{n}x\right)\right| \\ &= E_n\left(\Phi_r; -\frac{n}{1+\varepsilon}, \frac{n}{1+\varepsilon}\right). \end{aligned} \tag{7}$$

Let us denote by $g_1(x) = g_1(\Phi_r; x)$ the integral function of degree $\leqslant 1$ which deviates least from $\Phi_r(x)$ on $(-\infty, \infty)$. It is obvious that

$$\begin{aligned} &E_n\left(\Phi_r; -\frac{n}{1+\varepsilon}, \frac{n}{1+\varepsilon}\right) \\ &\quad\leqslant E_n\left(\Phi_r - g_1; -\frac{n}{1+\varepsilon}, \frac{n}{1+\varepsilon}\right) + E_n\left(g_1; -\frac{n}{1+\varepsilon}, \frac{n}{1+\varepsilon}\right) \\ &\quad\leqslant A_1(\Phi_r) + E_n\left(g_1; -\frac{n}{1+\varepsilon}, \frac{n}{1+\varepsilon}\right). \end{aligned} \tag{8}$$

Since $F_r(x) \in W^{(r)} H^{(\alpha)} M$, it follows that $\Phi_r(x) \in W^{(r)} H^{(\alpha)} M \left(\frac{1+\varepsilon}{n}\right)^{r+\alpha}$ and consequently

$$A_1(\Phi_r) \leqslant \left(\frac{1+\varepsilon}{n}\right)^{r+\alpha} M C_r^{(\alpha)}. \tag{9}$$

Moreover, we have

$$\sup_{-\infty<x<\infty} |g_1(x)| \leqslant 2 \max_x |\Phi_r(x)| = 2 \max_x |F_r(x)|. \tag{10}$$

From the definition of the function $F_r(x)$ (see section 5.6.1) we have for $r \geqslant 1$ (see section 3.5.5)

$$F_r(x) = \left(\frac{2}{\pi}\right)^r \sum_{\substack{k=-\infty \\ k\neq 0}}^{\infty} \frac{c_k}{(ik)^r} e^{i\frac{k\pi}{2}x} = \frac{2^{r-1}}{\pi^r} \int_{-2}^{2} F(t) D_0^{(r)}\left(\frac{x-t}{2}\pi\right) dt.$$

Hence

$$\max_x |F_r(x)| \leqslant \max_{|x|\leqslant 2} |F(x)| \frac{2^r}{\pi^{r+1}} \int_{-\pi}^{\pi} |D_0^{(r)}(t)|\,dt = \delta_r \max_{|x|\leqslant 2} |F(x)|.$$

Since the periodic function $F(x)$ is obtained as the continuation of the function $\phi(x) = f^{(r)}(x) - \frac{1}{2}\int_{-1}^{1} f^{(r)}(t)\,dt$ by the method indicated in the remark of section 3.5.71, the maximum of its modulus coincides with the maximum of the modulus of $\phi(x)$, i.e.

$$\max_x |F(x)| = \max_{|x|\leqslant 1} \left| f^{(r)}(x) - \frac{1}{2}\int_{-1}^{1} f^{(r)}(t)\,dt \right| \leqslant 2^\alpha M.$$

Hence $\max_x |F_r(x)| \leqslant 2^\alpha M\delta_r$ for any $r > 0$, and in virtue of 5.6(10)

$$\sup_{-\infty<x<\infty} |g_1(x)| \leqslant 2^{\alpha+1} M\delta_r.$$

Taking account of the last estimate, we have, by the theorem of section 5.4.4,

$$E_n\left(g_1; -\frac{n}{1+\varepsilon}, \frac{n}{1+\varepsilon}\right) < 2^{\alpha+2}\sqrt{\frac{1+\varepsilon}{\varepsilon}}\, M\delta_r\, e^{-\frac{2n}{3}}\left(\frac{\varepsilon}{1+\varepsilon}\right)^{\frac{3}{2}}. \tag{11}$$

Let us now choose N so large that for any $n > N$ the right-hand side of 5.6(11) will be less than $\frac{\varepsilon M}{n^{r+\alpha}}$. Then for all such n we obtain, in virtue of 5.6(8) and 5.6(9)

$$E_n\left(\Phi_r; -\frac{n}{1+\varepsilon}, \frac{n}{1+\varepsilon}\right) < M\left(\frac{1+\varepsilon}{n}\right)^{r+\alpha} C_r^{(\alpha)} + \frac{\varepsilon M}{n^{r+\alpha}},$$

or by 5.6(7)

$$n^{r+\alpha} E_n(f; -1, 1) < M[1+\varepsilon)^{r+\alpha} C_r^{(\alpha)} + \varepsilon]$$

for any function $f(x) \in W_1^{(r)} H^{(\alpha)} M$. It follows from this that for all n

$$\sup_{f\in W_1^{(r)}H^{(\alpha)}M} n^{r+\alpha} E_n(f; -1, 1) = MC_{r,n}^{(\alpha)} < M[(1+\varepsilon)^{r+\alpha} C_r^{(\alpha)} + \varepsilon],$$

i.e.

$$\overline{\lim_{n\to\infty}}\, C_{r,n}^{(\alpha)} < (1+\varepsilon)^{r+\alpha} C_r^{(\alpha)} + \varepsilon,$$

and in virtue of the arbitrariness of $\varepsilon > 0$,

$$\overline{\lim_{n\to\infty}}\, C_{r,n}^{(\alpha)} \leqslant C_r^{(\alpha)}. \tag{12}$$

The validity of 5.6(4) follows from the inequalities 5.6(6) and 5.6(12).

5.6.3. The numerical value of the constant $C_r^{(\alpha)}$ in the relation 5.6(4) for $0 < \alpha < 1$ is still unknown. However, in the case when $\alpha = 1$ this value can be determined from the considerations of section 5.5.1.

5.6.31. In the first place we shall show that *for any function* $f(x) \in W_1^{(r-1)}H^{(1)}M$ *there exists a sequence of algebraic polynomials* $P_n(x)$ *which possess the property that at each point* $x \in [-1, 1]$

$$\overline{\lim_{n\to\infty}}\, n^r |f(x)-P_n(x)| \leqslant MK_r\sqrt{1-x^2}, \tag{13}$$

where K_r *is the constant of* 2.11(10)**.

Let us consider the function $\phi(t) = f(\cos t)$ and let us note that if $f(x) \in W_1^{(r-1)}H^{(1)}M$ then

$$\phi^{(r)}(t) = \psi_r(t)+(-1)^r \sin^r t f^{(r)}(u), \tag{14}$$

where $u = \cos t$, $f^{(r)}(u) = f^{(r)}(\cos t)$, and $\psi_r(t) \in W_*^{(0)}H^{(1)}M_r$.

If

$$a_0 = \frac{1}{\pi}\int_{-\pi}^{\pi} f(\cos t)\,dt$$

then (see section 3.5.5)

$$\phi(y) = \frac{a_0}{2}+\frac{1}{\pi}\int_{-\pi}^{\pi} D_0^{(r)}(t)\phi^{(r)}(t+y)\,dt$$

and consequently

$$f(\cos y) = \frac{a_0}{2}+\frac{1}{\pi}\int_{-\pi}^{\pi} D_0^{(r+1)}\psi_r'(t+y)\,dt+$$

$$+\frac{(-1)^r}{\pi}\int_{-\pi}^{\pi} D_0^{(r)}(t)\sin^r(t+y)f^{(r)}[\cos(t+y)]\,dt.$$

Let us put

$$T_{n,r}(t) = \sum_{k=1}^{n} \lambda_r\left(\frac{k}{n+1}\right)\frac{\cos\left(kt-\frac{\nu\pi}{2}\right)}{k^r}, \tag{15}$$

where $\lambda_r\left(\frac{k}{n+1}\right)$ is the system of numbers 5.5(8) (if r is even) or 5.5(9) (if r is odd).

** A. F. Timan [18]. In the case $r = 1$ this result was obtained by S. M. Nikol'skii [4].

Since the function $\psi_r'(t)$ is of the form

$$\psi_r'(t) = \sum_{k=1}^{r} Q_{k,r}(t) f^{(k)}(\cos t),$$

where

$$Q_{k,r}(t) = \sum_{\nu=0}^{k} a_\nu^{(r)} \cos\left(\nu t - \frac{(\nu+1)\pi}{2}\right),$$

then the sum

$$\frac{a_0}{2} + \frac{1}{\pi} \int_{-\pi}^{\pi} T_{n,r+1}(t) \psi_r'(t+y) \, dt +$$

$$+ \frac{(-1)^r}{\pi} \int_{-\pi}^{\pi} T_{n,r}(t) \sin^r(t+y) f^{(r)}[\cos(t+y)] \, dt$$

represents a certain even trigonometric polynomial $P_n(\cos y)$ of order $\leqslant n$. For the difference $f(\cos y) - P_n(\cos y)$ we have the inequality

$$|f(\cos y) - P_n(\cos y)|$$

$$\leqslant \frac{1}{\pi} \int_{-\pi}^{\pi} |D_0^{(r+1)}(t) - T_{n,r+1}(t)| \, |\psi_r'(t+y)| \, dt +$$

$$+ \frac{1}{\pi} \int_{-\pi}^{\pi} |D_0^{(r)}(t) - T_{n,r}(t)| \, |\sin^r(t+y)| f^{(r)}[\cos(t+y)]| \, dt.$$

Consequently,

$$|f(\cos y) - P_n(\cos y)| \leqslant \frac{M}{\pi} \int_{-\pi}^{\pi} |D_0^{(r)}(t) - T_{n,r}(t)| \, |\sin^r(t+y) \, dt +$$

$$+ \frac{M_r}{\pi} \int_{-\pi}^{\pi} |D^{(r+1)}(t) - T_{n,r+1}(t)| \, dt. \qquad (16)$$

where with the condition $f^{(k)}(0) = 0$ $(k = 0, 1, \ldots, r-1)$ in virtue of 3.10(4) M_r depends only on r and on the value of the constant M. The following lemma can be used to obtain an estimate of the right-hand side of 5.6(16).

5.6.32. *For any* $\varepsilon > 0$,

$$\lim_{n\to\infty} n^r \int_{\varepsilon}^{\pi} |D_0^{(r)}(t) - T_{n,r}(t) \, dt = 0^{**}. \tag{17}$$

Since in the metric L the polynomial $T_{n,r}(t)$ deviates least from the function $D_0^{(r)}(t)$ (see section 5.5.1), it follows by theorem 2.11.4 that

$$\int_0^{\pi} |D_0^{(r)}(t) - T_{n,r}(t)| \, dt = \left| \int_0^{\pi} \{D_0^{(r)}(t) - T_{n,r}(t)\} \operatorname{sign}\cos\left[(n-1)t - \frac{r\pi}{2}\right] dt \right|.$$

Consequently, taking account of sections 2.8.7 and 2.11(12), we find that for every integral $\nu > 0$

$$\left| \int_0^{\pi} \{D_0^{(r)}(t) - T_{n,r}(t)\} \left\{ \operatorname{sign}\cos\left[(n+1)t - \frac{r\pi}{2}\right] - \right.\right.$$

$$\left.\left. - \operatorname{sign}\cos\left[(n+\nu+1)t - \frac{r\pi}{2}\right]\right\} dt \right|$$

$$= \int_0^{\pi} |D_0^{(r)}(t) - T_{n,r}(t)| \, dt - \int_0^{\pi} |D_0^{(r)}(t) - T_{n+\nu,r}(t)| \, dt$$

$$= \frac{\pi}{2} K_r \left\{ \frac{1}{(n+1)^r} - \frac{1}{(n+\nu+1)^r} \right\}$$

$$= \frac{\pi}{2} K_r \frac{r\nu}{n^{r+1}} + O\left(\frac{1}{n^{r+2}}\right). \tag{18}$$

Let us now note that for every $t \in [\varepsilon, \pi]$, where $\cos\left[(n+1)t - \frac{r\pi}{2}\right] \neq 0$, there exists at least one natural number $\nu \leqslant N = \left[\frac{\pi}{\varepsilon}\right] + 1$ for which

$$\operatorname{sign}\cos\left[(n+\nu+1)\,t - \frac{r\pi}{2}\right] = -\operatorname{sign}\cos\left[(n+1)t - \frac{r\pi}{2}\right]. \tag{19}$$

Hence if we denote by $G_\nu^{(n)}$ the set of all the t of $[0, \pi]$ where 5.6(19) holds, we have

$$\int_{\varepsilon}^{\pi} |D_0^{(r)}(t) - T_{n,r}(t)| \, dt$$

$$\leqslant \frac{1}{2} \sum_{\nu=1}^{N} \left| \int_{G_\nu^{(n)}} \{D_0^{(r)}(t) - T_{n,r}(t)\} \left\{ \operatorname{sign}\cos\left[(n+1)t - \frac{r\pi}{2}\right] - \right.\right.$$

** This lemma was communicated to the author by S. M. Nikol'skii. The proof given was proposed by V. K. Dzyadyk [1].

$$-\operatorname{sign}\cos\left[(n+\nu+1)t-\frac{r\pi}{2}\right]\Bigg\}\,dt$$

$$=\frac{1}{2}\sum_{\nu=1}^{N}\left|\int_0^{\pi}\{D_0^{(r)}(t)-T_{n,r}(t)\}\left\{\operatorname{sign}\cos\left[(n+1)t-\frac{r\pi}{2}\right]-\right.\right.$$

$$\left.\left.-\operatorname{sign}\cos\left[(n+\nu+1)t-\frac{r\pi}{2}\right]\right\}dt\right|.$$

Thus in virtue of 5.6(18) we have the inequality

$$\int_{\varepsilon}^{\pi}|D_0^{(r)}(t)-T_{n,r}(t)|\,dt\leqslant\frac{r\pi}{4}K_r\frac{N(N+1)}{2n^{r+1}}+O\left(\frac{N}{n^{r+2}}\right),$$

i.e.

$$\int_{\varepsilon}^{\pi}|D_0^{(r)}(t)-T_{n,r}(t)|\,dt=O\left(\frac{1}{\varepsilon^2n^{r+1}}\right). \tag{20}$$

This relation proves 5.6(17).

5.6.33. In virtue of theorem 2.11.4, the second of the integrals on the right-hand side of 5.6(16) is a quantity of the order $O\left(\frac{1}{n^{r+1}}\right)$, uniformly with respect to all $f\in W_1^{(r-1)}H^{(1)}M$ for which $f^{(k)}(0)=0\,(k=0,1,\dots,r-1)$. Hence it follows from the inequality 5.6(16) that

$$|f(\cos y)-P_n(\cos y)|\leqslant\frac{2M}{\pi}|\sin y|^r\int_0^{\pi}|D_0^{(r)}(t)-T_{n,r}(t)|\,dt+$$

$$+\frac{M}{\pi}\int_{-\pi}^{\pi}|D_0^{(r)}(t)-T_{n,r}(t)|\sum_{\nu=1}^{r-1}\binom{r}{\nu}|\sin t|^{r-\nu}\,dt+O\left(\frac{1}{n^{r+1}}\right).$$

For any $\varepsilon\,(0<\varepsilon<1)$ it is possible by the lemma of 5.6.32 to find an n_0 such that for all $n\geqslant n_0$

$$\frac{M}{\pi}\int_{\varepsilon}^{\pi}|D_0^{(r)}(t)-T_{n,r}(t)|\,dt<\frac{\varepsilon}{(2n)^r}.$$

Thus for all $n\geqslant n_0$

$$|f(\cos y)-P_n(\cos y)|<M|\sin y|^r\frac{K_r}{n^r}+\frac{2^{r+1}MK_r\varepsilon}{(n+1)^r}+\frac{2\varepsilon}{n^r}+O\left(\frac{1}{n^{r+1}}\right),$$

i.e. the difference between the function $f(x)$ and the algebraic polynomial $P_n(x)$ satisfies for $|x| \leqslant 1$ and $n \geqslant n_0$ the inequality

$$n^r|f(x)-P_n(x)| < MK_r(\sqrt{1-x^2})^r+2^{r+1}K_r M_\varepsilon+2\varepsilon+O\left(\frac{1}{n}\right).$$

Hence

$$\overline{\lim_{n\to\infty}}\, n^r|f(x)-P_n(x)| \leqslant MK_r(\sqrt{1-x^2})^r+2^{r+1}K_r M_\varepsilon+2\varepsilon,$$

and as $\varepsilon > 0$ is arbitrary, we obtain from this theorem 5.6.31. Since the condition $f^{(k)}(0) = 0$ $(k = 0, 1, \ldots, r-1)$ does not restrict the generality of this theorem, the inequality 5.6(13) is satisfied uniformly with respect to all $x \in [-1,1]$ and $f \in W_1^{(r-1)}H^{(1)}M$.

5.6.34. It follows from the inequality 5.5(5) and theorem 5.6.31 that

$$\lim_{n\to\infty} C_{r,n}^* = \lim_{n\to\infty} \sup_{f\in W_*^{(r-1)}H^{(1)}1} n^r E_n^*(f)$$

$$= K_r \geqslant \lim_{n\to\infty} C_{r-1,n}^{(1)}(-1,1) = C_{r-1}^{(1)}. \qquad (21)$$

On the other hand, let $f(x)$ be an arbitrary function of the class $W_*^{(r-1)}H^{(\alpha)}M$ and $\phi_n(x) = f\left(\frac{x}{n}\right)$. In virtue of theorem 2.6.21 (see 2.6(5) we have $E_n^*(f) = A_n(f)$ and, since $A_n(f) = A_1(\phi_n)$, we have

$$E_n^*(f) = A_1(\phi_n) \leqslant \frac{MC_{r-1}^{(\alpha)}}{n^{r+\alpha-1}}. \qquad (22)$$

Consequently, the inequality

$$C_{r,n}^* = \sup_{f\in W_*^{(r-1)}H^{(1)}1} (n+1)^r E_n^*(f) \leqslant \left(\frac{n+1}{n}\right)^r C_{r-1}^{(1)}$$

holds for all n, or

$$K_r \leqslant \left(\frac{n+1}{n}\right)^r C_{r-1}^{(1)},$$

i.e. $K_r \leqslant C_{r-1}^{(1)}$. Combining this inequality with 5.6(21), we find that *the constant* $C_{r-1}^{(1)}$ *is identical with the constant* K_r**. Hence in virtue of theorem 5.6.2 it follows that *it is impossible to decrease the coefficient* K_r *in the inequality* 5.6(13) *for the whole class* $W_1^{(r-1)}H^{(1)}M$. (See the end of section 5.6.33.)

5.6.35. Since for any $\sigma > 0$ and $\phi_\sigma(x) = f\left(\frac{x}{\sigma}\right)$

$$A_\sigma(f) = A_1(\phi_\sigma),$$

** M. G. Krein [2].

we find from theorem 5.6.2, taking into account that $C_{r-1}^{(1)} = K_r$, that for any natural number r

$$\sup_{f \in W^{(r-1)} H^{(1)} M} A_\sigma(f) = \frac{MK_r}{\sigma^r}. \tag{23}$$

Hence as in section 5.5.4, using the inequalities 3.12(11) for the functions of V. A. Steklov, it is possible to obtain

$$\begin{aligned} A_\sigma(f) &< \frac{3}{\sigma^r}\omega\left(f^{(r)}; \frac{1}{\sigma}\right), \\ A_\sigma(f) &< \frac{5}{2\sigma^r}\omega_2\left(f^{(r)}; \frac{1}{\sigma}\right) \end{aligned} \tag{24}$$

for the best approximation $A_\sigma(f)$ of bounded functions with bounded r-th derivative (r an integer) on $(-\infty, \infty)$, which is an improvement in precision on 5.1(17) for $k = 1$ and $k = 2$.

5.6.36. The inequality

$$A_\sigma(f) \leqslant \frac{K_r}{\sigma_r} \operatorname*{vrai\,sup}_{-\infty < x < \infty} |f^{(r)}(x),$$

which follows from the estimate 5.6(23), holds for *any* function $f(x)$, which for a given $r \geqslant 1$ belongs to one of the classes $W^{(r-1)} H^{(1)} M$. It follows from this that for the functions $f(x)$ considered we in fact always have

$$A_\sigma(f) \leqslant \frac{K_r}{\sigma^r} A_\sigma(f^{(r)}). \tag{25}$$

The estimate

$$A_\sigma(f^{(r)}) \leqslant C_r\left\{\sigma^r A_\sigma(f) + \sum_{\nu=[\sigma]}^{\infty} \nu^{r-1} A_\nu(f)\right\} \tag{26}$$

(C_r is a constant which depends only on r), which has a meaning if $\sum_{\nu=1}^{\infty} \nu^{r-1} A_\nu(f) < \infty$, is in a certain sense converse to 5.6(25). It can be obtained as follows. Since $f(x) = \lim_{\sigma \to \infty} g_\sigma(f; x)$, where $g_\sigma(f; x)$ is an integral function of degree $\leqslant \sigma$ for which

$$A_\sigma(f) = \sup_{-\infty < x < \infty} |f(x) - g_\sigma(f; x)|,$$

then, for any $\sigma > 0$,

$$f(x) - g_\sigma(f; x)$$
$$= g_{2^{p+2}}(f; x) - g_\sigma(f; x) + \sum_{k=p+2}^{\infty} \{g_{2^{k+1}}(f; x) - g_{2^k}(f; x)\}.$$

Moreover, thanks to the inequality of S. N. Bernstein for the derivative of integral functions of the class B_σ (see 4.8(5)),

$$|g^{(r)}_{2^{k+1}}(f;x)-g^{(r)}_{2^k}(f;x)| \leqslant 2^{r(k+1)+1}A_{2^k}(f) \leqslant 2^{2r+1} \sum_{\nu=2^{k-1}+1}^{2^k} \nu^{r-1}A_\nu(f).$$

Hence, choosing p such that $2^p \leqslant \sigma < 2^{p+1}$, we obtain

$$\begin{aligned} A_\sigma(f^{(\nu)}) &\leqslant \sup_{-\infty<x<\infty} |f^{(r)}(x)-g^{(r)}_\sigma(f;x)| \\ &\leqslant 2^{(p+2)r+1}A_\sigma(f)+2^{2r+1} \sum_{\nu=2^{p+1}+1}^{\infty} \nu^{r-1}A_\nu(f) \\ &\leqslant C_r\{\sigma^r A_\sigma(f)+\sum_{\nu=[\sigma]}^{\infty} \nu^{r-1}A_\nu(f)\}, \end{aligned}$$

i.e. 5.6(26) is satisfied.

In particular, if the function $f(x)$ is periodic with period 2π, then in virtue of theorem 2.6.21, the inequality 5.6(25) and the estimate 5.6(26) we obtain**

$$E^*_u(f) \leqslant \frac{K_r}{(n+1)^r} E^*_n(f^{(r)}) \tag{27}$$

and

$$E^*_n(f^{(r)}) \leqslant C_r\{(n+1)^r E^*_n(f)+\sum_{\nu=n=1}^{\infty} \nu^{r-1}E^*_\nu(f)\}^{***}. \tag{28}$$

5.6.4. As was shown by S. N. Bernstein, the limiting inequality

$$\lim_{n\to\infty} \sup_{f\in W_1^{(r)}H^{(1)}M} n^{r+1}E_n(f;-1,1) = \lim_{n\to\infty} \sup_{f\in W_*^{(r)}H^{(\alpha)}M} n^{r+1}E^*_n(f)^{****},$$

obtained in section 5.6.34 for the classes $W^{(r)}H^{(\alpha)}M$ for $\alpha=1$, holds for all $\alpha\,(0<\alpha\leqslant 1)$, i.e. for any $\alpha\,(0<\alpha\leqslant 1)$

$$\lim_{n\to\infty} \sup_{f\in W_1^{(r)}H^{(\alpha)}M} n^{r+\alpha}E_n(f;-1,1) = \lim_{n\to\infty} \sup_{f\in W_*^{(r)}H^{(\alpha)}M} n^{r+\alpha}E^*_n(f). \tag{29}$$

** It is also possible to obtain the inequality 5.6(27) directly from formula 3.5(15), if it is observed that for any trigonometric polynomials $U_n(x)$ and $V_n(x)$ of order $\leqslant n$

$$\frac{1}{\pi}\int_0^{2\pi} [f^{(r)}(t)-U_n(t)]\,[D_0^{(r)}(x-t)-V_n(x-t)]\,dt = f(x)-T_n(x),$$

where $T_n(x)$ is a certain trigonometric polynomial of order $\leqslant n$.

*** S. B. Stechkin [2].

**** For $r=0$ this equality was first obtained by S. M. Nikol'skii [1].

In the case $r = 0$ this can be verified if use is made of theorem 5.6.2 and it is noted that on the one hand, in virtue of the inequality 5.6(22) we have

$$n^\alpha E_n^*(f) \leqslant MC_0^{(\alpha)}, \tag{30}$$

for all functions $f(x) \in W_*^{(1)} H^{(\alpha)} M$, and consequently

$$\overline{\lim_{n\to\infty}} \sup_{f \in W_*^{(0)} H^{(\alpha)} M} n^\alpha E_n^*(f) \leqslant MC_0^{(\alpha)},$$

and on the other hand, to any function $f(x) \in W_1^{(0)} H^{(\alpha)} M$ there corresponds a function $F(t) \in W_*^{(0)} H^{(\alpha)} M$ $\big(F(t) = f(\cos t)\big)$ for which $E_n^*(F) = E_n(f; -1, 1)$ and consequently

$$\underline{\lim_{n\to\infty}} \sup_{f \in W_*^{(0)} H^{(\alpha)} M} n^\alpha E_n^*(f) \geqslant \lim_{n\to\infty} \sup_{f \in W_1^{(0)} H^{(\alpha)} M} n^\alpha E_n(f; -1, 1) = MC_0^{(\alpha)}.$$

5.7. The best approximation of functions analytic in a strip

Let us consider the problem of the character of the decrease of the best approximation (by integral functions of exponential type) of bounded functions $f(x)$ which are analytic in a certain strip containing the real axis $-\infty < x < \infty$. First and foremost the classes $A_{\infty,\delta} UM$ (or $A_{\infty,\delta} M$) defined in section 3.8.6 are of interest from this point of view.

One result which belongs here has already been derived in section 5.4.2 for the periodic case (see 5.4(6)). We now consider the wider classes indicated, and commencing with considerations similar to those given in section 5.5.1 for the classes $W_*^{(r)} M$, we give more precise estimates.

5.7.1. We make two preparatory remarks.

5.7.11. An integral of the form

$$\int_{-\infty}^{\infty} \phi_\sigma(x+t)\, g(t)\, dt,$$

where $\phi_\sigma(t)$ is an integral function of the class $B_\sigma(L)$, and $g(t)$ is an arbitrary bounded measurable function, is always some integral function of the class B_σ. This can be verified by expanding $\phi_\sigma(x+t)$ in a Taylor series in powers of x and applying the integral inequality of S. N. Bernstein (see 4.8(28), $q = 1$).

5.7.12.** *If the bounded function $C(t)$, defined on the semi-axis $t \geqslant 0$, is twice monotonic, i.e. $C'(t) \leqslant 0$, $C''(t) \geqslant 0$, and $\lim_{n\to\infty} C(t) = 0$, then for all $x \neq 0$*

$$b(x) = \int_0^\infty C(t) \cos xt\, dt \tag{1}$$

** E. C. Titchmarsh [2].

Here the function $b(x)$ is non-negative and integrable on the whole real axis. After integration by parts the first of these assertions follows from the integrability of the function $C'(t)$ on $[0, \infty)$, and the second from the identity

$$b(x)=\sum_{n=0}^{\infty}\int_{0}^{\frac{\pi}{2}}\left\{C\left(\frac{t+2n\pi}{x}\right)-C\left(\frac{\pi-t+2n\pi}{x}\right)-\right.$$
$$\left.-C\left(\frac{\pi-t+2n\pi}{x}\right)+C\left(\frac{2\pi-t+2n\pi}{x}\right)\right\}\frac{\cos t}{x}\,\mathrm{d}t.$$

In order to verify the integrability of the function $b(x)$ on $(-\infty, \infty)$, we remark that

$$xb(x)=-\int_{0}^{\infty}C'(t)\sin tx\,\mathrm{d}x. \tag{2}$$

Since $C'(t)$ is monotonic and integrable on $(0, \infty)$, it belongs to the class L_2 in this interval. By theorem 3.11.23 (see also section 3.11.21) the product $xb(x)$ also belongs to the same class. Hence it follows with the help of Hölder's inequality that the function $b(x)$ is integrable on any interval (ε, ∞) $(\varepsilon > 0)$. Applying Plancherel's theorem (see sections 3.11.21 and 3.11.23) to the transformation 5.7(2), we have

$$-C'(x)=\frac{2}{\pi}\frac{\mathrm{d}}{\mathrm{d}x}\int_{0}^{\infty}(1-\cos xt)b(t)\,\mathrm{d}t,$$

i.e.

$$C(0)-C(x)=\frac{2}{\pi}\int_{0}^{\infty}(1-\cos xt)b(t)\,\mathrm{d}t.$$

Since $b(t)\geqslant 0$, it follows that for any $\varepsilon > 0$

$$C(0)-C(x)\geqslant\frac{2}{\pi}\int_{\varepsilon}^{\infty}(1-\cos xt)b(t)\,\mathrm{d}t.$$

Passing to the limit as $x\to\infty$ and using the well-known Riemann-Lebesgue theorem, we obtain the inequality

$$\int_{\varepsilon}^{\infty}b(t)\,\mathrm{d}t\leqslant\frac{\pi}{2}C(0)$$

which shows that the function $b(t)$ belongs to the class L on the whole range $[0, \infty)$.

5.7.2. Let us consider an arbitrary function $f(x)$ of the class $A_{\infty,\delta}UM$ (see section 3.8.6). In virtue of the remark 5.7.11, for any $\sigma > 0$

$$A_\sigma(f) \leqslant \inf_{\substack{|g(t)|\leqslant M \\ \phi_\sigma(t)\in B_\sigma(L)}} \sup_{-\infty<x<\infty} \left| f(x) - \int_{-\infty}^{\infty} \phi_\sigma(x+t) g(t)\,\mathrm{d}t \right|.$$

Since $f(x) \in A_{\infty,\delta}UM$, the representation 3.8(17) holds, and using this we obtain from the above the inequality

$$A_\sigma(f) \leqslant \inf_{\phi_\sigma(t)\in B_\sigma(L)} \frac{M}{2\delta} \int_{-\infty}^{\infty} \left| \frac{1}{\operatorname{ch}\dfrac{\pi t}{2\delta}} - \phi_\sigma(t) \right| \mathrm{d}t. \tag{3}$$

Furthermore, in order to find an explicit expression for the right-hand side of 5.7(3) it turns out to be possible to apply considerations similar to those given in section 2.11.5.

We have the identity

$$\frac{1}{2\operatorname{ch}\dfrac{\pi x}{2\delta}} = \frac{1}{\pi} \int_0^{\infty} \frac{\cos xt}{\operatorname{ch}\delta t}\,\mathrm{d}t, \tag{4}$$

the validity of which is easily verified by expanding the function $a(t) = \dfrac{1}{\pi \operatorname{ch}\delta t}$ in a series of powers of $\mathrm{e}^{-\delta t}$ and integrating the corresponding series on the right-hand side of 5.7(4) term by term.

Here the function $a(t)$ plays the same part as was played in section 2.11.5 by the sequence of numbers a_k. By analogy with the sequence of numbers 2.11(17) let us consider for any $\sigma > 0$ the function

$$C(t) = \sum_{\nu=0}^{\infty} (-1)^\nu a[(2\nu+1)\sigma + t]. \tag{5}$$

Since for sufficiently great values of t the function $a(t)$ is triply monotonic, i.e. $a'(t) < 0$, $a''(t) > 0$, $a'''(t) < 0$, then for all σ beginning with a certain value $C(t)$ is doubly monotonic on the semi-axis $t \geqslant 0$.

As a result of this and in virtue of section 5.7.12 the product

$$2\cos \sigma x \int_0^{\infty} C(t) \cos tx\,\mathrm{d}x = \int_0^{\sigma} [C(\sigma-t) + C(\sigma+t)] \cos xt\,\mathrm{d}t + \\ + \int_{\sigma}^{\infty} [C(t-\sigma) + C(t+\sigma)] \cos xt\,\mathrm{d}t$$

has a sign which is identical with the sign of $\cos\sigma x$. On the other hand, since for $t\geqslant 0$

$$C(t+\sigma)+C(t-\sigma)=a(t),$$

it follows that this product can be written as the difference

$$2\cos\sigma x\int_0^\infty C(t)\cos tx\,dt$$

$$=\frac{1}{\pi}\int_0^\infty\frac{\cos tx}{\operatorname{ch}\delta t}\,dt-\int_0^\sigma[a(t)-C(\sigma-t)-C(\sigma+t)]\cos xt\,dt,$$

or in virtue of the identity 5.7(4)

$$2\cos\sigma x\int_0^\infty C(t)\cos tx\,dt=\frac{1}{2\delta}\left\{\frac{1}{\operatorname{ch}\frac{\pi x}{2\delta}}-\psi_\sigma(x)\right\},$$

where

$$\psi_\sigma(x)=2\delta\int_0^\sigma[a(t)-C(\sigma-t)-C(\sigma+t)]\cos xt\,dt.$$

Since $\psi_\sigma(x)$ is a certain integral function of degree $\leqslant\sigma$ ($\psi_\sigma(x)\in B_\sigma(L)$), it follows from 5.7(3) that

$$A_\sigma(f)\leqslant\frac{M}{2\delta}\int_{-\infty}^\infty\left|\frac{1}{\operatorname{ch}\frac{\pi t}{2\delta}}-\psi_\sigma(t)\right|dt$$

$$=\frac{M}{2\delta}\int_{-\infty}^\infty\left\{\frac{1}{\operatorname{ch}\frac{\pi t}{2\delta}}-\psi_\sigma(t)\right\}\operatorname{sign}\cos\sigma t\,dt.$$

Substituting here in place of $\operatorname{sign}\cos\sigma t$ the series

$$\operatorname{sign}\cos\sigma t=\frac{4}{\pi}\sum_{k=0}^\infty(-1)^k\frac{\cos(2k+1)\sigma t}{2k+1}$$

and integrating term by term, we obtain

$$A_\sigma(f)\leqslant\frac{2M}{\pi\delta}\left\{\sum_{k=0}^\infty\frac{(-1)^k}{2k+1}\int_{-\infty}^\infty\frac{\cos(2k+1)\sigma t}{\operatorname{ch}\frac{\pi t}{2\delta}}\,dt-\right.$$

$$\left.-\sum_{k=0}^\infty\frac{(-1)^k}{2k+1}\int_{-\infty}^\infty\psi_\sigma(t)\cos(2k+1)\sigma t\,dt\right\}.$$

If use is made of the identity 5.7(4) and the fact that the Fourier transform of the function $\psi_\sigma(t)$ for $t \geqslant \sigma$ is identically equal to zero, we finally obtain the inequality

$$A_\sigma(f) \leqslant \frac{4M}{\pi} \sum_{k=0}^{\infty} \frac{(-1)^k}{(2k+1)\,\mathrm{ch}\,(2k+1)\,\sigma\delta} \tag{6}$$

for the best approximation $A_\sigma(f)$ of the arbitrary function $f(x)$ of the class $A_{\infty,\delta}UM$ for sufficiently great values of σ.

Considering in the class $A_{\infty,\delta}UM$ the function

$$f_0(x) = \frac{M}{2\delta} \int_{-\infty}^{\infty} \frac{\mathrm{sign}\cos\sigma t}{\mathrm{ch}\dfrac{\pi(n-t)}{2\delta}}\,\mathrm{d}t,$$

which has period $\dfrac{2\pi}{\delta}$, and using the propositions of sections 2.6.21 and 2.7.32, we verify the fact that for it the quantity $A_{\sigma-0}(f_0)$ is identically equal to the right-hand side of 5.7(6). Consequently**, for all *sufficiently great values of σ the exact equality*

$$\sup_{f \in A_{\sigma,\delta}UM} A_{\sigma-0}(f) = \frac{4M}{\pi} \sum_{k=0}^{\infty} \frac{(-1)^k}{(2k+1)\,\mathrm{ch}\,(2k+1)\,\sigma\delta} \tag{7}$$

is satisfied.

5.7.21. Putting $\sigma = n+1$, we find that*** *for any periodic function $f(x) \in A_{\infty,\delta}UM$ of period 2π the inequality*

$$E_n^*(f) \leqslant \frac{8M}{\pi} \sum_{k=0}^{\infty} \frac{(-1)^k}{2k+1} \frac{\varrho^{(2k+1)(n+1)}}{1+\varrho^{2(2k+1)(n+1)}} < \frac{8M}{\pi} \varrho^{n+1}, \tag{8}$$

is satisfied for all sufficiently great values of n, where $\varrho = \mathrm{e}^{-\delta}$.

This result can also be obtained directly by using the representation 3.8(20).

5.7.22. The estimate 5.7(7) shows that if the real function $f(x)$, defined on $(-\infty, \infty)$, belongs to the class $A_{\infty,\varepsilon}M_\varepsilon$ for any positive $\varepsilon < \delta$ (see section 3.8.6), then

$$\overline{\lim_{\sigma\to\infty}} \sqrt[\sigma]{A_\sigma(f)} \leqslant \mathrm{e}^{-\delta}. \tag{9}$$

Thus the limiting relation 5.7(9) is valid for any function $f(x)$ of the class $A_{\infty,\delta}$ bounded on $(-\infty, \infty)$ (see section 3.8). This inequality, which

** N. I. Akhiezer [2]. See also N. I. Akhiezer [3], where the given result is obtained from more general considerations of the same type.

*** N. I. Akhiezer [3].

generalises 5.4(6), could also have been obtained directly from the estimate (see 5.6(23) and 5.5(14))

$$A_\sigma(f) \leqslant \frac{\pi M_n(f)}{2\sigma^n} \tag{10}$$

and the condition (see section 3.8)

$$\overline{\lim_{n\to\infty}} \sqrt[n]{\frac{M_n(f)}{n!}} \leqslant \frac{1}{\delta},$$

where $M_n(f) = \sup\limits_{-\infty<x<\infty} |f^{(n)}(x)|$, by making use of Stirling's formula (see 3.7(2)) and for arbitrary $\varepsilon > 0$ putting $\sigma = \frac{n(1+\varepsilon)}{\delta}$**.

Let us note that from the inequality 5.7(8) there immediately follows an estimate more precise than 5.4(2) of the best approximation by algebraic polynomials of a function $f(x)$ analytic on a given finite segment.

For definiteness, let us consider the segment $[-1,1]$ and a real function $f(x)$ which belongs to the corresponding class $A_R M$ (see section 3.7.21). Putting $z = \cos w$ let us consider the periodic function $F(w) = f(\cos w)$ of period 2π.

To the interior part of the ellipse C_R (see section 3.7.21) in the z-plane there then corresponds the interior part of the strip $|v| < \ln R$ in the plane of $w = u+iv$. Thus $F(W) \in A_{\infty, \ln R} M$, and as in this transformation to an even trigonometric polynomial there corresponds an algebraic polynomial of the same order, then on applying the inequality 5.7(8) to the function $F(w)$ we find that*** *if the real function $f(x)$ belongs to the class $A_R M(-1,1)$, then for sufficiently great n we have for its best approximation by algebraic polynomials on $[-1,1]$ the estimate*

$$E_n(f) \leqslant \frac{8M}{\pi} \sum_{k=0}^{\infty} \frac{(-1)^k}{2k+1} \frac{R^{(2k+1)(n+1)}}{1+R^{2(2k+1)(n+1)}} < \frac{8M}{\pi R^{n+1}}. \tag{11}$$

5.8. Constructive properties of some quasi-analytic classes of functions

The property of a sequence of best approximations of analytic functions established in section 5.4.1 is closely connected with the constructive properties considered in that section of functions which satisfy the conditions of analyticity indicated in section 3.10 (see 3.10(1) and 3.10(12)).

5.8.1. Let us begin with the quasi-analytic classes of functions characterised by condition 3.10(2), and prove that**** *if the infinitely differ-*

** S. N. Bernstein [45].
*** N. I. Akhiezer [3].
**** S. N. Bernstein [7].

entiable function $f(x)$, *defined on* $[a, b]$, *satisfies condition* 3.10(1) *on this segment, then the inequality* 3.10(13) *holds for its best approximation by algebraic polynomials, i.e.*

$$\overline{\lim_{n\to\infty}} \sqrt[n]{E_n(f; a, b)} < 1. \tag{1}$$

The inequality 5.8(1) can be obtained by using Jackson's theorem applied to the estimate 5.2(15), from which it follows that for any $m \geqslant n$

$$E_m(f; a, b) \leqslant C^n \frac{(b-a)^n (m-n)!}{m!} M_n(f) \leqslant \left(\frac{B}{m}\right)^n M_n(f),$$

where B is some constant. Hence it follows that if for some bounded sequence n_k we have $M_{n_k}(f) < (Mn_k)^{n_k}$, then for $m = 2BMn_k$ we have

$$\sqrt[m_k]{E_{m_k}(f; a, b)} \leqslant \left(\frac{1}{2}\right)^{\frac{1}{2BM}},$$ i.e. 5.8(1) holds**.

We give another proof of theorem 5.8.1 based on an idea of S. N. Bernstein, which is also interesting in itself. By Taylor's formula

$$f(x) = \sum_{\nu=0}^{n_k-1} \frac{1}{\nu!} f^{(\nu)}(c)\,(x-c)^\nu + \frac{f^{(n)_k}[c+\theta(x-c)]}{n_k!}(x-c)^{n_k}, \tag{2}$$

where $|\theta| < 1$, $c = \frac{1}{2}(a+b)$ and n_k is a subsequence of the natural numbers for which

$$\lim_{n_k\to\infty} \frac{1}{n_k} \sqrt[n_k]{M_{n_k}(f)} = \overline{\lim_{n\to\infty}} \frac{1}{n} \sqrt[n]{M_n(f)}. \tag{3}$$

As in section 3.10.1, it follows from formula 5.8(2) and condition 3.10(1) that for some constant M, depending only on the function f considered, and for

$$P_{n_k-1}(x) = \sum_{\nu=0}^{n_k-1} \frac{1}{\nu!} f^{(\nu)}(c)\,(x-c)^\nu$$

the inequality $|f(x)-P_{n_k-1}(x)| < (M|x-c|)^{n_k}$ is satisfied. Hence for all x for which $|x-c| < \frac{1}{2M}$,

$$|f(x)-P_{n_k-1}(x)| < \frac{1}{2^{n_k}}. \tag{4}$$

Then choosing instead of c the points $c \pm \frac{1}{4M}$ and repeating the same procedure for them, we obtain a similar inequality for two segments of the same length with centres

** For the best approximation $A_\sigma(f)$ of a function defined on $(-\infty, \infty)$ and satisfying there condition 3.10(1), the similar inequality $\overline{\lim_{\sigma\to\infty}} \sqrt[\sigma]{A_\sigma(f)} < 1$ holds. It follows directly from 5.7(10).

at $c \pm \frac{1}{4M}$. It is obvious that in this way we can cover $[a, b]$ by a finite system of consecutively intersecting intervals (a_ν, b_ν) $(\nu = 1, 2, \ldots, m)$ on each of which the inequality

$$E_{n_k}(f;\, a_\nu, b_\nu) < \frac{1}{2^{n_k}} \tag{5}$$

is satisfied.

It is now necessary to show that 5.8(1) follows from the finite system of inequalities 5.8(5). Let us consider the case $m = 2$. Then without loss of generality it may be supposed that $a = -1$, $b = 1$ and that the segments $[a_1, b_1]$, $[a_2, b_2]$ are $[-r, 1]$, $[-1, r]$, where $0 < r < 1$.

Thus it is required to prove that if for some infinite sequence of natural numbers n_k the function $f(x)$, considered on $[-1, 1]$, satisfies the conditions

$$\begin{gathered} E_{n_k}(f;\, -r, 1) < \frac{1}{2^{n_k}} \cdot \\ E_{n_k}(f;\, -1, r) < \frac{1}{2^{n_k}}, \end{gathered} \tag{6}$$

then

$$\lim_{n\to\infty} \sqrt[n]{E_n(f;\, -1, 1)} < 1. \tag{7}$$

For this purpose let us make use of the well-known binomial series, applied in chap. I (see section 1.2.1). After term-by-term differentiation of the binomial expansion

$$|x| = 1 - \frac{1}{2}(1-x^2) - \sum_{\nu=2}^{\infty} \frac{1\cdot 3 \ldots (2\nu-3)}{2^\nu \cdot \nu!}(1-x^2)^\nu$$

we find that for $r < |x| \leqslant 1$

$$1 + x + x\sum_{\nu=2}^{\infty} \frac{1\cdot 3 \ldots (2\nu-3)}{2^{\nu-1}(\nu-1)!}(1-x^2)^{\nu-1} = \begin{cases} 2 & \text{if} \quad x > 0, \\ 0 & \text{if} \quad x < 0. \end{cases}$$

The partial sums of the last series

$$S_{2n-1}(x) = 1 + x + x\sum_{\nu=2}^{n} \frac{1\cdot 3 \ldots (2\nu-3)}{2^{\nu-1}(\nu-1)!}(1-x^2)^{\nu-1} \tag{8}$$

possess the property that for all x on the segment $[-1, -r]$

$$|S_{2n-1}(x)| < \frac{1}{r}(1-r^2)^n \quad (-1 \leqslant x \leqslant -r) \tag{9}$$

and for all x on the segment $[r, 1]$

$$\left|1 - \frac{1}{2}S_{2n-1}(x)\right| < \frac{1}{r}(1-r^2)^n \quad (r \leqslant x \leqslant 1). \tag{10}$$

If, however, $|x| < r$, then

$$|S_{2n-1}(x)| < 1 + nr < 2n \quad (|x| < r). \tag{11}$$

Let us consider a sequence of polynomials $\sigma_n(x)$ such that $\sigma_{2n}(x) = \sigma_{2n-1}(x) = S_{2n-1}(x)$. Let $P_n(x)$ and $Q_n(x)$ be the best polynomial approximations of degree n to the function $f(x)$ on the segments $[-r, 1]$ and $[-1, r]$ respectively.

Let us form the polynomials

$$R_{2n}(x) = \tfrac{1}{2}\{P_n(x)\sigma_n(x)+Q_n(x)[2-\sigma_n(x)]\}. \tag{12}$$

From the fact that on the segment $[-r, r]$ the relations

$$|f(x)-P_{n_k}(x)| < \frac{1}{2^{n_k}}, \quad |f(x)-Q_{n_k}(x)| < \frac{1}{2^{n_k}}$$

are simultaneously satisfied, it follows in virtue of 5.8(11) that for $|x| \leqslant r$

$$|f(x)-R_{2n_k}(x)| < \frac{n_k+1}{2^{n_k}}.$$

In order to estimate the difference $f(x)-R_{2n_k}(x)$ for $|x| \geqslant r$, we make use of the inequality 4.8(39) for the segments $[-r, 1]$, $[-1, r]$ and for the polynomials $P_n(x)$, $Q_n(x)$ which correspond to them. If $\max\limits_{|x|\leqslant 1} |f(x)| = M$, then in virtue of 5.8(6)

$$|P_{n_k}(x)| < \left(M+\frac{1}{2^{n_k}}\right)\left[\frac{3-r+2\sqrt{2(1-r)}}{1+r}\right]^{n_k}, \quad |Q_{n_k}(x)| < M+\frac{1}{2^{n_k}}$$

for $-1 \leqslant x \leqslant -r$, and

$$|P_{n_k}(x)| < M+\frac{1}{2^{n_k}}, \quad |Q_{n_k}(x)| < \left(M+\frac{1}{2^{n_k}}\right)\left[\frac{3-r+2\sqrt{2(1-r)}}{1+r}\right]^{n_k}$$

for $r \leqslant x \leqslant 1$. Consequently, by 5.8(9) and 5.8(10),

$$|P_{n_k}(x)\sigma_{n_k}(x)| < \frac{1}{r}\left(M+\frac{1}{2^{n_k}}\right)\left\{[3-r+2\sqrt{2(1-r)}]\sqrt{\frac{1-r}{1+r}}\right\}^{n_k},$$

for $-1 \leqslant x \leqslant -r$ and

$$|Q_{n_k}(x)[2-\sigma_{n_k}(x)]| < \frac{2}{r}\left(M+\frac{1}{2^{n_k}}\right)\left\{3-r+2\sqrt{2(1-r)}\,\sqrt{\frac{1-r}{1+r}}\right\}^{n_k}$$

for $r \leqslant x < 1$. If the value of r is taken sufficiently close to unity, then for some $A > 0$ and $\delta < 1$ we find, that

$$|P_{n_k}(x)\sigma_{n_k}(x)| < A\,\delta^{n_k}$$

for $-1 \leqslant x \leqslant -r$ and

$$|Q_{n_k}(x)[2-\sigma_{n_k}(x)]| < A\,\delta^{n_k}$$

for $r \leqslant x \leqslant 1$. Hence if we take account of 5.8(6) and also the estimates 5.8(9), 5.8(10) and 5.8(11), it follows that for some $C > 0$ and $\varrho < 1$

$$|f(x)-R_{2n_k}(x)| < C\varrho^{2n_k} \quad (\varrho > 0) \tag{13}$$

for all x on the segment $[-1, 1]$.

Thus the assertion is proved if the ratio of the length of the common part of the two segments considered to the length of each of them is sufficiently close to unity. Let us note, then, that here the necessary lower bound of the value of this ratio is an absolute quantity. Hence in those cases where such a ratio does not satisfy the requirements indicated, it will be possible after repeating the given procedure a finite number of times to obtain for some natural number p a sequence of polynomials $R^*_{pn_k}(x)$ of degree $\leqslant pn_k$ which possess the property that for any x on the whole segment $[-1, 1]$ the inequality

$$|f(x)-R^*_{pn_k}(x)| < C\varrho^{pn_k} \tag{14}$$

is satisfied, where $C > 0$ and $0 < \varrho < 1$, i.e. 5.8(7) holds.

Similarly we conclude that from the validity of the proposition formulated above for $m = 2$ its validity follows for any finite value of m.

5.8.2. The inequality 5.8(1) could also have been obtained with less restrictive assumptions on $f(x)$. In particular, there is no necessity to suppose the function $f(x)$ infinitely differentiable in advance. If instead of Taylor's formula 5.8(2) use is made of its finite difference analogue 3.12(17) (see 3.12.11), then, using the same considerations as in section 5.8.1 for the proof of the inequality 5.8(1), we arrive at the following proposition**.

The assertion of theorem 5.8.1 *is valid for any function* $f(x)$ *continuous on* $[a, b]$ *for which there exists an infinite sequence of natural numbers* n_k *and an infinite null-sequence of values* h_k $(h_k \to 0$ *as* $k \to \infty)$ *such that*

$$\sqrt[n_k]{|\Delta_{h_k}^{n_k} f(x)|} < M n_k |h_k|, \tag{15}$$

where M is a constant which does not depend on x, n_k *and* h_k.

5.8.3. The following theorem** gives an idea of the character of the decrease of the sequence of best approximations of the functions $f(x)$ which satisfy the Denjoy condition 3.10.2. *If the function* $f(x)$, *defined on* $[a, b]$, *is infinitely differentiable and satisfies the condition* 3.10(12), *then*

$$\sum_{n=1}^{\infty} \frac{1}{\sup\limits_{k \geqslant 1} \sqrt[n]{E_k(f; a, b)}} = \infty. \tag{16}$$

In order to prove this proposition, we make use of the inequality

$$E_k(f; a, b) \leqslant C^n (b-a)^n \frac{M_n(f)}{(k-n+1)(k-n+2)(k-n+3)\dots k} \qquad (k \geqslant n), \tag{17}$$

which follows directly from 5.2(15). It follows from this that for any $k \geqslant n$

$$\sqrt[n]{M_n(f)} \geqslant \frac{k}{c(b-a)} \sqrt[n]{E_k(f; a, b)\left(1-\frac{1}{k}\right)\cdots\left(1-\frac{n-1}{k}\right)},$$

and as the lower bound of the expression

$$\sqrt[n]{\left(1-\frac{1}{k}\right)\cdots\left(1-\frac{n-1}{k}\right)}$$

for all integral n and for all $k \geqslant n$ is greater than zero, then, denoting it by δ, we obtain for $k \geqslant n$

$$\sqrt[n]{M_n(f)} \geqslant \frac{\delta}{c(b-a)} k \sqrt[n]{E_k(f; a, b)}. \tag{18}$$

** S. N. Bernstein [7].

Thus the conditions 3.10(12) and 5.8(18) imply the convergence of the series

$$\sum_{n=1}^{\infty} \frac{1}{\sup\limits_{k\geqslant n} \sqrt[n]{E_k(f;a,b)}}. \tag{19}$$

We shall show that the convergence of this series entails the convergence of the series 5.8(16).

Noting that for $|f(x)| \leqslant 1$ the terms of the series 5.8(16) do not increase as n increases (since $E_n(f;a,b) \leqslant 1$), we conclude that in the case when this series converges, by virtue of the lemma on numerical series given in section 3.8.3, there can exist not more than a finite number of values of n for which

$$\sup_{k\geqslant 1} k \sqrt[n]{E_k(f;a,b)} \leqslant n.$$

For all other values of n

$$\sup_{k\geqslant 1} k \sqrt[n]{E_k(f;a,b)} > n,$$

and consequently, since $E_k(f;a,b) \leqslant 1$, we have

$$\sup_{k\geqslant 1} k \sqrt[n]{E_k(f;a,b)} = \sup_{k\geqslant n} k \sqrt[n]{E_k(f;a,b)}. \tag{20}$$

Thus the series 5.8(16) differs from the series 5.8(19) only in a finite number of terms, and must converge with it.

5.8.31. If the infinitely differentiable function $f(x)$ considered is defined on the whole real axis and satisfies condition 3.10(12) there, then its best approximation $A_\sigma(f)$ by integral functions of exponential type possesses a property similar to 5.8(16), i.e.

$$\sum_{n=1}^{\infty} \frac{1}{\sup\limits_{k\geqslant 1} k \sqrt[n]{A_k(f)}} = \infty. \tag{21}$$

In particular, for the best approximations by trigonometric polynomials of periodic functions of period 2π for which 3.10(12) holds, we have

$$\sum_{n=1}^{\infty} \frac{1}{\sup\limits_{k\geqslant 1} k \sqrt[n]{E_k^*(f)}} = \infty. \tag{22}$$

The relation 5.8(21) follows directly from 3.10(12) after the application of the estimate (see section 5.6.35, and also 5.5(14))

$$A_\sigma(f) \leqslant \frac{\pi M_n(f)}{2\sigma^n}, \tag{23}$$

which holds for all integral n and for all $\sigma > 0$.

5.8.4. It directly follows from theorem 5.8.3. (or 5.8.31) with the help of S. N. Bernstein's criterion given in section 1.8, that** *if the infinitely differentiable function $f(x)$ defined on the finite segment $[a, b]$ (or on $(-\infty, \infty)$) is quasi-analytic in the sense of Denjoy, i. e. it satisfies condition* 3.10.12, *then the sequence of its best approximations $E_n(f;a, b)$ (or $A_n(f)$) satisfies the inequality*

$$E_n(f; a, b) < \frac{1}{\phi(n)} \qquad (n = 1, 2, \dots) \tag{24}$$

$\left(\text{or } A_n(f) < \dfrac{1}{\phi(n)}\right)$ *where $\phi(x)$ is some weight function.* In order to verify this it is sufficient to consider the function

$$\phi(x) = \frac{1}{6M_0(f)} + \sum_{\nu=1}^{\infty} c_\nu x^{2\nu},$$

where

$$c_\nu = \frac{1}{2\nu^2} \left\{\sup_{k\geq 1} k \sqrt[2\nu]{E_k(f; a, b)}\right\}^{-2\nu},$$

$\left(\text{or } c_\nu = \dfrac{1}{2\nu^2} \left\{\sup_{k\geq 1} k \sqrt[2\nu]{A_k(f)}\right\}^{-2\nu}\right)$, for which the series $\sum_{\nu=1}^{\infty} \sqrt[2\nu]{c_\nu}$ diverges in virtue of theorem 5.8.3. (see section 1.8.1), and note that $c_\nu n^{2\nu} E_n(f; a, b) \leq \dfrac{1}{2\nu^2}$.

5.9. Estimates of best approximations for some conjugate classes of functions

We will now consider the constructive properties of functions conjugate (see section 3.11) to the functions belonging to some of the classes studied above.

5.9.1. Let us commence with the classes $\tilde{W}_*^{(r)}M$ of periodic functions of period 2π conjugate to the functions belonging to the class $W_*^{(r)}M$. If $\tilde{f}(x) \in W_*^{(r)}M$ and

$$\frac{a_0}{2} + \sum_{k=1}^{\infty} (a_k \cos kx + b_k \sin kx) \tag{1}$$

is the Fourier series of the corresponding function $f(x)$, then by the definition of 3.11.1,

$$\tilde{f}(x) = \sum_{k=1}^{\infty} (a_k \sin kx - b_k \cos kx).$$

After r-fold integration by parts in the formulae for the Fourier coefficients a_k, b_k we obtain as in section 3.5.5,

$$\tilde{f}(x) = \frac{1}{\pi} \int_0^{2\pi} f^{(r)}(t) \tilde{D}_0^{(r)}(x-t) \, dt, \tag{2}$$

** S. N. Bernstein [7].

where

$$\tilde{D}_0^{(r)}(t) = \sum_{k=1}^{\infty} \frac{\sin\left(kt - \frac{r\pi}{2}\right)}{k^r}. \tag{3}$$

Putting

$$\tilde{T}_n(x) = \tilde{U}_n(f; x; \lambda) = \sum_{k=1}^{n} \lambda_k^{(n)} (a_k \sin kx - b_k \cos kx),$$

where $\lambda_k^{(n)}$ $(n = 1, 2, \ldots; k = 1, 2, \ldots, n)$ is an arbitrary system of numbers, let us consider the difference $\tilde{f}(x) - \tilde{T}_n(x)$. Since

$$\tilde{T}_n(x) = \tilde{U}_n(f; x; \lambda) = \frac{1}{\pi} \int_0^{2\pi} f^{(r)}(x-t) \sum_{k=1}^{n} \lambda_k^{(n)} \frac{\sin\left(kt - \frac{r\pi}{2}\right)}{k^r} \, dt,$$

we have, in the case where $|f^{(r)}(x)| \leqslant M$,

$$|\tilde{f}(x) - \tilde{T}_n(x)| = \frac{1}{\pi} \left| \int_0^{2\pi} \left\{ \tilde{D}_0^{(r)}(t) - \sum_{k=1}^{n} \frac{\lambda_k^{(n)}}{k^r} \sin\left(kt - \frac{r\pi}{2}\right) \right\} f^{(r)}(x-t) \, dt \right|$$

$$\leqslant \frac{M}{\pi} \int_0^{2\pi} \left| \tilde{D}_0^{(r)}(t) - \sum_{k=1}^{n} \frac{\lambda_k^{(n)}}{k^r} \sin\left(kt - \frac{r\pi}{2}\right) \right| dt.$$

Let us choose the polynomial $\lambda_k^{(n)}$, as in section 5.5.1, in such a way that the integral on the right-hand side of the last inequality will assume its least possible value. Then using the proposition of section 2.11.5 when r is odd, and 2.13.32 when r is even, we obtain the estimate

$$E_n^*(\tilde{f}) \leqslant \max_x |\tilde{f}(x) - \tilde{T}_n(x)| \leqslant \frac{M\tilde{K}_r}{(n+1)^r} \tag{4}$$

where

$$\tilde{K}_r = \frac{4}{\pi} \sum_{\nu=0}^{\infty} \frac{(-1)^{\nu r}}{(2\nu+1)^{r+1}}. \tag{5}$$

On the other hand, since the function

$$\tilde{f}(x) = \frac{4M}{\pi(n+1)^r} \sum_{k=0}^{\infty} \frac{\cos\left[(2k+1)(n+1)x - \frac{r\pi}{2}\right]}{(2k+1)^{r+1}}$$

belongs to the class $\tilde{W}_*^{(r)}M$ and at the $2n+2$ points

$$x_\nu = \frac{\nu\pi}{n+1} + \frac{\pi}{2(n+1)} [1 - (-1)^r] \quad (\nu = 0, 1, \ldots, 2n+1)$$

of the half-open interval $[0, 2\pi)$ assumes, with consecutive changes of sign, the value $\frac{\tilde{M}K_r}{(n+1)^r} = \max_x |\tilde{f}(x)|$, then $E_n^*(\tilde{f}) = \frac{M\tilde{K}_r}{(n+1)^r}$ (see section 2.7.32). Hence it follows from the inequality 5.9(4) that** *for any integers $r \geqslant 1$ and $n \geqslant 0$*

$$\sup_{f \in W_*^r M} E_n^*(f) = \sup_{f \in \tilde{W}_*^{(r)} M} E_n^*(\tilde{f}) = \frac{M\tilde{K}_r}{(n+1)^r}. \tag{6}$$

5.9.11. The result just obtained shows that for any integer $r \geqslant 1$ and for a function $f(x)$ of the class $\tilde{W}_*^{(r)}M$, just as for functions of the class $W_*^{(r)}M$,

$$E_n^*(f) = O\left(\frac{1}{n^r}\right).$$

From the estimates of the best approximations of conjugate functions given in section 5.9.2 (see 5.9(15)), and the remark in section 5.1.41, it follows *that this relation holds for any $r > 0$.*

5.9.12. The constants $\tilde{K}_r$ (see 5.9(5)) possess the property that $\tilde{K}_1 < \tilde{K}_3 < \tilde{K}_5 < \ldots < \frac{4}{\pi} < \ldots < \tilde{K}_6 < \tilde{K}_4 < \tilde{K}_2 < \frac{\pi}{2}$. Consequently, for any integer r

$$\tilde{K}_r < \frac{\pi}{2}. \tag{7}$$

Taking account of this and taking into consideration V. A. Steklov's function (see 3.12.1), we obtain from 5.9(6), as in section 5.5.4, that *if for some integer $r \geqslant 1$ we have $f(x) \in \tilde{W}_*^{(r)} H_\omega$* (see section 3.5.3) *then****

$$E_n^*(f) < \frac{3}{n^r}\,\omega\left(\tilde{f}^{(r)}; \frac{1}{n}\right) \tag{8}$$

and

$$E_n^*(f) < \frac{5}{2n^r}\,\omega_2\left(\tilde{f}^{(r)}; \frac{1}{n}\right). \tag{9}$$

5.9.13. Using the integral representation 5.9(2) and the estimate $E_n^*[\tilde{D}_0^{(r)}]_L = \frac{\tilde{K}_r}{(n+1)^r}$ (see sections 2.11.5 and 2.13.32), it is easy to obtain the inequality

$$E_n^*(f) \leqslant \frac{\tilde{K}_r}{(n+1)^r} E_n^*(\tilde{f}^{(r)}), \tag{10}$$

** N. I. Akhiezer and M. G. Krein [1].
*** N. I. Akhiezer [3].

similar to 5.6(27). It is sufficient to note that if $T_n(\tilde{f}^{(r)}; x)$ is the trigonometric polynomial giving the best uniform approximation of order n for $\tilde{f}^{(r)}(x)$, and $\tilde{T}_{n,r}(x)$ is the trigonometric polynomial of best approximation in the metric L for $\tilde{D}_0^{(r)}$, then

$$\frac{1}{\pi}\int_0^{2\pi}[\tilde{f}^{(r)}(t)-T_n(\tilde{f}^{(r)};t)][\tilde{D}_0^{(r)}(x-t)-\tilde{T}_{n,r}(x-t)]\,dt=f(x)-T_n(x),$$

where $T_n(x)$ is a certain trigonometric polynomial of order $\leqslant n$.

5.9.14. It is not difficult to see from the proof of the inequality 5.9(4) that *if* $f(x)\in W_*^{(r)}M(L_q)$ $(q\geqslant 1)$, i.e.

$$\left\{\int_0^{2\pi}|f^{(r)}(x)|^q\,dx\right\}^{\frac{1}{q}}\leqslant M,$$

then the best integral approximation of the conjugate function $\tilde{f}(x)$ *satisfies the estimate*

$$E_n^*(\tilde{f})_{L_q}\leqslant\frac{M\tilde{K}_r}{(n+1)^r}. \tag{11}$$

Hence as in section 5.9.12 it follows that

$$E_n^*(\tilde{f})_{L_q}<\frac{3}{n^r}\omega\left(\tilde{f}^{(r)};\frac{1}{n}\right)_{L_q^*} \tag{12}$$

and

$$E_n^*(\tilde{f})_{L_q}<\frac{5}{2n^r}\omega_2\left(\tilde{f}^{(r)};\frac{1}{n}\right)_{L_q^*}. \tag{13}$$

5.9.15. If the periodic function $f(x)$ of period 2π is quasi-analytic in the sense of Denjoy, i.e. it is infinitely differentiable and satisfies condition 3.10(12), then the conjugate function $\tilde{f}(x)$ possesses the property that

$$\sum_{n=1}^{\infty}\frac{1}{\sup\limits_{k\geqslant 1}k\sqrt[n]{E_n^*(\tilde{f})}}=\infty. \tag{14}$$

If the periodic function $f(x)$ of period 2π is infinitely differentiable and satisfies condition 3.10(1), then $\lim\limits_{n\to\infty}\sqrt[n]{E_n^*(\tilde{f})}<1$. These are obvious consequences of the inequality 5.9(4).

5.9.2. In section 3.11 (see section 3.11.3) it has already been mentioned that the continuity (integrability) of the function $f(x)$ is not sufficient to ensure the continuity (integrability) of the conjugate function $\tilde{f}(x)$ or the relation $E_n^*(\tilde{f})\to 0$ $(E_n^*(\tilde{f})\to 0)$ as $n\to\infty$. Hence an estimate of the value of $E_n^*(\tilde{f})$ for any classes of continuous (integrable) functions $f(x)$ only has meaning when certain limitations are imposed on the modulus of continuity $\omega(f;t)$ (integral modulus of continuity $\omega(f;t)_L$). If, for example, it is assumed that condition 3.11(23) is satisfied, then the following proposition is valid. *Let $f(x)$ be a function of period 2π which is continuous (integrable on $[0,2\pi]$). If the condition* 3.11(23) (*the condition* $\int_0^1 \frac{\omega(f;t)_L^*}{t}\,dt<\infty$) *is satisfied, the conjugate function $\tilde{f}(x)$ is continuous*** *(integrable) and for all $n\geqslant 0$*

$$E_n^*(\tilde{f})\leqslant C^*\left\{E_n^*(f)+\sum_{\nu=n+1}^{\infty}\frac{1}{\nu}E_\nu^*(f)\right\} \tag{15}$$

$$\left(\text{or respectively } E_n^*(\tilde{f})_L\leqslant C^*\left\{E_n^*(f)_L+\sum_{\nu=n+1}^{\infty}\frac{1}{\nu}E_\nu^*(f)_L\right\}\right), \tag{16}$$

*where C^* is a constant independent of f****.

Since in virtue of Jackson's theorem (see 5.1(10)) and the inequality 3.2(6)

$$\sum_{\nu=1}^{\infty}\frac{1}{\nu}E_\nu^*(f)\leqslant 4C\sum_{\nu=1}^{\infty}\nu\omega\left(f;\frac{1}{\nu}\right)\frac{1}{\nu(\nu+1)}<8C\int_0^1\frac{\omega(f;t)}{t}\,dt<\infty,$$

it is sufficient to prove 5.9(15).

Let $\tilde{f}_1(x)$ be the first integral of $\tilde{f}(x)$, i.e.

$$\tilde{f}_1(x)=-\sum_{k=1}^{\infty}\frac{a_k\cos kx+b_k\sin kx}{k}=\frac{1}{\pi}\int_0^{2\pi}f(x-t)\sum_{k=1}^{\infty}\frac{\cos kt}{k}\,dt.$$

Applying to $\tilde{f}_1(x)$ the estimate 5.6(28) (for $r=1$), we have

$$E_n^*(\tilde{f}_1')=E_n^*(\tilde{f})\leqslant C_1\{nE_n^*(\tilde{f}_1)+\sum_{\nu=n}^{\infty}E_\nu^*(\tilde{f}_1)\}.$$

But since in virtue of 5.9(10) (for $r=1$)

$$E_n^*(\tilde{f}_1)\leqslant\frac{\tilde{K}_1}{n+1}E_n^*(f),$$

** This has already been mentioned in section 3.11.3.

*** S. B. Stechkin [5].

it follows that

$$E_n^*(\tilde{f}) \leqslant C_1 \tilde{K}_1 \left\{ E_n^*(f) + \sum_{\nu=n}^{\infty} \frac{1}{\nu} E_\nu^*(f) \right\},$$

i.e. 5.9(15) is satisfied.

From similar considerations we may also verify the validity of 5.9(16).

5.9.21. It follows from the estimate 5.9(9) that *if for some integer* $r \geqslant 1$ *we have* $\omega_2(\tilde{f}^{(r)}; t) = O(t^\alpha)$ $(\alpha > 0)$, *then*

$$E_n^*(f) = O\left(\frac{1}{n^{r+\alpha}}\right). \tag{17}$$

The inequality 5.9(15), together with theorem 5.1.32, shows that this relation is also valid for $r = 0$. Similarly in the case of integral approximations. *For the functions* $f(x) \in \tilde{W}_*^{(r)} H_2^{(\alpha)}(L)$ *for any integer* $r \geqslant 0$

$$E_n^*(f)_L = O\left(\frac{1}{n^{r+\alpha}}\right). \tag{18}$$

5.9.22. If the periodic function $f(x)$ of period 2π is analytic, then from 5.9(15) and 5.4(6) it immediately follows that for the conjugate function $\tilde{f}(x)$

$$\overline{\lim_{n \to \infty}} \sqrt[n]{E_n^*(\tilde{f})} \leqslant e^{-\delta}, \tag{19}$$

where 2δ is the width of the strip of analyticity of $f(x+iy)$.

5.9.3. Theorems similar to 5.9.1 and 5.9.2 also hold for the best approximations $A_\sigma(\tilde{f})$ of conjugate functions by arbitrary integral functions of finite degree. The first of these establishes that** *if the function* $f(x)$ *defined on* $(-\infty, \infty)$ *has for some* $r \geqslant 1$ *an r-th derivative* $f^{(r)}(x)$ *which belongs to Wiener's class* (see section 3.11.4), *and the function* $\tilde{f}^{(r)}(x)$ *which is conjugate to* $f^{(r)}(x)$ *in the sense of* 3.11(20) *satisfies the condition* $\sup_{-\infty<x<\infty} |\tilde{f}^{(r)}(x)| \leqslant M$, *then for any* $\sigma > 0$

$$A_\sigma(\tilde{f}) \leqslant \frac{M\tilde{K}_r}{\sigma^r}. \tag{20}$$

We omit the proof of this assertion. It can be carried out by a method resembling that of section 5.7.2.

The second proposition establishes that *if the bounded uniformly continuous function* $f(x)$, *defined on* $(-\infty, \infty)$, *satisfies condition* 3.11(23),

** N. I. Akhiezer [3].

then the function $\tilde{f}(x)$ *conjugate in the sense of* 3.11(20) *is continuous and for any* $\sigma > 0$

$$A_\sigma(\tilde{f}) \leqslant C\left\{A_\sigma(f) + \sum_{\nu=[\sigma]+1}^{\infty} \frac{1}{\nu} A_\nu(f)\right\}, \tag{21}$$

where C is a constant independent of f.

The proof of 5.9(21) is based on the inequality 5.9(20) (for $r = 1$) and is carried out exactly as in the periodic case.

5.9.4. If the condition 3.11(23) imposed on the modulus of continuity of the function $f(x)$ and ensuring the continuity of the conjugate function $\tilde{f}(x)$ is replaced by the corresponding condition for the modulus of smoothness $\omega_k(f; t)$ $(k > 1)$ (see section 3.3), then when it is satisfied the function $\tilde{f}(x)$ is not only continuous but also differentiable.

Let the function $f(x)$, *defined on* $(-\infty, \infty)$, *be bounded and uniformly continuous and for some integer r let*

$$\int_0^1 \frac{\omega_{r+1}(f; t)}{t^{r+1}}\, dt < \infty. \tag{22}$$

Then the function $\tilde{f}(x)$, *conjugate to* $f(x)$ *in the sense defined in* 3.11(20) ($a = f(0)$, $b = 0$; see section 3.11.24), *has an r-th continuous derivative, and for any* $\sigma > 0$

$$A_\sigma(\tilde{f}^{(r)}) \leqslant C_r^*\{(\sigma+1)^r A_\sigma(f) + \sum_{\nu=[\sigma]+1}^{\infty} \nu^{r-1} A_\nu(f)\}, \tag{23}$$

where C_r^* *is a constant which depends only on r***.

In virtue of theorem 5.1.3 and 3.3(6)

$$\sum_{\nu=1}^{\infty} \nu^{r-1} A_\nu(f) \leqslant 2C_{r+1} \sum_{\nu=1}^{\infty} \nu^{r+1} \omega_{r+1}\left(f; \frac{1}{\nu}\right) \frac{1}{\nu(\nu+1)}$$

$$\leqslant 2^{r+2} C_{r+1} \int_0^1 \frac{\omega_{r+1}(f; t)}{t^{r+1}}\, dt < \infty. \tag{24}$$

Since $f(x) = \lim_{\sigma\to\infty} g_\sigma(f; x)$, where $g_\sigma(f; x)$ is an integral function of degree $\leqslant \sigma$ for which $A_\sigma(f) = \sup_{-\infty<x<\infty} |f(x) - g_\sigma(f; x)|$, it follows that

$$f(x) = g_1(f; x) + \sum_{k=0}^{\infty} \{g_{2^{k+1}}(f; x) - g_{2^k}(f; x)\}.$$

** For a similar proposition for periodic functions see the paper by N. K. Bari and S. B. Stechkin [1].

In virtue of the inequality 4.8(6),

$$\left|\tilde{g}^{(r)}_{2^{k+1}}(f;x)-\tilde{g}^{(r)}_{2^k}(f;x)\right| \leqslant 2^{r(k+1)+1}A_{2^k}(f)$$

$$\leqslant 2^{2r+1}\sum_{\nu=2^{k-1}+1}^{2^k}\nu^{r-1}A_\nu(f). \qquad (25)$$

Hence after the application of 5.9(24) and integration we verify that the conjugate series

$$\tilde{g}_1(f;x)+\sum_{k=0}^{\infty}\{\tilde{g}_{2^{k+1}}(f;x)-\tilde{g}_{2^k}(f;x)\}$$

converges on the whole real axis (uniformly on any finite segment). Its sum $\phi(x)=\lim\limits_{k\to\infty}\tilde{g}_{2^k}(f;x)$ is a continuous function possessing an r-th continuous derivative

$$\phi^{(r)}(x)=\tilde{g}_1^{(r)}(f;x)+\sum_{k=0}^{\infty}\{\tilde{g}^{(r)}_{2^{k+1}}(f;x)-\tilde{g}^{(r)}_{2^k}(f;x)\}. \qquad (26)$$

Moreover, since when the condition 5.9(22) is satisfied 3.11(23) holds (see 3.3(11)), it follows that the conjugate function $\tilde{f}(x)$ considered is continuous. In the definition of the conjugate function $\tilde{f}(x)$ (see section 3.11.24) taking $a=f(0)$, $b=0$, we have in virtue of Plancherel's theorem (see section 3.11.21)

$$\int_{-\infty}^{\infty}\frac{|\tilde{f}(x)-\tilde{g}_\sigma(f;x)|^2}{x^2}\,dx=\int_{-\infty}^{\infty}\frac{|f(x)-g_\sigma(f;x)-f(0)+g_\sigma(f;0)|^2}{x^2}\,dx. \qquad (27)$$

It follows that

$$\lim_{k\to\infty}\int_{-\infty}^{\infty}\frac{|\tilde{f}(x)-\tilde{g}_{2^k}(f;x)|^2}{x^2}\,dx=0.$$

Moreover for any $N>\varepsilon>0$

$$\lim_{k\to\infty}\int_{\varepsilon}^{N}\frac{|\phi(x)-\tilde{g}_{2^k}(f;x)|^2}{x^2}\,dx=\lim_{k\to\infty}\int_{-N}^{-\varepsilon}\frac{|\phi(x)-\tilde{g}_{2^k}(f;x)|^2}{x^2}\,dx=0.$$

Thus after passing to the limit we verify that

$$\int_{-N}^{-\varepsilon}\frac{|\tilde{f}(x)-\phi(x)|^2}{x^2}\,dx=\int_{\varepsilon}^{N}\frac{|\tilde{f}(x)-\phi(x)|^2}{x^2}\,dx=0,$$

or $\phi(x)=\tilde{f}(x)$.

Since in virtue of 5.9(26)

$$\tilde{f}^{(r)}(x)-\tilde{g}_{\sigma}^{(r)}(f;x) = \tilde{g}_{2^{p+2}}^{(r)}(f;x)-\tilde{g}_{\sigma}^{(r)}(f;x)+\sum_{k=p+2}^{\infty}\{\tilde{g}_{2^{k+1}}^{(r)}(f;x)-\tilde{g}_{2^{k}}^{(r)}(f;x)\}$$

for any integer $p \geqslant -2$ we have, choosing p such that $2^p \leqslant \sigma < 2^{p+1}$ and using the inequality 5.9(25),

$$A_{\sigma}(\tilde{f}^{(r)}) \leqslant \sup_{-\infty<x<\infty} |\tilde{f}^{(r)}(x)-\tilde{g}_{\sigma}^{(r)}(f;x)|$$

$$\leqslant C_r^{*}\Big\{(\sigma+1)^r A_{\sigma}(f)+\sum_{\nu=[\sigma]+1}^{\infty} \nu^{r-1}A_{\nu}(f)\Big\},$$

i.e. 5.9(23) holds.

5.10. Direct theorems in arbitrary Banach spaces

Using the concept of the ε-entropy of a compact set (see section 3.1.3), it is possible to make more precise the theorem derived in section 2.5.1, and for any linear normed Banach space to obtain an inequality of the same type as the estimates in the fundamental theorems of this chapter.

As in section 2.5 let $\mathfrak{X}\{x_0, x_1, x_2, \ldots\}$ be an arbitrary closed system of linearly independent elements in the space F, let W be some compact set in F, and

$$\mathcal{E}_n^{\mathfrak{X}}(W) = \sup_{x\in W} E_n^{\mathfrak{X}}(x) = \sup_{x\in W}\inf_{c_k}\|x-\sum_{k=0}^{n} c_k x_k\|. \tag{1}$$

The following theorem holds, which like the theorems derived in the preceding sections of this chapter gives an estimate of the upper bound 5.10(1)*.

The inequality

$$\mathcal{E}_{n-1}^{\mathfrak{X}}(W) \leqslant 4\Phi_{W_{\mathfrak{X}}}(n+1) \tag{2}$$

is satisfied for any integer n, where $W_{\mathfrak{X}}$ *is the set consisting of all elements* $x \in F$ *for which* $\|x\| \leqslant \mathcal{E}_0^{\mathfrak{X}}(W)$ *and* $\mathcal{E}_k^{\mathfrak{X}}(x) \leqslant \mathcal{E}_k^{\mathfrak{X}}(W)$ $(k = 1, 2, \ldots)$, *and* $\Phi_{W_{\mathfrak{X}}}(n+1)$ *is the function inverse to the ε-entropy* $I_{W_{\mathfrak{X}}}^{a}(\varepsilon)$ (see section 3.1.3), *i.e. the upper bound of all* ε *for which* $I_{W_{\mathfrak{X}}}^{a}(\varepsilon) \geqslant n+1$.

In order to prove the inequality 5.10(2) we consider the set $\overline{W}_{n+1}$ of all the polynomials $y = \sum_{k=0}^{n} c_k x_k$ which belong to $W_{\mathfrak{X}}$, and let us suppose that $\mathcal{E}_0^{\mathfrak{X}}(W) \neq 0$.

---

* Yu. A. Brudnyi and A. F. Timan [1].

It is not difficult to verify that $\overline{W}_{n+1}$ is a convex body centrally symmetrical about zero in the space of all the polynomials $\sum_{k=0}^{n} c_k x_k$, i.e. it is a closed convex set, possessing in this space interior points and containing together with any element y also the element $-y$. Since the null point θ belongs to $\overline{W}_{n+1}$ together with a certain $(n+1)$-dimensional neighbourhood of it, it follows that for any polynomial

$$y = \sum_{k=0}^{n} c_k x_k \tag{3}$$

a λ can be found such that $\frac{y}{\lambda} \in \overline{W}_{n+1}$.

In the space of all the polynomials 5.10(3) let us denote by $\|y\|^*$ the lower bound of all λ for which $\frac{y}{\lambda} \in \overline{W}_{n+1}$. It is obvious that if $\|y\| = \mathcal{E}_0^{\mathfrak{X}}(W)$ or for some value $k(1 \leqslant k \leqslant n-1)$ $E_k^{\mathfrak{X}}(y) = \mathcal{E}_k^{\mathfrak{X}}(W)$, then $\|y\| \geqslant \mathcal{E}_{n-1}^{\mathfrak{X}}(W)$ and $\|y\|^* = 1$, i.e. for such polynomials y the inequality

$$\|y\| \geqslant \mathcal{E}_{n-1}^{\mathfrak{X}}(W)\,\|y\|^* \tag{4}$$

holds. Since in virtue of the boundedness of $\overline{W}_{n+1}$, for any polynomial 5.10(3), there always exists a value of λ for which λy possesses this property, then the inequality 5.10(4) holds for any polynomial 5.10(3).

We now consider for some $\varepsilon > 0$ the maximal number of elements $y_1, \ldots, y_r$ in the set $\overline{W}_{n+1}$, which possess the property, that $\|y_k - y_\nu\|^* > \varepsilon$ if $k \neq \nu$, and the closed spheres $S^*(\varepsilon, y_k)(k = 1, \ldots, r)$ of radius ε with centres at these points, i.e. the set of all the polynomials 5.10(3) for which $\|y - y_k\|^* \leqslant \varepsilon$. It is obvious that $\overline{W}_{n+1}$ consists of the sum of all these spheres. Consequently, if mes $S^*(\varepsilon; y_k)$ is the Lebesgue measure of the set of all points $(c_0, c_1, \ldots, c_n)$ in $(n+1)$-dimensional Euclidean space which correspond to the polynomials contained in the sphere $S^*(\varepsilon; y_k)$, then, since for any k

$$\text{mes}\, S^*(\varepsilon; y_k) = \text{mes}\, S^*(\varepsilon; \theta) = \varepsilon^{n+1}\, \text{mes}\, S^*(1; \theta),$$

it follows that

$$r\varepsilon^{n+1}\, \text{mes}\, S^*(1; \theta) \geqslant \text{mes}\, S^*(1; \theta),$$

i.e. $r \geqslant \left(\frac{1}{\varepsilon}\right)^{n+1}$. Thus in $\overline{W}_{n+1}$ there exist at least 2^{n+1} polynomials $y_1, \ldots, y_{2^{n+1}}$ for which $\|y_k - y_\nu\|^* > \frac{1}{2}$, or in virtue of 5.10(4)

$$\|y_k - y_\nu\| > \tfrac{1}{2}\mathcal{E}_{n-1}^{\mathfrak{X}}(W)$$

if $k \neq \nu$. Since $\overline{W}_{n+1} \subset W_{\mathcal{X}}$, we obtain from this by applying the inequality 3.1(4),

$$n+1 \leqslant I^c_{\overline{W}_{n+1}}\{\tfrac{1}{2}\mathcal{E}^{\mathcal{X}}_{n-1}(W)\} \leqslant I^c_{W_{\mathcal{X}}}\{\tfrac{1}{2}\mathcal{E}^{\mathcal{X}}_{n-1}(W)\} \leqslant I^a_{W_{\mathcal{X}}}\{\tfrac{1}{4}\mathcal{E}^{\mathcal{X}}_{n-1}(W)\}.$$

The last inequality leads directly to the estimate 5.10(2).

5.11. Various problems and theorems

1. If the periodic function $f(x)$ of period 2π belongs on $[0, 2\pi]$ to the class $L_q (q \geqslant 1)$, then for any integral value of k any integer $n \geqslant 0$

$$E^*_n(f)_{L_q} \leqslant C_k \omega_k\left(f; \frac{1}{n+1}\right)_{L_q}, \tag{1}$$

where C_k is a constant depending only on k, and $\omega_k(f; t)_{L_q}$ is the corresponding integral modulus of smoothness.

See theorem 5.3.1 for $m = 1$.

2. If the periodic function $f(x)$ of period 2π possesses an $(r-1)$-th absolutely continuous derivative and a derivative $f^{(r)}(x)$ of order r which belongs to the class L_q on $[0, 2\pi]$, then for any integers k and $n \geqslant 0$

$$E^*_n(f)_{L_q} \leqslant \frac{C_{k,r}}{(n+1)^r}\omega_k\left(f^{(r)}; \frac{1}{n+1}\right)_{L_q}. \tag{2}$$

This inequality is a consequence of 5.11(1) and 3.3,(4). See 5.3.11.

3. The estimate 5.1(18) can be extended to the case of fractional derivatives. If the periodic function $f(x)$ of period 2π is representable in the form of the integral (see section 3.5.5)

$$f(x) = \frac{1}{\pi}\int_0^{2\pi} F(t) D_0^{(r)}(x-t)\, dt, \tag{3}$$

where $r > 0$, and

$$\int_0^{2\pi} F(t)\, dt = 0, \qquad \operatorname*{vrai\,sup}_{0 \leqslant t \leqslant 2\pi} |F(t)| < \infty,$$

then for any integer k

$$E^*_n(f) = O\left[\frac{1}{n^r}\omega_k\left(F; \frac{1}{n}\right)\right]. \tag{4}$$

Consider the polynomials 5.3(3) for $k_\nu = k$ and make use of the remark in section 5.1.41. A similar estimate holds in L^*_q $(q \geqslant 1)$.

4. The inequality 5.6(25) can be extended to the case of integral approximations. If the function $f(x)$, defined on $(-\infty, \infty)$, possesses an r-th (r an integer) derivative which belongs to the class L_q $(q \geqslant 1)$ on $(-\infty, \infty)$, then for any $\sigma > 0$

$$A_\sigma(f)_{L_q} \leqslant \frac{K_r}{\sigma_r} A_\sigma(f^{(r)})_{L_q}. \tag{5}$$

Similarly the inequality 5.6(27) can be extended to the case of integral approximations.

See N. I. Akhiezer [3].

5. Extend the estimates 5.6(26), 5.6(28) to the case of approximations in the metric $L_q (q \geqslant 1)$.

6. If $1 < q < \infty$ and the periodic function $f(x)$ of period 2π belongs to L_q on $[0, 2\pi]$, then for any $n \geqslant 0$

$$\left\{\int_0^{2\pi} |f(x)-S_n^*(f;x)|^q \mathrm{d}x\right\}^{\frac{1}{q}} \leqslant C_q E_n^*(f)_{L_q}, \tag{6}$$

where C_q is a constant depending only on q, and $S_n^*(f; x)$ is the partial sum of order n of the Fourier series for $f(x)$.

This assertion is a consequence of the inequality 3.12(20).

7. For any continuous function $f(x)$ of period 2π the inequality

$$|f(x)-S_n^*(f;x)| \leqslant \left[\frac{4}{\pi^2}\ln n + O(1)\right] E_n^*(f) \tag{7}$$

is satisfied for all $n \geqslant 1$. In addition

$$\left|f(x)-\frac{1}{2}\left[S_n^*\left(f;x+\frac{\pi}{2n}\right)+S_n^*\left(f;x-\frac{\pi}{2n}\right)\right]\right| \leqslant C\omega_2\left(f;\frac{1}{n}\right), \tag{8}$$

where C is a constant independent of f.

The inequality 5.11(7) with a less precise constant was first pointed out by Lebesgue [1]. For the inequality 5.11(8) see A. F. Timan [10]. See also S. B. Stechkin [3]. If $T_n(f; x)$ is a trigonometric polynomial of order $\leqslant n$ for which $|f(x)-T_n(f;x)| \leqslant E_n^*(f)$, then

$$\begin{aligned} |f(x)-S_n^*(f;x)| &\leqslant |f(x)-T_n(f;x)-S_n^*(f-T_n;x)| \\ &\leqslant E_n^*(f)\left[1+\frac{1}{\pi}\int_0^{2\pi} |D_n(t)|\,\mathrm{d}t\right], \end{aligned} \tag{9}$$

where $D_n(t)$ is Dirichlet's kernel. In order to obtain 5.11(7) we use the estimate 4.12(6). In a similar way 5.11(8) follows from 4.12(9).

8. If the periodic function $f(x)$ of period 2π is integrable over $[0, 2\pi]$, then

$$\int_0^{2\pi} |f(x)-S_n^*(f;x)|\,\mathrm{d}x \leqslant \left[\frac{4}{\pi^2}\ln n + O(1)\right] E_n^*(f)_L. \tag{10}$$

See 5.11.7.

9. If the continuous function $f(x)$ is periodic with period 2π and $T_n^*(f; x)$ is the trigonometric polynomial interpolating it at the points $y_k^{(n)} = \dfrac{2k\pi}{2n+1}$ $(k = 0, 1, \ldots, 2n)$, then for all $n \geqslant 1$ the inequality

$$|f(x)-T_n^*(f;x)| \leqslant \left[\frac{2}{\pi}\left|\sin\frac{2n+1}{2}x\right|\ln n + O(1)\right] E_n^*(f) \tag{11}$$

is satisfied.

Use formula 4.2(9) and the estimate 4.12(8).

10. If $f(x)$ is a continuous function, defined on $[-1, 1]$, and (see section 4.5.2)

$$S_n(f;x) = \frac{1}{\pi}\int_0^{\pi} f(\cos t)[D_n(t+y)+D_n(t-y)]\,\mathrm{d}t, \tag{12}$$

where $Dn(t)$ is Dirichlet's kernel, then

$$|f(x)-S_n(f;x)| \leqslant \left[\frac{4}{\pi^2}\ln n+O(1)\right] E_n(f;\ -1,1) \tag{13}$$

uniformly with respect to all $x \in [-1, 1]$.

See 5.11.7.

11. If $f(x)$ is a function which is bounded on $(-\infty, \infty)$, then for any $\sigma \geqslant 1$

$$\sup_{-\infty<x<\infty} \left| f(x)-\frac{1}{\pi}\int_{-\infty}^{\infty} f(t)\,\frac{\cos(\sigma-1)(x-t)-\cos\sigma(x-t)}{(x-t)^2}\,dt \right| \leqslant \left[\frac{4}{\pi^2}\ln\sigma+O(1)\right] A_{\sigma-1}(f). \tag{14}$$

See 4.12.7 and 4.12.11.

12. If $\int_{-\infty}^{\infty} |f(t)|\,dt < \infty$, then for any $\sigma \geqslant 1$

$$\int_{-\infty}^{\infty} \left| f(x)-\frac{1}{\pi}\int_{-\infty}^{\infty} f(t)\,\frac{\cos(\sigma-1)(x-t)-\cos\sigma(x-t)}{(x-t)^2}\,dt \right| dx \leqslant \left[\frac{4}{\pi^2}\ln\sigma+O(1)\right] A_{\sigma-1}(f)_L. \tag{15}$$

13. If $f(x_1, \ldots, x_n)$ is a continuous function of period 2π in each of the variables, then

$$E^*_{n_1,\ldots,n_m}(f) \leqslant C\min[\{E^*_{n_{\nu_1},\ldots,n_{\nu_i},\infty}(f)+E^*_{n_{\nu_{i+1}},\ldots,n_{\nu_m},\infty}(f)\}\times \times \ln(n_{\nu_1}+2)\ldots\ln(n_{\nu_i}+2)] \tag{16}$$

$$\left(\nu_r = 1,2,3,\ldots,m;\ \ r=1,2,\ldots,i;\ \ i \leqslant \left[\frac{m}{2}\right]\right),$$

where C is an absolute constant.

M. F. Timan [1,4]. Case $m = 2$, S. N. Bernstein [42]. Use the estimate 4.12(6).

14. If $\int_0^{2\pi}\ldots\int_0^{2\pi} |f(x_1,\ldots,x_m)|\,dx_1,\ldots dx_m < \infty$, then

$$E^*_{n_1,\ldots,n_m}(f)_L \leqslant C\min[\{E^*_{n_{\nu_1},\ldots,n_{\nu_i},\infty}(f)_L+E^*_{n_{\nu_{i+1}},\ldots,n_{\nu_m},\infty}(f)_L\}\times \times \ln(n_{\nu_1}+2)\ldots\ln(n_{\nu_i}+2)] \tag{17}$$

$$\left(\nu_r = 1,2,3,\ldots,m;\ \ r=1,2,\ldots,i;\ \ i \leqslant \left[\frac{m}{2}\right]\right),$$

where C is an absolute constant.

M. F. Timan [1,4]. See section 5.11.13.

15. Inequalities similar to 5.11(16) and 5.11(17) hold on passing to functions $f(x_1,\ldots, x_m)$ for which respectively

$$\sup_{-\infty<x_1,\ldots,x_m<\infty} |f(x_1,\ldots,x_m)| < \infty$$

and

$$\int_{-\infty}^{\infty} \cdots \int_{-\infty}^{\infty} |f(x_1, \ldots, x_m)| \, dx_1 \ldots dx_m < \infty,$$

if the best approximation by trigonometric polynomials is replaced by the corresponding best approximation by any integral functions of exponential type in the variables $x_1, \ldots, x_m$.

16. Show that the constant $C_0^{(\alpha)}$ in the inequality 5.6(30) is not less than $\frac{1}{2}\pi^{\alpha}$. Consider the function x^{α} and make use of theorem 2.7.32.

17. If the function $f(x)$ is representable in the form

$$f(x) = \frac{1}{\pi} \int_0^{2\pi} K(x-t) \phi(t) \, dt,$$

where $K(t)$ and $\phi(t)$ are functions integrable over $[0, 2\pi]$, then the upper bound $\underset{x}{\text{vrai sup}} |f(x)|$ for all possible bounded in modulus unit functions $\phi(t)$, which are orthogonal to any trigonometric polynomial of order $\leqslant n$, is identical with the upper bound $E_n^*(f)_L$ for all possible functions $\phi(t)$ for which $\int_0^{2\pi} |\phi(t)| \, dt \leqslant 1$.

Similarly the upper bound of $\int_0^{2\pi} |f(x)| \, dx$ for all possible functions $\phi(t)$ orthogonal to any trigonometric polynomial of order $\leqslant n$ and such that $\int_0^{2\pi} |\phi(t)| \, dt \leqslant 1$, is identical with the upper bound of $E_n^*(f)$ for all possible functions $\phi(t)$ for which $|\phi(t)| \leqslant 1$.

S. M. Nikol'skii [3].

18. If the periodic function $f(x)$ of period 2π belongs to the class L_q $(1 < q < \infty)$ over the period, then for the best approximations by trigonometric polynomials of the conjugate function $\tilde{f}(x)$ the inequality

$$E_n^*(\tilde{f})_{L_q} \leqslant M_q^* E_q^*(f)_{L_q} \tag{18}$$

is satisfied for all n, where M_q^* is a constant depending only on q.

See 3.11(5).

19. If $f(x)$ belongs to the class L_q $(1 < q < \infty)$ on $(-\infty, \infty)$, then for the function $\tilde{f}(x)$ conjugate in the sense of 3.11(7),

$$A_\sigma(\tilde{f})_{L_q} \leqslant M_q A_\sigma(f)_{L_q}, \tag{19}$$

for any $\sigma > 0$, where M_q is a constant depending only on q.

See 3.11(8).

20. If $f(x)$ is a function which is measurable and bounded on $(-\infty, \infty)$, then for any $\alpha > -1$ the series $\sum_{k=1}^{\infty} k^{\alpha} A_k(f)$ and $\sum_{k=1}^{\infty} k^{\alpha} A_k(\tilde{f})$, where $\tilde{f}(x)$ is the function conjugate in the sense of 3.11(20), converge or diverge together.

For the case where $f(x)$ is periodic see S. B. Stechkin [5]. Make use of the inequality 5.9(21).

21. Show that for $p < \gamma$, where γ is a root of equation 5.4(12),

$$\lim_{n\to\infty} E_n\left(e^x; -\frac{n}{p}, \frac{n}{p}\right) = \infty. \tag{20}$$

This example shows that the constant γ in theorem 5.4.5 cannot be decreased.

S. N. Bernstein [27]. Use the inequality for the best approximation given in 2.13.24, and the expansion of e^x in a Taylor series of powers of x.

22. If $f_q(x)$ is the function defined in 3.12.26, then

$$E_n^*(f_q)_{L_q} = O\left(\frac{1}{n}\right).$$

See 5.11(1).

23. If $1 \leqslant q \leqslant q'$ and $T_n(f; x)$ is the trigonometric polynomial which gives the best approximation of order n to the function $f(x)$ in the metric L_q, then

$$\left\{\int_0^{2\pi} |f(x) - T_n(f; x)|^{q'}\, dx\right\}^{\frac{1}{q'}} \leqslant Cn^{\frac{1}{q}-\frac{1}{q'}} E_n^*(f)_{L_{q'}}, \tag{21}$$

where C is some constant.

Use the inequality 4.9(7).

24. If $1 \leqslant q \leqslant q'$ and $P_n(f; x)$ is the algebraic polynomial giving the best approximation of order n to the function $f(x)$ in the metric L_q on $[-1, 1]$, then

$$\left\{\int_{-1}^{1} |f(x) - P_n(f; x)|^{q'}\, dx\right\}^{\frac{1}{q'}} \leqslant Cn^{2\left(\frac{1}{q}-\frac{1}{q'}\right)} E_n(f; -1, 1)_{L_{q'}}, \tag{22}$$

where C is some constant.

25. The inequalities 5.9(11) and 5.5(17) for $q = 1$ are precise for the whole class $W_*^{(r)}M(L)$ of periodic functions of period 2π which have an $(r-1)$-th absolutely continuous derivative and a derivative $f^{(r)}(x)$ for which $\int_0^{2\pi} |f^{(r)}(x)|\, dx \leqslant M$,

$$\sup_{f\in W_*^{(r)}M(L)} E_n^*(f)_L = \frac{MK_r}{(n+1)^r}$$

$$\sup_{f\in W_*^{(r)}M(L)} E_n^*(\tilde{f})_L = \frac{M\tilde{K}_r}{(n+1)^r}. \tag{23}$$

S. M. Nikol'skii [3].

26. A limiting relation, similar to 5.6(29) for $a = 1$, holds in an integral metric If $W_1^{(r)}M(L)$ is the class of all functions $f(x)$ defined on $[-1, 1]$ and having an absolutely continuous $(r-1)$-th derivative and a derivative $f^{(r)}(x)$ for which $\int_{-1}^{1} |f^{(r)}(x)|\, dx \leqslant M$, then

$$\lim_{n\to\infty} \sup_{f\in W_1^{(r)}M(L)} n^r E_n(f; -1, 1)_L = \lim_{n\to\infty} \sup_{f\in W_*^{(r)}M(L)} n^r E_n^*(f)_L = MK_r. \tag{24}$$

S. M. Nikol'skii [8].

27. The inequality 5.7(8) cannot be improved in the whole class $A_\delta^* \, UM$ of periodic functions $f(x) \in A_{\infty,\delta} \, UM$ of period 2π. Thus if $A_\delta^* \, UM(L)$ is the class of functions defined in section 3.8.8 ($q = 1$), then

$$\sup_{f \in A_\delta^* UM(L)} E_n^*(f)_L = \sup_{f \in A_\delta^* UM} E_n^*(f) = \frac{8M}{\pi} \sum_{k=0}^{\infty} \frac{(-1)^k}{2k+1} \frac{q^{(2k+1)(n+1)}}{1+q^{(2k+1)(n+1)}}, \tag{25}$$

where $q = e^{-\delta}$.

For the second of the inequalities 5.11(25) see N. I. Akhiezer [3], for the first see S. M. Nikol'skii [3].

CHAPTER VI

CONVERSE THEOREMS. CONSTRUCTIVE CHARACTERISTICS OF SOME CLASSES OF FUNCTIONS

6.1. Differential properties of functions with a given sequence of best approximations

Besides Weierstrass's theorem (see chap. I), one of the landmarks of the constructive theory of functions is theorem 2.5.

After the results given in chapter V, which represent a further development of the propositions of chapter I, it is of the utmost importance that theorem 2.5 on approximation by trigonometric polynomials and algebraic polynomials in the function spaces we are studying can be supplemented by indications of some of the differential properties of the functions which have a given sequence of best approximations.

Thanks to this, it is possible in some cases to give a complete description in terms of the best approximation of various structural properties of functions which are deeper than continuity, and to study them from the constructive point of view.

6.1.1. Let us commence with periodic functions of period 2π, and show that** *if for some* $q(1 \leqslant q \leqslant \infty)$ *the function* $f(x)$ *belongs to the class* L_q *over the period, and* $E_n^*(f)$ $(n=0, 1, 2, \ldots)$ *is the sequence of its best approximations by trigonometric polynomials* (see section 2.2.2), *then, whatever the natural numbers k and n may be, the inequality*

$$\omega_k\left(f; \frac{1}{n}\right)_{L_q} \leqslant \frac{c_k}{n^k} \sum_{\nu=0}^{n} (\nu+1)^{k-1} E_\nu^*(f)_{L_q} \tag{1}$$

is satisfied, in which c_k *is a constant depending only on* k *and* $\omega_k(f; t)_{L_q}$ *is the corresponding modulus of smoothness* (see section 3.3).

** A. F. Timan and M. F. Timan [1], $(q < \infty)$. The consideration of the inequality 6.1(1) in the case $q = \infty$, with the condition that the function $f(x)$ is continuous, is given in a paper by S. B. Stechkin [2].

In fact, if $T_n(f;x)$ is the trigonometric polynomial of best approximation of order n, i.e.

$$E_n^*(f)_{L_q}=\left\{\int_0^{2\pi}|f(x)-T_n(f;x)|^q\,dx\right\}^{\frac{1}{q}}, \tag{2}$$

then for any integer $m\geqslant 0$

$$\omega_k\left(f;\frac{1}{n}\right)_{L_q}\leqslant\omega_k\left(f-T_{2^{m+1}};\frac{1}{n}\right)_{L_q}+\omega_k\left(T_{2^{m+1}};\frac{1}{n}\right)_{L_q}$$

$$\leqslant 2^k E^*_{2^{m+1}}(f)_{L_q}+\omega_k\left(T_{2^{m+1}};\frac{1}{n}\right)_{L_q}. \tag{3}$$

Moreover (see 3.3,(4)), using Minkowski's inequality we have

$$\omega_k\left(T_{2^{m+1}};\frac{1}{n}\right)_{L_q}\leqslant\frac{1}{n^k}\left\{\int_0^{2\pi}|T^{(k)}_{2^{m+1}}(f;x)|^q\,dx\right\}^{\frac{1}{q}}$$

$$\leqslant\frac{1}{n^k}\left[\left\{\int_0^{2\pi}|T_1^{(k)}(f;x)|^q\,dx\right\}^{\frac{1}{q}}+\right.$$

$$\left.+\sum_{\nu=0}^{m}\left\{\int_0^{2\pi}|T^{(k)}_{2^{\nu+1}}(f;x)-T^{(k)}_{2^\nu}(f;x)|^q\,dx\right\}^{\frac{1}{q}}\right]$$

But since in virtue of 4.8(26)

$$\left\{\int_0^{2\pi}|T^{(k)}_{2^{\nu+1}}(f;x)-T^{(k)}_{2^\nu}(f;x)|^q\,dx\right\}^{\frac{1}{q}}$$

$$\leqslant 2^{(\nu+1)k}\left\{\int_0^{2\pi}|T_{2^{\nu+1}}(f;x)-T_{2^\nu}(f;x)|^q\,dx\right\}^{\frac{1}{q}}$$

$$\leqslant 2^{(\nu+1)k+1}E^*_{2^\nu}(f)_{L_q}. \tag{4}$$

and

$$\left\{\int_0^{2\pi}|T_1^{(k)}(f;x)|^q\,dx\right\}^{\frac{1}{q}}$$

$$=\left\{\int_0^{2\pi}|T_1^{(k)}(f;x)-T_0^{(k)}(f;x)|^q\,dx\right\}^{\frac{1}{q}}\leqslant 2E_0^*(f)_{L_q},$$

it follows that

$$\omega_k\left(T_{2^{m+1}};\frac{1}{n}\right)_{L_q}\leqslant\frac{2}{n^k}\left\{E_0^*(f)_{L_q}+\sum_{\nu=0}^{m}2^{(\nu+1)k}E^*_{2^\nu}(f)_{L_q}\right\}.$$

Taking account also of the fact that

$$2^{(\nu+1)k} E^*_{2^\nu}(f)_{L_q} \leqslant 2^{2k} \sum_{\mu=2^{\nu-1}+1}^{2^\nu} \mu^{k-1} E^*_\mu(f)_{L_q}, \tag{5}$$

we obtain the estimate

$$\omega_k\left(T_{2^{m+1}};\frac{1}{n}\right)_{L_q} \leqslant \frac{2^{2k+1}}{n^k}\Bigg\{E^*_0(f)_{L_q}+E^*_1(f)_{L_q}+$$

$$+\sum_{\nu=1}^{m}\sum_{\mu=2^{\nu-1}+1}^{2^\nu} \mu^{k-1}E^*_\mu(f)_{L_q}\Bigg\} \leqslant \frac{c^*_k}{n^k}\sum_{\nu=0}^{2^m}(\nu+1)^{k-1}E^*_\nu(f)_{L_q}.$$

Now choosing m so that $2^m \leqslant n < 2^{m+1}$, and substituting this estimate in 6.1(3), we arrive at the inequality 6.1(1).

6.1.2. It follows from the inequality 6.1(1), that** *if the sequence of best approximations* $E^*_n(f)L_q$ *of the periodic function* $f(x)$ *of period* 2π *of the class* L_q $(1 \leqslant q \leqslant \infty)$ *satisfies the relation*

$$E^*_n(f)_{L_q} = O\left(\frac{1}{n^\alpha}\right) \tag{6}$$

for some $\alpha > 0$, *then for any natural number* k

$$\omega_k(f;t)_{L_q} = \begin{cases} O(t^\alpha) & \text{when} \quad \alpha < k, \\ O(t^k|\ln t|) & \text{when} \quad \alpha = k, \\ O(t^k) & \text{when} \quad \alpha > k. \end{cases} \tag{7}$$

Thus for $0 < \alpha < k$ *the condition* 6.1(6) *is not only necessary* (see 5.1(16) and section 5.10.1), *but also sufficient in order that*

$$\omega_k(f;t)_{L_q} = O(t^\alpha).$$

The example of the continuous function 3.3(16) ($ab = 1$), which is not differentiable at a single point and, consequently, does not possess the property $\omega(f;t)_{L_q} = O(t)$ for any value $q \geqslant 1$ (see sections 3.6.1 and 3.12.13), but satisfies the condition $E^*_n(f)_{L_q} = O\left(\frac{1}{n}\right)$ (if b is an odd number; see section 2.11.2), shows that the last assertion would no longer be true for $\alpha = k$. It is obvious from theorem 5.1.32 (see also 5.11.1) and 6.1.1, that*** *for* $\alpha = k$ *the relation* 6.1(6) *is equivalent to the condition* $\omega_{k+1}(f;t)_{L_q} = O(t^k)$.

** S. N. Bernstein [1] for the case $k = 1$, $q = \infty$. Quade [1] for the case $k = 1$ $1 \leqslant q < \infty$.

*** A. Zygmund [2].

6.1.3. Theorem 6.1.1 can be supplemented by the following proposition**. *If for the periodic function $f(x)$ of period 2π*

$$\sum_{\nu=1}^{\infty} \nu^{r-1} E_\nu^*(f)_{L_q} < \infty \tag{8}$$

for some q $(1 \leqslant q < \infty)$ and some natural number r, then apart from values on a set of measure zero $f(x)$ is identical with a function which has an absolutely continuous derivative $f^{(r-1)}(x)$ and an r-th derivative $f^{(r)}(x)$ which belongs to L_q over the period (in the case where $q = \infty$ it is continuous). In addition, for any natural number k

$$\omega_k\left(f^{(n)}; \frac{1}{n}\right)_{L_q} \leqslant C_{k,r}\left\{\frac{1}{n_k}\sum_{\nu=0}^{n}(\nu+1)^{k+r-1}E_\nu^*(f)_{L_q} + \sum_{\nu=n+1}^{\infty}\nu^{r-1}E_\nu^*(f)_{L_q}\right\}, \tag{9}$$

where $C_{k,r}$ is a constant depending only on k and r.

For any natural number $p \leqslant r$ let us consider the series

$$T_1^{(p)}(f; x) + \sum_{\nu=0}^{\infty}\{T_{2^{\nu+1}}^{(p)}(f; x) - T_{2^\nu}^{(p)}(f; x)\},$$

where $T_n(f; x)$ is the best polynomial approximation (see 6.1(2)). The inequalities 6.1(4) and 6.1(5) show that in virtue of the assumptions the sequence of its partial sums

$$S_n^{(p)}(x) = T_1^{(p)}(f; x) + \sum_{\nu=0}^{n}\{T_{2^{\nu+1}}^{(n)}(f; x) - T_{2^\nu}^{(p)}(f; x)\}$$

converges in mean (in the sense of the metric L_q)*** on $[0, 2\pi]$. Consequently, by a well-known theorem there exists a subsequence $S_{n_m}^{(p)}(x)$, which for all $p = 1, 2, \ldots, r$ converges almost everywhere to a certain function $f_p(x)$ of the class L_q. Supposing that x_0 is one of the points of convergence for all $p = 1, 2, \ldots, r$, let us consider the difference

$$f_{p-1}(x) - f_{p-1}(x_0) - \int_{x_0}^{x} f_p(t)\,dt$$

$$= f_{p-1}(x) - S_{n_m}^{(p-1)}(x) - f_{p-1}(x_0) + S_{n_m}^{(p-1)}(x_0) - \int_{x_0}^{x}\{f_p(t) - S_{n_m}^{(p)}(t)\}\,dt.$$

** A. F. Timan [10]. See also S. B. Stechkin [2]. The first part of theorem 6.1.3 for $q = \infty$ is due to S. N. Bernstein [1].

*** In the case where $q = \infty$ it converges uniformly.

We see that

$$\left\{\int_0^{2\pi}\left|f_{p-1}(x)-f_{p-1}(x_0)-\int_{x_0}^{x} f_p(t)\,dt\right|^q dx\right\}^{\frac{1}{q}}$$

$$\leqslant\left\{\int_0^{2\pi}|f_{p-1}(t)-S_{n_m}^{(p-1)}(t)|^q\,dt\right\}^{\frac{1}{q}}+2\pi|f_{p-1}(x_0)-S_{n_m}^{(p-1)}(x_0)|+$$

$$+\left\{\int_0^{2\pi}\left|\int_{x_0}^{x}[f_p(t)-S_{n_m}^{(p)}(t)]\,dt\right|^q dx\right\}^{\frac{1}{q}}.$$

Since as m increases without limit the right-hand side tends to zero, it follows that for almost all x and for $p = 1, 2, \ldots, r$

$$f_{p-1}(x)-f_{p-1}(x_0)=\int_{x_0}^{x} f_p(t)\,dt.$$

Since $f_0(x) = f(x)$ almost everywhere, then almost everywhere $f(x)$ is identical with a function possessing an absolutely continuous derivative of the $(r-1)$-th order and an r-th derivative $f^{(r)}(x) = f_r(x) \in L_q^*$, which is the limit in mean of the sequence $S_n^{(r)}(x)$. In addition

$$\omega_k\left(f^{(r)};\frac{1}{n}\right)_{L_q}\leqslant\omega_k\left(f^{(r)}-S_m^{(r)};\frac{1}{n}\right)_{L_q}+\omega_k\left(S_m^{(r)};\frac{1}{n}\right)_{L_q}$$

and

$$\omega_k\left(f^{(r)}-S_m^{(r)};\frac{1}{n}\right)_{L_q}$$

$$\leqslant 2^k\left\{\int_0^{2\pi}|f^{(r)}(x)-S_m^{(r)}(x)|^q\,dx\right\}^{\frac{1}{q}}$$

$$\leqslant 2^k\sum_{\nu=m+1}^{\infty}\left\{\int_0^{2\pi}|T_{2^{\nu+1}}^{(r)}(f;x)-T_{2^\nu}^{(r)}(f;x)|^q\,dx\right\}^{\frac{1}{q}}$$

$$\leqslant 2^k\sum_{\nu=m+1}^{\infty}2^{(\nu+1)r+1}E_{2^\nu}^*(f)_{L_q}\leqslant 2^{k+1+2r}\sum_{\mu=2^m+1}^{\infty}\mu^{r-1}E_n^*(f)_{L_q}.$$

In addition to this (see 3.3,(4)),

$$\omega_k\left(S_n^{(r)};\frac{1}{n}\right)_{L_q}\leqslant\omega_k\left(T_1^{(r)};\frac{1}{n}\right)_{L_q}+\sum_{\nu=0}^{m}\omega_k\left(T_{2^{\nu+1}}^{(r)}-T_{2^\nu}^{(r)};\frac{1}{n}\right)_{L_q}$$

$$\leqslant \frac{1}{n^k}\left\{\int_0^{2\pi} |T_1^{(r+k)}(f; x)|^q \, dx\right\}^{\frac{1}{q}} +$$

$$+\frac{1}{n^k}\sum_{\nu=0}^{m}\left\{\int_0^{2\pi} |T_{2^{\nu+1}}^{(r+k)}(f; x)-T_{2^\nu}^{(r+k)}(f; x)|^q \, dx\right\}^{\frac{1}{q}}.$$

Now choosing m so that $2^m \leqslant n < 2^{m+1}$, and applying the estimates 6.1(4) and 6.1(5), we arrive at the inequality 6.1(9).

6.1.31. Condition 6.1(8), being sufficient for the existence for the function $f(x)$ of an r-th derivative $f^{(r)}(x)$, which belongs to L_q when $1 \leqslant q < \infty$, and is continuous for $q = \infty$, is not always necessary. Let us illustrate this by examples for the cases $q = 2$ and $q = \infty$.

(1) The function

$$f(x) = \sum_{k=2}^{\infty} \frac{\cos kx}{k^{3/2}\ln k}$$

has the derivative

$$f'(x) = -\sum_{k=2}^{\infty} \frac{\sin kx}{k^{1/2}\ln k},$$

which (in virtue of the Fischer–Riesz theorem) belongs to L_2^*. However, for this function

$$E_n^*(f)_{L_2} = \sqrt{\sum_{k=n+1}^{\infty} \frac{1}{k^3 \ln^2 k}}$$

and for some $C > 0$

$$E_n^*(f)_{L_2} > \frac{C}{(n+2)\ln(n+2)};$$

consequently, the series $\sum_{n=1}^{\infty} E_n^*(f)_{L_2}$ diverges.

(2) The function

$$f(x) = \sum_{k=2}^{\infty} \frac{\cos kx}{k^2 \ln k}$$

possesses the derivative

$$f'(x) = -\sum_{k=2}^{\infty} \frac{\sin kx}{k \ln \ln k},$$

which is continuous on account of the uniform convergence of the last series**. If, however use is made of the inequality 5.11(7) and it is noted that for some $C > 0$

$$\max_x \left| f(x) - \sum_{k=2}^{n} \frac{\cos kx}{k^2 \ln\ln k} \right| = \sum_{k=n+1}^{\infty} \frac{1}{k^2 \ln\ln k} > \frac{C}{(n+1)\ln\ln(n+1)},$$

we obtain

$$E_n^*(f) > \frac{C_1}{(n+2)\ln(n+2)\ln\ln(n+2)},$$

where C_1 is a positive constant.

Thus for this function $f(x)$ the series $\sum_{n=1}^{\infty} E_n^*(f)$ diverges.

6.1.32. As the series 6.1(8), where $r < \alpha$, converges for functions $f(x)$ which satisfy for some natural number l the condition $\omega_l(f; t) = O(t^\alpha)$ (see 5.1(16) and 5.11.1), we conclude from theorem 6.1.3 that *almost everywhere on $(-\infty, \infty)$ any such function is identical with a function possessing an absolutely continuous derivative $f^{(r-1)}(x)$ and a derivative $f^{(r)}(x)$ of order r $(r < \alpha)$, belonging to the class L_q over a period (for $q = \infty$ continuous), and whatever the natural number k may be*

$$\omega_k(f^{(r)}; t)_{L_q} = \begin{cases} O(t^{\alpha-r}), & \text{when} \quad \alpha - r < k, \\ O(t^k |\ln t|), & \text{when} \quad \alpha - r = k, \\ O(t^k), & \text{when} \quad \alpha - r > k. \end{cases}$$

6.1.4. It follows from 6.1(9) that*** *if the sequence of best approximations $E_n^*(f)_{L_q}$ of the periodic function $f(x) \in L_q$ $(1 \leqslant q \leqslant \infty)$ of period 2π satisfies for some natural number r and $\alpha > 0$ the relation*

$$E_n^*(f)_{L_q} = O\left(\frac{1}{n^{r+\alpha}}\right), \tag{10}$$

then almost everywhere on $(-\infty, \infty)$ $f(x)$ is identical with a function which possesses an absolutely continuous derivative $f^{(r-1)}(x)$ and an r-th derivative $f^{(r)}(x)$ which belongs to L_q over a period (for $q = \infty$ it is continuous), and for any natural number k

$$\omega_k(f_{-}^{(r)}; t)_{L_q} = \begin{cases} O(t^\alpha), & \text{when} \quad \alpha < k, \\ O(t^k |\ln t|), & \text{when} \quad \alpha = k, \\ O(t^k), & \text{when} \quad \alpha > k. \end{cases} \tag{11}$$

** The uniform convergence of this series can be verified by applying the Abel transformation and using the first of the relations 4.12(12).

*** S. N. Bernstein [1] for the case $q = \infty$, $k = 1$. Quade [1] for the case $k = 1$, $1 \leqslant q < \infty$. A. Zygmund [2] for the general case.

Thus *for* $0 < \alpha < k$ *the relation* 6.1(10) *is necessary* (see 5.1(18) *and* 5.11.2) *and sufficient for the periodic function* $f(x)$ *to be almost everywhere identical with a function possessing an absolutely continuous derivative* $f^{(r-1)}(x)$, *an r-th derivative* $f^{(r)}(x)$ *which belongs to* L_q (*for* $q = \infty$ *continuous*) *and satisfying the condition*

$$\omega_k(f^{(r)}; t)_{L_q} = O(t^{\alpha}).$$

In particular, a continuous function $f(x)$ of period 2π is infinitely differentiable when and only when for any r

$$\lim_{n\to\infty} n^r E_n^*(f) = 0.$$

It is obvious from theorems 5.1.4 (see also 5.11.2) and 6.1.3 that *the relation* 6.1(10) *in the case* $\alpha = k$ *is equivalent* (*in the same sense as in theorem* 6.1.3) *to the existence of an r-th derivative* $f^{(r)}(x) \in L_q$ *for which* $\omega_{k+1}(f^{(r)}; t)L_q = O(t^k)$.

6.1.5. The estimate 6.1(1), true for all the spaces L_q^* $(1 \leqslant q \leqslant \infty)$, can be improved for $1 < q < \infty$ and replaced by another estimate, the order of the right-hand side of which is in a number of cases more precise. The following theorem applies**, which we give here without proof. *If* $f(x) \in L_q^*$, *where* $1 < q < \infty$, *then for any natural number* k

$$\omega_k\left(f; \frac{1}{n}\right)_{L_q} \leqslant \begin{cases} \dfrac{M_{q,k}}{n^k} \left\{\sum\limits_{\nu=0}^{n} (\nu+1)^{kq-1} E_\nu^{*q}(f)_{L_q}\right\}^{\frac{1}{q}}, & \text{when } 1 < q \leqslant 2, \\ \dfrac{M_{q,k}}{n_k} \left\{\sum\limits_{\nu=0}^{n} (\nu+1)^{2k-1} E_\nu^{*2}(f)_{L_q}\right\}^{\frac{1}{2}}, & \text{when } 2 \leqslant q < \infty, \end{cases} \tag{12}$$

where $M_{q,k}$ is a constant depending only on q and k.

Let us compare the estimates 6.1(1) and 6.1(12) for $k = 1$. Considering the function x^q $(q > 1)$ and the value

$$x = \frac{1}{2}\left[\sum_{\mu=0}^{\nu} E_\mu^*(f)_{L_q} + (\nu-1) E_\nu^*(f)_{L_q}\right]$$

$$= \frac{1}{2}\left[\sum_{\mu=0}^{\nu-1} E_\nu^*(f)_{L_q} + \nu E_\nu^*(f)_{L_q}\right],$$

$$x - h = (\nu-1) E_\nu^*(f)_{L_q}, \qquad x + h = \sum_{\mu=0}^{\nu} E_\mu^*(f)_{L_q},$$

$$x - \delta = \nu E_\nu^*(f)_{L_q}, \qquad x + \delta = \sum_{\mu=0}^{\nu-1} E_\mu^*(f)_{L_q},$$

** M. F. Timan [2].

we obtain from considerations of convexity the inequality

$$[\nu E_\nu^*(f)_{L_q}]^q+\Big[\sum_{\mu=0}^{\nu-1} E_\mu^*(f)_{L_q}\Big]^q \leqslant [(\nu-1)E_\nu^*(f)_{L_q}]^q+\Big[\sum_{\mu=0}^{\nu} E_\mu^*(f)_{L_q}\Big]^q.$$

After summation with respect to ν we have

$$\sum_{\nu=1}^{n} [\nu^q-(\nu-1)^q]E_\nu^{*q}(f)_{L_q} \leqslant \sum_{\nu=1}^{n}\Big\{\Big[\sum_{\mu=0}^{\nu} E_\mu^*(f)_{L_q}\Big]^q-\Big[\sum_{\mu=0}^{\nu-1} E_\mu^*(f)_{L_q}\Big]^q\Big\},$$

i.e.

$$E_1^*(f)_{L_q}+\sum_{\nu=2}^{n}(\nu-1)^{q-1}E_\nu^{*q}(f)_{L_q} \leqslant \Big[\sum_{\mu=0}^{n} E_\mu^*(f)_{L_q}\Big]^q-E_0^{*q}(f)_{L_q}.$$

Thus for any $q>1$

$$\Big\{\sum_{\nu=0}^{n}(\nu+1)^{q-1}E_\nu^{*q}(f)_{L_q}\Big\}^{\frac{1}{q}} \leqslant 3\sum_{\nu=0}^{n} E_\nu^*(f)_{L_q} \tag{13}$$

Consequently the estimate 6.1(12) is never worse than 6.1(1) for large n. In addition, in a number of cases it leads to a more precise result. Thus, for example, for the function $f_q(x)$ of 3.12.26 we have $E_n^*(f_q)_{L_q}=O\left(\frac{1}{n}\right)$ (see 5.11.22) and the inequality 6.1(1) gives the estimate $\omega(f_q;t)_{L_q} = O(t|\ln t|)$, while in reality, by 6.1(12),

$$\omega(f_q;t)_{L_q}=\begin{cases} O\left(t|\ln t|^{\frac{1}{q}}\right) & \text{when} \quad 1<q\leqslant 2,\\ O\left(t|\ln t|^{\frac{1}{2}}\right) & \text{when} \quad 2\leqslant q<\infty. \end{cases} \tag{14}$$

The inequality 6.1(12) enables us to make 6.1(7) more precise in the case where $1<q<\infty$. It follows from it *that with the assumptions of theorem* 6.1.2

$$\omega_k(f;t)_{L_q}=\begin{cases} O\left(t^k|\ln t|^{\frac{1}{q}}\right) & \text{when} \quad 1<q\leqslant 2,\\ O\left(t^k|\ln t|^{\frac{1}{2}}\right) & \text{when} \quad 2\leqslant q<\infty, \end{cases} \tag{15}$$

for $\alpha=k$. Comparison of 6.1(15) and 3.12.26 in the case $k=1$ shows that this result is definitive in respect of order.

For uniformly best approximations ($q=\infty$), it is obvious from the function 3.3(16) (if $k=1$) that theorem 6.1.2 gives the precise order in the general case (see sections 2.11.2 and 9.16). Other examples show that the same also holds for $q=1$ (see 3.12.26 and 5.11.22).

Let us also note the following proposition**, which in some cases makes 6.1.3 more precise.

** M. F. Timan [2].

With the assumptions of theorem 6.1.3 *for* $1 < q < \infty$

$$\omega_k\left(f^{(r)}; \frac{1}{n}\right)_{L_q} \leqslant \begin{cases} C_{k,r,q}\left(\dfrac{1}{n_k}\left[\sum\limits_{\nu=0}^{n} (\nu+1)^{q(k+r)-1} E_n^{*q}(f)_{L_q}\right]^{\frac{1}{q}} + \right. \\ \qquad \left. + \sum\limits_{\nu=n+1}^{\infty} \nu^{r-1} E_\nu^{*}(f)_{L_q}\right), \quad \text{when } 1 < q \leqslant 2, \\ C_{k,r,q}\left(\dfrac{1}{n_k}\left[\sum\limits_{\nu=0}^{n} (\nu+1)^{2(k+r)-1} E_\nu^{*2}(f)_{L_q}\right]^{\frac{1}{2}} + \right. \\ \qquad \left. + \sum\limits_{\nu=n+1}^{\infty} \nu^{r-1} E_\nu^{*}(f)_{L_q}\right), \quad \text{when } 2 \leqslant q < \infty. \end{cases} \tag{16}$$

where $C_{k,r,q}$ *is a constant depending only on* k, r *and* q. It follows from this *that in theorem* 6.1.4 *for* $\alpha = k$ *and* $1 < q < \infty$ *we have instead of* 6.1(11) *the more precise relation*

$$\omega_k(f^{(r)}; t)_{L_q} = \begin{cases} O\left(t^k |\ln t|^{\frac{1}{q}}\right) & \text{when} \quad 1 < q \leqslant 2, \\ O\left(t^k |\ln t|^{\frac{1}{2}}\right) & \text{when} \quad 2 \leqslant q < \infty. \end{cases} \tag{17}$$

6.1.6. Propositions similar to those which were given in the periodic case hold for best approximations on the whole real axis by arbitrary integral functions of exponential type. *If the function $f(x)$ belongs to the class L_q $(1 \leqslant q \leqslant \infty)$ on $(-\infty, \infty)$, then, whatever the natural numbers k and n may be, the inequality*

$$\omega_k\left(f; \frac{1}{n}\right)_{L_q} < \frac{C_k}{n^k} \sum_{\nu=0}^{n} (\nu+1)^{k-1} A_\nu(f)_{L_q} \tag{18}$$

is satisfied, where C_k is a constant depending only on k, and $A_\sigma(f)_{L_q}$ *and* $\omega_k\left(f; \frac{1}{n}\right)_{L_q}$ *are respectively the best approximation by integral functions of degree $\leqslant \sigma$ and the modulus of smoothness of $f(x)$ in the metric L_q on $(-\infty, \infty)$.*

If for a function $f(x)$ which belongs to the class L_q $(1 \leqslant q \leqslant \infty)$ on $(-\infty, \infty)$ the series $\sum\limits_{\nu=1}^{\infty} \nu^{r-1} A_\nu(f)_{L_q}$ converges for some natural number r, then almost everywhere on $(-\infty, \infty)$ $f(x)$ is identical with a function possessing an absolutely continuous derivative $f^{(\nu-1)}(x)$ and an r-th derivative $f^{(\nu)}(x) \in L_q$ (for $q = \infty$ continuous), and for any natural number k

$$\omega_k\left(f^{(r}; \frac{1}{n}\right)_{L_q} \leqslant C_{k,r}\left\{\frac{1}{n^k} \sum_{\nu=0}^{n} (\nu+1)^{k+r-1} A_\nu(f)_{L_q} + \sum_{\nu=n+1}^{\infty} \nu^{r-1} A_\nu(f)_{L_q}\right\}, \tag{19}$$

where $C_{k,r}$ is a constant depending only on k and r.

The proofs of these assertions are similar to those given in sections 6.1.1 and 6.1.3 and are based on the inequality 4.8(28).

From this there immediately follow lemmas similar to those given in sections 6.1.2 and 6.1.4 for the periodic case.

6.1.61. In the general case, for functions defined on the whole real axis, these lemmas can be formulated as follows**. *If $E(t)$ is a function which is non-increasing on* $[0, \infty)$, *for which* $E(0) = 1$, $\lim_{t\to\infty} E(t) = 0$, *and for some natural number k*

$$u^k \int_u^1 \frac{1}{t^{k+1}} E\left(\frac{1}{t}\right) \mathrm{d}t = O\left[E\left(\frac{1}{u}\right)\right] \left(0 < u \leqslant \frac{1}{2}\right), \tag{20}$$

then for every function $f(x) \in L_q$ $(1 \leqslant q \leqslant \infty)$ *defined on* $(-\infty, \infty)$ *the condition* $\omega_k(f;\, t)_{L_q} = O\left[E\left(\frac{1}{t}\right)\right]$ *is equivalent to the condition* $A_\sigma(f)_{L_q} = O$ $[E(\sigma)]$. This assertion follows directly from the inequalities 5.1(15) and 6.1(18). It is only necessary to observe that in the case where $A_\sigma(f) = O[E(\sigma)]$:

$$\sum_{\nu=0}^{n} (\nu+1)^{k-1} A_\nu(f)_{L_q} = O\left\{\sum_{\nu=0}^{n} (\nu+1)^{k-1} E(\nu)\right\}$$

$$= O\left\{E(0) + E(1) + \sum_{\nu=2}^{n} E(\nu) \int_{\frac{1}{\nu}}^{\frac{1}{\nu-1}} \frac{\mathrm{d}t}{t^{k+1}}\right\}$$

$$= O\left\{E(0) + E(1) + \sum_{\nu=2}^{n} \int_{\frac{1}{\nu}}^{\frac{1}{\nu-1}} \frac{1}{t^{k+1}} E\left(\frac{1}{t}\right) \mathrm{d}t\right\}$$

$$= O\left\{E(0) + E(1) + \int_{\frac{1}{n}}^{1} \frac{1}{t^{k+1}} E\left(\frac{1}{t}\right) \mathrm{d}t\right\}.$$

Similarly, *if the function $E(t)$, in addition to the conditions enumerated above also satisfies the condition*

$$\int_0^u \frac{1}{t} E\left(\frac{1}{t}\right) \mathrm{d}t = O\left[E\left(\frac{1}{u}\right)\right] \quad (0 < u \leqslant 1), \tag{21}$$

** In a somewhat different form, equivalent to that given here (see section 6.9.5), this assertion was established for the periodic case by S. M. Lozinskii [3], who also proved the necessity of the corresponding conditions for the validity of the equivalence of the estimates of the best approximation and of the modulus of smoothness. See also A. F. Timan and M. F. Timan [1].

then the relation $A_\sigma(f)_{L_q} = O\left[\frac{1}{\sigma^r} E(\sigma)\right]$ *is equivalent* (*in the same sense as in theorem* 6.1.3) *to the existence for the function* $f(x)$ *of an absolutely continuous derivative of the* $(r-1)$*-th order and a derivative* $f^{(r)}(x)$ *for which* $\omega_k(f^{(r)}; t)_{L_q} = O\left[E\left(\frac{1}{t}\right)\right]$. In this case it is sufficient to apply the inequalities 5.1(19) and 6.1(19), and also to take into account that if $A_\sigma(f)_{L_q} = O\left[\frac{1}{\sigma^r} E(\sigma)\right]$, then

$$\sum_{\nu=n+1}^{\infty} \nu^{r-1} A_\nu(f)_{L_q} = O\left\{\sum_{\nu=n+1}^{\infty} \frac{1}{\nu} E(\nu)\right\}$$

$$= O\left\{\sum_{\nu=n+1}^{\infty} E(\nu) \int_{\frac{1}{\nu}}^{\frac{1}{\nu-1}} \frac{dt}{t}\right\} = O\left\{\int_0^{\frac{1}{n}} \frac{1}{t} E\left(\frac{1}{t}\right) dt\right\}.$$

6.1.7. In the case of uniform approximation it is also possible to obtain estimates of the structural characteristics of continuous functions, considered on the finite segment $[a, b]$, in terms of their best approximation $E_n(f; a, b)$ by algebraic polynomials. However, the corresponding results here, similar to theorems 6.1.1–6.1.4, are less complete and enable us to judge the differential properties of the functions studied not on the whole segment $[a, b]$ but only on segments lying entirely within it. Thus, for example, *for any fixed segment* $[a_1, b_1]$ *which lies within* $[a, b]$ $(a < a_1 < b_1 < b)$, *and any function* $f(x)$ *continuous on* $[a, b]$,

$$\sup_{\substack{|x_1|-x_2|\leq \frac{1}{n} \\ x_1, x_2 \in [a_1, b_1]}} |f(x_1)-f(x_2) \leq \frac{C(a_1, b_1)}{n} \sum_{\nu=0}^{n} E_\nu(f; a, b) \tag{22}$$

for all integral n, *where* $C(a_1, b_1)$ *is a constant depending only on the positions of* a_1 *and* b_1.

Such estimates follow from considerations similar to those for the periodic case, but in which, instead of the inequalities 4.8(26), use is made (for $a = -1$, $b = 1$) of the inequality 4.8(31) for the derivative of an algebraic polynomial. The coefficient $C(a_1, b_1)$ which is then obtained in relations of the type 6.1(22) increases without limit when a_1 approaches a or b_1 approaches b, on account of the degeneration of the inequality 4.8(31) at the ends of the segment considered. This circumstance is material and deprives us of the possibility of extending estimates of the type 6.1(22) for any of the differential characteristics of the corresponding functions to the whole of the basic segment. It is not difficult to verify that in the general case such estimates cannot hold for the whole segment $[a, b]$. If, for example, $f(x) = \sqrt[4]{1-x^2}$, then in virtue of 5.1(10)

$$E_n(f; -1, 1) = E_n^*(\sqrt{|\sin t|}) = O\left(\frac{1}{\sqrt{n}}\right).$$

and

$$\frac{1}{n}\sum_{\nu=0}^{n} E_\nu(f; -1, 1) = O\left(\frac{1}{\sqrt{n}}\right).$$

At the same time, as $n \to \infty$, the modulus of continuity $\omega\left(f; \frac{1}{n}\right)$ on the whole of the segment $[-1, 1]$ is of the exact order $\frac{1}{\sqrt[4]{n}}$.

Similar examples show that the theorems proved in section 5.1 on the best uniform approximation of continuous functions by algebraic polynomials on a finite segment do not admit of complete inversion. We have already encountered in chap. V the cause of this peculiarity of the case of approximations by algebraic polynomials on a finite segment, which distinguishes it from the case of approximations by integral functions of exponential type on the whole real axis, where such an inversion holds for some classes of functions (see sections 6.1.2, 6.1.4 and 6.6.6). It is due to the fact that the fundamental theorem of Jackson (see 5.1), although in some cases precise as regards order for approximations by integral functions of exponential type on the whole real axis, is crude for approximations by algebraic polynomials on a finite segment, as can be seen from the proposition of section 5.2.1. Theorem 5.2.1 fully reflects the constructive properties of functions defined on a finite segment, and this theorem is more natural in this case and for certain classes of functions it can be inverted.

6.2. Constructive characteristics of some classes of continuous functions defined on a finite segment

In this section we shall consider the question of the possibility of inversion of theorem 5.2.1, and at the same time we will give constructive characteristics of some differential properties of functions defined on a finite segment.

6.2.1. The following proposition** holds, which is the converse of theorem 5.2.1 for $r = 0$ and similar to 6.1.1 for $q = \infty$, $k = 1$. *If for the function $f(x)$ defined on $[-1, 1]$ it is possible to find a sequence of ordinary polynomials $P_n(x)$ $(n = 1, 2, \ldots)$ such that*

$$|f(x)-P_n(x)| \leqslant \omega\left[\frac{1}{n}\left(\sqrt{1-x^2}+\frac{1}{n}\right)\right], \quad x \in [-1,1], \tag{1}$$

where $\omega(t)$ is a certain modulus of continuity (see section 3.2), *then*

$$\omega(f; t) \leqslant Ct \int_t^1 \frac{\omega(u)}{u^2}\, du \quad (0 < t \leqslant \tfrac{1}{2}), \tag{2}$$

where C is a constant independent of f.

The proof is similar to that given in section 6.1.1 and is based on the inequality 4.8(36)***.

It follows from 6.2(1) that for any $x \in [-1, 1]$ and $n \geqslant 2$ there can be found a natural number

$$m = m(x, n) = \begin{cases} \left[\dfrac{\ln n}{2\ln 2}\right] & \text{if} \quad |x| \geqslant 1-\dfrac{2}{n}, \\ \left[\dfrac{\ln\left(n\sqrt{1-x^2}\right)}{\ln 2}\right] & \text{if} \quad |x| < 1-\dfrac{2}{n}, \end{cases} \tag{3}$$

such that for $0 \leqslant \delta \leqslant \dfrac{1}{n}$, $x+\delta \in [-1, 1]$,

$$|\Delta_\delta f(x)-\Delta_\sigma P_{2^{m+1}}(x)| \leqslant 4\omega\left(\frac{2}{n}\right). \tag{4}$$

Also, for any $x \in [-1, 1]$

$$|P'_{2^{s+1}}(x)| \leqslant |P'_1(x)| + \sum_{k=0}^{s} |P'_{2^{k+1}}(x)-P'_{k}(x)| \tag{5}$$

and

$$|P_{2^{k+1}}(x)-P_{2^k}(x)| \leqslant 2\omega\left[\frac{1}{2^k}\left(\sqrt{1-x^2}+\frac{1}{2^k}\right)\right].$$

** A. F. Timan [15]. For the case $\omega(t) = t^\alpha$ $(0 < \alpha < 1)$ see V. K. Dzyadyk [3].

*** The following more general assertion can be proved in the same way. With the assumptions of theorem 6.1.2, where $\omega(t)$ does not decrease and $\omega(t_1+t_2) \leqslant M[\omega(t_1)+\omega(t_2)]$, the following inequality holds for any natural number k;

$$\omega_k(f; t; -1, 1) \leqslant C_k t^k \int_t^1 \frac{\omega(u)}{u^{k+1}}\, du \quad (0 < t \leqslant \tfrac{1}{2}).$$

In the case where $k = 2$, we obtain a proposition converse to 5.2.3.

Applying theorem 4.8.72 to the function $\phi_{2^k}(x) = \omega\left[\frac{1}{2^k}\left(\sqrt{1-x^2}+\frac{1}{2^k}\right)\right]$, we see that for some constant $C_1 > 0$

$$|P'_{2^{k+1}}(x)-P'_{2^k}(x)| \leqslant C_1 2^k \frac{\omega\left[\frac{1}{2^k}\left(\sqrt{1-x^2}+\frac{1}{2^k}\right)\right]}{\sqrt{1-x^2}+\frac{1}{2^k}}. \tag{6}$$

Without loss of generality, we can take $x \geqslant 0$. Since $(1-x^2)^{-\frac{1}{2}} \geqslant 2^m-1$ for $x \geqslant 1-\frac{2}{n}$, there exists a positive constant C_2 with the property that for all such values of x and for $1 \leqslant k \leqslant m$ the inequality

$$|P'_{2^{k+1}}(x)-P'_{2^k}(x)| \leqslant C_2 2^{2k}\omega\left(\frac{1}{2^{2k}}\right)$$

is satisfied, and in virtue of 6.2(5)

$$|P'_{2^{m+1}}(x)| \leqslant C_3\sum_{k=0}^{m} 2^{2k}\omega\left(\frac{1}{2^{2k}}\right)$$

$$\leqslant C_4\sum_{k=0}^{2^{2m}}\omega\left(\frac{1}{k+1}\right) \leqslant C_4\sum_{k=0}^{n}\omega\left(\frac{1}{k+1}\right).$$

Now let $x < 1-\frac{2}{n}$ and $\nu = \left[\frac{-\ln(1-x^2)}{2\ln 2}\right]$. From the inequality 6.2(6) (see 3.2(6)) there follows the existence of a positive constant C_5 such that for all $k \leqslant \nu$ the inequality

$$|P'_{2^{k+1}}(x)-P'_{2^k}(x)| \leqslant C_5 2^{2k}\omega\left(\frac{1}{2^{2k}}\right)$$

is satisfied, and there exists a constant C_6 such that for all $k > \nu$

$$|P'_{2^{k+1}}(x)-P'_{2^k}(x)| \leqslant C_6\frac{2}{\sqrt{1-x^2}}\omega\left(\frac{\sqrt{1-x^2}}{2^k}\right).$$

Hence for $|x| < 1-\frac{2}{n}$ we obtain from 6.2(5)

$$|P'_{2^{m+1}}(x) \leqslant C_1\sum_{k=0}^{\nu} 2^{2k}\omega\left(\frac{1}{2^{2k}}\right)+\frac{C_6}{\sqrt{1-x^2}}\sum_{k=\nu+1}^{m} 2^k\omega\left(\frac{\sqrt{1-x^2}}{2^k}\right)$$

$$\leqslant C_8 \sum_{k=0}^{2^{2\nu}} \omega\left(\frac{1}{k+1}\right) + C_9 \sum_{k=\nu}^{m} 2^{k+\nu} \omega\left(\frac{1}{2^{k+\nu}}\right)$$

$$\leqslant C_8 \sum_{k=0}^{2^{2\nu}} \omega\left(\frac{1}{k+1}\right) + C_{10} \sum_{k=0}^{2^{m+\nu}} \omega\left(\frac{1}{k+1}\right),$$

or as $2^{m+\nu} \leqslant n$ $(n \geqslant 2)$ we have for all the values of x considered here,

$$|P'_{2^{m+1}}(x)| \leqslant C_{11} \sum_{k=0}^{n} \omega\left(\frac{1}{k+1}\right).$$

Thus from 6.2(4)

$$|f(x)-f(x+\delta)| \leqslant \frac{C_{12}}{n} \sum_{k=0}^{n} \omega\left(\frac{1}{k+1}\right) + 4\omega\left(\frac{2}{n}\right).$$

It remains to take into account that

$$\omega\left(\frac{2}{n}\right) \leqslant \frac{8}{n} \sum_{k=0}^{\left[\frac{n}{2}\right]} \omega\left(\frac{1}{k+1}\right) \leqslant \frac{8}{n} \sum_{k=0}^{n} \omega\left(\frac{1}{k+1}\right)$$

and

$$\sum_{k=0}^{n} \omega\left(\frac{1}{k+1}\right) \leqslant C_0 \int_{\frac{1}{n+1}}^{1} \frac{\omega(u)}{u^2}\, du,$$

where C_0 is a certain positive constant; after which we obtain the inequality 6.2(2).

6.2.2. Thus *if the function $f(x)$ satisfies the conditions of theorem* 6.2.1 *and*

$$t \int_{t}^{1} \frac{\omega(u)}{u^2}\, du = O\,[\omega(t)], \tag{7}$$

as $t \to 0$, then $\omega(f; t) = O[\omega(t)]$.

Combining this result with theorem 5.2.1, we see that *if the condition* 6.2(7) *is satisfied for $\omega(t)$, then the function $f(x)$, defined on* $[-1,1]$, *has there a modulus of continuity $\omega(f; t)$ of order $O\,[\omega(t)]$ if and only if there exists a sequence of algebraic polynomials $P_n(x)$ which satisfy the inequality* 6.2(1).

6.2.3. For the case when $r \geqslant 1$, the following proposition** is the converse of theorem 5.2.1. *If for a function $f(x)$ defined on $[-1,1]$, and for some modulus of continuity $\omega(t)$ satisfying the condition*

$$\int_0^1 \frac{\omega(u)}{u}\,du < \infty, \tag{8}$$

it is possible to find a sequence of ordinary polynomials $P_n(x)$ $(n = 1, 2, \ldots)$ such that

$$|f(x) - P_n(x)| \leqslant$$
$$\leqslant \frac{1}{n^r}\left(\sqrt{1-x^2} + \frac{1}{n}\right)^r \omega\left[\frac{1}{n}\left(\sqrt{1-x^2} + \frac{1}{n}\right)\right] \quad x \in [1, 1], \tag{9}$$

then $f(x)$ possesses on the given segment an r-th continuous derivative $f^{(r)}(x)$ and

$$\omega(f^{(r)}; t) \leqslant C_r\left\{t\int_t^1 \frac{\omega(u)}{u^2}\,du + \int_0^t \frac{\omega(u)}{u}\,du\right\} \quad \left(0 < t \leqslant \frac{1}{2}\right), \tag{10}$$

where C_r is a constant depending only on r.

The existance of the r-th continuous derivative $f^{(r)}(x)$ can be verified by considering the series

$$P_1^{(r)}(x) + \sum_{k=0}^{\infty}\left\{P_{2^{k+1}}^{(r)}(x) - P_{2^k}^{(r)}(x)\right\}. \tag{11}$$

It follows from 6.2(9), that for $|x| \leqslant 1$

$$\left|P_{2^{k+1}}(x) - P_{2^k}(x)\right| \leqslant \frac{2}{2^{kr}}\left(\sqrt{1-x^2} + \frac{1}{2^k}\right)^r \omega\left[\frac{1}{2^k}\left(\sqrt{1-x^2} + \frac{1}{2^k}\right)\right].$$

Hence in virtue of theorem 4.8.72

$$\left|P_{2^{k+1}}^{(r)}(x) - P_{2^k}^{(r)}(x)\right| \leqslant M_r\omega\left[\frac{1}{2^k}\left(\sqrt{1-x^2} + \frac{1}{2^k}\right)\right]$$
$$\leqslant 4M_r\omega\left(\frac{1}{2^{k+1}}\right) \leqslant 2M_r \sum_{\mu=2^k+1}^{2^{k+1}} \frac{1}{\mu}\omega\left(\frac{1}{\mu}\right), \tag{12}$$

where M_r is a constant depending only on r.

** A. F. Timan [15]. For the case $\omega(t) = t^\alpha$ $(0 < \alpha < 1)$ see V. K. Dzyadyk [3].

Since when condition 6.2(8) is satisfied

$$\sum_{k=0}^{\infty}\sum_{\mu=2^k+1}^{2^{k+1}}\frac{1}{\mu}\omega\left(\frac{1}{\mu}\right)=\sum_{\mu=2}^{\infty}\frac{1}{\mu}\omega\left(\frac{1}{\mu}\right)<4\int_0^1\frac{\omega(t)}{t}\,dt<\infty,$$

it follows that the series 6.2(11) converges uniformly on [—1,1], and consequently the function

$$f(x)=P_1(x)+\sum_{k=0}^{\infty}\{P_{2^{k+1}}(x)-P_{2^k}(x)\}$$

has the r-th continuous derivative

$$f^{(r)}(x)=P_1^{(r)}(x)+\sum_{k=0}^{\infty}\{P_{2^{k+1}}^{(r)}(x)-P_{2^k}^{(r)}(x)\}. \tag{13}$$

It follows from the inequality 6.2(12) that for $|x|\geqslant 1-\frac{2}{n}$, $m=m(x,n)$ (see 6.2(3)) and $k\geqslant m-2$ we have

$$|P_{2^{k+1}}^{(r)}(x)-P_{2^k}^{(r)}(x)|\leqslant 4M_r\omega\left[\frac{1}{2^{k+1}}\left(\frac{1}{2^{m-1}}+\frac{1}{2^{k+1}}\right)\right]$$

$$\leqslant 16M_r\omega\left(\frac{1}{2^{k+m+1}}\right)\leqslant 8M_r\sum_{\mu=2^{k+m}+1}^{2^{k+m+1}}\frac{1}{\mu}\omega\left(\frac{1}{\mu}\right).$$

Consequently, for such values of x

$$\sum_{k=m+1}^{\infty}|P_{2^{k+1}}^{(r)}(x)-P_{2^k}^{(r)}(x)|\leqslant 8M_r\sum_{k=m+1}^{\infty}\sum_{\mu=2^{k+m}+1}^{2^{k+m+1}}\frac{1}{\mu}\omega\left(\frac{1}{\mu}\right)$$

$$=8M_r\sum_{\mu=2^{2m+1}+1}^{\infty}\frac{1}{\mu}\omega\left(\frac{1}{\mu}\right)<8M_r\sum_{\mu=\left[\frac{n}{2}\right]+1}^{\infty}\frac{1}{\mu}\omega\left(\frac{1}{\mu}\right). \tag{14}$$

If $x<1-\frac{2}{n}$ $(n\geqslant 2)$ and $\nu=\left[\frac{-\ln(1-x^2)}{2\ln 2}\right]$, then $\frac{1}{1-x^2}<n$ and $2^{2\nu}\leqslant\frac{1}{1-x^2}<\left(n\sqrt{1-x^2}\right)^2\leqslant 2^{2m+2}$. Hence (see 6.2(12))

$$|P_{2^{k+1}}^{(r)}(x)-P_{2^k}^{(r)}(x)|\leqslant 4M_r\omega\left[\frac{1}{2^{k+1}}\left(\frac{1}{2^\nu}+\frac{1}{2^{k+1}}\right)\right],$$

i.e. for $k \geqslant \nu$

$$|P^{(r)}_{2^{k+1}}(x)-P^{(r)}_{2^k}(x)| \leqslant 8M_r\omega\left(\frac{1}{2^{\nu+k+1}}\right) \leqslant 8M_r \sum_{\mu=2^{\nu+k+1}+1}^{2^{\nu+k+2}} \frac{1}{\mu}\omega\left(\frac{1}{\mu}\right).$$

Thus

$$\sum_{k=m+1}^{\infty} |P^{(r)}_{2^{k+1}}(x)-P^{(r)}_{2^k}(x)| \leqslant 8M_r \sum_{k=m+1}^{\infty} \sum_{\mu=2^{\nu+k+1}+1}^{2^{\nu+k+2}} \frac{1}{\mu}\omega\left(\frac{1}{\mu}\right)$$

$$= 8M_r \sum_{\mu=2^{\nu+m+2}+1}^{\infty} \frac{1}{\mu}\omega\left(\frac{1}{\mu}\right) \leqslant 8M_r \sum_{\mu=n+1}^{\infty} \frac{1}{\mu}\omega\left(\frac{1}{\mu}\right). \tag{15}$$

Moreover, on applying theorem 4.8.7 to 6.2(12) we find that for all $x \in [-1, 1]$

$$|P^{(r+1)}_{2^{k+1}}(x)-P^{(r+1)}_{2^k}(x)| \leqslant M^*_r 2^k \frac{\omega\left[\frac{1}{2^k}\left(\sqrt{1-x^2}+\frac{1}{2^k}\right)\right]}{\sqrt{1-x^2}+\frac{1}{2^k}}.$$

Hence as in section 6.2.1 it follows that for $|x| \leqslant 1$

$$|P^{(r+1)}_{2^{m+1}}(x)| \leqslant B_r \sum_{k=0}^{n} \omega\left(\frac{1}{k+1}\right), \tag{16}$$

where B_r is a constant depending only on r.

If now $0 \leqslant \delta \leqslant \frac{1}{n}$ $(n \geqslant 2)$, then from 6.2(13) and the inequalities 6.2(14), 6.2(15), 6.2(16) we obtain the estimate

$$|f^{(r)}(x)-f^{(r)}(x+\delta)| \leqslant A_r\left\{\frac{1}{n}\sum_{k=0}^{n}\omega\left(\frac{1}{k+1}\right)+\sum_{k=n+1}^{\infty}\frac{1}{k}\omega\left(\frac{1}{k}\right)\right\},$$

which leads to 6.2(10).

6.2.4. *We see that if the function $f(x)$ satisfies the conditions of theorem 6.2.3 and*

$$t\int_t^1 \frac{\omega(u)}{u^2}\,du = O[\omega(t)], \qquad \int_0^t \frac{\omega(u)}{u}\,du = O[\omega(t)],$$

as $t \to 0$, then the derivative $f^{(r)}(x)$ exists, is continuous on [—1,1], and satisfies $\omega(f^{(r)}; t) = O[\omega(t)]$.

Thus *if* $\omega(t)$ *possesses the property* 6.2(17), *then, in order that the function* $f(x)$ *defined on* $[-1,1]$ *should have there an r-th continuous derivative with modulus of continuity* $\omega(f^{(r)};t) = O[\omega(t)]$, *it is necessary* (see section 5.2.1) *and sufficient that there should exist a sequence of algebraic polynomials* $P_n(x)$ *which satisfy the inequality* 6.2(9). Let us note that the condition 6.2(17), in particular, is satisfied for $\omega(t) = t^\alpha$ $(0 < \alpha < 1)$.

If account is taken of 5.4(1), it also follows from theorem 6.2.3 that a function $f(x)$ defined on the finite segment $[a, b]$ is infinitely differentiable if and only if for any r

$$\lim_{n\to\infty} n^r E_n(f; a,b) = 0.$$

6.2.5. In conclusion let us note that as in 6.2.1, commencing with the inequality for integral functions of exponential type given at the end of section 4.8.5, it is possible to obtain a converse of theorem 5.2.5 and indicate constructive characteristics of the corresponding classes of functions defined outside the finite interval, similar to those obtained in sections 6.2.2 and 6.2.4**.

6.3. Converse theorems for functions of many variables

Inequalities similar to those given in section 6.1 hold also for functions of many variables.

6.3.1. Let us begin with functions $f(x_1,\ldots,x_m)$ defined throughout the whole space R_m of the variables $x_1,\ldots,x_m$ and of period 2π in each of these variables. *If for some* q $(1 \leqslant q \leqslant \infty)$ $f(x_1,\ldots,x_m)$ *belongs to the class* L_q *on the cube* $0 \leqslant x_k \leqslant 2\pi$ $(k = 1, 2, 3,\ldots,m)$, *and* $E^*_{n},\ldots,_{n_r},_\infty(f)_{L_q}$ $(1 \leqslant r \leqslant m,\ n_k = 0, 1, 2, 3, \ldots)$ *is the sequence of its partial best approximations by trigonometric polynomials* (see section 2.2.6), *then whatever the natural numbers* $k_1,\ldots,k_r$ *may be the inequality****

$$\omega_{k_1,\ldots,k_r}\left(f; \frac{1}{n},\ldots,\frac{1}{n_r},0,\ldots,0\right)_{L_q}$$

$$\leqslant \frac{C_{k_1,\ldots,k_r}}{n_1^{k_1}\ldots n_r^{k_r}} \sum_{\nu_1=0}^{n_1}\ldots\sum_{\nu_r=0}^{n_r} (\nu_1+1)^{k_1-1}\ldots(\nu_r+1)^{k_r-1} E^*_{\nu_1,\ldots,\nu_r,\infty}(f)_{L_q}, \qquad (1)$$

is satisfied, where $C_{k_1,\ldots,k_r}$ *is a constant depending only on* $k_1,\ldots,k_r$,

** See Yu. A. Brudnyi [2].

*** M. F. Timan [3].

and $\omega_{k_1,\dots,k_r}(f\,;u_1,\dots,u_r,0,\dots,0)_{L_q}$ *is the corresponding mixed modulus of smoothness,* i.e.

$$\omega_{k_1,\dots,k_r}(f;u_1,\dots,u_r,0,\dots,0)_{L_q}$$

$$=\sup_{|h_k|\leqslant u_k}\left\{\int_0^{2\pi}\cdots\int_0^{2\pi}\left|\sum_{\nu_1=0}^{k_1}\cdots\sum_{\nu_r=0}^{k_r}(-1)^{k_1+\dots+k_r-\nu_1-\dots-\nu_r}\binom{k_1}{\nu_1}\cdots\times\right.\right.$$

$$\left.\left.\times\cdots\binom{k_r}{\nu_r}f(x_1+\nu_1h_1,\dots,x_r+\nu_rh_r,x_{r+1},\dots,x_m)\right|^q\mathrm{d}x_1\dots\mathrm{d}x_m\right\}^{\frac{1}{q}}. \qquad (2)$$

The inequality 6.3(1) can be obtained from the same considerations as in 6.1(1). We will give a proof of this for the case $r=m=2$. Let $T_{n_1,n_2}(f;x_1,x_2)$ be the best trigonometric polynomial approximation to $f(x_1,x_2)$, i.e.

$$E^*_{n_1,n_2}(f)_{L_q}=\|f-T_{n_1,n_2}\|_{L_q}$$

$$=\left\{\int_0^{2\pi}\int_0^{2\pi}|f(x_1,x_2)-T_{n_1,n_2}(x_1,x_2)|^q\mathrm{d}x_1\,\mathrm{d}x_2\right\}^{\frac{1}{q}}. \qquad (3)$$

For any integers $p_1\geqslant 0$ and $p_2\geqslant 0$

$$\omega_{k_1,k_2}\left(f;\frac{1}{n_1},\frac{1}{n_2}\right)_{L_q}$$

$$\leqslant\omega_{k_1,k_2}\left(f-T_{2^{p_1+1},2^{p_2+1}};\frac{1}{n_1},\frac{1}{n_2}\right)_{L_q}+\omega_{k_1,k_2}\left(T_{2^{p_1+1},2^{p_2+1}};\frac{1}{n_1},\frac{1}{n_2}\right)_{L_q}$$

$$\leqslant 2^{k_1+k_2}E^*_{2^{p_1+1},2^{p_2+1}}(f)_{L_q}+\omega_{k_1,k_2}\left(T_{2^{p_1+1},2^{p_2+1}};\frac{1}{n_1},\frac{1}{n_2}\right)_{L_q}. \qquad (4)$$

Since

$$T_{2^{p_1+1},2^{p_2+1}}(f;x_1,x_2)=T_{1,1}(f;x_1,x_2)+$$

$$+\sum_{\nu=0}^{p_1-1}\{T_{2^{\nu+1},1}(f;x_1,x_2)-T_{2^\nu,1}(f;x_1,x_2)\}+$$

$$+\sum_{\nu=0}^{p_2-1}\{T_{1,2^{\nu+1}}(f;x_1,x_2)-T_{1,2^\nu}(f;x_1,x_2)\}+$$

$$+\sum_{\nu_1=0}^{p_1-1}\sum_{\nu_2=0}^{p_2-1}\{T_{2^{\nu_1+1},2^{\nu_2+1}}(f;x_1,x_2)-T_{2^{\nu_1+1},2^{\nu_2}}(f;x_1,x_2)-$$

$$-T_{2^{\nu_1},2^{\nu_2+1}}(f;x_1,x_2)+T_{2^{\nu_1},2^{\nu_2}}(f;x_1,x_2)+$$

$$+T_{2^{p_1+1},2^{p_2+1}}(f;x_1,x_2)-T_{2^{p_1},2^{p_2}}(f;x_1,x_2),$$

we have, using Minkowski's inequality

$$\omega_{k_1,k_2}\left(T_{2^{p_1+1},2^{p_2+1}};\frac{1}{n_1},\frac{1}{n_2}\right)_{L_q}$$

$$\leqslant \frac{1}{n_1^{k_1}n_2^{k_2}}\left\|\frac{\partial^{k_1+k_2}T_{2^{p_1+1},2^{p_2+1}}}{\partial x_1^{k_1}\,\partial x_2^{k_2}}\right\|_{L_q}$$

$$\leqslant \frac{1}{n_1^{k_1}n_2^{k_2}}\left\{\left\|\frac{\partial^{k_1+k_2}T_{1,1}}{\partial x_1^{k_1}\,\partial x_2^{k_2}}\right\|_{L_q}\right.$$

$$+\sum_{\nu=0}^{p_1-1}\left\|\frac{\partial^{k_1+k_2}}{\partial x_1^{k_1}\,\partial x_2^{k_2}}[T_{2^{\nu+1},1}-T_{2^\nu,1}]\right\|_{L_q}+$$

$$+\sum_{\nu=0}^{p_2-1}\left\|\frac{\partial^{k_1+k_2}}{\partial x_1^{k_1}\,\partial x_2^{k_2}}[T_{1,2^{\nu+1}}-T_{1,2^\nu}]\right\|_{L_q}+$$

$$+\sum_{\nu_1=0}^{p_1-1}\sum_{\nu_2=0}^{p_2-1}\left\|\frac{\partial^{k_1+k_2}}{\partial x_1^{k_1}\,\partial x_2^{k_2}}[T_{2^{\nu_1+1},2^{\nu_2+1}}-T_{2^{\nu_1+1},2^{\nu_2}}-T_{2^{\nu_1},2^{\nu_2+1}}+T_{2^{\nu_1},2^{\nu_2}}]\right\|_{L_q}+$$

$$\left.+\left\|\frac{\partial^{k_1+k_2}}{\partial x_1^{k_1}\,\partial x_2^{k_2}}[T_{2^{p_1+1},2^{p_2+1}}-T_{2^{p_1},2^{p_2}}]\right\|_{L_q}\right\}.$$

But in virtue of 4.8(30) and 6.1(5)

$$\left\|\frac{\partial^{k_1+k_2}}{\partial x_1^{k_1}\,\partial x_2^{k_2}}[T_{2^{\nu+1},1}-T_{2^\nu,1}]\right\|_{L_q}$$

$$\leqslant 2^{k_1(\nu+1)+1}E^*_{2^\nu,1}(f)_{L_q}\leqslant 2^{2k_1+1}\sum_{\mu=2^{\nu-1}+1}^{2^\nu}\mu^{k-1}E^*_{\mu,1}(f)_{L_q},$$

$$\left\|\frac{\partial^{k_1+k_2}}{\partial x_1^{k_1}\,\partial x_2^{k_2}}[T_{1,2^{\nu+1}}-T_{1,2^\nu}]\right\|_{L_q}$$

$$\leqslant 2^{k_2(\nu+1)+1}E^*_{1,2^\nu}(f)_{L_q}\leqslant 2^{2k_2+1}\sum_{m=2^{\nu-1}+1}^{2^\nu}\mu^{k_2-1}E^*_{1,\mu}(f)_{L_q},$$

$$\left\|\frac{\partial^{k_1+k_2}}{\partial x_1^{k_1}\,\partial x_2^{k_2}}[T_{2^{\nu_1+1},2^{\nu_2+1}}-T_{2^{\nu_1+1},2^{\nu_2}}-T_{2^{\nu_1},2^{\nu_2+1}}+T_{2^{\nu_1},2^{\nu_2}}]\right\|_{L_q}$$

$$\leqslant 2^{k_1(\nu+1)+k_2(\nu_2+1)+2}E^*_{2^{\nu_1},2^{\nu_2}}(f)_{L_q}$$

$$\leqslant 4^{k_1+k_2+1}\sum_{\mu_1=2^{\nu_1-1}+1}^{2^{\nu_1}}\sum_{\mu_2=2^{\nu_2-1}}^{2^{\nu_2}}\mu_1^{k_1-1}\mu_2^{k_2-1}E^*_{\mu_1,\mu_2}(f)_{L_q},$$

$$\left\| \frac{\partial^{k_1+k_2}}{\partial x_1^{k_1}\,\partial x_2^{k_2}} [T_{2^{p_1+1},2^{p_2+1}} - T_{2^{p_1},2^{p_2}}] \right\|_{L_q}$$
$$\leqslant 2^{(p_1+1)k_1+(p_2+1)k_2+1} E^*_{2^{p_1},2^{p_2}}(f)_{L_q}$$
$$\leqslant 2^{2(k_1+k_2)+1} \sum_{\mu_1=2^{p_1-1}+1}^{2^{p_1}} \sum_{\mu_2=2^{p_2-1}+1}^{2^{p_2}} \mu_1^{k_1-1}\mu_2^{k_2-1} E^*_{\mu_1,\mu_2}(f)_{L_q},$$

$$\left\| \frac{\partial^{k_1+k_2}}{\partial x_1^{k_1}\,\partial x_2^{k_2}} T_{1,1} \right\|_{L_q} = \left\| \frac{\partial^{k_1+k_2}}{\partial x_1^{k_1}\,\partial x_2^{k_2}} [T_{1,1} - T_{0,0}] \right\|_{L_q} \leqslant 2E^*_{0,0}(f)_{L_q}.$$

Therefore

$$\left\| \frac{\partial^{k_1+k_2}}{\partial x_1^{k_1}\,\partial x_2^{k_2}} T_{2^{p_1+1},2^{p_2+1}} \right\|_{L_q}$$
$$\leqslant 2E^*_{0,0}(f)_{L_q} + \sum_{\nu=0}^{p_1-1} 2^{2k_1+1} \sum_{\mu=2^{\nu-1}+1}^{2^\nu} \mu^{k_1-1} E^*_{\mu,1}(f)_{L_q} +$$
$$+ \sum_{\nu=0}^{p_2-1} 2^{2k_2+1} \sum_{\mu=2^{\nu-1}+1}^{2^\nu} \mu^{k_2-1} E^*_{1,\mu}(f)_{L_q} +$$
$$+ 4^{k_1+k_2+1} \sum_{\nu_1=1}^{p_1-1} \sum_{\nu_2=1}^{p_2-1} \sum_{\mu_1=2^{\nu_1-1}+1}^{2^{\nu_1}} \sum_{\mu_2=2^{\nu_2-1}+1}^{2^{\nu_2}} \mu_1^{k_1-1}\mu_2^{k_2-1} E^*_{\mu_1,\mu_2}(f)_{L_q} +$$
$$+ 2^{2(k_1+k_2)+1} \sum_{\mu_2=2^{p-1}+1}^{2^{p_1}} \sum_{\mu=2^{p_2-1}+1}^{2^{p_2}} \mu_1^{k_1-1}\mu_2^{k_2-1} E^*_{\mu_1,\mu_2}(f)_{L_q} +$$
$$+ 2^{k_1+k_2+3} E^*_{1,1}(f)_{L_q} + 2^{k_1+2k_2+2} \sum_{\nu_2=1}^{p_2-1} \sum_{\mu=2^{\nu_2-1}+1}^{2^{\nu_2}} \mu^{k_2-1} E^*_{1,\mu}(f)_{L_q} +$$
$$+ 2^{2k_1+k_2+2} \sum_{\nu_1=1}^{p_1-1} \sum_{\mu=2^{\nu_1-1}+1}^{2^{\nu_1}} \mu_1^{k_1-1} E^*_{\mu,1}(f)_{L_q}.$$

Consequently

$$\omega_{k_1,k_2}\left(T_{2^{p_1+1},2^{p_2+1}}; \frac{1}{n_1}, \frac{1}{n_2}\right)_{L_q} \leqslant \frac{4^{k_1+k_2+2}}{n_1^{k_1} n_2^{k_2}} \Bigg\{ E^*_{0,0}(f)_{L_q} +$$
$$\sum_{\nu=2}^{2^{p_1-1}} \nu^{k_1-1} E^*_{\nu,1}(f)_{L_q} + \sum_{\nu=2}^{2^{p_2-1}} \nu^{k_2-1} E^*_{1,\nu}(f)_{L_q} +$$
$$\sum_{\nu_1=2}^{2^{p_1-1}} \sum_{\nu_2=2}^{2^{p_2-1}} \nu_1^{k_1-1}\nu_2^{k_2-1} E^*_{\nu_1,\nu_2}(f)_{L_q} \Bigg\}. \tag{5}$$

Now choosing p_1 and p_2 so that $2^{p_1-1} < n_1 \leqslant 2^{p_1}$, $2^{p_2-1} < n_2 \leqslant 2$ and substituting the last estimate in 6.3(4), we arrive at the inequality 6.3(1) for $r = m = 2$.

6.3.2. It follows from 6.3(1) for $r = m$ that *if the sequence of total best approximations* $E^*_{n_1,\ldots,n_m}(f)_{L_q}$ *of the function* $f(x_1, \ldots, x_m)$ *which is periodic with respect to each of the variables satisfies the relation*

$$E^*_{n_1,\ldots,n_m}(f)_{L_q} = O\left(\sum_{k=1}^{m} \frac{1}{n_k^{\alpha_k}}\right) \tag{6}$$

for some $\alpha_1 > 0, \ldots, \alpha_m > 0$, *then for any natural numbers* $k_1, \ldots, k_m$

$$\omega_{k_1,\ldots,k_m}(f; u_1, \ldots, u_m)_{L_q}$$

$$= \begin{cases} O\left(\sum\limits_{\nu=1}^{m} u_\nu^{\alpha_\nu}\right) \text{ when } \alpha_\nu < k_\nu (\nu = 1, \ldots, m), \\ O\left(\sum\limits_{\nu=1}^{r} u_\nu^{\alpha_\nu} + \sum\limits_{\nu=r+1}^{m} u_\nu^{k_\nu} |\ln u_\nu|\right) \\ \quad \text{when } \alpha_\nu < k_\nu (\nu = 1, \ldots, r);\ \alpha_\nu = k_\nu (\nu = r+1, \ldots, m), \\ O\left(\sum\limits_{\nu=1}^{m} u_\nu^{k_\nu}\right) \text{ when } \alpha_\nu > k_\nu (\nu = 1, \ldots, m), \\ O\left(\sum\limits_{\nu=1}^{r} u_\nu^{k_\nu} + \sum\limits_{\nu=r+1}^{m} u_\nu^{k_\nu} |\ln u|\right) \\ \quad \text{when } \alpha_\nu > k_\nu\ (\nu = 1, \ldots, r);\ \alpha_\nu = k_\nu\ (\nu = r+1, \ldots, m). \end{cases} \tag{7}$$

In addition, since

$$E^*_{n_1,\ldots,n_r,\infty}(f)_{L_q} = O\left(\sum_{\nu=1}^{r} \frac{1}{n_\nu^{\alpha_\nu}}\right) \tag{8}$$

in virtue of the inequality 2.2(10) when the relation 6.3(6) is satisfied, *it follows from* 6.3(6) *that*

$$\omega_{k_\nu}(f; 0, \ldots, 0, u_\nu, 0, \ldots, 0)_{L_q} = \begin{cases} O(u_\nu^{\alpha_\nu}) & \text{when } \alpha_\nu < k_\nu, \\ O(u_\nu^{k_\nu} |\ln u_\nu|) & \text{when } \alpha_\nu = k_\nu, \\ O(u_\nu^{k_\nu}) & \text{when } \alpha_\nu > k_\nu. \end{cases} \tag{9}$$

Thus *for* $0 < \alpha_\nu < k_\nu$ $(\nu = 1, \ldots, m)$ *condition* 6.3(6) *is necessary* (see 5.3(1)) *and sufficient in order that*

$$\omega_{k_\nu}(f; 0, \ldots, 0, u_\nu, 0, \ldots, 0)_{L_q} = O(u_\nu^{\alpha_\nu}) \tag{10}$$

for all $\nu = 1, 2, 3, \ldots, m$.

If $\alpha_\nu = k_\nu$ $(\nu = 1, \ldots, m)$, *then the relation* 6.3(6) *is equivalent to the set of conditions*

$$\omega_{k_\nu+1}(f; 0, \ldots, 0, u_\nu, 0, \ldots, 0)_{L_q} = O(u_\nu^{k_\nu}) \quad (\nu = 1, \ldots, m). \tag{11}$$

6.3.3. The following proposition can be proved by the same method as in section 6.1.3. *If for the function* $f(x_1, \ldots, x_m)$ *which is periodic of p eriol* 2π *in each of the variables, and for some* q $(1 \leqslant q \leqslant \infty)$ *and some naturad number* r_i *the sequence of partial best approximations* $E^*_{\nu_i, \infty}(f)_{L_q}$ *with respect to the variable* x_i *satisfies the condition*

$$\sum_{\nu_i=1}^{\infty} \nu_i^{r_i-1} E^*_{\nu_i, \infty}(f)_{L_q} < \infty, \tag{12}$$

then apart from values on a set of m-dimensional measure zero $f(x_1, \ldots, x_m)$ *is identical with a function which has for almost all* $x_1, \ldots, x_{i-1}, x_{i+1}, \ldots, x_m$ *an absolutely continuous (with respect to* x_i*) partial derivative* $\dfrac{\partial^{r_i-1} f}{\partial x_i^{r_i-1}}$, *and an* r_i*-th partial derivative* $\dfrac{\partial^{r_i} f}{\partial x_i^{r_i}}$ *with respect to* x_i *which belongs to the class* L_q *on the cube of periods (in the case, when* $q = \infty$ *it is continuous), while for any natural number* k_i

$$\omega_{k_i}\left(\frac{\partial^{r_i} f}{\partial x_i^{r_i}}; 0, \ldots, 0, \frac{1}{n_i}, 0, \ldots, 0\right)_{L_q}$$

$$\leqslant C_{k_i, r_i}\left\{\frac{1}{n_i^{k_i}} \sum_{\nu_i=0}^{n_i} (\nu_i+1)^{k_i+r_i-1} E^*_{\nu_i, \infty}(f)_{L_q} + \sum_{\nu_i=n_i+1}^{\infty} \nu_i^{r_i-1} E^*_{\nu_i, \infty}(f)_{L_q}\right\}, \tag{13}$$

where C_{k_i, r_i} *is a constant depending only* k_i, r_i.

In order to justify the possibility of the term-by-term differentiation of the series in the case where $q < \infty$, the following lemma is applied here**.

6.3.31. *If the sequence of functions* $S_n(x_1, \ldots, x_m)$ *which have partial derivatives* $\dfrac{\partial^{r_i} S_n}{\partial x_i^{r_i}}$ *and belong, together with these derivatives, to the class* $L_q(1 \leqslant q \leqslant \infty)$ *on the parallelepiped* $Q(-\infty < a_k \leqslant x \leqslant b_k < \infty;$ $k = 1, \ldots, m)$, *possesses the property that for some pair of functions* $f(x_1, \ldots, x_m)$ *and* $\phi(x_1, \ldots, x_m)$

$$\left\{\int \cdots \int_Q |f(x_1, \ldots, x_m) - S_n(x_1, \ldots, x_m)|^q \, dx_1, \ldots, dx_m\right\}^{\frac{1}{q}} \to 0 \tag{14}$$

$$\left\{\int \cdots \int_Q \left|\phi(x_1, \ldots, x_m) - \frac{\partial^{r_i} S_n(x_1, \ldots, x_m)}{\partial x_i^{r_i}}\right|^q dx_1, \ldots, dx_m\right\}^{\frac{1}{q}} \to 0 \tag{15}$$

** S. M. Nikol'skii [13].

as $n \to \infty$, then apart from values on a set of measure zero $f(x_1, \ldots, x_m)$ is identical with a function having for any x_i and for almost all $x_1, \ldots, x_{i-1}$, $x_{i+1}, \ldots, x_m$ an absolutely continuous partial derivative $\dfrac{\partial^{r_i-1} f}{\partial x_i^{r_i-1}}$ with respect to x_i and a partial derivative $\dfrac{\partial^{r_i} f}{\partial x_i^{r_i}}$ of order r_i equal to $\phi(x_1, \ldots, x_m)$.

We will give a proof of this lemma for the case $m = 2$. Without loss of generality it is possible to suppose that $i = 1$ and $a_1 = 0$. Since the partial derivative $\dfrac{\partial^{r_1} S_n(x_1, x_2)}{\partial x_1^{r_1}} \in L_q$ on Q, it is integrable with respect to x_1 for almost all x_2, and consequently

$$S_n(x_1, x_2) = \sum_{j=0}^{r_1-1} \frac{\partial^j S_n(0, x_2)}{\partial x_1} x_1^j + \\ + \frac{1}{\Gamma(r_1)} \int_0^{x_1} (x_1 - t)^{r_1-1} \frac{\partial^{r_1} S_n(t, x_2)}{\partial t^{r_1}} \, dt. \tag{16}$$

Moreover, if $0 \leqslant x_1 < b_1$, then in virtue of Hölder's inequality and 6.3(15)

$$\int_{a_2}^{b_2} \left| \int_0^{x_1} (x_1 - t)^{r_1-1} \frac{\partial^{r_1} S_n(t, x_2)}{\partial t^{r_1}} \, dt - \int_0^{x_1} (x_1 - t)^{r_1-1} \phi(t, x_2) \, dt \right| dx_2$$

$$\leqslant b_1^{r_1 - \frac{1}{q}} (b_2 - a_2)^{\frac{q-1}{q}} \left\{ \iint_Q \left| \frac{\partial^{r_1} S_n(t, x_2)}{\partial t^{r_1}} - \phi(t, x_2) \right|^q dt \, dx_2 \right\}^{\frac{1}{q}} \to 0$$

as $n \to \infty$. Hence there exists a subsequence of the natural numbers $n_l = n_l(x_1) \to \infty$ such that as $n_l \to \infty$

$$\int_0^{x_1} (x_1 - t)^{r_1-1} \frac{\partial^{r_1} S_{n_l}(t, x_2)}{\partial t^{r_1}} \, dt \to \int_0^{x_1} (x_1 - t)^{r_1-1} \phi(t, x_2) \, dt \tag{17}$$

for almost all x_2.

In addition, it follows from 6.3(14) that there exists a sequence of natural numbers μ_k such that as $\mu_k \to \infty$

$$S_{\mu_k}(x_1, x_2) \to f(x_1, x_2) \tag{18}$$

almost everywhere on Q. We may suppose that the chosen subsequence $n_l(x_1)$ is part of the subsequence $\{\mu_k\}$.

Considering r different values of x_1, for example $x_{1,1}, \ldots, x_{1,r}$, for each of which as $\mu_k \to \infty$ we have $S_{\mu_k}(x_{1,i}, x_2) \to f(x_{1,i}, x_2)$ $(i = 1, \ldots, r)$ almost everywhere with respect to x_2, we can, by what has just been said, choose a subsequence $\{\nu_k\}$ from the sequence $\{\mu_k\}$ so that for any $i = 1, 2, \ldots, r$ we have

$$\int_0^{x_{1,i}} (x_{1,i}-t)^{r_1-1} \frac{\partial^{r_1} S_{\nu_k}(t, x_2)}{\partial t^{r_1}} \, dt \to \int_0^{x_{1,i}} (x_{1,i}-t)^{r_1-1} \phi(t, x_2) \, dt$$

for almost all x_2 as $\nu_k \to \infty$. It then follows from 6.3(16) that for $i = 1, 2, \ldots, r$ the sequence

$$\sum_{j=0}^{r_1-1} \frac{\partial^j S_{\nu_k}(0, x_2)}{\partial x_1^j} x_{1,i}^j$$

has a finite limit for almost all x_2 when $\nu_k \to \infty$.

Since every algebraic polynomial of degree $r_1 - 1$ is uniquely determined by its values at r_1 distinct points, it follows that for any x_1

$$\sum_{j=0}^{r_1-1} \frac{\partial^j S_{\nu_k}(0, x_2)}{\partial x_1^j} x_1^j \to \sum_{j=0}^{r_1-1} C_j(x_2) x_1^j$$

for almost all x_2 when $\nu_k \to \infty$.

Almost all values $x_1 \in [0, b_1]$ possess the property that for each of them 6.3(18) is satisfied almost everywhere with respect to x_2. For any such x_1 it is possible to choose a subsequence $n_l(x_1)$ such that the relation 6.3(17) is satisfied. Here, it may be supposed that $\{n_l(x_1)\}$ is part of the subsequence $\{\nu_k\}$. Consequently, if x_1 possesses the given property, then, passing in 6.3(16) to the limit as $n = n_l(x_1) \to \infty$, we find that for almost all x_2

$$f(x_1, x_2) = \sum_{j=0}^{r_1-1} C_j(x_2) x_1^j + \int_0^{x_1} (x-t)^{r_1-1} \phi(t, x_2) \, dt \tag{19}$$

Thus the relation 6.3(19) holds almost everywhere on Q. But since its right-hand side is the r-th integral of the function $\phi(x_1, x_2)$ for almost all x_2, it follows that almost everywhere on Q the function $f(x_1, x_2)$ is identical with a function which has a partial derivative of order r_1 with respect to x_1, equal to $\phi(x_1, x_2)$.

6.3.4. It follows from 6.3.3 and 2.2(10) that *if the sequence of total best approximations $E^*_{n_1, \ldots, n_m}(f)_{L_q}$ of the function $f(x_1, \ldots, x_n)$ which is periodic with respect to each of the variables satisfies the condition*

$$E^*_{n_1, \ldots, n_m}(f)_{L_q} = O\left(\sum_{i=1}^{m} \frac{1}{n_r^{r_i+\alpha_i}}\right) \tag{20}$$

for some natural numbers $r_1, \ldots, r_m$ *and* $\alpha_1 > 0, \ldots, \alpha_m > 0$, *then, almost everywhere in the space* R_m, $f(x_1, \ldots, x_m)$ *is identical with a function which for almost all* $x_1, \ldots, x_{i-1}, x_{i+1}, \ldots, x_m$ *possesses a partial derivative* $\dfrac{\partial^{r_i-1} f}{\partial x_i^{r_i-1}}$ *absolutely continuous with respect to* x_i *and a partial derivative* $\dfrac{\partial^{r_i} f}{\partial x^{r_i}}$ *with respect to* x_i $(i = 1, \ldots, m)$, *belonging to the class* L_q *on the cube of periods, while for any natural number* k_i:

$$\omega_{k_i}\left(\frac{\partial^{r_i} f}{\partial x_i^{r_i}}; 0, \ldots, 0, u_i, 0, \ldots, 0\right)_{L_q} = \begin{cases} O(u_i^{\alpha_i}) & \text{when} \quad \alpha_i < k_i, \\ O(u_i^{k_i} |\ln u_i|) & \text{when} \quad \alpha_i = k_i, \\ O(u_i^{k_i}) & \text{when} \quad \alpha_i > k_i. \end{cases} \tag{21}$$

Thus *for* $0 < \alpha_i < k_i$ $(i = 1, \ldots, m)$ *the relation* 6.3(20) *is necessary* (see 5.3(1)) *and sufficient in order that the function* $f(x_1, \ldots, x_m)$ *should be almost everywhere in the space* R_m *identical with a function which possesses with respect to each of the variables* x_i $(i = 1, 2, \ldots, m)$, *for almost all* $x_1, \ldots, x_{i-1}, x_{i+1}, \ldots, x_m)$, *a partial derivative* $\dfrac{\partial^{r_i-1} f}{\partial x_i^{r_i-1}}$ *absolutely continuous with respect to* x_i, *and a partial derivative* $\dfrac{\partial^{r_i} f}{\partial x_i^{r_i}}$ *of order* r_i *belonging to the class* L_q *on the cube of periods, which satisfies the condition*

$$\omega_{k_i}\left(\frac{\partial^{r_i} f}{\partial x_i^{r_i}}; 0, \ldots, 0, u_i, 0, \ldots, 0\right)_{L_q} = O(u_i^{\alpha_i}) \quad (i = 1, \ldots, m). \tag{22}$$

In the case when for all $i = 1, 2, \ldots, m$ *we have* $a_i = k_i$, *the relation* 6.3(20) *is equivalent to the set of conditions*

$$\omega_{k_i+1}\left(\frac{\partial^{r_i} f}{\partial x_i^{r_i}}; 0, \ldots, 0, u_i, 0, \ldots, 0\right)_{L_q} = O(u_i^{k_i}) \quad (i = 1, 2, \ldots, m).$$

6.3.5. Theorem 6.3.3 is supplemented by the following assertion**. *If for the function* $f(x_1, \ldots, x_m)$ *which is periodic of period* 2π *with respect*

** M. F. Timan [3].

to each of the variables, the sequence of total best approximations $E^*_{n_1,\dots,n_m}(f)_{L_q}$ *satisfies the condition*

$$\sum_{\nu_i=1}^{\infty} \nu_i^{r-1} E^*_{n_1,\dots,\nu_i,\dots,n_m}(f)_{L_q} < \infty$$

$$(n_j = [\nu_i^{l_j}],\ l_j > 0;\ j = 1,\dots, m;\ l_i = 1), \tag{23}$$

for some $q(1 \leqslant q \leqslant \infty)$ *and some natural number* r, *then the function* $f(x_1,\dots, x_m)$ *is identical almost everywhere in the space* R_m *with a function which possesses any mixed derivative* $\dfrac{\partial^{r_1+\dots+r_m} f}{\partial x_1^{r_1}\dots \partial x_m^{r_m}}$ *of order* r_j *with respect to* $x_j (j = 1,\dots, m)$, *where* $\sum\limits_{\nu=1}^{\infty} l_\nu r_\nu \leqslant r$, *belonging to the class* L_q *on the cube of periods (in the case where* $q = \infty$ *it is continuous), and for any natural number* k_i *satisfying the inequality*

$$\omega_{k_i}\left(\frac{\partial^{r_1+\dots+r_m} f}{\partial x_1^{r_1}\dots \partial x_m^{r_m}}; 0,\dots, 0, \frac{1}{m_i}, 0,\dots, 0\right)_{L_q}$$

$$\leqslant C_{k_i,r}\left\{\frac{1}{m_i^{k_i}} \sum_{\nu_i=0}^{m_i} (\nu_1+1)^{k_i+r-1} E^*_{r_1,\dots,\nu_i,\dots,n_m}(f)_{L_q} + \right.$$

$$\left. + \sum_{\nu_i=m_i+1}^{\infty} \nu_i^{r-1} E^*_{n_1,\dots,\nu_i,\dots,n_m}(f)_{L_q}\right\}, \tag{24}$$

where $C_{k_i,r}$ *is a constant depending only on* r *and* k_i, *and* $n_j = [\nu_i^{l_j}] (j = 1,\dots, m)$.

The proof is similar to the proof given in section 6.1.3 for functions of a single variable, and as in the case of theorem 6.3.3, is based on the lemma 6.3.31, after term-by-term differentiation of the series

$$T_{1,\dots,1}(f; x_1,\dots, x_m) + \sum_{\nu_i=0}^{\infty} \{T_{n_1,\dots,2^{\nu_i+1},\dots,n_m}(f; x_1,\dots, x_m) -$$

$$- T_{n_1,\dots,2^{\nu_i},\dots,n_m}(f, x_1,\dots, x_m)\}, \tag{25}$$

where $T_{n_1,\dots,n_m}(f; x_1,\dots, x_m)$ is the sequence of trigonometric polynomial best approximations.

6.3.51. Let us consider specially the case when the function $f(x_1,\dots, x_m)$ possesses with respect to each of the variables $x_\nu (\nu = 1,\dots, m)$, for almost all $x_1,\dots, x_{\nu-1}, x_{\nu+1},\dots, x_m$, a $(p_\nu - 1)$-th ($p_\nu \geqslant 1$ an integer) absolutely continuous (with respect to x_ν) partial derivative, and a partial derivative of order p_ν which belongs to the class $L_q(1 \leqslant q \leqslant \infty)$ on the cube of periods, for which

$$\omega_2\left(\frac{\partial^{p_\nu} f}{\partial x_\nu^{p_\nu}}; 0,\dots, 0, u_\nu, 0,\dots, 0\right)_{L_q} = O(u_\nu^{\alpha_\nu}) \quad (0 < \alpha_\nu \leqslant 1). \tag{26}$$

By theorem 5.3.1 it follows from 6.3(26) that

$$E^*_{n_1,\dots,n_m}(f)_{L_q} = O\left(\sum_{\nu=1}^{m} \frac{1}{n_\nu^{p_\nu+\alpha_\nu}}\right).$$

In the present case the series 6.3(23) converges for $r = p_i$ $(i = 1,\dots, m)$ and $l_j = \dfrac{p_i+\alpha_i}{p_j+\alpha_j}$ $(j = 1,\dots, m)$, and hence the assertion of theorem 6.3.5 is satisfied, where

$$\sum_{\nu=1}^{m} \frac{r_\nu}{p_\nu+\alpha_\nu} + \frac{\alpha_i}{p_i+\alpha_i} \leqslant 1. \tag{27}$$

Thus

$$\omega_2\left(\frac{\partial^{r_1+\dots+r_m} f}{\partial x_1^{r_1}\dots\partial x_m^{r_m}}; 0,\dots, 0, u_i, 0,\dots, 0\right)_{L_q} = O(u_i^{a_i}). \tag{28}$$

*Hence*** *if the periodic function* $f(x_1,\dots, x_m)$ *satisfies the conditions enumerated above, then almost everywhere in the space* R_m *it is identical with a function, which possesses any mixed derivative* $\dfrac{\partial^{r_1+\dots+r_m} f}{\partial x_1^{r_1}\dots\partial x_m^{r_m}}$ *where* $r_1,\dots, r_m$ *are connected by the inequality* 6.3(27), *belonging to the class* L_q *and satisfying condition* 6.3(28). *It also follows from theorem* 6.3.5 *that if the periodic function* $f(x_1,\dots, x_m)$ *belongs to the class* $W^{c\,p,\dots,p_m}M_1,\dots,M_m(L_\infty)$ $(p_1 > 0,\dots, p_m > 0)$ (see section 3.5.8), *then for any* $r_1,\dots, r_m$ *which satisfy the condition*

$$\sum_{\nu=1}^{m} \frac{r_\nu}{p_\nu} < 1 \tag{29}$$

a continuous derivative $\dfrac{\partial^{r_1+\dots+r_m} f}{\partial x_1^{r_1}\dots\partial x_m^{r_m}}$ ****exists.*

6.3.52. Let us note also that the term-by-term differentiation of the series 6.3(25), and the estimates which were applied in order to obtain the theorems of 6.3.5 and 6.3.51, lead to the following proposition which is useful in some cases. *If the periodic function* $f(x_1, \dots, x_m)$ *of period* 2π *has for each of the variables* x_ν $(\nu = 1, \dots, m)$, *for almost all* $x_1, \dots, x_{\nu-1}$,

** Montel [1] for the case $q = \infty$, $m = 2$; S. M. Nikol'skii [11, 13] for the general case.

*** Examples exist which show that this assertion no longer applies when $\sum_{\nu=1}^{m} \dfrac{r_\nu}{p_\nu} = 1$. See B. S. Mityagin [1].

$x_{\nu+1}, \ldots, x_m$, *the absolutely continuous (with respect to* x_ν*) partial derivative* $\dfrac{\partial^{p_\nu-1} f}{\partial x_\nu^{p_\nu-1}}$ *and a partial derivative of order* p_ν *belonging to the class* $L_q(1 \leqslant q \leqslant \infty)$ *on the cube of the periods, then almost everywhere it is identical with a function possessing any mixed derivative* $\dfrac{\partial^{r_1+\ldots+r_m} f}{\partial x_1^{r_1} \ldots \partial x_m^{r_m}}$, *where*

$$\sum_{\nu=1}^{m} \frac{r_\nu}{p_\nu} < 1, \tag{30}$$

and the inequality

$$\left\| \frac{\partial^{r_1+\ldots+r_m} f(x_1, \ldots, x_m)}{\partial x_1^{r_1} \ldots \partial x_m^{r_m}} \right\|_{L_q} \leqslant C_{r_1, \ldots, r_m}^{(p_1, \ldots, p_m)} \sum_{\nu=0}^{m} \left\| \frac{\partial^{p_\nu} f(x_1, \ldots, x_m)}{\partial x_\nu^{p_\nu}} \right\|_{L_q}, \tag{31}$$

is satisfied, where $C_m^{(p_1, \ldots, p_m)}$ *is a constant depending only on* $r_1, \ldots, r_m$ *and* $p_1, \ldots, p_m$, *and* $p_0 = 0$**.

6.3.6. Theorems similar to 6.3.1–6.3.5, 6.3.51 and 6.3.52, hold for any measurable functions $f(x_1, \ldots, x_m)$ defined throughout the whole space of the variables $x_1, \ldots, x_m$ and satisfying the condition

$$\|f\|_{L_q} = \left\{ \int_{-\infty}^{\infty} \cdots \int_{-\infty}^{\infty} |f(x_1, \ldots, x_m)|^q \, dx_1 \ldots dx_m \right\}^{\frac{1}{q}} < \infty, \tag{32}$$

if their best approximation $A_{\sigma_1, \ldots, \sigma_m}(f)_{L_q}$ by arbitrary integral functions of exponential type are considered (see 5.3(11)).

Let us give the most fundamental ones.

6.3.61. If $f(x_1, \ldots, x_m)$ satisfies condition 6.3(32) and $A_{\sigma_1, \ldots, \sigma_r, \infty}(f)_{L_q}$ is the corresponding partial best approximation with respect to the variables $x_1, \ldots, x_r$, then for any natural numbers $k_1, \ldots, k_r$, $n_1, \ldots, n_r$ the inequality

$$\omega_{k_1, \ldots, k_r}\left(f; \frac{1}{n_1}, \ldots, \frac{1}{n_r} 0, \ldots, 0\right)_{L_q} \leqslant \frac{C_{k_1, \ldots, k_r}}{n_1^{k_1} \ldots n_r^{k_r}} \sum_{\nu_1=0}^{n_1} \cdots \sum_{\nu_r=0}^{n_r} (\nu_1+1)^{k_1-1} \ldots (\nu_r+1)^{k_r-1} \times A_{\nu_1, \ldots, \nu_r, \infty}(f)_{L_q} \tag{33}$$

** S. N. Bernstein [1] showed that for $q = 2$ the inequality 6.3(31) also holds in the case $m = 2$, $p_1 = p_2$, $r_1 = r_2 = 1$.

is satisfied where $C_{k_1,\dots,k_r}$ is a constant depending only on $k_1, \dots, k_r$, and

$$\omega_{k_1,\dots,k_r}(f; u_1,\dots, u_r, 0,\dots, 0)_{L_q}$$

$$= \sup_{|h_k|\leqslant u_k} \left\{ \int_{-\infty}^{\infty} \cdots \int_{-\infty}^{\infty} \left| \sum_{\nu_1=0}^{k_1} \cdots \sum_{\nu_r=0}^{k_r} (-1)^{k_1+\dots+k_r-\nu_1-\dots-\nu_r} \times \right.\right.$$

$$\left.\left. \times \binom{k_1}{\nu_1} \cdots \binom{k_r}{\nu_r} f(x_1+\nu_1 h_1,\dots, x_r+\nu_r h_r,\dots, x_{r+1},\dots, x_m) \right|^q dx_1\dots dx_m \right\}^{\frac{1}{q}} \quad (34)$$

6.3.62. If $f(x_1, \dots, x_m)$ satisfies condition 6.3(32) and for some natural number r_i the inequality

$$\sum_{\nu_i=1}^{\infty} \nu_i^{r_i-1} A_{\nu_i,\infty}(f)_{L_q} < \infty \quad (35)$$

holds for the partial best approximation $A_{\sigma_i,\infty}(f)$ with respect to the variable x_i, then almost everywhere in the space R_m the function $f(x_1,\dots, x_m)$ is identical with a function possessing, for almost all $(x_1,\dots, x_{i-1}, x_{i+1},\dots, x_m$, an absolutely continuous (with respect to x_i) partial derivative $\dfrac{\partial^{r_i-1} f}{\partial x_i^{r_i-1}}$ and an r-th partial derivative $\dfrac{\partial^{r_i} f}{\partial x_i^{r_i}}$ with respect to x_i for which

$$\left\{ \int_{-\infty}^{\infty} \cdots \int_{-\infty}^{\infty} \left| \frac{\partial^{r_i} f}{\partial x_i^{r_i}} \right|^q dx_1 \dots dx_m \right\}^{\frac{1}{q}} < \infty,$$

and, for any natural number k_i,

$$\omega_{k_i}\left(\frac{\partial^{r_i} f}{\partial x_i^{r_i}}; 0,\dots, 0, \frac{1}{n_i}, 0,\dots, 0 \right)_{L_q}$$

$$\leqslant C_{k_i, r_i} \left\{ \frac{1}{n_i^{k_i}} \sum_{\nu_i=0}^{n_i} (\nu_i+1)^{k_i+r_i-1} A_{\nu_i,\infty}(f)_{L_q} + \right.$$

$$\left. + \sum_{\nu_i=n_i+1}^{\infty} \nu_i^{r_i-1} A_{\nu_i,\infty}(f)_{L_q} \right\}, \quad (36)$$

where C_{k_i, r_i} is a constant depending only on k_i, r_i.

6.3.63. *If* $f(x_1,\dots, x_m)$ *satisfies condition* 6.3(32) *and for some natural number* r

$$\sum_{\nu_i=1}^{\infty} \nu_i^{r-1} A_{\sigma_1,\dots,\nu_i,\dots,\sigma_m}(f)_{L_q} < \infty \quad (37)$$

$$(\sigma_j = \nu_i^{l_j}, l_j > 0;\ j = 1,\dots, m;\ l_i = 1),$$

then almost everywhere in the space R_m *the function* $f(x_1, \ldots, x_m)$ *is identical with a function which has any mixed derivative* $\dfrac{\partial^{r_1+\cdots+r_m} f}{\partial x_1^{r_1} \ldots \partial x_m^{r_m}}$ *of order* r_j *with respect to* x_j $(j = 1,\ldots,m)$, *where* $\sum_{\nu=1}^{m} l_\nu r_\nu \leqslant r$, *belonging to the class* L_q *on the whole space* $-\infty < x_1,\ldots,x_m < \infty$ *(in the case where* $q = \infty$ *it is continuous), and for any natural number* k_i *satisfying*

$$\omega_{k_i}\left(\frac{\partial^{r_1+\cdots+r_m} f}{\partial x_1^{r_1}\ldots \partial x_m^{r_m}}; 0,\ldots,0,\frac{1}{n},0,\ldots,0\right)_{L_q}$$

$$\leqslant C_{k_i,r}\left\{\frac{1}{n_i^{k_i}}\sum_{\nu_i=0}^{n_i}(\nu_i+1)^{k_i+r-1}A_{\sigma_1,\ldots,\nu_i,\ldots,\sigma_m}(f)_{L_q}+\right.$$

$$\left.+\sum_{\nu_i=n_i+1}^{\infty}\nu^{r-1}A_{\sigma_1,\ldots,\nu_i,\ldots,\sigma_m}(f)_{L_q}\right\}, \qquad (38)$$

where $C_{k_i,r}$ *is a constant depending only on* r, k_i *and* $\sigma_j = \nu_j$ $(j = 1,\ldots, m)$.

The proof of each of these propositions is based on the application of the inequality 4.2(28) and is derived as in the periodic case. It can be shown that in important individual cases the estimate 6.3(38) in a certain sense cannot be improved**.

6.3.7. For functions $f(x_1, \ldots, x_m)$ defined on the parallelepiped Q $(-\infty < a_\nu \leqslant x_\nu \leqslant b_\nu < \infty;\ \nu = 1, \ldots, m)$, inequalities of the type 6.1(22) hold which enable us from the best approximations of such functions by algebraic polynomials to judge their differential properties in any m-dimensional domain which together with its boundaries lies within Q. They can be obtained in the same way as the corresponding estimates in the periodic case if use is made of the inequality for the derivative of an algebraic polynomial which here plays a fundamental part (in the case $q = \infty$, $a = -1$, $b = 1$: see 4.8(31)).

The special feature of such estimates, for functions defined in the finite part of the space, have already been indicated above in the case of a single variable (see section 6.1.7).

Omitting a general formulation of the propositions which are relevant here, we will note only one particular case for continuous functions and the lemma similar to 6.3.51 which follows from it.

6.3.71. *If for some natural numbers* $p_1,\ldots,p_m$ *and* $0 < \alpha_1,\ldots,\alpha_m < 1$ *the sequence of best uniform approximations* $E_{n_1,\ldots,n_m}(f; Q)$ *of the function*

** S. M. Nikol'skii [13].

$f(x_1,\dots,x_m)$ by algebraic polynomials on Q satisfies the relation

$$E_{n_1,\dots,n_m}(f;Q)=O\left(\sum_{k=1}^{m}\frac{1}{n_k^{p_k+\alpha_k}}\right), \tag{39}$$

then $f(x_1,\dots,x_m)$ possesses everywhere within Q a p_k-th partial derivative $\dfrac{\partial^{p_k} f}{\partial x_k^{p_k}}$ with respect to x_k, which satisfies the condition

$$\omega_2\left(\frac{\partial^{p_k} f}{\partial x_k^{p_k}};0,\dots,0,u_k,0,\dots,0\right)\leqslant M(G)u_k^{\alpha_k}, \tag{40}$$

in every domain G which together with its boundaries lies within Q. Here the constant $M(G)$ in condition 6.3(40) depends on G.

6.3.72. *If the function $f(x_1,\dots,x_m)$ has in the parallelepiped Q a partial derivative of order $p_\nu>0$ with respect to each of the variables $x_\nu(\nu=1,\dots,m)$, which satisfies in Q condition* 6.3(40), *then it has within Q any mixed derivative $\dfrac{\partial^{r_1+\dots+r_m} f}{\partial x_1^{r_1}\dots\partial x_m^{r_m}}$ where $r_1,\dots,r_m$ are connected by the relation* 6.3(27), *and*

$$\omega_2\left(\frac{\partial^{r_1+\dots+r_m} f}{\partial x_1^{r_1}\dots\partial x_m^{r_m}};0,\dots,0,u_k,0,\dots,0\right)\leqslant M(G)u_k^{\alpha_k} \tag{41}$$

in every domain G which together with its boundaries lies within Q.

Here, as in 6.3(71), the constant $M(G)$ depends on G.

6.3.73. Let us note that *the assertions of sections* 6.3.71, 6.3.72 *remain valid for other m-dimensional domains.* This follows from the fact that if Q is an arbitrary bounded m-dimensional domain, then every domain G, which together with its boundaries lies within Q, can be covered by a finite number of parallelepipeds located within Q.

6.4. Differential properties and best approximations of functions in various metrics. On inclusion theorems for some classes of functions

Using the inequalities for the integral norms of integral functions of finite degree established in section 4.9, it is possible to obtain estimates of the structural characteristics of functions in a given metric $L_{q'}$ from their best approximations in some other metric L_q.

6.4.1. Let $1\leqslant q\leqslant q'\leqslant\infty$ and let the function $f(x)$, defined on the whole real axis $-\infty<x<\infty$, belong to the class L_q, i.e.

$$\|f\|_{L_q}=\left\{\int_{-\infty}^{\infty}|f(x)|^q\,dx\right\}^{\frac{1}{q}}<\infty. \tag{1}$$

If the best approximation $A_\sigma(f)_{L_q}$ possesses the property that for some integer $\varrho \geqslant 0$

$$\sum_{\nu=1}^{\infty} \nu^{\rho-1+\frac{1}{q}-\frac{1}{q'}} A_\nu(f)_{L_q} < \infty . \tag{2}$$

then almost everywhere on $(-\infty, \infty)$ the function $f(x)$ is identical with a function having an absolutely continuous derivative $f^{(\rho-1)}(x)$, a derivative $f^{(\rho)}(x)$ of order ϱ which belongs to the class $L_{q'}$ (in the case $q' = \infty$ continuous), and for any natural number k satisfies

$$\omega_k\left(f^{(\rho)}; \frac{1}{n}\right)_{L_{q'}} \leqslant C_{k,\rho}\left\{\frac{1}{n^k}\sum_{\nu=0}^{n}(\nu+1)^{k+\rho-1+\frac{1}{q}-\frac{1}{q'}} A_\nu(f)_{L_q} + \right.$$

$$\left. + \sum_{\nu=n+1}^{\infty} \nu^{\rho-1+\frac{1}{q}-\frac{1}{q'}} A_\nu(f)_{L_q}\right\}, \tag{3}$$

where the constant $C_{k,\rho}$ depends only on ϱ and k.

To prove this it is necessary to consider the series

$$g_1^{(p)}(f; x) + \sum_{\nu=0}^{\infty} \{g_{2^{\nu+1}}^{(p)}(f; x) - g_{2^\nu}^{(p)}(f; x)\}, \tag{4}$$

where $g_\sigma(f; x)$ is an integral function of degree $\leqslant \sigma$ for which

$$A_\sigma(f) = \|f(x) - g_\sigma(f; x)\|_{L_q},$$

and the integer $p \leqslant \varrho$. Since, by 4.8(28) and 4.9 (29),

$$\|g_{2^{\nu+1}}^{(p)}(f; x) - g_{2^\nu}^{(p)}(f; x)\|_{L_{q'}} \leqslant 2^{p(\nu+1)}\|g_{2^{\nu+1}}(f; x) - g_{2^\nu}(f; x)\|_{L_{q'}}$$

$$< 2^{(\nu+1)\left(p+\frac{1}{q}-\frac{1}{q'}\right)+1}\|g_{2^{\nu+1}}(f, x) - g_{2^\nu}(f; x)\|_{L_q}$$

$$\leqslant 2^{(\nu+1)\left(p+\frac{1}{q}-\frac{1}{q'}\right)+2} A_{2^\nu}(f)_{L_q}$$

$$\leqslant 2^{2\left(p+\frac{1}{q}-\frac{1}{q'}\right)+2} \sum_{\mu=2^{\nu-1}+1}^{2^\nu} \mu^{p-1+\frac{1}{q}-\frac{1}{q'}} A_\mu(f)_{L_q},$$

then in virtue of condition 6.4(2) the series 6.4(4) converges in mean in the sense of the metric $L_{q'}$ on $(-\infty, \infty)$. Moreover, reasoning exactly as in section 6.1.3, we see that the derivative $f^{(\rho)}(x)$ exists almost everywhere and is the limit in mean (in the sense of the metric $L_{q'}$ on $(-\infty, \infty)$) of the partial sums of the series 6.4(4) for $p = \varrho$. Then repeating the same estimates as in section 6.1.3, we obtain the inequality 6.4(3) for the modulus of smoothness $\omega_k\left(f^{(\rho)}; \frac{1}{n}\right)_{L_{q'}}$.

6.4.2. Since the series 6.4(2), where $0 \leqslant \varrho < \alpha - \left(\frac{1}{q} - \frac{1}{q'}\right)$, converges (see 5.1(15)) for a function $f(x)$ which satisfies for some natural l the condition $\omega_l(f; t)_{L_q} = O(t^\alpha)$ $(q \geqslant 1)$, it follows that from theorem 6.4.1. it can be asserted *that almost everywhere on* $(-\infty, \infty)$ *any such function* $f(x)$ *is identical with a function having an absolutely continuous derivative* $f^{(\rho-1)}(x)$ (ϱ *an integer*), *and a derivative* $f^{(\rho)}(x)$ *of order* ϱ *which belongs to the class* $L_{q'}$ $(q \leqslant q' \leqslant \infty)$ *on* $(-\infty, \infty)$ *and for any natural number* k *satisfies*

$$\omega_k(f^{(\rho)}, t)_{L_{q'}} = \begin{cases} O\left[t^{\alpha-\left(\frac{1}{q}-\frac{1}{q'}\right)-\rho}\right] & \text{when} \quad \alpha - \left(\frac{1}{q} - \frac{1}{q'}\right) - \varrho < k, \\ O(t^k|\ln t|) & \text{when} \quad \alpha - \left(\frac{1}{q} - \frac{1}{q'}\right) - \varrho = k, \\ O(t^k) & \text{when} \quad \alpha - \left(\frac{1}{q} - \frac{1}{q'}\right) - \varrho > k. \end{cases} \tag{5}$$

In particular, if use is made of the notation adopted in section 3.5.3, it follows directly from this (see 3.3.4), that** the class $W^{(r)} H_2^{(\alpha)} M(L_q)$ (r an integer, $0 < \alpha \leqslant 1$, $1 \leqslant q \leqslant \infty$) is contained in the class $W^{(\rho)} H_2^{(\beta)} M^*(L_{q'})$. for $q' \geqslant q$ and $\varrho + \beta = r + \alpha - \left(\frac{1}{q} - \frac{1}{q'}\right)$, where ϱ is an integer and $0 < \beta \leqslant 1$.

6.4.3. A theorem similar to 6.4.1, is valid also for functions of many variables. *If the function* $f(x, \ldots, x_m)$, *defined throughout the whole of the space* $-\infty < x_1, \ldots, x_m < \infty$, *satisfies condition* 6.3(32) $(q \geqslant 1)$ *and for some integer* $\varrho_i \geqslant 0$ *the best approximation* $A_{\sigma_1, \ldots, \sigma_m}(f)_{L_q}$ satisfies

$$\sum_{\nu_i=0}^{\infty} \nu_i^{\rho_i - 1 + \left(\frac{1}{q} - \frac{1}{q'}\right) \sum_{k=1}^{m} l_k} A_{\sigma_1, \ldots, \nu_i, \ldots, \sigma_m}(f)_{L_q} < \infty \qquad (q' \geqslant q) \tag{6}$$

when $\sigma_j = \sigma_i^{l_j}$ $(j = 1, \ldots, m;\ l_j > 0,\ l_i = 1)$, *then almost everywhere in the space* R_m *the function* $f(x_1, \ldots, x_m)$ *is identical with a function possessing, for almost all* $x_1, \ldots, x_{i-1}, x_{i+1}, \ldots, x_m$, *a partial derivative* $\frac{\partial^{\rho_i - 1} f}{\partial x_i^{\rho_i - 1}}$ *absolutely continuous with respect to* x_i, and *a partial derivative* $\frac{\partial^{\rho_i} f}{\partial x_i^{\rho_i}}$ *of order* ϱ_i *with respect to* x_i *for which*

$$\left\{ \int_{-\infty}^{\infty} \cdots \int_{-\infty}^{\infty} \left| \frac{\partial^{\rho_i} f}{\partial x_i^{\rho_i}} \right|^{q'} dx_1, \ldots, dx_m \right\}^{\frac{1}{q'}} < \infty \tag{7}$$

** S. M. Nikol'skii [13].

and for any natural number k_i

$$\omega_{k_i}\left(\frac{\partial^{\rho_i} f}{\partial x_i^{\rho_i}}; 0, \ldots, 0, \frac{1}{n_i}, 0, \ldots, 0\right)_{L_{q'}}$$

$$\leqslant C_{k_i, \rho_i}^{(m)} \left\{ \frac{1}{n_i^{k_i}} \sum_{\nu_i=0}^{n_i} (\nu_i+1)^{k_i+\rho_i-1+\left(\frac{1}{q}-\frac{1}{q'}\right)\sum_{j=1}^{m} l_j} A_{\sigma_1, \ldots, \nu_i, \ldots, \sigma_m}(f)_{L_q} + \right.$$

$$\left. + \sum_{\nu_i=n_i+1}^{\infty} \nu_i^{\rho_i-1+\left(\frac{1}{q}-\frac{1}{q'}\right)\sum_{j=1}^{m} l_j} A_{\sigma_1, \ldots, \nu_i, \ldots, \sigma_m}(f)_{L_q} \right\} \tag{8}$$

where the constant $C_{k_i, \rho_i}^{(m)}$ *depends only on* m, k_i, ϱ_i, *and* $\sigma_j = \nu_i^{l_j}$ $(j = 1, \ldots$ $\ldots, m; l_i = 1)$.

The proof is carried out in exactly the same way as in the case when $m = 1$ (see 6.4.1). It is based on the lemma of 6.3.31 and on the inequalities 4.8(28), 4.9(34).

6.4.4. In the case where the function $f(x, \ldots, x_m)$ has an absolutely continuous partial derivative $\dfrac{\partial^{r_\nu - 1} f}{\partial x_\nu^{r_\nu - 1}}$ with respect to each of the variables $x_\nu (\nu = 1, \ldots, m)$ and

$$\omega_2\left(\frac{\partial^{r_\nu} f}{\partial x_\nu^{r_\nu}}; 0, \ldots, 0, u_\nu, 0, \ldots, 0\right)_{L_q} = O\left(u_\nu^{\alpha_\nu}\right) \qquad (0 < \alpha_\nu \leqslant 1), \tag{9}$$

the series 6.4(6) converges in virtue of 5.3.3 if $l_j = \dfrac{r_i + \alpha_i}{r_j + \alpha_j}$ subject to the condition that $0 \leqslant \varrho_i < (r_i + \alpha_i)\left[1 - \left(\dfrac{1}{q} - \dfrac{1}{q'}\right) \sum_{j=1}^{m} \dfrac{1}{r_j + \alpha_j}\right]$. Hence by 6.4.3 in this case $f(x_1, \ldots, x_m)$ is identical almost everywhere in the space R_m with a function having with respect to each variable x_i, for almost all $x_1, \ldots, x_{i-1}, x_{i+1}, \ldots, x_m$, an absolutely continuous partial derivative $\dfrac{\partial^{\rho_i - 1} f}{\partial x_i^{\rho_i - 1}}$ (ϱ_i an integer), and a partial derivative $\dfrac{\partial^{\rho_i} f}{\partial x_i^{\rho_i}}$ of order ϱ_i which satisfies condition 6.4.(7) and for any natural number k_i

$$\omega_{k_i}\left(\frac{\partial^{\rho_i} f}{\partial x_i^{\rho_i}}; 0, \ldots, 0, u_i, 0, \ldots, 0\right)_{L_{q'}}$$

$$= \begin{cases} O\left\{u_i^{(r_i+\alpha_i)\left[1-\left(\frac{1}{q}-\frac{1}{q'}\right)\sum_{j=1}^{m}\frac{1}{r_j+\alpha_j}\right]-\rho_i}\right\}, \\ O\left(u_i^{k_i}\right)|\ln u_i|\right), \\ O\left(u_i^{k_i}\right), \end{cases} \tag{10}$$

when $(r_i+\alpha_i)\left[1-\left(\frac{1}{q}-\frac{1}{q'}\right)\sum_{j=1}^{m}\frac{1}{r_j+\alpha_j}\right]-\varrho_i$ is respectively less than, equal to or greater than k_i.

This implies immediately the validity of the first of the inclusion theorems for the classes $H_2^{(r_1,\dots,r_m)}M_1\dots M_m(L_q)$ formulated in section 3.5.81.

The second of these theorems can be obtained in a similar way from the following proposition.

6.4.5. *If the function $f(x_1,\dots,x_m)$, defined throughout the whole of the space $-\infty < x_1,\dots,x_m < \infty$, satisfies the condition 6.3(32) $(q \geqslant 1)$ and for some integer $\varrho_1 \geqslant 0$ the best approximation $A_{\sigma_1,\dots,\sigma_m}(f)_{L_q}$ satisfies*

$$\sum_{\nu=1}^{\infty} \nu^{\rho_i-1+\left(\frac{1}{q}-\frac{1}{q'}\right)\sum_{k=1}^{s} l_k+\frac{1}{q}\sum_{k=s+1}^{m} l_k} A_{\sigma_1,\dots,\nu_i,\dots,\sigma_m}(f)_{L_q} < \infty \quad (q' \geqslant q) \tag{11}$$

when $\sigma_j = \sigma_i^{l_j}(j=1,\dots,m;\ l_j>0)$, $1 \leqslant i \leqslant s \leqslant m$, then almost everywhere in the space R_m the function $f(x_1,\dots,x_m)$ is identical with a function which, for any fixed values $x_{s+1},\dots,x_m$, as a function of the variables $x_1,\dots,x_s$, has, for almost all $x_1,\dots,x_{i-1},x_{i+1},\dots,x_s$, an absolutely continuous (with respect to x_i) partial derivative $\frac{\partial^{\rho_i-1}f}{\partial x_i^{\rho_i-1}}$, and a partial derivative $\frac{\partial^{\rho_i}f}{\partial x_i^{\rho_i}}$ with respect to x_i such that

$$\left\{\int_{-\infty}^{\infty}\cdots\int_{-\infty}^{\infty}\left|\frac{\partial^{\rho_i}f}{\partial x_i^{\rho_i}}\right|^{q'}\mathrm{d}x_1\dots\mathrm{d}x_s\right\}^{\frac{1}{q'}} < \infty, \tag{12}$$

and, for every natural number k:

$$\omega_{k_i}\left(\frac{\partial^{\rho_i}f}{\partial x_i^{\rho_i}};0,\dots,0,\frac{1}{n_i},0,\dots,0\right)_{L_{q'}}$$
$$\leqslant C_{k_i,\rho_i}^{(m)}\left\{\frac{1}{n_i^{k_i}}\sum_{\nu_i=0}^{n_i}(\nu_i+1)^{k_i+\rho_i-1+\left(\frac{1}{q}-\frac{1}{q'}\right)\sum_{j=1}^{s} l_j+\frac{1}{q}\sum_{j=s+1}^{m} l_j}\times\right.$$
$$\times A_{\sigma_1,\dots,\nu_i,\dots,\sigma_m}(f)_{L_q}+$$
$$\left.+\sum_{\nu_1=n_i+1}^{\infty}\nu_i^{\rho_i-1+\left(\frac{1}{q}-\frac{1}{q'}\right)\sum_{j=1}^{s} l_j+\frac{1}{q}\sum_{j=s+1}^{m} l_j} A_{\sigma_1,\dots,\nu_i,\dots,\sigma_m}(f)_{L_q}\right\}, \tag{13}$$

where $1 \leqslant i \leqslant s$, $\sigma_i = \nu_i^{l_j}(j=1,\dots,m;\ l_i=1)$, and $C_{k_i,\rho_i}^{(m)}$ is a constant depending only on m, k_i, ϱ_i.

The proof here is the same as in the case when $s = m$ (section 6.4.3). For $s < m$, besides the lemma 6.3.31 and the inequalities 4.8(28), 4.9(34), the inequality 4.9(35) is also required.

In order to obtain from this the theorem given in section 3. 5. 81, it is necessary in virtue of 5.3.3 to put $l_j = \frac{r_i}{r_j}$ and take account of the inequalities

$$0 \leqslant \varrho_i < r_i \left[1 - \left(\frac{1}{q} - \frac{1}{q'} \right) \sum_{j=1}^{s} \frac{1}{r_j} - \frac{1}{q} \sum_{j=s+1}^{m} \frac{1}{r_j} \right]. \tag{14}$$

As was shown by S. M. Nikol'skii [13], in a certain sense this estimate for ϱ_i cannot be improved.

6.4.6. In conclusion let us note that theorems similar to 6.4.1–6.4.5 hold for functions periodic with respect to each of the variables, if all the integrals are taken over a period and instead of the best approximations $A_{\sigma_1, \dots, \sigma_m}(f)_{L_q}$ we consider best approximations by trigonometric polynomials. In this case on passing from the metric L_q to the metric $L_{q'}$, $(q' \geqslant q)$ the corresponding differential properties of the functions are not improved (see 2.13.3).

6.5. The analyticity of functions and their best approximations

In chap. V in the study of the constructive properties of bounded functions, analytic in a strip containing the real axis (see section 5.7), or on some finite segment $[a, b]$ (see section 5.4), it was established that for their best approximations by transcendental integral functions of exponential type or by algebraic polynomials, the limiting inequalities 5.7(9) and 5.4(2) are satisfied. Here it will be shown that the corresponding classes of functions $A_{\infty, \delta}$ (see section 3.8) and $A_R(a, b)$ (see section 3.7.31) are completely described by these properties.

6.5.1. Let the function $f(x)$ be continuous on the finite segment $[a, b]$. Let us prove *that if for the sequence of best uniform approximations* $E_n(f; a, b)$ *the inequality* 5.4(2) *is satisfied, where* $R > \frac{b-a}{2}$, *then the function* $f(x)$ *is analytic on* $[a, b]$ *and belongs to the class* $A_R(a, b)$**.

It is sufficient to consider the case when $a = -1$, $b = 1$. It follows from 5.4(2) that for any $\varrho > \frac{1}{R}$ $(\varrho < 1)$ for all values of $n \geqslant N(\varrho)$

$$E_n(f; -1, 1) < \varrho^n \tag{1}$$

** S. N. Bernstein [1].

Considering the series

$$P_1(f; z)+\sum_{n=1}^{\infty}\{P_{n+1}(f; z)-P_n(f; z)\}, \tag{2}$$

where $P_n(f; x)$ is the best uniform algebraic polynomial approximation of order n to the function $f(x)$ on $[-1, 1]$, i.e.

$$E_n(f; -1, 1) = \max_{-1\leqslant x\leqslant 1} |f(x)-P_n(f; x)|,$$

we see that in virtue of 6.5(1) its terms satisfy for all $n \geqslant N(\varrho)$ the inequality

$$\max_{-1\leqslant x\leqslant 1} |P_{n+1}(f; x)-P_n(f; x)| < 2\varrho^n. \tag{3}$$

Since the polynomial $P_{n+1}(f; x)-P_n(f; x)$ satisfies the inequality 6.5(3) on the segment $[-1,1]$, it follows that at points z lying on the ellipse C_r with foci at the ends of the segment and with the sum of the semi-axes equal to r (see section 2.13.27),

$$|P_{n+1}(f; z)-P_n(f; z)| < 2\varrho^n r^{n+1}. \tag{4}$$

Hence for $r < \dfrac{1}{p}$ the series 6.5(2) converges uniformly within the domain bounded by the ellipse C_r, and consequently its sum is there an analytic function, which on the segment $[-1, 1]$ is identical with the function $f(x)$ considered. Hence in virtue of the arbitrariness of the choice of $\varrho > \dfrac{1}{R}$ and $r < \dfrac{1}{\varrho} < R$ it follows that $f(x) \in A_R(-1, 1)$.

6.5.2.. Comparing theorems 6.5.1 and 5.4.1, we see that *a function $f(x)$ continuous on the finite segment $[a, b]$ belongs to the class $A_R(a, b)$ if and only if the sequence of its best approximations by algebraic polynomials on $[a, b]$ satisfies the limiting relation* 5.4(2). *In particular, f is an integral function if and only if*

$$\lim_{n\to\infty} \sqrt[n]{E_n(f; a, b)} = 0. \tag{5}$$

6.5.3. *If the function $f(x)$, bounded and continuous on $(-\infty, \infty)$, possesses the property that for its best uniform approximation $A_\sigma(f)$ by integral functions of finite degree the inequality* 5.7(9) *is satisfied, then $f(x)$ belongs to the class $A_{\infty,\delta}$***.

** S. N. Bernstein [22].

It follows from 5.7(9) that for any $\varepsilon < \delta$ $(\varepsilon > 0)$

$$A_\sigma(f) < e^{-\sigma\varepsilon} \tag{6}$$

for all $\sigma > N(\varepsilon)$. The terms of the series

$$g_1(f; z) + \sum_{n=1}^{\infty} \{g_{n+1}(f; z) - g_n(f; z)\}, \tag{7}$$

where $g_\sigma(f; z)$ is an integral function of degree $\leqslant \sigma$ for which $A_\sigma(f) = \sup_{-\infty<x<\infty} |f(x) - g_\sigma(f; x)|$, satisfy the inequality

$$\sup_{-\infty<x<\infty} |g_{n+1}(f; x) - g_n(f; x)| \leqslant 2A_n(f) < 2e^{-n\varepsilon} \tag{8}$$

for $n \geqslant N(\varepsilon)$ consequently (see section 4.8.3),

$$|g_{n+1}(f, x+iy) - g_n(f; x+iy)| < 2e^{|y|} e^{n(|y|-\varepsilon)}.$$

Thus for any $\tau < \varepsilon$ $(\tau > 0)$ the series 6.5(7) converges uniformly in the strip $|y| < \tau$, $-\infty < x < \infty$, and its sum in this strip is a bounded analytic function, which is identical with $f(x)$ on the real axis. Hence in virtue of the arbitrariness of the choice of $\varepsilon < \delta$ and $\tau < \varepsilon < \delta$ (if use is made of Cauchy's integral formula) it follows that the function $f(x)$ belongs to the class $A_{\infty,\delta}$.

6.5.4. Thus**, *the continuous and bounded function $f(x)$ belongs to the class $A_{\infty,\delta}$ if and only if* (see section 5.7.22) *its best approximation by integral functions of exponential type satisfies the inequality* 5.7(9). In particular, *a necessary and sufficient condition in order that $f(x) \in A_{\infty,\delta}$ for any δ is the equality*

$$\lim_{\delta\to\infty} \sqrt[\sigma]{A_\sigma(f)} = 0. \tag{9}$$

6.5.41. It follows immediately from theorem 6.5.4 (see also sections 3.8 and 5.7.22) that *for any function $f(x)$ continuous and bounded on $(-\infty < x < \infty)$ the condition* 5.7(9) *is equivalent to its infinite differentiability and the inequality*

$$\overline{\lim_{n\to\infty}} \sqrt[n]{\frac{M_n(f)}{n!}} \leqslant \frac{1}{\delta}, \tag{10}$$

and 6.5(9) *is equivalent to the equality*

$$\lim_{n\to\infty} \sqrt[n]{\frac{M_n(f)}{n!}} = 0. \tag{11}$$

** S. N. Bernstein [45].

6.6. On constructive characteristics of quasi-analytic classes of functions

Theorems 5.8.1, 5.8.3 and 5.8.4 on the properties of the best approximation of infinitely differentiable functions which satisfy the condition 3.10(1) and 3.10(12) form the starting points of the study of the nature of quasi-analyticity from the constructive point of view and lead to the determination of new quasi-analytic classes of functions.

6.6.1. Let the function $f(x)$ be continuous on $[a, b]$. The condition

$$\overline{\lim_{n\to\infty}} \sqrt[n]{E_n(f; a, b)} < 1, \tag{1}$$

which describes the class of all functions analytic on $[a, b]$ (see section 6.5.2), completely determines such a function $f(x)$ from its values in any sufficiently small interval. If the function $f(x)$ which satisfies it is equal to zero at every point x of some interval $(\alpha, \beta) \subset [a, b]$, then it is identically zero on the whole segment $[a, b]$.

It is not difficult to see that the *last proposition remains valid if the condition* 6.6(1) *is replaced by the weaker condition*

$$\underline{\lim_{n\to\infty}} \sqrt[n]{E_n(f; a, b)} < 1. \tag{2}$$

In fact, if 6.6(2) holds for the function $f(x)$, there exists an infinite sequence of natural numbers n_k and a certain positive number $\varrho < 1$ such that

$$E_{n_k}(f; a, b) < \varrho^{nk} \qquad (k = 1,2,3,\ldots). \tag{3}$$

Hence for the algebraic polynomials $P_{n_k}(f; x)$ of order n_k which deviate least from $f(x)$ on $[a,b]$, subject to the condition that $f(x) = 0$ on (α,β), the inequality

$$\max_{\alpha\leqslant x\leqslant\beta} |P_{n_k}(f; x)| \leqslant \varrho^{nk} \qquad (k = 1,2,3,\ldots). \tag{4}$$

is satisfied. Hence in virtue of section 2.9.11 (see 2.9(9)) it follows that at points x outside the segment $[\alpha, \beta]$

$$|P_{n_k}(f; x)| \leqslant \varrho^{nk} \left| \cos n_k \arccos \frac{2x-\alpha-\beta}{\beta-\alpha} \right|, \tag{5}$$

or

$$|P_{n_k}(f; x)| \leqslant \varrho^{nk} \left(\frac{2R}{\beta-\alpha} \right)^{nk}, \tag{6}$$

where R is the sum of the semi-axes of the ellipse which passes through x with foci at the points β and α (see 4.8(39)). Taking, for example, the value $R = \frac{\beta-\alpha}{4}\left(1+\frac{1}{\varrho}\right)$, we find that for all x belonging to a certain interval (α_1,β_1) $(\alpha_1 < \alpha < \beta < \beta_1;\ \alpha-\alpha_1 = \beta_1-\beta)$,

$$|P_{n_k}(f;\,x)| < \varrho_1^{n_k} \qquad (\varrho < \varrho_1 < 1).$$

Consequently, thanks to the inequality 6.6(3),

$$\max_{\alpha_1 \leqslant x \leqslant \beta_1} |f(x)| \leqslant \varrho^{n_k} + \varrho_1^{n_k} < 2\varrho_1^{n_k} \qquad (k = 1,2,3,\ldots),$$

i.e. $f(x) = 0$ everywhere on the segment $[\alpha_1, \beta_1]$. Since the difference $\alpha-\alpha_1 = \beta_1-\beta$ depends only on the value of ϱ and on the length of the segment $[\alpha,\beta]$, it follows by repetition of the argument a finite number of times that $f(x) = 0$ for all $x \in [a, b]$.

From the above considerations it is obvious that** *for any infinite sequence of natural numbers $n_1 < n_2 < n_3 < \ldots$, the class of functions $f(x)$ continuous on $[a, b]$, for each of which*

$$E_{n_k}(f\,;a,b) < \varrho^{n_k}(f)_1 \tag{7}$$

for some $\varrho = \varrho(f) < 1$, is quasi-analytic, i.e. if two functions $f_1(x)$ and $f_2(x)$ which belong to this class assume identical values at the points of an arbitrarily small interval $(\alpha,\beta) \subset [a,b]$, then they are identically equal everywhere on $[a,b]$. It is sufficient merely to consider the function $f(x) = f_1(x)-f_2(x)$ for which

$$E_{n_k}(f\,;a,b) < \varrho^{n_k}(f_1) + \varrho^{n_k}(f_2),$$

and consequently for all sufficiently great values of k and for some $r < 1$

$$E_{n_k}(f\,;a,b) < r^{n_k}.$$

6.6.2. Thus any sequence of natural numbers $\{n_k\}$ which increases without limit determines some quasi-analytic (in the sense of S. N. Bernstein) class of functions, continuous on $[a,b]$ and satisfying condition 6.6(7).

It follows immediately from theorem 6.5.1 that *in the case where for the sequence $\{n_k\}$ the ratio $\frac{n_{k+1}}{n_k}$ remains bounded as $k \to \infty$ $\left(\frac{n_{k+1}}{n_k} \leqslant c < \infty\right)$, the corresponding quasi-analytic class contains only functions analytic on $[a,b]$. In this case, for any $n(n_k \leqslant n < n_{k+1})$,*

$$E_n(f\,;a,b) \leqslant E_{n_k}(f\,;a,b) < \left(\varrho^{\frac{n_k}{n_{k+1}}}\right)^n \leqslant \left(\varrho^{\frac{1}{c}}\right)^n.$$

** S. N. Bernstein [7].

If, however, the given ratio is unbounded, then the quasi-analytic class determined by the sequence $\{n_k\}$ necessarily contains functions which are not analytic. In order to verify this, let us consider the monotonic null sequence of numbers $E_1 \geqslant E_2 \geqslant E_3 \geqslant \ldots \geqslant E_n \geqslant \ldots$ such that for some $\varrho < 1$ and any value n_k of a given subsequence of the natural series $E_n = \varrho^{n_k}$ when $n_k \leqslant n < n_{k+1}$. By theorem 2.5 there exists a function $f(x)$ for which $E_n(f; a,b) = E_n$ for any $n \geqslant 1$. Such a function satisfies the condition 6.6(7) and, consequently, belongs to the given quasi-analytic class. However, by theorem 5.4.1, this function cannot be analytic on $[a, b]$, since $E_{n_{k+1}-1}(f; a, b) = \varrho^{n_k}$, and under the assumption that $\varliminf_{k\to\infty} \frac{n_k}{n_{k+1}} = 0$, we have

$$\varlimsup_{n\to\infty} \sqrt[n]{E_n(f;a,b)} = \varlimsup_{k\to\infty} \sqrt[n_{k+1}-1]{E_{n_{k+1}-1}(f; a, b)} = \varlimsup_{k\to\infty} \varrho^{\frac{n_k}{n_{k+1}}} = 1.$$

According to the character of the sequence $\{n_k\}$ the quasi-analytic classes of functions satisfying conditions 6.6(7) may turn out to be sufficiently large. Functions quasi-analytic in the sense of S. N. Bernstein (i.e. satisfying condition 6.6(7) need not be infinitely differentiable. Moreover, one cannot even guarantee that such functions will have the first derivative or a fractional derivative of an arbitrary small order $r > 0$. This can be seen from the following theorem of A. N. Markushevich**.

6.6.3. *Every function $f(x)$ continuous in the interval $[a, b]$ is the difference of two functions quasi-analytic in the sense of S. N. Bernstein.* Let $P_n(f; x)$ be an algebraic polynomial of order $\leqslant n$, for which $E_n(f; a, b) = \max_{a\leqslant x\leqslant b} |f(x)-P_n(f; x)|$. It is evident that for every sequence of numbers ν_i $(i = 0, 1, 2, \ldots)$ increasing sufficiently fast the series

$$P_{\nu_0}(f; x) + \sum_{i=0}^{\infty} [P_{\nu_{i+1}}(f; x) - P_{\nu_i}(f; x)] \tag{8}$$

is absolutely and uniformly convergent, i.e. the function $f(x)$ can be represented as the sum of the absolutely and uniformly convergent series

$$f(x) = \sum_{n=0}^{\infty} Q_n(x) \tag{9}$$

whose terms $Q_n(x)$ are algebraic polynomials of order $\leqslant n$.

** A. N. Markushevich [1]. The proof presented here is due to S. N. Bernstein (Coll. papers, vol. I, p. 549).

We define two unbounded sequences of integers $\{n_k\}$ and $\{m_k\}$ in such a way that $0 = m_0 < n_1 < m_1 < n_2 < m_2 < n_3 < \ldots$ and for some positive $\varrho < 1$

$$\sum_{n=n_1+1}^{\infty} |Q_n(x)| < 1, \qquad \sum_{n=m_1+1}^{\infty} |Q_n(x)| < \varrho^{n_1},$$
$$\cdots\cdots\cdots\cdots\cdots\cdots \tag{10}$$
$$\sum_{n=n_{k+1}+1}^{\infty} |Q_n(x)| < \varrho^{mk}, \qquad \sum_{n=m_{k+1}+1}^{\infty} |Q_n(x)| < \varrho^{nk+1}.$$

Let

$$R_{n_1}(x) = \sum_{n=0}^{n_1} Q_n(x), \qquad R_{m_1}(x) = \sum_{n=n_1+1}^{m_1} Q_n(x),$$
$$\cdots\cdots\cdots\cdots\cdots\cdots \tag{11}$$
$$R_{n_{k+1}}(x) = \sum_{n=m_k+1}^{n_{k+1}} Q_n(x), \qquad R_{m_{k+1}}(x) = \sum_{n=n_{k+1}+1}^{m_{k+1}} Q_n(x)$$
$$\cdots\cdots\cdots\cdots\cdots\cdots$$

and

$$f_1(x) = \sum_{k=1}^{\infty} R_{n_k}(x), \qquad f_2(x) = \sum_{k=1}^{\infty} R_{m_k}(x). \tag{12}$$

Then

$$f_1(x)+f_2(x) = \sum_{k=1}^{\infty} [R_{n_k}(x)+R_{m_k}(x)] = \sum_{n=0}^{\infty} Q_n(x) = f(x).$$

Besides, since $R_{n_k}(x)$ (resp. $R_{m_k}(x)$) are algebraic polynomials of order not exceeding n_k (respectively m_k), we have

$$E_{n_k}(f_1; a, b) \leqslant \max_{a \leqslant x \leqslant b} \Big|f_1(x) - \sum_{j=1}^{k} R_{n_j}(x)\Big| = \max_{a \leqslant x \leqslant b} \Big|\sum_{j=k+1}^{\infty} R_{n_j}(x)\Big|$$
$$\leqslant \max_{a \leqslant x \leqslant b} \sum_{n=m_k+1}^{\infty} |Q_n(x)| < \varrho^{nk}. \tag{13}$$

and

$$E_{m_k}(f_2; a, b) \leqslant \max_{a \leqslant x \leqslant b} \Big|f_2(x) - \sum_{j=1}^{k} R_{m_j}(x)\Big| = \max_{a \leqslant x \leqslant b} \Big|\sum_{j=k+1}^{\infty} R_{m_j}(x)\Big|$$
$$\leqslant \max_{a \leqslant x \leqslant b} \sum_{n=n_{k+1}+1}^{\infty} |Q_n(x)| < \varrho^{mk}. \tag{14}$$

Thus, the function $f(x)$ is the sum of two functions, and each of them, in view of 6.6(13) and 6.6(14) is quasi-analytic in the sense of S. N. Bernstein.

6.6.31. We mention here that in the case when the function $f(x)$ does not possess a derivative, at least one of the quasi-analytic functions $f_1(x)$ and $f_2(x)$ in the representation $f(x) = f_1(x)+f_2(x)$ does not have a derivative on a set of positive measure. If, however, the function $f(x)$ has a bounded r-th ($r \geqslant 1$) derivative on $[a, b]$, then, by theorem 5.1.5 and the inequality for the derivative of an algebraic polynomial (see 4.8.(31) where $a = -1, b = 1$) the sequence of numbers ν_i ($i = 0, 1, 2, \ldots$) can be chosen in such a way that the series 6.6(8) will converge absolutely and uniformly on every segment $[a_1, b_1] \subset (a, b)$ after differentiation $(r-1)$ times; in other words, in this case the function $f(x)$ represented in the form of the series 6.6(9) possesses this property. It is evident that both functions $f_1(x)$ and $f_2(x)$ represented as sums of series 6.6(12) will have an $(r-1)$-th continuous derivative at each point inside the interval (a, b). Similar considerations based on the application of the inequality 4.8(32) show, that if the function $f(x)$ is infinitely differentiable on $[a, b]$, then it may be represented as a difference of two functions infinitely differentiable on this interval and quasi-analytic in the sense of S. N. Bernstein.

6.6.4. We arrived at the concept of the quasi-analytic classes of S. N. Bernstein by considering the infinitely differentiable functions satisfying condition 3.10(1). We now see that these classes may also include functions which are differentiable only a finite number of times, or even functions which are not differentiable at all.

These differential properties of functions satisfying condition 6.6(7) depend upon how rapidly the sequences $\{n_k\}$ defining particular classes increase to infinity. Thus, if

$$\lim_{k\to\infty} \frac{\ln n_{k+1}}{n_k} = 0, \tag{15}$$

then every continuous function $f(x)$ satisfying 6.6(7) is infinitely differentiable**. In fact, for the terms of the series

$$f(x) = P_{n_0}(f; x)+ \sum_{k=0}^{\infty} \{P_{n_{k+1}}(f; x)-P_{n_k}(f; x)\}, \tag{16}$$

where $P_n(f; x)$ is the algebraic polynomial of the n-th order best approximation of the function $f(x)$ in the given interval, we have for every $r \geqslant 0$ the inequality (see 4.8(32)):

$$|P^{(r)}_{n_{k+1}}(f; x)-P^{(r)}_{n_k}(f; x)| \leqslant 2\left(\frac{2n^2_{k+1}}{b-a}\right)^r E_{n_k}(f; a, b).$$

** S. N. Bernstein [7].

Besides, if $f(x)$ satisfies condition 6.6(7), and the sequence $\{n_k\}$ satisfies 6.6(15), then

$$n_{k+1}^{2r} E_{n_k}(f; a, b) \leqslant e^{2r \varepsilon_k n_k} \varrho^{n_k},$$

where $\varepsilon_k \to 0$ as $k \to \infty$. Thus, for every integer r, the series 6.6(16) converges uniformly after r-fold term-by-term differentiation, i.e. $f(x)$ has the continuous r-th derivative $f^{(r)}(x)$.

6.6.5. In spite of the fact, that, in general, a function quasi-analytic in the sense of S. N. Bernstein may turn out to be non differentiable, *such a function $f(x)$ defined on $[a,b]$ has always the property that for every $r > 0$ there exists a non-negative null-sequence of numbers $t_k < 1$, for which the r-th modulus of smoothness* $\omega_r(f; t) = \omega r\ (f; \alpha, \beta; t)$ (see 3.3) *where* $[\alpha, \beta] \subset [a, b]$, *satisfies, as $t_k \to 0$ the relation*

$$\sqrt[r]{\omega_r(f; \alpha, \beta; t_k)} = O\left\{\frac{r t_k |\ln t_k|}{\sqrt{(b-\beta)(\alpha-a)} + \dfrac{b-a}{r|\ln t_k|}}\right\} \tag{17}$$

*uniformly in α, β and r***. In fact, for every x and h, for which $a \leqslant \alpha \leqslant x \leqslant x+rh \leqslant \beta \leqslant b$, we have

$$|\Delta_h^r f(x)| = \left|\sum_{\nu=0}^{r} (-1)^\nu \binom{r}{\nu} f(x+\nu h)\right| \leqslant |\Delta_h^r P_n(f; x)| + 2^r E_n(f; a, b)$$

$$\leqslant h^r \max_{\alpha \leqslant u \leqslant \beta} |P_n^{(r)}(f; u)| + 2^r E_n(f; a, b),$$

Thus, if $f(x)$ satisfies condition 6.6(7), then for every sufficiently small $t > 0$

$$\omega_r(f; \alpha, \beta; t) \leqslant t^r \max_{\alpha \leqslant u \leqslant \beta} |P_{nk}^{(r)}(f; u)| + 2^r \varrho^{n_k}, \tag{18}$$

Applying the inequality 4.8(51) for the derivative of an algebraic polynomial, we obtain, passing to the segment $[a, b]$:

$$\max_{\alpha \leqslant u \leqslant \beta} |P_{nk}^{(r)}(f; u)| \leqslant \frac{C_r n_k^r}{\left[\dfrac{2}{b-a}\sqrt{(b-\beta)(\alpha-a)} + \dfrac{1}{n_k}\right]^r} \times$$

$$\times \left(\frac{2}{b-a}\right)^r \max_{a \leqslant u \leqslant b} |P_{n_k}(f; u)|, \tag{19}$$

where C_r is a constant depending only on r, and, as may be seen from

** S. N. Bernstein [7], the case when α, β $(a < \alpha < \beta < b)$ are fixed.

the proof of theorem 4.8.72, satisfies the inequality $C_r < B^r$ for some $B > 0$. If we substitute this estimate into 6.6(18), and use the fact that

$$\max_{a \leqslant u \leqslant b} |P_{n_k}(f; u)| \leqslant \max_{a \leqslant u \leqslant b} |f(u)| + E_{n_k}(f; a, b) \leqslant 2 \max_{a \leqslant u \leqslant b} |f(u)|,$$

we get 6.6(17) for $t_k = \varrho^{\frac{n_k}{r}}$.

6.6.6. We now pass to certain other properties of best approximation, connected with quasi-analyticity. Let $f(x)$ be an arbitrary function defined for all real x, bounded and uniformly continuous, whose optimal approximation with entire functions of exponential type satisfies 5.8(21). We shall see, that *every function satisfying the conditions stated above is infinitely differentiable and the relation* 3.10(12) *holds for upper bounds of its derivatives***.

Indeed, if $g_\sigma(f; x)$ is an entire function of order $\leqslant \delta$, for which $A_\sigma(f) = \sup_{-\infty < x < +\infty} |f(x) - g_\sigma(f; x)|$, then (see 2.6.2):

$$f(x) = g_\sigma(f; x) + \sum_{k=1}^{\infty} \{g_k(f; x) - g_{k-1}(f; x)\};$$

without loss of generality we may assume that $|f(x)| \leqslant 1$. Under this condition, the divergence of the series 5.8(21) implies that for every n

$$\sup_{k \geqslant 1} k^n A_k(f) < \infty.$$

Thus, since by theorem 4.8.2 for all integers n and $k > 1$:

$$|g_k^{(n)}(f; x) - g_{k-1}^{(n)}(f; x)| \leqslant k^n \sup_{-\infty < k < +\infty} |g_k(f; x) - g_{k-1}(f; x)|$$
$$\leqslant 2k^n A_{k-1}(f) \leqslant 2^{n+1}(k-1)^n A_{k-1}(f),$$

we have

$$\sum_{k=2}^{\infty} |g_k^{(n)}(f; x) - g_{k-1}^{(n)}(f; x)| \leqslant 2^{n+1} \sup_{k \geqslant 1} k^{n+2} A_k(f) \sum_{\nu=1}^{\infty} \frac{1}{\nu^2}$$
$$= \frac{2^n \pi^2}{3} \sup_{k \geqslant 1} k^{n+2} A_k(f) < \infty.$$

Hence the function $f(x)$ is infinitely differentiable, and

$$M_n(f) = \sup_{-\infty < x < \infty} |f^{(n)}(x)| \leqslant \frac{2^n \pi^2}{3} \max \left\{ \sup_{k \geqslant 1} k^{n+2} A_k(f), \frac{A_0(f)}{2^n} \right\},$$

** S. N. Bernstein, the periodic case.

i.e. for all sufficiently large n

$$\sqrt[n]{M_n(f)} \leqslant 2 \sqrt[n]{\frac{\pi^2}{3}} \{\sup_{k \geqslant 1} k^{n+2} \sqrt[n+2]{A_k(f)}\}^{1+\frac{2}{n}}.$$

Thus the series $\sum_{n=1}^{\infty} \frac{1}{\sqrt[n]{M_n(f)}}$ diverges together with the series 5.8(21). We come to the conclusion, that *in the class of functions bounded and uniformly continuous on the whole axis, the condition* 5.8(21) *is not only necessary* (see 5.8(31)), *but also sufficient for quasi-analyticity in the sense of Denjoy.*

6.6.7. *The condition* 5.8(16), *similar to* 5.8(21), *for best approximation by algebraic polynomials of functions $f(x)$ defined on a finite segment $[a, b]$, is sufficient for infinite differentiability of $f(x)$ in the interval (a, b) and for the divergence of the series*

$$\sum_{n=1}^{\infty} \frac{1}{\sqrt[n]{M_n(f; a_1, b_1)}}, \qquad M_n(f; a, b) = \max_{a_1 \leqslant x \leqslant b_1} |f^{(n)}(x)|, \tag{20}$$

for every segment $[a_1, b_1]$ contained in (a, b). This statement follows from estimates similar to these presented in 6.6.6, if, instead of theorem 4.8.2 we use the inequality for the derivative of an algebraic polynomial (see 4.8(31)). There exist examples** showing that condition 5.8(16) is not sufficient for the divergence of the series 6.6(20) for $a_1 = a$ or $b_1 = b$. We have already met such singularities of converse theorems in the case of a finite segment in 6.1.7. These singularities can be removed by replacing the best uniform approximation by approximations of the type discussed in 6.2 (see also 5.2).

Let $f(x)$ be a continuous function on $[-1, 1]$. For arbitrary integers n and r let

$$E_n^{(r)}(f) = \inf_{P_n(x)} \max_{|x| \leqslant 1} \frac{|f(x) - P_n(x)|}{\left(\sqrt{1-x^2} + \frac{1}{n}\right)^r}, \tag{21}$$

where the inf is taken over all algebraic polynomials $P_n(x)$ of degree $\leqslant n$. *If*

$$\sum_{n=1}^{\infty} \frac{1}{\sup_{k \geqslant 1} k \sqrt[n]{E_k^{(n)}(f)}} = \infty, \tag{22}$$

then the function $f(x)$ is infinitely differentiable and satisfies 3.10(12) *on the whole segment* $[-1,1]$.

** See S. N. Bernstein [7] and Carleman [1].

For what follows it is sufficient to assume that for every positive integer n

$$\sup_{k\geq 1} k^n E_k^{(n)}(f) < \infty. \tag{23}$$

Further, the sequence of polynomials $P_{k,n}(f; x)$ for which

$$E_k^{(n)}(f) = \max_{|x|\leq 1} \frac{|f(x)-P_{k,n}(f;x)|}{\left(\sqrt{1-x^2}+\frac{1}{k}\right)^n}$$

has the property that

$$|P_{k+1,n+2}(f;x)-P_{k,n+2}(f;x)|$$
$$\leq E_k^{(n+2)}(f)\left(\sqrt{1-x^2}+\frac{1}{k}\right)^{n+2}+E_{k+1}^{(n+2)}(f)\left(\sqrt{1-x^2}+\frac{1}{k}\right)^{n+2}$$
$$\leq \{E_k^{(n+2)}(f)+E_{k+1}^{(n+2)}(f)\}\left(\sqrt{1-x^2}+\frac{1}{k}\right)^{n+2}.$$

Thus, by theorem 4.8(73)

$$|P_{k+1,n+2}^{(n)}(f;x)-P_{k,n+2}^{(n)}(f;x)| \leq C^n k^n \{E_k^{(n+2)}(f)+E_{k+1}^{(n+2)}(f) \tag{24}$$

where C is a constant.

Since

$$f(x) = P_{1,n+2}(f;x)+\sum_{k=1}^{\infty}\{P_{k+1,n+2}(f;x)-P_{k,n+2}(f;x)\},$$

it follows from inequalities 6.6(24) and 6.6(23), that the function $f(x)$ is infinitely differentiable, and

$$M_n(f) = \max_{|x|\leq 1}|f^{(n)}(x)|$$
$$\leq \sum_{k=n-1}^{\infty}\max_{|x|\leq 1}|P_{k+1,n+2}^{(n)}(f;x)-P_{k,n+2}^{(n)}(f;x)|$$
$$\leq 2C^n \sum_{k=n-1}^{\infty} k^n E_k^{(n+2)}(f) \leq \frac{C^n\pi^2}{3}\sup_{k\geq 1} k^{n+2}E_k^{(n+2)}(f)$$

Hence, by the condition 6.6(22) the series

$$\sum_{n=1}^{\infty}\frac{1}{\sqrt[n]{M_n(f)}}$$

diverges.

6.6.71. If we consider the class of functions $f(x)$ defined on the segment $[a, b]$ and satisfying for every integer n the condition

$$\sup_{k\geqslant 1} k\sqrt[n]{E_k(f; a, b)} \leqslant M(f)\varrho_n, \tag{25}$$

where $\{\varrho_n\}$ is some monotonically increasing sequence of numbers, and $M(f)$ is a constant depending on f, then it can be shown, using the theorem of Carleman presented in 3.10.3, that** *a necessary and sufficient condition for quasi-analyticity of this class in the Denjoy and Carleman sense, is the divergence of the series* $\sum_{n=1}^{\infty} \frac{1}{\varrho_n}$. We can arrive at the necessity of this condition by considering the class R of functions $f(x)$ defined on $[a, b]$ and possessing the property that for every integer n

$$\sqrt[n]{M_n(f)} \leqslant M^*(f)\varrho_n.$$

If the series $\sum_{n=1}^{\infty} \frac{1}{\varrho_n}$ converges, then by the theorem of Carleman mentioned above (see 3.10.3) there exists a function $f(x) \in R$ not vanishing identically $(0 < M_0(f) < 1)$ and assuming the value 0 together with all its derivatives at some point $x_0 \in [a, b]$. The inequality 5.8(18) shows that for this function

$$\sup_{k\geqslant n} k\sqrt[n]{E_R(f; a, b)} \leqslant B\sqrt{M_n(f)} \leqslant BM\varrho_n.$$

In addition, from section 5.8.4 and theorem 6.6.8 presented below it follows that for such a function the series 5.8(16) converges. On the other hand, using the lemma for numerical series presented in section 3.10.3, we conclude that there exists a number N such that for all $n \geqslant N$ (see 5.8.3)

$$\sup_{k\geqslant 1} k\sqrt[n]{E_k(f; a, b)} = \sup_{k\geqslant n} k\sqrt[n]{E_k(f; a, b)} \leqslant BM^*(f)\varrho_n.$$

Thus, the function under consideration satisfies the inequality 6.6(25) for all $n > 0$, hence the class defined by this inequality is not quasi-analytic.

The sufficiency follows from section 5.8.4 and theorem 6.6.8.

6.6.8. In chapter V (see 5.8.4) we mentioned that the best uniform approximation of infinitely differentiable functions $f(x)$ satisfying 3.10(12) decreases to zero at least as fast as the sequence of numbers $\frac{1}{\phi(n)}$ where $\phi(x)$ is some even weight function non- decreasing together with $|x|$. Now, we consider the class of all functions $f(x)$ defined on a given finite segment $[a, b]$ and such that for every $n = 1,2, \ldots$

$$E_n(f; a, b)\frac{M(f)}{\phi(n)}, \tag{26}$$

** S. N. Bernstein, Coll. Papers vol. I., p. 552.

where $\phi(x) \geqslant 1$ $(-\infty < x < +\infty)$ is a certain even weight function non-decreasing together with $|x|$ (see section 1.8) and $M(f)$ is a constant depending on f.

In the case when $\phi(x) = e^{\delta|x|}$ $(\delta > 0)$ this class consists of all the functions analytic on $[a, b]$ (see section 6.5.1). If, however,

$$\varliminf_{n\to\infty} \sqrt[n]{\phi(n)} = 1,$$

then, as may be seen from theorems 2.5.1 and 5.4.1, this class must contain non-analytic functions. Besides, *for every weight function $\phi(x)$, the corresponding class defined by* 6.6(26) *is quasi-analytic* (see section 3.10.3) *and consists only of infinitely differentiable functions***.

We may assume that $a = -1$, $b = 1$. If the function $f(x)$, defined on $[-1,1]$ satisfies 6.6(26), and $P_n(f; x)$ is an algebraic polynomial of degree $\leqslant n$ for which

$$E_n(f) = E_n(f; -1,1) = \max_{-1\leqslant x\leqslant 1} |f(x) - P_n(f; x)|,$$

then

$$f(x) = P_0(f; x) + \sum_{n=0}^{\infty} \{P_{n+1}(f; x) - P_n(f; x)\},$$

and, by the inequality 4.8.(43) for every positive integer r

$$|P_{n+1}^{(r)}(f; x) - P_n^{(r)}(f; x)| \leqslant (n+1)^{2r} \max_{|x|\leqslant 1} |P_{n+1}(f; x) - P_n(f; x)|$$
$$\leqslant 2^{2r+1} n^{2r} E_n(f) \leqslant 2^{2r+1} M(f) \frac{n^{2r}}{\phi(n)}.$$

Since however the function $\phi(x)$ is a weight function, we have

$$\sup_{n\geqslant 1} \frac{n^{2r+2}}{\phi(n)} < \infty$$

i.e. the series $\sum_{n=1}^{\infty} \frac{n^{2r}}{\phi(n)}$ converges. If follows that $f(x)$ is infinitely differentiable. Now we shall show that if $f(x)$ satisfies 6.6(26), and at some point $x_0 = \cos t_0 \in [-1,1]$ its derivatives $f^{(r)}(x_0)$ $(r = 0,1,2,\dots)$ are equal to zero, then $f(x) \equiv 0$ on the whole segment. For this, let us consider the Fourier series

$$f(\cos t) = \sum_{k=0}^{\infty} a_k \cos kt$$

** See S. N. Bernstein [7].

and note that for every integer $r \geqslant 0$

$$\sum_{k=0}^{\infty} k^{2r} a_k \cos kt_0 = \sum_{k=1}^{\infty} k^{2r+1} a_k \sin kt_0 = 0.$$

Thus for every even (odd) algebraic polynomial $P_n(x)$ $Q_n(x)$) we have:

$$\sum_{k=0}^{\infty} P_n(k) a_k \cos kt_0 = \sum_{k=1}^{\infty} Q_n(k) a_k \sin kt_0.$$

Let us denote by $C(x)$ $S(x)$) a continuous function, whose graph is a polygonal line with vertices at the points $(k, a_{|k|} \cos kt_0)$ $(k, a_{|k|} \sin kt_0)$ respectively $(k = 0, \pm 4, \pm 2, \ldots)$. Since $\phi(x)$ and hence $\dfrac{\phi(x/2)}{x^2-1/4}$ is a weight function (see section 1.8.6), we can for every $\varepsilon > 0$, select polynomials $P_n(x)$ and $Q_n(x)$ so that for all $x(-\infty < x < \infty)$

$$|C(x)-P_n(x)| < \varepsilon \frac{\phi\left(\dfrac{x}{2}\right)}{\left|x^2-\dfrac{1}{4}\right|}$$

and

$$|S(x)-Q_n(x)| < \varepsilon \frac{\phi\left(\dfrac{x}{2}\right)}{\left|x^2-\dfrac{1}{4}\right|}.$$

We then have:

$$\sum_{k=0}^{\infty} a_k^2 \cos^2 kt_0 = \sum_{k=0}^{\infty} \{C(k)-P_n(k)\} a_k \cos kt_0 < \varepsilon \sum_{k=0}^{\infty} \phi\left(\frac{k}{2}\right) \frac{|a_k|}{\left|k^2-\dfrac{1}{4}\right|},$$

$$\sum_{k=1}^{\infty} a_k^2 \sin^2 kt_0 = \sum_{k=1}^{\infty} \{S(k)-Q_n(k)\} a_k \sin kt_0 < \varepsilon \sum_{k=0}^{\infty} \phi\left(\frac{k}{2}\right) \frac{|a_k|}{\left|k^2-\dfrac{1}{4}\right|},$$

i.e.

$$\sum_{k=0}^{\infty} a_k^2 < 2\varepsilon \sum_{k=0}^{\infty} \phi\left(\frac{k}{2}\right) \frac{|a_k|}{\left|k^2-\dfrac{1}{4}\right|}$$

$$\leqslant 8\varepsilon \left\{ |a_0|\phi(0) + |a_1|\phi(1) + \frac{1}{2} \sum_{k=2}^{\infty} \phi(k-1) \frac{|a_k|}{\left|k^2-\dfrac{1}{4}\right|} \right\}.$$

Since for $k \geqslant 1$

$$E_{k-1}(f; -1, 1) \geqslant \frac{1}{2\pi} \int_{-\pi}^{\pi} |f(\cos t) - P_n(f; \cos t)|\, |\cos kt|\, dt$$

$$\geqslant \frac{1}{\pi} \left| \int_{0}^{\pi} f(\cos t) \cos kt\, dt \right| = \frac{1}{2} |a_k|,$$

we have, by 6.6(26):

$$\sum_{k=0}^{\infty} a_k^2 < 8\varepsilon \left\{ |a_0|\, \phi(0) + |a_1| \phi(1) + \sum_{k=2}^{\infty} \frac{\phi(k-1) E_{k-1}(f)}{\left| k^2 - \frac{1}{4} \right|} \right\}$$

$$< 8\varepsilon \left\{ |a_0|\, \phi(0) + |a_1| \phi(1) + \frac{\pi^2}{6} \mathrm{M}(f) \right\}.$$

It follows that $a_k = 0$ for every $k \geqslant 0$, i.e. $f(x) \equiv 0$.

6.6.81.** *In the cases when the weight function* $\varphi(x)$ *is an integral function of type* 1.8(3) ($c_k \geqslant 0$), *the corresponding quasi-analytical class defined by* 6.6(26) *coincides with the class of functions satisfying* 5.8(16). This statement follows from the criterion for weight functions given in 1.8 (see 1.8.3), by using the results of section 5.8.4 and the inequality

$$\frac{1}{\sup\limits_{k \geqslant 1} k \sqrt[n]{E_k(f; a, b)}} \geqslant \min_{x \geqslant 1} \frac{1}{x} \sqrt[n]{\phi(x)},$$

valid under the assumption 6.6(26).

6.6.82. Similaıly we obtain, that *in the class of functions bounded und uniformly continuous for all* $-\infty < x < +\infty$ *the condition*

$$\mathrm{A}_\sigma(f) \leqslant \frac{M(f)}{\phi(\sigma)} \qquad (\sigma > 0), \tag{27}$$

where $\phi(x)$ *is an integral weight function of type* 1.8(3) ($c_k \geqslant 0$) *is equivalent to* 5.8(21), *i.e.* (see section 6.6.6) *to the condition* 3.10(12). It may, therefore, be used as a constructive characterisation of functions quasi-analytic in the Denjoy sense for $(-\infty, +\infty)$.

6.6.83. Theorem 6.6.8 shows the close connection between quasi-analyticity and weight functions. We mention here, that under certain additional assumptions the converse theorem is also true: *for the class of all functions* $f(x)$ *satisfying* 6.6(26) *where* $\phi(x)$ *is even non-decreasing on* $(0, \infty)$ *to be quasi-analytic* (see section 3.10.3) *it is sufficient* (see 6.6.8), *and, in the case when* $\varphi(x)$ *increases normally, i.e.* $x \frac{\phi'(x)}{\phi(x)}$ *is non-negative and strictly increasing to infinity together with* $|x|$ *— also necessary, that* $\phi(x)$ *is a weight function**.*

** S. N. Bernstein [43].

We have to verify only the second part of this theorem. If the function $\phi(x)$ in the inequality 6.6(26) is not a weight function, then by a theorem of S. N. Bernstein given in section 1.8 (see section 1.8.1), it cannot satisfy condition 1.8(4), i.e.

$$\sum_{n=1}^{\infty} \min_{x>0} \frac{1}{x} \sqrt[n]{\phi(x)} < \infty. \tag{28}$$

Thus, by 6.6(71), among functions $f(x)$ for which

$$\sup_{k \geqslant 1} k \sqrt[n]{E_k(f; a, b)} \leqslant \frac{1}{2 \min\limits_{x>0} \frac{1}{x} \sqrt[n]{\phi(x)}}$$

there exists a function $f_0(x) \not\equiv 0$, which is equal zero together with all its derivatives at some point $x_0 \in [a, b]$. The function $\phi_0(x)$ defined in such a way that

$$\phi_0(n) = \frac{1}{E_n(f_0; a, b)} \qquad (n = 0, 1, 2, \ldots),$$

cannot, by theorem 6.6.8 be a weight function. Besides

$$\sup_{k \geqslant 1} k \sqrt[n]{\frac{1}{\phi_0(k)}} \leqslant \frac{1}{2 \min\limits_{x>0} \frac{1}{x} \sqrt[n]{\phi(x)}},$$

i.e.

$$\frac{1}{2} \min_{k \geqslant 1} \frac{1}{k} \sqrt[n]{\phi_0(k)} \geqslant \min_{x>0} \frac{1}{x} \sqrt[n]{\phi(x)}. \tag{29}$$

If $\phi(x)$ increases normally, then the function

$$n(x) = x \frac{\phi'(x)}{\phi(x)}$$

is not bounded and strictly increases as $|x| \to \infty$. Thus, for all sufficiently large values of n there exists one and only one value $x = x(n)$ at which the function $\frac{1}{t} \sqrt[n]{\phi(t)}$ achieves its minimum on $(0, \infty)$. Conversely, to each $x > 0$ there corresponds some $n = n(x)$ for which such a function achieves its minimum at the point x. Putting $n = n(x)$ we find, using the inequality 6.6(29), that for every $x \geqslant 1$

$$\frac{1}{x} \sqrt[n]{\phi(x)} \leqslant \frac{1}{x} \sqrt[n]{\phi_0(x)},$$

i.e. $\phi(x) \leqslant \phi_0(x)$. It follows therefore, that under the condition 6.6(28) the function $\phi(x)$ cannot be a weight function, hence in the class of normally increasing functions condition 1.8(4) is not only sufficient, but also necessary for $\phi(x)$ to be a weight function (see section 1.8.2).

Moreover, since

$$E_n(f; a, b) = \frac{1}{\phi_0(n)} \leqslant \frac{1}{\phi(n)}$$

the class of functions satisfying 6.6(26) contains $f_0(x)$, and hence is not quasi-analytic.

6.6.9. Theorem 6.6.8 shows that quasi-analyticity may hold even in the case when condition 6.6(2) is not satisfied.

We see, however, that in these cases the corresponding quasi-analytic classes are already defined by the condition which makes use not of some part of the sequence of best approximations, but of the whole sequence (see sections 6.6.6.–6.6.8). It turns out that such a singularity of classes of functions, for which $\lim_{n\to\infty}\sqrt[n]{E_n(f;a,b)}=1$ is essential here, and is closely connected with their quasi-analyticity. *For every sequence of numbers E_n decreasing monotonically to zero and such that $\lim_{n\to\infty}\sqrt[n]{E_n}=1$ there exists a function $f(x)$ not vanishing identically on the segment $[a,b]$ and vanishing everywhere on the part $[a,c]$ $(a<c<b)$ of $[a,b]$ with the best approximations $E_n(f;a,b)$ satisfying the inequality $E_n(f;a,b)<E_n$ for an infinite number of subscripts an chosen from any previously specified increasing sequence of integers***.

We assume for simplicity that $a=-1$, $c=0$, $b=1$. We shall show that if $\lim_{n\to\infty}\sqrt[n]{E_n}=1$, then *for a fixed m, for every algebraic polynomial $R_m(x)$ of degree $\leqslant m$ there exists a sequence of algebraic polynomials $Q_n(x)$ of degree $\leqslant n$ tending uniformly to $R_m(x)$ on $[0,1]$ and satisfying the condition*

$$Q_n(0)=0, \qquad \max_{-1\leqslant x\leqslant 0}|Q_n(x)|<E_n, \tag{30}$$

for all $n=0,1,2,\ldots$

In the case $R_m(x)\equiv 1$ we consider the binomial series

$$\frac{1}{2}\left\{x-\frac{1}{r}+\left|x-\frac{1}{r}\right|\right\}=\frac{1}{4}\left(x+\frac{r-1}{r}\right)^2-$$
$$-\frac{1}{2}\sum_{k=2}^{\infty}\frac{1.3\ldots(2k-3)}{2.4\ldots 2k}\left\{1-\left(x-\frac{1}{r}\right)^2\right\}^k, \tag{31}$$

where $r=r(n)$ is an arbitrary integer-valued function increasing to infinity with n. Thanks to the uniform convergence of the expansion of $\sqrt{1-t}$ in a power series of t in $-1\leqslant t\leqslant 1$, the partial sums of the series 6.6(31)

$$P_{2n}(x)=\frac{1}{4}\left(x+\frac{r-1}{r}\right)^2-\frac{1}{2}\sum_{k=2}^{n}\frac{1.3\ldots(2k-3)}{2.4\ldots 2k}\left\{1-\left(x-\frac{1}{r}\right)^2\right\}^k \tag{32}$$

** S. N. Bernstein [7].

converge uniformly to x on [0,1] as $n \to \infty$, and for $-1 \leqslant x \leqslant 0$ they satisfy the inequality

$$|P_{2n}(x)| \leqslant \frac{1}{2} \sum_{k=n+1}^{\infty} \frac{1.3 \ldots (2k-3)}{2.4 \ldots 2k} \left(1-\frac{1}{r^2}\right)^k. \tag{33}$$

The right-hand side of 6.6(33) is, as $n \to \infty$, of order not exceeding $\frac{r^2}{n^{\frac{5}{2}}}\left(1-\frac{1}{r^2}\right)^n$, i.e. $\frac{r^2}{n^{\frac{5}{2}}} e^{-\frac{n}{r^2}}$. Thus, if $\alpha_n < 1$ is any arbitrary numerical sequence for which $\lim\limits_{n\to\infty} \sqrt[n]{\alpha_n} = 1$, then for

$$r^2 = r^2(n) < \min\left\{-\frac{n}{\ln \alpha_n}, n^{\frac{5}{2}}\right\} \tag{34}$$

we have, for all sufficiently large n ($n \geqslant N$) and $-1 \leqslant x \leqslant 0$:

$$\max_{-1\leqslant x\leqslant 0} |P_{2n}(x)| < \alpha_n.$$

Putting $\alpha_n = \frac{1}{2} E_{2n+1}$, $Q_n(x) \equiv 0$ if $0 \leqslant n < 2N$ and $Q_{2n+1}(x) = Q_{2n}(x) = P_{2n}(x) - P_{2n}(0)$ ($n \geqslant N$), we obtain the sequence of polynomials, for which 6.6(30) holds, and $\lim\limits_{n\to\infty} \max\limits_{0\leqslant x\leqslant 1} |Q_n(x)-x| = 0$. Taking an arbitrary fixed polynomial $R_m(x) \not\equiv 0$, in order to construct the sequence $Q_n(x)$, let us consider the polynomials $Q_{m+2n}(x) = P_{2n}(x)\, R_m(x)$, where the $P_{2n}(x)$ are the partial sums of 6.6(32) for $r = r(n)$ satisfying 6.6(34) with

$$\alpha_n = \frac{E_{m+2n+1}}{2\max\limits_{|x|\leqslant 1}|R_m(x)|}.$$

We put $Q_n(x) \equiv 0$ for $0 \leqslant n < m+2N$ and $Q_{m+2n+1}(x) = Q_{m+2n}(x) = Q_{m+2n}(x) - Q_{m+2n}(0)$ for $n \geqslant N$, and thus we obtain a sequence of polynomials for which 6.6(30) holds and $\lim\limits_{n\to\infty} \max\limits_{0\leqslant x\leqslant 1} |Q_n(x) - x R_m(x)| = 0$.

From the above-proved theorem it follows that, having a sequence of integers, one can choose consecutively from it an unbounded subsequence $n_0 < n_1 < n_2 < n_3 < \ldots$ and construct the corresponding sequence of algebraic polynomials $S_{n_k}(x)$ of the degree n_k so that

$$\left.\begin{aligned} E_{n_0} < \frac{1}{2}, \quad S_{n_0}(x) = x, \quad E_{n_{k+1}} < \frac{1}{2} E_{n_k}, \quad S_{n_k}(0) = 0, \\ \max_{-1\leqslant x 0} |S_{n_k}(x)| < E_{n_k}, \\ \max_{0\leqslant x\leqslant 1} |S_{n_k}(x) - S_{n_{k-1}}(x)| < \frac{1}{2} E_{n_{k-1}} \quad (k \geqslant 1). \end{aligned}\right\} \tag{35}$$

By the inequalities 6.6(35) the function

$$f(x) = S_{n_0}(x) + \sum_{k=1}^{\infty} \{S_{n_k}(x) - S_{n_{k-1}}(x)\}$$

satisfies the condition: for every $x \in [-1, 0]$

$$f(x) = \lim_{n_k \to \infty} S_{n_k}(x) = 0,$$

and for every $x \in [0,1]$

$$|f(x) - S_{n_k}(x)| \leqslant \sum_{\nu=k}^{\infty} |S_{n_{\nu+1}}(x) - S_{n_\nu}(x)|$$

$$< \frac{1}{2} \sum_{\nu=k}^{\infty} E_{n_\nu} < \frac{1}{2} E_{n_k} \sum_{\nu=0}^{\infty} \frac{1}{2^\nu} = E_{n_k}.$$

Thus, for every $k = 0, 1, 2, \ldots$, and for all $x \in [-1,1]$ we have:

$$|f(x) - S_{n_k}(x)| < E_{n_k},$$

i.e. $E_{n_k}(f; -1, 1) < E_{n_k}$ and $f(x)$ vanishes identically on $[-1, 0]$ and does not vanish identically on $[0, 1]$.

6.6.10. The singularity of classes of functions for which $\lim_{n \to \infty} \sqrt[n]{E_n(f; a, b)} = 1$, mentioned in section 6.6.9, leads to the following important characterization of analytic functions, selecting them from among all possible quasi-analytic classes**. *A function is analytic on* $[a, b]$ *if and only if each modification of it on any part of an arbitrary interval contained in* (a, b) *implies the lowering of the order of the best approximation* $E_n(f; a, b)$ *for all sufficiently large values of* n.

On the one hand, if $f(x)$ is analytic on $[a, b]$ then, by theorem 5.4.1 we have $\overline{\lim_{n \to \infty}} \sqrt[n]{E_n(f; a, b)} < 1$. Changing $f(x)$ on a part of any interval contained in (a, b) we obtain a certain new function $F(x)$ which coincides with $f(x)$ on a sufficiently small interval. If we had $\underline{\lim_{n \to \infty}} \sqrt[n]{E_n(f; a, b)} < 1$ then by theorem 6.6.1 these functions would have to coincide everywhere on $[a, b]$. Consequently, $\lim_{n \to \infty} \sqrt[n]{E_n(f; a, b)} = 1$, i.e. for all sufficiently large n

$$E_n(f; a, b) = o[E_n(F; a, b)].$$

On the other hand, if a bounded function $F(x)$ is not analytic then by theorem 6.5.1 $\overline{\lim_{n \to \infty}} \sqrt[n]{E_n(F; a, b)} = 1$, i.e. there exists an unbounded

** S. N. Bernstein [3].

sequence of numbers $\nu_1 < \nu_2 < \nu_3 < \dots$ for which $\lim_{\nu_k \to \infty} \sqrt[\nu_k]{E_{\nu_k}(F; a, b)} = 1$. For a subsequence of numbers E_n such that $E_n = E_{\nu_k}(F; a, b)$ for $\nu_k \leqslant n < \nu_{k+1}$ ($\nu_0 = 1$)

$$\lim_{n \to \infty} \sqrt[n]{E_n} = 1.$$

Thus, from the considerations of section 6.6.9 it follows that there exists a function $f(x)$ not vanishing identically, which satisfies the property that $f(x) = 0$ everywhere on the segment $\left[a, \frac{a+b}{2}\right]$ and with the best approximation $E_n(f; a, b)$, satisfying the inequality $E_n(f; a, b) < E_n$ for an infinite subsequence $\{n_k\}$ of values n from the sequence $\{\nu_k\}$.

For any $\delta > 0$ the function $\phi(x) = F(x) + \delta f(x)$ obtained from $F(x)$ by changing it at the points of the interval $\left(\frac{a+b}{2}, b\right)$ has the best approximation $E_n(\phi; a, b)$ which for $n = n_k (k = 0, 1, 2, \dots)$ satisfies the inequality

$$E_{n_k}(\phi; a, b) \leqslant E_{n_k}(F; a, b) + \delta E_{n_k}(f; a, b)$$
$$< E_{n_k}(F; a, b) + \delta E_{n_k} = (1+\delta) E_{n_k}(F; a, b).$$

Thus, after the change of function $F(x)$ given above, the order of its best approximation is not worse, at least for values of n chosen from some unbounded subsequence of natural numbers.

6.7. Certain converse theorems for conjugate classes of functions

Now we shall briefly deal with the problem of differential properties of functions conjugate to functions from a given sequence of best approximations, and we shall prove certain inequalities similar to those given in section 6.1.

6.7.1. If $f(x)$ is an essentially bounded measurable function of period 2π, then the condition $\lim_{n \to \infty} E_n^*(f) = 0$ for $f(x)$, as has been already mentioned in sections 3.11.3 and 5.9.2, does not imply the continuity of the conjugate function $\tilde{f}(x)$. Thus the estimate of the number $\omega(\tilde{f}, t)$ for $t \to 0$ is meaningful only under certain restrictions imposed upon the sequence of best approximations $E_n^*(f)$. The considerations of sections 5.9.2 and 6.1.1 show that such a restriction could be, for instance, the convergence of the series

$$\sum_{\nu=1}^{\infty} \frac{1}{\nu} E_\nu^*(f). \tag{1}$$

In those cases when the series (1) *converges the conjugate function* $\tilde{f}(x)$ *is continuous, and for all integers n and k*

$$\omega_k\left(\tilde{f};\frac{1}{n}\right)\leqslant \tilde{C}_k\left\{\frac{1}{n^k}\sum_{\nu=0}^{n}(\nu+1)^{k-1}E_\nu^*(f)+\sum_{\nu=n+1}^{\infty}\frac{1}{\nu}E_\nu^*(f)\right\}, \tag{2}$$

where $\tilde{C}_k$ *is a constant, depending only on* k**.

The inequality 6.7(2) is a consequence of theorems 6.1.1 and 5.9.2. By theorem 6.1.1

$$\omega_k\left(\tilde{f};\frac{1}{n}\right)\leqslant\frac{C_k}{n^k}\sum_{\nu=0}^{n}(\nu+1)^{k-1}E_\nu^*(\tilde{f}).$$

Using the estimate 5.9(15), we get

$$\omega_k\left(\tilde{f};\frac{1}{n}\right)\leqslant 2^kC_kC^*\left\{\frac{1}{n^k}\sum_{\nu=0}^{n}(\nu+1)^{k-1}E_\nu^*(f)+\right.$$

$$\left.+\frac{1}{n^k}\sum_{\nu=0}^{n}(\nu+1)^{k-1}\sum_{\mu=\nu+1}^{n}\frac{1}{\mu}E_\mu^*(f)+\sum_{\mu=n+1}^{\infty}\frac{1}{\mu}E_\mu^*(f)\right\} \tag{3}$$

and after changing the order of summation in the second term of the right-hand side of 6.7(3) we get 6.7(2).

6.7.11. The condition of convergence of the series 6.7(1) is sufficient for the continuity of the conjugate function $\tilde{f}(x)$, in general however, it is not necessary. Let us consider, for instance the function

$$f(x)=\sum_{k=2}^{\infty}a_k\cos kx, \tag{4}$$

where

$$a_k=\frac{1}{\ln\ln k}-\frac{1}{\ln\ln(k+1)},$$

Since $a_k=O\left(\dfrac{1}{k\ln k\ln^2\ln k}\right)$, the series 6.7(4), and also the series

$$\sum_{k=2}^{\infty}a_k\sin kx$$

are uniformly convergent. Consequently, in this case, the conjugate function $\tilde{f}(x)$ is continuous together with $f(x)$. Moreover, since

$$\max_x\left|f(x)-\sum_{k=2}^{n}a_k\cos kx\right|=\sum_{k=n+1}^{\infty}a_k=\frac{1}{\ln\ln(n+1)},$$

** N. K. Bari and S. B. Stechkin [1].

it follows from the inequality 5.11(7), for some constant $C > 0$

$$E_n^*(f) > \frac{C}{\ln(n+1)\ln\ln(n+1)},$$

i.e. the series 6.7(1) diverges.

6.7.2. Theorem 6.7.1 admits the following generalisation**. *If for some integer $r \geqslant 0$ the sequence of best approximations $E_\nu^*(f)$ of a function $f(x)$ with period 2π satisfies the condition*

$$\sum_{\nu=1}^{\infty} \nu^{r-1} E_\nu^*(f) < \infty \tag{5}$$

then the conjugate function $\tilde{f}(x)$ has an r-th continuous derivative, and for all integers n and k

$$\omega_k\left(\tilde{f}^{(r)}; \frac{1}{n}\right) \leqslant \tilde{C}_{k,r}\left\{\frac{1}{n^k}\sum_{\nu=0}^{n}(\nu+1)^{k+r-1}E_\nu^*(f) + \sum_{\nu=n+1}^{\infty}\nu^{r-1}E_\nu^*(f)\right\} \tag{6}$$

where $\tilde{C}_{k,r}$ is a constant depending only on k and r. This generalisation, like to that of section 6.7.1 for $r = 0$, follows immediately from theorem 5.9.4 and 6.1.1.

6.7.3. From the inequality 5.1(16) and 6.7.1 it follows at once that*** *if $E(t)$ is a function non-increasing on* $[0, \infty)$ $(E(0) = 1, \lim_{t\to\infty} E(t) = 0)$, *satisfying* 6.1(20) *and* 6.1(21), *and for a periodic function $f(x)$ of period 2π we have $\omega_k(f, t) = O\left[E\left(\frac{1}{t}\right)\right]$, then $\omega_k(\tilde{f}, t) = O\left[E\left(\frac{1}{t}\right)\right]$. In particular putting for $t \geqslant 1$ $E(t) = t^{-a}(0 < a < k)$ we obtain***** from the condition $\omega_k(f; t) = O(t^a)(0 < a < k)$ *the relation $\omega_k(\tilde{f}, t) = O(t^a)$.*

6.7.4. In a similar way, using theorems 5.1.4 and 6.7.2 we see, that** *if a periodic function $f(x)$ of period 2π has, for a certain integer r a continuous r-th derivative $f^{(r)}(x)$ with $\omega_k(f^{(r)}, t) = O\left[E\left(\frac{1}{t}\right)\right]$, then under the conditions* 6.1(20) *and* 6.1(21), *the conjugate function $\tilde{f}(x)$ has the same property, i.e. it has continuous r-th derivative, and $\omega_k(\tilde{f}^{(r)}; t) = O\left[E\left(\frac{1}{t}\right)\right]$.*

In particular, *if $\omega_k(f^{(r)}, t) = O(t^a)$ $(0 < a < k)$, then $\tilde{f}(x)$ has a continuous r-th derivative for which $\omega_k(\tilde{f}^{(r)}, t) = O(t^a)$.*

** N. K. Bari and S. B. Stechkin [1].

*** In a slightly different form, equivalent to ours (see section 6.9.5) this result was obtained by N. K. Bari [1].

**** I. I. Privalov [1], case $k = 1$; A Zygmund [2], case $k = 2$.

6.7.5. We mention also, that since condition 3.11(24) implies, by the inequality 5.1(16), the convergence of the series 6.7(5), then from 6.7.2 we obtain the theorem presented in section 3.11.3, if only in the estimate 6.7(6) we use the relation 5.1(16) (for $k = l$) and replace sums by integrals (see for instance section 6.1.61).

6.7.6. It is understood, that estimates similar to those given in 6.7.1–6.7.4, and also the inequalities on which these estimates are based, remain valid if we pass to the integral metric L_q ($q \geqslant 1$). Here, obviously, besides the case $q = \infty$ considered above, owing to the inequality 3.11(5) of M. Riesz (see also 5.11(18)) only is of interest.

6.8. Converse theorems and some constructive characteristics of compact sets in Banach spaces

In this section we shall consider the problem of theorems converse to the general inequality 5.10(2) and we shall present some estimations concerning this inequality, supplementing to theorem 2.5.1.

6.8.1. For every normed linear space F we have the following theorem which is, in a certain sense, converse to 5.10: *For every $\varepsilon > 0$, every compact set $W \subset F$ and every linearly independent system $\mathcal{X}\,\{x_0, x_1, x_2, \ldots\}$ we have the following inequality*

$$I^a_W(4\varepsilon) \leqslant \{n^{\mathcal{X}}_W(\varepsilon)+1\} \log_2 \frac{4M}{\varepsilon} + \{n^{\mathcal{X}}_W(\varepsilon)+1\} \log_2 (n^{\mathcal{X}}_W(\varepsilon)+2), \tag{1}$$

where $I^a_W(\varepsilon)$ is the ε-entropy of the set W (see section 3.1.3), $M = \sup_{x \in W} \|x\|$ *and $n = n^{\mathcal{X}}_W(\varepsilon)$ is the smallest non-negative number for which*

$$\mathcal{E}^{\mathcal{X}}_n(W) \leqslant \varepsilon. \tag{2}$$

To prove the estimate 6.8(1), it suffices to assume that $\varepsilon \leqslant M$ and consider the set of all polynomials

$$y_n(x) = \sum_{\nu=0}^{n^{\mathcal{X}}_W(\varepsilon)} c^*_\nu x_\nu, \tag{3}$$

whose distance from elements $x \in W$ is smallest. Since $\|y_n(x)\| \leqslant 2M$ for all $x \in W$, the coefficients of the polynomials 6.8(3) satisfy for every $x \in W$ the inequality (see 2.13.1)

$$\sup_{0 \leqslant \nu \leqslant n} \|c^*_\nu x_\nu\| \leqslant \frac{2M}{d^{\mathcal{X}}_n}, \tag{4}$$

where

$$d_n^{\mathfrak{X}} = \inf_{0 \leqslant \nu \leqslant n} \frac{1}{\|x_\nu\|} \inf_{c_k} \Big\| x_\nu - \sum_{\substack{k=0 \\ k \neq \nu}} c_k x_k \Big\|.$$

If

$$c_\nu^{(k_\nu)} = \pm \frac{k_\nu \varepsilon}{(n+1)\|x_\nu\|} \quad \left(k_\nu = 0, 1, 2, \ldots, \left[\frac{2M(n+1)}{\varepsilon d_n^{\mathfrak{X}}}\right]\right),$$

then for every polynomial 6.8(3) one can find a polynomial of the form $\sum_{\nu=0}^{n} c_\nu^{(k_\nu)} x_\nu$ such that

$$\Big\| \sum_{\nu=0}^{n} c_\nu^* x_\nu - \sum_{\nu=0}^{n} c_\nu^{(k_\nu)} x_\nu \Big\| \leqslant \sum_{\nu=0}^{n} |c_\nu^* - c_\nu^{(k_\nu)}| \, \|x_\nu\| \leqslant \varepsilon.$$

This means that for any $x \in W$, for a given $n = \mathrm{n}_W^{\mathfrak{X}}(\varepsilon)$ there exists a certain polynomial $\sum_{\nu=0}^{n} c_\nu^{(k_\nu)} x_\nu$ such that

$$\Big\| x - \sum_{\nu=0}^{n} c_\nu^{(k_\nu)} x_\nu \Big\| \leqslant \|x - y_n(x)\| + \Big\| y_n(x) - \sum_{\nu=0}^{n} c_\nu^{(k_\nu)} x_\nu \Big\| < 2\varepsilon.$$

Choosing any element $x \in W$ from each sphere of radius 2ε with the centre $\sum_{\nu=0}^{n} c_\nu^{(k_\nu)} x_\nu$ we obtain a 4ε-net for the set W. Since however the number of those elements does not exceed the number of polynomials of the form $\sum_{\nu=0}^{n} c_\nu^{(k_\nu)} x_\nu$ and $d_n^{\mathfrak{X}} \leqslant 1$, the minimal 4ε-net for W consists of not more than

$$\left(2\left[\frac{2M(n+1)}{\varepsilon d_n^{\mathfrak{X}}}\right] + 1\right)^{n+1} \leqslant \left\{\frac{4M(n+2)}{\varepsilon d_n^{\mathfrak{X}}}\right\}^{n+1}$$

elements. Thus

$$I_W^a(4\varepsilon) \leqslant (n+1)\left(\log_2 \frac{4M}{\varepsilon} + \log_2 \frac{n+2}{d_n^{\mathfrak{X}}}\right).$$

It remains to use the fact that *for each* $(n+1)$*-dimensional Banach space there exists a system of* $n+1$ *elements* $z_0, z_1, \ldots, z_n$ *for which* $\|z_k\| = 1$ $(k = 0, 1, \ldots, n)$ *and* $d_n^{\mathfrak{Z}} = 1$**, and also use the fact that for any $\varepsilon > 0$ and

** In the $(n+1)$-dimensional space under consideration, such a property is possessed by the solution of the system $x_k = \sum_{\nu=0}^{n} f_\nu(x_k) z_\nu$ $(k = 0, 1, \ldots, \mathrm{n})$ where $f_0, f_1, \ldots, f_n$ are linear functionals in F with $\|f_\nu\| = 1$ $(\nu = 0, 1, \ldots, n)$ for which the determinant of the system achieves its maximal possible value. This fact, due to G. Auerbach (see S. Banach [1], p. 205) was pointed out to the author by M. I. Kadets.

$n = n_W^{\mathcal{Z}}(\varepsilon)$, for the system $\mathcal{Z}(z_0, z_1, \ldots, z_n, x_{n+1}, \ldots)$ the inequality $n_W^{\mathcal{Z}}(\varepsilon) \leqslant n_W^{\mathcal{X}}(\varepsilon)$ holds.

6.8.2. The inequality 6.8(1), and also 5.10(2) may be used to estimate the increase of ε-entropy of compact sets in given specific cases.

6.8.21. As an example, let us consider in the space C of functions $f(t)$ continuous on $[-1,1]$ with $\|f\| = \sup_{-1 \leqslant t \leqslant 1} |f(t)|$, the class $A_R(K)(-1, 1)$ (see section 3.7.21) and as a class $\mathcal{X}\{x_0, x_1, x_2, \ldots\}$ let us take the system $1, t, t^2, \ldots$. It follows from 5.4(4) that in this case

$$n_{A_R K(-1,1)}(\varepsilon) \leqslant \frac{1}{\log_2 R} \log_2 \frac{1}{\varepsilon} + O(1). \tag{5}$$

Thus, it follows from 6.8(1) that**

$$I^a_{A_R K(-1,1)}(\varepsilon) \leqslant \frac{1}{\log_2 R} \left(\log_2 \frac{1}{\varepsilon}\right)^2 + O\left(\log_2 \frac{1}{\varepsilon} \log_2 \log_2 \frac{1}{\varepsilon}\right). \tag{6}$$

6.8.22. Let $KH_2^{(r+a)} M(L_q^{*(m)})$ be the set of all periodic functions $f(t_1, \ldots, t_m)$ with period 2π from the class

$$H_2^{(r_1+a_1, \ldots, r_m+a_m)} M_1, \ldots, M_m(L_q^{*(m)}),$$

$$r_1 = \ldots = r_m = r \text{ integer}; \quad 0 < \alpha_1 = \ldots = \alpha_m = \alpha < 1,$$

$$M_1 = \ldots = M_m = M.$$

(see section 3.8.5), for which

$$\|f\|_{L_q^{*(m)}} = \left\{\int_0^{2\pi} \cdots \int_0^{2\pi} |f(t_1, \ldots, t_m)|^q \, dt_1 \ldots dt_m\right\}^{\frac{1}{q}} \leqslant K. \tag{7}$$

It follows from theorem 6.3.4 that the set of all functions $f \in L_q^{*(m)}$ with $\|f\|_{L_q^{*(m)}} \leqslant K$ for which

$$E^*_{n_1, \ldots, n_m}(f)_{L_q^{*(m)}} \leqslant C \sum_{\nu=1}^{m} \frac{1}{(n_\nu + 1)^{r+\alpha}}$$

is, for a sufficiently small value of the constant $C > 0$, contained in the set $KH_2^{(r+a)} M(L_q^{*(m)})$. Thus, from the inequality 5.10(2) we get:

$$\Phi_{KH_2^{(r+a)} M(L_q^{*(m)})} \{(2n+1)^m\} \geqslant \frac{Cm}{4(n+1)^{r+\alpha}},$$

** In fact, as it was shown by A. G. Vitushkin [1], we have:

$$I^a_{A_R K(-1,1)}(\varepsilon) = \frac{1}{2 \log_2 R} \left(\log_2 \frac{1}{\varepsilon}\right)^2 + O\left(\log_2 \frac{1}{\varepsilon} \log_2 \log_2 \frac{1}{\varepsilon}\right).$$

i.e. as $\varepsilon \to 0$:

$$I^{\alpha}_{KH_2^{(r+\alpha)}M(L_q^{*(m)})}(\varepsilon) \geqslant B(K, M, m, r\alpha)\left(\frac{1}{\varepsilon}\right)^{\frac{m}{r+\alpha}} \tag{8}$$

On the other hand, for functions $f \in KH_2^{(r+\alpha)}M(L_q^{*(m)}$ we have, by be inequality 5.3(7):

$$E^*_{n_1, \dots, n_m}(f)_{L_q^{*(m)}} \leqslant CM \sum_{\nu=1}^{m} \frac{1}{(n_\nu+1)^{r+\alpha}},$$

where C is a constant. Consequently, for every integer s

$$\mathcal{E}_{(2s+1)^m}\left(KH_2^{(r+\alpha)}M(L_q^{*(m)})\right) = \sup_{f \in KH_2^{(r+\alpha)}M(L_q^{*(m)})} E^*_{s, \dots, s}(f) \leqslant \frac{mMC}{(s+1)^{r+\alpha}}$$

and for $s = \left[\left(\frac{mMC}{\varepsilon}\right)^{\frac{1}{r+\alpha}}\right]$:

$$\mathcal{E}_{(2s+1)^m}\left(KH_2^{(r+\alpha)}M(L_q^{*(m)})\right) \leqslant \varepsilon.$$

Thus, in this case

$$n_W(\varepsilon) \leqslant (2s+1)^m \leqslant \left\{2\left(\frac{mMC}{\varepsilon}\right)^{\frac{1}{r+\alpha}} + 1\right\}^m$$

i.e. for $\varepsilon \to 0$ the function $n_W(\varepsilon)$ increases no faster than $\left(\frac{1}{\varepsilon}\right)^{\frac{m}{r+\alpha}}$. Thus, by the basic inequality 6.8(1), we get the estimate from above equal to $\left(\frac{1}{\varepsilon}\right)^{\frac{m}{r+\alpha}} \log_2 \frac{1}{\varepsilon}$ ** for the order of increase of the ε-entropy $I^{a}_{KH_2^{(r+a)}M(L_q^{*(m)})}(\varepsilon)$ as $\varepsilon \to 0$. Combining this result with the inequality 6.8(8) we get the asymptotic estimate

$$\log_2 I^{a}_{KH_2^{(r+a)}M(L_q^{*(m)})}(\varepsilon) = \frac{m}{r+a}\log_2\frac{1}{\varepsilon} + O\left(\log_2\log_2\frac{1}{\varepsilon}\right). \tag{9}$$

6.8.23. If $K\Lambda_\alpha M(L_q^{*(m)})$ is the set of all periodic functions $f(t_1, \dots, t_m)$ satisfying condition 6.8(7) and the condition $\omega_2(f; 0, \dots, h_\nu, \dots, 0)$

** For $q = \infty$, $0 < \alpha < 1$, A. N. Kolmogorov [6] has obtained an estimate which does not contain the factor $\log_2 \frac{1}{\varepsilon}$, i.e. whose order coincides with the right-hand side of 6.8.(8). Kolmogorov's result shows that in this case the inequality 5.10(2) becomes an equality in the sense of order of decrease.

$\leqslant M|\log_2 h_\nu|^{-\alpha}$ $(\nu = 1, \ldots, m;\ \alpha > 0)$, then from considerations similar to those of section 6.8.22 it follows that in this case, as $n \to \infty$ the order of decrease of the function $\Phi_{KA_\alpha M(L_q^{*(m)}),}(n)$, inverse to the entropy, equals $\dfrac{1}{(\log_2 n)^\alpha}$.

6.8.3. Theorems 5.10 and 2.5 show that in an arbitrary infinite-dimensional Banach space F there exist sets, whose ε-entropy increases arbitrarily fast as $\varepsilon \to 0$**. This indicates the need for the introduction of a scale of increase of entropy $I_W^a(\varepsilon)$ as $\varepsilon \to 0$. Let us denote by $\mathfrak{M}_k^\beta(F)$ the class of sets $W \subset F$ for which

$$A\left(\log_2 \frac{1}{\varepsilon}\right)^\beta \leqslant (\log_2)_k I_W^a(\varepsilon) \leqslant B\left(\log_2 \frac{1}{\varepsilon}\right)^\beta,$$

where $(\log_2)_k I_W^a(\varepsilon)$ is the iterated logarithm of order k $(k = 0, 1, 2, \ldots)$ of $I_W^a(\varepsilon)$, B and A are some positive constants and $\beta > 0$.

L. S. Pontryagin and L. G. Shnirel'man [1] have shown that the inequality $\inf\limits_\varepsilon \dfrac{\log_2 \frac{1}{\varepsilon}}{I_W^b(\varepsilon)} > 0$ holds only in the case when the set W is finite-dimensional, i.e. only in the case when $\inf\limits_{\mathfrak{X}} \mathcal{E}_n^{\mathfrak{X}}(W) = 0$ for all sufficiently large n. In the general case the lower bound of $\inf\limits_{\mathfrak{X}} \mathcal{E}_n^{\mathfrak{X}}(W)$ taken over all possible linearly independent systems $\mathfrak{X}$ in F, may serve as a characteristic of the degree of deviation of this set from all possible sets of given dimension n.

We mention some properties which follow from the inequality 6.8(1), giving the estimate of $\inf\limits_{\mathfrak{X}} \mathcal{E}_n^{\mathfrak{X}}(W)$ for sets $W \in \mathfrak{M}_k^\beta(F)$.

6.8.31. If $W \in \mathfrak{M}_0^\beta(F)$ $(\beta > 1)$, then for all sufficiently large n

$$\inf_{\mathfrak{X}} \mathcal{E}_n^{\mathfrak{X}}(W) \geqslant \varrho^{\sqrt[\beta-1]{n}}$$

where $\varrho < 1$ is a positive constant independent of n.

6.8.32. If $W \in \mathfrak{M}_1^\beta(F)$ $(0 < \beta \leqslant 1)$, then for all sufficiently large n

$$\inf_{\mathfrak{X}} \mathcal{E}_n^{\mathfrak{X}}(W) \geqslant \Phi_W\left[C_n(\log_2 n)^{\frac{1}{\beta}}\right],$$

where C is a positive constant independent of n.

** A. N. Kolmogorov [7].

6.8.33. If $W \in \mathfrak{M}_1^\beta(F)$ $(\beta > 1)$, then for all sufficiently large n

$$\inf_{\mathcal{X}} \mathcal{E}_n^{\mathcal{X}}(W) \geqslant \varrho^{\sqrt[\beta]{\ln n}},$$

where $\varrho < 1$ is a constant independent of n.

6.8.34. If $W \in \mathfrak{M}_k^\beta(F)$, where $k \geqslant 2$, then for all sufficiently large n

$$\inf_{\mathcal{X}} \mathcal{E}_n^{\mathcal{X}}(W) \geqslant \Phi_W(n^2).$$

6.9 Various problems and theorems

1. A function $f(x)$ from the class $L_q (1 \leqslant q \leqslant \infty)$ on $(-\infty, +\infty)$ satisfies the generalised condition of Dini-Lipschitz: $\omega(f; t)_{L_q} \ln t \to 0$ as $t \to 0$ if and only if $\lim\limits_{\sigma\to\infty} A_\sigma (f)_{L_q} \ln \sigma = 0$ (see 6.1(18)) and 5.1(15)).

2. If a functions f(x) defined on $(-\infty, +\infty)$ belongs to the class L_q and $\alpha > 0$, the series $\sum\limits_{n=1}^{\infty} n^{-\alpha} \omega\left(f; \frac{1}{n}\right)_{L_q}$ and $\sum\limits_{n=1}^{\infty} n^{-\alpha} A_n(f)_{L_q}$ are either both convergent or both divergent. The same holds in the periodic case if we replace $A_n(f)_{L_q}$ by $E_n^*(f)_{L_q}$.

For the cases $q = 2$ and $q = \infty$ and $f(x)$ continuous and periodic see S. B. Stechkin [6]. Use the inequalities 5.1(9) and 6.1(18) $(k = 1)$ and apply Abel's transformation.

3. If the function $f(x)$ defined on $(-\infty, +\infty)$ belongs to the class L_q and has the property that for a certain integer k

$$A_\sigma(f)_{L_q} = O\left(\frac{1}{\sigma^k}\right),$$

then

$$\omega_k(f; t)_{L_q} = O\,(t^k |\ln t|).$$

See 6.1(18).

4. Theorems 6.1.61 (see also 6.1.2) remain valid if we replace O by o.

5. Conditions 6.1(30) and 6.1(21) are, respectively, equivalent to the existence of a constant $C > 1$ such that

$$\overline{\lim_{t\to\infty}} \frac{E\left(\frac{c}{t}\right)}{E\left(\frac{1}{t}\right)} < C^k \tag{1}$$

and

$$\underline{\lim_{t\to\infty}} \frac{E\left(\frac{c}{t}\right)}{E\left(\frac{1}{t}\right)} > 1. \tag{2}$$

N. K. Bari and S. B. Stechkin [1].

6. Let $f(x)$ be a periodic function of period 2π and let $\sigma_n(f; x)$ denote the arithmetic means of partial sums of its Fourier series (Fejér sums). The function $f(x)$ satisfies the Lipschitz condition $\omega(f, t) = O(t^\alpha)$ $(0 < \alpha \leqslant 1)$ if and only if $\omega(\sigma_n; t) = O(t^\alpha)$ uniformly in n.

A. Zygmund [1].

7. The function $f(x)$ continuous and bounded on $(-\infty, +\infty)$ satisfies the Lipschitz condition $\omega(f; t) = O(t^\alpha)$ $(0 < \alpha \leqslant 1)$ if and only if the entire functions $G_\sigma(x) = G_\sigma(f; x)$ of the best uniform approximation satisfy on $(-\infty, +\infty)$ the condition $\omega(G_\sigma; t) = O(t^\alpha)$ uniformly in σ.

8. If a sequence of positive numbers $E_1 \geqslant E_2 \geqslant E_3 \geqslant \ldots$ is such that the series $\sum_{n=1}^{\infty} n^{r-1} E_n$ is divergent, then there exists a function $f(x)$ with a discontinuous r-th derivative, for which for all n the best approximation by algebraic polynomials satisfies the inequality $E_n(f; -1, +1) < E_n$.

S. N. Bernstein [1].

9. If the sequence of best approximations $E_n^*(f)_{L_q}$ of a periodic function $f(x)$ of period 2π satisfies the condition $E_n^*(f)_{L_q} = O\left(\frac{1}{n^\alpha}\right)$ $(0 < \alpha \leqslant 1)$, then for every positive $\beta < \alpha$ there exists the derivative $f^{(\beta)}(x)$ (see section 3.5.6) and $E_n^*(f^{(\beta)})_{L_q} = O\left(\frac{1}{n^{\alpha-\beta}}\right)$.

D. Kralik [1].

10. If a function $f(x, y)$ defined in some bounded plane set G has the property that for the sequence of best uniform approximations by algebraic polynomials on G: $\lim_{m\to\infty} E_{k,m}(f; G) = 0$ and $\lim_{n\to\infty} E_{n,l}(f; G) = 0$, then

$$f(x, y) = \sum_{\nu=0}^{k} \sum_{\mu=0}^{l} a_{\nu\mu} x^\nu y^\mu.$$

Use the inequality 2.2(10).

11. If a function $f(x, y)$ is periodic with period 2π with respect to each variable, and has the partial derivatives $\frac{\partial^2 f}{\partial x^2}$ and $\frac{\partial^2 f}{\partial y^2}$ such that

$$\int_0^{2\pi}\int_0^{2\pi} \left(\frac{\partial^2 f}{\partial x^2}\right)^2 dx\,dy < \infty, \quad \int_0^{2\pi}\int_0^{2\pi} \left(\frac{\partial^2 f}{\partial y^2}\right)^2 dx\,dy < \infty, \tag{3}$$

then the mixed derivative $\frac{\partial^2 f}{\partial x \partial y}$ exists and

$$\int_0^{2\pi}\int_0^{2\pi} \left(\frac{\partial^2 f}{\partial x \partial y}\right)^2 dx\,dy = \int_0^{2\pi}\int_0^{2\pi} \frac{\partial^2 f}{\partial x^2}\frac{\partial^2 f}{\partial y^2}\,dx\,dy. \tag{4}$$

S. N. Bernstein [1]. Use the Parseval equality.

12. The equation of elliptic type:

$$A\frac{\partial^2 u}{\partial x^2}+2B\frac{\partial^2 u}{\partial x\,\partial y}+C\frac{\partial^2 u}{\partial y^2}=0 \qquad (AC-B^2>0)$$

where A, B, C are functions of $x, y, u, \dfrac{\partial u}{\partial x}, \dfrac{\partial u}{\partial y}$ does not have solutions periodic in x and y with finite derivatives of the first two orders, except the solution identically constant.

S. N. Bernstein [1]. See section 6.9.11.

13. The function $f(x)$ continuous on $[a, b]$ has the r-th derivative $f^{(r)}(x)$ satisfying the Lipschitz condition $\omega(f^{(r)}; t) = O(t^{\alpha})$ $(0<\alpha\leqslant 1)$ if and only if for every two points $x_1, x_2 \in [a, b]$

$$E_r(f; x_1, x_2) \leqslant C_r|x_1-x_2|^{r+\alpha}, \tag{5}$$

where C_r is a constant depending on r.

D. A. Raikov [1].

14. A function $f(x)$ continuous on $[a, b]$ has a continuous r-th derivative $(r\geqslant 1)$ if and only if

$$\frac{E_{r-1}(f; \alpha, \beta)}{(\beta-\alpha)^r} \to \frac{1}{r!\,2^{2r-1}}\,|f^{(r)}(x)| \tag{6}$$

uniformly with respect to x as $\beta-\alpha\to 0$, $\alpha<x<\beta$.

S. N. Bernstein [21].

15. Let $f(x)$ be defined on $[a, b]$,

$$\left\{\int_a^b |f(x)|^q\,dx\right\}^{\frac{1}{q}}<\infty,$$

and let $E_n(f; a, b)_{L_q}$ be its best mean approximation by algebraic polynomials of order $\leqslant n$. The condition $\overline{\lim_{n\to\infty}}\sqrt[n]{E_n(f; a, b)_{L_q}}<1$ holds if and only if $f(x)$ coincides almost everywhere on $[a, b]$ with an analytic function.

S. N. Bernstein [7].

16. (continuation). If $\underline{\lim_{n\to\infty}}\sqrt[n]{E_n(f; a, b)_{L_q}}<1$ and $f(x)$ equals zero for almost all points of an interval $(\alpha, \beta)\subset[a, b]$ then $f(x)=0$ almost everywhere on $[a, b]$.

S. N. Bernstein [7].

17. A function $f(x)$ continuous and bounded on $(-\infty, +\infty)$ is infinitely differentiable if and only if $\lim_{n\to\infty}\sigma^r A_\sigma(f)=0$ for every r.

18. If a function $f(x)$ bounded and continuos on $(-\infty, +\infty)$ equals zero on some interval and has the property that $\underline{\lim_{\sigma\to\infty}}\sqrt[\sigma]{A_\sigma(f)}<1$, then $f(x)\equiv 0$.

19. If a function $f(x)$ is quasi-analytic in the sense of S. N. Bernstein then its indefinite integral satisfies the same condition.

S. N. Bernstein [7].

20. The function

$$f(x) = \sum_{n=1}^{\infty} \frac{\cos n!\arccos x}{2^{(n-1)!}} \tag{7}$$

is not analytic on $[-1, +1]$ and is infinitely differentiable and quasi-analytic in the sense of S. N. Bernstein.

S. N. Bernstein [7]. See sections 6.6.2 and 6.6.4.

21. The function

$$f(x) = \sum_{n=0}^{\infty} \frac{\cos F(n)\arccos x}{F(n)} \tag{8}$$

where $F(0) = 1, F(n+1) = 2^{F(n)}$ $(n = 0, 1, 2, \ldots)$ is not differentiable at any point of the interval $-1 \leqslant x \leqslant 1$ and is quasi-analytic in the sense of S. N. Bernstein.

S. N. Bernstein [7]. See section 6.6.2.

22. If a periodic function $f(x)$ has the property that for a certain indefinitely increasing sequence n_k and a certain null-sequence h_k

$$|\Delta_{h_k}^{n_k} f(x)| \leqslant (Mn_k h_k)^{n_k}$$

uniformly in x, then

$$|\Delta_h^n f(x)| \leqslant (M^* nh|\ln h|)^n$$

uniformly in x for all n and a certain infinite set of values of h tending to zero.

S. N. Bernstein [7]. See sections 5.8.2 and 6.6.5.

23. If a periodic function f(x) satisfies the property that

$$E_{n_k}^*(f) < \frac{C}{n_k^r}$$

for a certain indefinitely increasing sequence n_k, then for every n

$$\omega_n(f; t) \leqslant Mt^{\frac{rn}{r+n}}$$

for a certain infinite set of values of t tending to zero.

S. N. Bernstein [11]. See section 6.6.5.

24. The function

$$f(x) = \sum_{m=1}^{\infty} \frac{\cos 2^m!\, x}{2^m!} \tag{9}$$

satisfies for every r the condition of the preceding theorem and does not have any derivatives.

S. N. Bernstein [7].

25. If a sequence of positive numbers $E_1 \geqslant E_2 \geqslant E_3 \geqslant \ldots$ has the property that $\sum_{n=1}^{\infty} \frac{1}{n} E_n$ diverges, then there exists a continuous function $f(x)$ with period 2π, with discontinuous conjugate function $\tilde{f}(x)$ and $E_n^*(f) < E_n$ for every n.

N. K. Bari [1]. Consider the function $f(x) = \sum_{n=1}^{\infty} E_n \frac{\sin nx}{n}$.

26. Let a bounded function $f(x)$ defined on $(-\infty, +\infty)$ satisfy the condition $\omega_2(f; t) = O(t^\alpha)$ $(0 < \alpha \leqslant 1)$, and let $\tilde{f}(t)$ be the function conjugate in the sense 3.11(20). If there exists a constant C such that the integral

$$\int_0^x \{\tilde{f}(t) - C\}\, dt$$

is bounded for all x, then the function $\tilde{f}(x)$ is also bounded and $\omega_2(\tilde{f}; t) = O(t^\alpha)$.

N. I. Akhiezer [3].

CHAPTER VII

FURTHER THEOREMS CONNECTING THE BEST APPROXIMATIONS OF FUNCTIONS WITH THEIR STRUCTURAL PROPERTIES

7.1. On the exact order of decrease of best approximations

The results of the preceding chapter (see section 6.1) show that in all the cases when the best approximation $A_\sigma(f)_{L_q}$ of a function $f(x)$ from the class L_q $(1 \leqslant q \leqslant \infty)$ defined on $(-\infty, +\infty)$ satisfies, for a certain integer k, the condition

$$\frac{1}{n^k}\sum_{\nu=0}^{n}(\nu+1)^{k-1}A_\nu(f)_{L_q} = O[A_n(f)_{L_q}],$$

then the modulus of smoothness $\omega_k\left(f; \frac{1}{\sigma}\right)_{L_q}$ has, as $\sigma \to \infty$, the same order of tendency to 0 as $A_\sigma(f)_{L_q}$. Owing to this fact, in the study of differential characteristics of functions with a given sequence of best approximations E_n the special role is played by functions, for which the sequence $\{E_n\}$ satisfies, for a certain integer $k > 0$ the property

$$\frac{1}{n^k}\sum_{\nu=0}^{n}(\nu+1)^{k-1}E_\nu = O(E_n). \tag{1}$$

We shall describe the class of such functions in structural terms.

7.1.1. *A periodic function of period* 2π *of the class* L_q *has the sequence* $E_n^*(f)_{L_q} = E_n$ *of best approximations satisfying condition* 7.1(1) *if and only if as* $t \to 0$**

$$t^k\int_t^1 \frac{\omega_k(f; u)_{L_q}}{u^{k+1}}\,du = O[\omega_k(f; t)_{L_q} \tag{2}$$

** For $q = \infty$, in the case when $f(x)$ is continuous, an equivalent theorem can be found in papers of S. M. Lozinskii [3] and N. K. Bari and S. B. Stechkin [1].

If for the sequence $E_n = E_n^*(f)_{L_q}$ we have 7.1(1), then by the inequality 5.1(16) (see also 5.11(1) and 6.1(1))

$$M_k^* E_n^*(f)_{L_q} \leqslant \omega_k\left(f; \frac{1}{n+1}\right)_{L_q} \leqslant C_k^* E_n^*(f)_{L_q}, \tag{3}$$

where M_k^* and C_k^* are positive constants depending only on k. Thus, using the fact that for $\frac{1}{n+1} < t \leqslant \frac{1}{n}$

$$t^k \int_t^1 \frac{\omega_k(f; u)_{L_q}}{u^{k+1}} du \leqslant \frac{1}{n^k} \int_{\frac{1}{n+1}}^1 \frac{\omega_k(f; u)_{L_q}}{u^{k+1}} du = \frac{1}{n^k} \sum_{\nu=1}^n \int_{\frac{1}{\nu+1}}^{\frac{1}{\nu}} \frac{\omega_k(f; u)_{L_q}}{u^{k+1}} du$$

$$\leqslant \frac{1}{n^k} \sum_{\nu=1}^n \omega_k\left(f; \frac{1}{\nu}\right)_{L_q} \int_{\frac{1}{\nu+1}}^{\frac{1}{\nu}} \frac{du}{u^{k+1}} \leqslant \frac{B_k}{n^k} \sum_{\nu=1}^n \nu^{k-1} \omega_k\left(f; \frac{1}{\nu}\right)_{L_q},$$

we have

$$t^k \int_t^1 \frac{\omega_k(f; u)_{L_q}}{u^{k+1}} du \leqslant \frac{B_k C_k^*}{n^k} \sum_{\nu=0}^n (\nu+1)^{k-1} E_\nu^*(f)_{L_q}$$

$$= O[E_n^*(f)_{L_q}] = O\left[\omega_k\left(f; \frac{1}{n+1}\right)_{L_q}\right] = O[\omega_k(f; t)_{L_q}].$$

To prove the converse statement, we prove first the following lemma.

7.1.11. Every modulus of smoothness $\omega_k(t) = \omega_k(f; t)_{L_q}$ satisfying condition 7.1(2) has the property that

$$\omega_k(t) = O\left[\frac{1}{v^k |\ln v|} \omega_k(vt)\right] \tag{4}$$

uniformly with respect to all positive $v \leqslant \frac{1}{2}$ and $t \leqslant \frac{1}{2}$.

Let $\frac{1}{n+1} < t \leqslant \frac{1}{n}$ and $\frac{1}{m+1} < v \leqslant \frac{1}{m}$. It is evident that

$$t^k \int_t^1 \frac{\omega_k(u)}{u^{k+1}} du > \frac{1}{(n+1)^k} \int_{\frac{1}{n}}^1 \frac{\omega_k(u)}{u^{k+1}} du = \frac{1}{(n+1)^k} \sum_{\nu=1}^{n-1} \int_{\frac{1}{\nu+1}}^{\frac{1}{\nu}} \frac{\omega_k(u)}{u^{k+1}} du$$

$$\geqslant \frac{1}{(n+1)^k} \sum_{\nu=1}^{n-1} \omega_k\left(\frac{1}{\nu+1}\right) \int_{\frac{1}{\nu+1}}^{\frac{1}{\nu}} \frac{du}{u^{k+1}} \geqslant \frac{a_k}{(n+1)^k} \sum_{\nu=1}^{n-1} \nu^{k-1} \omega_k\left(\frac{1}{\nu}\right), \tag{5}$$

where a_k is a positive constant.

It follows that when 7.1(2) is satisfied

$$\frac{1}{(m+1)^k(n+1)^k}\sum_{\nu=1}^{mn}\nu^{k-1}\omega_k\left(\frac{1}{\nu}\right)\leqslant b_k\omega_k\left(\frac{1}{mn}\right). \tag{6}$$

Moreover

$$\frac{1}{(n+1)^k}\sum_{\nu=1}^{mn}\nu^{k-1}\omega_k\left(\frac{1}{\nu}\right)=\frac{1}{(n+1)^k}\sum_{j=1}^{m}\sum_{\nu=(j-1)n+1}^{jn}\nu^{k-1}\omega_k\left(\frac{1}{\nu}\right)$$

$$\geqslant\frac{1}{(n+1)^k}\sum_{j=1}^{m}\omega_k\left(\frac{1}{jn}\right)\sum_{\nu=(j-1)n+1}^{jn}\nu^{k-1}\geqslant\alpha_k\sum_{j=1}^{m}j^{k-1}\omega_k\left(\frac{1}{jn}\right),$$

where α_k is a positive constant. By 3.3(5), for every integer $j\geqslant 1$ we have the inequality:

$$j^k\omega_k\left(\frac{1}{jn}\right)\geqslant\omega_k\left(\frac{1}{n}\right).$$

Thus

$$\frac{1}{(n+1)^k}\sum_{\nu=1}^{mn}\nu^{k-1}\omega_k\left(\frac{1}{\nu}\right)\geqslant\alpha_k\omega_k\left(\frac{1}{n}\right)\sum_{j=1}^{m}\frac{1}{j},$$

i.e.

$$\omega_k\left(\frac{1}{n}\right)=O\left\{\frac{1}{(n+1)^k\ln m}\sum_{\nu=1}^{mn}\nu^{k-1}\omega_k\left(\frac{1}{\nu}\right)\right\}.$$

Combining the last relation with the inequality 7.1(6) we obtain

$$\omega_k\left(\frac{1}{n}\right)=O\left\{\frac{m_k}{\ln m}\omega_k\left(\frac{1}{mn}\right)\right\}, \tag{7}$$

or

$$\omega_k(t)=O\left\{\frac{1}{v^k|\ln v|}\omega_k(vt)\right\}.$$

7.1.12. Using lemma 7.11.1 we find easily that *condition* 7.1(2) *implies* 7.1(3). Indeed, by inequality 5.1(16) (see also 5.11(1)):

$$\frac{1}{n^k}\sum_{\nu=0}^{n}(\nu+1)^{k-1}E_\nu^*(f)_{L_q}\leqslant\frac{C_k}{n^k}\sum_{\nu=0}^{n}(\nu+1)^{k-1}\omega_k\left(f;\frac{1}{\nu+1}\right)_{L_q}$$

$$=O\left\{\frac{1}{n^k}\int_{\frac{1}{n}}^{1}\frac{\omega_k(f;u)_{L_q}}{u^{k+1}}\,du\right\}.$$

Thus, if 7.1(2) holds, then there exists a constant β_k, such that for all $n\geqslant 1$ we have

$$\frac{1}{n^k}\sum_{\nu=0}^{n}(\nu+1)^{k-1}E_\nu^*(f)_{L_q}\leqslant\beta_k\omega_k\left(f;\frac{1}{n}\right)_{L_q}. \tag{8}$$

Besides, by theorem 6.1.1 for every integer $m \geqslant 1$ we have

$$\omega_k\left(f; \frac{1}{mn}\right)_{L_q} \leqslant \frac{C_k}{m^k n^k} \sum_{\nu=0}^{mn} (\nu+1)^{k-1} E_\nu^*(f)_{L_q}.$$

Consequently

$$\sum_{\nu=n+1}^{mn} (\nu+1)^{k-1} E_\nu^*(f)_{L_q} \geqslant \frac{m^k n^k}{C_k} \omega_k\left(f; \frac{1}{mn}\right)_{L_q} - \beta_k n^k \omega_k\left(f; \frac{1}{n}\right)_{L_q},$$

i.e.

$$E_n^*(f)_{L_q} \sum_{\nu=n+1}^{mn} (\nu+1)^{k-1} \geqslant \frac{m^k n^k}{C_k} \omega_k\left(f; \frac{1}{mn}\right)_{L_q} - \beta_k n^k \omega_k\left(f; \frac{1}{n}\right)_{L_q},$$

or

$$\begin{aligned} E_n^*(f)_{L_q} &\geqslant \frac{m^k n^k}{C_k(m+1)^k (n+1)^k} \omega_k\left(f; \frac{1}{mn}\right)_{L_q} - \\ &\quad - \beta_k \frac{n^k}{(m+1)^k (n+1)^k} \omega_k\left(f; \frac{1}{n}\right)_{L_q} \\ &> \frac{1}{4^k C_k} \omega_k\left(f; \frac{1}{mn}\right)_{L_q} - \frac{\beta_k}{m^k} \omega_k\left(f; \frac{1}{n}\right)_{L_q}. \end{aligned} \tag{9}$$

On the other hand, 7.1(2) implies 7.1(7), and, consequently, as $m \to \infty$ the second term of the right-hand side of 7.1(9) becomes infinitely small in comparison with $\omega_k\left(f; \frac{1}{mn}\right)_{L_q}$. It follows that m may be chosen in such a way that (see section 3.3.5):

$$E_n^*(f)_{L_q} \geqslant \frac{1}{8^k C_k} \omega_k\left(f; \frac{1}{mn}\right)_{L_q} \geqslant \frac{1}{8^k C_k m^k} \omega_k\left(f; \frac{1}{n}\right)_{L_q},$$

which, together with 5.1(16) (see also 5.11(1)) proves 7.1(3).

7.1.13. To complete the proof of theorem 7.1.1 it remains to notice that 7.1(18) and 7.1(3) imply 7.1(1).

Thus, *in all cases when for some k the modulus of smoothness $\omega_k(f; t)_{L_q}$ satisfies* 7.1(2), *the exact order of decrease of best approximations $E_n^*(f)_{L_q}$ coincides with the order of $\omega_k\left(f; \frac{1}{n}\right)_{L_q}$.*

7.1.2. In connection with theorem 6.1.3 (see also section 6.1.6) among the functions, whose sequence of best approximations satisfies a condition of the type 7.1(1), the functions for which for some positive integer k

$$\sum_{\nu=1}^{\infty} \nu^{r-1} E_\nu < \infty$$

and

$$\frac{1}{n^r} \sum_{\nu=n+1}^{\infty} \nu^{r-1} E_\nu = O(E_n) \tag{10}$$

are of special interest.

A periodic function $f(x)$ of period 2π, with the r-th derivative $f^{(r)}(x) \in L_q (1 \leqslant q \leqslant \infty)$ has the sequence of best approximations $E_n^*(f)_{L_q} = E_n$ $(n = 1,2,\ldots)$ satisfying 7.1(10) and

$$\frac{1}{n^{k+r}} \sum_{\nu=0}^{n} (\nu+1)^{k+r-1} E_\nu = O(E_n), \quad (k > 0 \text{ integer}) \tag{11}$$

if and only if as $t \to 0^+$

$$\int_0^t \frac{\omega_k(f^{(r)}; u)_{L_q}}{u} \, du = O[\omega_k(f^{(r)}; t)_{L_q}] \tag{12}$$

and

$$t^k \int_t^1 \frac{\omega_k(f^{(r)}; u)_{L_q}}{u^{k+1}} \, du = O[\omega_k(f^{(r)}; t)_{L_q}]^{**}. \tag{13}$$

If the sequence $E_n = E_n^*(f)_{L_q}$ satisfies 7.1(10), then by 5.6(27) and 5.6(28) (see also 5.11.4)

$$M'_{k,r} n^r E_n^*(f)_{L_q} \leqslant E_n^*(f^{(r)})_{L_q} \leqslant C'_{k,r} n^r E_n^*(f)_{L_q}, \tag{14}$$

where $M'_{k,r}$ and $C'_{k,r}$ are positive constants. In this case 7.1(11) is equivalent to

$$\frac{1}{n^k} \sum_{\nu=0}^{n} (\nu+1)^{k-1} E_\nu^*(f^{(r)})_{L_q} = O[E_n^*(f^{(r)})_{L_q}]. \tag{15}$$

By theorem 7.1.1 the last equality is equivalent to 7.1(13). Also from the inequalities 5.1(18) (see also 5.11(2)) and 6.1(9) it follows that

$$M^*_{k,r} n^r E_n^*(f)_{L_q} \leqslant \omega_k\left(f^{(r)}; \frac{1}{n}\right)_{L_q} \leqslant C^*_{k,r} n^r E_n^*(f)_{L_q}. \tag{16}$$

** For $q = \infty$ an equivalent theorem can be found in papers of S. M. Lozinskii [3], N. K. Bari and S. B. Stechkin [1].

where $M_{k,r}^*$ and $C_{k,r}^*$ are positive constants. Thus, since for $\frac{1}{n+1} < t \leqslant \frac{1}{n}$ we have:

$$\int_0^t \frac{\omega_k(f^{(r)}; u)_{L_q}}{u} du \leqslant \int_0^{\frac{1}{n}} \frac{\omega_k(f^{(r)}; u)_{L_q}}{u} du$$

$$= \sum_{\nu=n}^{\infty} \int_{\frac{1}{\nu+1}}^{\frac{1}{\nu}} \frac{\omega_k(f^{(r)}; u)_{L_q}}{u} du$$

$$\leqslant \sum_{\nu=n}^{\infty} \omega_k\left(f^{(r)}; \frac{1}{\nu}\right)_{L_q} \int_{\frac{1}{\nu+1}}^{\frac{1}{\nu}} \frac{du}{u} \leqslant \sum_{\nu=n}^{\infty} \frac{1}{\nu} \omega_k\left(f^{(r)}; \frac{1}{\nu}\right)_{L_q},$$

we obtain:

$$\int_0^t \frac{\omega_k(f^{(r)}; u)_{L_q}}{u} du \leqslant C_{k,r}^* \sum_{\nu=n}^{\infty} \nu^{-1} E_\nu^*(f)_{L_q}$$

$$= O\left[n^r E_n^{*}(f)_{L_q}\right] = O\left[\omega_k\left(f^{(r)}; \frac{1}{n}\right)_{L_q}\right].$$

To prove the second part of the theorem we use the following lemma.

7.1.21. *If* $\omega_k(t) = \omega_k(f^{(r)}; t)_{L_q}$ *satisfies condition* 7.1(12), *then*

$$\omega_k(vt) = O\left[\frac{1}{|\ln v|} \omega_k(t)\right] \tag{17}$$

uniformly for all positive $v \leqslant \frac{1}{2}$ and $t \leqslant \frac{1}{2}$.

Let $\frac{1}{n+1} < t \leqslant \frac{1}{n}$ and $\frac{1}{m+1} < v \leqslant \frac{1}{m}$. Then

$$\int_0^t \frac{\omega_k(u)}{u} du \geqslant \int_0^{\frac{1}{n+1}} \frac{\omega_k(u)}{u} du$$

$$= \sum_{\nu=n+1}^{\infty} \int_{\frac{1}{\nu+1}}^{\frac{1}{\nu}} \frac{\omega_k(u)}{u} du \geqslant a_k^* \sum_{\nu=n+1}^{\infty} \frac{1}{\nu} \omega_k\left(\frac{1}{\nu}\right),$$

where a_k^* is a positive constant. It follows that when condition 7.1(12) is satisfied

$$\sum_{\nu=n+1}^{mn} \frac{1}{\nu}\omega_k\left(\frac{1}{\nu}\right) \leqslant b_k^*\omega_k\left(\frac{1}{n}\right),$$

i.e.

$$\omega_k\left(\frac{1}{mn}\right) = O\left[\frac{1}{\ln m}\omega_k\left(\frac{1}{n}\right)\right].$$

7.1.22. We shall show that 7.1(12) and 7.1(13) imply 7.1(16). On one hand, by the inequality 5.1(18) (see also 5.11(2)) we have

$$\sum_{\nu=n+1}^{\infty} \nu^{r-1}E_\nu^*(f)_{L_q} \leqslant C_{k,r}\sum_{\nu=n+1}^{\infty}\frac{1}{\nu}\omega_k\left(f^{(r)};\frac{1}{\nu}\right)_{L_q}$$

$$= O\left\{\int_0^{\frac{1}{n}} \frac{\omega_k(f^{(r)};u)_{L_q}}{u}\,du\right\},$$

and, consequently, by condition 7.1(12) there exists a constant β_k^* such that

$$\sum_{\nu=mn+1}^{\infty} \nu^{r-1}E_\nu^*(f)_{L_q} \leqslant \beta_k^*\omega_k\left(f^{(r)};\frac{1}{mn}\right)_{L_q}, \tag{18}$$

where m and n are arbitrary natural numbers.

On the other hand, since by theorem 7.1.1, condition 7.1(13) implies that

$$\frac{1}{n^k}\sum_{\nu=0}^{n}(\nu+1)^{k-1}E_\nu^*(f^{(r)})_{L_q} = O\left[E_n^*(f^{(r)})_{L_q}\right]$$

we have, by inequalities 5.6(27) and 5.6(28) (see also 5.11.4)

$$\frac{1}{n^k}\sum_{\nu=0}^{n}(\nu+1)^{k+r-1}E_\nu^*(f)_{L_q} = O\left\{\frac{1}{n^k}\sum_{\nu=0}^{n}(\nu+1)^{k-1}E_\nu^*(f^{(r)})_{L_q}\right\}$$

$$= O\left[E_n^*(f^{(r)})_{L_q}\right] = O\left\{n^rE_n^*(f)_{L_q} + \sum_{\nu=n+1}^{\infty}\nu^{r-1}E_\nu^*(f)_{L_q}\right\}, \tag{19}$$

and according to theorem 6.1.3 we get

$$\omega_k\left(f^{(r)};\frac{1}{n}\right)_{L_q} \leqslant B_{k,r}\left\{n^rE_n^*(f)_{L_q} + \sum_{\nu=n+1}^{\infty}\nu^{r-1}E_\nu^*(f)_{L_q}\right\}. \tag{20}$$

The inequalities 7.1(18) and 7.1(20) imply, that for some values of the positive constants $A^*_{k,r}$ and $B^*_{k,r}$

$$n^r E^*_n(f)_{L_q} + \sum_{\nu=n+1}^{mn} \nu^{r-1} E^*_\nu(f)_{L_q}$$
$$\geqslant A^*_{k,r}\omega_k\left(f^{(r)}; \frac{1}{n}\right)_{L_q} - B^*_{k,r}\omega_k\left(f^{(r)}; \frac{1}{mn}\right)_{L_q}. \qquad (21)$$

Since, however, 7.1(12) implies 7.1(17), the second term on the right-hand side of 7.1(21) becomes negligible in comparison with $\omega_k\left(f^{(r)}; \frac{1}{n}\right)_{L_q}$ as m increases.

Thus, we can choose m in such a way that for all $n \geqslant 1$

$$\omega_k\left(f^{(r)}; \frac{1}{n}\right)_{L_q} \leqslant A_{k,r} n^r(1+m^r) E^*_n(f)_{L_q},$$

where $A_{k,r}$ is a constant. The first of the inequalities 7.1(16) has already been obtained in Chapter V (see 5.1(8) and 5.11(2)).

7.1.23. To complete the proof of theorem 7.1.2 it remains to note that 7.1(18), 7.1(19) and 7.1(16) imply 7.1(10) and 7.1(11).

Thus, in all cases when the function $f(x)$ has the r-th derivative $f^{(r)}(x) \in L_q$ and for some k the modulus of smoothness $\omega_k(f^{(r)}; t)_{L_q}$ satisfies 7.1(12) *and* 7.1(13), *the exact order of decrease of best approximations $E^*_n(f)_{L_q}$ coincides with the order of* $\frac{1}{n^r}\omega_k\left(f^{(r)}; \frac{1}{n}\right)_{L_q}$.

7.1.3. We see easily that the theorems similar to those found in sections 7.1.1 and 7.1.2 for the periodic case, remain valid also in the case when we consider functions $f(x)$ of the class L_q on $(-\infty, +\infty)$ and their best approximations $A_\sigma(f)_{L_q}$ by arbitrary entire functions of exponential type.

7.1.4. Theorems 7.1.13 and 7.1.23 allow us, in certain cases, to find the exact order of decrease of best approximations for some specific functions.

7.1.41. Let us consider, for instance, the function $f(t) = |\cos t|^\alpha$, where $0 < \alpha \leqslant 1$. Since in this case $\omega_2(f; t)$ is, for $t \to 0$ of order t^α, and consequently, it satisfies 7.1(2) $(q = \infty, k = 2)$, we find by 7.1.13 that the exact order of decrease of $E^*_n(f)$ as $n \to \infty$ equals $\frac{1}{n^\alpha}$.

Similarly, if $f(t)$ is an even function of period 2π, which coincides with the function $\omega\left(\frac{t}{\pi}\right)$ on $[0, \pi]$ (where $\omega(t)$ is the modulus of continuity considered in section 3.2.4.), then, since in this case $\omega(f; t) = \omega\left(\frac{t}{\pi}\right)$, $t < \pi$, and $\omega(t)$ is of order $t^{\log_3 2}$ for $t \to 0$, the exact order of decrease of best approximations $E_n^*(f)$ for this function (differentiable almost everywhere), equals $\frac{1}{n^{\log_3 2}}$.

The function $f(t) = |\cos t|^{1+\alpha}$, where $0 < \alpha < 1$ has the derivative

$$f'(t) = \begin{cases} -(1+\alpha)|\cos t|^{\alpha} \sin t & \text{for} \quad 0 \leqslant t \leqslant \frac{\pi}{2}, \\ (1+\alpha)|\cos t|^{\alpha} \sin t & \text{for} \quad \frac{\pi}{2} \leqslant t \leqslant \pi, \end{cases}$$

for which $\omega_2(f'; t)$ as $t \to 0$, is of order t^{α}, and consequently, satisfies 7.1(12) and 7.1(13) ($k = 2$). Thus, by 7.1.23, the exact order of decrease of best approximations $E_n^*(f)$ as $n \to \infty$ for this function equals $\frac{1}{n^{1+\alpha}}$.

If $f(t) = (\cos t)^{r-1} |\cos t|^{1+\alpha}$, where $r > 0$, is an integer, and $-1 < \alpha < 1$, then the derivative $f^{(r-1)}(t)$ exists, and it can be written in the form of a sum

$$f^{(r-1)}(t) = \pm (r+\alpha) \dots (\alpha+2) |\cos t|^{1+\alpha} + \psi_r(t),$$

where $\psi_r(t)$ is a function with a bounded second derivative for $\alpha \geqslant 0$, and bounded first derivative if $\alpha < 0$.

Since the function $\varphi(t) = |\cos t|^{1+\alpha}$ has the modulus of smoothness $\omega_2(\varphi'; t)$ of order $t^{1+\alpha}$ as $t \to 0$, then $\omega_2(f^{(r-1)}, t)$ has the same order of decrease. Thus, by 7.1.23, *the exact order of decrease of the sequence of best approximations* $E_n^*(f)$ *for* $f(t) = (\cos t)^{r-1}.\ |\cos t|^{1+\alpha}$ *for any positive integer* r *and* $-1 < \alpha < 1$ *equals* $\frac{1}{n^{r+\alpha}}$.

The same considerations show that *for any number* c *with* $|c| < 1$, *the exact order of decrease of the sequence of best approximations* $E_n^*(f)$ *of the function* $f(t) = (\cos t - c)^{r-1} |\cos t - c|^{1+\alpha}$, *for any positive integer* r *and* $-1 < \alpha < 1$ *equals* $\frac{1}{n^{r+\alpha}}$ *as* $n \to \infty$.

It follows directly that *for any positive integer* r *and* $-1 < \alpha < 1$ *the exact order of decrease of best approximations* $E_n(f, -1, 1)$ *by algebraic polynomials of the function* $f(x) = (x-c)^{r-1} |x-c|^{1+\alpha}$ *for* $|c| < 1$ *and* $n \to \infty$ *equals* $\frac{1}{n^{r+\alpha}}$**.

** In the case $r = 1$, $\alpha = 0$, this result has been obtained by S. N. Bernstein (see [15]).

By the substitution $y^2 = \frac{1 \pm x}{2}$, we easily find that *in the case* $c = \pm 1,\ \alpha \neq 0,$ *this order equals* $\frac{1}{n^{2r+2\alpha}}$.

7.1.42. For some positive integer m and $0 < \alpha \leqslant 1$ let

$$f(t) = |\cos t|^{\alpha} (\ln |\cos t|)^m. \tag{22}$$

Here $\omega_2(f;\ t)$ is of order $t^{\alpha} |\ln t|^m$ as $t \to 0$, and consequently, it satisfies 7.1(2) ($q = \infty$, $k = 2$) the exact order of decrease of best approximations $E_n^*(f)$ equals, as $n \to \infty$, $\frac{(\mathrm{lu}\, n)^m}{n}$. The function

$$f(t) = \cos t (\ln |\cos t|)^m \tag{23}$$

has the modulus of smoothness $\omega_2(f;\ t)$ of order $t|\ln|^{m-1}$** as $t \to 0$. Thus we conclude that in this case the exact order of decrease of $E_n^*(f)$ as $n \to \infty$, equals $\frac{(\mathrm{lu}\, n)^{m-1}}{n}$. If

$$f(t) = |\cos t|^s (\ln |\cos t|)^m, \tag{24}$$

where s is an arbitrary positive number, then (if s is an even integer), the function 7.1(24) has an $(s-1)$-th derivative which can be written in the form of the sum

$$f^{(s-1)}(t) = \pm s! \cos t (\ln |\cos t|)^m + \psi_s(t) + \sum_{k=1}^{m} a_k \cos t (\ln |\cos t|)^k,$$

where $\psi_s(t)$ has a bounded first derivative. Thus, the modulus of smoothness $\omega_2(f^{(s-1)},\ t)$, as $t \to 0$, is of order $t|\ln t|^{m-1}$, and satisfies 7.1(12) and 7.1(13). Hence *for even values of* s, *by* 7.1.23, *the exact order of decrease of best approximations* $E_n^*(f)$ *for the function* 7.1(24) *equals* $\frac{(\ln n)^{m-1}}{n^s}$

It follows from similar considerations that *for all remaining values of* s, *this order equals* $\frac{(\ln n)^m}{n^s}$.

Since the best approximation of function $|x|^s (\ln |x|)^m$ by algebraic polynomials on $[-1,1]$ coincide with the best approximations of the function 7.1(24) by trigonometric polynomials of the same order, we have in consequence: *the exact order of decrease of best approximations*

** Using the remark made in section 3.5.71 it suffices to verify that the function 7.1(23) satisfies this property on the segment $\left[-\frac{\pi}{2}, \frac{\pi}{2}\right]$.

$E_n[|x|^s(\ln|x|)^m; -1,1]$ *($s>0$, $m\geqslant 0$ an integer), as $n\to\infty$ equals* $\dfrac{(\ln n)^{m-1}}{n^s}$ *if s is an even integer and* $\dfrac{(\ln n)^m}{n^s}$ *in all remaining cases.*

7.1.43. The integral modulus of smoothness $\omega_2(f_0; t)_{L_q} (q\geqslant 1)$ of an odd periodic function of period 2π:

$$f_0(t)=\begin{cases} -\sin t, & 0\leqslant t\leqslant \dfrac{\pi}{2},\\ \sin t, & \dfrac{\pi}{2}\leqslant t\leqslant \pi, \end{cases} \tag{25}$$

is, as $t\to 0$, of the order $t^{\frac{1}{q}}$.

Thus, by 7.1.13 the exact order of decrease of the sequence $E_n^*(f_0)_{L_q}$ is, as $n\to\infty$, equal to $\dfrac{1}{n^{\frac{1}{q}}}$.

If for some integer $r\geqslant 0$

$$f(t)=(\cos t)^{r-1}|\cos t|, \tag{26}$$

then there exists the derivative $f^{(r)}(t)$, and it can be written as the sum

$$f^{(r)}(t)=\pm r!\, f_0(t)+\psi_r(t),$$

where $\psi_r(t)$ is a function which has a first derivative, with integral modulus of continuity $\omega(\psi_r'; t)_{L_q}$ of the order $t^{\frac{1}{q}}$. Consequently the integral modulus of continuity $\omega_2(f^{(r)}; t)_{L_q}$ is, as $t\to 0$, of the order $t^{\frac{1}{q}}$, i.e. it satisfies conditions 7.1(12) and 7.1(13). Thus, by 7.1.23, *the exact order of decrease of best approximations $E_n^*(f)_{L_q}$ for the function* 7.1(26) *equals to* $\dfrac{1}{n^{r+\frac{1}{q}}}$, *as $n\to\infty$.*

From the same considerations it follows that *for every c with $|c|<1$, the exact order of decrease of the sequence of best integral approximations $E_n^*(f)_{L_q}$ for the function $f(t)=(\cos t-c)^{r-1}|\cos t-c|$ for every integer $r\geqslant 0$ and $n\to\infty$ equals* $\dfrac{1}{n^{r+\frac{1}{q}}}$.

Similarly it can be shown that, *for every integer $r\geqslant 0$ and $|c|<1$ the exact order of decrease of best approximations $E_n^*[(x-c)^{r-1}|x-c|: -1,1]_{L_q}$ equals* $\dfrac{1}{n^{r+\frac{1}{q}}}$ *as $n\to\infty$.*

7.1.44. If, instead of the function 7.1(25), we consider the derivative of the finite sum $\sum_{\nu=1}^{m} A_\nu |\cos - c_\nu|$, where $|c_\nu| < 1$, $(\nu = 1, 2, \ldots, m)$, then, just as in section 7.1.43, we show that *the exact order of decrease of best approximations* $E_n^*(f)_{L_q}$ *of the function*

$$f(t) = \sum_{\nu=1}^{m} A_\nu (\cos t - c_\nu)^{r-1} |\cos t - c_\nu| \tag{27}$$

or of the best approximations $E_n(F; -1, 1)$ *of the function*

$$F(x) = \sum_{\nu=1}^{m} A_\nu (x - c_\nu)^{r-1} |x - c_\nu| \tag{28}$$

is, for every fixed m and $|c_\nu| < 1 (\nu = 1, 2, \ldots, m)$ *equal to* $\dfrac{1}{n^{r+1/q}}$

7.1.45. The examples considered, despite their specialised character, may be used to investigate the exact order of decrease of the best approximations of functions with this type of singularity. This remark is based on the obvious fact, that for every function $f(x)$ of the form

$$f(x) = \phi(x) + \psi(x) \tag{29}$$

where $E_n(\psi) = o[E_n(\phi)]$, the best approximation of $f(x)$, as $n \to \infty$, asymptotically coincides with the best approximation of $\phi(x)$.

The last remark is, in some cases, useful also for investigating of the order of decrease of best approximations of functions with various singularities of another character. Thus if for instance

$$f(x) = \frac{1}{x - c} + \psi(x) \quad (c > 1) \tag{30}$$

where $\psi(x)$ is an arbitrary function, analytic in some elliptical domain containing the point c, then by 5.4.1, as $n \to \infty$, the best approximation $E_n(f; -1, 1)$ equals $\dfrac{1}{(c^2-1)\left(c + \sqrt{(c^2-1)}\right)^n}$ (see 2.11(4)), neglecting terms of higher order.

7.1.5. To complete this section, we shall consider another convenient criterion for the exact order of decrease of best approximations of periodic functions by trigonometric polynomials, in the cases when these functions are defined by their Fourier series**. *For an arbitrary function* $f(x)$ *with the Fourier series*

$$f(x) = \sum_{k=1}^{\infty} (a_k \cos n_k x + b_k \sin n_k x) \tag{31}$$

** See also section 2.11.2.

where $\frac{n_{k+1}}{n_k} \geqslant \lambda > 1$ $(k = 1, 2, \ldots)$** *the order of decrease of the sequence of best uniform approximations* $E_n^*(f)$, *as* $n \to \infty$, *coincides with the order of* $R_n(f) = \sum_{k=\nu+1}^{\infty} \varrho_k$, *where* $n_\nu \leqslant n < n_{\nu+1}$ and $\varrho_k = \sqrt{a_k^2 + b_k^2}$***.

Since for $n_\nu \leqslant n < n_{\nu+1}$ we have

$$E_n^*(f) \leqslant r_n(f) = \max_x |f(x) - S_n^*(f; x)| \leqslant R_n(f), \tag{32}$$

where $S_n(f; x)$ is the partial sum of n-th order of the Fourier series of the function $f(x)$, it follows that for every such function the partial sums of the Fourier series have the deviation of the same order as the polynomials of the best uniform approximation.

The proof of the above statement is based on the fact**** that *for every function* $f(x)$ *with Fourier series* 7.1(31) *satisfying the condition* $\frac{n_{k+1}}{n_k} \geqslant \lambda > 1$ $(k = 1, 2, \ldots)$ *we have*

$$R_1(f) \leqslant C(\lambda) \max_x |f(x)| \tag{33}$$

for some constant $C(\lambda)$ *depending only on* λ.

Inequality 7.1(33) shows that for every n

$$R_n(f) \leqslant C(\lambda) r_n(f). \tag{34}$$

Besides, since

$$r_{n_\nu}(f) = \max_x |f(x) - S_{n_\nu}(f; x)|$$

$$= \max_x \left| f(x) - \frac{S_{n_\nu}^*(f; x) + \ldots + S_{n_{\nu+1}-1}^*(f : x)}{n_{\nu+1} - n_\nu} \right|,$$

we have for $\nu = 1, 2, \ldots$ (see inequality 8.4(5)):

$$r_{n_\nu}(f) \leqslant \left[\frac{4}{\pi^2} \ln \frac{n_{\nu+1}}{n_{\nu+1} - n_\nu} + O(1)\right] E_{n_\nu}^*(f)$$

$$\leqslant \left[\frac{4}{\pi^2} \ln \frac{\lambda}{\lambda+1} + O(1)\right] E_{n_\nu}^*(f) \leqslant C_1(\lambda) E_{n_\nu}^*(f)$$

and, consequently,

$$R_{n_\nu}(f) \leqslant C(\lambda) C_1(\lambda) E_{n_\nu}^*(f).$$

In the case $n_\nu < n < n_{\nu+1}$:

$$R_n(f) = \varrho_{\nu+1} + R_{n_{\nu+1}}(f).$$

** Such trigonometric series are called lacunary (or trigonometric series with gaps. Translator).

*** S. B. Stechkin [4].

**** S. Szidon [1]. The proof of inequality 7.1(33) may be found in the well-known monograph of A. Zygmund [1].

Thus, if some $\gamma > 1$, $\varrho_{\nu+1} \leqslant \gamma R_{n_{\nu+1}}(f)$, then

$$R_n(f) \leqslant (1+\gamma) R_{n_{\nu+1}}(f) \leqslant (1+\gamma) C(\lambda) C_1(\lambda) E^*_{n_{\nu+1}}(f) \leqslant (1+\gamma) C(\lambda) C_1(\lambda) E^*_n(f).$$

If, however, $\varrho_{\nu+1} > \gamma R_{n_{\nu+1}}(f)$, then the difference

$$f(x) - \sum_{k=1}^{\nu} (a_k \cos n_k x + b_k \sin n x) = \sum_{k=\nu+1}^{\infty} (a_k \cos n_k x + b_k \sin n_k x) = \sum_{k=\nu+1}^{\infty} \varrho_k \cos(n_k x - \alpha_k)$$

at $2n_{\nu+1} \geqslant 2n+2$ points $x_r = \dfrac{r\pi + \alpha_{\nu+1}}{n_{\nu+1}}$ $(r = 1, 2, \ldots, 2n_{\nu+1})$ inside the period of the function, assumes with the consecutive change of sign values greater than $\dfrac{\gamma-1}{\gamma} \varrho_{\nu+1}$ in absolute value. Consequently, by theorem 2.10, the best approximation $E^*_n(f)$ of the function $f(x)$ by trigonometric polynomials of order $\leqslant n$ satisfies the inequality

$$E^*_n(f) > \frac{\gamma-1}{\gamma} \varrho_{\nu+1} > (\gamma-1) R_{n_{\nu+1}}(f) = (\gamma-1)[R_n(f) - \varrho_{\nu+1}] > (\gamma-1) R_n(f) - \gamma E^*_n(f)$$

i.e.

$$R_n(f) < \frac{\gamma+1}{\gamma-1} E^*_n(f).$$

Thus, in both cases for all n

$$R_n(f) \leqslant C_2(\lambda) E^*_n(f). \tag{35}$$

The inequality obtained, together with 7.1(32), proves the theorem.

7.2. Asymptotic properties of the best uniform approximation of some simple functions with singularities

The basic idea of the proof of the classical theorem of Weierstrass, presented in section 1, 1, is based on replacing the graph of the continuous function approximated on $[a, b]$ by a polygonal line inscribed in it, i.e. by the graph of some function of the type

$$Ax + B + \sum_{k=1}^{m} A_k |x - c_k|, \tag{1}$$

where $a < c_1 < c_2 < c_3 < \ldots < c_m < b$.

Since the sum 7.2(1) represents a simple continuous function which is not a polynomial, it is natural, that in the constructive theory of functions an important problem** arises, concerning the character of the behaviour of the sequence of its best uniform approximations by polynomials. We have already seen (see sections 7.1.41 and 7.1.44) that this sequence decreases fairly regularly and in respect of order always behaves like the sequence of numbers $\frac{1}{n}$ $(n = 1, 2, 3, \ldots)$. It turns out that the n-th best approximation of the function 7.2(1) is identical with the quantity $\frac{1}{n}$ not only in order, but apart from a constant factor it is asymptotically (as $n \to \infty$) equal to it. The sequences of best approximations of the other elementary functions considered in section 7.1.4 also possess similar asymptotic properties.

7.2.1.*** *Whatever the natural number r and $-1 < \alpha < 1$, for the best approximation of the function $x^{r-1}|x|^{1+\alpha}$ by algebraic polynomials on $[-1, 1]$ the limit*

$$\lim_{n\to\infty} n^{r+\alpha} E_n(x^{r-1}|x|^{1+\alpha}; -1, 1) = \mu(r, \alpha) > 0, \tag{2}$$

exists, i.e. $E_n(x^{r-1}|x|^{1+\alpha}; -1, 1)$ is asymptotically equal to the quantity $\dfrac{\mu(r, \alpha)}{n^{r+\alpha}}$ as $n \to \infty$.

Let us first note that if for some $k > 0$ the continuous function $f(x)$ defined on $(-\infty, \infty)$, satisfies the condition $f(x) = O(|x|^k)$, then for any $\sigma > 0$ there exists an integral function $g_\sigma(x)$ of degree $\leqslant \sigma$, which possesses the property

$$\sup_{-\infty<x<\infty} |f(x)-g_\sigma(x)| < \infty,$$

i.e. $A_\sigma(f) < \infty$ for any $\sigma > 0$. Such a function $g_\sigma(x)$ is, for example, the function $Q_\sigma(f; x)$ defined in section 5.1.3, where $r \geqslant \frac{1}{2}(k+2)$.

As a result of this, for all $\sigma > 0$ the relation 2.6(8) (see section 5.4.6) is applicable to the function $f(x) = x^{r-1}|x|^{1+\alpha}$ considered here. In particular,

$$\lim_{\varepsilon\to 0} \overline{\lim_{n\to\infty}}\, E_n\left(x^{r-1}|x|^{1+\alpha}; -\frac{n}{1+\varepsilon}, \frac{n}{1+\varepsilon}\right) = A_1(x^{r-1}|x|^{1+\alpha}). \tag{3}$$

But as

$$E_n(x^{r-1}|x|^{1+\alpha}; -\lambda, \lambda) = |\lambda|^{r+\alpha} E_n(x^{r-1}|x|^{1+\alpha}; -1, 1),$$

** This problem, in the case $m = 1$, $c_1 = 0$, was first considered by Vallée-Poussin in 1908.

*** S. N. Bernstein [18] for r odd. See also S. N. Bernstein [29], I. I. Ibragimov [2].

it follows that

$$\lim_{n\to\infty} n^{r+\alpha}E_n(x^{r-1}|x|^{1+\alpha}; -1, 1) = A_1(x^{r-1}|x|^{1+\alpha}). \tag{4}$$

This proves the existence of the limit 7.2(2) and also establishes that $\mu(r, \alpha) = A_1(x^{r-1}|x|^{1+\alpha})$.

7.2.2. Theorem 7.2.1 can be generalised as follows**. *If* $-1 < c < 1$, *then whatever the natural number r and* $-1 < \alpha < 1$, *for the best approximation of the function* $f(x) = (x-c)^{r-1}|x-c|^{1+\alpha}$ *by algebraic polynomials on* $[-1, 1]$ *as* $n\to\infty$ *the limit of the product*

$$n^{r+\alpha}E_n[(x-c)^{r-1}|x-c|^{1+\alpha}; -1, 1]$$

exists and

$$\lim_{n\to\infty} n^{r+\alpha}E_n[(x-c)^{r-1}|x-c|^{1+\alpha}; -1, 1] = (\sqrt{1-c^2})^{r+\alpha}\mu(r, \alpha), \tag{5}$$

where $\mu(r, \alpha)$ *is the constant of* 7.2(2).

Following S. N. Bernstein [20], let us divide the proof into several stages.

7.2.21. *For any* $\delta > 0$ *the inequality*

$$E_{m-1}(x^{r-1}|x|^{1+\alpha}; a, b) \leqslant \frac{(1+\delta)^m+1}{(1+\delta)^{m-r-\alpha}-1}E_m(x^{r-1}|x|^{1+\alpha}; a, b) \tag{6}$$

is satisfied for all integral m on the segments $[a, b]$, *where* $a \leqslant 0 \leqslant b$. *If*

$$\max_{a\leqslant x\leqslant b}\left|x^{r-1}|x|^{1+\alpha}-\sum_{k=0}^{m} a_k x^k\right| = E_m(x^{r-1}|x|^{1+\alpha}; a, b),$$

then for any $\delta > 0$

$$\max_{a\leqslant x\leqslant b}\left|\left(\frac{x}{1+\delta}\right)^{r-1}\left|\frac{x}{1+\delta}\right|^{1+\alpha} - \sum_{k=0}^{m} a_k\left(\frac{x}{1+\delta}\right)^k\right| \leqslant E_m(x^{r-1}|x|^{1+\alpha}; a, b),$$

i.e.

$$\max_{a\leqslant x\leqslant b}\left|x^{r-1}|x|^{1+\alpha}(1+\delta)^{m-r-\alpha}-\sum_{k=0}^{m}\frac{a_k}{(1+\delta)^{k-m}}x^k\right|$$
$$\leqslant (1+\delta)^m E_m(x^{r-1}|x|^{1+\alpha}; a, b).$$

Consequently there exists a polynomial $\sum_{k=0}^{m-} d_k x^k$ such that

$$\max_{a\leqslant x\leqslant b}\left|x^{r-1}|x|^{1+\alpha}\{(1+\delta)^{m-r-\alpha}-\}-\sum_{k=0}^{m-1} d_k x^k\right|$$
$$\leqslant \{(1+\delta)^m+1\}E_m(x^{r-1}|x|^{1+\alpha}; a, b)$$

** S. N. Bernstein [20], *r* odd.

Hence we obtain 7.2(6). It is also obvious that if $-1 < c < 1$ then

$$E_{m-1}[(x-c)^{r-1}|x-c|^{1+\alpha}; -1, 1] = E_{m-1}(x^{r-1}|x|^{1+\alpha}; -1-c, 1-c)$$

$$\leqslant \frac{(1+\delta)^m+1}{(1+\delta)^{m-r-\alpha}-1} E_m(x^{r-1}|x|^{1+\alpha}; -1-c, 1-c)$$

$$= \frac{(1+\delta)^m+1}{(1+\delta)^{m-r-\alpha}-1} E_m[(x-c)^{r-1}|x-c|^{1+\alpha}; -1, 1]. \qquad (7)$$

7.2.22. Let us denote by $E_{n,q(x)}(f; a, b)$ the best uniform approximation of the function $f(x)$ by algebraic polynomials $P_n(x)$ of degree $\leqslant n$ on $[a, b]$ with weight $q(x)$, i.e.

$$E_{n,q(x)}(f; a, b) = \inf_{P_n(x)} \max_{a\leqslant x\leqslant b} |q(x)|\,|f(x)-P_n(x)|, \qquad (8)$$

and let us prove that if $-1 \leqslant a \leqslant 0 < b \leqslant 1$, *then*

$$E_{n,n^{x^2}}(x^{r-1}|x|^{1+\alpha}; a, b) < (1+\tau_n)E_n(x^{r-1}|x|^{1+\alpha}; a, b), \qquad (9)$$

where $\tau_n \to 0$ *as* $n\to\infty$.

For this purpose, putting $h = \ln n$ and $k = [\ln^2 n]$, let us consider the polynomials

$$R_{2k}(x) = \sum_{\nu=0}^{k} \frac{h^\nu x^{2\nu}}{\nu!}, \qquad Q_{2k}(x) = \sum_{\nu=0}^{k} (-1)^\nu \frac{h^\nu x^{2\nu}}{\nu!}.$$

It is obvious that

$$n^{x^2} = R_{2k}(x)+\varepsilon_k(x), \qquad n^{-x^2} = Q_{2k}(x)+\varepsilon_k^*(x),$$

where for $|x| \leqslant 1$ and $n > e^2$ (see 3.7(2))

$$|\varepsilon_k(x)| = \sum_{\nu=k+1}^{\infty} \frac{h^\nu x^{2\nu}}{\nu!} \leqslant h^{k+1}\sum_{\nu=1}^{\infty} \frac{h^{\nu-1}}{(k+\nu)!}$$

$$< \frac{h^{k+1}}{(k+1)!}\sum_{\nu=0}^{\infty}\left(\frac{1}{\ln n}\right)^\nu < \frac{2h^{k+1}}{(k+1)!} < B\left(\frac{he}{k+1}\right)^{k+1},$$

and similarly

$$|\varepsilon_k^*(x)| < B\left(\frac{he}{k+1}\right)^{k+1},$$

where B is some constant. Consequently

$$|1-R_{2k}(x)\,Q_{2k}(x)| = |\,\varepsilon_k(x)\,e^{-hx^2}+\varepsilon_k^*(x)\,R_{2k}(x)|$$

$$< |\varepsilon_k(x)|\,e^{-hx^2}+|\varepsilon_k^*(x)|\,e^{hx^2} < 2Be^h\left(\frac{he}{k+1}\right)^{k+1} < 2Bn\left(\frac{e}{\ln n}\right)^{\ln^2 n} = \beta_n,$$

where $\beta_n \to 0$ as $n \to \infty$. Hence

$$E_{n, R_{2k}(x)Q_{2k}(x)}(x^{r-1}|x|^{1+\alpha}; a, b) < (1+\beta_n)\, E_n(x^{r-1}|x|^{1+\alpha}; a, b). \tag{10}$$

But if $m = n+2k$, then

$$E_{n, R_{2k}(x)Q_{2k}(x)}(x^{r-1}|x|^{1+\alpha}; a, b) > E_{m, R_{2k}(x)}(x^{r-1}|x|^{1+\alpha}Q_{2k}(x); a, b) \tag{11}$$

and, in addition,

$$E_{m, R_{2k}(x)}(x^{r-1}|x|^{1+\alpha}; a, b) < E_{m, R_{2k}(x)}(x^{r-1}|x|^{1+\alpha}Q_{2k}(x); a, b) + \\ + \sum_{\nu=1}^{k} \frac{h^\nu}{\nu!} E_{m, R_{2k}(x)}(x^{r+2\nu-1}|x|^{1+\alpha}; a, b). \tag{12}$$

Since on the segment considered $|R_{2k}(x)| < n$, it follows that

$$E_{m, R_{2k}(x)}(x^{r+2\nu-1}|x|^{1+\alpha}; a, b) < nE_m(x^{r+2\nu-1}|x|^{1+\alpha}; a, b).$$

The function $f(x) = x^{r+2\nu-1}|x|^{1+\alpha}$ has a derivative of order $p_\nu = r+2\nu+[\alpha]$, which on $[-1, 1]$ satisfies the condition

$$\omega(f^{(p_\nu)}; t) < B_0^{r+2\nu}(r+2\nu)^{r+2\nu}t^{\alpha-[\alpha]}, \tag{13}$$

where B_0 is some constant not depending on ν. Therefore, if use is made of the inequality 5.2(15), we obtain

$$E_m(f; a, b) < \frac{(2C)^{p_\nu}}{(m-p_\nu+1)(m-p_\nu+2)\dots m} E_{m-p_\nu}(f^{(p_\nu)}; a, b) \\ < \frac{2^{p_\nu}C^{p_\nu+1}p_p^{p_\nu}}{p_\nu!\, m^{p_\nu}} B_0^{r+2\nu}(r+2\nu)^{r+2\nu}\frac{1}{(m-p_\nu)^{\alpha-[\alpha]}} \tag{14}$$

i.e.

$$E_m(x^{r+2\nu-1}|x|^{1+\alpha}; a, b) < \frac{(2B_0^* C)^{r+2\nu+1}(r+2\nu)^{r+2\nu}}{n^{r+2\nu+\alpha}}\left(\frac{n}{n-r}\right)^{\alpha-[\alpha]}.$$

Consequently, for sufficiently great values of n

$$E_m(x^{r+2\nu-1}|x|^{1+\alpha}; a, b) < \frac{B_1^{r+2\nu+1}(r+2\nu)^{r+2\nu}}{n^{r+2\nu+\alpha}}$$

where B_1 is a constant not depending on ν.

Thus

$$E_{m, R_{2k}(x)}(x^{r+2\nu-1}|x|^{1+\alpha}; a, b) < \frac{B_1^{r+2\nu+1}(r+2\nu)^{r+2\nu}}{n^{r+2\nu+\alpha-1}},$$

and hence, taking into account that $h \ln n$ and $k = [\ln^2 n]$, we have

$$\sum_{\nu=1}^{k} \frac{h^\nu}{\nu!} E_{m, R_{2k}(x)}(x^{r+2\nu-1}|x|^{1+\alpha}; a, b) < \sum_{\nu=1}^{k} \frac{h^\nu}{\nu!} \frac{B_1^{r+2\nu+1}(\nu+2)^{r+2\nu}}{n^{r+2\nu+\alpha-1}}$$

$$= \frac{h}{n^{r+\alpha+1}} \sum_{\nu=1}^{k} \left(\frac{h}{n^2}\right)^{\nu-1} \frac{B_1^{r+2\nu+1}(r+2\nu)^{r+2\nu}}{\nu!}$$

$$< \frac{B_2 \ln n}{n^{r+\alpha+1}} \sum_{\nu=1}^{k} \left(\frac{B_3 \ln^2 n}{n}\right)^{2\nu-2} < \frac{B_4 \ln n}{n^{r+\alpha+1}},$$

where B_2, B_3 and B_4 are certain constants. On substituting these estimates in 7.2(12), we obtain the inequality

$$E_{m, R_{2k}(x)}(x^{r-1}|x|^{1+\alpha}; a, b)$$
$$< E_{m, R_{2k}(x)}[x^{r-1}|x|^{1+\alpha}Q_{2k}(x); a, b] + \frac{B_4 \ln n}{n^{r+\alpha+1}},$$

from which in virtue of 7.2(10) and 7.2(11)

$$E_{m, R_{2k}(x)}(x^{r-1}|x|^{1+\alpha}; a, b)$$
$$< (1+\beta_n) E_n(x^{r-1}|x|^{1+\alpha}; a, b) + \frac{B_4 \ln n}{n^{r+\alpha+1}}, \qquad (15)$$

where $\beta_n \to 0$ as $n \to \infty$. Since for $a < 0 < b$ by theorem 7.2.1

$$\frac{\ln n}{n^{r+\alpha+1}} = o\{E_n(x^{r-1}|x|^{1+\alpha}; a, b)\},$$

when $n \to \infty$, the inequality 7.2(15) indicates that

$$E_{m, R_{2k}(x)}(x^{r-1}|x|^{1+\alpha}; a, b) < (1+\gamma_n) E_n(x^{r-1}|x|^{1+\alpha}; a, b), \qquad (16)$$

where $\gamma_n \to 0$ as $n \to \infty$.

In order to obtain 7.2.(9) from this, let us make use of the inequality 7.2(6), substituting in it $(1+\delta)^{n-r-\alpha} = 2m$. Then for sufficiently great m we obtain

$$E_{m-1}(x^{r-1}|x|^{1+\alpha}; a, b) < \frac{(2m)^{\overline{\frac{m}{m-r-\alpha}}}+1}{2m-1} E_m(x^{r-1}|x|^{1+\alpha}; a, b)$$

$$< \frac{2m+1}{2m-1}\left[1+\frac{r+a}{m-r-a} 2\ln(2m)\right] E_m(x^{r-1}|x|^{1+\alpha}; a, b).$$

Hence, since $k = [\ln^2 n]$, we obtain the estimate

$$\begin{aligned} &E_n(x^{r-1}|x|^{1+\alpha}; a, b) \\ &\quad < \frac{2m+1}{2n+1} \prod_{\nu=1}^{2k} \left[1+\frac{r+\alpha}{n+\nu-r-\alpha} 2\ln 2(n+\nu)\right] E_n(x^{r-1}|x|^{1+\alpha}; a, b) \\ &\quad < \frac{2m+1}{2n+1}\left[1+\frac{r+\alpha}{n-r-\alpha} 2\ln(2m)\right]^{2k} E_m(x^{r-1}|x|^{1+\alpha}; a, b) \\ &\quad = (1+\eta_n) E_m(x^{r-1}|x|^{1+\alpha}; a, b), \end{aligned}$$

where $\eta_k \to 0$ as $n > \infty$. If we compare this inequality with 7.2(16), then for sufficiently great n we shall have

$$E_{m, R_{2k}(x)}(x^{r-1}|x|^{1+\alpha}; a, b) < (1+\eta_n)(1+\gamma_n) E_m(x^{r-1}|x|^{1+\alpha}; a, b).$$

Thus

$$\begin{aligned} &E_{m, m^{x^2}}(x^{r-1}|x|^{1+\alpha}; a, b) \\ &\qquad \leqslant E_{m, R_{2k}(x)}(x^{r-1}|x|^{1+\alpha}; a, b) \max_{|x|\leqslant 1} \frac{m^{x^2}}{R_{2k}(x)} \\ &\qquad < (1+\eta_n)(1+\gamma_n) E_m(x^{r-1}|x|^{1+\alpha}; a, b) \frac{m}{n} \max_{|x|\leqslant 1} \frac{n^{x^2}}{n^{x^2}-\varepsilon_k(x)} \\ &\qquad = (1+\tau_n) E_m(x^{r-1}|x|^{1+\alpha}; a, b), \end{aligned}$$

i.e. 7.2(9) is satisfied.

7.2.23. *For any arbitrarily small values of $\beta > 0$ and $\varepsilon > 0$ it is possible to find an N, such that for every $n \geqslant N$ there exists an algebraic polynomial $P_n(x)$ of degree $\leqslant n$ which satisfies the inequality*

$$\left|P_n(x)-(x-c)^{r-1}|x-c|^{1+\alpha}\right| < (1+\beta) E_n[(x-c)^{r-1}|x-c|^{1+\alpha}; -1, 1] \quad (17)$$

on the whole of the segment $[-1, 1]$ and the inequality

$$|P_n(x)-(x-c)^{r-1}|x-c|^{1+\alpha}| < \beta E_n[(x-c)^{r-1}|x-c|^{1+\alpha}; -1, 1] \quad (18)$$

if $|x-c| > \varepsilon$.

This assertion follows directly from section 7.2.22. In the case where $c > 0$, let us consider the segment $[-1, b]$, where $b = \dfrac{1-c}{1+c}$, and put $\delta = \dfrac{\varepsilon}{1+c}$.

From the inequality 7.2(9) follows the existence of polynomials $P_n^*(y)$ such that

$$|y^{r-1}|y|^{1+\alpha}-P_n^*(y)| < (1+\tau_n) E_n(y^{r-1}|y|^{1+\alpha}; -1, b)$$

on the whole of the segment $-1 \leqslant y \leqslant b$ and

$$|y^{r-1}|y|^{1+\alpha} - P_n^*(y)| < \frac{1+\tau_n}{n^{\delta^2}} E_n(y^{r-1}|y|^{1+\alpha}; -1, b)$$

for $|y| \geqslant \delta$. After the substitution $y = \dfrac{x-c}{1+c}$, we obtain the inequality

$$\left|(x-c)^{r-1}|x-c|^{1+\alpha} - (1+c)^{r+\alpha} P_n^*\left(\frac{x-c}{1+c}\right)\right|$$
$$< (1+\tau_n) E_n[(x-c)^{r-1}|x-c|^{1+\alpha}; -1, 1]$$

on the segment $-1 \leqslant x \leqslant 1$ and the inequality

$$\left|(x-c)^{r-1}|x-c|^{1+\alpha} - (1+c)^{r+\alpha} P_n^*\left(\frac{x-c}{1+c}\right)\right|$$
$$< \frac{1+\tau_n}{n^{\delta^2}} E_n[(x-c)^{r-1}|x-c|^{1+}; -1, 1]$$

for $|x-c| > \delta(1+c) = \varepsilon$. It remains to choose N so that for $n \geqslant N$ we have $\tau_n < \beta$ and $1+\beta < \beta n^{\left(\frac{\varepsilon}{1+c}\right)^2}$.

If, however, $c < 0$, then in exactly the same way we apply the proposition of 7.2.22 to the segment $[a, 1]$, where $a = -\dfrac{1+c}{1-c}$, and put $\delta = \dfrac{\varepsilon}{1-c}$ and $y = \dfrac{x-c}{1-c}$.

7.2.24. Let $-1 < a < c < b < 1$. Denote by $E_n(f; a, l; M)$ the best approximation of the function $f(x)$ on $[a, b]$ by the algebraic polynomials $P_n(x)$ of degree $\leqslant n$, which for $-1 \leqslant x \leqslant a$ and $b \leqslant n \leqslant 1$ satisfy the condition $|P_n(x)| \leqslant M$.

Let us prove that *if* $c-a \geqslant \dfrac{1}{n^\sigma}$, $b-c \geqslant \dfrac{1}{n^\sigma}$, $0 < \sigma < \frac{1}{2}$, *then, whatever* $h > 0$ *may be, for any arbitrarily small* $\varepsilon > 0$ *it is possible to find an* N *such that for all* $n \geqslant N$

$$E_n[(x-c)^{r-1}|x-c|^{1+\alpha}; a, b; n^h]$$
$$> (1-\varepsilon) E_n[(x-c)^{r-1}|x-c|^{1+\alpha}; -1, 1]. \tag{19}$$

Let us put $\lambda = [n^\varrho]$, where $2\delta < \varrho < 1$, and

$$q(x) = \left\{1 - \left(\frac{x-c}{1+|c|}\right)^2\right\}^\lambda.$$

It is obvious that

$$E_{n,q(x)}[(x-c)^{r-1}|x-c|^{1+\alpha};-1,1] \geqslant E_m[(x-c)^{r-1}|x-c|^{1+\alpha}q(x);\ -1,1], \quad (20)$$

where $m = n+2\lambda$, and

$$E_m[(x-c)^{r-1}|x-c|^{1+\alpha}q(x);\ -1,1]$$

$$\geqslant E_m[(x-c)^{r-1}|x-c|^{1+\alpha};\ -1,1]-$$

$$-\sum_{\nu=1}^{\lambda}\binom{\lambda}{\nu}E_m\left[(x-c)^{r-1}|x-c|^{1+\alpha}\left(\frac{x-c}{1+|c|}\right)^{2\nu};\ -1,1\right]. \quad (21)$$

The function $f(x) = (x-c)^{r+2\nu-1}|x-c|^{1+\alpha}$ has a derivative of order $p_\nu = r+2\nu-[\alpha]$, which on $[-1,1]$ satisfies condition 7.2(13).

Hence in virtue of 7.2(14) for $n \geqslant r$ we have

$$E_m\left[(x-c)^{r-1}|x-c|^{1+\alpha}\left(\frac{x-c}{1+|c|}\right)^{2\nu};\ -1,1\right] < \frac{(2B_0c)^{r+2\nu+1}(r+2\nu)^{2\nu+4\nu}}{(r+2\nu)!\,m^{p_\nu}}\,\frac{1}{(m-p_\nu)^{\alpha-[\alpha]}}.$$

Consequently, for sufficiently great values of n

$$E_m\left[(x-c)^{r-1}|x-c|^{1+\alpha}\left(\frac{x-c}{1+|c|}\right)^{2\nu};\ -1,1\right] < \frac{B_1^{r+2\nu}(r+2\nu)^{r+2\nu}}{m^{r+2\nu+\alpha}},$$

where B_1 is a constant which does not depend on ν.

Thus

$$\sum_{\nu=1}^{\lambda}\binom{\lambda}{\nu}E_m\left[(x-c)^{r-1}|x-c|^{1+\alpha}\left(\frac{x-c}{1+|c|}\right)^{2\nu};\ -1,1\right] < \frac{1}{m^{r+\alpha}}\sum_{\nu=1}^{\lambda}\binom{\lambda}{\nu}\frac{B_1^{r+2\nu}(r+2\nu)^{r+2\nu}}{m^{2\nu}}.$$

Here

$$\sum_{\nu=1}^{\lambda}\binom{\lambda}{\nu}\frac{B_1^{r+2\nu}(r+2\nu)^{r+2\nu}}{m^{2\nu}} < \sum_{\nu=1}^{\lambda}\binom{\lambda}{\nu}\left(\frac{C_1\nu}{m}\right)^{2\nu}$$

$$< C_0\sum_{\nu=1}^{\lambda-1}\frac{\lambda^\lambda}{\nu^\nu(\lambda-\nu)^{\lambda-\nu}}\left(\frac{C_1\nu}{m}\right)^{2\nu}+\left[O\left(\frac{\lambda}{m}\right)\right]^{2\lambda},$$

where C_0 and C_1 are certain constants, and

$$\sum_{\nu=1}^{\lambda-1} \frac{\lambda^\lambda}{\nu^\nu(\lambda-\nu)^{\lambda-\nu}} \left(\frac{C_1\nu}{m}\right)^{2\nu} = \sum_{\nu=1}^{\lambda-1} \left(\frac{\lambda}{\lambda-\nu}\right)^{\lambda-\nu} \left(\frac{C_1^2\lambda\nu}{m^2}\right)^{\nu}$$

$$< \sum_{\nu=1}^{\infty} \left(\frac{C_2\lambda^2}{m^2}\right)^{\nu} = O\left(\frac{\lambda^2}{m^2}\right).$$

Hence, since $\lambda = [n^\varrho]$ $(0 < \varrho < 1)$ and $m = n+2\lambda$,

$$\sum_{\nu=1}^{\lambda} \binom{\lambda}{\nu} E_m\left[(x-c)^{r-1}|x-c|^{1+\alpha}\left(\frac{x-c}{1+|c|}\right)^{2\nu}; -1, 1\right] = o\left(\frac{1}{m^{r+\alpha}}\right)$$

as $n \to \infty$. Since the exact order of decrease of $E_m[(x-c)^{r-1}|x-c|^{1+\alpha}; -1, 1]$, when $-1 < c < 1$, is equal to $\frac{1}{m^{r+\alpha}}$ (see section 7.1.41), then in virtue of 7.2(21)

$$E_m[(x-c)^{r-1}|x-c|^{1+\alpha}q(x); -1, 1]$$
$$> (1+\gamma_n)E_m[(x-c)^{r-1}|x-c|^{1+\alpha}; -1, 1], \quad (22)$$

where $\gamma_n \to 0$ as $n \to \infty$.

Let us now make use of the inequality 7.2(7), in which we put $(1+\delta)^{m-r+\alpha} = 2m$. Then as in section 7.2.22

$$E_{m-1}[(x-c)^{r-1}|x-c|^{1+\alpha}; -1, 1]$$

$$< \frac{(2m)^{\frac{m}{m-r-\alpha}}+1}{2m-1} E_m[(x-c)^{r-1}|x-c)|^{1+\alpha}; -1, 1]$$

$$< \frac{2m+1}{2m-1}\left[1+\frac{r+\alpha}{m-r-\alpha} 2\ln(2m)\right] E_m[(x-c)^{r-1}|x-c|^{1+\alpha}; -1,1]$$

and consequently

$$E_n[(x-c)^{r-1}|x-c|^{1+\alpha}; -1, 1]$$

$$< \frac{2m+1}{2m-1} \prod_{\nu=1}^{2\lambda} \left[1+\frac{r+\alpha}{n+\nu-r-\alpha} 2\ln 2(n+\nu)\right] \times$$
$$\times E_m[(x-c)^{r-1}|x-c|^{1+\alpha}; -1,1]$$
$$= (1+\tau_n)E_m[(x-c)^{r-1}|x-c|^{1+\alpha}; -1, 1],$$

where $\tau_n \to 0$ as $n \to \infty$.

The last estimate together with 7.2(20) and 7.2(22) leads to the inequality

$$E_{n,q(x)}[(x-c)^{r-1}|x-c|^{1+\alpha}; -1, 1]$$
$$> (1-\varepsilon_n)E_n[(x-c)^{r-1}|x-c|^{1+\alpha}; -1, 1], \quad (23)$$

in which $\varepsilon_n \to 0$ as $n \to \infty$.

If the polynomial $P_n(x)$ of degree $\leqslant n$ satisfies the condition $|P_n(x)| \leqslant n^h$ on the whole of the segment $[-1, 1]$, then when $|x-c| \geqslant \frac{1}{n^\sigma}$ for sufficiently great values of n

$$q(x)\left|P_n(x)-(x-c)^{r-1}|x-c|^{1+\alpha}\right| \leqslant \left[1-\frac{1}{(1+|c|)^2 n^{2\sigma}}\right]^\lambda (n^h+2^{r+\alpha}) < \frac{1}{n^{r+\alpha+1}}. \quad (24)$$

But since the exact order of decrease of the sequence $E_n[(x-c)^{r-1}|x-c|^{1+\alpha}; -1, 1]$ is equal to $\frac{1}{n^{r+\alpha}}$ (see section 7.1.41), then from 7.2(23) and 7.2(24), for sufficiently great n, at the points where $|x-c| \geqslant \frac{1}{n^\sigma}$ we obtain the inequality

$$q(x)|P_n(x)-(x-c)^{r-1}|x-c|^{1+\alpha}| < E_{n,\,q(x)}[(x-c)^{r-1}|x-c|^{1+\alpha}; -1, 1],$$

as a result of which, since $|q(x)| \leqslant 1$, we have for any polynomial $P_n(x)$ considered

$$\begin{aligned}(1-\varepsilon_n)E_n[(x-c)^{r-1}|x-c|^{1+\alpha}; -1; 1]& \\ < E_{n,\,q(x)}[(x-c)^{r-1}|x-c|^{1+\alpha}; -1, 1]& \\ \leqslant \max_{-1\leqslant x\leqslant 1} q(x)\left|(x-c)^{r-1}|x-c|^{1+\alpha}-P_n(x)\right|& \\ = \max_{a\leqslant x\leqslant b}\left|(x-c)^{r-1}|x-c|^{1+\alpha}-P_n(x)\right|q(x)& \\ \leqslant \max_{a\leqslant x\leqslant b}\left|(x-c)^{r-1}|x-c|^{1+\alpha}-P_n(x)\right|,&\end{aligned}$$

i.e.

$$(1-\varepsilon_n)E_n[(x-c)^{r-1}|x-c|^{1+\alpha}; -1, 1] < E_n[(x-c)^{r-1}|x-c|^{1+\alpha}; a, b; n^h].$$

7.2.25. Now let $f(t)$ be a continuous function of period 4π. Let us denote by $E_n^{**}(f)$ its best uniform approximation by trigonometric polynomials of the form

$$T_n^*(t) = \sum_{k=0}^{n}\left(a_k \cos k\frac{t}{2} + b_k \sin k\frac{t}{2}\right), \quad (25)$$

i.e. $E_n^{**}(f) = \inf_{T_n^*(t)} \max |f(t)-T_n^*(t)|$. It is obvious that in the case where $f(x)$ has period 2π we have $E_n^*(f) = E_{2n}^{**}(f)$ (see section 2.6.21).

Let us also denote by $E_n^{**}(f; a, b; M)$ the best uniform approximation on $[a, b](-2\pi < a < b < 2\pi)$ of the function $f(t)$ by trigonometric polynomials $T_n^*(t)$ of the type 7.2(25), which for $-2\pi \leqslant t \leqslant a$ and $b \leqslant t \leqslant 2\pi$

satisfy the condition $|T_n^*(t)| \leqslant M$. It follows from 7.2.24 after the substitution $x = \sin\frac{t}{2}$ that, *whatever* $h > 0$ *may be, for* $0 < \sigma < \frac{1}{2}$ *and* $a \geqslant \frac{1}{n^\sigma}$ $(a < 2\pi)$ *we have*

$$E_n^{**}\left[\left(\sin\frac{t}{2}\right)^{r-1}\left|\sin\frac{t}{2}\right|^{1+\alpha}; -a, a; n^h\right]$$
$$> (1-\varepsilon_n)\, E_n^{**}\left[\left(\sin\frac{t}{2}\right)^{r-1}\left|\sin\frac{t}{2}\right|^{1+\alpha}\right], \tag{26}$$

where $\varepsilon_n \to 0$ *when* $n \to \infty$.

In addition, we see from 7.2.23 *that for any arbitrarily small values of* $\beta > 0$ *and* $\varepsilon > 0$ *it is possible to find an* $N(\varepsilon, \beta)$ *such that for every* $n \geqslant N$ *there exists a polynomial* $T_n^*(t)$ *of the type* 7.2(25) *which satisfies the inequality*

$$\left|\left(\sin\frac{t}{2}\right)^{r-1}\left|\sin\frac{t}{2}\right|^{1+\alpha} - T_n^*(t)\right| < (1+\beta)\, E_n^{**}\left[\left(\sin\frac{t}{2}\right)^{r-1}\left|\sin\frac{t}{2}\right|^{1+\alpha}\right] \tag{27}$$

for all t and the inequality

$$\left|\left(\sin\frac{t}{2}\right)^{r-1}\left|\sin\frac{t}{2}\right|^{1+\alpha} - T_n^*(t)\right| < \beta E_n^{**}\left[\left(\sin\frac{t}{2}\right)^{r-1}\left|\sin\frac{t}{2}\right|^{1+\alpha}\right], \tag{28}$$

if $\left|\sin\frac{t}{2}\right| > \varepsilon$.

We make use of the last two remarks in order to prove theorem 7.2.2. In the first place we note that if $c = \cos t_0$ $\left(0 < t_0 < \frac{\pi}{2}\right)$, then

$$\begin{aligned} &E_n[(x-c)^{r-1}|x-c|^{1+\alpha}; -1, 1] \\ &\quad = E_n^*[(\cos t-\cos t_0)^{r-1}|\cos t-\cos t_0|^{1+\alpha}] \\ &\quad = E_{2n}^{**}[(\cos t-\cos t_0)^{r-1}|\cos t-\cos t_0|^{1+\alpha}] \\ &\quad = 2^{r+\alpha}E_{2n}^{**}\left[\left(\sin\frac{t-t_0}{2}\sin\frac{t+t_0}{2}\right)^{r-1}\left|\sin\frac{t-t_0}{2}\sin\frac{t+t_0}{2}\right|^{1+\alpha}\right]. \end{aligned}$$

Let r be odd and $s = r+\alpha$. Consider the identity

$$\left|\sin\frac{t-t_0}{2}\sin\frac{t+t_0}{2}\right|^s = (\sin t_0)^s\left[\left|\sin\frac{t-t_0}{2}\right|^s + \left|\sin\frac{t+t_0}{2}\right|^s\right] + \psi(t),$$

where

$$\psi(t) = \left[\left|\sin\frac{t-t_0}{2}\right|^s - (\sin t_0)^s\right]\left[\left|\sin\frac{t+t_0}{2}\right|^s - (\sin t_0)^s\right] - (\sin t_0)^{2s}.$$

Since the function $\psi(t)$ has a bounded derivative of order $r+1+[\alpha]$, then by Jackson's theorem (see section 5.1.41), $E_n^*(\psi) = O\left(\frac{1}{n^{s+1-\alpha+[\alpha]}}\right)$. Therefore

$$E_n[|x-c|^s; -1, 1]$$
$$= (2\sin t_0)^s E_n^*\left[\left|\sin\frac{t-t_0}{2}\right|^s + \left|\sin\frac{t+t_0}{2}\right|^s\right] + O\left(\frac{1}{n^{s+1-\alpha+[\alpha]}}\right), \tag{29}$$

$$E_n[(x-c)|x-c|^s; -1, 1]$$
$$= 2^{s+1}(\sin t_0)^s E_n^*\left[\sin\frac{t-t_0}{2}\sin\frac{t+t_0}{2}\left\{\left|\sin\frac{t-t_0}{2}\right|^s + \right.\right.$$
$$\left.\left. + \left|\sin\frac{t+t_0}{2}\right|^s\right\}\right] + O\left(\frac{1}{n^{s+2-\alpha+[\alpha]}}\right).$$

Let us verify the fact that as $n \to \infty$

$$E_n^*\left[\left|\sin\frac{t-t_0}{2}\right|^s + \left|\sin\frac{t+t_0}{2}\right|^s\right] = [1+o(1)]\, E_{2n}^{**}\left[\left|\sin\frac{t}{2}\right|^s\right], \tag{30}$$

$$E_n^*\left[\sin\frac{t-t_0}{2}\sin\frac{t+t_0}{2}\left\{\left|\sin\frac{t-t_0}{2}\right|^s + \left|\sin\frac{t+t_0}{2}\right|^s\right\}\right]$$
$$= \sin t_0[1+o(1)]\, E_{2n}^{**}\left[\sin\frac{t}{2}\left|\sin\frac{t}{2}\right|^s\right].$$

In fact, if $T_n^*(t) = T_{n,r,\alpha}^*(t)$ are the trigonometric polynomials 7.2(25), which satisfy the inequalities 7.2(27) and 7.2(28) for any arbitrarily small $\beta > 0$ and $\varepsilon < t_0 < t_0+\varepsilon < \frac{\pi}{2}$, when $n \geqslant N(\varepsilon, \beta)$, then for all t

$$\left|\left|\sin\frac{t-t_0}{2}\right|^s + \left|\sin\frac{t+t_0}{2}\right|^s - T_{2n,r,\alpha}^*(t-t_0) - T_{2n,r,\alpha}^*(t+t_0)\right|$$
$$< (1+2\beta)\, E_{2n}^{**}\left[\left|\sin\frac{t}{2}\right|^s\right],$$

and

$$\left|\sin\frac{t-t_0}{2}\sin\frac{t+t_0}{2}\left\{\left|\sin\frac{t-t_0}{2}\right|^s + \left|\sin\frac{t+t_0}{2}\right|^s\right\} - \right.$$
$$\left. - \sin\frac{t+t_0}{2}\, T_{2n,r+1,\alpha}^*(t-t_0) - \sin\frac{t-t_0}{2}\, T_{2n,r+1,\alpha}^*(t+t_0)\right|$$
$$< [\sin(t_0+\varepsilon)(1+\beta)+\beta]\, E_{2n}^{**}\left[\sin\frac{t}{2}\left|\sin\frac{t}{2}\right|^s\right],$$

i.e. as $n \to \infty$

$$E_n^*\left[\left|\sin\frac{t-t_0}{2}\right|^s+\left|\sin\frac{t+t_0}{2}\right|^s\right] \leqslant [1+o(1)]E_{2n}^{**}\left[\left|\sin\frac{t}{2}\right|^s\right],$$

$$E_n^*\left[\sin\frac{t-t_0}{2}\sin\frac{t+t_0}{2}\left\{\left|\sin\frac{t-t_0}{2}\right|^s+\left|\sin\frac{t+t_0}{2}\right|^s\right\}\right]$$

$$\leqslant \sin t_0[1+o(1)]E_{2n}^{**}\left[\sin\frac{t}{2}\left|\sin\frac{t}{2}\right|^s\right]. \quad (31)$$

On the other hand, if there existed an infinite sequence of ordinary trigonometric polynomials $\overline{T}_{n_k}(t)$ such that for some $\beta > 0$

$$\max_t\left|\left|\sin\frac{t-t_0}{2}\right|^s+\left|\sin\frac{t+t_0}{2}\right|^s-\overline{T}_{n_k}(t)\right| < (1-2\beta)\,E_{2n_k}^{**}\left[\left|\sin\frac{t}{2}\right|^s\right],$$

or

$$\max_t\left|\sin\frac{t-t_0}{2}\sin\frac{t+t_0}{2}\left\{\left|\sin\frac{t-t_0}{2}\right|^s+\left|\sin\frac{t+t_0}{2}\right|^s\right\}-\overline{T}_{n_k}(t)\right|$$

$$< \{\sin t_0(1-\beta)-\beta\}\,E_{2n_k}^{**}\left[\sin\frac{t}{2}\left|\sin\frac{t}{2}\right|^s\right],$$

then in a sufficiently small interval $t_0-\varepsilon_0 < t < t_0+\varepsilon_0$ we should have the inequality

$$\left|\overline{T}_{n_k}(t)-T^*_{2n_k,r,\alpha}(t+t_0)-\left|\sin\frac{t-t_0}{2}\right|^s\right| < (1-\beta)E_{2n_k}^{**}\left[\left|\sin\frac{t}{2}\right|^s\right]$$

or the inequality

$$\left|\overline{T}_{n_k}(t)-\sin\frac{t-t_0}{2}\,T^*_{2n_k,r+1,\alpha}(t+t_0)-\right.$$

$$\left.-\sin\frac{t-t_0}{2}\sin\frac{t+t_0}{2}\left|\sin\frac{t+t_0}{2}\right|^s\right| < \sin t_0(1-\beta)\,E_{2n_k}^{**}\left[\sin\frac{t}{2}\left|\sin\frac{t}{2}\right|^s\right].$$

Since

$$|\overline{T}_{n_k}(t)-T^*_{2n_k,r,\alpha}(t+t_0)| \leqslant 4$$

and

$$\left|\overline{T}_{n_k}(t)-\sin\frac{t-t_0}{2}\,T^*_{2n_k,r+1,\alpha}(t+t_0)\right| \leqslant 4,$$

it would follow from this that

$$E_{2n_k}^{**}\left[\left|\sin\frac{t-t_0}{2}\right|^s;\, t_0-\varepsilon_0,\, t_0+\varepsilon_0;\, 4\right] < (1-\beta)\,E_{2n_k}^{**}\left[\left|\sin\frac{t}{2}\right|^s\right],$$

or for sufficiently small values of ε_0

$$E^{**}_{2n_k+1}\left[\sin\frac{t-t_0}{2}\left|\sin\frac{t-t_0}{2}\right|^s;\ t_0-\varepsilon_0;\ t_0+\varepsilon_0;\ 4\right]$$
$$<\left(1-\frac{\beta}{2}\right)E^{**}_{2n_k}\left[\sin\frac{t}{2}\left|\sin\frac{t}{2}\right|^s\right];$$

but this is incompatible with 7.2(26).

Thus in virtue of the estimate 7.2(31)

$$E^*_n\left[\left|\sin\frac{t-t_0}{2}\right|^s+\left|\sin\frac{t+t_0}{2}\right|^s\right]=[1+o(1)]E^{**}_{2n}\left[\left|\sin\frac{t}{2}\right|^s\right]$$

and

$$E^*_n\left[\sin\frac{t-t_0}{2}\sin\frac{t+t_0}{2}\left\{\left|\sin\frac{t-t_0}{2}\right|^s+\left|\sin\frac{t+t_0}{2}\right|^s\right\}\right]$$
$$=\sin t_0[1+o(1)]E^{**}_{2n}\left[\sin\frac{t}{2}\left|\sin\frac{t}{2}\right|^s\right],$$

i.e. 7.2(30) is satisfied.

Thus by 7.2(29), as $n\to\infty$ and for any integer $r>0$

$$E_n[(x-c)^{r-1}|x-c|^{1+\alpha};\ -1,1]=|2\sin t_0|^{r+\alpha}[1+o(1)]\times$$
$$\times E^{**}_{2n}\left[\left(\sin\frac{t}{2}\right)^{r-1}\left|\sin\frac{t}{2}\right|^{1+\alpha}\right]+O\left(\frac{1}{n^{r+1[\alpha]}}\right). \tag{32}$$

Putting here $c=0$ and using theorem 7.2.1, we find that as $n\to\infty$

$$2^{r+\alpha}E^{**}_{2n}\left[\left(\sin\frac{t}{2}\right)^{r-1}\left|\sin\frac{t}{2}\right|^{1+\alpha}\right]=\frac{\mu(r,\alpha)}{n^{r+\alpha}}+o\left(\frac{1}{n^{r+\alpha}}\right). \tag{33}$$

If we substitute this estimate in 7.2(32), we obtain the limiting inequality 7.2(5).

7.2.26. Here let us note a remarkable corollary of theorem 7.2.2 in the case when $r=1$, $\alpha=0$. *If the vertex c of an angle is located on the segment $[-1,1]$, and its arms pass through given points with co-ordinates $(-1,a)$ and $(+1,b)$ $(a>0,\ b>0)$, then for all sufficiently great n the minimum of the best uniform approximation of the zig-zag line $a\,c\,b$ by polynomials of degree $\leqslant n$ on $[-1,1]$ is obtained when the point c is such that the angle of incidence is equal to the angle of reflection***.

In the general case the tangents of these angles are equal to $\dfrac{a}{1+c}$ and $\dfrac{b}{1-c}$.

** S. N. Bernstein [22].

The best approximation considered for $n \geqslant 1$ is not changed if to the function with a given graph is added the arbitrary linear function $\lambda(x-c)$. As a result of this** the best approximation considered of order n for $n \geqslant 1$ is identical with the best approximation of the function $\frac{a(1-c)+b(1+c)}{2(1-c^2)}|x-c|$, and consequently, when $n \to \infty$, it is asymptotically equal to the quantity

$$\frac{a(1-c)+b(1+c)}{2\sqrt{1-c^2}}\,\frac{\mu(1,0)}{n},$$

which assumes its least value when $c = \frac{a-b}{a+b}$, i.e. when $\frac{a}{1+c} = \frac{b}{1-c}$.

7.2.3. It is now not difficult to verify that for every function 7.2(1) the best approximation by algebraic polynomials of degree $\leqslant n$ on any finite segment $[a, b]$ is apart from a constant factor, asymptotically (as $n \to \infty$) equal to $\frac{1}{n}$.

Moreover, *if*

$$f(x) = \sum_{k=1}^{m} A_k(x-c_k)^{r-1}|x-c_k|^{1+\alpha} \tag{34}$$

where r is a natural number, $-1 < \alpha < 1$ and $-1 = c_0 < c_1 < c_2 < c_3 \dots$ $\dots < c_m < 1$, then as $n \to \infty$ the limit of the product $n^{r+\alpha}E_n(f; -1, 1)$ always exists and

$$\lim_{n\to\infty} n^{r+\alpha}E_n(f; -1, 1) = \max_{1\leqslant k\leqslant m} |A_k|\left(\sqrt{1-c_k^2}\right)^{r+\alpha}\mu(r, \alpha). \tag{35}$$

Let $P_{n,k}(x)(k = 1, 2, \dots, m)$ be an algebraic polynomial of degree $\leqslant n$ which satisfies inequalities 7.2(17) and 7.2(18), where $c = c_k$, for arbitrarily small $\beta > 0$ and $\varepsilon < \frac{1}{2}\min\limits_{0\leqslant\nu\leqslant n}(c_{\nu+1}-c_\nu)$, when $n \geqslant N_k(\varepsilon\beta)$. It is obvious that at any point $x \in [-1, 1]$

$$|f(x)-\sum_{k=1}^{m} A_kP_{n,k}(x)| < (1+m\beta)\max_{1\leqslant k\leqslant m}|A_k|E_n[(x-c_k)^{r-1}|x-c_k|^{r-1}|^{1+\alpha}, -1, 1],$$

i.e.

$$E_n(f; -1, 1) < [1+o(1)]\max_{1\leqslant k\leqslant m}|A_k|E_n[(x-c_k)^{r-1}|x-c_k|^{1+\alpha}; -1, 1]. \tag{36}$$

** Put $\lambda = \frac{a(1-c)-b(1+c)}{2(1-c^2)}$.

Contrary to the above assertion, suppose that there exists for some $\beta > 0$ an infinite sequence of polynomials $P_{n_\nu}(x)$ such that

$$\max_{-1\leqslant x\leqslant 1} |f(x)-P_{n_\nu}(x)| < (1-2\beta) \max_{1\leqslant k\leqslant m} |A_k| E_{n_\nu}[(x-c_k)^{r-1}|x-c_k|^{1+\alpha}; -1, 1].$$

Let us take into consideration the existence in virtue of 7.2.23 of the polynomials $Q_{n,j}(x)$ of degree $\leqslant n$ such that at the points $x \in (c_j-\varepsilon, c_j+\varepsilon)$ for sufficiently great values of n,

$$|f(x)-A_j(x-c_j)^{r-1}|x-c_j|^{1+\alpha}-Q_{n,j}(x)| < \beta \max_{1\leqslant k\leqslant m} |A_k| E_n[(x-c_k)^{r-1}|x-c_k|^{1+\alpha}; -1, 1],$$

and for these sufficiently great n let us choose j such that**

$$|A_j| E_n[(x-c_j)^{r-1}|x-c_j|^{1+\alpha}; -1, 1] = \max_{1\leqslant k\leqslant m} |A_k| E_n[(x-c_k)^{r-1}|x-c_k|^{1+\alpha}; -1, 1].$$

Then the polynomials $R_{n_\nu,j}(x) = P_{n_\nu}(x)-Q_{n_\nu,j}(x)$ possess the property that for $c_j-\varepsilon < x < c_j+\varepsilon$

$$\begin{aligned} &|R_{n_\nu,j}(x)-A_j(x-c_j)^{r-1}|x-c_j|^{1+\alpha}| \\ &\quad \leqslant |P_{n_\nu}(x)-f(x)| + |f(x)-A_j(x-c_j)^{r-1}|x-c_j|^{1+\alpha}-Q_{n_\nu,j}(x)| \\ &\quad \leqslant (1-\beta)\, |A_j|\, E_{n_\nu}[(x-c_j)^{r-1}\, |x-c_j|^{1+\alpha}; -1, 1], \end{aligned}$$

i.e. for sufficiently great n_ν

$$E_{n_\nu}[A_j(x-c_j)^{r-1}|x-c_j|^{1+\alpha}; c_j-\varepsilon, c_j+\varepsilon; M] < (1-\beta)\, |A_j|\, E_{n_\nu}[(x-c_j)^{r-1}|x-c_j|^{1+\alpha}; -1, 1],$$

where $M = 4 \sum_{k=1}^{m} |A_k|$. We have arrived at a contradiction of proposition 7.2.24.

Consequently, as $n \to \infty$

$$E_n(f; -1, 1) > [1-o(1)] \max_{1\leqslant k\leqslant m} |A_k|\, E_n[(x-c_k)^{r-1}\,|x-c_k|^{1+\alpha}; -1, 1],$$

which together with 7.2(36) and 7.2(5) leads to 7.2(34).

7.3. The best uniform approximation of functions which have a discontinuous derivative of bounded variation

This limit theorems established in section 7.2 for the best approximation of the simple functions considered there, play a fundamental part in the study of the asymptotic properties of the best approximation by polynomials of other functions with singularities of the same type.

** In virtue of theorem 7.2.2, for sufficiently great values of n the given number j does not depend on n.

In particular, starting from them it is possible to show that all functions which have a discontinuous derivative of bounded variation possess similar asymptotic properties.

7.3.1. In the first place let us consider from this point of view functions of the type

$$f(x) = \sum_{k=1}^{\infty} A_k (x-c_k)^{r-1} |x-c_k|^{1+\alpha}, \tag{1}$$

where r is a natural number, $-1 < \alpha < 1$, and

$$\sum_{k=1}^{\infty} |A_k| < \infty. \tag{2}$$

We will prove that** *if* $|c_k| < 1$ $(k = 1, 2, \ldots)$, *then as* $n \to \infty$ *the limit of the product* $n^{r+\alpha} E(f; -1, 1)$ *exists and*

$$\lim_{n\to\infty} n^{r+\alpha} E_n(f; -1, 1) = \max_{1\leqslant k<\infty} |A_k| (\sqrt{1-c_k^2})^{r+\alpha} \mu(r, \alpha), \tag{3}$$

where $\mu(r, \alpha)$ *is the constant* of 7.2(2).

Let

$$|A_j| (\sqrt{1-c_j^2})^{r+\alpha} = \max_{1\leqslant k\leqslant\infty} |A_k| (\sqrt{1-c_k^2})^{r+\alpha},$$

and for a given $\varepsilon > 0$ choose $m \geqslant j$ so that

$$\sum_{k=m+1}^{\infty} |A_k| < \varepsilon.$$

Then

$$f(x) = \sum_{k=1}^{m} A_k (x-c_k)^{r-1} |x-c_k|^{1+\alpha} + \psi(x),$$

where

$$\psi(x) = \sum_{k=m+1}^{\infty} A_k (x-c_k)^{r-1} |x-c_k|^{1+\alpha}.$$

Since

$$\begin{aligned} E_n[(x-c)^{r-1}|x-c|^{1+\alpha}; -1, 1] &\leqslant E_n[(x-c)^{r-1}|x-c|^{1+\alpha}; c-2, c+2] \\ &= E_n(x^{r-1}|x|^{1+\alpha}; -2, 2) = 2^{r+\alpha} E_n(x^{r-1}|x|^{1+\alpha}; -1, 1), \end{aligned}$$

** S. M. Nikol'skii [5] in the case where r is odd.

then in virtue of theorem 7.2.1 there exists a constant a such that for all $n \geqslant 1$

$$E_n[(x-c)^{r-1}|x-c|^{1+\alpha}; -1, 1] < \frac{a}{n^{r+\alpha}}.$$

Therefore

$$E_n(\psi) \leqslant \sum_{k=m+1}^{\infty} |A_k[E_n(x-c_k)^{r-1}|x-c_k|^{1+\alpha}; -1, 1] < \frac{a\varepsilon}{n^{r+\alpha}}$$

and

$$E_n\left[\sum_{k=1}^{m} A_k(x-c_k)^{r-1}|x-c_k|^{1+\alpha}; -1, 1\right] - \frac{a\varepsilon}{n^{r+\alpha}} < E_n(f; -1, 1)$$

$$< E_n\left[\sum_{k=1}^{m} A_k(x-c_k)^{r-1}|x-c_k|^{1+\alpha}; -1, 1\right] + \frac{a\varepsilon}{n^{r+\alpha}}.$$

But, by theorem 7.2.3, for all sufficiently great values of n

$$\frac{\mu(r, \alpha)}{n^{r+\alpha}} (1-\varepsilon)\, |A_j| (\sqrt{1-c_j^2})^{r+\alpha}$$

$$< E_n\left[\sum_{k=1}^{m} A_k(x-c_k)^{r-1}\, |x-c_k|^{1+\alpha}; -1, 1\right]$$

$$< \frac{\mu(r, \alpha)}{n^{r+\alpha}} (1+\varepsilon)|A_j| \left(\sqrt{1-c_j^2}\right)^{r+\alpha}.$$

Consequently, for all such n

$$(1-\varepsilon)|A_j| \left(\sqrt{1-c_j^2}\right)^{r+\alpha} \mu(r, \alpha) - a\varepsilon < n^{r+\alpha} E_n(f; -1, 1)$$

$$< (1+\varepsilon)|A_j| (\sqrt{1-c_j^2})^{r+\alpha} \mu(r, \alpha) + a\varepsilon,$$

i.e.

$$(1-\varepsilon)|A_r| \left(\sqrt{1-c_j^2}\right)^{r+\alpha} \mu(r, \alpha) - a\varepsilon$$

$$\leqslant \varliminf_{n\to\infty} n^{r+\alpha} E_n(f; -1, 1) \leqslant \varlimsup_{n\to\infty} n^{r+\alpha} E_n(l; -1, 1)$$

$$\leqslant (1+\varepsilon)|A_j| \left(\sqrt{1-c_j^2}\right)^{r+\alpha} \mu(r, \alpha) + a\varepsilon.$$

In view of the arbitrariness of $\varepsilon > 0$, 7.3(3) follows.

7.3.2. Let $f(x)$ have on $[-1, 1]$ an r-th ($r \geqslant 1$ an integer) derivative $f^{(r)}(x) = \phi(x)$, which is a discontinuous function of bounded variation. It is known that such a function $\phi(x)$ can have only discontinuities of the first kind, and not more than an enumerable number of these. Let us denote the points of discontinuity of the function $\phi(x)$ by $c_k (k = 1, 2, \ldots)$ and let us suppose that at each of these points $\phi(x)$ is continuous on the right, that is, $\phi(c_k) = \phi(c_k+0)$. We shall prove that** *as* $n \to \infty$ *the limit of the product* $n^r E_n(f: -1, 1)$ *exists and the equality*

$$\lim_{n\to\infty} n^r E_n(f, -1, 1) = \frac{\mu(r, 0)}{2r!} \max_{-1<x<1} |f^{(r)}(x+0)-f^{(r)}(x-0)|(\sqrt{1-x^2})^r, \quad (4)$$

is satisfied, in which $\mu(r, 0)$ *is the constant of* 7.2(2).

Since as $k \to \infty$

$$A_k = \phi(c_k+0)-\phi(c_k-0) \to \infty$$

there exists a j such that

$$\max_{-1<x<1} |\phi(x+0)-\phi(x-0)|(\sqrt{1-x^2})^r = \max_{1\leqslant k<\infty} |A_k|(\sqrt{1-c_k^2})^r = |A_j|(\sqrt{1-c_j^2})^r, \quad (5)$$

and for a given $\varepsilon > 0$ an $m \geqslant j$ such that $|A_k| < \varepsilon$ for all $k \geqslant m$.

Let us consider the simplest step functions

$$F_x(t) = \begin{cases} 0, & -1 \leqslant t < x, \\ 1 & x \leqslant t \leqslant 1. \end{cases} \quad (6)$$

and represent the function $\phi(t)$ in the form of the sum

$$\phi(t) = \sum_{k=1}^{m} A_k F_{c_k}(t) + \psi(t). \quad (7)$$

In virtue of the choice of m the function $\psi(t)$ possesses the property that for all points of the segment $[-1, 1]$

$$|\psi(t+0)-\psi(t-0)| < \varepsilon$$

Hence each point $t \in [-1, 1]$ has a neighbourhood*** $(t-\eta, t+\eta)$ $(\eta > 0)$, for all points x of which $|\psi(x)-\psi(t)| < \varepsilon$. Selecting (as is possible in virtue of the Borel-Lebesgue lemma) from the system of segments $\left[t-\frac{\eta}{2}, t+\frac{\eta}{2}\right]$ a finite number which cover the segment $[-1, 1]$ and denoting by 2δ the length of the least of them, we obtain $\omega(\psi; \delta) \leqslant 2\varepsilon$.

** S. M. Nikol'skii [5]. See also I. I. Ibragimov [6].

*** At the ends of the segment $[-1, 1]$ a one-sided neighbourhood is considered.

Consequently, by Jackson's theorem (see 5.1(1)), if $n \geqslant \frac{2}{\delta}$,

$$E_n(\psi; -1, 1) \leqslant C\omega\left(\psi; \frac{2}{n}\right) \leqslant 2\varepsilon C. \tag{8}$$

Since the function $f(x)$ can be represented in the form (see 3.5(5))

$$f(x) = \frac{1}{(r-1)!} \int_{-1}^{x} (x-t)^{r-1}\phi(t)\, dt + P_{r-1}(x),$$

where $P_{r-1}(x)$ is an algebraic polynomial of degree $\leqslant r-1$, then, taking account of 7.3(7), we have

$$f(x) = \frac{1}{(r-1)!} \sum_{k=1}^{m} A_k \int_{-1}^{x} (x-t)^{r-1} F_{c_k}(t)\, dt +$$

$$+ \frac{1}{(r-1)!} \int_{-1}^{x} (x-t)^{r-1}\psi(t)\, dt + P_{r-1}(x).$$

But

$$\int_{-1}^{x} (x-t)^{r-1} F_{c_k}(t) dt = \frac{(x-c_k)^{r-1}|x-c_\kappa| + (x-c_k)^r}{2r} \begin{cases} \frac{(x-c_k)^r}{r}, & x \geqslant c_k, \\ 0 & x \leqslant c_k. \end{cases}$$

Hence

$$f(x) = \frac{1}{2r!} \sum_{k=1}^{m} A_k (x-c_k)^{r-1}|x-c_k|$$

$$+ \frac{1}{(r-1)!} \int_{-1}^{x} (x-t)^{r-1}\psi(t) dt + Q_r(x), \tag{9}$$

where $Q_r(x)$ is some algebraic polynomial of degree $\leqslant r$.

The second expression on the right-hand side of 7.3(9) is a function $\psi_r(x)$, the r-th derivative of which, $\psi(t)$, has a best approximation which satisfies the inequality 7.3(8). Hence if use is made of the inequality 5.2(15), we see that for $n-r > \frac{2}{\delta}$

$$E_n(\psi_r; -1, 1) < \frac{(2C)^r}{(n-r+1) \dots (n+1)n} E_{n-r}(\psi; -1, 1) < \frac{C_r \varepsilon}{n^r},$$

where C_r is a constant depending only r.

Thus

$$\frac{1}{2r!}E_n\Big[\sum_{k=1}^{m}A_k(x-c_k)^{r-1}|x-c_k|;\,-1,1\Big]-\frac{C_r\varepsilon}{n^r}<E_n(f;\,-1,1)$$

$$<\frac{1}{2r!}E_n\Big[\sum_{k=1}^{m}A_k(x-c_k)^{r-1}|x-c_k|;\,-1,1\Big]+\frac{C_r\varepsilon}{n^r},$$

and by theorem 7.2.3, for all sufficiently great n,

$$\frac{\mu(r,0)}{2r!\,n^r}\max_{1\leqslant k\leqslant m}|A_k|\big(\sqrt{1-c_k^2}\big)^r-\frac{\varepsilon}{n^r}-\frac{C_r\varepsilon}{n^r}<E_n(f;\,-1,1)$$

$$<\frac{\mu(r,0)}{2r!\,n^r}\max_{1\leqslant k\leqslant m}|A_k|\big(\sqrt{1-c_k^2}\big)^r+\frac{\varepsilon}{n^r}+\frac{C_r\varepsilon}{n^r},$$

or

$$\frac{\mu(r,0)}{2r!}|A_j|\big(\sqrt{1-c_j^2}\big)^r-\varepsilon-C_r\varepsilon\leqslant\varliminf_{n\to\infty}n^rE_n(f;\,-1,1)$$

$$\leqslant\varlimsup_{n\to\infty}n^rE_n(f;\,-1,1)\leqslant\frac{\mu(r,0)}{2r!}|A_j|\big(\sqrt{1-c_j^2}\big)^r+\varepsilon+C_r\varepsilon.$$

It follows from this in virtue of the arbitrariness of $\varepsilon>0$ that

$$\lim_{n\to\infty}n^rE_n(f;\,-1,1)=\frac{\mu(r,0)}{2r!}|A_j|\big(\sqrt{1-c_j^2}\big)^r,$$

i.e. 7.3(4) is satisfied.

7.3.3. Theorem 7.3.2 can be extended to the case of a fractional derivative of any order $r>0$. For this it is merely necessary, reasoning exactly as in sections 7.2.1–7.2.3, to obtain similar results for the best approximation of the function

$$(x-c+|x-c|)\,|x-c|^{r-1}$$

We will limit ourselves here to the formulation of the final results. If on $[-1,1]$ the function $f(x)$ can be represented in the form

$$f(x)=\frac{1}{\Gamma(r)}\int_{-1}^{x}(x-t)^{r-1}\phi(t)\,dt, \tag{10}$$

where $r>0$, and $\phi(t)$ is a discontinuous function of bounded variation, then as $n\to\infty$ the limit of the product $n^rE_n(f;\,-1\ 1)$ exists and the equality

$$\lim_{n\to\infty}n^rE_n(f,\,-1,1)=\frac{\mu_r}{2\Gamma(r+1)}\max_{-1<x<1}|\phi(x+0)-\phi(x-0)|\big(\sqrt{1-x^2}\big)^r \tag{11}$$

is satisfied, in which

$$\mu_r=A_1[(x+|x|)\,|x|^{r-1}]. \tag{12}$$

7.4. On the best mean approximation of the simpler functions with singularities

A number of extremal problems, formulated in a uniform metric (see, for example, sections 5.5.1, 5.7.2, 5.9.1), reduce to corresponding problems on the best integral approximation of individual functions possessing certain singularities. As in the case of uniform approximation, in the study of the asymptotic properties of the best approximation of such functions a fundamental part is frequently played by the simplest of them.

We will consider here from this point of view only periodic functions which have derivatives with discontinuities of the first kind.

7.4.1. The simplest periodic function whose $(r-1)$-th $(r \geqslant 1)$ derivative has a discontinuity of the first kind, is the function $D_0^{(r)}(x)$ defined in section 3.5.5 (see 3.5(16)), which has in a period $[0,2\pi)$ only one such singularity, at the point $x = 0$. Knowing the exact value of the best approximation of this function by trigonometric polynomials in the metric L (see section 2.11.4 or 2.11.5), it is possible to determine the asymptotic behaviour of the best approximation $E_n^*(f)_L$ also for other functions $f(x)$ with singularities of the same type. A fundamental part is played here by the lemma of 5.6.32 (see 5.6(17)) and by the following proposition, which is similar to 7.2.24.

If we denote by $E_n^*(f; a, b; M)_L$ the best integral approximation (in the metric L) of the function $f(x)$ on $[a, b]$ $(-\pi < a < b < \pi)$ by trigonometric polynomials $T_n(x)$ of order $\leqslant n$ for which

$$\int_{-\pi}^{a} |T_n(x)|\,dx + \int_{b}^{\pi} |T_n(x)|\,dx \leqslant M, \tag{1}$$

then *whatever* $h > 0$ *and* $0 < \sigma < \frac{1}{2}$ *may be, for any arbitrarily small* $\varepsilon > 0$ *it is possible to find an* N *such that for all* $n \geqslant N$

$$E_n^*\left[D_0^{(r)}; -\frac{1}{n^\sigma}, \frac{1}{n^\sigma}; n^h\right]_L > (1-\varepsilon)\, E_n^*[D_0^{(r)}]_L. \tag{2}$$

Let us put $\lambda = [n^\varrho]$, where $2\sigma < \varrho < 1$, and let

$$q(x) = \left(1 - \sin^2\frac{x}{2}\right)^\lambda.$$

It is obvious that for the given best approximation by trigonometric polynomials

$$E^*_{n,q(x)}[D_0^{(r)}]_L = \inf_{T_n} \int_{-\pi}^{\pi} q(x)|D_0^{(r)}(x)-T_n(x)|\,dx$$

the inequality

$$E^*_{n,q(x)}[D_0^{(r)}]_L \geqslant E^*_m[q(x)D_0^{(r)}(x)]_L \tag{3}$$

is satisfied, where $m = n+\lambda$. Thus

$$E^*_m[q(x)D_0^{(r)}(x)]_L \geqslant E^*_m[D_0^{(r)}]_L - \sum_{\nu=1}^{\lambda} \binom{\lambda}{\nu} E^*_m\left[\left(\sin\frac{x}{2}\right)^{2\nu} D_0^{(r)}(x)\right]_L. \tag{4}$$

Everywhere, except at the points $x = 2k\pi$ (k an integer), the function $f(x) = \left(\sin\frac{x}{2}\right)^{2\nu} D_0^{(r)}(x)$ has a bounded derivative of order $r+2\nu-1$ for any $\nu = 1, 2, \ldots, \lambda$, and

$$|f^{(r+2\nu-1)}(x)| \leqslant B_0^{2\nu+r}\nu^{2\nu+r},$$

where B_0 is some constant not depending on ν. Hence (see 5.5(11) and 5.5(14))

$$E^*_m\left[\left(\sin\frac{x}{2}\right)^{2\nu} D_0^{(r)}(x)\right]_L \leqslant 2\pi E^*_m\left[\left(\sin\frac{x}{2}\right)^{2\nu} D_0^{(r)}(x)\right] < \frac{\pi^2 B_0^{2\nu+r}\nu^{2\nu+r}}{m^{r+2\nu+1}},$$

and

$$\sum_{\nu=1}^{\lambda} \binom{\lambda}{\nu} E^*_m\left[\left(\sin\frac{x}{2}\right)^{2\nu} D_0^{(r)}(x)\right]_L < \frac{C_0}{m^{r-1}} \sum_{\nu=1}^{\lambda} \binom{\lambda}{\nu}\left(\frac{C_1\nu}{m}\right)^{2\nu},$$

where C_0 and C_1 are certain constants. But since $\lambda = [n^\varrho]$ ($\varrho < 1$) it follows that as $n \to \infty$

$$\sum_{\nu=1}^{\lambda} \binom{\lambda}{\nu}\left(\frac{C_1\nu}{m}\right)^{2\nu} \leqslant C_2 \sum_{\nu=1}^{\lambda-1} \frac{\lambda^\lambda}{\nu^\nu(\lambda-\nu)^{\lambda-\nu}}\left(\frac{C_1\nu}{m}\right)^{2\nu} + \left[o\left(\frac{\lambda}{m}\right)\right]^{2\lambda}$$

$$< \sum_{\nu=1}^{\lambda}\left(\frac{C_3\lambda\nu}{m^2}\right)^{\nu} = o\left(\frac{1}{m}\right).$$

Consequently, we find from 7.4(4) (see section 2.11.4) that

$$E^*_m[q(x)D_0^{(r)}(x)]_L \geqslant E^*_m[D_0^{(r)}]_L - o\left(\frac{1}{m^r}\right) = \frac{\pi K_r}{(n+1)^r} - o\left(\frac{1}{n^r}\right), \tag{5}$$

where K_r is the constant of 2.11(10).

If $T_n(x)$ is an arbitrary trigonometric polynomial of order $\leqslant n$ which satisfies condition 7.4(1), where $M = n^h$, then, since $|D_0^{(r)}(x)| \leqslant \frac{\pi^2}{6}$, we have

$$\int_{-\pi}^{-\frac{1}{n^\sigma}} q(x)|D_0^{(r)}(x)-T_n(x)|\,dx + \int_{\frac{1}{n^\sigma}}^{\pi} q(x)|D_0^{(r)}(x)-T_n(x)|\,dx$$

$$\leqslant \max_{\frac{1}{n^\sigma}\leqslant x\leqslant\pi}\left(1-\sin^2\frac{x}{2}\right)^\lambda \left\{\int_{-\pi}^{-\frac{1}{n^\sigma}} |D_0^{(r)}(x)-T_n(x)|\,dx + \right.$$

$$\left. + \int_{\frac{1}{n^\sigma}}^{\pi} |D_0^{(r)}(x)-T_n(x)|\,dx\right\} \leqslant C\left(1-\frac{1}{\pi^2 n^{2\sigma}}\right)^{n^\varrho} n^h = o\left(\frac{1}{n^r}\right),$$

where C is some constant. Therefore in virtue of 7.4(3) and 7.4(4) the inequality

$$\int_{-\frac{1}{n^\sigma}}^{\frac{1}{n^\sigma}} q(x)|D_0^{(r)}(x)-T_n(x)|\,dx > \frac{\pi K r}{(n+1)^r} - o\left(\frac{1}{n^r}\right),$$

is satisfied for such polynomials, and *a fortiori*

$$\int_{-\frac{1}{n^\sigma}}^{\frac{1}{n^\sigma}} |D_0^{(r)}(x)-T_n(x)\,dx > \frac{\pi K r}{(n+1)^r} - o\left(\frac{1}{n^r}\right),$$

i.e.

$$E_n^*\left[D_0^{(r)}; -\frac{1}{n^\sigma}, \frac{1}{n^\sigma}\, n^h\right]_L > (1-\varepsilon_n)\, E_n^*[D_0^{(r)}]_L, \tag{6}$$

where $\varepsilon \to 0$ as $n \to \infty$.

7.4.2. Now let

$$f(x) = \sum_{\nu=1}^{m} A_\nu D_0^{(r)}(x-c_\nu), \tag{7}$$

where $0 \leqslant c_1 < c_2 < c_3 < \ldots < c_m < 2\pi$. We will prove that *as* $n \to \infty$ *the limit of the product* $n^r E_n^*(f)_L$ *exists and*

$$\lim_{n\to\infty} n^r E_n^*(f)_L = \pi K_r \sum_{\nu=1}^{m} |A_\nu|, \tag{8}$$

where K_r *is the constant of* 2.11(10).

It follows from 2.11(12) that for any $n \geqslant 0$

$$E_n^*(f)_L \leqslant \sum_{\nu=1}^{m} |A_\nu| E_n^*[D_0^{(r)}(x-c_\nu)]_L = \frac{\pi K_r}{(n+r)} \sum_{\nu=1}^{m} |A_\nu|. \tag{9}$$

On the other hand, if $T_n(x)$ and $T_{n,r}(x)$ are the corresponding trigonometric polynomials of order $\leqslant n$ for which

$$E_n^*(f)_L = \int_0^{2\pi} |f(x) - T_n(x)| \mathrm{d}x,$$

$$E_n^*[D_0^{(r)}]_L = \int_0^{2\pi} |D_0^{(r)}(x) - T_{n,r}(x)| \, \mathrm{d}x,$$

then for sufficiently small $\delta > 0$

$$\begin{aligned} E_n^*(f)_L &\geqslant \sum_{\nu=1}^{m} \int_{c_\nu-\delta}^{c_\nu+\delta} \Big|\sum_{k=1}^{m} A_k D_0^{(r)}(x-c_k) - T_n(x)\Big| \, dx \\ &= \sum_{\nu=1}^{m} \Big\{ \int_{c_\nu-\delta}^{c_\nu+\delta} \Big|A_\nu D_0^{(r)}(x-c_\nu) - \Big[T_n(x) + A_\nu T_{n,r}(x-c_\nu) - \\ &\quad - \sum_{k=1}^{m} A_k T_{n,r}(x-c_k)\Big] + \sum_{k=1}^{m} A_k [D_0^{(r)}(x-c_k) - T_{n,r}(x-c_k)] - \\ &\quad - A_\nu [D_0^{(r)}(x-c_\nu) - T_{n,r}(x-c_\nu)]\Big| \, \mathrm{d}x \Big\} \\ &\geqslant \sum_{\nu=1}^{m} \Big\{ \int_{c_\nu-\delta}^{c_\nu+\delta} \Big|A_\nu D_0^{(r)}(x-c_\nu) - \Big[T_n(x) - \sum_{\substack{k=1 \\ k\neq\nu}}^{m} A_k T_{n,r}(x-c_k)\Big]\Big| \, dx - \\ &\quad - \int_{c_\nu-\delta}^{c_\nu+\delta} \Big| \sum_{\substack{k=1 \\ k\neq\nu}}^{m} A_k [D_0^{(r)}(x-c_k) - T_{n,r}(x-c_k)]\Big| \, \mathrm{d}x \Big\}. \end{aligned}$$

But since for any $\nu = 1, 2, \ldots, m$ the sequence of polynomials

$$T_n(x) = \sum_{\substack{k=1 \\ k\neq\nu}}^{m} A_k T_{n,r}(x-c_k)$$

possesses the property that

$$\int_{-\pi}^{\pi} \Big|T_n(x) - \sum_{\substack{n=1\\k\neq\nu}}^{m} A_k T_{n,r}(x-c_k)\Big|\, dx$$

$$\leqslant \int_{-\pi}^{\pi} |T_n(x)|\, dx + \sum_{k=1}^{m} |A_k| \int_{-\pi}^{\pi} |T_{n,r}(x)|\, dx$$

$$\leqslant 2\int_{-\pi}^{\pi} |f(x)|\, dx + 2\sum_{k=1}^{m} |A_k| \int_{-\pi}^{\pi} |D_0^{(r)}(x)|\, dx = M,$$

then in virtue of the assertion of 7.4.1, whatever $\varepsilon > 0$, $h > 0$ and $0 < \delta < \frac{1}{2}$ may be, for all sufficiently great n

$$\int_{c_\nu-\delta}^{c_\nu+\delta} \Big|A_\nu D_0^{(r)}(x-c_\nu) - \Big[T_n(x) - \sum_{\substack{k=1\\k\neq\nu}}^{m} A_k T_{n,r}(x-c_k)\Big]\Big|\, dx$$

$$\geqslant E_n^*[A_\nu D_0^{(r)}(x); -\delta, \delta; M]_L \geqslant |A_\nu| E_n^*\Big[D_{0c}^{(r)}; -\frac{1}{n^\sigma}, \frac{1}{n^\sigma}; n^h\Big]_L$$

$$> (1-\varepsilon)\, |A_\nu| E^*[D_0^{(r)}]_L.$$

Moreover, by the lemma of 5.6.32

$$\int_{c_\nu-\delta}^{c_\nu+\delta} \Big|\sum_{\substack{k=1\\k\neq\nu}} A_k [D_0^{(r)}(x-c_k) - T_{n,r}(x-c_k)]\Big|\, dx = o\Big(\frac{1}{n^r}\Big) = o\{E_n^*[D_0^{(r)}]_L\}.$$

Consequently, for any $\varepsilon > 0$,

$$E_n^*(f)_L > (1-2\varepsilon) E_n^*[D_0^{(r)}]_L \sum_{\nu=1}^{m} |A_\nu| = \frac{(1-2\varepsilon)\pi K_r}{(n+1)^r} \sum_{\nu=1}^{m} |A_\nu|,$$

for all sufficiently great n, or

$$\varliminf_{n\to\infty} n^r E_n^*(f)_L \geqslant (1-2\varepsilon)\pi K_r \sum_{\nu=1}^{m} |A_\nu|.$$

Thus in virtue of the arbitrariness of $\varepsilon > 0$ we have

$$\varliminf_{n\to\infty} n^r E_n^*(f)_L \geqslant \pi K_r \sum_{\nu=1}^{m} |A_\nu|.$$

The relation 7.4(8) follows from this inequality and 7.4(9).

7.4.3.. Let us consider an arbitrary periodic function $f(x)$ of period 2π of the form

$$f(x) = \sum_{\nu=1}^{\infty} A_\nu D_0^{(r)}(x-c_\nu) + \psi(x), \tag{10}$$

where $0 \leqslant c_\nu < 2\pi (\nu = 1, 2, \ldots)$, $c_k \neq c_\nu (k \neq \nu)$,

$$0 < \sum_{\nu=1}^{\infty} |A_\nu| < \infty,$$

and $\psi(x)$ has an absolutely continuous derivative of the $(r-1)$-th $(r \geqslant 1)$ order. Any such function $f(x)$ has an $(r-1)$-th discontinuous (at the point c_ν) derivative of bounded variation.

Here since

$$D_0^{(1)}(x) = \begin{cases} \dfrac{\pi - x}{2}, & 0 < x \leqslant \pi, \\ \dfrac{-\pi - x}{2}, & -\pi \leqslant x < 0, \end{cases}$$

it follows that

$$A_\nu = \frac{1}{\pi}[f^{(r-1)}(c_\nu + 0) - f^{(r-1)}(c_\nu - 0)].$$

Let us verify that *for the function* 7.4(10) *the limit of the product* $n^r E_n^*(f)_L$ *exists as* $n \to \infty$ *and*

$$\lim_{n\to\infty} n^r E_n^*(f)_L = \pi K_r \sum_{\nu=1}^{\infty} |A_\nu|, \tag{11}$$

where K_r *is the constant of* 2.11(10).

Since for the function $\psi(x)$ there exists an absolutely continuous derivative $\psi^{(r-1)}(x)$, the derivative $\psi^{(r)}(x)$ is integrable over $[0, 2\pi]$, and as $n \to \infty$ (see 5.11.2) we have

$$E_n^*(\psi)_L = o\left(\frac{1}{n^r}\right). \tag{12}$$

Also, given an arbitrary $\varepsilon > 0$, let us choose m so that

$$\sum_{\nu=m+1}^{\infty} |A_\nu| < \varepsilon.$$

Then we will have the inequality

$$E_n^*\Big[\sum_{\nu=1}^{m} A_\nu D_0^{(r)}(x - c_\nu)\Big]_L - \varepsilon E_n^*[D_0^{(r)}]_L - E_n^*(\psi)_L$$
$$\leqslant E_n^*(f)_L \leqslant E_n^*\Big[\sum_{\nu=1}^{m} A_\nu D_0^{(r)}(x - c_\nu)\Big]_L + \varepsilon E_n^*[D_0^{(r)}]_L + E_n^*(\psi)_L.$$

It follows from this in virtue of theorem 7.4.2, 2.11(12) and 7.4(12), that for all sufficiently great n

$$\frac{\pi K_r}{n^r}\sum_{\nu=1}^{m}|A_\nu| - \frac{\varepsilon\pi K_r}{(n+1)^r} - \frac{\varepsilon}{n^r} \leqslant E_n^*(f)_L \leqslant \frac{\pi K_r}{n^r}\sum_{\nu=1}^{m}|A_\nu| + \frac{\varepsilon\pi K_r}{(n+1)^r} + \frac{\varepsilon}{n^r},$$

i.e.

$$\pi K_r \sum |A_\nu| - \varepsilon\pi K_r - \varepsilon \leqslant \varliminf_{n\to\infty} n^r E_n^*(f)_L$$

$$\leqslant \varlimsup_{n\to\infty} n^r E_n^*(f)_L \leqslant \pi K_r \sum_{\nu=1}^{m} |A_\nu| + \varepsilon\pi K_r + \varepsilon.$$

Proceeding to the limit when $\varepsilon \to 0$, we obtain 7.4(11).

7.4.4. The propositions derived in this section were announced without proof by S.M. Nikol'skii**, who considered similar problems for the best mean approximations by algebraic polynomials of functions defined on a finite segment. In the latter case the simplest function possessing an r-th derivative with discontinuities of the first kind is the function $(x-c)^{r-1}$ $|x-c|$, which plays the same part as is played in the periodic case by the function $D_0^{(r+1)}(x)$.

To S.M. Nikols'skii*** is due also a series of results which establish the corresponding asymptotic properties of the best approximation by algebraic polynomials in the metric L of the function $|x-c|^{r+\alpha}$ ($r \geqslant 0$ an integer, $-1 < \alpha < 1$) and other functions with singularities of the same type.

7.5. On the asymptotic behaviour of the best uniform approximation of some analytic functions

The limit theorems established in sections 7.2 and 7.3 refer to functions whose the singularities are located on the segment where their best approximation is studied. Here we will consider from the same point of view the best approximation of functions analytic on this segment, but possessing singularities elsewhere on the real axis. In some cases it is possible, for such functions, to obtain an improvement of theorem 5.4.1. We have already indicated one of these cases in section 7.1.45.

7.5.1. Let us consider the function

$$f(x) = \frac{1}{(x-c)^r}, \tag{1}$$

where $c > 1$ and r is a natural number, which is the simplest function with a unique pole of order r at the point c.

** The proofs were given by V. K. Dzyadyk [1].

*** S. M. Nikol'skii [6, 7]. Note also the paper I. I. Ibragimov [4], and also the work of V. I. Gukevich [1], in which the case of singularities of logarithmic type was studied.

By considering the investigation of the quantity $E_n\left(\frac{1}{x-c};\ -1,\ 1\right)$ carried out in section 2.11.1, it is not difficult to determine an asymptotic expression for $E_n(f;\ -1,\ 1)$ for any $r > 1$. *For any natural number r,*

$$E_n\left[\frac{1}{(x-c)^r};\ -1,\ 1\right] = (1-\varepsilon_n)\frac{n^{r-1}}{(r-1)!\left(\sqrt{c^2-1}\right)^{r+1}\left(c+\sqrt{c^2-1}\right)^n}, \tag{2}$$

where $\varepsilon_n \to 0$ *as* $n \to \infty$**.

If $P_n(x)$ is the polynomial 2.11(5), then on differentiation with respect to the parameter c we obtain

$$\frac{d^{r-1}}{dc^{r-1}}P_n(x) = \frac{(r-1)!}{(x-c)^r} - \frac{d^{r-1}}{dc^{r-1}}\frac{\cos[n\arccos x+\phi(x)]}{(c^2-1)\left(c+\sqrt{c^2-1}\right)^n},$$

where

$$\phi(x) = \arccos\frac{cx-1}{x-c}.$$

Now

$$\frac{d^{r-1}}{dc^{r-1}}\frac{\cos[n\arccos x+\phi(x)]}{(c^2-1)\left(c+\sqrt{c^2-1}\right)^n} = \frac{\cos[n\arccos x+\phi(x)]}{c^2-1}\frac{d^{r-1}}{dc^{r-1}}\frac{1}{\left(c+\sqrt{c^2-1}\right)^n}+$$

$$+(r-1)\frac{d}{dc}\left\{\frac{\cos[n\arccos x+\phi(x)]}{c^2-1}\right\}\frac{d^{r-2}}{dc^{r-2}}\frac{1}{\left(c+\sqrt{c^2-1}\right)^n}+\dots$$

But for any $k \geqslant 0$,

$$\frac{d^k}{dc^k}\frac{1}{\left(c+\sqrt{c^2-1}\right)^n} = [1+o(1)]\frac{(-n)^k}{\left(\sqrt{c^2-1}\right)^k\left(c+\sqrt{c^2-1}\right)^n}$$

when $n \to \infty$ and

$$\frac{d^k}{dc^k}\frac{\cos[n\arccos x+\phi(x)]}{c^2-1} = O(1)$$

uniformly with respect to x and n. Consequently

$$\frac{d^{r-1}}{dc^{r-1}}\frac{\cos[n\arccos x+\phi(x)]}{(c^2-1)\left(c+\sqrt{c^2-1}\right)^n} = \frac{(-n)^{r-1}\{\cos[n\arccos x+\phi(x)]+\delta_n(x)\}}{\left(\sqrt{c^2-1}\right)^{r+1}\left(c+\sqrt{c^2-1}\right)^n},$$

where $\delta_n(x) \to 0$ uniformly with respect to x as $n \to \infty$.

Thus the difference

$$F_n(x) - \frac{d^{r-1}}{dc^{r-1}}P_n(x),$$

** S. N. Bernstein [15]. In the same work S. N. Bernstein obtained an asymptotic expression for $E_n\left[\frac{1}{(c-x)^r};\ -1,\ 1\right]$ for any value of r.

where

$$F_n(x) = \frac{(r-1)!}{(x-c)^r} - \frac{(-n)^{r-1}\delta_n(x)}{\left(\sqrt{c^2-1}\right)^{r+1}\left(c+\sqrt{c^2-1}\right)^n},$$

assumes its maximum value in absolute magnitude

$$\frac{n^{r-1}}{\left(\sqrt{c^2-1}\right)^{r+1}\left(c+\sqrt{c^2-1}\right)^n}$$

with consecutive alternations of sign at the $n+2$ points of the segment $[-1, 1]$ at which $\cos[n\arccos x+\phi(x)] = \neq 1$. Since $\frac{d^{r-1}}{dc^{r-1}} P_n(x)$ is an algebraic polynomial of degree $\leqslant n$, then, by theorem 2.7.2,

$$E_n(F_n; -1, 1) = \frac{n^{r-1}}{\left(\sqrt{c^2-1}\right)^{r+1}\left(c+\sqrt{c^2-1}\right)^n}.$$

It follows that

$$E_n\left[\frac{(r-1)!}{(x-c)^r}; -1, 1\right] = (1+\varepsilon_n)\frac{n^{r-1}}{\left(\sqrt{c^2-1}\right)^{r+1}\left(c+\sqrt{c^2-1}\right)^n},$$

where $\varepsilon_n \to 0$ as $n \to \infty$, i.e. 7.5(2) holds.

7.5.2. In virtue of the remark in section 7.1.45 it immediately follows from 7.5(2) that *if*

$$f(x) = \sum_{\nu=1}^{r} \frac{A_\nu}{(x-c)^\nu} + \psi(x) \tag{3}$$

where $c > 1$, r *is a natural number,* $\psi(x)$ *is any function analytic in a certain ellipse which contains within it the point* c *and with foci at the ends of the segment* $[-1, 1]$, *the equality*

$$E_n(f; -1, 1) = [1+o(1)]\frac{|A_r|\, n^{r-1}}{(r-1)!\left(\sqrt{c^2-1}\right)^{r+1}\left(c+\sqrt{c^2-1}\right)^n}, \tag{4}$$

is satisfied as $n \to \infty$.

7.5.3. Now let

$$f(x) = (c-x)^r \ln(c-x), \tag{5}$$

where $c > 1$ and r is an arbitrary integral non-negative number.

*The best approximation by algebraic polynomials of this function, which has a unique singularity of logarithmic type at the point c, satisfies the asymptotic equality***

$$E_n[(c-x)^r \ln(c-x); -1, 1] = [1+o(1)]\frac{r!\left(\sqrt{c^2-1}\right)^{r-1}}{n^{r+1}\left(c+\sqrt{c^2-1}\right)^n}. \tag{6}$$

** S. N. Bernstein [15].

Choosing some value $b > c$, let us integrate 2.11(5) with respect to the parameter a from c to b. We then obtain

$$\int_c^b P_n(x)\,\mathrm{d}a = \ln(c-x) - \ln(b-x) - \int_c^b \frac{\cos[n \arccos x + \phi(x)]}{(a^2-1)\left(a+\sqrt{a^2-1}\right)^n}\,\mathrm{d}a, \tag{7}$$

where $\phi(x) = \arccos \dfrac{ax-1}{x-a}$, or after integration by parts

$$\int_c^b P_n(x)\,\mathrm{d}a = \ln(c-x) - \ln(b-x) + \frac{\cos[n \arccos x + \phi(x)]}{n\sqrt{a^2-1}\left(a+\sqrt{a^2-1}\right)^n}\Bigg|_{a=c}^{a=b} -$$

$$-\frac{1}{n}\int_c^b \frac{\mathrm{d}}{\mathrm{d}a}\left\{\frac{\cos[n \arccos x + \phi(x)]}{\sqrt{a^2-1}}\right\}\frac{\mathrm{d}a}{\left(a+\sqrt{a^2-1}\right)^n}$$

$$= \ln(c-x) - \ln(b-x) + \frac{\cos[n \arccos x + \phi(x)}{n\sqrt{a^2-1}\left(a+\sqrt{a^2-1}\right)^n}\Bigg|_{a=c}^{a=b} +$$

$$+\frac{1}{n^2}\frac{\sqrt{a^2-1}}{\left(a+\sqrt{a^2-1}\right)}\frac{\mathrm{d}}{\mathrm{d}a}\frac{\cos[n \arccos x + \phi(x)]}{\sqrt{a^2-1}}\Bigg|_{a=c}^{a=b} -$$

$$-\frac{1}{n^2}\int_c^b \frac{1}{\left(a+\sqrt{a^2-1}\right)^n}\frac{\mathrm{d}}{\mathrm{d}a}\left\{\sqrt{a^2-1}\,\frac{\mathrm{d}}{\mathrm{d}a}\frac{\cos[n \arccos x + \phi(x)]}{\sqrt{a^2-1}}\right\}\mathrm{d}a. \tag{8}$$

But since for any k

$$\frac{\mathrm{d}}{\mathrm{d}a}\frac{\cos[n \arccos x + \phi(x)]}{\left(\sqrt{a^2-1}\right)^k} = O(1),$$

$$\frac{\mathrm{d}}{\mathrm{d}a}\left\{\sqrt{a^2-1}\,\frac{\mathrm{d}}{\mathrm{d}a}\frac{\cos[n \arccos x + \phi(x)]}{\left(\sqrt{a^2-1}\right)^k}\right\} = O(1) \tag{9}$$

uniformly with respect to $x \in [-1, 1]$, $a \in [c, b]$ and n, then the last two terms on the right-hand side of 7.5(8) are quantities of the order

$$O\left[\frac{1}{n^2\left(c+\sqrt{c^2-1}\right)^n}\right]$$

uniformly with respect to x and n. Moreover, as $n \to \infty$

$$\frac{\cos[n \arccos x + \phi(x)]}{n\sqrt{a^2-1}\left(a+\sqrt{a^2-1}\right)^n}\Bigg|_{a=b} = o\left[\frac{1}{n\left(c+\sqrt{c^2-1}\right)^n}\right]$$

uniformly with respect to $x \in [-1, 1]$.

Thus equation 7.5(7) can be written in the form

$$\ln(a-x)-\ln(b-x)-\int_a^b P_n(x)\,\mathrm{d}a = \frac{\cos[n\arccos x+\phi(x)]+\varepsilon_n(a,x)}{n\sqrt{a^2-1}\left(a+\sqrt{a^2-1}\right)}, \tag{10}$$

where $\varepsilon_n(a, x)\to 0$ uniformly with respect to $x\in[-1, 1]$ and $a\in[c, b]$ as $n\to\infty$.

Let us suppose that for some $r\geqslant 0$ and for $c\leqslant a<b$ the equality

$$(a-x)^r\ln(a-x)+Q_r(x)\ln(b-x)-R_n(x) = \frac{r!\left(\sqrt{a^2-1}\right)^{r-1}\cos[n\arccos x+\phi(x)]+\varepsilon_n(a,x)}{n^{r+1}\left(a+\sqrt{a^2-1}\right)^n} \tag{11}$$

has already been established, in which $Q_r(x)$ and $R_n(x)$ are algebraic polynomials depending on a, respectively of degree not higher than r and n, and $\varepsilon_n(a, x)$ is a sequence of functions possessing the property that $\varepsilon_n(a, x)\to 0$ uniformly with respect to $x\in[-1, 1]$ and $a\in[c, b]$ as $n\to\infty$. Integrating the left- and right-hand sides of 7.5(11) with respect to the parameter a from c to b, we obtain

$$\begin{aligned}
&-\frac{(c-x)^{r+1}}{r+1}\ln(c-x)+Q^*_{r+1}(x)\ln(b-x)-\int_c^b R_n(x)\,\mathrm{d}a-\frac{(c-x)^{r+1}-(b-x)^{r+1}}{(r+1)^2}\\
&=\frac{r!}{n^{r+1}}\int_c^b\frac{\left(\sqrt{a^2-1}\right)^{r+1}\cos[n\arccos x+\phi(x)]}{\left(a+\sqrt{a^2-1}\right)^n}\,\mathrm{d}a+o\left[\frac{1}{n^{r+1}}\int_c^b\frac{\mathrm{d}a}{\left(a+\sqrt{a^2-1}\right)^n}\right]\\
&=-\frac{r!}{n^{r+2}}\frac{\left(\sqrt{a^2-1}\right)^{r}\cos[n\arccos x+\phi(x)]}{\left(a+\sqrt{a^2-1}\right)^n}\Bigg|_{a=c}^{a=b}+\\
&+\frac{r!}{n^{r+1}}\int_c^b\frac{\mathrm{d}}{\mathrm{d}a}\left\{\left(\sqrt{a^2-1}\right)\cos[n\arccos x+\phi(x)]\right\}\frac{\mathrm{d}a}{\left(a+\sqrt{a^2-1}\right)^n}+\\
&\qquad+o\left[\frac{1}{n^{r+2}\left(c+\sqrt{c^2-1}\right)^n}\right].
\end{aligned}$$

Here in virtue of 7.5(9)

$$\begin{aligned}
&\int_c^b\frac{\mathrm{d}}{\mathrm{d}a}\left\{\left(\sqrt{a^2-1}\right)^r\cos[n\arccos x+\phi(x)]\right\}\frac{\mathrm{d}a}{\left(a+\sqrt{a^2-1}\right)^n}\\
&\qquad=-\frac{\sqrt{a^2-1}}{n\left(a+\sqrt{a^2-1}\right)^r}\frac{\mathrm{d}}{\mathrm{d}a}\left\{\left(\sqrt{a^2-1}\right)^r\cos[n\arccos x+\phi(x)]\right\}\Bigg|_{a=c}^{a=b}+
\end{aligned}$$

$$+\frac{1}{n}\int_c^b \frac{1}{\left(a+\sqrt{a^2-1}\right)^n}\frac{d}{da}\left[\sqrt{a^2-1}\frac{d}{da}\left\{\left(\sqrt{a^2-1}\right)^r\cos\left[n\arccos x+\phi(x)\right]\right\}\right]da$$

$$= o\left[\frac{1}{n\left(c+\sqrt{c^2-1}\right)^n}\right].$$

Consequently

$$(c-x)^{r+1}\ln(c-x)+Q_{r+1}(x)\ln(b-x)-R^*(x)$$

$$= \left.\frac{(r+1)!\left(\sqrt{a^2-1}\right)^r\cos\left[n\arccos x+\phi(x)\right]+\delta_n(a,x)}{n^{r+2}\left(a+\sqrt{a^2-1}\right)^n}\right|_{a=c}$$

where $\delta_n(a, x) \to 0$ uniformly with respect to $x \in [-1, 1]$ and $a \in [c, b]$ as $n \to \infty$.

Thus the equality 7.5(11), if valid for a given r, also holds for the value $r+1$. But this relation has already been established for $r = 0$ (see 7.5(10)). Hence it is true for any $r \geqslant 0$. It follows that the difference $F_n(x)-R_n(x)$, where

$$F_n(x) = (c-x)^r\ln(c-x)+Q_r(x)\ln(b-x)-\frac{\varepsilon_n(c, x)}{n^{r+1}\left(c+\sqrt{c^2-1}\right)^n},$$

and $R_n(x)$ is some algebraic polynomial of degree $\leqslant n$, assumes its maximum absolute value

$$\frac{r!\left(\sqrt{c^2-1}\right)^{r-1}}{n^{r+1}\left(c+\sqrt{c^2-1}\right)^n}$$

at the $n+2$ points of the segment $[-1, 1]$ where $\cos\left[n\arccos x+\phi(x)\right]_{a=c} = \pm 1$, with consecutive alternations of sign. By theorem 2.7.2 it follows that

$$E_n(F_n; -1, 1) = \frac{r!\left(\sqrt{c^2-1}\right)^{r-1}}{n^{r+1}\left(c+\sqrt{c^2-1}\right)^n}. \tag{12}$$

If we now take into account the remark in section 7.1.45, then from 7.5(12) we obtain 7.5(6).

7.5.4. It should be noted that the estimate 7.5(6) is not uniform with respect to c as $c \to 1$. Hence the case when $c = 1$ (the singularity of the function occurs at the end of the segment) requires separate consideration. Reasoning exactly as in section 7.2.1, it is not difficult to show that *whatever the natural number r may be, as $n \to \infty$ the limit of the product $n^{2r}E_n[(1-x)^r\ln(1-x); -1, 1]$ exists and*

$$\lim_{n\to\infty} n^{2r}E_n[(1-x)^r\ln(1-x); -1, 1] = \frac{1}{2^{r-1}}A_1(x^{2r}\ln|x|), \tag{13}$$

where $A_1(x^{2r}\ln|x|)$ *is the best approximation of the function* $x^{2r}\ln|x|$ *by integral functions of degree* $\leqslant 1$ *on* $(-\infty, \infty)$.

In fact, after the substitution $2y^2 = 1-x$ we have

$$E_n[(1-x)^r\ln(1-x); -1, 1] = 2^{r+1}E_{2n}(y^{2r}\ln|y|; -1, 1)$$
$$= 2^{r+1}E_{2n}\left[\left(\frac{x}{2n}\right)^{2r}\ln\frac{|x|}{2n}; -2n, 2n\right]$$
$$= \frac{2^{r+1}}{(2n)^{2r}}E_{2n}(x^{2r}\ln|x|; -2n, 2n). \tag{14}$$

But the relation 2.6(8) is applicable to the function $x^{2r}\ln|x|$ for all $\sigma > 0$ (see sections 7.2.1 and 5.4.6). Therefore

$$\lim_{n\to\infty} E_{2n}(x^{2r}\ln|x|; -2n, 2n) = A_1(x^{2r}\ln|x|),$$

and consequently 7.5(13) follows from 7.5(14).

Similarly, applying theorem 7.2.1, it is possible to obtain an asymptotic estimate for the best approximation on $[-1, 1]$ of the function $(1-x)^{r+\alpha}$, where r is a non-negative integer and $0 < \alpha < 1$ (see 7.8.2).

7.5.5. Let us make a comparison of the best approximation 7.5(2) with the value of the uniform deviation of the function $\frac{1}{(c-x)^r}$ $(c > 1)$ from the partial sums of its expansion

$$\frac{1}{(c-x)^r} = \sum_{k=0}^{\infty} c_k^{(r)}\cos k\arccos x \tag{15}$$

in a Fourier series of Chebyshev polynomials, i.e. with the quantity

$$r_n\left[\frac{1}{(c-x)^r}\right] = \max_{-1\leqslant x\leqslant 1}\left|\sum_{k=n+1}^{\infty} c_k^{(r)}\cos k\arccos x\right|.$$

Let us start with the formula (see 3.7(8))

$$c_k^{(1)} = \begin{cases} \dfrac{2}{\pi}\displaystyle\int_0^{\pi}\frac{\cos kt}{c-\cos t}\,dt & \text{if} \quad k \geqslant 1, \\[2ex] \dfrac{1}{\pi}\displaystyle\int_0^{\pi}\frac{dt}{c-\cos t} & \text{if} \quad k = 0. \end{cases} \tag{16}$$

Thus

$$c_0^{(1)} = \frac{1}{\sqrt{c^2-1}}, \qquad c_1^{(1)} = 2c\cdot c_0^{(1)} - 2,$$

and since for any $k \geqslant 1$

$$c_{k+1}^{(1)} = \frac{2}{\pi}\int_0^{\pi} \frac{\cos(k+1)t}{c-\cos t}\,\mathrm{d}t = 2c\cdot c_k^{(1)} - c_{k-1}^{(1)},$$

then

$$c_k^{(1)} = \frac{2}{\sqrt{c^2-1}\,(c+\sqrt{c^2-1})^k},$$

and

$$\frac{1}{c-x} = \frac{1}{\sqrt{c^2-1}} + 2\sum_{k=1}^{\infty} \frac{\cos k \arccos x}{\sqrt{c^2-1}\,(c+\sqrt{c^2-1})^k}. \tag{17}$$

Differentiating the left- and right-hand sides of 7.5(17) $r-1$ times with respect to the parameter c, we see that for any $k \geqslant 1$

$$c_k^{(r)} = \frac{\mathrm{d}^{r-1}}{\mathrm{d}c^{r-1}} \frac{2}{\sqrt{c^2-1}\,(c+\sqrt{c^2-1})^k}. \tag{18}$$

Formula 7.5(18) shows that

$$c_k^{(r)} = [1+\varepsilon_k^{(r)}] \frac{2k^{r-1}}{(r-1)!(\sqrt{c^2-1})^r\,(c+\sqrt{c^2-1})^k}, \tag{19}$$

where $\varepsilon_k^{(r)} \to 0$ as $k \to \infty$. Therefore, when n increases indefinitely,

$$\max_{-1\leqslant x\leqslant 1} \left| \sum_{k=n+1}^{\infty} c_k^{(r)} \cos k \arccos x \right|$$
$$= [1+o(1)] \frac{2}{(r-1)!(\sqrt{c^2-1})^r} \sum_{k=n+1}^{\infty} \frac{k^{r-1}}{(c+\sqrt{c^2-1})^k}.$$

But

$$\varrho_n = \sum_{k=n+1}^{\infty} \frac{k^{r-1}}{(c+\sqrt{c^2-1})^k} = \sum_{k=n+1}^{n+[\sqrt{n}]} \frac{k^{r-1}}{(c+\sqrt{c^2-1})^k} + \varrho_{n+[\sqrt{n}]}$$
$$= [1+o(1)]n^{-1} \sum_{k=n+1}^{n+[\sqrt{n}]} \frac{1}{(c+\sqrt{c^2-1})^k} + \varrho_{n+[\sqrt{n}]}$$
$$= [1+o(1)]n^{r-1} \sum_{k=n+1}^{\infty} \frac{1}{(c+\sqrt{c^2-1})^k} + \varrho_{n+[\sqrt{n}]}$$

and $\varrho_{n+[\sqrt{n}]} = o(\varrho_n)$. Consequently, as $n \to \infty$

$$r_n\left[\frac{1}{(c-x)^r}\right] = \max_{-1\leqslant x\leqslant 1} \left| \sum_{k=n+1}^{\infty} c_k^{(r)} \cos k \arccos x \right|$$
$$= [1+o(1)] \frac{2n^{r-1}}{(r-1)!(\sqrt{c^2-1})^r(c-1+\sqrt{c^2-1})(c+\sqrt{c^2-1})^n}. \tag{20}$$

Comparing equation 7.5(20) with 7.5(2), we obtain the relation

$$E_n\left[\frac{1}{(x-c)^r};-1,1\right]=\frac{1}{2}[1+o(1)]\left(1+\sqrt{\frac{c-1}{c+1}}\right)r_n\left[\frac{1}{(c-x)^r}\right]. \tag{21}$$

It is not difficult to understand that in virtue of the inequality 3.7(9) such a relationship between $E_n(f;-1,1)$ and the remainder $r_n(f)$ of the Fourier series should hold for any function of the type 7.5(3).

Thus *as the singularity c recedes to infinity the partial sums of order n of the Fourier series of the function* 7.5(3) *give a uniform approximation, asymptotically identical with the quantity* $E_n(f;-1,1)$.

Moreover, it follows from 7.5(2) and 7.5(19) that *for every such function* $f(x)$

$$E_n(-f;-1,1)=[1+o(1)]\frac{c_n}{2\sqrt{c^2-1}} \tag{22}$$

as $n\to\infty$, *where*

$$c_n=\frac{2}{\pi}\int_0^\pi f(\cos t)\cos nt\,dt$$

is the n-th coefficient of the Fourier series of the function $f(\cos t)$.

7.5.6. The relations 7.5(21) and 7.5(22) were established by S. N. Bernstein [15], who also indicated their validity in certain other cases. In particular, he proved that 7.5(22) and the equality

$$E_n(f;-1,1)=\frac{1}{2}[1+o(1)]\left(1+\sqrt{\frac{c-1}{c+1}}\right)r_n(f), \tag{23}$$

where $r_n(f)=\max_t\left|\sum_{k=n+1}^{\infty}c_k\cos kt\right|$, are valid for any function $f(x)$ of the form

$$f(x)=\sum_{\nu=1}^{\infty}\frac{A_\nu}{(c-x)^\nu}\qquad (A_\nu>0;\ \nu=1,2,\ldots), \tag{24}$$

which has an essential singularity at the point $c>1$.

7.6. On a constructive property of regularly monotonic functions

The functions 7.5(1), 7.5(5) and 7.5(24) ($A_\nu\geqslant 0$) are the simplest examples of functions regularly monotonic on $[-1,1]$ (see section 3.9.1). Such functions possess a remarkable property, which lies at the basis of the investigation carried out in the preceding section**. *Any function* $f(x)$

** S. N. Bernstein [15].

regularly monotonic on $[-1, 1]$ *possesses the property that if* $P_n(f; x)$ *is an algebraic polynomial of degree* $\leqslant n$ *for which*

$$E_n(f; -1, 1) = \max_{-1\leqslant x\leqslant 1} |f(x)-P_n(f; x)|,$$

then for any natural number n the number of points $\xi_k^{(n)}$ in $[-1, 1]$ *at which*

$$|f(\xi_k^{(n)})-P_n(f; \xi_k^{(n)})| = E_n(f; -1, 1) > 0 \tag{1}$$

is equal to $n+2$, *and*

$$-1 = \xi_{n+1}^{(n)} < \xi_n^{(n)} < \cos\frac{n-1}{n}\pi < \xi_{n-1}^{(n)} < \dots < \xi_{k+1}^{(n)} < \cos\frac{k\pi}{n} < \xi_k^{(n)} < \dots$$
$$\dots < \cos\frac{\pi}{n} < \xi_1^{(n)} < \xi_0^{(n)} = 1. \tag{2}$$

7.6.1. If $f(x)$ is regularly monotonic, then by Rolle's theorem, for any n, the system of functions $1, x, x^2, \dots, x^n, f(x)$ satisfies Haar's condition (see sections 2.3 and 2.3.2, (5)) and there are not more than n points $\xi_k^{(n)}$ within the interval $(-1, 1)$ at which 7.6(1) holds.

On the other hand, by theorem 2.7.2, there are not less than $n+2$ such points. Consequently there are exactly $n+2$ of them, of which two coincide with the ends of the given segment.

7.6.2. Let

$$-1 = \xi_{n+1}^{(n)} < \xi_n^{(n)} < \xi_{n-1}^{(n)} < \dots < \xi_1^{(n)} < \xi_0^{(n)} = 1 \tag{3}$$

and

$$\begin{aligned} \phi_n(x) &= f(x)-P_n(f; x)-E_n \cos n \arccos x, \\ \Phi_n(x) &= f(x)-P_n(f; x)+E_n \cos n \arccos x, \end{aligned} \tag{4}$$

where $E_n = E_n(f; -1, 1)$.

If it is considered that $f(1)-P_n(f; 1) = E_n$, then (see 7.6.1)

$$f(\xi_k^{(n)})-P_n(f; \xi_k^{(n)}) = (-1)^k E_n, \tag{5}$$

i.e. $\Phi_n(-1) = \phi_n(-1) = 0$, and

$$(-1)^k \phi_n(\xi_k^{(n)}) \geqslant 0, \qquad (-1)^k \Phi_n(\xi_k^{(n)}) \geqslant 0. \tag{6}$$

Moreover,

$$(-1)^k \phi_n\left(\cos\frac{k\pi}{n}\right) \leqslant 0, \qquad (-1)^k \Phi_n\left(\cos\frac{k\pi}{n}\right) \geqslant 0. \tag{7}$$

In virtue of the regular monotonicity of the function $f(x)$, the function $\phi_n(x)$ (correspondingly $\Phi_n(x)$) has on $[-1, 1]$ exactly $n+1$ zeros** (taking

** Consider the relationship between the graphs of the functions $f(x)-P_n(f; x)$ and $\pm E_n \cos n \arccos x$, and also use the fact that the derivative $f^{(n+1)}(x)$ does not have a zero anywhere within $(-1, 1)$.

into account the multiplicity of those which lie within $(-1, 1)$). It is obvious, that if any of the points $\xi_k^{(n)}$ $(k = 0, 1, \ldots, n+1)$ is a zero of the function $\phi_n(x)$ $(\Phi_n(x))$, then $\xi_k^{(n)} = \cos\frac{\nu\pi}{n}$ where ν is an integer, and conversely, if $\phi_n\left(\cos\frac{\nu\pi}{n}\right) = 0$ (correspondingly $\Phi_n\left(\cos\frac{\nu\pi}{n}\right) = 0$), then $\cos\frac{\nu\pi}{n}$ is identical with one of the points $\xi_k^{(n)}$. In this case, within the interval $(-1, 1)$ the points $\xi_k^{(n)}$ where $\phi_n(\xi_k^{(n)}) = 0$ $(\Phi_n(\xi_k^{(n)}) = 0)$, are double zeros of the function $\phi_n(x)$ $(\Phi_n(x))$.

Let us note also that *in the case when at any of the points* 7.6(3) *we have* $\phi_n(\xi_k^{(n)}) = 0$ (correspondingly $\Phi_n(\xi_k^{(n)}) = 0$), *then at the neighbouring points* $\xi_{k-1}^{(n)}$ and $\xi_{k+1}^{(n)}$ *the function* $\phi_n(x)$ $(\Phi_n(x))$ *does not vanish*. Assuming the contrary, we would be able to find two adjacent points $\xi_k^{(n)}$ and $\xi_{k+1}^{(n)}$ $(0 \leqslant k \leqslant n)$, at which (see 7.6(5))

$$f(\xi_k^{(n)}) - P_n(f;\, \xi_k^{(n)}) = E_n \cos n \arccos \xi_k^{(n)} = (-1)^k E_n$$

and

$$f(\xi_{k+1}^{(n)}) - P_n(f;\, \xi_{k+1}^{(n)}) = E_n \cos n \arccos \xi_{k+1}^{(n)} = (-1)^{k+1} E_n.$$

Considering the relationship between the graphs of the functions $f(x) - P_n(f;\, x)$ and $E_n \cos n \arccos x$, we see that with such an assumption their difference $\phi_n(x)$ would have to have on $[-1, 1]$, at least $(n+2)$ zeros (taking account of the multiplicity of those which lie within $(-1, 1)$). But this is impossible, since $\phi_n(x)$ has only $n+1$ zeros altogether.

Hence it follows that (see the footnote on page 452)

$$\cos\frac{\pi}{n} < \xi_1^{(n)} \quad \text{and} \quad \xi_n^{(n)} < \cos\frac{n-1}{n}\pi.$$

7.6.3. Let us now verify that *if neither of a given pair of neighbouring points* $\xi_k^{(n)}$ *and* $\xi_{k+1}^{(n)}$ *is a point of the type* $\cos\frac{i\pi}{n}$ (*i an integer*), *then within the interval* $(\xi_{k+1}^{(n)}, \xi_k^{(n)})$ *there cannot be more than one value of* $\cos\frac{\nu\pi}{n}$.

In fact, in this case $\phi_n(\xi_k^{(n)}) \neq 0$ and $\phi_n(\xi_{k+1}^{(n)}) \neq 0$ (see section 7.6.2). Therefore (see 7.6(6)) in the interval $(\xi_{k+1}^{(n)}, \xi_k^{(n)})$ there occurs one and only one (in virtue of the regular monotonicity of $f(x)$; see the footnote on page 452) zero of the function $\phi_n(x)$. Hence in $(\xi_{k+1}^{(n)}, \xi_k^{(n)})$ there cannot fall more than two points of the type $\cos\frac{\nu\pi}{n}$ (ν an integer) (see 7.6(7)).

But if for some integer ν we had

$$\xi_{k+1}^{(n)} < \cos\frac{\nu+1}{n}\pi < \cos\frac{\nu\pi}{n} < \xi_k^{(n)}$$

then, since between $\cos\dfrac{\nu+1}{n}\pi$ and $\cos\dfrac{\nu\pi}{n}$ there is a zero of the function $\phi_n(x)$ (see 7.6(7)), we would have the inequalities

$$\phi_n(\xi_k^{(n)})\,\phi_n\left(\cos\frac{\nu\pi}{n}\right) > 0, \qquad \phi_n(\xi_{k+1}^{(n)})\,\phi_n\left(\cos\frac{\nu+1}{n}\pi\right) > 0.$$

Hence in virtue of 7.6(6) and 7.6(7) it would follow that

$$\Phi_n(\xi_k^{(n)})\,\Phi_n\left(\cos\frac{\nu\pi}{n}\right) \geqslant 0, \qquad \Phi(\xi_{k+1}^{(n)})\,\Phi_n\left(\cos\frac{\nu+1}{n}\pi\right) \leqslant 0,$$

the function $\Phi_n(x)$ on $[\xi_{k+1}^{(n)}, \xi_k^{(n)}]$ would have not less than two, and consequently on $[-1, 1]$ not less than $n+2$, zeros (see the footnote on page 452), which is impossible (see section 7.6.2).

7.6.4. Let us also show that *within* $(-1, 1)$ *there are no points* $\xi_k^{(n)}$ *which coincide with* $\cos\dfrac{\nu\pi}{n}$ *for any integer* ν. If such a point $\xi_k^{(n)}$ were to exist, then it would be a zero either of $\phi_n(x)$ or of $\Phi_n(x)$. Without restriction of generality it is possible to suppose that $\phi_n(\xi_k^{(n)}) = 0$. But then (see section 7.6.2)

$$\phi_n(\xi_{k+1}^{(n)}) \neq 0, \qquad \phi_n(\xi_{k-1}^{(n)}) \neq 0 \tag{8}$$

and (see 7.6(6)) $\phi_n(\xi_{k-1}^{(n)})\,\phi_n(\xi_{k+1}^{(n)}) > 0$.

Moreover, since

$$f(\xi_k^{(n)}) - P_n(f;\, \xi_k^{(n)}) = E_n \cos n \arccos \xi_k^{(n)} = E_n \cos n \arccos\frac{\nu\pi}{n},$$

then

$$[f(\xi_k^{(n)}) - P_n(f;\, \xi_k^{(n)})] \cos n \arccos\frac{\nu\pi}{n} > 0.$$

Hence (see 7.6(6)),

$$[f(\xi_{k-1}^{(n)}) - P_n(f;\, \xi_{k-1}^{(n)})] \cos n \arccos\frac{\nu-1}{n}\pi > 0,$$

$$[f(\xi_{k+1}^{(n)}) - P_n(f;\, \xi_{k+1}^{(n)})] \cos n \arccos\frac{\nu+1}{n}\pi > 0.$$

As a result of this it is possible to write the inequalities

$$\phi_n(\xi_{k-1}^{(n)})\phi_n\left(\cos\frac{\nu-1}{n}\pi\right) \leqslant 0, \quad \phi_n(\xi_{k+1}^{(n)})\phi_n\left(\cos\frac{\nu+1}{n}\pi\right) \leqslant 0, \tag{9}$$

each of which can become an equality only when

$$\phi_n\left(\cos\frac{\nu-1}{n}\pi\right) = 0 \quad \text{or} \quad \phi_n\left(\cos\frac{\nu+1}{n}\pi\right) = 0$$

(see 7.6(8)).

Since the point $\xi_k^{(n)}$ is a double zero of $\phi_n(x)$ (see 7.6.2), there cannot be other zeros of the function $\phi_n(x)$ on the segment $[\xi_{k+1}^{(n)}, \xi_{k-1}^{(n)}]$ (see the footnote on page 452). Hence in virtue of 7.6(9) it follows that the points $\cos\frac{\nu-1}{n}\pi$ and $\cos\frac{\nu+1}{n}\pi$ are located outside this segment.

Thus, if a point $\xi_k^{(n)}$ existed which coincided with $\cos\frac{\nu\pi}{n}$ (ν an integer), then the two half-open intervals $[\xi_{k+1}^{(n)}, \xi_k^{(n)})$ and $(\xi_k^{(n)}, \xi_{k-1}^{(n)}]$ abutting on it would not contain values of $\cos\frac{i\pi}{n}$ for integral i.

Comparing this result with the assertions of 7.6.2 and 7.6.3, we reach the conclusion that there cannot be within $(-1, 1)$ points $\xi_k^{(n)} = \cos\frac{\nu\pi}{n}$, and all the $n-1$ values of $\cos\frac{i\pi}{n}$ $(i = 1, 2, \ldots n-1)$ are situated each in one of the intervals $(\xi_{k+1}^{(n)}, \xi_k^{(n)})$. This proves theorem 7.6.

7.7. The asymptotic behaviour of the best uniform approximation by polynomials of some transcendental integral functions

In sections 7.2, 7.3 and 7.5 we studied the asymptotic behaviour of the best uniform approximation by polynomials of functions which on the segment considered or outside it have some singularity. Let us now consider the case when the function $f(x)$ being approximated has no singularities (i.e. it is an integral function) and let us prove the following theorem*.

7.7.1. *If*

$$f(x) = \sum_{n=0}^{\infty} c_n x^n \tag{1}$$

is an integral function which possesses the property that for some sequence of the natural numbers $n_1 < n_2 < n_3 < \dots$ *increasing without limit*

$$\lim_{n_k\to\infty} \sqrt[m]{\frac{(n_k+m)^{\frac{1}{2}(n_k+m)}|c_{n_k+m}|}{n_k^{\frac{1}{2}n_k}|c_{n_k}|}} = 0 \quad (c_{n_k} \neq 0), \tag{2}$$

uniformly with respect to $m \geqslant 1$, *then the best approximation* $E_{n_k-1}(f; -1, 1)$ *satisfies the limiting equality*

$$\lim_{n_k\to\infty} \frac{2^{n_k-1}E_{n_k-1}(f; -1, 1)}{|c_{n_k}|} = 1, \tag{3}$$

i.e. the sequence $E_{n_k-1}(f; -1, 1)$ *is asymptotically (as* $n_k\to\infty$*) identical with the sequence of numbers* $\frac{1}{2^{n_k-1}}|c_{n_k}|$.

7.7.2. The proof of this theorem is based on a similar proposition for integral functions of period 2π**.

Let

$$F(t) = \sum_{n=0}^{\infty} (a_n \cos nt + b_n \sin nt) = \sum_{n=0}^{\infty} \varrho_n \cos(nt+\alpha_n) \tag{4}$$

be an integral function and $\{n_k\}$ *a sequence of natural numbers* $n_1 < n_2 < n_3 < \dots$ *increasing without limit, for which*

$$\lim_{n_k\to\infty} \sqrt[m]{\frac{\varrho_{n_k+m}}{\varrho_{n_k}}} = 0 \quad (\varrho_{n_k} \neq 0) \tag{5}$$

uniformly with respect to $m \geqslant 1$***. *Then*

$$\lim_{n_k\to\infty} \frac{E^*_{n_k-1}(F)}{\varrho_{n_k}} = 1. \tag{6}$$

Let us consider the sequence of partial sums

$$S_{n_k-1}(F; t) = \sum_{\nu=0}^{n_k-1} \varrho_\nu \cos(\nu t - \alpha\nu). \tag{7}$$

The difference

$$F(t) - S_{n_k-1}(F; t) = \sum_{n=n_k}^{\infty} \varrho_\nu \cos(\nu t - \alpha_\nu)$$

** S. N. Bernstein [15].

*** It is not difficult to verify that such a sequence always exists. This follows directly from the fact that for any integral function $F(t)$ of period 2π the equation $\lim_{n\to\infty} \sqrt[n]{\varrho_n} = 0$ is satisfied (see 3.12.15).

at the $2n_k$ points $t_j = \dfrac{j\pi+\alpha_k}{n_k}$ $(j=0,1,\ldots,2n_k-1)$ on a period assumes the values

$$F(t_j)-S_{n_k-1}(F;\, t_j) = (-1)^j\varrho_{n_k}+\sum_{\nu=n_k+1}^{\infty}\varrho_\nu\cos(\nu t_j-\alpha_\nu).$$

But for any $\varepsilon>0$ there exists in virtue of 7.7(5) an N such that for all $n_k\geqslant N$ and $\nu>n_k$

$$\varrho_\nu<\varepsilon^{\nu-n_k}\varrho_{n_k}. \tag{8}$$

Hence, supposing that $n_k\geqslant N$ and $\varepsilon<\frac{1}{2}$, we have the inequality

$$\left|\sum_{\nu=n_k+1}^{\infty}\varrho_\nu\cos(\nu t_j-\alpha_\nu)\right|\leqslant\sum_{\nu=n_k+1}^{\infty}\varrho_\nu<\varrho_{n_k}\sum_{\nu=n_k+1}^{\infty}\varepsilon^{\nu-n_k}=\frac{\varepsilon}{1-\varepsilon}\varrho_{n_k},$$

as a result of which

$$(-1)^j[F(t_j)-S_{n_k-1}(F;\, t_j)]>\left[1-\frac{\varepsilon}{1-\varepsilon}\right]\varrho_{n_k}.$$

From theorem 2.10 it follows from this that

$$E^*_{n_k-1}(F)>\left(1-\frac{\varepsilon}{1-\varepsilon}\right)\varrho_{n_k}. \tag{9}$$

On the other hand, we have in virtue of 7.7(8)

$$E^*_{n_k-1}(F)\leqslant\max_t|F(t)-S_{n_k-1}(F;\, t)|\leqslant\sum_{\nu=n_k}^{\infty}\varrho_\nu<\left(1+\frac{\varepsilon}{1-\varepsilon}\right)\varrho_{n_k}.$$

Thus

$$1-\frac{\varepsilon}{1-\varepsilon}<\frac{E^*_{n_k-1}(F)}{\varrho_{n_k}}<1+\frac{\varepsilon}{1-\varepsilon},$$

i.e.

$$1-\frac{\varepsilon}{1-\varepsilon}\leqslant\varliminf_{n_k\to\infty}\frac{E^*_{n_k-1}(F)}{\varrho_{n_k}}\leqslant\varlimsup_{n_k\to\infty}\frac{E^*_{n_k-1}(F)}{\varrho_{n_k}}\leqslant1+\frac{\varepsilon}{1-\varepsilon}.$$

Since the value of $\varepsilon>0$ is arbitrary, this is equivalent to 7.7(6).

7.7.3. Let us now consider the integral function $F(t)=f(\cos t)$ of period 2π, where $f(x)$ is the function 7.7(1), which satisfies the conditions of theorem 7.7.1, and let

$$F(t)=\sum_{n=0}^{\infty}a_n\cos nt. \tag{10}$$

The Fourier coefficients of the series 7.7(10) can be expressed in terms of the coefficients c_n of the Taylor series 7.7(1). We have

$$F(t) = f(\cos t) = \sum_{n=0}^{\infty} c_n \cos^n t.$$

But as

$$\cos^n t = \frac{1}{2^n}(e^{it}+e^{-it})^n = \begin{cases} \dfrac{1}{2^{2p-1}}\left[\sum_{\nu=1}^{p}\binom{2p}{p-\nu}\cos 2\nu t+\dfrac{1}{2}\binom{2p}{p}\right] & \text{if} \quad n=2p, \\ \dfrac{1}{2^{2p}}\sum_{\nu=0}^{p}\binom{2p+1}{p-\nu}\cos(2\nu+1)\,t & \text{if} \quad n=2p+1, \end{cases}$$

where $p \geqslant 0$ is an integer, then

$$F(t) = c_0+\sum_{p=1}^{\infty}\frac{c_{2p}}{2^{2p-1}}\left\{\sum_{\nu=1}^{p}\binom{2p}{p-\nu}\cos 2\nu t+\frac{1}{2}\binom{2p}{p}\right\}+$$
$$+\sum_{p=0}^{\infty}\frac{c_{2p+1}}{2^{2p}}\sum_{\nu=0}^{p}\binom{2p+1}{p-\nu}\cos(2\nu+1)\,t.$$

Consequently

$$F(t) = \sum_{p=0}^{\infty}\frac{1}{2^{2p}}\binom{2p}{p}c_{2p}+\sum_{\nu=1}^{\infty}\cos 2\nu t\sum_{p=\nu}^{\infty}\frac{1}{2^{2p-1}}\binom{2p}{p-\nu}c_{2p}$$
$$+\sum_{\nu=0}^{\infty}\cos(2\nu+1)\,t\sum_{p=\nu}^{\infty}\frac{1}{2^{2p}}\binom{2p+1}{p-\nu}c_{2p+1}.$$

Thus for any $n \geqslant 1$

$$a_n = \begin{cases} \dfrac{1}{2^{n-1}}\sum_{p=\nu}^{\infty}\dfrac{1}{2^{2(p-\nu)}}\binom{2p}{p-\nu}c_{2p} & \text{if} \quad n=2\nu, \\ \dfrac{1}{2^{n-1}}\sum_{p=\nu}^{\infty}\dfrac{1}{2^{2(p-\nu)}}\binom{2p+1}{p-\nu}c_{2p+1} & \text{if} \quad n=2\nu+1, \end{cases}$$

or

$$a_n = \frac{1}{2^{n+1}}\left\{c_n+\sum_{j=1}^{\infty}\frac{(n+j+1)\dots(n+2j)}{j!\,4^j}c_{n+2j}\right\}. \tag{11}$$

If for a given sequence of indices $n_1 < n_2 < n_3 < \dots$ increasing to infinity the coefficients c_n satisfy condition 7.7(2), then, putting

$m = \mu+2j$ ($\mu \geqslant 0$ an integer) and choosing an arbitrarily small $\varepsilon > 0$, we have for any $j = 1, 2, \ldots$ and for all sufficiently great n_k,

$$|c_{n_k+\mu+2j}| < \varepsilon^{\mu+2j} \frac{|c_{n_k}|}{(n_k+\mu+2j)^j}. \tag{12}$$

Hence for $\varepsilon^2 < \frac{1}{2}$

$$\left| \sum_{j=1}^{\infty} \frac{(n_k+\mu+j+1) \ldots (n_k+\mu+2j)}{j!\, 4^j} c_{n_k+\mu+2j} \right| < \frac{|c_{n_k}|}{n_k+\mu} \cdot \frac{\varepsilon^{\mu+2}}{1-\varepsilon^2},$$

and in virtue of 7.7(11), for any integer $\mu \geqslant 0$,

$$\frac{1}{2^{n_k+\mu-1}} \left\{ |c_{n_k+\mu}| - \frac{|c_{n_k}|}{n_k+\mu} \frac{\varepsilon^{\mu+2}}{1-\varepsilon^2} \right\} < |a_{n_k+\mu}|$$
$$< \frac{1}{2^{n_k+\mu-1}} \left\{ |c_{n_k+\mu}| + \frac{|c_{n_k}|}{n_k+\mu} \frac{\varepsilon^{\mu+2}}{1-\varepsilon^2} \right\}. \tag{13}$$

Dividing the second of these inequalities by the first for $\mu = 0$ and taking account of the fact that by 7.7(2)

$$|c_{n_k+\mu}| < \frac{\varepsilon^\mu n_k^{\frac{1}{2}n_k}}{(n_k+\mu)^{\frac{1}{2}(n_k+\mu)}} |c_{n_k}|$$

for all sufficiently great values of n_k, we obtain the inequality

$$\left| \frac{a_{n_k}+\mu}{a_{n_k}} \right| < \frac{\varepsilon^\mu(1-\varepsilon^2)}{2^\mu(1-2\varepsilon^2)} \left\{ \frac{n_k^{\frac{1}{2}n_k}}{(n_k+\mu)^{\frac{1}{2}(n_k+\mu)}} + \frac{1}{n_k+\mu} \cdot \frac{\varepsilon^2}{1-\varepsilon^2} \right\},$$

which is thus, valid for all sufficiently great values of n_k and for any integer $\mu > 1$. By theorem 7.7.2 it follows that

$$\lim_{n_k \to \infty} \frac{E^*_{n_k-1}(F)}{|a_{n_k}|} = 1. \tag{14}$$

If we now observe that in virtue of 7.7(13)

$$\lim_{n_k \to \infty} \left| \frac{2^{n_k-1} a_{n_k}}{c_{n_k}} \right| = 1$$

and $E^*_{n_k-1}(F) = E_{n_k-1}(f; -1, 1)$, then from 7.7(14) we obtain 7.7(3).

7.7.4. In those cases when there is no sequence $\{n_k\}$ for which 7.7(2) holds uniformly with respect to $m \geqslant 1$, it is possible to find some $\varepsilon > 0$

and for every arbitrarily great N a sequence of values $m_1 < m_2 < m_3 < \ldots$, increasing to infinity, such that

$$\sqrt[m_k]{\frac{(N+m_k)^{\frac{1}{2}(N+m_k)}\,|c_{N+m_k}|}{N^{\frac{1}{2}N}\,|c_N|}} \geqslant \varepsilon,$$

i.e. in these cases $\overline{\lim_{n\to\infty}}\, n\sqrt[n]{|c_n|} > 0$.

Consequently theorem 7.7.1 is always applicable to functions $f(x)$ for which, as $n \to \infty$,

$$\sqrt[n]{|c_n|} = o\left(\frac{1}{\sqrt{n}}\right). \tag{15}$$

Frequently the sequence $\{n_k\}$ indicated in this theorem is identical with the whole sequence of natural numbers or with a sequence of all integers with the same parity. In these cases it is possible for us to judge the asymptotic value of the quantity $E_n(f; -1, 1)$ for any arbitrarily great n. We will limit ourselves to a single example.

Let

$$f(x) = \cos\lambda x = \sum_{k=0}^{\infty} (-1)^k \frac{\lambda^{2k}}{(2k)!} x^{2k}.$$

Here

$$c_n = \begin{cases} (-1)^k \dfrac{\lambda^{2k}}{(2k)!} & \text{if} \quad n = 2k, \\ 0, & \text{if} \quad n = 2k+1 \end{cases}$$

and the condition 7.7(2) is satisfied for the sequence of numbers $n_k = 2k$. Hence, taking into account the evenness of the function considered, we see in virtue of theorem 8.7.1, that

$$E_{2k+1}(\cos\lambda x; -1, 1) = E_{2k}(\cos\lambda x; -1, 1) = [1+o(1)]\frac{2\lambda^{2k}}{4^k(2k)!}$$

as $k \to \infty$.

7.8. Various problems and theorems

1. If $-1 < c < 1$, then for any natural numbers m and $s > 0$ the exact order of decrease of the sequence of best approximations $E_n[|x-c|^s (\ln|x-c|)^m; -1, 1]$ as $n \to \infty$ is equal to $\dfrac{(\ln n)^{m-1}}{n^s}$ when s is an even integer, and $\dfrac{(\ln n)^m}{n^s}$ in all the remaining cases. Consider also the case, when $c = \pm 1$.

See section 7.1.42.

2. If r is a non-negative integer and $0 < \alpha < 1$, then as $n \to \infty$

$$E_n[(1-x)^{r+\alpha};\ -1,1] = [1+o(1)]\,2^{r+\alpha} A_1(|x|^{2r+2\alpha})\,\frac{1}{n^{2r+2\alpha}}, \tag{1}$$

where $A_1(|x|^{2r+2\alpha})$ is the best approximation of the function $|x|^{2r+2\alpha}$ by integral functions of degree $\leqslant 1$ on $(-\infty, \infty)$.

See sections 7.2.1 and 7.5.4.

3. When r increases without limit, assuming odd values, and $-1 < \alpha < 1$, the constant $\mu(r, \alpha)$ (see section 7.2.1) is asymptotically equal to $\frac{1}{\pi}\left|\sin\frac{r+\alpha}{2}\pi\right|\Gamma(r+\alpha)$.

S. N. Bernstein [18].

4. If $-1 < \alpha < 1$, then as $\alpha \to 1$ the constant $\mu(1, \alpha)$ is asymptotically identical with the quantity $\frac{1}{\pi}\left|\sin\frac{1+\alpha}{2}\pi\right|\Gamma(1+\alpha)$.

S. N. Bernstein [18].

5. If $-1 < c_k < 1$ $(k = 1, \ldots, m)$, $c_\nu \neq c_k$ $(\nu \neq k)$, $0 < s_1 < s_2 < s_3 < \ldots < s_m$, $A_1 \neq 0$ and s_1 is not an even number, then as $n \to \infty$

$$E_n\Big[\sum_{k=1}^{m} A_k|x-c_k|^{s_k};\ -1,1\Big] = [1+o(1)]|A_1|\big(\sqrt{1-c_1^2}\big)^{s_1} E_n(|x|^{s_1};\ -1,1). \tag{2}$$

See sections 7.1.41, 7.1.45 and 7.2.2.

6. For any periodic function $f(x)$ of period 2π with Fourier series 7.1(31), in which $\lim \frac{n_{k+1}}{n_k} = \infty$, the asymptotic equality

$$E_n^*(f) = [1+o(1)]R_n(f), \tag{3}$$

is satisfied, where $R_n(f)$ is defined just as in section 7.1.5.

S. B. Stechkin [4].

7. If $c > 0$, then as $n \to \infty$

$$E_n\left[\frac{1}{c-\ln|x|};\ -1,1\right] = [1+o(1)]\frac{1}{2\ln n}. \tag{4}$$

S. N. Bernstein [32].

8. If m is a natural number and the value $s > 0$ is not even, then

$$E_n[|x|^s(\ln|x|)^m;\ -1,1] = [1+o(1)]\frac{(\ln n)^m}{n^s}A_1(|x|^s), \tag{5}$$

where $A_1(|x|^s)$ is the best approximation of the function $|x|^s$ by integral functions of degree $\leqslant 1$ on $(-\infty, \infty)$.

S. N. Bernstein [29].

9. If m and k are natural numbers, then as $n \to \infty$

$$E_n[x^{2k}(\ln|x|)^m;\ -1,1] = [1+o(1)]\,mA_1(x^{2k}\ln|x|)\frac{(\ln n)^{m-1}}{n^{2k}} \tag{6}$$

where $A_1(x^{2k}\ln|x|)$ is the best approximation of the function $x^{2k}\ln|x|$ by integral functions of degree $\leqslant 1$ on $(-\infty, \infty)$.

S. N. Bernstein [29].

10. If $|c| < 1$ and $s > -1$, then as $n \to \infty$

$$E_n[|x-c|^s; -1, 1]_L = [1+o(1)]\, M_s(\sqrt{1-c^2})^{s+1} \frac{1}{n^{s+1}} \tag{7}$$

where

$$M_s = \frac{4\left|\sin\frac{s\pi}{2}\right|}{\pi} \int_{-\infty}^{\infty} |\cos t| \int_0^{\infty} \frac{u^{s+1}\, du}{(t^2+u^2)(e^u+e^{-u})}\, dt. \tag{8}$$

For any $s > 0$ the constant M_s can be written in the form

$$M_s = \frac{8}{\pi}\left|\sin\frac{s\pi}{2}\right| \Gamma(s+1) \sum_{k=0}^{\infty} \frac{(-1)^k}{(2k+1)^{s+2}}. \tag{9}$$

S. M. Nikol'skii [7]. See also S. N. Bernstein [20].

11. Let $-1 < c_1 < c_2 < c_3 < \dots < c_m < 1$ and $s > -1$.

If $q(x) > 0$ is a function continuous on $[-1, 1]$, then for the best weighted mean approximation by algebraic polynomials, $E_{n, q(x)}[f; -1, 1]_L$, of the function

$$f(x) = \sum_{k=1}^{m} A_k |x-c_k|^s \tag{10}$$

we have, as $n \to \infty$, the asymptotic equality

$$E_{n, q(x)}(f; -1, 1)_L = \{1+o(1)\} \frac{M_s}{n^{s+1}} \sum_{k=1}^{m} |A_k| (\sqrt{1-c_k^2})^{s+1} q(c_k), \tag{11}$$

where M_s is the constant 7.8(8).

S. M. Nikol'skii [6].

12. If $s > -1$, then as $n \to \infty$

$$E_n[(1-x)^s; -1, 1]_L = \frac{2}{\pi}[1+o(1)] \frac{|\sin s\pi|}{2^{s-1}} \frac{\Gamma(2s+2)}{n^{2s+2}} \sum_{k=0}^{\infty} \frac{1}{(2k+1)^{2s+3}}. \tag{12}$$

N. I. Akhiezer [3].

13. As $n \to \infty$ the asymptotic equality

$$E_n[\ln(1-x); -1, 1]_L = [1+o(1)] \frac{4}{n^2} \sum_{k=0}^{\infty} \frac{1}{(2k+1)^3}. \tag{13}$$

is satisfied.

N. I. Akhiezer [3]. See section 2.12.33.

14. If $c > 1$ and $r \neq 0, -1, -2, -3, \dots$, then as $n \to \infty$

$$E_n\left[\frac{1}{(c-x)^r}; -1, 1\right] = [1+o(1)] \frac{n^{r-1}}{\Gamma(r)(\sqrt{c^2-1})^{r+1}(c+\sqrt{c^2-1})^n}. \tag{14}$$

Formula 7.8(14) is a generalisation of 7.5(2).

S. N. Bernstein [29].

15. The asymptotic relation

$$E_n(f'; -1, 1) = [1+o(1)]\frac{n}{\sqrt{c^2-1}}E_n(f; -1, 1)$$

$$= [1+o(1)]\frac{d}{dc}E_n(f; -1, 1) \qquad (15)$$

is satisfied for the function 7.5(3) and its derivative, as $n \to \infty$.

S. N. Bernstein [15].

16. If $A_\sigma(f)$ is the best approximation of the function $f(x)$ by integral functions of degree $\leqslant \sigma$ on $(-\infty, \infty)$, then whatever the real value of c, in the case where $r \neq 0, -1, -2, -3, \ldots$, the asymptotic equality

$$A_\sigma\left[\frac{1}{(x^2+c^2)^r}\right] = [1+o(1)]\frac{e^{-\sigma c}\sigma^{r-1}}{2^r\Gamma(r)c^{r+1}}. \qquad (16)$$

is satisfied for $\sigma \to \infty$.

S. N. Bernstein [29].

17. If the function $f(x)$ defined on $[-1, 1]$ is infinitely differentiable, then for some unbounded sequence of values of n there exist algebraic polynomials $P_n(x)$ for which

$$\max_{-1\leqslant x\leqslant 1} |f(x)-P_n(x)| = [1+o(1)]E_n(f; -1, 1)$$

and such that the points $\xi_k^{(n)}$, where the maximum modulus of the difference $|f(x)-P_n(x)|$ is attained, satisfy the limiting equality

$$\lim_{n\to\infty}\left(\xi_k^{(n)} - \cos\frac{k\pi}{n}\right) = 0. \qquad (17)$$

S. N. Bernstein [15].

18. If $c > 1$ and the function $\Phi(x)$ is analytic in a certain ellipse containing the point c within it and with foci at the ends of the segment $[-1, 1]$, then, subject to the conditions that $\Phi(c) \neq 0$ and m is not an integer, the asymptotic equality

$$E_n[(c-x)^m\Phi(x); -1, 1] = [1+o(1)]|\Phi(c)|\,E_n[(c-x)^m; -1, 1] \qquad (18)$$

is always satisfied as $n \to \infty$.

S. N. Bernstein [33].

19. Show that the sequence of best approximations $E_n^*(f)$ of the function

$$f(x) = \sum_{n=1}^{\infty}\frac{\cos nx}{n!} \qquad (19)$$

always satisfies the asymptotic equation

$$E_n^*(f) = \frac{1+o(1)}{(n+1)!} \qquad (20)$$

as $n \to \infty$.

See section 7.7.2.

20. Show that as $k\to\infty$

$$E_{2k}(\sin\lambda x;\,-1,1)=E_{2k-1}(\sin\lambda x;\,-1,1)$$
$$=[1+o(1)]\frac{\lambda^{2k+1}}{4^k(2k+1)!}. \tag{21}$$

See section 7.7.4.

21. Show that as $n\to\infty$

$$E_n(e^{\lambda x};\,-1,1)=[1+o(1)]\frac{|\lambda|^{n+1}}{2^n(n+1)!}. \tag{22}$$

See section 7.7.4.

22. Show that as $k\to\infty$

$$E_{2k-1}(e^{x^2};\,-1,1)=E_{2k-2}(e^{x^2};\,-1,1)=[1+o(1)]\frac{\sqrt{e}}{2^{2k-1}k!}. \tag{23}$$

See section 7.8.21.

23. Show that as $k\to\infty$

$$E_{2k-1}(e^{-x^2};\,-1,1)=E_{2k-2}(e^{-x^2};\,-1,1)=\frac{1+o(1)}{\sqrt{e}\,2^{2k-1}k!}. \tag{24}$$

See section 7.8.21.

CHAPTER VIII

LINEAR PROCESSES OF APPROXIMATION OF FUNCTIONS BY POLYNOMIALS AND SOME ESTIMATES CONNECTED WITH THEM

8.1. On the convergence of linear processes of approximation of functions by polynomials

In connection with Weierstrass's theorem (see Chap. I) and the results of Chap. V which form an extension of it, there arose the problem of the study of the methods most convenient, in this or that respect, of constructing for each function f considered sequences of polynomials approximating to it. If use is made of the terminology of functional analysis, then to every such method there corresponds a certain sequence of finite-dimensional (polynomial) operators $U_n(f)$, defined in the function space F considered and mapping it into the finite-dimensional subspaces F_n $(F_n \subset F_{n+1})$ of polynomials of order $\leqslant n$.

From the point of view of the rapidity with which $U_n(f)$ tends to f (in norm in F) as $n \to \infty$, the most advantageous is the sequence of operators which establishes a correspondence between each function f and its best polynomial approximation (see section 2.2). Such operators are homogeneous, i.e. $U_n(\lambda f) = \lambda U_n(f)$, and as a result of the inequality $\|U_n(f)\|_F \leqslant 2\|f\|_F$ bounded. However, not being in general additive, they do not belong to the category of linear operators**, and this circumstance in the general case materially impedes their study and the more effective construction of the corresponding polynomials. The exact determination of such polynomials is successfully achieved when F is a Hilbert space (here the given operators $U_n(f)$ are additive) and for individual concrete functions in other spaces (see sections 2.9 and 2.11).

** Let us recall that the operator $U(f)$ which maps the space F onto part of itself is said to be linear if it is bounded, i.e. $\|U\| = \sup\limits_{\|f\|_F \leqslant 1} \|U(f)\|_F < \infty$ and for any pair of functions f_1, f_2 of F

$$U(f_1+f_2) = U(f_1)+U(f_2).$$

If however we do not make it our aim to obtain for each individual function $f \in F$ the best rapidity of convergence of the sequence $U_n(f)$ and restrict ourselves to the class of linear polynomials operators**, then the investigation of the problem in which we are interested becomes more accessible.

An important part is played here by the following well-known theorem of Banach.

In order that the sequence of linear operators $U_n(f)$ which maps the complete linear normed space F into part of itself should possess the property

$$\lim_{n\to\infty} \|f - U_n(f)\|_F = 0, \tag{1}$$

whatever the function $f \in F$ may be, it is necessary that the condition

$$\|U_n\| = \sup_{\|f\|_F \leqslant 1} \|U_n(f)\|_F = O(1) \tag{2}$$

should be satisfied when $n \to \infty$.

8.1.1. The sequence of operators $U_n(f)$, which establishes a correspondence between each function f and its best polynomial approximation of order n, possesses in addition to the property 8.1(1) the property that for any n every polynomial of order $\leqslant n$ is mapped by the operator U_n into itself, i.e. if f is such a polynomial, then $U_n(f) = f$.

Is it possible that any sequence whatever of linear finite-dimensional operators in the spaces C and L_q possesses these two properties? In the spaces L_q for $1 < q < \infty$, as is well known from the theory of trigonometric series***, the sequence of partial sums of the expansion in a Fourier series possesses the properties indicated.

It is not difficult to show that in the classical case, when F is the space C of functions $f(x)$ continuous on the given finite segment $[a, b]$ ($\|f\| = \max\limits_{a \leqslant x \leqslant b} |f(x)|$), and F_n is the subspace of algebraic polynomials

$$P_n(x) = \sum_{k=0}^{n} c_k x^k \tag{3}$$

of degree $\leqslant n$ or the space C^* of continuous functions of period 2π, where $\|f\| = \max\limits_{t} |f(t)|$, and the subspace F_n consists of all trigonometric polynomials

** The papers of S. M. Lozinski [2, 4, 5] and D. L. Berman [1, 2] are devoted to the study of different properties of such operators.

*** See A. Zygmund [1].

$$T_n(t) = \sum_{k=0}^{n} (a_k \cos kt + b_k \sin kt) \tag{4}$$

of order $\leqslant n$, the answer must be negative**.

Let $U_n(f)(n = 0, 1, 2, \ldots)$ be a sequence of linear operators defined in the space C^* and such that for any function $f(t) \in C^*$ $U_n(f)$ is some trigonometric polynomial 8.1(4), and if $T(t)$ is a trigonometric polynomial of order $\leqslant n$, then $U_n(T_n) \equiv T_n(t)$. We shall show that in this case condition 8.1(2) is violated.

Let us consider the sequence of trigonometric polynomials

$$T_{2n-1}(t) = 2 \sin nt \sum_{k=1}^{n-1} \frac{\sin kt}{k} \tag{5}$$

of order $2n-1$, which is uniformly bounded with respect to t and n (see 4.12.12).

For any value of the parameter α let us put

$$T_{2n-1}^{(\alpha)}(t) = T_{2n-1}(t+\alpha) + T_{2n-1}(t-\alpha). \tag{6}$$

Since

$$T_{2n-1}^{(\alpha)}(t) = 2 \sum_{k=1}^{n-1} \frac{\cos kt \cos k\alpha}{n-k} - 2 \sum_{k=1}^{n-1} \frac{\cos (k+n)\, t \cos (k+n)\, \alpha}{k},$$

then in virtue of the properties of the operator $U_n(f)$

$$\begin{aligned} U_{n-1}[T_{2n-1}^{(\alpha)}] &= 2 \sum_{k=1}^{n-1} \frac{\cos kt \cos k\alpha}{n-k} - 2 \sum_{k=1}^{n-1} V_k(t) \frac{\cos (n+k)\, \alpha}{k} \\ &= \sum_{k=1}^{n-1} \frac{\cos k(t-\alpha)}{n-k} + \sum_{k=1}^{n-1} \left[\frac{\cos k(t+\alpha)}{n-k} - 2V_k(t) \frac{\cos (n+k)\, \alpha}{k} \right], \end{aligned} \tag{7}$$

where $V_k(t) = U_{n-1}[\cos (n+k)\, t]$ $(k = 1, 2, \ldots, n-1)$ are trigonometric polynomials of order $\leqslant n-1$.

We also note that for any integer $k \geqslant 1$

$$\int_{-\pi}^{\pi} \left[\frac{\cos 2kt}{n-k} - 2V_k(t) \frac{\cos (n+k)\, t}{k} \right] dt = 0.$$

Consequently, there exists a value t_0, for which

$$\sum_{k=1}^{n-1} \left[\frac{\cos 2kt_0}{n-k} - 2V_k(t_0) \frac{\cos (n+k)\, t_0}{k} \right] = 0$$

** This result is due to S. M. Lozinskii and F. I. Kharshiladza. See S. M. Lozinskii [2], I. P. Natanson [2].

and in virtue of 8.1(7) the trigonometric polynomial $W_{n-1}(t) = U_{n-1}[T_{2n-1}^{(t_0)}]$ assumes the value

$$W_{n-1}(t_0) = U_{n-1}[T_{2n-1}^{(t_0)}] = \sum_{k=1}^{n-1} \frac{1}{k},$$

at the point $t = t_0$. But

$$|W_{n-1}(t_0)| \leqslant \max_t |W_{n-1}(t)| = \|U_{n-1}[T_{2n-1}^{(t_0)}]\|_{C^*} \leqslant \|U_{n-1}\| \max_t |T_{2n-1}^{(t_0)}(t)|.$$

Therefore, taking into account that $\max_t |T_{2n-1}^{(t_0)}(t)| = O(1)$, we obtain $\lim_{n\to\infty} \|U_{n-1}\| = \infty$.

The case when $U_n(f)$ $(n = 0, 1, 2, \ldots)$ is a sequence of linear operators defined in the space C of functions continuous on the finite segment $[a, b]$, is easily reduced to that considered. Without restriction of generality it is possible to suppose that $a = -1$, $b = 1$. After the substitution $x = \cos t$ we obtain the sequence of operators $U_n[f(\cos t)]$, defined in the space of continuous even functions $\phi(t) = f(\cos t)$ of period 2π. If for any function $f(x) \in C$, $U_n(f)$ is some polynomial 8.1(3), then $U_n(\phi) = U_n[f(\cos t)]$ is the even trigonometric polynomial 8.1(4). If for any polynomial $P_n(x)$ of degree $\leqslant n$, $U_n(P_n) \equiv P_n(x)$, then whatever the even trigonometric polynomial $T_n(t)$ of order $\leqslant n$, $U_n(T_n) = U_n[P_n(\cos t)] \equiv T_n(t)$. It remains to note that for any value of the parameter α the trigonometric polynomials 8.1(6) are even.

The given theorem, if account had been taken of the estimate 4.12.6, could also have been obtained from the identity**

$$\frac{1}{2\pi} \int_0^{2\pi} U_n(f_t; x-t)\, dt = \dot{S}_n(f; x),$$

in which $f(x) \in C^*$, $f_t(x) = f(x+t)$, $S_n(f; x)$ is the partial sum of order n of the Fourier series of $f(x)$, and $U_n(f; x)$ are the operators considered here. In the case when $f(x)$ is a trigonometric polynomial it is verified immediately, and for arbitrary continuous functions by means of Weierstrass's theorem (see 1.3.1). From the same considerations it follows that this identity is valid for the linear polynomial operators $U_n(f; x)$ defined in the space L_1^* of all integrable functions $f(x)$ of period 2π. It shows that in this case

$$\int_0^{2\pi} |S_n(f; x)|\, dx \leqslant \frac{1}{2\pi} \int_0^{2\pi} dt \int_0^{2\pi} |U_n(f_t; x-t)|\, dt \leqslant \|U_n\| \cdot \|f\|_{L_1^*},$$

** D. L. Berman [3]. See also Marcinkiewicz and Zygmund [1].

i.e. (see 4.12.6),

$$||U_n|| \geqslant \sup_{||f||_{L_1^*} \leqslant 1} \int_0^{2\pi} |S_n(f; x)| \, dx = \frac{1}{\pi} \int_0^{2\pi} |D_n(t)| \, dt = \frac{4}{\pi} \ln n + O(1) \quad (n > 0).$$

Thus, as in the case of the space C^*, there does not exist a sequence of linear polynomial operators satisfying condition 8.1(1), where F is the space L_1^*, for all functions f of L_1^*.

8.1.2. The simplest and at the same time most natural example of a linear process of approximation of continuous functions by polynomials is the approximation by the sequences of partial sums of their Fourier series expansions (the trigonometric system, in the periodic case, and the system of Chebyshev polynomials $\cos n \arccos \frac{2x-a-b}{b-a}$ in the case of the finite segment $[a, b]$, or interpolation by Lagrange's formula (see sections 4.1 and 4.2).

However, as follows from section 8.1.1, *the sequence of partial sums of a Fourier series is not uniformly convergent in the whole class of continuous functions***. For this reason, if $P_n(f; x)$ is a sequence of algebraic polynomials of degree $\leqslant n$ which at the $n+1$ points $x_0^{(n)} < x_1^{(n)} < \dots < x_n^{(n)}$ on $[a, b]$ are identical with the function $f(x)$, *then whatever the system of points* $x_k^n (k = 0, 1, \dots, n; n = 0, 1, 2, \dots)$, *the sequence* $P_n(f; x)$ *cannot converge uniformly for all continuous functions***. Similarly, if $T_n(f; t)$ is a sequence of trigonometric polynomials of order $\leqslant n$, which at the $2n+1$ points $t_0^{(n)} < t_1^{(n)} < \dots < t_{2n}^{(n)}$ on $(0, 2\pi]$ is identical with the function $T_n(f; t)$, *then whatever the system of points* $t_k^{(n)} (k = 0, 1, \dots, 2n;\ n = 0, 1, 2, \dots)$, *the sequence* $T_n(f; t)$ *cannot converge uniformly for all continuous functions****.

This shows that the partial sums of Fourier series expansions and Lagrange interpolation polynomials are not of themselves, for any system of points whatever, suitable as a means of uniform approximation in the whole class of continuous functions. In this connection there arises the question of how it is possible, starting with a Fourier series expansion or interpolation formula of Lagrange, to achieve by means of some variation of the given processes uniform convergence of the resulting polynomials for any continuous function.

8.1.3. One of the best-known methods for achieving this purpose is the so-called summation of Fourier series. In the periodic case many

** The first example showing this was due to Du Bois Reymond [1].

*** Faber [1], S. N. Bernstein [4].

methods of summation consists in specifying some infinite triangular matrix of numbers $\lambda_k^{(n)}(k = 0, 1, \ldots, n;\ n = 0, 1, 2, \ldots)$, by means of which the partial sums of the Fourier series of any continuous function of period 2π are transformed into the form

$$U_n^*(f;\, x;\, \lambda) = \frac{a_0}{2}\lambda_0^{(n)} + \sum_{k=1}^{n} \lambda_k^{(n)}(a_k \cos kx + b_k \sin kx), \tag{8}$$

where

$$a_k = \frac{1}{\pi}\int_0^{2\pi} f(t)\cos kt\,\mathrm{d}t, \qquad b_k = \frac{1}{\pi}\int_0^{2\pi} f(t)\sin kt\,\mathrm{d}t \tag{9}$$

are the Fourier coefficients.

Similarly for the function $f(x)$, defined on $[-1, 1]$, the corresponding sequence of algebraic polynomials

$$U_n(f;\, x;\, \lambda) = \frac{c_0}{2}\lambda^{(n)} + \sum_{k=1}^{n} \lambda_k^{(n)} c_k \cos k \arccos x, \tag{10}$$

is considered, where the

$$c_k = \frac{2}{\pi}\int_0^{\pi} f(\cos t)\cos kt\,\mathrm{d}t \tag{11}$$

are obtained from the expansion of $f(x)$ in a Fourier series of Chebyshev polynomials.

In the study of linear processes of approximation of continuous functions by the trigonometric polynomials 8.1(8) or by the algebraic polynomials 8.1(10), it is in the first place natural to consider the problem of the conditions which must be satisfied by the system of numbers $\lambda_k^{(n)}$ in order that, as $n \to \infty$,

$$U_n^*(f;\, x\colon \lambda) \to f(x)$$

or

$$U_n(f;\, x;\, \lambda) \to f(x)$$

for all continuous functions.

It is easily understood that the solution of this problem follows directly from the general propositions of functional analysis. In fact, if the point x is fixed, then the sum 8.1(8) for any n represents a linear functional defined in the complete linear space C^* of all continuous functions of period 2π, in which in virtue of Weierstrass's theorem (see section 1.3.1) the sequence of functions cos kx, sin kx $(k = 0, 1, 2, \ldots)$ is closed (see section 2.2). And correspondingly, if the point x is fixed, then the sum 8.1(10) for any n represents a linear functional defined in the space C of all functions continuous on $[-1, 1]$, in which in virtue of Weierstrass's theorem (see section 1.3.1) the sequence $1, x, x^2, \ldots$ or the sequence of polynomials $\cos k \arccos x$ $(k = 0, 1, 2, \ldots)$ is closed (see section 2.2). Thus it is a question of the conditions which must be satisfied by the system of numbers $\lambda_k^{(n)}$ in order that the given sequence of linear functionals should converge weakly, as $n \to \infty$, in the space C^* (or in the space C respectively). It is known that *a necessary and sufficient condition for this is the boundedness of the sequence of norms of these functionals and the convergence of the sequence of their values for every function of some system closed* (see section 2.2) *in the given space.*

In our case these norms are respectively

$$\sup_{|f(t)|\leqslant 1} |U_n^*(f; x; \lambda)|, \qquad \sup_{|f(t)|\leqslant 1} |U_n(f; x; \lambda)|.$$

Since

$$U_n^*(f; x; \lambda) = \frac{1}{\pi}\int_0^{\pi} f(t)\left\{\frac{\lambda_0^{(n)}}{2} + \sum_{k=1}^{n} \lambda_k^{(n)} \cos k(x-t)\right\} dt \tag{12}$$

and

$$U_n(f; x; \lambda)$$

$$= \frac{2}{\pi}\int_0^{\pi} f(\cos t)\left\{\frac{\lambda_0^{(n)}}{2} + \sum_{k=1}^{n} \lambda_k^{(n)} \cos k \arccos x \cdot \cos kt\right\} dt, \tag{13}$$

they may be expressed in the form of the integrals

$$\sup_{|f(t)|\leqslant 1} |U_n^*(f; x; \lambda)| = L_n = \frac{2}{\pi}\int_0^{\pi} \left|\frac{\lambda_0^{(n)}}{2} + \sum_{k=1}^{n} \lambda_k^{(n)} \cos kt\right| dt \tag{14}$$

and

$$\sup_{|f(t)|\leqslant 1} |U_n(f; x; \lambda)| = L_n(x)$$

$$= \frac{2}{\pi}\int_0^{\pi} \left|\frac{\lambda_0^{(n)}}{2} + \sum_{k=1}^{n} \lambda_k^{n} \cos k \arccos x \cdot \cos kt\right| dt. \tag{15}$$

Consequently, the first condition reduces to the boundness of the sequence of numbers L_n (or correspondingly $L_n(x)$).

If the second condition of the above theorem on weak convergence in the periodic case is applied to the sequence of functions $\cos kx$, $\sin kx$ ($k = 0, 1, 2, \ldots$), then we find that for any fixed value of $k = 0, 1, 2, \ldots$ the limit equality

$$\lim_{n\to\infty} \lambda_k^{(n)} = 1 \tag{16}$$

must be satisfied.

Whatever $\varepsilon > 0$ may be, it is possible by theorem 1.3.1, for any continuous function $f(x)$ of period 2π, to find a trigonometric polynomial $T_m(f; x)$ such that $\max_x |f(x) - T_m(f; x)| < \varepsilon$. Hence for any n

$$\begin{aligned}\max_x |f(x) - U_n^*(f; x; \lambda)| &\leqslant \max_x |f(x) - T_m(f; x)| + \\ &+ \max_x |T_m(f; x) - U_n^*(T_m; x; \lambda)| + \max_x |U_n^*(T_m - f; x; \lambda)| \\ &< (1 + L_n)\varepsilon + \max_x |T_m(f; x) - U_n^*(T_m; x; \lambda)|. \end{aligned} \tag{17}$$

It is obvious from this that when condition 8.1(16) is satisfied, then when the sequence of norms L_n is bounded the sequence of polynomials $U_n^*(f; x; \lambda)$ converges uniformly to $f(x)$ as $n \to \infty$ for any function $f \in C^*$.

Thus *condition* 8.1(16) *together with condition*

$$L_n = \frac{2}{\pi}\int_0^\pi \left| \frac{\lambda_0^{(n)}}{2} + \sum_{k=1}^n \lambda_k^{(n)} \cos kt \right| dt = O(1) \tag{18}$$

is necessary and sufficient for the convergence at every point and for the uniform convergence to $f(x)$ *of the sequence of trigonometric polynomials* 8.1(8) *in the whole class of continuous functions of period* 2π**.

We see that in the periodic case, in distinction from the case of a finite segment, the norms of the corresponding functionals do not depend on the position of the point x. Hence the quantity L_n is at the same time also the norm of the operator 8.1(10).

In the non-periodic case the norms of the operators 8.1(10) are defined by the equation

$$\|U_n\| = \max_{-1 \leqslant x \leqslant 1} L_n(x).$$

The formula (see 8.1(13))

$$U_n(f; x; \lambda) = \frac{1}{\pi}\int_{-\pi}^{\pi} f(\cos t)\left\{\frac{\lambda_0^{(n)}}{2} + \sum_{k=1}^n \lambda_k^{(n)} \cos k(t - \arccos x)\right\} dt$$

** S. M. Nikol'skii [9].

shows that, subject to the condition $|f(\cos t)| \leqslant 1$,

$$|U_n(f; x; \lambda)| \leqslant \frac{1}{\pi} \int_{-\pi}^{\pi} \left| \frac{\lambda_0^{(n)}}{2} + \sum_{k=1}^{n} \lambda_k^{(n)} \cos k(t - \arccos x) \right| dt = L_n .$$

Consequently

$$L_n(x) \leqslant L_n,$$

and since $L_n(1) = L_n$, then

$$\max_{-1 \leqslant x \leqslant 1} L_n(x) = L_n, \tag{19}$$

i.e. the quantity L_n is also a norm of the operator 8.1(10). Hence by Banach's theorem (see section 8.1, condition 8.1(2)), the condition 8.1(18) is necessary for the uniform convergence of the sequence of polynomials 8.1(10) in the whole class of functions continuous in $[-1, 1]$. Since $L_n(1) = L_n$, then in virtue of the first condition of the theorem on weak convergence of a sequence of linear functionals given above it is also necessary for the convergence of these polynomials at each point.

If the second condition of the given theorem is applied to the system of polynomials $\cos k \arccos x$ $(k = 0, 1, 2, \ldots)$ at the point $x = 1$, then we obtain the necessity of condition 8.1(16).

On the other hand, reasoning as in the periodic case (see 8.1(17)), we find that when conditions 8.1(16) and 8.1(18) are satisfied the sequence of polynomials $U_n(f; x; \lambda)$ converges uniformly to $f(x)$ as $n \to \infty$ for any function continuous on $[-1, 1]$.

Thus *these two conditions are necessary and sufficient both for convergence at each point and for uniform convergence to* $f(x)$ *of the sequence of algebraic polynomials* 8.1(10) *on the whole class* C *of functions continuous on* $[-1, 1]$.

We also note that, as is obvious from the inequality

$$|\lambda_\nu^{(n)}| = \frac{2}{\pi} \left| \int_0^{\pi} \cos \nu t \left[\frac{\lambda_0^{(n)}}{2} + \sum_{k=1}^{n} \lambda_k^{(n)} \cos kt \right] dt \right| \leqslant L_n, \tag{20}$$

condition 8.1(18) *can hold only when*

$$\lambda_\nu^{(n)} = O(1) \tag{21}$$

uniformly with respect to all n *and* ν $(\nu = 0, 1, 2, \ldots, n)$.

8.1.4. To the transformations 8.1(8) and 8.1(10) for the partial sums of expansions in Fourier series there correspond similar transformations of the polynomials obtained by interpolation by Lagrange's formula for the system of equally spaced points 4.2(7) or for the Chebyshev system 4.1(7).

If the trigonometric polynomial $T_n(x)$ of order $\leqslant$ interpolates the periodic function $f(x)$ of period 2π at the points $y_k^{(n)} = \dfrac{2k\pi}{2n+1}$ $(k = 0, 1, \ldots, 2n)$, that is $T_n(y_k^{(n)}) = f(y_k^{(n)})$, then by formula 4.2(9)

$$T_n(x) = \frac{2}{2n+1}\sum_{k=0}^{2n} f(y_k^{(n)})\,D_n(x-y_k^{(n)})$$

$$= \frac{a_0^{(n)}}{2} + \sum_{\nu=1}^{n} (a_\nu^{(n)} \cos \nu x + b_\nu^{(n)} \sin \nu x), \tag{22}$$

where $D_n(t)$ is Dirichlet's kernel (see 4.1(10)) and

$$a_\nu^{(n)} = \frac{2}{2n+1}\sum_{k=0}^{2n} f(y_k^{(n)}) \cos \nu y_k^{(n)}, \qquad b_\nu^{(n)} = \frac{2}{2n+1}\sum_{k=0}^{2n} f(y_k^{(n)}) \sin \nu y_k^{(n)}. \tag{23}$$

Similarly, if the algebraic polynomial $P_n(x)$ of degree $\leqslant n$ interpolates the function $f(x)$, defined on $[-1, 1]$, at the points $x_k^{(n+1)} = \cos \dfrac{2k+1}{2(n+1)}\pi$ $(k = 0, 1, \ldots, n)$, i.e. $P_n(x_k^{(n+1)}) = f(x_k^{(n+1)})$, then by formula 4.1(11)

$$P_n(x) = \frac{1}{n+1}\sum_{k=0}^{n} f(x_k^{(n+1)})\,[D_n(y+t_k^{(n+1)}) + D_n(y-t_k^{(n+1)})]$$

$$= \frac{c_0^{(n)}}{2} + \sum_{\nu=1}^{n} c_\nu^{(n)} \cos \nu \arccos x, \tag{24}$$

where $y = \arccos x$, $t_k^{(n+1)} = \dfrac{2k+1}{2(n+1)}\pi$ and

$$c_\nu^{(n)} = \frac{1}{n+1}\sum_{k=0}^{n} f(x_k^{(n+1)}) \cos \nu \arccos x_k^{(n+1)} \qquad (\nu = 0, 1, \ldots, n). \tag{25}$$

Every infinite triangular matrix of numbers $\lambda_\nu^{(n)} (\nu = 0, 1, \ldots, n; 0, 1, 2, \ldots)$ defines a sequence of trigonometric polynomials

$$V_n^*(f; x; \lambda) = \frac{a_0^{(n)}}{2}\lambda_0^{(n)} + \sum_{\nu=1}^{n} \lambda_\nu^{(n)} (a_\nu^{(n)} \cos \nu x + b_\nu^{(n)} \sin \nu x) \tag{26}$$

or a sequence of algebraic polynomials

$$V_n(f; x; \lambda) = \frac{c_0^{(n)}}{2}\lambda_0^{(n)} + \sum_{\nu=1}^{n} \lambda_\nu^{(n)} c_\nu^{(n)} \cos \nu \arccos x \tag{27}$$

obtained from 8.1(22) or 8.1(24) by the same method as the polynomials 8.1(8) or the polynomials 8.1(10) from the partial sums of Fourier series.

Since

$$V_n^*(f; x; \lambda) = \frac{2}{2n+1}\sum_{k=0}^{2n} f(y_k^{(n)})\left\{\frac{\lambda_0^{(n)}}{2} + \sum_{\nu=1}^{n} \lambda_\nu^{(n)} \cos \nu (x - y_k^{(n)})\right\} \tag{28}$$

and

$$V_n(f; x; \lambda) = \frac{1}{n+1}\sum_{k=0}^{n} f(x_k^{(n+1)})\left\{\frac{\lambda_0^{(n)}}{2} + \right.$$

$$\left. + \sum_{\nu=1}^{n} \lambda_\nu^{(n)} \cos \nu \arccos x \cdot \cos \nu \arccos x_k^{(n+1)}\right\}, \tag{29}$$

the question of the conditions which the system of numbers $\lambda_\nu^{(n)}$ must satisfy in order that $V_n^*(f; x; \lambda) \to f(x)$ or $V_n(f; x; \lambda) \to f(x)$ as $n \to \infty$, for all continuous functions, just as in section 8.1.3 reduces to the investigation of the upper bounds

$$M_n^*(x) = \sup_{|f(t)| \leqslant 1} |V_n^*(f; x; \lambda)|$$

$$= \frac{2}{2n+1}\sum_{k=0}^{2n}\left|\frac{\lambda_0^{(n)}}{2} + \sum_{\nu=1}^{n} \lambda_\nu^{(n)} \cos \nu (x - y_k^{(n)})\right| \tag{30}$$

and

$$M_n(x) = \sup_{|f(t)| \leqslant 1} |V_n(f; x; \lambda)|$$

$$= \frac{1}{n+1}\sum_{k=0}^{n}\left|\frac{\lambda_0^{(n)}}{2} + \sum_{\nu=1}^{n} \lambda_\nu^{(n)} \cos \nu \arccos x \cdot \cos \nu \arccos x_k^{(n+1)}\right|. \tag{31}$$

8.1.5. The two methods considered of construction of convergent linear processes of approximation by polynomials can also be applied to functions of many variables. Thus, for example, in the study of the class of continuous functions $f(x, y)$ of period 2π in each of the variables, any infinite matrix of numbers $\lambda_{k, l}^{(m, n)}(k = 0, 1, \ldots, m;\ l = 0, 1, \ldots n;$ $m, n = 0, 1, 2, \ldots)$ defines the sequence of trigonometric polynomials

$$U_{m, n}^*(f; x; y; \lambda)$$

$$= \lambda_{0, 0}^{(m, n)} \frac{a_{0, 0}}{4} + \frac{1}{2}\sum_{k=1}^{m} \lambda_{k, 0}^{(m, n)}(a_{k, 0} \cos kx + b_{k, 0} \sin kx) +$$

$$+ \frac{1}{2}\sum_{l=1}^{m} \lambda_{0, l}^{(m, n)}(a_{0, l} \cos ly + b_{0, l} \sin ly) +$$

$$+ \sum_{k=1}^{m}\sum_{l=1}^{n} \lambda_{k, l}^{(m, n)}(a_{k, l} \cos kx \cos ly + b_{k, l} \sin kx \cos ly +$$

$$+ c_{k, l} \cos kx \sin ly + d_{k, l} \sin kx \sin ly), \tag{32}$$

where $a_{k, l}$, $b_{k, l}$, $c_{k, l}$, $d_{k, l}$ are the Fourier coefficients of $f(x, y)$.

From the general considerations of functional analysis given in section 8.1.3, it follows that necessary and sufficient conditions that $U^*_{m,n}(f; x; y; \lambda) \to f(x, y)$ as $m, n \to \infty$ at every point (x, y) or uniformly with respect to (x, y), are the conditions

$$\lim_{m, n\to\infty} \lambda_{k, l}^{(m, n)} = 1 \tag{33}$$

for any fixed values of k, l and

$$L_{m,n} = \frac{4}{\pi^2}\int_0^\pi\int_0^\pi \left| \frac{\lambda_{0,0}^{(m,n)}}{4} + \frac{1}{2}\sum_{k=1}^{m}\lambda_{k,0}^{(m,n)}\cos ku + \frac{1}{2}\sum_{l=1}^{n}\lambda_{0,l}^{(m,n)}\cos lv + \right.$$

$$\left. + \sum_{k=1}^{m}\sum_{l=1}^{n}\lambda_{k,l}^{(m,n)}\cos ku\cos lv\right| du\, dv = O(1) \tag{34}$$

uniformly with respect to m, n.

It is clear (see, for example, 8.1(20)) that 8.1(34) can only hold when

$$\lambda_{k,l}^{(m,n)} = O(1) \tag{35}$$

uniformly with respect to m, n and k, l ($k = 0, 1, \ldots, m$; $l = 0, 1, \ldots, n$).

8.2. Lebesgue constants and functions

It is customary to term the upper bounds 8.1(14) and 8.1(34) the Lebesgue constants of the given method of summation of the Fourier series, and the upper bounds 8.1(15), 8.1(30) and 8.1(31), obtained for a finite segment or as a result of interpolation by Lagrange's formula, which depend on the position of the point x, the corresponding Lebesgue functions.

As has already been mentioned in sections 8.1.3–8.1.5, in deciding the question of convergence of any one of the linear processes of approximation by polynomials considered, a fundamental part belongs to the investigation of the behaviour of a sequence of these quantities. Such an investigation is connected with the study of the character of the dependence of the given upper bounds on the structure of the corresponding numerical matrix λ which determines the process of approximation, and has for its aim the finding of effective conditions which the system of numbers $\lambda_k^{(n)}$ must satisfy for the satisfaction of conditions of the type 8.1(18). Hence (see 8.1(20)) we can consider only bounded matrices λ, i.e. matrices possessing the property 8.1(21).

8.2.1. In the first place we shall show that in the general case *for any* $n = 0, 1, 2, \ldots$ *the upper bounds* 8.1(14) *always satisfy the inequality***

** An equivalent inequality was established by Fejes [1]. See also A. F. Timan [17].

$$L_n \geqslant \frac{8}{\pi^2}\left|\sum_{\nu=0}^{\infty}\frac{1}{2\nu+1}\sum_{k=(2n+1)\nu}^{(2n+1)(\nu+1)-1}\frac{\lambda^{(n)}_{|(2n+1)\nu+n-k|}}{2k+1}\right|, \tag{1}$$

which it is impossible to improve for any n in the class of all triangular matrices λ.

In order to obtain the estimate 8.2(1), it is necessary to consider the Fourier series

$$\operatorname{sign}\sin x = \frac{4}{\pi}\sum_{\nu=0}^{\infty}\frac{\sin(2\nu+1)x}{2\nu+1} \tag{2}$$

and make use of the inequality

$$L_n \geqslant \frac{2}{\pi}\left|\int_0^{\pi}\operatorname{sign}\sin\left(n+\frac{1}{2}\right)t\left\{\frac{\lambda_0^{(n)}}{2}+\sum_{k=1}^{n}\lambda_k^{(n)}\cos kt\right\}\mathrm{d}t\right|$$
$$= \frac{8}{\pi^2}\left|\int_0^{\pi}\left\{\frac{\lambda_0^{(n)}}{2}+\sum_{k=1}^{n}\lambda_k^{(n)}\cos kt\right\}\sum_{\nu=0}^{\infty}\frac{\sin(2\nu+1)(n+\frac{1}{2})t}{2\nu+1}\mathrm{d}t\right|$$
$$= \frac{4}{\pi^2}\left|\sum_{\nu=0}^{\infty}\frac{1}{2\nu+1}\sum_{k=0}^{n}{}'\lambda_k^{(n)}\left[\frac{1}{(2\nu+1)(n+\frac{1}{2})-k}+\frac{1}{(2\nu+1)(n+\frac{1}{2})+k}\right]\right|, \tag{3}$$

in which the dash indicates that the term which corresponds to the index $k = 0$ is divided by two.

Since in the case of the identical matrix, i.e., when for any n and k ($k = 0, 1, \ldots, n$; $n = 0, 1, 2, \ldots$) $\lambda_k^{(n)} = 1$, in virtue of 4.1(10) the inequality 8.2(3) becomes an equality it follows that the estimate 8.2(1) is precise.

We note two corollaries which follow from this.

8.2.11. In the case where the matrix λ is non-negative, i.e. $\lambda_k^{(n)} \geqslant 0$ ($k = 0, 1, \ldots, n$; $n = 0, 1, 2, \ldots$), the right-hand side of 8.2(1) is not less than one of its terms corresponding to the index $\nu = 0$. Hence in this case the inequality

$$L_n \geqslant \frac{8}{\pi^2}\sum_{k=0}^{2n}\frac{\lambda^{(n)}_{|n-k|}}{2k+1}. \tag{4}$$

is always satisfied.

8.2.12. In the case where the matrix λ is bounded, i.e. 8.1.(21) holds, then

$$\sum_{\nu=1}^{\infty}\frac{1}{2\nu+1}\sum_{k=(2n+1)\nu}^{(2n+1)(\nu+1)-1}\frac{\lambda^{(n)}_{|(2n+1)\nu+n-k|}}{2k+1} = O\left(\sum_{\nu=1}^{\infty}\frac{1}{\nu^2}\right) = O(1) \tag{5}$$

and

$$\sum_{k=n+1}^{2n} \frac{\lambda_{|n-k|}^{(n)}}{2k+1} = O(1). \tag{6}$$

Hence *in this case the inequality* 8.2.(1) *leads to the estimate*

$$L_n \geqslant \frac{8}{\pi^2} \sum_{k=0}^{n} \frac{\lambda_{n-k}^{(n)}}{2k+1} + O(1) \tag{7}$$

*or***, what is the same thing,

$$L_n \geqslant \frac{4}{\pi^2} \sum_{k=0}^{n} \frac{\lambda_k^{(n)}}{n-k+1} + O(1). \tag{8}$$

8.2.13. The inequality 8.2(8) together with 8.1(21) shows that in the general case *the sequence of numbers L_n can be bounded only when****

$$\sum_{k=0}^{n} \frac{\lambda_k^{(n)}}{n-k+1} = O(1). \tag{9}$$

8.2.14. It is not difficult to see that *condition* 8.2(9), *although necessary for the validity of* 8.1(18), *is not in the general case sufficient*. Let us consider, for example, a matrix of the form

$$\lambda_k^{(n)} = \begin{cases} 1 & \text{when} \quad 0 \leqslant k \leqslant \left[\frac{n}{2}\right], \\ 0 & \text{when} \quad \left[\frac{n}{2}\right] < k \leqslant n. \end{cases} \tag{10}$$

For this matrix 8.2(9) is satisfied, and for $m = \left[\frac{n}{2}\right]$ (see 8.2(3))

$$\begin{aligned} L_n &= \frac{2}{\pi} \int_0^{\pi} \left| \frac{1}{2} + \sum_{k=1}^{n} \cos kt \right| dt \\ &= \frac{2}{\pi} \int_0^{\pi} \operatorname{sign} \sin\left(m+\frac{1}{2}\right) t \left\{ \frac{1}{2} + \sum_{k=1}^{m} \cos kt \right\} dt \\ &= \frac{4}{\pi^2} \sum_{\nu=0}^{\infty} \frac{1}{2\nu+1} {\sum_{k=0}^{m}}' \left[\frac{1}{(2\nu+1)(m+\frac{1}{2})-k} + \frac{1}{(2\nu+1)(m+\frac{1}{2})+k} \right] \\ &= \frac{8}{\pi^2} \sum_{k=0}^{m} \frac{1}{2m-2k+1} + O(1) = \frac{4}{\pi^2} \ln m + O(1), \end{aligned} \tag{11}$$

i.e. L_n increases indefinitely with n.

** A. F. Timan [17]. If it is not assumed that the matrix λ is bounded, then the second term in 8.2(8) must be replaced by a quantity of order $O[\max_{0 \leqslant k \leqslant n} |\lambda_k^{(n)}|]$.

*** S. M. Nikol'skii [9].

8.2.2. Passing to an estimate for the Lebesgue functions 8.1(15), we consider the case where the rows of the matrix λ are of uniformly bounded variation, i.e.

$$\sum_{k=0}^{n} |\Delta\lambda_k^{(n)}| = O(1), \tag{12}$$

where $\Delta\lambda_k^{(n)} = \lambda_k^{(n)} - \lambda_{k+1}^{(n)}$ $(\lambda_{n+1}^{(n)} = 0,\ \lambda_0^{(n)} = 1)$.

We shall show that** *subject to this condition the inequality*

$$L_n(x) \geqslant \frac{4}{\pi^2}\left|\sum_{k=1}^{n} \frac{\lambda_{n-k}^{(n)}}{k} - \right.$$
$$\left. -\{1-|\cos n \arccos |x||\} \sum_{k=1}^{n} \frac{\lambda_{n-k}^{(n)}}{k} \cos k \arccos |x| \right| + O(1) \tag{13}$$

is always satisfied uniformly with respect to $x \in [-1, 1]$ *and* n.

If $y = \arccos x$, then

$$L_n(x) = \frac{2}{\pi}\int_0^{\pi} \left|\frac{1}{2} + \sum_{k=1}^{n} \lambda_k^{(n)} \cos ky \cos kt\right| \mathrm{d}t$$
$$= \frac{1}{\pi}\int_0^{\pi} |K_n(t+y) + K_n(t-y)| \,\mathrm{d}t, \tag{14}$$

where

$$K_n(u) = \frac{1}{2} + \sum_{\nu=1}^{n} \lambda_\nu^{(n)} \cos \nu u. \tag{15}$$

It is obvious from 8.2(14) that, without loss of generality, it is possible to take $0 \leqslant y \leqslant \frac{\pi}{2}$.

Let

$$\phi_n(t) = \begin{cases} \operatorname{sign} \sin n(y-t), & 0 \leqslant t \leqslant y, \\ \operatorname{sign} \sin n(t-y), & y \leqslant t \leqslant 2y, \\ \operatorname{sign} [\cos ny \sin nt], & 2y \leqslant t \leqslant \pi. \end{cases} \tag{16}$$

** A. F. Timan [13].

Then by 8.2(14)

$$L_n(x) \geqslant \frac{1}{\pi}\left|\int_0^{\pi} \phi_n(t)\{K_n(t+y)+K_n(t-y)\}\,\mathrm{d}t\right| = |I_1+I_2+I_3|, \tag{17}$$

where

$$\left.\begin{aligned} I_1 &= \frac{1}{\pi}\int_0^{y} \phi_n(t)\{K_n(t+y)+K_n(t-y)\}\,\mathrm{d}t, \\ I_2 &= \frac{1}{\pi}\int_y^{2y} \phi_n(t)\{K_n(t+y)+K_n(t-y)\}\,\mathrm{d}t, \\ I_3 &= \frac{1}{\pi}\int_{2y}^{\pi} \phi_n(t)\{K_n(t+y)+K_n(t-y)\}\,\mathrm{d}t. \end{aligned}\right\} \tag{18}$$

In virtue of 8.2(2) we have

$$\begin{aligned} I_1 = {} & \frac{4}{\pi^2}\int_0^{y}\sum_{\nu=0}^{\infty}\frac{\sin(2\nu+1)n(y-t)}{2\nu+1}\Big\{1+\sum_{k=1}^{n}\lambda_k^{(n)}[\cos k(y+t)+ \\ & +\cos k(y-t)]\Big\}\,\mathrm{d}t \\ = {} & \frac{2}{\pi^2}\int_0^{y}\sum_{k=1}^{n}\lambda_k^{(n)}[\sin(n+k)(y-t)+\sin(n-k)(y-t)]\,\mathrm{d}t+ \\ & +\frac{2}{\pi^2}\int_0^{y}\sum_{k=1}^{n}\lambda_k^{(n)}\{\sin[(n+k)y-(n-k)t]+ \\ & +\sin[(n-k)y-(n+k)t]\}\,\mathrm{d}t+ \\ & +\frac{4}{\pi^2}\int_0^{y}\sum_{\nu=1}^{\infty}\frac{\sin(2\nu+1)n(y-t)}{2\nu+1}\sum_{k=1}^{n}\lambda_k^{(n)}[\cos k(y+t)+ \\ & +\cos k(y-t)]\,\mathrm{d}t+O(1), \end{aligned}$$

$$I_2 = \frac{4}{\pi^2}\int\limits_y^{2y}\sum_{\nu=0}^{\infty}\frac{\sin(2\nu+1)n(t-y)}{2\nu+1}\Big\{1+\sum_{k=1}^{n}\lambda_k^{(n)}[\cos k(y+t)+$$

$$+\cos k(y-t)]\Big\}\,dt$$

$$=\frac{2}{\pi^2}\int\limits_y^{2y}\sum_{k=1}^{n}\lambda_k^{(n)}[\sin(n+k)(t-y)+\sin(n-k)(t-y)]\,dt+$$

$$+\frac{2}{\pi^2}\int\limits_y^{2y}\sum_{k=1}^{n}\lambda_k^{(n)}\{\sin[(n+k)t-(n-k)y]+$$

$$+\sin[(n-k)t-(n+k)y]\}\,dt+$$

$$+\frac{4}{\pi^2}\int\limits_y^{2y}\sum_{\nu=1}^{\infty}\frac{\sin(2\nu+1)n(t-y)}{2\nu+1}\sum_{k=1}^{n}\lambda_\nu^{(n)}[\cos k(y+t)+$$

$$+\cos k(y-t)]\,dt+O(1),$$

$$I_3 = \frac{4\,\mathrm{sign}\cos ny}{\pi^2}\int\limits_{2y}^{\pi}\sum_{\nu=0}^{\infty}\frac{\sin(2\nu+1)nt}{2\nu+1}\Big\{1+\sum_{k=1}^{n}\lambda_k^{(n)}[\cos k(y+t)+$$

$$+\cos k(y-t)]\Big\}\,dt$$

$$=\frac{2\,\mathrm{sign}\cos ny}{\pi^2}\int\limits_{2y}^{\pi}\sum_{k=1}^{n}\lambda_k^{(n)}\{\sin[(n+k)t+ky]+$$

$$+\sin[(n-k)t--ky]\}\,dt+$$

$$+\frac{2\,\mathrm{sign}\cos ny}{\pi^2}\int\limits_{2y}^{\pi}\sum_{k=1}^{n}\lambda_k^{(n)}\{\sin[(n+k)t-ky]+$$

$$+\sin[(n-k)t+ky]\}\,dt+$$

$$+\frac{4\,\mathrm{sign}\cos ny}{\pi^2}\int\limits_{2y}^{\pi}\sum_{\nu=1}^{\infty}\frac{\sin(2\nu+1)nt}{2\nu+1}\sum_{k=1}^{n}\lambda_k^{(n)}[\cos k(y+t)+$$

$$+\cos k(y-t)]\,dt+O(1).$$

Evaluating the integrals involved, we obtain

$$\left.\begin{aligned}
I_1 &= \frac{2}{\pi^2}\sum_{k=1}^{n}\lambda_{n-k}^{(n)}\frac{1-\cos ky}{k}+\\
&\qquad+\frac{2}{\pi^2}\sum_{k=1}^{n-1}\lambda_k^{(n)}\frac{\cos 2ky-\cos(n+k)y}{n-k}+O(1),\\
I_2 &= \frac{2}{\pi^2}\sum_{k=1}^{n}\lambda_{n-k}^{(n)}\frac{1-\cos ky}{k}+\\
&\qquad+\frac{2}{\pi^2}\sum_{k=1}^{n-1}\lambda_k^{(n)}\frac{\cos 2ky-\cos(n-3k)y}{n-k}+O(1),\\
I_3 &= \frac{2\operatorname{sign}\cos ny}{\pi^2}\left\{\sum_{k=1}^{n}\lambda_k^{(n)}\frac{\cos(2n-3k)y+(-1)^{n-k}\cos ky}{n-k}+\right.\\
&\qquad\left.+\sum_{k=1}^{n-1}\lambda_k^{(n)}\frac{\cos(2n-k)y+(-1)^{n-k}\cos ky}{n-k}\right\}+O(1),
\end{aligned}\right\} \tag{19}$$

where $O(1)$ is a quantity which is uniformly bounded with respect to y and n.

But subject to the condition 8.2(12) the following relations are satisfied (see 4.12(13)):

$$\left.\begin{aligned}
&\sum_{k=1}^{n-1}\lambda_k^{(n)}\frac{(-1)^{n-k}\cos ky}{n-k}=O(1),\\
&\sum_{k=1}^{n-1}\lambda_k^{(n)}\frac{\cos 2ky-\cos(n+k)y}{n-k}=O(1),\\
&\sum_{k=1}^{n-1}\lambda_k^{(n)}\frac{\cos 2ky-\cos(n-3k)y}{n-k}=O(1),\\
&\sum_{k=1}^{n-1}\lambda_k^{(n)}\frac{\cos(2n-3k)y+\cos(2n-k)y}{n-k}\\
&\qquad\qquad=2\cos ny\sum_{k=1}^{n}\lambda_{n-k}^{(n)}\frac{\cos ky}{k}+O(1).
\end{aligned}\right\} \tag{20}$$

As a result of this the estimates 8.2(19) lead to the equation

$$I_1+I_2+I_3=\frac{4}{\pi^2}\sum_{k=1}^{n}\lambda_{n-k}^{(n)}\frac{1-\cos ky}{k}+$$
$$+\frac{4}{\pi^2}|\cos ny|\sum_{k=1}^{n}\lambda_{n-k}^{(n)}\frac{\cos ky}{k}+O(1), \qquad (21)$$

from which, in virtue of 8.2(17), 8.2(13) follows.

8.2.21. We note that for any fixed $x(-1\leqslant x\leqslant 1)$ *the right-hand side of* 8.2(13) *is bounded for the matrix* λ *considered when and only when* 8.2(9) *holds*. This follows from the fact that for any fixed $y(0<y\leqslant\pi)$ the inequality

$$\sum_{k=1}^{n}\frac{\lambda_{n-k}^{(n)}}{k}\cos ky=-\sum_{k=1}^{n-1}\Delta\lambda_{n-k-1}^{(n)}\sum_{\nu=1}^{k}\frac{\cos\nu y}{\nu}+\sum_{\nu=1}^{n}\frac{\cos\nu y}{\nu}$$
$$=O\left[\sum_{k=1}^{n-1}|\Delta\lambda_{n-k-1}^{(n)}|\right]=O(1)$$

is satisfied.

Thus if condition 8.2(9) is not satisfied, then for any $x\in[-1,1]$, by the theorem on weak convergence of the sequence of linear functionals (see section 8.1.3), a function $f(x)$ continuous on $[-1,1]$ can be found for which the sequence of polynomials 8.1(10) diverges at this point.

8.2.22. In those cases when the sum $\sum_{k=1}^{n}\frac{\lambda_k^{(n)}}{n-k+1}$ is unbounded, it is possible to raise the question of the choice of a sequence of points x_n for which the right-hand side of 8.2(13) remains bounded. It follows from the inequality 8.2(13) *that for a bounded sequence of the numbers* $L_n(x_n)$ *the satisfaction of the condition*

$$[1-|\cos n\arccos|x_n||]\sum_{k=1}^{n}\frac{\lambda_{n-k}^{(n)}}{k}\cos k\arccos|x_n|=\sum_{k=1}^{n}\frac{\lambda_{n-k}^{(n)}}{k}+O(1) \qquad (22)$$

is necessary.

8.2.23. In particular, if only non-negative matrices $(\lambda_k^{(n)}\geqslant 0)$ are considered, it then follows from this that *for weak convergence of the corresponding sequence of linear functionals* $U_n(f;x_n;\lambda)$ (see 8.1(13)) *it is necessary that there should exist a sequence of odd numbers* $p=p(n)$ $(n=1,2,\ldots)$ *for which*

$$|x_n|=\cos\left\{\frac{p\pi}{2n}+O\left[\frac{1}{n\sum_{k=1}^{n}\frac{\lambda_k^{(n)}}{n-k+1}}\right]\right\} \qquad (23)$$

and

$$\sum_{k=1}^{n} \frac{\lambda_{n-k}^{(n)}}{k} \cos k \arccos |x_n| = \sum_{k=1}^{n} \frac{\lambda_{n-k}^{(n)}}{k} + O(1). \tag{24}$$

In fact, let

$$y_n = \arccos |x_n| = \frac{p\pi}{2n} + \varepsilon_n,$$

where $p = p(n)$ assumes only odd values, and $|\varepsilon_n| \leqslant \frac{\pi}{n}\cdot$ It is obvious from 8.2(22) that in our case, for any sequence of values of n increasing without limit for which $\sum_{k=1}^{n} \frac{\lambda_{n-k}^{(n)}}{k} \to \infty$, we have

$$\sum_{k=1}^{n} \frac{\lambda_{n-k}^{(n)}}{k} \cos ky_n \geqslant 0.$$

Moreover, since

$$\sum_{k=1}^{n} \frac{\lambda_{n-k}^{(n)}}{k} \cos ky_n \leqslant \sum_{k=1}^{n} \frac{\lambda_{n-k}^{(n)}}{k},$$

then by 8.2(22) the relations

$$|\cos ny_n| \sum_{k=1}^{n} \frac{\lambda_{n-k}^{(n)}}{k} \cos ky_n = O(1),$$

$$\sum_{k=1}^{n} \frac{\lambda_{n-k}^{(n)}}{k} \cos ky_n = \sum_{k=1}^{n} \frac{\lambda_{n-k}^{(n)}}{k} + O(1) \tag{25}$$

must be satisfied. It follows that

$$\sin n\varepsilon_n = O\left[\frac{1}{\sum_{k=1}^{n} \frac{\lambda_k^{(n)}}{n-k+1}}\right],$$

i.e. 8.2(24) and 8.2(23) are satisfied.

8.2.3. It is impossible to improve the estimate 8.2(13) in the class of all the matrices λ considered. The following theorem holds**.

If the matrix of numbers $\lambda_k^{(n)}$ *is bounded, i.e. it satisfies condition* 8.1(21), *and for any* n

$$\Delta^2\lambda_k^{(n)} = \lambda_k^{(n)} - 2\lambda_{k+1}^{(n)} + \lambda_{k+2}^{(n)} \leqslant 0 \quad (k = 0, 1, \ldots, n-1;\ \lambda_{n+1}^{(n)} = 0), \tag{26}$$

or

$$\Delta^2\lambda_k^{(n)} \geqslant 0 \quad (k = 0, 1, \ldots, n-1;\ \lambda_{n+1}^{(n)} = 0), \tag{27}$$

** A. F. Timan [13].

then

$$L_n(x) = \frac{4}{\pi^2}\left| \sum_{k=1}^{n} \frac{\lambda_{n-k}^{(n)}}{k} - \right.$$

$$\left. -\{1-|\cos n \arccos|x||\} \sum_{k=1}^{n} \frac{\lambda_{n-k}^{(n)}}{k} \cos k \arccos|x| \right| + O(1) \qquad (28)$$

uniformly with respect to all $x \in [-1, 1]$ *and* n.

8.2.31. For the proof we need the following lemma**. *Let the function* $f(x)$ *be continuous on* $[-1, 1]$ *and let*

$$S_n(f;\, x) = \frac{c_0}{2} + \sum_{k=1}^{n} c_k \cos k \arccos x \qquad (29)$$

be the partial sum of order n of its Fourier-Chebyshev series (see 8.1(11)). *If* $0 \leqslant k \leqslant n$ *and*

$$\sigma_n(f;\, k;\, x) = \frac{1}{k+1} \sum_{\nu=0}^{k} S_{n-\nu}(f;\, x), \qquad (30)$$

$$L_n(k;\, x) = \sup_{|f(x)| \leqslant 1} |\sigma_n(f;\, k;\, x)|, \qquad (31)$$

then

$$L_n(k;\, x) = \frac{4}{\pi^2}\left\{ \ln \frac{1+ny}{1+(k+1)y} + |\cos ny| \ln \frac{y+\dfrac{1}{k+1}}{y+\dfrac{1}{n+1}} \right\} + O(1) \qquad (32)$$

uniformly with respect to $y = \arccos|x|$ $\left(0 \leqslant y \leqslant \dfrac{\pi}{2}\right)$, *n and* $k\,(0 \leqslant k \leqslant n)$.

Since

$$\sigma_n(f;\, k;\, x) = \frac{1}{(k+1)\pi} \int_0^{\pi} f(\cos t) \sum_{\nu=0}^{k} \left\{ 1 + 2\sum_{j=1}^{n-\nu} \cos jy \cos jt \right\} dt, \qquad (33)$$

then

$$L_n(k;\, x) = \frac{1}{(k+1)\pi} \int_0^{\pi} \left| \sum_{\nu=0}^{k} \left\{ 1 + 2\sum_{j=1}^{n-\nu} \cos jy \cos jt \right\} \right| dt$$

$$= \frac{1}{(k+1)\pi} \int_0^{\pi} \left| \sum_{\nu=0}^{k} \left\{ D_{n-\nu}(t+y) + D_{n-\nu}(t-y) \right\} \right| dt, \qquad (34)$$

** L. M. Abramov [1]. For the case $x = \pm 1$ see S. M. Nikol'skii [1]; for the case $k = 0$ see A. F. Timan [2].

where $D_n(u)$ is Dirichlet's kernel (see 4.1(10)). The following formula holds:

$$\sum_{\nu=0}^{k} D_{n-\nu}(u) = \frac{1}{2\sin\frac{u}{2}} \sum_{\nu=0}^{k} \sin\left(n-\nu+\frac{1}{2}\right)u$$

$$= \frac{\sin\frac{2n+1-k}{2}u \sin\frac{k+1}{2}u}{2\sin^2\frac{u}{2}}. \tag{35}$$

Therefore

$$L_n(k;x) = \frac{1}{2(k+1)\pi}\int_0^\pi \left| \frac{\sin\frac{2n+1-k}{2}(t+y)\sin\frac{k+1}{2}(t+y)}{\sin^2\frac{t+y}{2}} + \right.$$

$$\left. + \frac{\sin\frac{2n+1-k}{2}(t-y)\sin\frac{k+1}{2}(t-y)}{\sin^2\frac{t-y}{2}} \right| dt. \tag{36}$$

Formula 8.2(36) shows that without restriction of generality it is possible to suppose that $0 \leqslant y \leqslant \frac{\pi}{2}$.

Let us consider the three integrals

$$I_1 = \frac{1}{2\pi(k+1)}\int_0^y \left| \frac{\sin\frac{2n+1-k}{2}(t+y)\sin\frac{k+1}{2}(t+y)}{\sin^2\frac{t+y}{2}} + \right.$$

$$\left. + \frac{\sin\frac{2n+1-k}{2}(t-y)\sin\frac{k+1}{2}(t-y)}{\sin^2\frac{t-y}{2}} \right| dt,$$

$$I_2 = \frac{1}{2\pi(k+1)}\int_y^{2y} \left| \frac{\sin\frac{2n+1-k}{2}(t+y)\sin\frac{k+1}{2}(t+y)}{\sin^2\frac{t+y}{2}} + \right.$$

$$\left. + \frac{\sin\frac{2n+1-k}{2}(t-y)\sin\frac{k+1}{2}(t-y)}{\sin^2\frac{t-y}{2}} \right| dt,$$

$$I_3 = \frac{1}{2\pi(k+1)}\int_{2y}^{\pi}\left|\frac{\sin\frac{2n+1-k}{2}(t+y)\sin\frac{k+1}{2}(t+y)}{\sin^2\frac{t+y}{2}}+\right.$$

$$\left.+\frac{\sin\frac{2n+1-k}{2}(t-y)\sin\frac{k+1}{2}(t-y)}{\sin^2\frac{t-y}{2}}\right|dt.$$

In the first place, we note that

$$\int_0^b \frac{|\sin lu|}{u}\,du = \frac{2}{\pi}\ln(1+bl)+O(1) \tag{37}$$

uniformly with respect to all $b > 0$ and $l > 0$. In fact, if $m, \pi \leqslant bl < (m+1)\pi$, then

$$\int_0^b \frac{|\sin lu|}{u}\,du = \sum_{\nu=0}^{m-1}(-1)^\nu \int_{\frac{\nu\pi}{l}}^{\frac{\nu+1}{l}\pi} \frac{\sin lu}{u}\,du + \int_{\frac{m\pi}{l}}^{b}\frac{|\sin lu|}{u}\,du$$

$$= \int_0^{\frac{\pi}{l}} \sin lu \sum_{\nu=0}^{m-1}\frac{1}{u+\frac{\nu\pi}{l}}\,du + O(1) = \int_0^{\frac{\pi}{l}} \sin lu \sum_{\nu=1}^{m-1}\frac{l}{\nu\pi}\,du + O(1)$$

$$= \frac{2}{\pi}\ln(m+1)+O(1) = \frac{2}{\pi}\ln(1+bl)+O(1).$$

Also for any $a > 0$,

$$\int_{\frac{(k+1)\frac{a}{2}}{1+(k+1)\frac{a}{2}}}^{(k+1)\frac{a}{2}} \frac{\left|\sin\frac{2n+1-k}{k+1}t\sin t\right|}{t^2}\,dt < 1$$

uniformly with respect to n, k and a. Consequently, since the difference $\frac{\sin t}{t^2}-\frac{1}{t}$ is bounded, we have

$$\frac{1}{k+1}\int_0^a \frac{\left|\sin\frac{2n+1-k}{2}t\sin\frac{k+1}{2}t\right|}{t^2}\,dt$$

$$=\frac{1}{2}\int_0^{(k+1)\frac{a}{2}} \frac{\left|\sin\frac{2n+1-k}{k+1}t\sin t\right|}{t^2}\,dt$$

$$=\frac{1}{2}\int_0^{\frac{(k+1)\frac{a}{2}}{1+(k+1)\frac{a}{2}}} \frac{\left|\sin\frac{2n+1-k}{k+1}t\sin t\right|}{t^2}\,dt+O(1)$$

$$=\frac{1}{2}\int_0^{\frac{(k+1)\frac{a}{2}}{1+(k+1)\frac{a}{2}}} \frac{\left|\sin\frac{2n+1-k}{k+1}t\right|}{t}\,dt+O(1)$$

$$=\frac{1}{\pi}\ln\frac{1+na}{1+(k+1)a}+O(1). \tag{38}$$

Taking into account now that the difference $\frac{1}{\sin^2 u}-\frac{1}{u^2}$ is bounded in the interval $0<u<\frac{3}{4}\pi$, we obtain by 8.2(38)

$$I_1=\frac{2}{\pi^2}\ln\frac{1+ny}{1+(k+1)y}+O(1) \tag{39}$$

and similarly

$$I_2=\frac{2}{\pi^2}\ln\frac{1+ny}{1+(k+1)y}+O(1). \tag{40}$$

In order to estimate the integral I_3 we note that

$$I_3 = \frac{2}{\pi(k+1)} \int_{2y}^{\pi} \left| \frac{\sin\frac{2n+1-k}{2}(t+y)\sin\frac{k+1}{2}(t+y)}{(t+y)^2} + \right.$$

$$\left. + \frac{\sin\frac{2n+1-k}{2}(t-y)\sin\frac{k+1}{2}(t-y)}{(t-y)^2} \right| dt + O(1)$$

$$= \frac{2}{\pi(k+1)} \int_{2y}^{\pi} \left| \frac{\sin\frac{2n+1-k}{2}(t+y)\sin\frac{k+1}{2}t\cos\frac{k+1}{2}y}{t^2} + \right.$$

$$\left. + \frac{\sin\frac{2n+1-k}{2}(t-y)\sin\frac{k+1}{2}t\cos\frac{k+1}{2}y}{t^2} \right| dt + O(1)$$

$$= \frac{4}{\pi(k+1)} \left| \cos\frac{2n+1-k}{2}y\cos\frac{k+1}{2}y \right| \times$$

$$\times \int_{2y}^{\pi} \frac{\left| \sin\frac{2n+1-k}{2}t\sin\frac{k+1}{2}t \right|}{t^2} dt + O(1).$$

If we again take account of 8.2(38), we then find that

$$I_3 = \frac{4}{\pi^2} \left| \cos\frac{2n+1-k}{2}y\cos\frac{k+1}{2}y \right| \ln\frac{y+\frac{1}{k+1}}{y+\frac{1}{n}} + O(1)$$

$$= \frac{4}{\pi^2} |\cos ny| \ln\frac{y+\frac{1}{k+1}}{y+\frac{1}{n}} + O(1). \qquad (41)$$

Considering 8.2(39), 8.2(40), 8.2(41) and 8.2(36), we obtain 8.2(32).

8.2.32. If now $U_n(f; x; \lambda)$ are the polynomials 8.1(10), then

$$U_n(f; x; \lambda) = \sum_{k=0}^{n} [\lambda_k^{(n)} - \lambda_{k+1}^{(n)}] S_k(f; x) = \sum_{k=0}^{n} \Delta\lambda_{n-k}^{(n)} S_{n-k}(f; x).$$

Hence after a second application of Abel's transformation we obtain

$$U_n(f;x;\lambda) = -\sum_{k=0}^{n-1}(k+1)\sigma_n(f;k;x)^2\Delta\lambda_{n-k-1}^{(n)}+$$

$$+(n+1)\sigma_n(f;n;x)\Delta\lambda_0^{(n)},$$

or

$$U_n(f;x;\lambda) = -\sum_{k=0}^{\left[\frac{n}{2}\right]}(k+1)\sigma_n(f;k,x)\Delta^2\lambda_{n-k-1}^{(n)}+$$

$$+\sum_{k=\left[\frac{n}{2}\right]+1}^{n-1}(n-k)\sigma_{n-k-1}(f;n-k-1;x)\Delta^2\lambda_{n-k-1}^{(n)}-$$

$$-\sum_{k=\left[\frac{n}{2}\right]+1}^{n-1}(n+1)\sigma_n(f;n;x)\Delta^2\lambda_{n-k-1}^{(n)}+$$

$$+(n+1)\sigma_n(f;n;x)\Delta\lambda_0^{(n)},$$

$$U_n(f;x;\lambda) = -\sum_{k=0}^{\left[\frac{n}{2}\right]}(k+1)\sigma_n(f;k;x)\Delta^2\lambda_{n-k-1}^{(n)}+$$

$$+\sum_{k=\left[\frac{n}{2}\right]+1}^{n-1}(n-k)\sigma_{n-k-1}(f;n-k-1;x)\Delta^2\lambda_{n-k-1}^{(n)}-$$

$$-(n+1)\sigma_n(f;n;x)\Delta\lambda_{n-\left[\frac{n}{2}\right]-1}^{(n)}. \tag{42}$$

But in the case when the matrix λ considered is bounded and satisfies the conditions 8.2(26) or 8.2(27), then

$$\lambda_{n-\left[\frac{n}{2}\right]-1}^{(n)}-\lambda_{n-\left[\frac{n}{2}\right]}^{(n)} = O\left(\frac{1}{n}\right),$$

$$\sum_{k=\left[\frac{n}{2}\right]+1}^{n-1}(n-k)\Delta^2\lambda_{n-k-1}^{(n)}$$

$$= \sum_{k=\left[\frac{n}{2}\right]+1}^{n-1}(n-k)\left[\Delta\lambda_{n-k-1}^{(n)}-\Delta\lambda_{n-k}^{(n)}\right] = O(1). \tag{43}$$

Also the sums

$$\sigma_n(f;n;x) = \frac{1}{n+1}\sum_{\nu=0}^{n}S_\nu(f;x) \tag{44}$$

in virtue of formulae 8.2(33) and 8.2(35) possess the property that

$$L_n(n;x) = \sup_{|f(\cos t)|\leqslant 1} |\sigma_n(f;n\,x)| = 1. \tag{45}$$

Therefore

$$L_n(x) = \sup_{|f(\cos t)|\leqslant 1} |U_n(f;x;\lambda)| \leqslant \left|\sum_{k=0}^{\left[\frac{n}{2}\right]} (k+1)\,L_n(k;x)\,\Delta^2\lambda_{n-k-1}^{(n)}\right| + O(1).$$

Substituting the estimate 8.2(32) in this, and taking account of the fact that

$$\sum_{k=0}^{\left[\frac{n}{2}\right]} (k+1)\Delta^2\lambda_{n-k-1}^{(n)} = O(1), \tag{46}$$

we obtain the inequality

$$L_n(x) \leqslant \frac{4}{\pi^2}\left|\sum_{k=0}^{\left[\frac{n}{2}\right]}(k+1)\left\{\ln\frac{n}{k+1} - (1-|\cos ny|)\ln\frac{y+\dfrac{1}{k+1}}{y+\dfrac{1}{n+1}}\right\}\Delta^2\lambda_{n-k-1}^{(n)}\right| + O(1). \tag{47}$$

Here

$$\sum_{k=0}^{\left[\frac{n}{2}\right]}(k+1)\ln\frac{n}{k+1}\Delta^2\lambda_{n-k-1}^{(n)} = \sum_{k=0}^{\left[\frac{n}{2}\right]}(k+1)\Delta^2\lambda_{n-k-1}^{(n)}\sum_{\nu=k+1}^{n}\frac{1}{\nu} + O(1)$$

and (see 4.12(13))

$$\sum_{k=0}^{\left[\frac{n}{2}\right]}(k+1)\ln\frac{y+\dfrac{1}{k+1}}{y+\dfrac{1}{n+1}}\Delta^2\lambda_{n-k-1}^{(n)} = \sum_{k=0}^{\left[\frac{n}{2}\right]}(k+1)\Delta^2\lambda_{n-k-1}^{(n)}\sum_{\nu=k+1}^{n}\frac{\cos\nu y}{\nu} + O(1).$$

Consequently

$$L_n(x) \leqslant \frac{4}{\pi^2}\left|\sum_{k=0}^{n-1}(k+1)\left\{\sum_{\nu=k+1}^{n}\frac{1}{\nu} - (1-|\cos ny|)\sum_{\nu=k+1}^{n}\frac{\cos\nu y}{\nu}\right\}\Delta^2\lambda_{n-k-1}^{(n)}\right| + O(1). \tag{48}$$

A twofold application of Abel's transformation to the right-hand side of 8.2(48) shows that for the matrix considered

$$L_n(x) \leqslant \frac{4}{\pi^2}\left|\sum_{k=1}^{n}\frac{\lambda_{n-k}^{(n)}}{k}-(1-|\cos ny|)\sum_{k=1}^{n}\frac{\lambda_{n-k}^{(n)}}{k}\cos ky\right|+O(1).$$

The resulting inequality together with 8.2(13) assures us of the validity of the estimate 8.2(28).

8.2.33. It follows from theorem 8.2.3 that *if the matrix λ is bounded and for any n satisfies the condition* 8.2(26) *or* 8.2(27), *then*

$$L_n(1) = \frac{4}{\pi^2}\sum_{k=1}^{n}\frac{\lambda_{n-k}^{(n)}}{k}+O(1). \tag{49}$$

This result shows that the estimate 8.2(8) for the Lebesgue constant $L_n = L_n(1)$ cannot be improved in the general case.

8.2.34. Summarising the results obtained in 8.1.3 (see 8.1(16), 8.1(18), 8.1(21)), 8.2.13 and 8.2.33, we obtain the following theorem**.

In order that the sequence of trigonometric polynomials 8.1(8) *should converge at every point x, or uniformly with respect to x, to $f(x)$ on the whole class of continuous functions of period 2π, it is necessary, and in the case when* $\Delta^2\lambda_k^{(n)} \geqslant 0\,(k = 0, 1, \ldots, \; n-1; \;\; \lambda_{n+1}^{(n)} = 0; \;\; n = 0, 1, 2, \ldots)$ or $\Delta^2\lambda_k^{(n)} \leqslant 0 \;(k = 0, 1, \ldots, n-1; \lambda_{n+1}^{(n)} = 0; \; n = 0, 1, 2, \ldots)$, *it is also sufficient, that the conditions* 8.1(16), 8.1(21) *and* 8.2(9) *should be satisfied.*

8.2.35. The last assertion remains valid if the class C^* of all continuous functions of period 2π and the sequence of trigonometric polynomials 8.1(18) is replaced by the class C of all functions continuous on $[-1, 1]$ and the sequence of algebraic polynomials 8.1(10) see section 8.1.3).

8.2.36. Let α_n be an arbitrary sequence of numbers $\left(0 \leqslant \alpha_n \leqslant \frac{\pi}{2}\right)$ and

$$\lambda_k^{(n)} = l_k^{(n)}\cos k\alpha_n, \tag{50}$$

where $\Delta^2 l_k^{(n)} \geqslant 0 \;\; (k = 0, 1, \ldots, n-1; \; l_{n+1}^{(n)} = 0; \; n = 0, 1, 2, \ldots)$ or $\Delta^2 l_k^{(n)} \leqslant 0 \;\; (k = 0, 1, \ldots, n-1; l_{n+1}^{(n)} = 0; \; n = 0, 1, 2, \ldots)$.

The Lebesgue constants for the method 8.1(8), corresponding to the matrix λ of 8.2(50), are identical with the numbers L_n (cos α_n) (see 8.1(15)) corresponding to the matrix l. Taking into account that the system of numbers $l_k^{(n)}$ satisfies for any n a condition of convexity (concavity), it is not difficult to verify the fact that if 8.1(16) and 8.1(21) hold, then

$$l_k^{(n)} = O(1), \quad \lim_{n\to\infty}\alpha_n = 0 \tag{51}$$

** S.M. Nikol'skii [9].

and for any fixed value of k

$$\lim_{n\to\infty} l_k^{(n)} = 1. \tag{52}$$

Hence it is possible to apply theorem 8.2.3 to the matrix l, from which it follows that in the case of boundedness of the corresponding sequence of numbers L_n ($\cos \alpha_n$) the condition

$$(1-|\cos n\alpha_n|) \sum_{k=1}^{n} \frac{l_{n-k}^{(n)}}{k} \cos k\alpha_n = \sum_{k=1}^{n} \frac{l_{n-k}^{(n)}}{k} + O(1) \tag{53}$$

is not only necessary (see section 8.2.22), but also sufficient.

Thus on the basis of the considerations given in section 8.1.3 we arrive at the following proposition, which generalises the criteria 8.2.34**.

If the matrix λ *is defined by the equations* 8.2(50), *where the system of numbers* $l_k^{(n)}$ *for any* n *is convex or concave, then in order that the sequence of trigonometric polynomials* 8.1(8) *should converge at every point* x, *or uniformly in* x, *to* $f(x)$ *on the whole class* C^* *of continuous functions of period* 2π, *it is necessary and sufficient that conditions* 8.2(51)–8.2(53) *should be satisfied.*

8.2.37. In particular, if the system of numbers $l_k^{(n)}$ is convex upwards, i.e. $\Delta^2 l^{(n)} \leqslant 0$, then condition 8.2(53) is equivalent to the following two requirements:

(a) There exists a sequence of odd numbers $p = p(n)$ ($n = 1, 2, \ldots$) for which

$$\alpha_n = \frac{p\pi}{2n} + O\left\{\frac{1}{n\sum_{k=1}^{n} \frac{l_{n-k}^{(n)}}{k}}\right\}. \tag{54}$$

(b) The relation

$$\sum_{k=1}^{n} \frac{l_{n-k}^{(n)}}{k} \cos k\alpha_n = \sum_{k=1}^{n} \frac{l_{n-k}^{(n)}}{k} + O(1). \tag{55}$$

is satisfied.

We note that in the case when $l_k^{(n)} = 1$ for all n and $k (0 \leqslant k \leqslant n)$, condition 8.2(55) is equivalent to the boundedness of the integer-valued function $p = p(n)$.

** Timan [13]. See also A. F. Timan [1,5], where the case $l_k^{(n)} = l$ ($0 \leqslant k \leqslant n$, $n = 0, 1, 2, \ldots$) is considered.

8.2.4. Let us pass on to consideration of the Lebesgue interpolation functions 8.1(30) and show that in all cases where the matrix λ is bounded and satisfies condition 8.2(26) or 8.2(27), the relation**

$$M_n^*(x) = \frac{\pi}{2}\left|\sin\frac{2n+1}{2}x\right| L_n + O(1). \tag{56}$$

holds.

8.2.41. For this purpose as a preliminary we establish the following lemma. If $T_n(x)$ is the trigonometric polynomial 8.1(22), where $y_k^{(n)} = \frac{2k\pi}{2n+1}$, and for $0 \leqslant k \leqslant n$

$$\tau_n^*(f; k; x) = \frac{1}{k+1}\sum_{\nu=0}^{k} T_{n-\nu}(x), \tag{57}$$

then, *whatever the periodic function $f(x)$ of period 2π for which $|f(x)| \leqslant 1$,*

$$\tau_n^*(f; k; x) = \frac{2}{2n+1}\sum_{|x-y_\nu^{(n)}|<\frac{2\pi}{k+1}} f(y_\nu^{(n)})\frac{\sin\frac{2n+1}{2}(x-y_\nu^{(n)})}{x-y_\nu^{(n)}} + O(1) \tag{58}$$

uniformly with respect to all z, n, k $(0 \leqslant k \leqslant n)$ and f.

From the definition of the polynomials $T_n(x)$ and formula 8.2(35) it follows by the periodicity of the function $f(x)$ that

$$\tau_n^*(f; k; x) = \frac{1}{(2n+1)(k+1)}\times$$

$$\times \sum_{-\pi<x-y_\nu^{(n)}\leqslant\pi} f(y_\nu^{(n)})\frac{\sin\frac{2n+1-k}{2}(x-y_\nu^{(n)})\sin\frac{k+1}{2}(x-y_\nu^{(n)})}{\sin^2\frac{x-y_\nu^{(n)}}{2}}. \tag{59}$$

Hence, since in the interval $0 < t < \frac{\pi}{2}$ the difference $\frac{1}{\sin^2 t} - \frac{1}{t^2}$ is bounded and $|f(x)| \leqslant 1$,

$$\tau_n^*(f; k; x) = \frac{4}{(2n+1)(k+1)}\times$$

$$\times \sum_{|x-y_\nu^{(n)}|<\pi} f(y_\nu^{(n)}\frac{\sin\frac{2n+1-k}{2}(x-y_\nu^{(n)})\sin\frac{k+1}{2}(x-y_\nu^{(n)})}{x-y_\nu^{(n)})^2} + O(1).$$

** N. A. Pogodicheva and A. F. Timan [1].

Also

$$\frac{4}{(2n+1)(k+1)}\left|\sum_{|x-y_\nu^{(n)}|\geqslant\pi}{}_{\nu} f(y_\nu^{(n)})\frac{\sin\frac{2n+1-k}{2}(x-y_\nu^{(n)})\sin\frac{k+1}{2}(x-y_\nu^{(n)})}{(x-y_\nu^{(n)})^2}\right|$$

$$\leqslant\frac{8(2n+1)}{\pi^2(k+1)}\sum_{\nu=0}^{\infty}\frac{1}{(2n+1+2\nu)^2}=O(1)$$

and similarly

$$\frac{4}{(2n+1)(k+1)}\left|\sum_{|y_\nu^{(n)}-x|\geqslant\frac{2\pi}{k+1}}{}_{\nu} f(y_\nu^{(n)})\frac{\sin\frac{2n+1-k}{2}(x-y_\nu^{(n)})\sin\frac{k+1}{2}(x-y_\nu^{(n)})}{(x-y_\nu^{(n)})^2}\right|$$

$$\leqslant\frac{2(k+1)}{\pi^2(2n+1)}\sum_{\nu=0}^{\infty}\frac{1}{\left(1+\frac{k+1}{2n+1}\nu\right)^2}=\frac{2}{\pi^2}\int_0^{\infty}\frac{dx}{1+x^2}+O(1)=O(1).$$

Therefore

$$\tau_n^*(f;k;x)=\frac{4}{(2n+1)(k+1)}\times$$

$$\times\sum_{|x-y_\nu^{(n)}|<\frac{2\pi}{k+1}}{}_{\nu} f(y_\nu^{(n)})\frac{\sin\frac{2n+1-k}{2}(x-y_\nu^{(n)})\sin\frac{k+1}{2}(x-y_\nu^{(n)})}{(x-y_\nu^{(n)})^2}+O(1), \quad (60)$$

and since the function $\frac{\sin t}{t^2}-\frac{1}{t}$ is bounded $|f(x)|\leqslant 1$, and the number of terms in the sum 8.2(60) is a quantity of order $O(n/k)$, we have

$$\tau_n^*(f;k;x)=\frac{2}{2n+1}\sum_{|x-y_\nu^{(n)}|<\frac{2\pi}{k+1}}{}_{\nu} f(y_\nu^{(n)})\frac{\sin\frac{2n+1-k}{2}(x-y_\nu^{(n)})}{x-y_\nu^{(n)}}+O(1).$$

But

$$\frac{2}{2n+1}\sum_{|x-y_\nu^{(n)}|<\frac{2\pi}{k+1}}{}_{\nu}\left|\frac{\sin\frac{k}{2}(x-y_\nu^{(n)})}{x-y_\nu^{(n)}}\right|=O(1)$$

and

$$\frac{2}{2n+1}\sum_{|x-y_\nu^{(n)}|<\frac{2\pi}{k+1}} f(y_\nu^{(n)})\frac{\sin\frac{2n+1}{2}(x-y_\nu^{(n)})\cos\frac{k}{2}(x-y_\nu^{(n)})}{x-y_\nu^{(n)}}$$

$$=\frac{2n+1}{2}\sum_{|x-y_\nu^{(n)}|<\frac{2\pi}{k+1}} f(y_\nu^{(n)})\frac{\sin\frac{2n+1}{2}(x-y_\nu^{(n)})}{x-y_\nu^{(n)}}+O(1)$$

Consequently

$$\tau_n^*(f;k;x)=\frac{2n+1}{2}\sum_{|x-y_\nu^{(n)}|<\frac{2\pi}{k+1}} f(y_\nu^{(n)})\frac{\sin\frac{2n+1}{2}(x-y_\nu^{(n)})}{x-y_\nu^{(n)}}+O(1),$$

i.e. 8.2(58) is satisfied.

8.2.42. It follows from 8.2.41 that *for the upper bound*

$$M_n^*(k;x)=\sup_{|f(t)|\leqslant 1}\tau_\nu^*(f;k;x) \tag{61}$$

*the asymptotic equality***

$$M_n^*(k;x)=\frac{2}{\pi}\left|\sin\frac{2n+1}{2}x\right|\ln\frac{n}{k+1}+O(1) \tag{62}$$

holds uniformly with respect to x and $k(0\leqslant k\leqslant n)$.

In fact, we obtain from 8.2(58)

$$M_n^*(k;x)\leqslant\frac{2}{2n+1}\sum_{|x-y_\nu^{(n)}|<\frac{2\pi}{k+1}}\left|\frac{\sin\frac{2n+1}{2}(x-y_\nu^{(n)})}{x-y_\nu^{(n)}}\right|+O(1)$$

$$=\frac{2}{\pi}\left|\sin\frac{2n+1}{2}x\right|\sum_{\nu=1}^{\left[\frac{2n+1}{k+1}\right]}\frac{1}{\nu}+O(1)$$

$$=\frac{2}{\pi}\left|\sin\frac{2n+1}{2}x\right|\ln\frac{n}{k+1}+O(1).$$

On the other hand, it is obvious that for the function

$$f_x(t)=\operatorname{sign}\sin\frac{2n+1}{2}(x-t) \tag{63}$$

** S. M. Nikol'skii [1].

we have by 8.2(58)

$$\tau_n^*(f_x; k; x) = \frac{2}{2n+1} \sum_{|x-y_\nu^{(n)}| < \frac{k\pi}{k+1}} {}_{\nu} \left| \frac{\sin\frac{2n+1}{2}x}{x-y_\nu^{(n)}} \right| + O(1)$$

$$= \frac{2}{\pi} \left| \sin\frac{2n+1}{2}x \right| \ln\frac{n}{k+1} + O(1). \quad (64)$$

8.2.43. Now let $V_n^*(f; x; \lambda)$ be the polynomials 8.1(28). Then applying Abel's transformation we have

$$V_n^*(f; x; \lambda) = \frac{2}{2n+1} \sum_{\nu=0}^{2n} f(y_\nu^{(n)}) \sum_{k=0}^{n} \Delta\lambda_{n-k}^{(n)} D_{n-k}(x-y_\nu^{(n)}),$$

where $D_n(u)$ is Dirichlet's kernel (see 4.1(10)).

After a second application of Abel's transformation we obtain

$$V_n^*(f; x; \lambda)$$
$$= -\sum_{k=0}^{n-1} (k+1)\Delta^2\lambda_{n-k-1}^{(n)} \tau_n^*(f; k; x) + (n+1)\Delta\lambda_0^{(n)} \tau_n^*(f; n; x).$$

Moreover, as in section 8.2.32,

$$V_n^*(f; x; \lambda) = -\sum_{k=0}^{\left[\frac{n}{2}\right]} (k+1)\Delta^2\lambda_{n-k-1}^{(n)} \tau_n^*(f; k; x) -$$
$$- \sum_{k=\left[\frac{n}{2}\right]+1}^{n-1} (n+1)\Delta^2\lambda_{n-k-1}^{(n)} \tau_n^*(f; n; x) +$$
$$+ \sum_{k=\left[\frac{n}{2}\right]+1}^{n-1} (n-k)\Delta^2\lambda_{n-k-1}^{(n)} \tau_{n-k-1}^*(f; n-k-1; x) +$$
$$+ (n+1)\Delta\lambda_0^{(n)} \tau_n^*(f; n; x),$$

i.e.

$$V_n^*(f; x; \lambda) = -\sum_{k=0}^{\left[\frac{n}{2}\right]} (k+1)\Delta^2\lambda_{n-k-1}^{(n)} \tau_n^*(f; k; x) +$$
$$+ \sum_{k=\left[\frac{n}{2}\right]+1}^{n-1} (n-k)\Delta^2\lambda_{n-k-1}^{(n)} \tau_{n-k-1}^*(f; n-k-1; x) -$$
$$- (n+1)\tau_n^*(f; n; x)\Delta\lambda_{n-\left[\frac{n}{2}\right]-1}^{(n)}. \quad (65)$$

If now the matrix λ is bounded and satisfies condition 8.2(26) or 8.2(27), then the relations 8.2(43) and 8.2(46) hold. In addition, in virtue of the formula 8.2(59) (see also 4.2(9) and 8.2(35)) the sums $\tau_n{}^*(f; n; x)$ possess the property that

$$M_n^*(n; x) = \sup_{|f(t)|\leqslant 1} |\tau_n^*(f; n; x)| = 1. \tag{66}$$

Therefore

$$M_n^*(x) = \sup_{|f(t)\leqslant 1} |V_n^*(f; x; \lambda)| \leqslant$$

$$\leqslant \left|\sum_{k=0}^{\left[\frac{n}{2}\right]} (k+1) M_n^* (k; x) \Delta^2 \lambda_{n-k-1}^{(n)}\right| + O(1)$$

$$= \frac{2}{\pi} \left|\sin \frac{2n+1}{2} x\right| \left|\sum_{k=0}^{\left[\frac{n}{2}\right[} (k+1) \Delta^2 \lambda_{n-k-1}^{(n)} \ln \frac{n}{k+1}\right| + O(1).$$

On the other hand, for the function 8.2.(63), by 8.2(64) and 8.2(66)

$$|V_n^*(f_x; x; \lambda)| = \frac{2}{\pi} \left|\sin \frac{2n+1}{2} x\right| \left|\sum_{k=0}^{\left[\frac{n}{2}\right]} (k+1) \Delta^2 \lambda_{n-k-1}^{(n)} \ln \frac{n}{k+1}\right| + O(1).$$

Thus

$$M_n^*(x) = \frac{2}{\pi} \left|\sin \frac{2n+1}{2} x\right| \left|\sum_{k=0}^{\left[\frac{n}{2}\right]} (k+1) \Delta^2 \lambda_{n-k-1}^{(n)} \ln \frac{n}{k+1}\right| + O(1)$$

$$= \frac{2}{\pi} \left|\sin \frac{2n+1}{2} x\right| \left|\sum_{k=0}^{\left[\frac{n}{2}\right]} (k+1) \Delta^2 \lambda_{n-k-1}^{(n)} \sum_{\nu=k+1}^{n} \frac{1}{\nu}\right| + O(1).$$

But since (see 8.2(43))

$$\left|\sum_{k=\left[\frac{n}{2}\right]+1}^{n-1} (k+1) \Delta^2 \lambda_{n-k-1}^{(n)} \sum_{\nu=k+1}^{n} \frac{1}{\nu}\right| \leqslant$$

$$\leqslant \left|\sum_{k=\left[\frac{n}{2}\right]+1}^{n-1} (n-k) \Delta^2 \lambda_{n-k-1}^{(n)}\right| = O(1),$$

it follows that

$$M_n^*(x) = \frac{2}{\pi} \left|\sin \frac{2n+1}{2} x\right| \left|\sum_{k=0}^{n-1} (k+1) \Delta^2 \lambda_{n-k-1}^{(n)} \sum_{\nu=k+1}^{n} \frac{1}{\nu}\right| + O(1).$$

From this, after a twofold application of Abel's transformation, we obtain

$$M_n^*(x) = \frac{2}{\pi}\left|\sin\frac{2n+1}{2}x\right|\left|\sum_{k=1}^{n}\frac{\lambda_k^{(n)}}{n-k+1}\right|+O(1). \tag{67}$$

Comparing this result with 8.2(49) we arrive at the relation 8.2(56).

8.2.44. It follows from 8.2(56) that in the class of bounded matrices λ which satisfy a convexity condition (see 8.2(26), 8.2(27)), the sequence of norms $M_n^*(n)$ is bounded at every point x, or uniformly with respect to x, when and only when 8.2(9) holds. As a result of this, *for such matrices theorem* 8.2.34 *remains valid if the polynomials* 8.1(8) *are replaced by the polynomials* 8.1(26).

However, let us note that because of the factor $\left|\sin\frac{2n+1}{2}x\right|$ in formula 8.2(56) the sequence $M_n^*(x)$, in distinction from the sequence L_n or $L_n(x)$ (see section 8.2.21), may turn out to be bounded at isolated points even when condition 8.2(9) is not satisfied.

8.2.45. The condition of convexity (concavity) of a system of numbers, although not necessary, generally speaking, is in a certain sense essential for the validity of the relation 8.2(56). Such a relation does not hold in the general case. This can be verified by considering, for example, the system of numbers

$$\lambda_k^{(n)} = \begin{cases} (-1)^{\frac{k}{2}} & \text{if } k \text{ is even,} \\ 0 & \text{if } k \text{ is odd} \end{cases} \tag{68}$$

for which** $M_n^*\left(\frac{2\pi}{2n+1}\right) \neq O(1)$.

Moreover, it must also be noted that the relation 8.2(56) may also sometimes be satisfied even in cases not covered by theorem 8.2.4, i.e. it is in reality valid for a wider class of matrices (see 8.7(18).

8.2.5. With respect to the Lebesgue functions 8.1(31), which correspond to interpolation on $[-1, 1]$ at the zeros of the Chebyshev polynomials, the following theorem*** is valid, which we will state here without proof. *If the bounded matrix λ satisfies the condition* 8.2(26) *or the condition* 8.2(27), *then*

$$M_n(x) = \frac{2}{\pi}|\cos n \arccos x|\left|\sum_{k=1}^{n}\frac{\lambda_k^{(n)}}{n-k+1}\right|+O(1), \tag{69}$$

uniformly with respect to all $x \in [-1, 1]$ *and* n.

** See for example, A. F. Timan [1].

*** N. A. Pogodicheva [1].

Comparison of this estimate with 8.2(28) for the function $L_n(x)$, which corresponds to an expansion in a Fourier-Chebyshev series, shows that between $M_n(x)$ and $L_n(x)$ there is, generally speaking, no relation of dependence similar to 8.2(56), where in place of the factor $\left|\sin\frac{2n+1}{2}x\right|$ it would be natural to have the factor $|\cos n \arccos x|$. Moreover, we see that, as in the previous cases, in the class of bounded matrices λ which satisfy a condition of convexity (concavity) the sequence $M_n(x)$ is bounded at each point x or uniformly with respect to x when and only when 8.2(9) is satisfied.

8.2.6. In concluding this section we note that results similar to those stated above do hold also in the many-dimensional case. Thus, for example, starting from inequalities of the type 8.2(3) it can be shown that** *for boundedness of the sequence of constants* $L_{m,n}$ (see 8.1(34)) *in the general case, it is necessary for the following conditions to be satisfied besides* 8.1(35):

$$\sum_{k=1}^{m} \frac{\lambda_{k,l}^{(m,n)}}{m-k+1} = O(1), \quad \sum_{l=1}^{n} \frac{\lambda_{k,l}^{(m,n)}}{n-l+1} = O(1),$$
$$\sum_{k=1}^{m} \sum_{l=1}^{n} \frac{\lambda_{k,l}^{(m,n)}}{(m-k+1)(n-l+1)} = O(1), \tag{70}$$

uniformly with respect to all m, n *and* k, l $(0 \leqslant k \leqslant m, 0 \leqslant l \leqslant n)$.

8.3. On linear methods of approximation by polynomials which give the best order of approximation.

In the sections 8.1 and 8.2 we studied linear processes of approximation only from the point of view of their convergence in the whole class of continuous functions. Let us now pass to questions connected with the rapidity of their convergence, and first of all consider the elucidation of conditions which must be satisfied by linear methods of approximation which give the best possible order of approximation.

8.3.1. In the Hilbert space L_2 the best linear method of approximation with respect to a given orthonormal system of polynomials is the Fourier series expansion. Whatever the function $f \in L_z$ may be, the partial sums of its Fourier series form a sequence which realises the best approximation in the class of polynomials considered (see 2.8.2).

** I. V. Matveev [1].

As regards order of approximation the Fourier series expansion also possesses a similar property in other spaces L_q, when $1 < y < \infty$ (see 5.11.6).

However, as regards uniform approximation, as mentioned above (see section 8.1.2), the partial sums of the Fourier series are unsuitable as a means of approximation for the whole class of continuous functions. Uniform approximation by them can be identical with the best approximation as regards order only for functions possessing certain special properties (see, for example, sections 7.5.5 and 7.5.6). In this respect the process of expansion in a Fourier series is not an exception among all other linear processes of approximation by polynomials. No linear process of approximation by polynomials can, for any continuous function, accomplish uniform approximation identical in order of decrease with the best approximation. This follows directly from the considerations advanced in section 8.1.1. If, for example, in the periodic case there existed a sequence of operators $U_n(f)$ which established a correspondence between each continuous function f of period 2π and a trigonometric polynomial 8.1(4), which possessed the property that in the whole class C*

$$\|f-U_n(f)\|_{C^*} \leqslant C(f)E_n^*(f) \qquad (n = 0, 1, 2, \ldots), \tag{1}$$

where $C(f)$ is a constant which depends only on f, then whatever the trigonometric polynomial $T_n(t)$ of order $\leqslant n$, we should have $U_n(T_n) \equiv T_n(t)$. Consequently (see section 8.1.1) condition 8.1(2) is violated, and there exists a continuous function f for which the deviation $\|f-U_n(f)\|_{C^*}$ does not tend to zero as $n \to \infty$. This is incompatible with 8.3(1).

The situation is similar also in the case of approximation by algebraic polynomials on a finite segment.

Thus the study of linear processes of approximation which accomplish the best possible order of uniform approximation has a meaning only in certain special classes of continuous functions. If $\mathfrak{M}$ is some class of continuous functions $f(x)$ and $U_n(f)$ $(n = 0, 1, 2, \ldots)$ is a linear method of approximation which establishes a relation between each function f and a sequence of polynomials $U_n(f; x)$, then a natural characteristic of the approximative properties of this process on the whole class $\mathfrak{M}$ is the sequence of upper bounds

$$\mathcal{E}(\mathfrak{M}; x; U_n) = \sup_{f \in \mathfrak{M}} |f(x) - U_n(f; x)|. \tag{2}$$

When $\mathfrak{M}$ consists of all continuous functions $f(t)$ for which $\max\limits_t |f(t)| \leqslant 1$, the upper bounds 8.3(2) are, apart from a bounded quantity, identical with the norms of the functionals $U_n(f; x)$, and in this case the convergence of the corresponding sequence of polynomials for any function $f \in \mathfrak{M}$ (see section 8.1.3) depends on their behaviour as $n \to \infty$.

In other cases, when $\sup_x \mathcal{E}(\mathfrak{M}; x; U_n) \to 0$ as $n \to \infty$, the order of decrease of the sequence 8.3(2) determines the rapidity of convergence of the given process on the whole class $\mathfrak{M}$. It follows from theorem 2.5.1 that in the case of the class $\mathfrak{M}$, it must always be compact if bounded. We obtain the best possible order of approximation on a given compact class $\mathfrak{M}$ when as $n \to \infty$ $\sup_x \mathcal{E}(\mathfrak{M}; x; U_n)$ has the same order of decrease as the sequence of upper bounds of the corresponding best approximations.

Let us consider the problem of the behaviour of the sequence 8.3(2) for certain compact classes of continuous periodic functions and for methods of approximation of the type 8.1(8).

We have already seen (see section 8.1.3), that for the convergence of the sequence of trigonometric polynomials 8.1(8) on the whole class C^* of continuous functions of period 2π it is necessary that for any fixed value of k the difference $1-\lambda_k^{(n)}$ should tend to zero when $n \to \infty$ (see 8.1(16)). Now let us note that the rapidity of convergence of the sequence $U_n^*(f; x; \lambda)$ to $f(x)$ is connected in a known manner with the order of decrease to zero of this difference. In particular, it is not difficult to verify the fact that if $f(x) \not\equiv 0$ then the deviation $\max_x |f(x)-U_n^*(f; x; \lambda)|$ cannot decrease to zero as $n \to \infty$, more rapidly (in respect of order) than any of the differences

$$1-\lambda_k^{(n)} \quad (k = 0, 1, 2, \ldots).$$

In fact, since

$$(1-\lambda_k^{(n)})(a_k+ib_k) = \frac{1}{\pi} \int_0^{2\pi} \{f(x)-U_n^*(f; x; \lambda)\}\, e^{ikx}\, dx,$$

it follows that

$$|1-\lambda_k^{(n)}| \sqrt{a_k^2+b_k^2} \leqslant 2 \max_x |f(x)-U_n^*(f; x; \lambda)|. \tag{3}$$

It follows that in those cases when for any $k = 0, 1, 2, \ldots$ and as $n \to \infty$

$$\max_x |f(x)-U_n^*(f; x; \lambda)| = o\{|1-\lambda_k^{(n)}|\} \quad (\lambda_k^{(n)} \neq 1)$$

all the Fourier coefficients of the function $f(x)$ must be equal to zero, i.e. $f(x) \equiv 0$.

Let the class $\mathfrak{M}$ of continuous functions of period 2π possess the property that for any $k = 0, 1, 2, \ldots$

$$\sup_{f \in \mathfrak{M}} \left| \int_0^{2\pi} f(t)\, e^{ikt}\, dt \right| \neq 0. \tag{4}$$

It follows from the inequality 8.3(3) that *on such a compact class* $\mathfrak{M}$ *the process of approximation* 8.1(8), *defined by the infinite triangular matrix* λ, *can accomplish the best order of approximation*, i.e. *satisfy the condition*

$$\mathcal{E}_n(\mathfrak{M}; x; \lambda) = \sup_{f \in \mathfrak{M}} |f(x) - U_n^*(f; x; \lambda)| \leqslant C \sup_{f \in \mathfrak{M}} E_n^*(f) \tag{5}$$

where C is some constant, only when

$$|1-\lambda_k^{(n)}| \sup_{f \in \mathfrak{M}} \left| \int_0^{2\pi} f(t) \mathrm{l}\, e^{ikt}\, dt \right| = O\,\{\sup_{f \in \mathfrak{M}} E_n^*(f)\} \tag{6}$$

uniformly with respect to n and k $(0 \leqslant k \leqslant n)$.

8.3.2. Many frequently encountered classes of periodic functions satisfy condition 8.3(4). In particular, it is satisfied by the classes H_ω^* of all periodic functions whose modulus of continuity does not exceed the given modulus of continuity $\omega(t)$ (see section 3.2.7), of the class $W_*^{(r)} H_\omega$ (see 3.5.3) for any $r > 0$, and also by the classes $\tilde{H}_\omega^*$ and $W_*^{(r)} \tilde{H}_\omega$ conjugate to them (see section 3.11.1).

As an example we consider the class H_ω^* and we show that *for any natural number k*

$$\sup_{f \in H_\omega^*} \int_0^{2\pi} f(t) \sin kt\, dt \geqslant 2 \int_0^{\frac{\pi}{2}} \omega\left(\frac{t}{k}\right) \sin t\, dt, \tag{7}$$

and in the case when the modulus of continuity $w(t)$ is convex, i.e. $\frac{1}{2}[\omega(t_1)+\omega(t_2)] \leqslant \omega\left(\frac{t_1+t_2}{2}\right)$ *we have***

$$\sup_{f \in H_\omega^*} \int_0^{2\pi} f(t) \sin kt\, dt = 2 \int_0^{\frac{\pi}{2}} \omega\left(\frac{2t}{k}\right) \sin t\, dt. \tag{8}$$

In order to obtain 8.3(7), it is sufficient to consider the odd function of period $\frac{2\pi}{k}$

$$f_k^*(t) = \begin{cases} \frac{1}{2}\omega(t) & \text{when} \quad 0 \leqslant t \leqslant \frac{\pi}{2k}, \\ \frac{1}{2}\omega\left(\frac{\pi}{k} - t\right) & \text{when} \quad \frac{\pi}{2k} \leqslant t \leqslant \frac{\pi}{k}, \end{cases} \tag{9}$$

** See Lebesgue [3].

which, as it is not difficult to verify, belongs to the class H_ω^*, and satisfies

$$\int_0^{2\pi} f_k^*(t)\sin kt\,dt = 4k\int_0^{\frac{\pi}{2k}} f_k^*\sin kt\,dt = 2\int_0^{\frac{\pi}{2}} \omega\left(\frac{t}{k}\right)\sin t\,dt \tag{10}$$

If, however, $\omega(t)$ satisfies the condition of convexity, then in order to obtain a more precise estimate from below we may take, instead of the function 8.3(9) in the class H_ω^*, the odd function of period $\frac{2\pi}{k}$

$$\phi_k^*(t) = \begin{cases} \frac{1}{2}\omega(2t) & \text{when} \quad 0 \leqslant t \leqslant \frac{\pi}{2k}, \\ \frac{1}{2}\omega\left(\frac{2\pi}{k} - 2t\right) & \text{when} \quad \frac{\pi}{2k} \leqslant t \leqslant \frac{\pi}{k}. \end{cases} \tag{11}$$

Then we obtain

$$\sup_{f\in H_\omega^*}\int_0^{2\pi} f(t)\sin k(t)\,dt \geqslant \int_0^{2\pi} \phi_k^*(t)\sin kt\,dt$$

$$= 4k\int_0^{\frac{\pi}{2k}} \phi_k^*(t)\sin kt\,dt = 2\int_0^{\frac{\pi}{2}} \omega\left(\frac{2t}{k}\right)\sin t\,dt. \tag{12}$$

On the other hand, if

$$F_k(t) = \frac{1}{k}\sum_{\nu=0}^{k-1} f\left(t+\frac{2\nu\pi}{k}\right), \tag{13}$$

then

$$\sup_{f\in H_\omega^*}\int_0^{2\pi} f(t)\sin kt = \sup_{f\in H_\omega^*}\int_0^{2\pi} F_k(t)\sin kt\,dt.$$

But $F_k(t)$ has period $\frac{2\pi}{k}$. Therefore

$$\sup_{f\in H_\omega^*}\int_0^{2\pi} f(t)\sin kt\,dt = k\sup_{f\in H_\omega^*}\int_{-\frac{\pi}{2k}}^{\frac{3\pi}{2k}} F_k(t)\sin kt$$

$$= \frac{k}{2}\sup_{f\in H_\omega^*}\int_{-\frac{\pi}{2k}}^{\frac{3\pi}{2k}} \left[F_k(t)+F_k\left(\frac{\pi}{k}-t\right)\right]\sin kt\,dt$$

$$= k \sup_{f \in H_\omega^*} \int_{-\frac{\pi}{2k}}^{\frac{\pi}{2k}} \left[F_k(t)+F_k\left(\frac{\pi}{k}-t\right)\right] \sin kt\, dt$$

$$= k \sup_{f \in H_\omega^*} \int_0^{\frac{\pi}{2k}} \left[F_k(t)-F_k(-t)+F_k\left(\frac{\pi}{k}-t\right)-\right.$$

$$\left.-F_k\left(\frac{\pi}{k}+t\right)\right] \sin kt\, dt$$

$$\leqslant 2k \int_0^{\frac{\pi}{2k}} \omega(2t) \sin kt\, dt = 2 \int_0^{\frac{\pi}{2}} \omega\left(\frac{2t}{k}\right) \sin t\, dt.$$

Similarly we may establish the inequality

$$\sup_{f \in H_\omega^*} \int_0^{2\pi} f(t) \cos kt\, dt \geqslant 2 \int_0^{\frac{\pi}{2}} \omega\left(\frac{t}{k}\right) \sin t\, dt \tag{14}$$

in the general case or the equality

$$\sup_{f \in H_\omega^*} \int_0^{2\pi} f(t) \cos kt\, dt = 2 \int_0^{\frac{\pi}{2}} \omega\left(\frac{t}{k}\right) \sin t\, dt \tag{15}$$

in the case when $\omega(t)$ satisfies the condition of convexity.

After integration by parts we obtain from 8.3(7) and 8.3(14) the estimate

$$\sup_{f \in W_*^{(r)} H_\omega} \left|\int_0^{2\pi} f(t)\, e^{ikt}\, dt\right| \geqslant \frac{2}{(k+1)^r} \int_0^{\frac{\pi}{2}} \omega\left(\frac{t}{k+1}\right) \sin t\, dt \quad (k = 0, 1, 2, \ldots) \tag{16}$$

for every natural number r and any modulus of continuity $\omega(t)$.

8.3.3. In order to verify condition 8.3(6) it is important to know the character of the behaviour of the sequence of upper bounds $\sup_{f \in \mathfrak{M}} E_n^*(f)$ as $n \to \infty$. Let us consider from this point of view the classes H_w^*, $W_*^{(r)} H_\omega$ and $\tilde{W}_*^{(r)} H_w$ ($r \geqslant 0$ an integer).

The function 8.3(9) for any k belongs to the class H_ω^*. In addition, $f_{n+1}(t)$ assumes the value $\frac{1}{2}\omega\left(\frac{\pi}{2n+2}\right)$ with consecutively alternating signs at $2n+2$ points of $(0, 2\pi)$. Therefore (see section 2.10)

$$E_n^*(f_{n+1}) \geqslant \frac{1}{2}\omega\left(\frac{\pi}{2n+2}\right).$$

Consequently, whatever the modulus of continuity $\omega(t)$ may be, the inequality

$$\sup_{f \in H_\omega^*} E_n^*(f) \geqslant \frac{1}{2}\omega\left(\frac{\pi}{2n+2}\right), \tag{17}$$

supplementing 5.1(10), is satisfied.

This same kind of estimate can be obtained for the classes $W_*^{(r)}H_\omega$ for any natural number r if use is made of the Fourier expansion

$$f_n^*(x) = \sum_{\nu=0}^{\infty} b_{2\nu+1}^{(n)}(\omega)\sin(2\nu+1)nx \tag{18}$$

and the function

$$\Phi_n(x) = \frac{1}{n^r}\sum_{\nu=0}^{\infty}\frac{b_{2\nu+1}^{(n)}(\omega)}{(2\nu+1)^r}\sin\left[(2\nu+1)nx-\frac{r\pi}{2}\right] \tag{19}$$

is considered.

Since $\Phi_n^{(r)}(x) = f_n^*(x)$, it follows that $\Phi_n(x) \in W_*^{(r)}H_\omega$. In addition, at the $2n$ points $x_k = \frac{k\pi}{n}+\frac{\pi}{4n}[1+(-1)^r]$ $(k = 0, 1, \ldots, 2n-1)$ of the half-open interval $[0, 2\pi)$ the function $\Phi_n(x)$ assumes with consecutive changes of sign the value

$$\frac{1}{n^r}\sum_{\nu=0}^{\infty}(-1)^{\nu(r+1)}\frac{b_{2\nu+1}^{(n)}(\omega)}{(2\nu+1)^r}.$$

Therefore (see section 2.10)

$$E_{n-1}^*(\Phi_n) \geqslant \frac{1}{n^r}\sum_{\nu=0}^{\infty}(-1)^{\nu(r+1)}\frac{b_{2\nu+1}^{(n)}(\omega)}{(2\nu+1)^r}$$

But since

$$b_{2\nu+1}^{(n)}(\omega) = \frac{2}{(2\nu+1)\pi}\int_0^{\frac{2\nu+1}{2}\pi}\omega\left[\frac{t}{(2\nu+1)n}\right]\sin t\,dt, \tag{20}$$

then

$$\sum_{\nu=0}^{\infty}(-1)^{\nu(r+1)}\frac{b_{2\nu+1}^{(n)}(\omega)}{(2\nu+1)^r}$$

$$=\frac{2}{\pi}\sum_{\nu=0}^{\infty}\frac{(-1)^{\nu(r+1)}}{(2\nu+1)^r}\int_0^{\frac{\pi}{2}}\omega\left(\frac{t}{n}\right)\sin(2\nu+1)t\,dt$$

$$=\frac{2}{\pi}\int_0^{\frac{\pi}{2}}\omega\left(\frac{t}{n}\right)\sum_{\nu=0}^{\infty}(-1)^{\nu(r+1)}\frac{\sin(2\nu+1)t}{(2\nu+1)^r}\,dt.$$

Applying Rolle's theorem, it is easily verified that the sum

$$\sum_{\nu=0}^{\infty}(-1)^{\nu(r+1)}\frac{\sin(2\nu+1)t}{(2\nu+1)^r}$$

retains its sign on $(0, \pi)$. Therefore

$$\int_0^{\frac{\pi}{2}}\omega\left(\frac{t}{n}\right)\sum_{\nu=0}^{\infty}(-1)^{\nu(r+1)}\frac{\sin(2\nu+1)t}{(2\nu+1)^r}\,dt$$

$$>\omega\left(\frac{\pi}{4n}\right)\int_{-\frac{\pi}{2}}^{\frac{\pi}{2}}\sum_{\nu=0}^{\infty}(-1)^{\nu(r+1)}\frac{\sin(2\nu+1)t}{(2\nu+1)^r}\,dt.$$

Consequently

$$E_{n-1}^*(\Phi_n)\geqslant C_r\frac{\omega\left(\frac{1}{n}\right)}{n^r},$$

where C_r is some constant, and

$$\sup_{f\in W_*^{(r)}H_\omega} E_{n-1}^*(f)\geqslant C_r\frac{\omega\left(\frac{1}{n}\right)}{n^r}. \tag{21}$$

It is similarly verified that whatever the modulus of continuity $\omega(t)$ may be

$$\sup_{f\in \tilde{W}_*^{(r)}H_\omega} E_{n-1}^*(f)\geqslant C_r\frac{\omega\left(\frac{1}{n}\right)}{n^r} \tag{22}$$

for any integer $r\geqslant 0$

Comparing the inequalities 8.3(17), 8.3(21) and 8.3(22) with 5.5(16) and 5.9(8), we obtain the result that *whatever the modulus of continuity* $\omega(t)$ *may be, the exact order of decrease of the upper bounds* $\sup_{f\in W_*^{(r)}H_\omega} E_n^*(f)$ (*or correspondingly* $\sup_{f\in \tilde{W}_*^{(r)}H_\omega} E_n^*(f)$ *for any integer* $r\geqslant 0$, *when* $n\to\infty$, *is equal to* $\dfrac{1}{n^r}\omega\left(\dfrac{1}{n}\right)$.

Thus in the case when $\mathfrak{M}$ is the class $W_*^{(r)}H_\omega$ or $\tilde{W}_*^{(r)}H_\omega$, the condition 8.3(6) reduces as a result of the inequality 8.3(16) to the fact that

$$1-\lambda_k^{(n)} = O\left[\left(\frac{k+1}{n+1}\right)^r \frac{\omega\left(\dfrac{1}{n+1}\right)}{\left(\omega\dfrac{1}{k+1}\right)}\right] \tag{23}$$

uniformly with respect to n and k $(0\leqslant k\leqslant n)$.

Only when this is satisfied can the process of approximation by polynomials 8.1(18) produce the best order of approximation in these classes.

8.3.4. Since condition 8.2(23) entails the boundedness of the matrix λ, we may in what follows, in studying methods of approximation of the type 8.1(8) which give the best order of approximation in the classes of functions indicated, suppose that $\lambda_k^{(n)} = O(1)$ uniformly with respect to n and k $(0\leqslant k\leqslant n)$.

We will consider the classes H_ω^* and $\tilde{H}_\omega^*$ separately, and will prove that *whatever the modulus of continuity* $\omega(t)$ *may be, the inequality***

$$\left.\begin{aligned}\mathcal{E}_n(H_\omega^*; x; \lambda) &= \sup_{f\in H_\omega^*}|f(x)-U_n^*(f; x; \lambda)|\\ \mathcal{E}_n(\tilde{H}_\omega^*; x; \lambda) &= \sup_{f\in \tilde{H}_\omega^*}|f(x)-U_n^*(f; x; \lambda)|\end{aligned}\right\}$$

$$\geqslant \frac{2}{\pi^2}\int_0^{\frac{\pi}{2}}\omega\left(\frac{1}{n}\right)\sin t\,dt\left|\sum_{k=1}^{n}\frac{\lambda_k^{(n)}}{n-k+1}\right| + O\left[\omega\left(\frac{1}{n}\right)\right], \tag{24}$$

is satisfied for any bounded matrix λ, *and in the case when the function* $\omega(t)$ *is convex* (see section 8.3.2),

$$\left.\begin{aligned}\mathcal{E}_n(H_\omega^*; x; \lambda)\\ \mathcal{E}_n(\tilde{H}_\omega^*; x; \lambda)\end{aligned}\right\}$$

$$\geqslant \frac{2}{\pi^2}\int_0^{\frac{\pi}{2}}\omega\left(\frac{2t}{n}\right)\sin t\,dt\left|\sum_{k=1}^{n}\frac{\lambda_k^{(n)}}{n-k+1}\right| + O\left[\omega\left(\frac{1}{n}\right)\right]. \tag{25}$$

** A. F. Timan [12].

In order to obtain the inequalities 8.3(24) and 8.3(25) we note that for any a there corresponds to any function $F(x)$ of the class H^*_ω (or correspondingly $\tilde{H}^*_\omega$) the function $F_a(x) = F(x+a)$, and

$$F(a) - U^*_n(F;\ a;\ \lambda) = F_a(0) - U^*_n(F;\ 0;\ \lambda).$$

Therefore the upper bounds $\mathcal{E}\,(H^*_\omega\,;\ x;\ \lambda)$, $\mathcal{E}_n(\tilde{H}^*_\omega\,;\ x;\ \lambda)$, as also in the case of the Lebesgue constant (see section 8.3.1), must not depend on x, i.e.

$$\mathcal{E}_n(H^*_\omega;\ x;\ \lambda) = \mathcal{E}_n(H^*_\omega;\ 0;\ \lambda) = \sup_{f\in H^*_\omega} |f(0) - U^*_n(f;\ 0;\ \lambda)|,$$

$$\mathcal{E}_n(\tilde{H}^*_\omega;\ x;\ \lambda) = \mathcal{E}_n(\tilde{H}^*_\omega;\ 0;\ \lambda) = \sup_{f\in \tilde{H}^*_\omega} |f(0) - U^*_n(f;\ 0;\ \lambda)|.$$

It follows that

$$\mathcal{E}_n(H^*_\omega;\ x;\ \lambda) \geqslant |f_n(0) - U^*_n(f_n;\ 0;\ \lambda)| = |U^*_n(f_n;\ 0;\ \lambda)|, \tag{26}$$

where

$$f_n(t) = \begin{cases} f^*_n(t), & 0 \leqslant t \leqslant \pi, \\ f^*_n(-t), & -\pi \leqslant t \leqslant 0, \end{cases}$$

and $f^*_n(t)$ is the periodic function of period $\dfrac{2\pi}{n}$ of 8.3(9).

But (see 8.1(12) and 8.3(18))

$$|U^*_n(f_n;\ 0;\ \lambda)|$$

$$= \frac{2}{\pi}\left|\int_0^\pi \left\{\frac{\lambda_0^{(n)}}{2} + \sum_{k=1}^n \lambda_k^{(n)} \cos kt\right\} \sum_{\nu=0}^\infty b_{2\nu+1}^{(n)}(\omega) \sin(2\nu+1)\, nt\, dt\right|$$

$$= \frac{2}{\pi}\left|\int_0^\pi \sum_{k=1}^n \lambda_k^{(n)} \cos kt \sum_{\nu=0}^\infty b_{2\nu+1}^{(n)}(\omega) \sin(2\nu+1)\, nt\, dt\right| + O\left[\omega\left(\frac{1}{n}\right)\right]$$

$$= \left|\frac{2b_1^{(n)}(\omega)}{\pi} \int_0^\pi \sum_{k=1}^n \lambda_k^{(n)} \cos kt \sin n\, t\, dt + \right.$$

$$\left. + \frac{2}{\pi} \int_0^\pi \sum_{k=1}^n \lambda_k^{(n)} \cos kt \sum_{k=1}^\infty b_{2\nu+1}^{(n)}(\omega) \sin(2\nu+1)\, nt\, dt\right| + O\left[\omega\left(\frac{1}{n}\right)\right], \tag{27}$$

since we are supposing that $\lambda_k^{(n)} = O(1)$.

With this condition

$$\frac{2b_1^{(n)}(\omega)}{\pi}\int\limits_0^\pi\sum_{k=1}^n \lambda_k^{(n)}\cos kt\sin nt\,\mathrm{d}t$$

$$=\frac{b_1^{(n)}(\omega)}{\pi}\int\limits_0^\pi\sum_{k=1}^n \lambda_k^{(n)}\sin(n-k)\,t\,\mathrm{d}t+$$

$$+\frac{b_1^{(n)}(\omega)}{\pi}\int\limits_0^\pi\sum_{k=1}^n \lambda_k^{(n)}\sin(n+k)\,t\,\mathrm{d}t$$

$$=\frac{b_1^{(n)}(\omega)}{\pi}\sum_{k=1}^{n-1}\frac{1+(-1)^{n-k+1}}{n-k}\lambda_k^{(n)}+$$

$$+\frac{b_1^{(n)}(\omega)}{\pi}\sum_{k=1}^{n}\frac{1+(-1)^{n-k+1}}{n+k}\lambda_k^{(n)}$$

$$=\frac{b_1^{(n)}(\omega)}{\pi}\sum_{k=1}^{n-1}\frac{1+(-1)^{n-k+1}}{n-k}\lambda_k^{(n)}+O\left[\omega\left(\frac{1}{n}\right)\right]. \tag{28}$$

Moreover, since by 8.3(20)

$$|b_{2\nu+1}^{(n)}(\omega)|=\frac{2}{(2\nu+1)\pi}\left|\int\limits_0^{\frac{2\nu+1}{2}\pi}\cos t\,\mathrm{d}\omega\left[\frac{t}{(2\nu+1)n}\right]\right|$$

$$\leqslant\frac{2}{(2\nu+1)\pi}\omega\left(\frac{\pi}{2n}\right), \tag{29}$$

we have

$$\frac{2}{\pi}\sum_{\nu=1}^\infty b_{2\nu+1}^{(n)}(\omega)\int\limits_0^\pi\sum_{k=1}^n\lambda_k^{(n)}\cos kt\sin(2\nu+1)nt\,\mathrm{d}t$$

$$=\frac{1}{\pi}\sum_{\nu=1}^\infty b_{2\nu+1}^{(n)}(\omega)\int\limits_0^\pi\sum_{k=1}^n\lambda_k^{(n)}\sin[(2\nu+1)n-k]t\,\mathrm{d}t+$$

$$+\frac{1}{\pi}\sum_{\nu=1}^\infty b_{2\nu+1}^{(n)}(\omega)\int\limits_0^\pi\sum_{k=1}^n\lambda_k^{(n)}\sin[(2\nu+1)n+k]t\,\mathrm{d}t+$$

$$+O\left[\omega\left(\frac{1}{n}\right)\right]. \tag{30}$$

Consequently, if we substitute the estimates 8.3(28) and 8.3(29) in 8.3(27), then by 8.3(26) we obtain the inequality

$$\mathcal{E}_n(H_w^*;\, x;\, \lambda) \geqslant \frac{b_1^{(n)}(\omega)}{\pi}\left|\sum_{k=1}^{n-1} \frac{1+(-1)^{n-k+1}}{n-k}\lambda_k^{(n)}\right| + O\left[\omega\left(\frac{1}{n}\right)\right]. \tag{31}$$

Considering instead of $f_n(t)$ the function $f_{n-1}(t)$ and observing that $b_1^{(n-1)}(\omega) - b_1^{(n)}(\omega)$

$$= \frac{2(n-1)}{\pi}\int_0^{\frac{\pi}{2(n-1)}} \omega(2t)\sin(n-1)t\,\mathrm{d}t - \frac{2n}{\pi}\int_0^{\frac{\pi}{2n}} \omega(2t)\sin nt\,\mathrm{d}t$$

$$= \frac{4n}{\pi}\int_0^{\frac{\pi}{2n}} \omega(2t)\cos\frac{2n-1}{2}t\sin t\,\mathrm{d}t + O\left[\frac{1}{n}\omega\left(\frac{1}{n}\right)\right]$$

$$\leqslant \frac{4n}{\pi}\int_0^{\frac{\pi}{2n}} t\omega(2t)\,\mathrm{d}t + O\left[\frac{1}{n}\omega\left(\frac{1}{n}\right)\right] = O\left[\frac{1}{n}\omega\left(\frac{1}{n}\right)\right], \tag{32}$$

in addition to the inequality 8.3(31) we obtain

$$\mathcal{E}_n(H_\omega^*;\, x;\, \lambda) \geqslant \frac{b_1^{(n)}(\omega)}{\pi}\left|\sum_{k=1}^{n-2} \frac{1+(-1)^{n-k}}{n-k-1}\lambda_k^{(n)}\right| + O\left[\omega\left(\frac{1}{n}\right)\right]. \tag{33}$$

From 8.3(31) and 8.3(33) it immediately follows as a result of 8.3(32) that for any modulus of continuity $\omega(t)$

$$\mathcal{E}_n(H_\omega^*;\, x;\, \lambda) > \frac{b_1^{(n)}(\omega)}{\pi}\left|\sum_{k=1}^{n} \frac{\lambda_k^{(n)}}{n-k+1}\right| + O\left[\omega\left(\frac{1}{n}\right)\right],$$

i.e. (see 8.3(20)) the first of the inequalities 8.3(24) is satisfied.

In order to obtain the first of the inequalities 8.3(25) subject to the condition that the modulus of continuity $\omega(t)$ is convex, it is necessary instead of the function $f_n^*(t)$ to consider the function $\phi_n^*(t)$ (see 8.3(11)) and repeat the same estimate.

In passing to the conjugate classes $\tilde{H}_\omega^*$ we reason similarly, considering instead of the functions $f_n(t)$ the odd functions of period 2π

$$f_{n,1}(t) = \begin{cases} 0, & 0 \leqslant t \leqslant \dfrac{3\pi}{2n},\quad \pi - \dfrac{\pi}{2n} \leqslant t \leqslant \pi, \\[2ex] f_n^*\left(t + \dfrac{\pi}{2n}\right), & \dfrac{3\pi}{2n} \leqslant t \leqslant \pi - \dfrac{\pi}{2n}, \end{cases}$$

if n is even, and

$$f_{n,1}(t)=\begin{cases}0, & 0\leqslant t\leqslant\frac{\pi}{2n},\ \pi-\frac{\pi}{2n}\leqslant t\leqslant\pi,\\ f_n^*\left(t+\frac{\pi}{2n}\right), & \frac{\pi}{2n}\leqslant t\leqslant\pi-\frac{\pi}{2n},\end{cases}$$

if n is odd, or the functions $\phi_{n,1}(t)$ defined also in terms of $\phi_n^*(t)$ (see 8.3(11)).

Since (see 3.11(3))

$$\tilde{f}_{n,1}(0)=-\frac{2}{\pi}\int_{\frac{\pi}{2n}}^{\pi}\frac{f_{n,1}(t)}{2\tan\frac{t}{2}}\,dt=O\left[\omega\left(\frac{1}{n}\right)\right],$$

the estimation of the quantity $\mathcal{E}(\tilde{H}_\omega^*;0;\lambda)$, as in the case of the class H_ω^*, reduces to the evaluation of the integral

$$\frac{2}{\pi}\int_0^{\pi}f_{n,1}(t)\sum_{k=1}^{n}\lambda_k^{(n)}\sin kt\,dt.$$

8.3.5. It follows from the inequality 8.3(24) that the *condition* 8.2(9) *is necessary* not only for the convergence of the sequence of polynomials 8.1(18) on the *whole* class C^* of continuous functions of period 2π (see 8.2.34), but also *in order that this process of approximation should produce the best possible order of approximation by trigonometric polynomials on the class H_ω^* of functions which have a given modulus of continuity $\omega(t)$, or on the class $\tilde{H}_\omega^*$ conjugate to it.*

It is necessary to bear in mind that in the case considered here, in addition to 8.2(9), it is also necessary for the condition

$$1-\lambda_k^{(n)}=O\left[\frac{\omega\left(\frac{1}{n+1}\right)}{\omega\left(\frac{1}{k+1}\right)}\right] \tag{34}$$

to be satisfied uniformly with respect to all n and k $(0\leqslant k\leqslant n)$.

8.3.6. A theorem similar to 8.3.5 holds also in the many-dimensional case. We state here without proof the following result for continuous periodic functions of two variables**. Let H_{ω_1,ω_2}^* be the class of continuous

** V. G. Ponomarenko [1].

functions $f(x, y)$, of period 2π in each of the variables, the partial moduli of continuity of which respectively do not exceed the moduli of continuity $\omega_1(t)$ and $\omega_2(t)$ (see section 3.4.33). *In order that the process of approximation* 8.1(32) *should produce on the class $H^*_{\omega_1, \omega_2}$ the best order of approximation by trigonometric polynomials, it is necessary that the conditions*

$$\lambda_{k,l}^{(m,n)} = O(1),$$

$$\sum_{k=1}^{m} \frac{\lambda_{k,l}^{(m,n)}}{m-k+1} = O(1), \qquad \sum_{l=1}^{n} \frac{\lambda_{k,l}^{(m,n)}}{n-l+1} = O(1),$$

$$\min\left[\omega_1\left(\frac{1}{m}\right), \omega_2\left(\frac{1}{n}\right)\right] \sum_{k=1}^{m}\sum_{l=1}^{n} \frac{\lambda_{k,l}^{(m,n)}}{(m-k+1)(n-l+1)} = O\left[\omega_1\left(\frac{1}{m}\right)+\omega_2\left(\frac{1}{n}\right)\right]$$

should be satisfied uniformly with respect to all m, n, k, l $(0 \leqslant k \leqslant m;$ $0 \leqslant l \leqslant n)$.

8.3.7. For the classes $W_*^{(r)}H_\omega$ and $W_*^{(r)}H_\omega$ for natural values of r, the following theorem holds**. *Whatever the modulus of continuity $\omega(t)$, for any bounded matrix λ the inequalities*

$$\left.\begin{aligned} \mathcal{E}_n(\tilde{W}_*^{(r)}H_\omega; x; \lambda) &= \sup_{f \in W_*^{(r)}H_\omega} |f(x) - U_n^*(f; x; \lambda)| \\ \mathcal{E}_n(W_*^{(r)}H_\omega; x; \lambda) &= \sup_{f \in \tilde{W}_*^{(r)}H_\omega} |f(x) - U_n^*(f; x; \lambda)| \end{aligned}\right\}$$

$$\geqslant \frac{2}{\pi^2}\int_0^{\frac{\pi}{2}} \omega\left(\frac{t}{n}\right)\sin t\,dt \left|\sum_{k=1}^{n} \frac{1-\lambda_k^{(n)}}{k^r(n-k+1)} - \frac{\ln n}{n^r}\right| + O\left[\frac{\omega\left(\frac{1}{n}\right)}{n}\sum_{k=1}^{n}\frac{|1+\lambda_k^{(n)}|}{k^r}\right] + O\left[\frac{\omega\left(\frac{1}{n}\right)}{n^r}\right], \tag{35}$$

are satisfied, and in the case when the function $\omega(t)$ is convex,

$$\left.\begin{aligned} &\mathcal{E}_n(W_*^{(r)}H_\omega; x; \lambda) \\ &\mathcal{E}_n(\tilde{W}_*^{(r)}H_\omega; x; \lambda) \end{aligned}\right\}$$

$$\geqslant \frac{2}{\pi^2}\int_0^{\frac{\pi}{2}} \omega\left(\frac{2t}{n}\right)\sin t\,dt \left|\sum_{k=1}^{n} \frac{1-\lambda_k^{(n)}}{k^r(n-k+1)} - \frac{\ln n}{n^r}\right| + O\left[\frac{\omega\left(\frac{1}{n}\right)}{n}\sum_{k=1}^{n}\frac{|1-\lambda_k^{(n)}|}{k^r}\right] + O\left[\frac{\omega\left(\frac{1}{n}\right)}{n^r}\right] \tag{36}$$

** A. F. Timan [12].

We conclude from the same considerations as in section 8.3.4 that the upper bounds $\mathcal{E}_n(W_*^{(r)}H_\omega; x; \lambda)$ and $\mathcal{E}_n(\tilde{W}_*^{(r)}H_\omega; x; \lambda)$ do not depend on x.

Let us also consider the sequence of functions (see 8.3(9))

$$f_n(t) = f_{n,r}(t) = \begin{cases} f_n^*\left(t+\dfrac{r\pi}{2n}\right), & 0 \leqslant t \leqslant \pi, \\ f_n(-t), & -\pi \leqslant t \leqslant 0, \end{cases}$$

when n is even and r is even;

$$f_n(t) = f_{n,r}(t) = \begin{cases} 0, & 0 \leqslant t \leqslant \dfrac{\pi}{n}, \\ f_n^*\left(t+\dfrac{r\pi}{2n}\right), & \dfrac{\pi}{n} \leqslant t \leqslant \pi, \\ f_n(-t), & -\pi \leqslant t \leqslant 0, \end{cases}$$

when n is odd and r is even;

$$f_n(t) = f_{n,r}(t) = \begin{cases} 0, & 0 \leqslant t \leqslant \dfrac{3\pi}{2n}, \quad \pi-\dfrac{\pi}{2n} \leqslant t \leqslant \pi, \\ f_n^*\left(t+\dfrac{r\pi}{2n}\right), & \dfrac{3\pi}{2n} \leqslant t \leqslant \pi-\dfrac{\pi}{2n}, \\ -f_n(-t), & -\pi \leqslant t \leqslant 0, \end{cases}$$

when n is even and r is odd;

$$f_n(t) = f_{n,r}(t) = \begin{cases} 0, & 0 \leqslant t \leqslant \dfrac{\pi}{2n}, \quad \pi-\dfrac{\pi}{2n} \leqslant t \leqslant \pi, \\ f_n^*\left(t+\dfrac{r\pi}{2n}\right), & \dfrac{\pi}{2n} \leqslant t \leqslant \pi-\dfrac{\pi}{2n}, \\ -f_n(-t), & -\pi \leqslant t \leqslant 0, \end{cases}$$

when n is odd and r is odd.

It is easily verified that the sequence of functions $f_0(t)$ defined in this way belongs to the class H_ω^*. Also, for all $t \in \left[\dfrac{3\pi}{2n}, \ \pi-\dfrac{\pi}{2n}\right]$ (see 8.3(18))

$$f_n(t) = \sum_{\nu=0}^{\infty} b_{2\nu+1}^{(n)}(\omega) \sin\left[(2\nu+1)nt+(2\nu+1)\frac{r\pi}{2}\right]. \tag{37}$$

The function

$$\psi_n(x) = \frac{1}{\pi}\int_{-\pi}^{\pi} f_n(t) \sum_{k=1}^{\infty} \frac{\cos\left[k(t-x)+\dfrac{r\pi}{2}\right]}{k^r}\, dt$$

possesses the property that its derivative $\psi^{(r)}(x)$ (see 3.5(15)) belongs to the class H_ω^*, i.e. $\psi_n(x) \in W_*^{(r)} H_\omega$. The polynomials $U_n^*(\psi_n; x; \lambda)$ for it can be represented in the form

$$U_n^*(\psi_n; x; \lambda) = \frac{1}{\pi} \int_{-\pi}^{\pi} f_n(t) \sum_{k=1}^{n} \frac{\lambda_k^{(n)}}{k^r} \cos\left[k(t-x)+\frac{r\pi}{2}\right] dt.$$

Thus

$$\psi_n(x) - U_n^*(\psi_n; x; \lambda) = \frac{1}{\pi} \int_{-\pi}^{\pi} f_n(t) \left\{ \sum_{k=1}^{n} \frac{1-\lambda_k^{(n)}}{k^r} \times \right.$$

$$\left. \times \cos\left[k(t-x)+\frac{r\pi}{2}\right] + \sum_{k=n+1}^{\infty} \frac{\cos\left[k(t-x)+\frac{r\pi}{2}\right]}{k^r} \right\} dt.$$

Therefore

$$\mathcal{E}_n(W_*^{(r)} H_\omega; x; \lambda) = \mathcal{E}_n(W_*^{(r)} H_\omega; 0; \lambda) \geqslant |\psi_n(0) - U_n^*(\psi_n; 0; \lambda)|$$

$$= \frac{2}{\pi} \left| \int_0^{\pi} f_n(x) \left\{ \sum_{k=1}^{n} \frac{1-\lambda_k^{(n)}}{k^r} \cos\left(kt+\frac{r\pi}{2}\right) + \right.\right.$$

$$\left.\left. + \sum_{k=n+1}^{\infty} \frac{\cos\left(kt+\frac{r\pi}{2}\right)}{k^r} \right\} dt \right|. \tag{38}$$

But since for any constant $C > 0$

$$\int_0^{\frac{C}{n}} \left| \sum_{k=1}^{n} \frac{1-\lambda_k^{(n)}}{k^r} \cos\left(kt+\frac{r\pi}{2}\right) \right| dt = O\left\{ \frac{1}{n} \sum_{k=1}^{n} \frac{|1-\lambda_k^{(n)}|}{k^r} \right\}$$

and also

$$\int_0^{\frac{C}{n}} \left| \sum_{k=n+1}^{\infty} \frac{\cos\left(kt+\frac{r\pi}{2}\right)}{k^r} \right| dt = O\left(\frac{1}{n^r}\right),$$

then

$$\int_0^{\frac{C}{n}} f_n^*\left(t+\frac{r\pi}{2n}\right) \left\{ \sum_{k=1}^{n} \frac{1-\lambda_k^{(n)}}{k^r} \cos\left(kt+\frac{r\pi}{2}\right) + \sum_{k=n+1}^{\infty} \frac{\cos\left(kt+\frac{r\pi}{2}\right)}{k^r} \right\} dt$$

$$= O\left[\frac{\omega\left(\frac{1}{n}\right)}{n} \sum_{k=1}^{n} \frac{|1-\lambda_k^{(n)}|}{k^r} \right] + O\left[\frac{\omega\left(\frac{1}{n}\right)}{n^r} \right].$$

Consequently, from 8.3(18), 8.3(37) and 8.3(38) it follows that

$$\mathcal{E}_n(W_*^{(r)}H_\omega;\, x;\, \lambda)$$

$$\geqslant \frac{2}{\pi}\left|\int_0^\pi \left\{\sum_{k=1}^{n} \frac{1-\lambda_k^{(n)}}{k^r}\cos\left(kt+\frac{r\pi}{2}\right)+\sum_{k=n+1}^{\infty}\frac{\cos\left(kt+\dfrac{r\pi}{2}\right)}{k^r}\right\}\times\right.$$

$$\left.\times\sum_{\nu=0}^{\infty} b_{2\nu+1}^{(n)}(\omega)\sin\left[(2\nu+1)nt+(2\nu+1)\frac{r\pi}{2}\right]\mathrm{d}t\right|+$$

$$+O\left[\frac{\omega\left(\dfrac{1}{n}\right)}{n}\sum_{k=1}^{n}\frac{|1-\lambda_k^{(n)}|}{k^r}\right]+O\left[\frac{\omega\left(\dfrac{1}{n}\right)}{n^r}\right],$$

i.e.

$$\mathcal{E}_n(W_*^{(r)}H_\omega;\, x;\, \lambda)\geqslant\left|\frac{2b_1^{(n)}(\omega)}{\pi}\int_0^\pi\left\{\sum_{k=1}^{n}\frac{1-\lambda_k^{(n)}}{k^r}\cos\left(kt+\frac{r\pi}{2}\right)+\right.\right.$$

$$\left.+\sum_{k=n+1}^{\infty}\frac{\cos\left(kt+\dfrac{r\pi}{2}\right)}{k^r}\right\}\sin\left(nt+\frac{r\pi}{2}\right)\mathrm{d}t+$$

$$+\frac{2}{\pi}\int_0^\pi\left\{\sum_{k=1}^{n}\frac{1-\lambda_k^{(n)}}{k^r}\cos\left(kt+\frac{r\pi}{2}\right)+\right.$$

$$\left.\left.+\sum_{k=n+1}^{\infty}\frac{\cos\left(kt+\dfrac{r\pi}{2}\right)}{k^r}\right\}\sum_{\nu=1}^{\infty}b_{2\nu+1}^{(n)}(\omega)\sin\left[(2\nu+1)nt+(2\nu+1)\frac{r\pi}{2}\right]\mathrm{d}t\right|+$$

$$+O\left[\frac{\omega\left(\dfrac{1}{n}\right)}{n}\sum_{k=1}^{n}\frac{|1-\lambda_k^{(n)}|}{k^r}\right]+O\left[\frac{\omega\left(\dfrac{1}{n}\right)}{n^r}\right]$$

$$=|I_n'+I_n''|+O\left[\frac{\omega\left(\dfrac{1}{n}\right)}{n}\sum_{k=1}^{n}\frac{|1-\lambda_k^{(n)}|}{k^r}\right]+O\left[\frac{\omega\left(\dfrac{1}{n}\right)}{n^r}\right]. \tag{39}$$

In order to estimate the integrals occurring here, we require the following lemma.

8.3.71. *For any natural numbers r, n and h, the relations*

$$\sum_{k=n+1}^{(2p+1)n-1} \frac{1+(-1)^{k-n+1}}{k^r[(2p+1)n-k]} = \frac{1}{(2p+1)^r}\frac{\ln n}{n^r} + O\left[\frac{\ln(2p+1)}{(2p+1)n^r}\right], \tag{40}$$

$$\sum_{k=(2p+1)n+1}^{\infty} \frac{1+(-1)^{k+n-1}}{k^r[k-(2p+1)n]} = \frac{1}{(2p+1)^r}\frac{\ln n}{n^r} + O\left[\frac{\ln(2p+1)}{(2p+1)^r n^r}\right], \tag{41}$$

$$\sum_{k=n+1}^{\infty} \frac{1+(-1)^{k-n+1}}{k^r(k-n)} = \frac{\ln n}{n^r} + O\left(\frac{1}{n}\right) \tag{42}$$

hold uniformly with respect to all p and n.

We make use of the identity

$$\frac{1}{x^r(c-x)} = \frac{d}{dx}\left[\frac{1}{c^r}\ln\frac{x}{c-x} - \sum_{\nu=1}^{r-1}\frac{1}{(r-\nu)c^\nu x^{r-\nu}}\right], \tag{43}$$

and note that if $0 < a < b < c$ then in the interval $\left[a, \frac{cr}{r+1}\right]$ the function $\frac{1}{x^r(c-x)}$ decreases monotonically, and for $x > \frac{cr}{r+1}$ increases monotonically. Therefore

$$\sum_{k=n+1}^{(2p+1)n-1} \frac{1+(-1)^{n-k+1}}{k^r[(2p+1)n-k]} = \sum_{k=n+1}^{(2p+1)n-1} \frac{1}{k^r[(2p+1)n-k]} + O\left[\frac{1}{(2p+1)n^r}\right]$$

$$= \int_{n+1}^{(2p+1)n-1} \frac{dx}{x^r[(2p+1)n-x]} + O\left[\frac{1}{(2p+1)n^r}\right]$$

$$= \frac{1}{(2p+1)^r n^r}\ln\frac{[(2p+1)n-1](2pn-1)}{n+1} +$$

$$+\sum_{\nu=1}^{r-1}\frac{1}{(r-\nu)[(2p+1)n]^\nu(n+1)^{r-\nu}} -$$

$$-\sum_{\nu=1}^{r-1}\frac{1}{(r-\nu)[(2p+1)n]^\nu[(2p+1)n-1]^{r-\nu}} + O\left[\frac{1}{(2p+1)n^r}\right]$$

$$= \frac{\ln n}{[2p+1)n]^r} + O\left[\frac{[\ln(2p+1)}{(2p+1)n^r}\right]$$

uniformly for all p and n.

The validity of the estimates 8.3(42) and 8.3(41) is verified similarly. Since the function $\frac{1}{x^r(x-c)}$ decreases monotonically for $x > c$, then in virtue of 8.3(43)

$$\sum_{k=(2p+1)n+1}^{\infty} \frac{1+(-1)^{k-n+1}}{k^r[k-(2p+1)n]}$$

$$= \sum_{k=(2p+1)n+1}^{\infty} \frac{1}{k^r[k-(2p+1)n]} + O\left[\frac{1}{(2p+1)^r n^r}\right]$$

$$= \int\limits_{(2p+1)n+1}^{\infty} \frac{dx}{x^r[x-(2p+1)n]} + O\left[\frac{1}{(2p+1)^r n^r}\right]$$

$$= \frac{1}{[(2p+1)n]^r} \ln[(2p+1)n+1] -$$

$$-\sum_{\nu=1}^{r-1} \frac{1}{(r-\nu)[(2p+1)n]^\nu[(2p+1)n+1]^{r-\nu}} + O\left[\frac{1}{(2p+1)^r n^r}\right]$$

$$= \frac{\ln n}{[(2p+1)n]^r} + O\left[\frac{\ln(2p+1)}{(2p+1)^r n^r}\right]$$

uniformly with respect to p and n.

8.3.72. We now continue the proof of theorem 8.3.6. We have

$$I_n' = \frac{b_1^{(n)}(\omega)}{\pi} \int\limits_0^\pi \sum_{k=1}^{n} \frac{1-\lambda_k^{(n)}}{k^r} \sin(n-k)t\,dt +$$

$$+ \frac{b_1^{(n)}(\omega)}{\pi} \int\limits_0^\pi \left\{\sum_{k=1}^{n} \frac{1-\lambda_k^{(n)}}{k^r} \sin[(n+k)t+r\pi] +\right.$$

$$\left.+ \sum_{k=n+1}^{\infty} \frac{1}{k^r}[\sin(n-k)t + \sin\{(n+k)t+r\pi\}]\right\} dt$$

$$= \frac{b_1^{(n)}(\omega)}{\pi} \left\{\sum_{k=1}^{n-1} \frac{1+(-1)^{n-k+1}}{n-k} \cdot \frac{1-\lambda_k^{(n)}}{k^r} +\right.$$

$$\left.+(-1)^r \sum_{k=1}^{n-1} \frac{1+(-1)^{n-k+1}}{n+k} \cdot \frac{1-\lambda_k^{(n)}}{k^r}\right\} +$$

$$+ \frac{b_1^{(n)}(\omega)}{\pi} \left\{\sum_{k=n+1}^{\infty} \frac{1+(-1)^{n-k+1}}{k^r(n-k)} + (-1)^r \sum_{k=n+1}^{\infty} \frac{1+(-1)^{n-k+1}}{k^r(n+k)}\right\}.$$

Taking account of the relation 8.3(42), we obtain

$$I_n' = \frac{b_1^{(n)}(\omega)}{\pi}\left\{\sum_{k=1}^{n-1}\frac{1+(-1)^{n-k+1}}{n-k}\cdot\frac{1-\lambda_k^{(n)}}{k^r}-\frac{\ln n}{n^r}\right\}+$$
$$+O\left[\frac{\omega\left(\frac{1}{n}\right)}{n}\sum_{k=1}^{n}\frac{1+(-1)^{n-k+1}}{k^r}(1-\lambda_k^{(n)})\right]+O\left[\frac{\omega\left(\frac{1}{n}\right)}{n^r}\right]. \tag{44}$$

Moreover,

$$I_n'' = \frac{1}{\pi}\sum_{\nu=1}^{\infty} b_{2\nu+1}^{(n)}(\omega)\int_0^{\pi}\sum_{k=1}^{n}\frac{1-\lambda_k^{(n)}}{k^r}\sin\{[(2\nu+1)n-k]t+\nu r\pi\}\,dt+$$
$$+\frac{1}{\pi}\sum_{\nu=1}^{\infty} b_{2\nu+1}^{(n)}(\omega)\int_0^{\pi}\sum_{k=1}^{n}\frac{1-\lambda_k^{(n)}}{k^r}\sin\{[(2\nu+1)n+k]t+(\nu+1)r\pi\}\,dt+$$
$$+\frac{1}{\pi}\sum_{\nu=1}^{\infty} b_{2\nu+1}^{(n)}(\omega)\int_0^{\pi}\sum_{k=n+1}^{\infty}\frac{1}{k^r}\sin\{[(2\nu+1)n-k]t+\nu r\pi\}\,dt+$$
$$+\frac{1}{\pi}\sum_{\nu=1}^{\infty} b_{2\nu+1}^{(n)}(\omega)\int_0^{\pi}\sum_{k=n+1}^{\infty}\frac{1}{k^r}\sin\{[(2\nu+1)n+k]t+(\nu+1)\nu\pi\}\,dt$$
$$=\frac{1}{\pi}\sum_{\nu=1}^{\infty} b_{2\nu+1}^{(n)}(\omega)(-1)^{\nu r}\sum_{k=1}^{n}\frac{[1+(-1)^{n-k+1}](1-\lambda_k^{(n)})}{k^r[(2\nu+1)n-k]}+$$
$$+\frac{1}{\pi}\sum_{\nu=1}^{\infty} b_{2\nu+1}^{(n)}(\omega)(-1)^{(\nu+1)r}\sum_{k=1}^{n}\frac{1+(-1)^{n-k+k}}{(2\nu+1)n+k}\cdot\frac{1-\lambda_k^{(n)}}{k^r}+$$
$$+\frac{1}{\pi}\sum_{\nu=1}^{\infty} b_{2\nu+1}^{(n)}(\omega)(-1)^{\nu r}\left\{\sum_{k=n+1}^{(2\nu+1)n-1}\frac{1+(-1)^{n-k+1}}{k^r[(2\nu+1)n-k]}+\right.$$
$$\left.+\sum_{k=(2\nu+1)n+1}^{\infty}\frac{1+(-1)^{n-k+1}}{k^r[(2\nu+1)n-k]}\right\}+$$
$$+\frac{1}{\pi}\sum_{\nu=1}^{\infty} b_{2\nu+1}^{(n)}(\omega 1)(-1)^{(\nu+1)r}\sum_{k=n+1}^{\infty}\frac{1+(-1)^{n-k+1}}{k^r[(2\nu+1)n+k]}.$$

Therefore, taking account of 8.3(29) and also of 8.3(40)–8.3(42), we find

$$I_n'' = O\left[\frac{\omega\left(\frac{1}{n}\right)}{n}\sum_{k=1}^{n}\frac{|1-\lambda_k^{(n)}|}{k^r}\right]+O\left[\frac{\omega\left(\frac{1}{n}\right)}{n^r}\right]. \tag{45}$$

If we substitute the estimates 8.3(44) and 8.3(45) in 8.3.(39), we obtain the inequality

$$\mathcal{E}_n(W_*^{(r)}H_\omega; x; \lambda) \geqslant \frac{b_1^{(n)}(\omega)}{\pi}\left|\sum_{k=1}^{n-1}\frac{1+(-1)^{n-k+1}}{n-k}\cdot\frac{1-\lambda_k^{(n)}}{k^r}-\frac{\ln n}{n^r}\right|+$$
$$+O\left[\frac{\omega\left(\frac{1}{n}\right)}{n}\sum_{k=1}^{n}\frac{|1-\lambda_k^{(n)}|}{k^r}\right]+O\left[\frac{\omega\left(\frac{1}{n}\right)}{n}\right]. \tag{46}$$

Considering instead of the function $f_n^*(t)$ the function $f_{n-1}^*(t)$ and taking account of 8.3(32), we obtain as well as the inequality 8.3(46) the inequality

$$\mathcal{E}_n(W_*^{(r)}H_\omega; x; \lambda) \geqslant \frac{b_1^{(n)}(\omega)}{\pi}\left|\sum_{k=1}^{n-2}\frac{1+(-1)^{n-k}}{n-k-1}\cdot\frac{1-\lambda_k^{(n)}}{k^r}-\frac{\ln n}{n^r}\right|+$$
$$+O\left[\frac{\omega\left(\frac{1}{n}\right)}{n}\sum_{k=1}^{n}\frac{|1-\lambda_k^{(n)}|}{k^r}\right]+O\left[\frac{\omega\left(\frac{1}{n}\right)}{n^r}\right]. \tag{47}$$

Since

$$\sum_{k=1}^{n}\frac{1-\lambda_k^{(n)}}{k^r(n-k+1)^2}=\sum_{1\leqslant k\leqslant\frac{n}{2}}\frac{1-\lambda_k^{(n)}}{k^r(n-k+1)^2}+\sum_{\frac{n}{2}<k\leqslant n}\frac{1-\lambda_k^{(n)}}{k^r(n-k+1)^2}$$
$$=O\left[\frac{1}{n^2}\sum_{k=1}^{n}\frac{|1-\lambda_k^{(n)}|}{k^r}\right]+O\left(\frac{1}{n^r}\right),$$

then, adding 8.3(46) and 8.3(47), we verify the validity of the first of the inequalities 8.3(35). In order to obtain the first inequality of 8.3(36) subject to the condition that the modulus of continuity $\omega(t)$ is convex, it is necessary to replace the functions $f_n^*(t)$ everywhere by the functions $\varphi_n^*(t)$ (see 8.3(11)) and repeat the same estimates.

8.3.73. We note that the calculations are not changed, and lead to the same result, if instead of the function $\psi_n(x)$ we consider the function

$$\frac{1}{\pi}\int_{-\pi}^{\pi}f_n(t)\sum_{k=1}^{\infty}\frac{\sin\left[[k(t-x)+\frac{r\pi}{2}\right]}{k^r}$$

and in the definition of $f_n(t)$ replace r by $r+1$. Therefore theorem 8.3.7 is valid also for the conjugate classes $\tilde{W}_*^{(r)}H_\omega$.

8.3.8. It follows from the inequality 8.3(35) that** *the process of approximation* 8.1(8) *can achieve the best order of uniform approximation by trigonometric polynomials on the class* $W_*^{(r)}H_\omega$ *or on the conjugate class* $\tilde{W}_*^{(r)}H_\omega$ *only when*

$$\sum_{k=1}^{n} \frac{1-\lambda_k^{(n)}}{k^r(n-k+1)} = \frac{\ln n}{n^r} + O\left(\frac{1}{n^r}\right). \tag{48}$$

In fact, if the method 8.1(8) gives the best order of approximation on the class $W_*^{(r)}\, H_\omega$ or $\tilde{W}_*^{(r)}\, H_\omega$, then the system of numbers $\lambda_k^{(n)}$ must satisfy the condition 8.2(23) (see section 8.3.3), from which it follows that $\lambda_k^{(n)} = O(1)$ and

$$\sum_{k=1}^{n} \frac{|1-\lambda_k^{(n)}|}{k^r} = O\left(\frac{1}{n^{r-1}}\right). \tag{49}$$

Therefore the inequality 8.3(35) must hold, the right-hand side of which can be a quantity of order $O\left[\frac{1}{n^r}\omega\left(\frac{1}{n}\right)\right]$ only when 8.3(48) is satisfied.

8.4. Approximation of functions by arithmetic means of the partial sums of Fourier series

It has been mentioned in the preceding section that there do not exist linear methods of approximation by polynomials which for any continuous function $f(x)$ would give an approximation satisfying an inequality of the type 8.3(1). It happens that this situation can be changed if the requirements are somewhat weakened by replacing, on the right-hand side of 8.3(1), the best approximation $E_n^*(f)$ by $E_{[\theta n]}^*\ (f)$ (and correspondingly $E_n(f;\, a, b)$ by $E_{[\theta n]}^*(f;\, a, b)$), where $0 < \theta < 1$. Whatever the fixed value of $\theta (0 < \theta < 1)$, there exists a linear process of approximation by algebraic polynomials $U_n(f;\, x)$ of degree $\leqslant n$ $(n = 0, 1, 2, \ldots)$, which for every function $f(x)$ continuous on the segment $[a, b]$ produces an approximation satisfying the inequality

$$\max_{a \leqslant x \leqslant b} |f(x) - U_n(f;\, x)| \leqslant B_\theta E_{[\theta n]}(f;\, a, b), \tag{1}$$

where B_θ is a constant depending only on θ.

If for definiteness we take $a = -1$, $b = 1$, then it is easily shown that the sequence of polynomials $U_n(f;\, x) = \sigma_n(f;\, k;\, x)$ (see 8.2(30)) possesses

** A. F. Timan [12].

this property for $k = n-[\theta n]$. In fact, let $P_n(f; x)$ $(n = 0, 1, 2, \ldots)$ be an algebraic polynomial of degree $\leqslant n$ for which

$$E_n(f; -1, 1) = \max_{-1\leqslant x\leqslant 1} |f(x)-P_n(f; x)|.$$

Since for any $\nu > m$

$$c_\nu = \frac{2}{\pi}\int_0^\pi P_m(f; \cos t) \cos \nu t \, dt = 0,$$

then in the case where $n > m$ (see section 8.2.31)

$$\sigma_n(P_m; n-m; x) = \frac{1}{n-m+1}\sum_{\nu=0}^{n-m} S_{n-\nu}(P_m; x)$$

$$= \frac{1}{n-m+1}\sum_{\nu=0}^{n-m} P_m(f; x) = P_m(f; x).$$

Hence (see 8.2.31)

$$\begin{aligned} |f(x)-\sigma_n(f; n-[\theta n]; x)| & \\ \leqslant |f^{(x)}-P_{[\theta n]}(f; x)| &+ |\sigma_n(f; n-[\theta_n]; x)-P_{[\theta n]}(f; x)| \\ \leqslant E_{[\theta n]}(f; -1, 1) &+ |\sigma_n(f-P_{[\theta n]}; n-[\theta n]; x)| \\ \leqslant \{1+L_n(n-[\theta n]; x)\} & E_{[\theta n]}(f; -1, 1). \end{aligned}$$

But (see 8.2(32))

$$L_n(n-[\theta n]; x)+O(1) \tag{2}$$

uniformly with respect to x and n.

Consequently

$$\max_{-1\leqslant x\leqslant 1} |f(x)-\sigma_n(f; n-[\theta n]; x)| \leqslant B_\theta E_{[\theta n]}(f; -1, 1), \tag{3}$$

where the constant B_θ depends only on the value of θ.

Similarly, commencing with 8.4(2) for $x = 1$, it is possible to verify that the process of approximation by trigonometric polynomials**

$$\sigma_n^*(f; k; x) = \frac{1}{k+1}\sum_{\nu=0}^{k} S_{n-\nu}^*(f; x), \tag{4}$$

where $S_m^*(f; x)(m = 0, 1, 2, \ldots)$ are the partial sums of order m of the Fourier series of the periodic function $f(x)$ of period 2π, satisfies the inequality

$$\max_x |f(x)-\sigma_n^*(f; n-[\theta n]; x)| \leqslant B_\theta E_{[\theta n]}(f) \tag{5}$$

on the whole class C^*.

** The method of approximation of periodic functions by polynomials 8.4(4) was first considered by De la Valiée Poussin [3].

Thus for any continuous function $f(x)$ of period 2π for which, for some $\theta(0 < \theta < 1)$, we have $E^*_{[\theta n]}(f) = O\,[E^*_n(f)]$, the sequence of sums 8.4(4) gives the best approximation, as regards order, by trigonometric polynomials.

The sums 8.4(4) and 8.2(30) can be written respectively in the form 8.1(8) and 8.1(10), where the numbers $\lambda_\nu^{(n)}$ are determined by the equations

$$\lambda_\nu^{(n)} = \begin{cases} 1, & 0 \leqslant \nu \leqslant n-k, \\ \dfrac{n-\nu+1}{k+1}, & n-k \leqslant \nu \leqslant n. \end{cases} \tag{6}$$

We shall see that the process of approximation 8.1(8) corresponding to the matrix 8.4(6), where $k = n-[\theta n]$ $(0 < \theta < 1)$, achieves the best order of approximation by trigonometric polynomials on any compact class $\mathfrak{M}$ of continuous functions of period 2π which possess the property

$$\sup_{f\in\mathfrak{M}} E^*_{[\theta n]}(f) = O\,[\sup_{f\in\mathfrak{M}} E^*_n(f)], \tag{7}$$

and in particular on the classes $W_*^{(r)}H_\omega$, $\tilde{W}_*^{(r)}H_\omega$ considered in section 8.3 (see section 8.3.3).

In the investigation of other methods of the type 8.1(8) which possess this property, an important part is played by the study of the behaviour of the upper bounds $\mathcal{E}_n(\mathfrak{M}; x; \lambda)$ (see 8.3(5)) for the arbitrary matrices 8.4(6). We will here consider from this point of view the classes $W_*^{(r)}H_\omega$ and $\tilde{W}_*^{(r)}H_\omega$ for any $r \geqslant 0$ in the case when $\omega(t) = Mt^\alpha(0 < \alpha < 1)$, i.e. the classes $W_*^{(r)}H^{(\alpha)}M$ and $\tilde{W}_*^{(r)}H^{(\alpha)}M$

Before passing to the estimation of the quantities

$$\mathcal{E}_{\sigma_n^*}(W_*^{(r)}H^{(\alpha)}M; x) = \sup_{f\in W_*^{(r)}H^{(\alpha)}M} |f(x)-\sigma_n^*(f; k; x)| \tag{8}$$

and

$$\mathcal{E}_{\sigma_n^*}(\tilde{W}_*^{(r)}H^{(\alpha)}M; x) = \sup_{f\in \tilde{W}_*^{(r)}H^{(\alpha)}M} |f(x)-\sigma_n^*(f; k; x)|, \tag{9}$$

where $k = k(n)$ $(0 \leqslant k \leqslant n)$ are arbitrary integer-valued functions of n, let us pause for some preparatory discussion of an auxiliary character.

8.4.1. For $r > 0$ let

$$D_n^{(r)}(t) = \sum_{\nu=n-1}^{\infty} \frac{\cos\left(\nu t+\dfrac{r\pi}{2}\right)}{\nu^r}. \tag{10}$$

We show that for any constant $c > 0$

$$\int_{-\frac{c}{n}}^{\frac{c}{n}} |D_n^{(r)}(t)|\,dt = O\left(\frac{1}{n^r}\right). \tag{11}$$

In the case when $r \geqslant 1$, this relation follows directly from the fact that

$$D_n^{(r)}(t) = O\left(\frac{1}{n^{r-1}}\right)$$

uniformly with respect to t and n.

If, however, $0 < r < 1$, then it can be obtained after the application of Abel's** transformation to the sum 8.4(10). In fact, since

$$\sum_{\nu=n+1}^{\infty} \frac{\cos\nu t}{\nu^r} = -\frac{D_n(t)}{(n+1)^r} - \left[\frac{1}{n^r} - \frac{1}{(n+1)^r}\right]\sum_{j=0}^{n} D_j(t) +$$
$$+\sum_{\nu=n+1}^{\infty}\left[\frac{1}{\nu^r} - \frac{2}{(\nu+1)^r} + \frac{1}{(\nu+2)^r}\right]\sum_{j=0}^{\nu} D_j(t), \tag{12}$$

where $D_n(t)$ is Dirichlet's kernel (see 4.1(10)), and

$$\sum_{\nu=n+1}^{\infty} \frac{\sin\nu t}{\nu^r}$$
$$= -\frac{1}{(n+1)^r}\sum_{j=1}^{n} \sin jt + \sum_{\nu=n+1}^{\infty}\left[\frac{1}{\nu^r} - \frac{1}{(\nu+1)^r}\right]\sum_{j=1}^{\nu} \sin jt, \tag{13}$$

$$\sum_{j=0}^{\nu} D_j(t) = \frac{\sin^2\frac{\nu+1}{2}t}{2\sin^2\frac{t}{2}}, \qquad \sum_{j=1}^{\nu} \sin jt = \frac{\cos\frac{t}{2} - \cos\frac{2\nu+1}{2}t}{2\sin\frac{t}{2}}, \tag{14}$$

then in virtue of the relations $(\nu \geqslant n)$

$$\int_0^{\frac{c}{n}} |D_n(t)|\,dt = \int_0^{\frac{c}{n}} \left|\frac{\sin\frac{2n+1}{2}t}{2\sin\frac{t}{2}}\right| dt = O(1),$$

$$\int_0^{\frac{c}{n}} \left|\sum_{j=0}^{\nu} D_j(t)\right| dt \leqslant \int_0^{2\pi} \sum_{j=0}^{\nu} D_j(t)\,dt = (\nu+1)\pi,$$

** V. T. Pinkevich [1].

$$\int_0^{\frac{c}{n}} \left| \frac{\cos\frac{t}{2} - \cos\frac{2\nu+1}{2}t}{2\sin\frac{t}{2}} \right| dt = \int_0^{\frac{c}{n}} \left|\cot\frac{t}{2}\right| \sin^2\frac{\nu t}{2} dt + O(1)$$

$$= 2\int_0^{\frac{c}{n}} \frac{1}{t}\sin^2\frac{\nu t}{2} dt + O(1) \leqslant 2\int_c^{\frac{c\nu}{n}} \frac{dt}{t} + O(1) = 2\ln\frac{\nu}{n} + O(1)$$

we have

$$\left.\begin{aligned} \int_{-\frac{c}{n}}^{\frac{c}{n}} \left| \sum_{\nu=n+1}^{\infty} \frac{\cos\nu t}{\nu^r} \right| dt &= O\left(\frac{1}{n^r}\right) + \sum_{\nu=n+1}^{\infty} O\left(\frac{1}{\nu^{r+1}}\right) = O\left(\frac{1}{n^r}\right), \\ \int_{-\frac{c}{n}}^{\frac{c}{n}} \left| \sum_{\nu=n+1}^{\infty} \frac{\sin\nu t}{\nu^r} \right| dt &= O\left(\frac{1}{n^r}\right) + \sum_{\nu=n+1}^{\infty} O\left(\frac{1}{\nu^{r+1}}\ln\frac{\nu}{n}\right). \end{aligned}\right\} \quad (15)$$

But

$$\sum_{\nu=n+1}^{\infty} \frac{\ln\nu}{\nu^{r+1}} = \ln(n+1)\sum_{\nu=n+1}^{\infty}\frac{1}{\nu^{r+1}} + \sum_{\nu=n+1}^{\infty}\ln\left(1+\frac{1}{\nu}\right)\sum_{j=\nu+1}^{\infty}\frac{1}{j^{r+1}}$$

$$= \ln(n+1)\sum_{\nu=n+1}^{\infty}\frac{1}{\nu^{r+1}} + O\left(\frac{1}{n^r}\right).$$

Therefore the second integral of 8.4(15) is also a quantity of order $O\left(\frac{1}{n^r}\right)$, i.e. 8.4(11) holds.

8.4.2. For any $t \neq 2k\pi$ (k an integer) the sums 8.4(10) can be represented in the form**

$$D_n^{(r)}(t) = -\frac{A_n^{(r)}(t)}{(n+1)^r} - G_n^{(r)}(t), \quad (16)$$

where

$$G_n^{(r)}(t) = \Delta\left[\frac{1}{(n+1)^r}\right][B_n^{(r)}(t) + C_n^{(r)}(t)] -$$

$$- \sum_{\nu=n+1}^{\infty} \Delta^2\left(\frac{1}{\nu^r}\right)[B_\nu^{(r)}(t) + C_\nu^{(r)}(t)], \quad (17)$$

** The case where r is an integer was considered by A. N. Kolmogorov [2], and the general case by V. T. Pinkevich [1].

and

$$\left.\begin{aligned} A_n^{(r)}(t) &= \frac{\sin\left(\frac{2n+1}{2}t+\frac{r\pi}{2}\right)}{2\sin\frac{t}{2}}, \\ B_\nu^{(r)}(t) &= \frac{\cos\frac{r\pi}{2}\sin^2\frac{\nu+1}{2}t}{2\sin^2\frac{t}{2}}, \quad C_\nu^{(r)}(t) = \frac{\sin\frac{r\pi}{2}\sin(\nu+1)t}{4\sin^2\frac{t}{2}}, \\ \Delta\left(\frac{1}{\nu^r}\right) &= \frac{1}{\nu^r}-\frac{1}{(\nu+1)^r}, \quad \Delta^2\left(\frac{1}{\nu^r}\right) = \frac{1}{\nu^r}-\frac{2}{(\nu+1)^r}+\frac{1}{(\nu+2)^r}. \end{aligned}\right\} \tag{18}$$

Here, for any periodic function $\phi(x)$ of period 2π which satisfies the conditions

$$\phi(0) = 0, \quad |\phi(x_1)-\phi(x_2)| \leqslant M|x_1-x_2|^\alpha \quad (0 \leqslant \alpha < 1), \tag{19}$$

the relations**

$$\int_{\frac{1}{n}}^{\pi} G_n^{(r)}(t)\,\phi(t)\,\mathrm{d}t = O\left(\frac{1}{n^{r+\alpha}}\right), \quad \int_{-\pi}^{-\frac{1}{n}} G_n^{(r)}(t)\,\phi(t)\,\mathrm{d}t = O\left(\frac{1}{n^{r+\alpha}}\right) \tag{20}$$

hold uniformly with respect to n and ϕ.

We obtain the identity 8.4(16) if we apply Abel's transformation to the partial sums of the series 8.4(10) and pass to the limit. In fact, putting

$$a^{(r)}(t) = \frac{\sin\frac{r\pi-t}{2}}{2\sin\frac{t}{2}}$$

and taking account of the identity

$$\sum_{j=0}^{\nu} \cos\left(jt+\frac{r\pi}{2}\right) = A_\nu^{(r)}(t)-a^{(r)}(t),$$

we have for any $m > 0$

$$\sum_{\nu=n+1}^{n+m+1} \frac{\cos\left(\nu t+\frac{r\pi}{2}\right)}{\nu^r} = \sum_{\nu=n+1}^{n+m} \Delta\left(\frac{1}{\nu^r}\right)[A_\nu^{(r)}(t)-a^{(r)}(t)] -$$

$$-\frac{A_n^{(r)}(t)}{(n+1)^r}+\frac{a^{(r)}(t)}{(n+1)^r}+\frac{A_{n+m}^{(r)}(t)}{(n+m+1)^r}-\frac{a^{(r)}(t)}{(n+m+1)^r}.$$

** S. M. Nikol'skii [2].

Passing to the limit as $m \to \infty$, we obtain for any $t \neq 2k\pi$ (k an integer) the equation

$$D_n^{(r)}(t) = -\frac{A_n^{(r)}(t)}{(n+1)^r} + \sum_{\nu=n+1}^{\infty} \Delta\left(\frac{1}{\nu^r}\right) A_\nu^{(r)}(t). \tag{21}$$

Moreover, since $\sum_{j=0}^{\nu} A_j^{(r)}(t) = B_\nu^{(r)}(t) + C_\nu^{(r)}(t)$ it follows that

$$\sum_{\nu=n+1}^{n+m+1} \Delta\left(\frac{1}{\nu^r}\right) A_\nu^{(r)}(t) = -\Delta\left[\frac{1}{(n+1)^r}\right][B_n^{(r)}(t) + C_n^{(r)}(t)] +$$

$$+ \Delta\left[\frac{1}{(n+m+1)^r}\right][B_{n+m+1}^{(r)}(t) + C_{n+m+1}^{(r)}(t)] +$$

$$+ \sum_{\nu=n+1}^{n+m} \Delta^2\left(\frac{1}{\nu^r}\right)[B_\nu^{(r)}(t) + C_\nu^{(r)}(t)].$$

Therefore after passing to the limit as $m \to \infty$ we find that for all $t \neq 2\,k\pi$(k an integer)

$$\sum_{\nu=n+1}^{\infty} \Delta\left(\frac{1}{\nu^r}\right) A_\nu^{(r)}(t) = -\Delta\left[\frac{1}{(n+1)^r}\right][B_n^{(r)}(t) + C_n^{(r)}(t)] +$$

$$+ \sum_{\nu=n+1}^{\infty} \Delta^2\left(\frac{1}{\nu^r}\right)[B_\nu^{(r)}(t) + C_\nu^{(r)}(t)],$$

i.e. in virtue of 8.4(21) and 8.4(17) we arrive at the identity 8.4(16).

The validity of the relation 8.4(20) follows from the fact that for any $\nu \geqslant n$, for any function $\phi(t)$ satisfying the condition 8.4(19),

$$\int_{\frac{1}{\nu}}^{\pi} |C_\nu^{(r)}(t)\phi(t)|\,dt \leqslant M \int_{\frac{1}{\nu}}^{\pi} \frac{t^\alpha\,dt}{2\sin^2\frac{t}{2}} < M\frac{\pi^2}{2}\int_{\frac{1}{\nu}}^{\pi} t^{\alpha-2}\,dt = O(\nu^{1-\alpha}),$$

and

$$\int_{\frac{1}{\nu}}^{\pi} |B_\nu^{(r)}(t)\phi(t)|\,dt \leqslant \int_0^{\pi} |\phi(t)| \frac{\sin^2\frac{\nu+1}{2}t}{2\sin^2\frac{t}{2}}\,dt \leqslant M\int_0^{\pi} t^\alpha \frac{\sin^2\frac{\nu+1}{2}t}{2\sin^2\frac{t}{2}}\,dt$$

$$\leqslant M\pi^2 \int_0^{\pi} \frac{\sin^2\frac{\nu+1}{2}t}{t^{2-\alpha}}\,dt \leqslant M\,\frac{\pi^2(\nu+1)^{1-\alpha}}{2^{1-\alpha}}\int_0^{\infty}\frac{\sin^2 t}{t^{2-\alpha}}\,dt = O(\nu^{1-\alpha}).$$

$$\left|\int_{\frac{1}{n}}^{\pi} \sum_{\nu=n+1}^{\infty} \Delta^2\left(\frac{1}{\nu^r}\right)[B_\nu^{(r)}(t)+C_\nu^{(r)}(t)]\phi(t)\,\mathrm{d}t\right|$$

$$\leqslant \sum_{\nu=n+1}^{\infty}\left|\Delta^2\left(\frac{1}{\nu^r}\right)\right|\int_{\frac{1}{\nu}}^{\pi}|B_\nu^{(r)}(t)+C_\nu^{(r)}(t)|\,|\phi(t)|\,\mathrm{d}t$$

$$=\sum_{\nu=n+1}^{\infty} O\left(\frac{1}{\nu^{r+\alpha+1}}\right)=O\left(\frac{1}{n^{r+a}}\right),$$

i.e. in virtue of 8.4(17) the relations 8.4(20) are satisfied.

8.4.3. It follows from 8.4(15) that if

$$\tilde{D}_n^{(r)}(t)=\sum_{\nu=n+1}^{\infty}\frac{\sin\left(\nu t+\dfrac{r\pi}{2}\right)}{\nu^r}, \tag{22}$$

then whatever the constant $c>0$

$$\int_{-\frac{c}{n}}^{\frac{c}{n}}|\tilde{D}_n^{(r)}(t)|\,\mathrm{d}t=O\left(\frac{1}{n^r}\right). \tag{23}$$

Moreover, considerations similar to those of 8.4.2 show that for any $t\neq 2k\pi$ (k an integer)

$$\tilde{D}_n^{(r)}(t)=\frac{A_n^{(r+1)}(t)}{(n+1)^r}+\tilde{G}_n^{(r)}(t), \tag{24}$$

where

$$\tilde{G}_n^{(r)}(t)=\Delta\left[\frac{1}{(n+1)^r}\right][B_n^{(r+1)}(t)+C_n^{(r+1)}(t)]-$$
$$-\sum_{\nu=n+1}^{\infty}\Delta^2\left(\frac{1}{\nu^r}\right)[B_\nu^{(r+1)}(t)+C_\nu^{(r+1)}(t)]$$

and

$$\int_{\frac{1}{n}}^{\pi}\tilde{G}_n^{(r)}(t)\phi(t)\,\mathrm{d}t=O\left(\frac{1}{n^{r+\alpha}}\right),\qquad \int_{-\frac{1}{n}}^{-\pi}\tilde{G}_n^{(r)}(t)\phi(t)\,\mathrm{d}t=O\left(\frac{1}{n^{r+\alpha}}\right) \tag{25}$$

uniformly with respect to n and for all functions $\phi(t)$ which satisfy the condition 8.4(19).

8.4.4. We now pass to the estimation of the upper bounds 8.4(8), 8.4(9) and show that** *for any numbers* r $(r \geqslant 0)$ *and* $\alpha(0 < \alpha \leqslant 1)$, *and subject to the condition* $\lim\limits_{n\to\infty} \dfrac{k(n)}{n} = 0$, *the asymptotic equalities*

$$\left.\begin{array}{l}\mathcal{E}_{\sigma_n^*}(W_*^{(r)}H^{(\alpha)}M;\, x)\\ \mathcal{E}_{\sigma_n^*}(\tilde{W}_*^{(r)}H^{(\alpha)}M;\, x)\end{array}\right\} = \frac{2^{\alpha+1}M}{\pi^2 n^{r+\alpha}} \ln\frac{n}{k+1} \int\limits_{2}^{\frac{\pi}{2}} t^{\alpha}\sin t\, dt + O\left(\frac{1}{n^{r+\alpha}}\right), \tag{26}$$

are satisfied, where $O(1)$ *is a quantity which is uniformly bounded with respect to all the functions* $k = k(n)$ $(0 \leqslant k < n)$, *for which* $k \leqslant (1-\theta)n$ $(0 < \theta \leqslant 1)$. *In the case where* $r = 0$, *the first of the equations* 8.4(26) *is satisfied uniformly with respect to all the functions* $k = k(n)$ $(0 \leqslant k < n)$ *for any* $\alpha(0 \leqslant \alpha < 1)$ *and the second is satisfied uniformly for all the functions* $k = k(n)$ $(0 \leqslant k < n)$ *for any* $\alpha(0 < \alpha < 1)$***.

Since the upper bounds 8.4(8) and 8.4(9) do not depend on the position of the point (see section 8.3.4), we shall suppose that $x = 0$. We consider first the case where $r > 0$.

The class $W_*^{(r)}H^{(\alpha)}M$ consists of all functions $f(x)$ representable in the form of the integral

$$f(x) = \frac{a_0}{2} + \frac{1}{\pi}\int\limits_{-\pi}^{\pi} \phi(t) D_0^{(r)}(x-t)\, dt, \tag{27}$$

where $D_0^{(r)}(u)$ is the kernel 3.5(16) and $\phi(x)$ is an arbitrary periodic function of period 2π which satisfies the conditions

$$\int\limits_{-\pi}^{\pi} \phi(x)\, dx = 0, \qquad \omega(\phi;\, t) \leqslant Mt^{\alpha}. \tag{28}$$

Therefore

$$S_n^*(f;\, x) = \frac{1}{\pi}\int\limits_{-\pi}^{\pi} \phi(t) \sum_{\nu=1}^{n} \frac{\cos\left[\nu(x-t) - \dfrac{r\pi}{2}\right]}{\nu^r}\, dt, \tag{29}$$

and

$$f(0) - S_n^*(f;\, 0) = \frac{1}{\pi}\int\limits_{-\pi}^{\pi} \phi(t) D_n^{(r)}(t)\, dt, \tag{30}$$

** A. F. Timan [8, 12]. For the case $k = 0$, S. M. Nikol'skii [2].
*** For the case $r = 0$, $\alpha = 0$ see section 8.2.31.

i.e.

$$f(0)-\sigma_n^*(f;k;0) = \frac{1}{k+1}\sum_{\nu=0}^{k}\frac{1}{\pi}\int_{-\pi}^{\pi}\phi(t)D_{n-\nu}^{(r)}(t)\,\mathrm{d}t \tag{31}$$

Thus the finding of the quantity $\mathcal{E}_{\sigma_n^*}(W_*^{(r)}H^{(\alpha)}M;x)$ reduces to the investigation of the upper bound

$$\frac{1}{(k+1)\pi}\sup_{\phi}\int_{-\pi}^{\pi}\phi(t)\sum_{\nu=0}^{k}D_{n-\nu}^{(r)}(t)\,\mathrm{d}t, \tag{32}$$

extended over all functions $\phi(t)$ satisfying conditions 8.4(28). Since

$$\int_{-\pi}^{\pi}D_n^{(r)}(t)\,\mathrm{d}t = 0,$$

the upper bound 8.4(32) is not decreased if it extends only over the subclass $H_0^{(\alpha)}M$ of those functions $\phi(t)$ for which $\phi(0) = 0$. Consequently, in virtue of 8.4(31)

$$\begin{aligned}\mathcal{E}_{\sigma_n^*}(W_*^{(r)}H^{(\alpha)}M;x) &= \mathcal{E}_{\sigma_n^*}(W_*^{(r)}H^{(\alpha)}M;0)\\ &= \frac{1}{(k+1)\pi}\sup_{\phi\in H_0^{(\alpha)}M}\int_{-\pi}^{\pi}\phi(t)\sum_{\nu=n-k}^{n}D_\nu^{(r)}(t)\,\mathrm{d}t.\end{aligned} \tag{33}$$

We consider separately the cases $0<\alpha<1$ and $\alpha = 1$.

8.4.41. Let $0<\alpha<1$. If $\phi(x)\in H_0^{(\alpha)}M$, then subject to the condition $k\leqslant(1-\theta)n$ $(0<\theta\leqslant 1)$ we have in virtue of 8.4(11) and 8.4(20)

$$\frac{1}{k+1}\int_{-\frac{1}{n}}^{\frac{1}{n}}\phi(t)\sum_{\nu=n-k}^{n}D_\nu^{(r)}(t)\,\mathrm{d}t = O\left(\frac{1}{n^{r+\alpha}}\right) \tag{34}$$

and

$$\left.\begin{aligned}\frac{1}{k+1}\int_{\frac{1}{n}}^{\pi}\phi(t)\sum_{\nu=n-k}^{n}G_\nu^{(r)}(t)\,\mathrm{d}t &= O\left(\frac{1}{n^{r+\alpha}}\right),\\ \frac{1}{k+1}\int_{-\pi}^{-\frac{1}{n}}\phi(t)\sum_{\nu=n-k}^{n}G_\nu^{(r)}(t)\,\mathrm{d}t &= O\left(\frac{1}{n^{r+\alpha}}\right)\end{aligned}\right\} \tag{35}$$

uniformly with respect to n, k $(0 \leqslant k \leqslant (1-\theta)n)$ and ϕ. Therefore (see section 8.4.2)

$$\frac{1}{(k+1)\pi}\left|\int_{-\pi}^{\pi} \phi(t) \sum_{\nu=n-k}^{n} D_\nu^{(r)}(t)\,\mathrm{d}t\right|$$

$$= \frac{1}{(k+1)\pi}\left|\int_{-\pi}^{-\frac{1}{n}} \phi(t) \sum_{\nu=n-k}^{n} \frac{A_\nu^{(r)}(t)}{(\nu+1)^r}\,\mathrm{d}t + \right.$$

$$\left. + \int_{\frac{1}{n}}^{\pi} \phi(t) \sum_{\nu=n-k}^{n} \frac{A_\nu^{(r)}(t)}{(\nu+1)^r}\,\mathrm{d}t\right| + O\left(\frac{1}{n^{r+\alpha}}\right),$$

or

$$\frac{1}{(k+1)\pi}\left|\int_{-\pi}^{\pi} \phi(t) \sum_{\nu=n-k}^{n} D_\nu^{(r)}(t)\,\mathrm{d}t\right|$$

$$= \frac{1}{(k+1)\pi}\left|\frac{1}{(n+1)^r}\int_{-\pi}^{-\frac{1}{n}} \phi(t) \sum_{\nu=n-k}^{n} A_\nu^{(r)}(t)\,\mathrm{d}t + \right.$$

$$+ \frac{1}{(n+1)^r}\int_{\frac{1}{n}}^{\pi} \phi(t) \sum_{\nu=n-k}^{n} A_\nu^{(r)}(t)\,\mathrm{d}t +$$

$$+ \int_{-\pi}^{-\frac{1}{n}} \phi(t) \sum_{\nu=n-k}^{n} \left\{\frac{1}{(\nu+1)^r} - \frac{1}{(n+1)^r}\right\} A_\nu^{(r)}(t)\,\mathrm{d}t +$$

$$\left. + \int_{\frac{1}{n}}^{\pi} \phi(t) \sum_{\nu=n-k}^{n} \left\{\frac{1}{(\nu+1)^r} - \frac{1}{(n+1)^r}\right\} A_\nu^{(r)}(t)\,\mathrm{d}t\right| + O\left(\frac{1}{n^{r+\alpha}}\right). \tag{36}$$

For any pair of values of n, k considered, and for any function $\phi(t) \in H_0^{(\alpha)} M$, let us denote by N the integer for which

$$\left|\sum_{\nu=n-k}^{N} \int_{\frac{1}{n}}^{\pi} \phi(t) A_\nu^{(r)}(t)\,\mathrm{d}t\right| = \max_{n-k \leqslant \nu \leqslant n} \left|\sum_{j=n-k}^{\nu} \int_{\frac{1}{n}}^{\pi} \phi(t) A_j^{(r)}(t)\,\mathrm{d}t\right|$$

$$(n-k \leqslant N \leqslant n).$$

Applying Abel's transformation and taking into account the positiveness of the difference $\frac{1}{(\nu+1)^r}-\frac{1}{(n+1)^r}$ $(n-k \leqslant \nu \leqslant n)$, we obtain the inequality

$$\frac{1}{(k+1)\pi}\left|\int_{\frac{1}{n}}^{\pi} \phi(t) \sum_{\nu=n-k}^{n}\left\{\frac{1}{(\nu+1)^r}-\frac{1}{(n+1)^r}\right\} A_\nu^{(r)}(t)\,\mathrm{d}t\right|$$

$$\leqslant\left\{\frac{1}{(n-k+1)^r}-\frac{1}{(n+1)^r}\right\}\frac{1}{(k+1)\pi}\left|\int_{\frac{1}{n}}^{\pi}\phi(t)\sum_{\nu=n-k}^{N} A_\nu^{(r)}(t)\,\mathrm{d}t\right|$$

$$\leqslant \frac{B}{n^{r+1}\pi}\left|\int_{\frac{1}{n}}^{\pi}\phi(t)\sum_{\nu=n-k}^{N} A_\nu^{(r)}(t)\,\mathrm{d}t\right| = \frac{B(k+1)}{n^{r+1}} I_N^{(n)}(\phi), \tag{37}$$

where B is a certain positive constant not depending on n and ϕ. In virtue of the identity

$$\sum_{\nu=0}^{p} A_\nu^{(r)}(t) = \frac{\sin(p+1)\frac{t}{2}\sin\left[(p+1)\frac{t}{2}+\frac{r\pi}{2}\right]}{2\sin^2\frac{t}{2}}$$

we have

$$\sum_{\nu=n-k}^{N} A_\nu^{(r)}(t) = \frac{1}{2\sin^2\frac{t}{2}}\left\{\left[\sin^2(N+1)\frac{t}{2}-\sin^2(n-k)\frac{t}{2}\right]\cos\frac{r\pi}{2}+\right.$$

$$\left.+\frac{1}{2}\left[\sin(N+1)t-\sin(n-k)t\right]\sin\frac{r\pi}{2}\right\}$$

$$=\frac{\sin(N-n+k+1)\frac{t}{2}\sin\left[(N+n-k+1)\frac{t}{2}+\frac{r\pi}{2}\right]}{2\sin^2\frac{t}{2}}.$$

Thus

$$I_N^{(n)}(\phi) = \frac{1}{(k+1)\pi}\left|\int_{\frac{1}{n}}^{\pi}\phi(t)\frac{\sin\left[(N+n-k+1)\frac{t}{2}+\frac{r\pi}{2}\right]}{2\sin^2\frac{t}{2}}\times\right.$$

$$\left.\times\sin(N-n+k+1)\frac{t}{2}\,\mathrm{d}t\right|. \tag{38}$$

Denote by l the least integer satisfying the inequality $l \geqslant \frac{r}{2}$, and put $t_\nu = \dfrac{2\left(\nu+l-\frac{r}{2}\right)}{N+n-k+1}\pi$.

For $\frac{1}{n} \leqslant t \leqslant \pi$, let

$$K_{\nu,N}(t) = \begin{cases} 0, & t \overline{\in} [t_{\nu-\frac{1}{2}}, t_{\nu+\frac{1}{2}}], \\ \dfrac{\sin(N-n+k+1)\frac{t}{2}\sin\left[(N+n-k+1)\frac{t}{2}+\frac{r\pi}{2}\right]}{2\sin^2\frac{t}{2}}, \\ t \in [t_{\nu-\frac{1}{2}}, t_{\nu+\frac{1}{2}}]. \end{cases}$$

Put

$$\psi_{\nu,N}(t_\nu+t) = \begin{cases} K_{\nu,N}(t_\nu+t), & t \geqslant 0, \\ -K_{\nu,N}(t_\nu-t), & t \leqslant 0, \end{cases}$$

and consider the difference

$$r_{\nu,N}(t) = K_{\nu,N}(t) - \psi_{\nu,N}(t).$$

It is not difficult to verify that if $\phi(x) \in H_0^{(\alpha)}M$ and $|t_\nu| \leqslant \pi$, then

$$\int_{t_{\nu-\frac{1}{2}}}^{t_{\nu-\frac{1}{2}}} \phi(t) r_{\nu,N}(t)\,dt = O\left(\frac{N-n+k+1}{n^\alpha \nu^{2-\alpha}}\right) \tag{39}$$

uniformly with respect to n, $\nu(\nu \neq 0)$, $k(0 \leqslant k \leqslant (1-\theta)n)$ and ϕ.

The validity of this relation follows from the estimate

$$\left|\int_{t_{\nu-\frac{1}{2}}}^{t_{\nu+\frac{1}{2}}} \phi(t) r_{\nu,N}(t)\,dt\right| = \left|\int_{t_{-\frac{1}{2}}}^{t_{\frac{1}{2}}} \phi(t_\nu+t)\, r_{\nu,N}(t_\nu+t)\,dt\right|$$

$$= \left|\int_0^{t_{\frac{1}{2}}} \phi(t_\nu-t)\{K_{\nu,N}(t_\nu-t) - K_{\nu,N}(t_\nu+t)\}\,dt\right|$$

$$= O\left(\frac{\nu^\alpha}{n^\alpha}\right) \int_0^{t_{\frac{1}{2}}} \sin\frac{N+n-k+1}{2}t \times$$

$$\times \left|\frac{\sin\frac{N-n+k-1}{2}(t_\nu+t)}{\sin^2\frac{t_\nu+t}{2}} - \frac{\sin\frac{N-n+k+1}{2}(t_\nu-t)}{\sin^2\frac{t_\nu-t}{2}}\right| dt$$

$$\leqslant \Bigg\{ \int_0^{t_{\frac{1}{2}}} \frac{\sin \frac{N+n-k+1}{2} t}{\sin^2 \frac{(t_\nu+t)}{2}} \times$$

$$\times \left| \sin \frac{N-n+k+1}{2}(t_\nu+t) - \sin \frac{N-n+k+1}{2}(t_\nu-t) \right| \mathrm{d}t +$$

$$+ \int_0^{t_{\frac{1}{2}}} \sin \frac{N+n-k+1}{2} t \left| \sin \frac{N-n+k+1}{2}(t_\nu-t) \right| \times$$

$$\times \left| \frac{1}{\sin^2 \frac{t_\nu+t}{2}} - \frac{1}{\sin^2 \frac{t_\nu-t}{2}} \right| \mathrm{d}t \Bigg\} O\left(\frac{\nu^\alpha}{n^\alpha}\right)$$

$$= O\left(\frac{\nu^\alpha}{n^\alpha} \cdot \frac{N-n+k+1}{\nu^2}\right).$$

Moreover, since $\frac{1}{n} \leqslant \frac{t_1}{2}$, $t_0 > 0$, $t_\nu < \pi$ for $\nu \geqslant \frac{1}{2}(N+n-k)$ and $t_\nu > \pi$ for $\nu > \frac{1}{2} N+n-k+1)$, then in virtue of the relations

$$\frac{1}{k+1} \int_{\frac{1}{n}}^{t_{\frac{1}{2}}} \phi(t) \frac{\sin\left(\frac{N+n-k+1}{2} t + \frac{r\pi}{2}\right) \sin \frac{N-n+k+1}{2} t}{\sin^2 \frac{t}{2}} \mathrm{d}t = O\left(\frac{1}{n^\alpha}\right),$$

$$\frac{1}{k+1} \int_{t_{N_*+\frac{1}{2}}}^{\pi} \phi(t) \frac{\sin\left(\frac{N+n-k+1}{2} t + \frac{r\pi}{2}\right) \sin \frac{N-n+k+1}{2} t}{\sin^2 \frac{t}{2}} \mathrm{d}t = O\left(\frac{1}{n}\right),$$

where $N_* = \left[\frac{N+n-k}{2}\right]$, we obtain

$$I_N^{(n)}(\phi) = \frac{1}{(k+1)\pi} \left| \sum_{\nu=1}^{N_*} \int_{t_{\nu-\frac{1}{2}}}^{t_{\nu+\frac{1}{2}}} \phi(t) K_{\nu,N}(t) \mathrm{d}t \right| + O\left(\frac{1}{n^\alpha}\right),$$

i.e.

$$I_N^{(n)}(\phi) = \frac{1}{(k+1)\pi} \left| \int_{-\frac{\pi}{N+n-k+1}}^{\frac{\pi}{N+n-k+1}} \sum_{\nu=1}^{N_*-1} \phi(t_\nu+t) K_{\nu,N}(t_\nu+t) \mathrm{d}t \right| + O\left(\frac{1}{n^\alpha}\right),$$

or

$$I_N^{(n)}(\phi) = \frac{1}{(k+1)\pi}\left| \int_0^{\frac{\pi}{N+n-k+1}} \sum_{\nu=1}^{N*} \{\phi(t_\nu+t)K_{\nu,N}(t_\nu+t) - \phi(t_\nu-t)K_{\nu,N}(t_\nu-t)\}\,dt \right|$$

$$= \frac{1}{(k+1)\pi}\left| \int_0^{\frac{\pi}{N+n-k+1}} \sum_{\nu=1}^{N*} \{\phi(t_\nu+t)-\phi(t_\nu-t)\}\psi_{\nu,N}(t_\nu+t)\,dt + \int_0^{\frac{\pi}{N+n-k+1}} \sum_{\nu=1}^{N*} \phi(t_\nu-t)r_{\nu,N}(t_\nu-t)\,dt \right| + O\left(\frac{1}{n^\alpha}\right).$$

But since, as a consequence of 8.4(39),

$$\frac{1}{(k+1)\pi} \int_0^{\frac{\pi}{N+n-k+1}} \sum_{\nu=1}^{N*} \phi(t_\nu-t)r_{\nu,N}(t_\nu-t)\,dt = \frac{1}{n^\alpha}\sum_{\nu=1}^{N*} O\left(\frac{1}{\nu^{2-\alpha}}\right) = O\left(\frac{1}{n^\alpha}\right),$$

it follows that

$$I_N^{(n)}(\phi) = \frac{1}{(k+1)\pi}\left| \int_0^{\frac{\pi}{N+n-k+1}} \sum_{\nu=1}^{N*} \{\phi(t_\nu+t) - \phi(t_\nu-t)\}\psi_{\nu,N}(t_\nu+t)\,dt \right| + O\left(\frac{1}{n^\alpha}\right). \tag{40}$$

From this relation, since $\phi(t) \in H_0^{(\alpha)}M$, the inequality

$$I_N^{(n)}(\phi) \leqslant \frac{2^\alpha M}{(k+1)\pi} \sum_{\nu=1}^{N*} \int_0^{\frac{\pi}{N+n-k+1}} t^\alpha |\psi_{\nu,N}(t_\nu+t)|\,dt + O\left(\frac{1}{n^\alpha}\right), \tag{41}$$

follows, i.e.

$$I_N^{(n)}(\phi) \leqslant \frac{2^{\alpha-1} M}{(k+1)\pi} \sum_{\nu=1}^{N*} \int_0^{\frac{\pi}{N+n-k+1}} t^\alpha \times \frac{\sin\frac{N+n-k+1}{2}t \left|\sin\frac{N-n+k+1}{2}(t_\nu+t)\right|}{\sin^2\frac{t_\nu+t}{2}}\,dt + O\left(\frac{1}{n^\alpha}\right).$$

From this we obtain the estimate

$$I_N^{(n)}(\phi) \leqslant \frac{2^{\alpha+1}M}{(k+1)\pi} \sum_{\nu=1}^{N*} \int_0^{\frac{\pi}{N+n-k+1}} t^\alpha \times$$

$$\times \frac{\sin\frac{N+n-k+1}{2}t\left|\sin\frac{N-n+k+1}{2}t_\nu\right|}{t_\nu^2}\,dt + O\left(\frac{1}{n^\alpha}\right)$$

$$= \frac{2^{\alpha-1}(N+n-k+1)^2 M}{(k+1)\pi^3} \sum_{\nu=1}^{N*} \frac{\left|\sin(N-n+k+1)\frac{\nu\pi}{N+n-k+1}\right|}{\nu^2} \times$$

$$\times \int_0^{\frac{\pi}{N+n-k+1}} t^\alpha \sin\frac{N+n-k+1}{2}t\,dt + O\left(\frac{1}{n^\alpha}\right)$$

$$= \frac{2^{2\alpha}M}{(N+n-k+1)^\alpha}\,\frac{N+n-k+1}{\pi^3(k+1)} \sum_{\nu=1}^{N*} \frac{\left|\sin\nu\frac{N-n+k+1}{N+n-k+1}\pi\right|}{\nu^2} \times$$

$$\times \int_0^{\frac{\pi}{2}} t^\alpha \sin t\,dt + O\left(\frac{1}{n^\alpha}\right). \tag{42}$$

Further, we note that

$$\frac{N+n-k+1}{\pi(N-n+k-1)} \sum_{\nu=1}^{N*} \frac{\left|\sin\nu\frac{N-n+k+1}{N+n-k+1}\pi\right|}{\nu^2}$$

$$= \frac{N+n-k+1}{\pi(N-n+k-1)} \sum_{\nu=1}^{\left[\frac{N+n-k+1}{N-n+k+1}\right]} \frac{\sin\nu\frac{N-n+k+1}{N+n-k+1}\pi}{\nu^2} +$$

$$+ \frac{N+n-k+1}{\pi(N-n+k+1)} \sum_{\nu=\left[\frac{N+n-k+1}{N-n+k+1}\right]+1}^{N*} \frac{\left|\sin\nu\frac{N-n+k+1}{N+n-k+1}\pi\right|}{\nu^2}$$

$$= \frac{N+n-k+1}{\pi(N-n+k+1)} \sum_{\nu=1}^{\left[\frac{N+n-k+1}{N-n+k+1}\right]} \frac{\sin\nu\frac{N-n+k+1}{N+n-k+1}\pi}{\nu^2} + O(1)$$

$$= \frac{N+n-k+1}{\pi(N-n+k+1)} \sum_{\nu=1}^{\infty} \frac{\sin\nu\frac{N-n+k+1}{N+n-k+1}\pi}{\nu^2} + O(1).$$

But for $0 \leqslant x \leqslant \pi$**

$$\sum_{\nu=1}^{\infty} \frac{\sin \nu x}{\nu^2} = -\int_0^x \ln\left(2\sin\frac{t}{2}\right) dt$$

and

$$-\int_0^x \ln\left(2\sin\frac{t}{2}\right) dt = -\int_0^x \ln\,[t+O(t^3)]\,dt$$

$$= -\int_0^x \ln t\,dt - \int_0^x \ln\,[1+O(t^2)]\,dt,$$

i.e.

$$\sum_{\nu=1}^{\infty} \frac{\sin \nu x}{\nu^2} = -\int_0^x \ln\left(2\sin\frac{t}{2}\right) dt = -\int_0^x \ln t\,dt + O(x) = x\ln\frac{1}{x} + O(x).$$

Therefore, putting $x = \dfrac{N-n+k+1}{N+n-k+1}\pi$, we find

$$\frac{N+n-k+1}{\pi(N-n+k+1)} \sum_{\nu=1}^{N*} \frac{\left|\sin \nu \dfrac{N-n+k+1}{N+n-k+1}\pi\right|}{\nu^2} = \ln\frac{N+n-k+1}{N-n+k+1} + O(1)$$

Hence in virtue of 8.4(42), for any function $\phi(t) \in H_0^{(\alpha)}M (0 < \alpha < 1)$

$$I_N^{(n)}(\phi) \leqslant \frac{N-n+k+1}{k+1}\left(\frac{2}{N+n-k+1}\right)^{\alpha} \frac{2^\alpha M}{\pi^2} \times$$

$$\times \ln\frac{N+n-k+1}{N-n+k+1} \int_0^{\frac{\pi}{2}} t^\alpha \sin t\,dt + O\left(\frac{1}{n^\alpha}\right),$$

or

$$I_N^{(n)}(\phi) \leqslant \frac{2^\alpha M}{\pi^2 n^\alpha} \ln\frac{n}{k+1} \int_0^{\frac{\pi}{2}} t^\alpha \sin t\,dt + O\left(\frac{1}{n^\alpha}\right). \tag{43}$$

Reasoning in a similar way, we obtain a similar inequality for the integral

$$I_{N'}^{(n)}(\phi) = \frac{1}{(k+1)\pi}\left|\int_{-\pi}^{-\frac{1}{n}} \phi(t) \sum_{\nu=n-k}^{N'} A_\nu^{(r)}(t)\,dt\right|,$$

where N' is determined from the same considerations as N.

** See G. M. Fikhtengol'ts [1], v. III.

By virtue of 8.4(42) and 8.4(37) we find from 8.4(36) that for the functions $\phi(t) \in H_0^{(\alpha)}M$

$$\frac{1}{(k+1)\pi}\left|\int_{-\pi}^{\pi} \phi(t) \sum_{\nu=n-k}^{n} D_\nu^{(r)}(t)\,dt\right|$$

$$= \frac{1}{(n+1)^r(k+1)\pi}\left|\int_{-\pi}^{-\frac{1}{n}} \phi(t) \sum_{\nu=n-k}^{n} A_\nu^{(r)}(t)\,dt + \right.$$

$$\left. + \int_{\frac{1}{n}}^{\pi} \phi(t) \sum_{\nu=n-k}^{n} A_\nu^{(r)}(t)\,dt\right| + O\left(\frac{1}{n^{r+\alpha}}\right) \tag{44}$$

uniformly in n, $k(0 \leqslant k \leqslant (1-\theta)n)$ and ϕ.

Applying to the two integrals on the right-hand side of 8.4(44) the inequality 8.4(43) for $N = n$, and the similar inequality $\left(\text{for the interval } \left(-\pi, -\frac{1}{n}\right)\right)$ for $N' = n$, we obtain by 8.4(33), for $0 < \alpha < 1$ and $0 \leqslant k \leqslant (1-\theta)n$, the inequality

$$\mathscr{E}_{\sigma_n}^{*}(W_*^{(r)}H^{(\alpha)}M;\, x;\, \lambda)$$

$$\leqslant \frac{2^{\alpha+1}M}{\pi^2 n^{r+\alpha}} \ln\frac{n}{k+1} \int_0^{\frac{\pi}{2}} t^\alpha \sin t\,dt + O\left(\frac{1}{n^{r+\alpha}}\right). \tag{45}$$

On the other hand, for every n a function $\Phi_n(x) \in W_*^{(r)}H^{(\alpha)}M$ can be found such that $|\Phi_n(0) - \sigma_n^*(\Phi_n;\, k;\, 0)|$, apart from an additive term of order $O\left(\frac{1}{n^{r+\alpha}}\right)$, is identical with the right-hand side of 8.4(45). For this it is sufficient to take the function whose r-th derivative $\phi_n(x)$ is determined on $[0, \pi]$ by the equations

$$\phi_n(x) = \begin{cases} 0, \ 0 \leqslant x \leqslant t_1, \ t_{n*} \leqslant x \leqslant \pi, \\ (-1)^\nu 2^{\alpha-1} M (x - t_\nu)^\alpha, \ t_\nu \leqslant x \leqslant t_{\nu+\frac{1}{2}} \\ (-1)^\nu 2^{\alpha-1} M (t_{\nu+1} - x)^\alpha, \ t_{\nu+\frac{1}{2}} \leqslant x \leqslant t_{\nu+1}, \\ \nu = 1, 2, \ldots, n^* - 1. \end{cases} \tag{46}$$

and by similar equations on the segment $[-\pi, 0]$.

The fact that $\phi_n(x)$ belongs to the class $H_0^{(\alpha)}M$ is verified directly. From 4.8(40), for $N = n$, it is easily seen that for the function $\phi_n(x)$ the inequality 8.4(41) becomes an equality. If we now take into account the fact that all transformations for obtaining 8.4(43) which follow the inequality 8.4(41) are asymptotic in character, we see that $\Phi_n(x)$ is an extremal function, i.e. the sign of equality occurs in 8.4(45).

The last remark also follows from 8.3(36).

8.4.42. Now let $\alpha = 1$. In virtue of the known properties of functions which satisfy a Lipschitz condition of the first order, the class $W_*^{(p)}H^{(1)}M$ is identical with the class $W_*^{(p+1)}M$ of functions representable by equation 8.4(27), in which $r = p+1$ and $\phi(t)$ is any measurable function of period 2π satisfying the conditions

$$\int_{-\pi}^{\pi} \phi(t)\mathrm{d}t = 0, \qquad \operatorname*{vrai\,sup}_t |\phi(t)| \leqslant M. \tag{47}$$

Hence in this case (see 8.4(31))

$$\mathcal{E}_{\sigma_n^*}(W_*^{(r)}H^{(1)}M; x) = \mathcal{E}_{\sigma_n^*}(W_*^{(r+1)}M; 0)$$

$$= \frac{1}{(k+1)\pi} \sup_{|\phi(t)|\leqslant 1} \int_{-\pi}^{\pi} \phi(t) \sum_{\nu=n-k}^{n} D_\nu^{(r+1)}(t)\mathrm{d}t.$$

From the same considerations as in section 8.4.41 we obtain

$$\mathcal{E}_{\sigma_n^*}(W^{(r)}H^{(1)}M; x) = \frac{1}{(n+1)^{r+1}(k+1)\pi} \times$$

$$\times \sup_{|\phi(t)|\leqslant 1} \left| \int_{-\frac{1}{n}}^{-\frac{1}{n}} \phi(t) \sum_{\nu=n-k}^{n} A_\nu^{(r+1)}(t)\mathrm{d}t + \int_{\frac{1}{n}}^{n} \phi(t) \sum_{\nu=n-k}^{n} A_\nu^{(r+1)}(t)\mathrm{d}t + \right.$$

$$+ \int_{-\pi}^{-\frac{1}{n}} \phi(t) \sum_{\nu=n-k}^{n} \left\{ \frac{1}{(\nu+1)^{r+1}} - \frac{1}{(n+1)^{r+1}} \right\} A_\nu^{(r+1)}(t)\mathrm{d}t +$$

$$\left. + \int_{\frac{1}{n}}^{\pi} \phi(t) \sum_{\nu=n-k}^{n} \left\{ \frac{1}{(\nu+1)^{r+1}} - \frac{1}{(n+1)^{r+1}} \right\} A_\nu^{(r+1)}(t)\mathrm{d}t \right| + O\left(\frac{1}{n^{r+1}}\right). \tag{48}$$

Thus, since $0 \leqslant k \leqslant (1-\theta)n$ and $\operatorname{vrai\,sup} |\phi(t)| \leqslant M$, we have (see section 8.4.41)

$$\frac{1}{(k+1)\pi}\left| \int_{\pm\frac{1}{n}}^{\pm\pi} \phi(t) \sum_{\nu=n-k}^{n} \left\{ \frac{1}{(\nu+1)^{r+1}} - \frac{1}{(n+1)^{r+1}} \right\} A_\nu^{(r+1)}(t)\,dt \right|$$

$$\leqslant \frac{BM}{n^{r+1}\pi} \int_{\pm\frac{1}{n}}^{\pm\pi} \left| \frac{\sin(N-n+k+1)\frac{t}{2}\sin\left[(N+n-k+1)\frac{t}{2}+\frac{r+1}{2}\pi\right]}{2\sin^2\frac{t}{2}} \right| dt, \quad (49)$$

where $N(n-k \leqslant N \leqslant n)$ is defined as in obtaining 8.4(37).

Calculations similar to those given in section 8.2.31 show that

$$\frac{1}{(k+1)\pi} \int_{\pm\frac{1}{n}}^{\pm\pi} \left| \frac{\sin(N-n+k+1)\frac{t}{2}\sin\left[(N+n-k+1)\frac{t}{2}+\frac{r+1}{2}\pi\right]}{2\sin^2\frac{t}{2}} \right| dt$$

$$= \pm\frac{2}{\pi^2}\ln\frac{n}{k+1} + O(1). \quad (50)$$

Therefore from 8.4(48) it follows (see section 8.4.41) that

$$\mathcal{E}_{\sigma_n^*}(W^{(r)}H^{(1)}M; x) = \frac{1}{(n+1)^{r+1}(k+1)\pi} \times$$

$$\times \sup_{|\phi(t)|\leqslant 1} \left| \int_{-\pi}^{-\frac{1}{n}} \phi(t) \frac{\sin\frac{k+1}{2}t\sin\left[\frac{2n+1-k}{2}t+\frac{r+1}{2}\pi\right]}{2\sin^2\frac{t}{2}}\,dt + \right.$$

$$\left. + \int_{\frac{1}{n}}^{\pi} \phi(t) \frac{\sin\frac{k+1}{2}t\sin\left[\frac{2n+1-k}{2}t+\frac{r+1}{2}\pi\right]}{2\sin^2\frac{t}{2}}\,dt \right| + O\left(\frac{1}{n^{r+1}}\right).$$

i.e.

$$\mathcal{E}_{\sigma_n^*}(W_*^{(r)}H^{(1)}M; x) = \frac{M}{(n+1)^{r+1}(k+1)\pi} \times$$

$$\times\left\{\int_{-\pi}^{-\frac{1}{n}}\left|\frac{\sin\frac{k+1}{2}t\sin\left(\frac{2n+1-k}{2}t+\frac{r+1}{2}\pi\right)}{2\sin^2\frac{t}{2}}\right|dt+\right.$$

$$\left.+\int_{\frac{1}{n}}^{\pi}\left|\frac{\sin\frac{k+1}{2}t\sin\left(\frac{2n+1-k}{2}t+\frac{r+1}{2}\pi\right)}{2\sin^2\frac{t}{2}}\right|dt\right\}+O\left(\frac{1}{n^{r+1}}\right).$$

Substituting in this the estimate 8.4(50) for $N=n$, we obtain the first of the inequalities 8.4(26) in the case when $r>0$, $\alpha=1$.

8.4.43. In order to obtain the first equation of 8.4(26) for $r>0$, we make use of the fact that the class $W_*^{(r)}H^\alpha M$ consists of functions $f(x)$ representable in the form of the integral

$$f(x)=\frac{1}{\pi}\int_{-\pi}^{\pi}\phi(t)\tilde{D}_0^{(r)}(x-t)\,dt, \tag{51}$$

where $\tilde{D}_n^{(r)}(t)$ is the kernel 8.4(22) and $\phi(x)$ satisfies the conditions 8.4(28). Hence, as in section 8.4.41,

$$\mathcal{E}_{\sigma_n^*}(W_*^{(r)}H^{(\alpha)}M;x)=\mathcal{E}_{\sigma_n^*}(\tilde{W}_*^{(r)}H^{(\alpha)}M;0)$$

$$=\frac{1}{(k+1)\pi}\sup_{\phi\in H_0^{(\alpha)}M}\int_{-\pi}^{\pi}\phi(t)\sum_{\nu=n-k}^{n}\tilde{D}_\nu^{(r)}(t)\,dt. \tag{52}$$

Moreover, using the identity 8.4(24) and the relation 8.4(25), we arrive at the estimates given in sections 8.4.41 and 8.4.42.

8.4.44. The case when $r=0$ is reduced to those considered previously. Since for any value of the constant C

$$f(x)+C-\sigma_n^*(f+C;k;x)=f(x)-\sigma_n^*(f;k;x),$$

it is possible, without restriction of generality, to suppose that $f(0)=0$. Thus

$$\mathcal{E}_{\sigma_n^*}(W^{(0)}H^{(\alpha)}M;x)=\sup_{f(x)}\sigma_n^*(f;k;0),$$

where the upper bound is taken for $f(x) \in W^{(0)}H^{(\alpha)}M$ with $f(0) = 0$. But (see 8.4(4) and 8.2(35))

$$\sigma_n^*(f; k; 0) = \frac{1}{(k+1)\pi} \int_{-\pi}^{\pi} f(t) \sum_{\nu=n-k}^{n} D_\nu(t)\, dt$$

$$= \frac{1}{2(k+1)\pi} \int_{-\pi}^{\pi} f(t) \frac{\sin \frac{2n+1-k}{2} t \sin \frac{k+1}{2} t}{\sin^2 \frac{t}{2}}\, dt. \tag{53}$$

Therefore, taking into account that the integral 8.4(53) is asymptotically $\left(\text{to within the order } O\left(\frac{1}{n^\alpha}\right)\right)$ identical with the quantity $2I_n^{(n)}(f)$ (see 8.4(38)) when $r = 0$, we verify on the basis of the considerations of sections 8.4.41 and 8.4.42 the validity of the first of the relations 8.4(26) when $r = 0$.

In order to obtain the first of the equalities 8.4(26) in the case $r = 0$, $0 < \alpha < 1$, we note that

$$S_n^*(\tilde{f}; x) = -\frac{1}{\pi} \int_{-\pi}^{\pi} f(x+t) \sum_{k=1}^{n} \sin kt\, dt$$

$$= -\frac{1}{\pi} \int_{-\pi}^{\pi} f(x+t) \frac{\cos \frac{t}{2} - \cos \frac{2n+1}{2} t}{2 \sin \frac{t}{2}}\, dt. \tag{54}$$

Hence it is possible, without loss of generality, to take $f(0) = 0$. Since

$$\tilde{f}(0) = -\frac{1}{\pi} \int_{-\pi}^{\pi} f(t) \frac{\cos \frac{t}{2}}{2 \sin \frac{t}{2}}\, dt, \tag{55}$$

we have

$$\tilde{f}(0) - S_n^*(\tilde{f}; 0) = -\frac{1}{\pi} \int_{-\pi}^{\pi} f(t) \frac{\cos \frac{2n+1}{2} t}{2 \sin \frac{t}{2}}\, dt$$

and consequently

$$\tilde{f}(0)-\sigma_n^*(\tilde{f}; k; 0) = -\frac{1}{(k+1)\pi}\int_{-\pi}^{\pi}\frac{f(t)}{2\sin\frac{t}{2}}\sum_{\nu=n-k}^{n}\cos\frac{2\nu+1}{2}t$$

$$= -\frac{1}{2(k+1)\pi}\int_{-\pi}^{\pi} f(t)\frac{\cos\frac{2n+1-k}{2}t\sin\frac{k+1}{2}t}{\sin^2\frac{t}{2}}\,dt. \tag{56}$$

It remains to be noted also that the estimation of the integral 8.4(56) reduces to the investigation of the quantity $2I_n^{(n)}(f)$ for $r = 1$.

8.4.5. We also show that if $|f(t)| \leqslant 1$ *and*

$$\tilde{f}_n(0) = -\frac{1}{2\pi}\int_{\frac{1}{n}}^{\pi}[f(t)-f(-t)]\cot\frac{t}{2}\,dt \tag{57}$$

*then***

$$|\tilde{f}_n(0)-\sigma_n^*(\tilde{f}; k; 0)| \leqslant \frac{4}{\pi^2}\ln\frac{n}{k+1}+O(1). \tag{58}$$

Let us represent $\tilde{\sigma}^*(f; k; x)$ in the form of the sum

$\tilde{\sigma}_n^*(f; k; x)$

$$= -\frac{1}{2(k+1)\pi}\int_0^{\pi}[f(t)-f(-t)]\sum_{\nu=n-k}^{n}\frac{\cos\frac{t}{2}-\cos\frac{2\nu+1}{2}t}{\sin\frac{t}{2}}\,dt$$

$$= -\frac{1}{2(k+1)\pi}\int_0^{\frac{1}{n}}[f(t)-f(-t)]\sum_{\nu=n-k}^{n}\frac{\cos\frac{t}{2}-\cos\frac{2\nu+1}{2}t}{\sin\frac{t}{2}}\,dt$$

$$-\frac{1}{2(k+1)\pi}\int_{\frac{1}{n}}^{\pi}[f(t)-f(-t)]\sum_{\nu=n-k}^{n}\frac{\cos\frac{t}{2}-\cos\frac{2\nu+1}{2}t}{\sin\frac{t}{2}}\,dt$$

and note that

$$\frac{1}{2(k+1)\pi}\int_0^{\frac{1}{n}}\left|\sum_{\nu=n-k}^{n}\frac{\cos\frac{t}{2}-\cos\frac{2\nu+1}{2}t}{\sin\frac{t}{2}}\right|dt$$

** A. D. Shcherbina [1].

$$= \frac{1}{2(k+1)\pi} \int_0^{\frac{1}{n}} \frac{\left| (k+1)\cos\frac{t}{2}\sin\frac{t}{2} - \cos\frac{2n+1-k}{2}t\sin\frac{k+1}{2}t \right|}{\sin^2\frac{t}{2}} dt$$

$$\leqslant \frac{1}{2(k+1)\pi} \int_0^{\frac{1}{n}} \frac{\left| (k+1)\sin\frac{t}{2} - \sin(k+1)\frac{t}{2} \right|}{\sin^2\frac{t}{2}} dt +$$

$$+ \frac{1}{2(k+1)\pi} \int_0^{\frac{1}{n}} \frac{\sin\frac{k+1}{2}t \left| \cos\frac{t}{2} - \cos\frac{2n+1-k}{2}t \right|}{\sin^2\frac{t}{2}} dt = O(1).$$

From this it follows that

$$\tilde{f}_n(0) - \tilde{\sigma}_n^*(f; k; 0)$$

$$= -\frac{1}{2(k+1)\pi} \int_{\frac{1}{n}}^{\pi} [f(t) - f(-t)] \sum_{\nu=n-k}^{n} \frac{\cos\frac{2\nu+1}{2}t}{\sin\frac{t}{2}} dt + O(1),$$

and from this we obtain

$$\sup_{|f(t)|\leqslant 1} |\tilde{f}_n(0) - \tilde{\sigma}_n^*(f; k; 0)|$$

$$\leqslant \frac{1}{(k+1)\pi} \int_{\frac{1}{n}}^{\pi} \frac{\left| \cos\frac{2n+1-k}{2}t\sin\frac{k+1}{2}t \right|}{\sin^2\frac{t}{2}} dt + O(1). \tag{59}$$

Estimating the right-hand side of 8.4(59) in the same way as the similar integrals in section 8.2.31, we obtain

$$\frac{1}{(k+1)\pi} \int_{\frac{1}{n}}^{\pi} \frac{\left| \cos\frac{2n+1-k}{2}t\sin\frac{k+1}{2}t \right|}{\sin^2\frac{t}{2}} dt$$

$$= \frac{4}{(k+1)\pi} \int_{\frac{1}{n}}^{\pi} \frac{\left| \cos\frac{2n+1-k}{2}t\sin\frac{k+1}{2}t \right|}{t^2} dt + O(1)$$

$$= \frac{2}{\pi} \int\limits_{\frac{k+1}{2n}}^{\frac{k+1}{2}\pi} \frac{\left| \cos \frac{2n+1-k}{k+1} t \sin t \right|}{t^2} dt + O(1)$$

$$= \frac{2}{\pi} \int\limits_{\frac{k+1}{n}}^{\frac{\pi}{2}} \frac{\left| \cos \frac{2n+1-k}{k+1} t \sin t \right|}{t^2} dt + O(1)$$

$$= \frac{2}{\pi} \int\limits_{\frac{k+1}{n}}^{\frac{\pi}{2}} \frac{\left| \cos \frac{2n+1-k}{k+1} t \right|}{t} dt + O(1)$$

$$= \frac{4}{\pi^2} \ln \frac{n}{k+1} + O(1).$$

8.4.6. We indicate an application of the inequality 8.4(5) to the investigation of the approximative properties of the derivatives of best approximation polynomials and establish the following theorem**. *If the periodic function $f(x)$ of period 2π has a bounded derivative $f^{(r)}(x)$ of order r and $T_n(f; x)$ is the trigonometric polynomial of order $\leqslant n$ for which* $E_n^*(f) = \max\limits_x |f(x) - T_n(f; x)|$, *then as* $n \to \infty$

$$\sup_x |f^{(r)}(x) - T_n^{(r)}(f; x)| = O\{E_n^*(f^{(r)})\}. \tag{60}$$

Since $\sigma_{2n}^*(f^{(r)}; n; x) = \sigma_{2n}^{*(r)}(f; n; x)$, then in virtue of 8.4(5) and the inequalities 4.8(7) we have

$$\begin{aligned}
&\sup_x |f^{(r)}(x) - T_n^{(r)}(f; x)| \\
&\quad \leqslant \sup_x |f^{(r)}(x) - \sigma_{2n}^*(f^{(r)}; n; x)| + \\
&\qquad + \sup |T_n^{(r)}[\sigma_{2n}^*(f; n); x] - T_n^{(r)}(f; x)| + \\
&\qquad + \sup |\sigma_{2n}^{*(r)}(f; n; x) - T_n^{(r)}[\sigma_{2n}^*(f; n); x)| \\
&\quad = O\{E_n^*(f^{(r)})\} + n^r \sup_x |T_n[\sigma_{2n}^*(f; n); x] - T_n(f; x)| + \\
&\qquad + 2^r n^r E_n^*[\sigma_{2n}^*(f; n)].
\end{aligned}$$

Also applying 8.4(5) and 8.4(2), we obtain

$$\sup_x |T_n[\sigma_{2n}^*(f; n); x] - T_n(f; x)| = O\{E_n^*[\sigma_{2n}^*(f; n)]\} + O\{E_n^*(f)\}.$$

** G. Freud [1]. See also A. L. Garvaki [1]. If we take 4.12.17 into account, it is obvious from the proof given here that this result remains valid also for derivatives of fractional order.

It remains to take into account that since

$$\frac{1}{(n+1)\pi}\int_{-\pi}^{\pi}(\{f(x+t)-T_n(f;x+t)\}\sum_{k=n}^{2n}D_k(t)dt = \sigma_{2n}^*|(f;n;x)-T_n(x),$$

where $D_k(t)$ is Dirichlet's kernel (see 4.1(10)) and $T_n(x)$ is some trigonometric polynomial of order $\leqslant n$, then by 8.4(2)

$$E_n^*[\sigma_{2n}^*(f;n)] = O\{E_n^*(f)\},$$

and to apply the inequality 5.6(27).

In virtue of 3.10(8) and 8.4(60) the inequality

$$E_n^*(f^{(p)}) \leqslant C_{r,p}\{E_n^*(f)\}^{1-\frac{p}{r}}\{E_n^*(f^{(r)})\}^{\frac{p}{r}}, \tag{61}$$

is valid for any $n \geqslant 0$ and $0 \leqslant p \leqslant r$ for every periodic function $f(x)$ of period 2π, where $C_{r,p}$ is a constant depending only on p and r**.

8.4.7. In concluding this section we note that a theorem of the type 8.4.4 also holds in the case where we consider an approximation to functions $f(x)$ belonging to the class $W^{(r)}H^{(\alpha)}M$ on $[-1, 1]$, by the algebraic polynomials 8.2(30). Here we will limit ourselves to its formulation***. *For any numbers r ($r \geqslant 0$ an integer), $\alpha (0 < \alpha \leqslant 1)$, subject to the condition $\lim_{n\to\infty}\frac{k(n)}{n} = 0$ the asymptotic equation*

$$\sup_{f\in W^{(r)}H^{(\alpha)}M[-1,1]} |f(x)-\sigma_n(f;k;x)|$$
$$= \frac{2^{\alpha+1}M}{\pi^2 n^{r+\alpha}}\left(\sqrt{1-x^2}\right)^{r+\alpha}\ln\frac{n}{k+1}\int_0^{\frac{\pi}{2}} t^\alpha \sin t\,dt+O\left(\frac{1}{n^{r+\alpha}}\right) \tag{62}$$

is satisfied uniformly with respect to $x \in [-1, 1]$ and for all functions $k = k(n)$ $(0 \leqslant k < n)$ for which $k \leqslant (1-\theta)n$ $(0 < \theta \leqslant 1)$. In the case when $r = 0$, the relation 8.4(62) is satisfied uniformly with respect to all functions $k = k(n)$ $(0 \leqslant k < n)$ for any α $(0 < \alpha < 1)$.

As in other estimates, relating to approximation by algebraic polynomials on a finite segment, here, in distinction from the periodic case, the corresponding deviation is materially dependent on the position of the point x.

** It can be shown that this property is possessed by the best approximations $A_\delta(f)$ of arbitrary bounded functions $f(x)$ which have a bounded derivative $f^{(r)}(x)$ on $(-\infty, \infty)$.

*** For the case $k = 0$ see A. F. Timan and L. I. Tuchinskii [1]; for the case $r = 0$ see L. M. Ganzburg [2]; for the case $r = 0$, $k = 0$, $\alpha = 1$ see S. M. Nikol'skii [4]; for the case $k = 0$, $\alpha = 1$ see S. G. Selivanova [1]; for the case $k = 0$, $0 < \alpha < 1$ see A. F. Timan [6]; for the general case see S. I. Rabinovich [1].

8.5. Further estimates for linear methods of approximation of functions by polynomials

In section 8.3 we indicated a necessary condition which an infinite triangular matrix of numbers must satisfy in order that the process of approximation by the polynomials 8.1(8) corresponding to it should realise the best uniform approximation on the classes $W_*^{(r)} H_\omega$ or $\tilde{W}_*^{(r)} H_\omega$ (see sections 8.3.5 and 8.3.8). It should be noted that in the general case this condition (see 8.3(48)) is of itself not sufficient. Let us consider, for example, the classes H_ω^* or $\tilde{H}_\omega^*$, when $\omega(t) = Mt^\alpha (0 < \alpha < 1)$. The matrix 8.2(10) obviously satisfies the relation 8.3(48) (for $r = 0$ it is equivalent to 8.2(9)). At the same time since the polynomials 8.1(8) corresponding to it are of the form

$$U_n^*(f; x; \lambda) = \frac{a_0}{2} + \sum_{k=1}^{m} (a_k \cos kx + b_k \sin kx),$$

where $m = \left[\frac{n}{2}\right]$, then by theorem 8.4.4 in the present case the exact order of decrease of the upper bounds $\mathcal{E}_n(H_*^{(\alpha)} M; x; \lambda)$ and $(\tilde{H}_*^{(\alpha)} M; x; \lambda)$ as $n \to \infty$ is equal to $\frac{\ln n}{n^\alpha}$, and hence is not identical with the best.

However, in this section it is established that if we consider a matrix λ which satisfies some limitation of convexity type, then condition 8.3(48) becomes not only necessary but also sufficient for the method of approximation 8.1(8) to give the best order of approximation on the classes $W_*^{(r)} H_\omega$ or $\tilde{W}_*^{(r)} H_\omega$ for integral r and for $\omega(t) = Mt^\alpha$, i.e. on the classes $W_*^{(r)} H^{(\alpha)} M$, $\tilde{W}_*^{(r)} H^{(\alpha)} M$.

In those cases where the condition 8.3(48) is not satisfied, the order of the corresponding approximation on the classes $W_*^{(r)} H^{(\alpha)} M$, $\tilde{W}_*^{(r)} H^{(\alpha)} M$ will not be identical with the best. In such cases an estimate can be given for the magnitude of this order, and for the supplementary limitations on the matrix λ to obtain also asymptotic behaviour of the upper bounds $\mathcal{E}_n(W_*^{(r)} H^{(\alpha)} M; x; \lambda)$ and $\mathcal{E}_n(\tilde{W}_*^{(r)} H^{(\alpha)} M; x; \lambda)$ as $n \to \infty$. In what follows, for brevity in writing, we will use the notations

$$\mu_k^{(n)} = \begin{cases} 0, & k = 0, \\ \dfrac{1 - \lambda_k^{(n)}}{k^r}, & 1 \leqslant k \leqslant n, \\ \dfrac{1}{(n+1)^r}, & k = n+1, \end{cases}$$

$$\Delta \mu_k^{(n)} = \mu_k^{(n)} - \mu_{k+1}^{n}, \quad \Delta^2 \mu_k^{(n)} = \mu_k^{(n)} - 2\mu_{k+1}^{(n)} + \mu_{k+2}^{(n)}.$$

8.5.1. *For any bounded matrix λ, and for any integer $r \geqslant 0$ and $0 < \alpha < 1$, the inequality***

$$\left.\begin{array}{l}\mathcal{E}_n(W_*^{(r)}H^{(\alpha)}M; x; \lambda)\\ \mathcal{E}_n(\tilde{W}^{(r)}H^{(\alpha)}M; x; \lambda)\end{array}\right\}$$

$$\leqslant \frac{2^{\alpha+1}M}{\pi^2 n^\alpha}\int_0^{\frac{\pi}{2}} t^\alpha \sin t\,dt \sum_{k=0}^{n-1}\left(\sum_{\nu=n-k}^{n}\frac{n-k}{\nu}\right)|\Delta^2\mu_k^{(n)}|+$$

$$+O\left\{\sum_{k=0}^{[\theta n]}(k+1)^{1-\alpha}|\Delta^2\mu_k^{(n)}|+\right.$$

$$\left.+\frac{1}{n^\alpha}\sum_{k=[\theta n]+1}^{n-1}(n-k)|\Delta^2\mu_k^{(n)}|\right\}+O\left(\frac{1}{n^{r+\alpha}}\right) \tag{1}$$

*is satisfied, in which $\theta > 0$ is an arbitrary fixed number less than unity. In the case when r is even (odd), the first (second) of these estimates also holds for $\alpha = 0$ if its right-hand side is doubled****.*

For other integral values of r,

$$\left.\begin{array}{l}\mathcal{E}_n(W_*^{(r)}H^{(0)}M; x; \lambda)\\ \mathcal{E}_n(\tilde{W}_*^{(r)}H^{(0)}M; x; \lambda)\end{array}\right\}\leqslant \frac{4M}{\pi^2}\sum_{k=0}^{n-1}\left(\sum_{\nu=n-k}^{n}\frac{n-k}{\nu}\right)|\Delta^2\mu_k^{(n)}|+$$

$$+O\left\{\sum_{k=0}^{n-1}\left(\sum_{\nu=k+1}^{n}\frac{k+1}{\nu}\right)|\Delta^2\mu_k^{(n)}|\right\}+O\left(\frac{1}{n^r}\right). \tag{2}$$

We consider the four cases separately.

8.5.11. Let $r > 0$ be even and $0 \leqslant \alpha < 1$. For functions $f(x) \in W_*^{(r)}H^{(\alpha)}M$ the integral representation 8.4(27) holds, in which $\phi(x)$ satisfies the conditions 8.4(28). Therefore

$$f(x) - U_n^*(f; x; \lambda) = \frac{1}{\pi}\int_{-\pi}^{\pi}\phi(t)\sum_{k=0}^{n+1}\mu_k^{(n)}\cos\left[k(t-x)+\frac{r\pi}{2}\right]dt+$$

$$+\frac{1}{\pi}\int_{-\pi}^{\pi}\phi(t)D_{n+1}^{(r)}(t-x)\,dt, \tag{3}$$

where the kernel $D_n^{(r)}(t)$ is defined by 8.4(10). Since

$$\int_{-\pi}^{\pi}\sum_{k=0}^{n+1}\mu_k^{(n)}\cos\left[k(t-x)+\frac{r\pi}{2}\right]dt = 0, \qquad \int_{-\pi}^{\pi}D_{n+1}^{(r)}(t-x)\,dt = 0,$$

** A. F. Timan [12].

*** $W_*^{(r)}H^{(0)}M$ is the class of functions of period 2π with an $(r-1)$-th absolutely continuous derivative and a derivative of the r-th order which almost everywhere does not exceed M in absolute value.

then, without restriction of generality, we may suppose that $\phi(0) = 0$ Moreover, since the upper bounds considered do not depend on x (see section 8.3.4), it is sufficient to put $x = 0$.

Applying Abel's transformation to the sum on the right-hand side of 8.5(3), we obtain

$$|f(0)-U_n^*(f;0;\lambda)| = \left| \sum_{k=0}^{n} \Delta\mu_k^{(n)} S_k^*(\phi;0) + \right.$$

$$\left. + \frac{1}{\pi(n+1)^r} \int_{-\pi}^{\pi} \phi(t) \frac{\sin\frac{2n+3}{2}}{2\sin\frac{t}{2}} \mathrm{d}t + \frac{1}{\pi} \int_{-\pi}^{\pi} \phi(t) \sum_{k=n+2}^{\infty} \frac{\cos kt}{k^r} \mathrm{d}t \right|,$$

where $S_k^*(\phi;0)$ is the partial sum of order k of the Fourier series of the function $\phi(t)$ at the point $t = 0$. In virtue of the relations 8.4(11), 8.4(16) and 8.4(20) we find from this that

$$\begin{aligned} |f(0)-U_n^*(f;0;\lambda)| &= \left| \sum_{k=0}^{n} \Delta\mu_k^{(n)} S_k^*(\phi;0) \right| + O\left(\frac{1}{n^{r+\alpha}}\right) \\ &= \left| \sum_{k=0}^{n} \Delta\mu_{n-k}^{(n)} S_{n-k}^*(\phi;0) \right| + O\left(\frac{1}{n^{r+\alpha}}\right) \end{aligned} \tag{4}$$

uniformly with respect to all the functions $\phi(t)$ considered.

If Abel's transformation is again applied to the sum on the right-hand side of 8.5(4), we arrive at the equality

$$\begin{aligned} |f(0)-U_n^*(f;0;\lambda)| &= \left| \sum_{k=0}^{n-1} (k+1)\Delta^2\mu_k^{(n)} \sigma_k^*(\phi;k;0) + \right. \\ &\quad \left. + (n+1)\mu_n^{(n)} \sigma_n^*(\phi;n;0) \right| + O\left(\frac{1}{n^{r+\alpha}}\right) \\ &= \left| \sum_{k=0}^{[\theta n]} (k+1)\Delta^2\mu_k^{(n)} \sigma_k^*(\phi;k;0) + \right. \\ &\quad + \sum_{k=[\theta n]+1}^{n-1} (n-k)\Delta^2\mu_k^{(n)} \sigma_n^*(\phi;n-k;0) + \\ &\quad \left. + (n+1)\Delta\mu_{[\theta n]}^{(n)} \sigma_n(\phi;n;0) \right| + O\left(\frac{1}{n^{r+\alpha}}\right), \end{aligned}$$

where $\sigma_n^*(\phi;k;x)$ is the sum 8.4(4).

Hence in virtue of theorem 8.4.4 or the lemma 8.2.31 for $x=1$ (if $\alpha=0$) and the identity

$$[\theta n]\Delta\mu_{[\theta n]}^{(n)}=\mu_0^{(n)}-\mu_{[\theta n]}^{(n)}+\sum_{k=0}^{[\theta n]-1}(k+1)\Delta^2\mu_k^{(n)}$$

it follows that

$$|f(0)-U_n^*(f;0;\lambda)|=\Big|\sum_{k=[\theta n]}^{n-1}(n-k)\Delta^2\mu_k^{(n)}\sigma_n^*(\phi;n-k;0)\Big|+$$
$$+O\Big\{\sum_{k=0}^{[\theta n]}(k+1)^{1-\alpha}|\Delta^2\mu_k^{(n)}|\Big\}+O\left(\frac{1}{n^{r+\alpha}}\right).$$

After using once more theorem 8.4.4 or the lemma 8.2.31 (if α=0) we obtain the inequality**

$$|f(0)-U_n^*(f;0;\lambda)|$$
$$\leqslant\frac{2^{\alpha+1}}{\pi^2n^\alpha}\int_0^{\frac{\pi}{2}}t^\alpha\sin t\,dt\sum_{k=[\theta n]}^{n-1}(n-k)\ln\frac{n}{n-k}|\Delta^2\mu_k^{(n)}|+$$
$$+O\Big\{\sum_{k=0}^{[\theta n]}(k+1)^{1-\alpha}|\Delta^2\mu_k^{(n)}|\Big\}+$$
$$+O\Big\{\sum_{k=[\theta n]}^{n-1}(n-k)|\Delta^2\mu_k^{(n)}|\Big\}+O\left(\frac{1}{n^{r+\alpha}}\right),$$

from which follows the first estimate of 8.5(1) when $r>0$ is even and $0\leqslant\alpha<1$.

8.5.12. Let r be odd and $0<\alpha<1$. In virtue of 8.5(3), where, as in section 8.5.11, we suppose $\phi(0)=0$, we find after applying Abel's transformation

$$|f(0)-U_n^*(f;0;\lambda)|$$
$$=\Bigg|-\sum_{k=0}^{n}\Delta\mu_k^{(n)}\tilde{S}_n^*(\phi;0)+\frac{1}{\pi(n+1)^r}r\int_{-\pi}^{\pi}\phi(t)\frac{\cos\frac{t}{2}-\cos\frac{2n+3}{2}t}{2\sin\frac{t}{2}}dt+$$
$$+\frac{1}{\pi}\int_{-\pi}^{\pi}\phi(t)\sum_{k=n+2}^{\infty}\frac{\sin kt}{k^r}dt\Bigg|, \tag{5}$$

where $\tilde{S}_k^*(\phi;0)$ is the partial sum of order k of the series conjugate to the Fourier series of the function $\phi(t)$, taken at the point t=0.

** Where the right-hand side has to be doubled when $\alpha=0$.

Using the relations 8.4(11), 8.4(16) and 8.4(20), and also taking into account that

$$\tilde{\phi}(0) = -\frac{1}{2\pi}\int_{-\pi}^{\pi}\phi(t)\cot\frac{t}{2}\,dt, \qquad \mu_{n+1}^{(n)} = \frac{1}{(n+1)^r} = -\sum_{k=0}^{n}\Delta\mu_k^{(n)},$$

it is not difficult to obtain the equation

$$|f(0) - U_n^*(f; 0; \lambda)| = \Big|\sum_{k=0}^{n}\Delta\mu_k^{(n)}[\tilde{\phi}(0) - \tilde{S}_k^*(\phi; 0)]\Big| + O\left(\frac{1}{n^{r+\alpha}}\right)$$

$$= \Big|\sum_{k=0}^{n}\Delta\mu_{n-k}^{(n)}[\tilde{\phi}(0) - \tilde{S}_{n-k}^*(\phi; 0)]\Big| + O\left(\frac{1}{n^{r+\alpha}}\right), \tag{6}$$

which is satisfied uniformly with respect to all the functions $\varphi(t)$ considered.

From this by the same transformations which were applied in 8.5.11 after 8.5(4), we establish the validity of the first of the estimates 8.5(1) when r is odd and $0 < \alpha < 1$.

8.5.13. Let r be odd and $\alpha = 0$. In this case from 8.5(5), in virtue of 8.4(11), 8.4(16) and 8.4(20), follows the relation

$$|f(0) - U_n^*(f; 0; \lambda)| = \Big|\sum_{k-0}^{n}\Delta\mu_k^{(n)}[\tilde{\phi}_n(0) - \tilde{S}_k^*(\phi; 0)]\Big| + O\left(\frac{1}{n^r}\right)$$

$$= \Big|\sum_{k=0}^{n}\Delta\mu_{n-k}^{(n)}[\tilde{\phi}_n(0) - \tilde{S}_{n-k}^*(\phi; 0)\Big| + O\left(\frac{1}{n^r}\right), \tag{7}$$

where

$$\tilde{\phi}_n(0) = -\frac{1}{2\pi}\int_{\frac{1}{n}}^{\pi}[\phi(t) - \phi(-t)]\cot\frac{t}{2}\,dt,$$

which is satisfied uniformly for all bounded functions.

Further, we apply the same reasoning which was applied in section 8.5.11 after 8.5(4), using the estimate

$$\tilde{\phi}_n(0) - \tilde{\phi}_k(0) = O\left[\ln\frac{n}{k+1}\right]$$

and the inequality 8.4(58).

8.5.14. Thus the first part of theorem 8.5.1 for the case when $r > 1$, is proved. The proof of the second part (i.e. for the conjugate classes) for $r > 0$ is based on the integral representation 8.4(51) and reduces to similar estimates (see sections 8.4.43 and 8.4.3).

8.5.15. Now let $r = 0$. Since

$$U_n^*(f; 0; \lambda) = \frac{1}{\pi} \int_{-\pi}^{\pi} f(t) \left\{ \frac{1}{2} + \sum_{k=1}^{n} \lambda_k^{(n)} \cos kt \right\} dt, \tag{8}$$

then in estimating the difference $f(0) - U_n^*(f; 0; \lambda)$ it is possible to take $f(0) = 0$. If for the sum

$$U_n^*(f; 0; \lambda) = \sum_{k=0}^{n} \Delta\lambda_k^{(n)} S_k^*(f; 0) = \sum_{k=0}^{n} \Delta\lambda_{n-k}^{(n)} S_{n-k}^*(f; 0) \tag{9}$$

the same reasoning is repeated which was carried through in section 8.5.11 after 8.5(4), and use is made of theorem 8.4.4, then we obtain the first of the inequalities 8.5(1) for $r = 0$, $0 < \alpha < 1$.

The case when $r = 0$, $\alpha = 0$, was investigated in section 8.2.31.

In order to verify the validity of the second inequality of 8.5(1) for $r = 0$, $0 < \alpha < 1$, we consider the difference

$$\tilde{f}(0) - U_n^*(\tilde{f}; 0; \lambda) = \tilde{f}(0) - \frac{1}{\pi} \int_{-\pi}^{\pi} f(t) \sum_{k=1}^{n} \lambda_k^{(n)} \sin kt \, dt,$$

which after Abel's transformation assumes the form

$$\tilde{f}(0) - U_n^*(\tilde{f}; 0; \lambda) = \tilde{f}(0) - \sum_{k=0}^{n} \Delta\lambda_k^{(n)} \tilde{S}_k^*(f; 0)$$

$$= \sum_{k=0}^{n} \Delta\lambda_{n-k}^{(n)} [\tilde{f}(0) - \tilde{S}_{n-k}^*(f; 0)]. \tag{10}$$

We also repeat the reasoning of section 8.5.11 after 8.5(4), and use theorem 8.4.4.

8.5.2. The inequalities 8.5(1) show that *if for the bounded matrix* λ

$$\sum_{k=0}^{n-1} \left(\sum_{\nu=n-k}^{n} \frac{n-k}{\nu} \right) |\Delta\mu_k^{(n)}| = O\left(\frac{1}{n^r}\right) \tag{11}$$

and there exists a value of $\theta (0 < \theta < 1)$ *for which*

$$\sum_{k=0}^{[\theta n]} (k+1)^{1-\alpha} |\Delta^2 \mu_k^{(n)}| = O\left(\frac{1}{n^{r+\alpha}}\right), \tag{12}$$

then the corresponding method of approximation 8.1(8) *gives on the classes* $W_*^{(r)} H^{(\alpha)} M$ *and* $\tilde{W}_*^{(r)} H^{(\alpha)} M$ $(r \geqslant 0,\ 0 < \alpha < 1)$ *the best uniform approximation, as regards order, by trigonometric polynomials.*

8.5.3. Taking into account that for every fixed $\theta (0 < \theta < 1)$

$$\sum_{k=0}^{[\theta n]} (k+1)|\Delta^2 \mu_k^{(n)}| = o\left\{\sum_{k=0}^{n-1}\left(\sum_{\nu=n-k}^{n} \frac{n-k}{\nu}\right)|\Delta^2 \mu_k^{(n)}|\right\},$$

by theorem 8.5.1 in the case $\alpha=0$ we obtain the following result. For any *integer* $r > 0$,

$$\mathcal{E}_n(W_*^{(r)}M; x; \lambda) = \begin{cases} O\left[\sum_{k=0}^{n-1}\left(\sum_{\nu=n-k}^{n} \frac{n-k}{\nu}\right)|\Delta^2 \mu_k^{(n)}|\right] + O\left(\frac{1}{n^r}\right), \\ \quad \text{when } r \text{ is even}, \\ O\left[\sum_{k=0}^{n-1}\left(\sum_{\nu=n-k}^{n} \frac{n-k}{\nu}\right)|\Delta^2 \mu_k^{(n)}| + \right. \\ \quad \left. + \sum_{k=0}^{n-1}\left(\sum_{\nu=k+1}^{n} \frac{k+1}{\nu}\right)|\Delta^2 \mu_k^{(n)}|\right] + O\left(\frac{1}{n^r}\right), \\ \quad \text{when } r \text{ is odd}, \end{cases} \tag{13}$$

and

$$\mathcal{E}_n(\tilde{W}_*^{(r)}M; x; \lambda) = \begin{cases} O\left[\sum_{k=0}^{n-1}\left(\sum_{\nu=n-k}^{n} \frac{n-k}{\nu}\right)|\Delta^2 \mu_k^{(n)}|\right] + O\left(\frac{1}{n^r}\right), \\ \quad \text{when } r \text{ is odd}, \\ O\left[\sum_{k=0}^{n-1}\left(\sum_{\nu=n-k}^{n} \frac{n-k}{\nu}\right)|\Delta^2 \mu_k^{(n)}| + \right. \\ \quad \left. + \sum_{k=0}^{n-1}\left(\sum_{\nu=k+1}^{n} \frac{k+1}{\nu}\right)|\Delta^2 \mu_k^{(n)}|\right] + O\left(\frac{1}{n^r}\right) \\ \quad \text{when } r \text{ is even}. \end{cases} \tag{14}$$

Thus if the matrix λ *satisfies the condition* 8.5(11), *the process of approximation* 8.1(8) *produces the best uniform approximation, as regards order, by trigonometric polynomials on the class* $W_*^{(r)}M$ *if* r *is even, and on the class* $\tilde{W}_*^{(r)}M$ *if* r *is odd***.

8.5.4. In those cases where the bounded matrix λ is such that for any n the system of numbers $\mu_k^{(n)} (k=0, 1, \ldots, n+1)$ is convex or concave, i.e.

$$\Delta^2 \mu_k^{(n)} \leqslant 0 \quad (k = 0, 1, \ldots, n-1) \tag{15}$$

** For $r = 0$ the condition 8.5 (11) ensures the boundedness of the sequence of Lebesgue constants L_n(see 8.1(14)). This case was considered by B. Sz. Nagy [3].

or

$$\Delta^2\mu_k^{(n)} \geqslant 0 \quad (k = 0,\ 1, \dots, n-1), \tag{16}$$

the relations 8.5(13) and 8.5(14) can be reduced to a simpler form. For this it is necessary to make use of the equalities

$$\begin{gathered}\sum_{k=0}^{n-1}\left(\sum_{\nu=n-k}^{n}\frac{n-k}{\nu}\right)\Delta^2\mu_k^{(n)} = \mu_0^{(n)}-\mu_n^{(n)}-\sum_{k=1}^{n}\frac{\mu_k^{(n)}}{n-k+1}+\mu_{n+1}^{(n)}\sum_{k=1}^{n}\frac{1}{k},\\ \sum_{k=0}^{n-1}\left(\sum_{\nu=k+1}^{n}\frac{k+1}{\nu}\right)\Delta^2\mu_k^{(n)} = \mu_0^{(n)}\sum_{k=1}^{n}\frac{1}{k}-\mu_1^{(n)}+\mu_{n+1}^{(n)}-\sum_{k=1}^{n}\frac{\mu_k^{(n)}}{k},\end{gathered} \tag{17}$$

which we obtain by two Abel transformations.

Thus *in these cases*

$$\mathcal{E}_n(W_*^{(r)}M;x;\lambda) = \begin{cases} O\left[\left|\sum_{k=1}^{n}\frac{1-\lambda_k^{(n)}}{k^r(n-k+1)}-\frac{\ln n}{n^r}\right|\right]+O\left(\frac{1}{n^r}\right) & \text{when } r \text{ is even},\\ O\left[\left|\sum\frac{1-\lambda_k^{(n)}}{k^r(n-k+1)}-\frac{\ln n}{n^r}\right|+\right. & \\ \qquad\left.+\left|\sum_{k=1}^{n}\frac{1-\lambda_k^{(n)}}{k^{r+1}}\right|\right]+O\left(\frac{1}{n^r}\right) & \text{when } r \text{ is odd},\end{cases} \tag{18}$$

and

$$\mathcal{E}_n(\tilde{W}_*^{(r)}M;x;\lambda) = \begin{cases} O\left[\left|\sum_{k=1}^{n}\frac{1-\lambda_k^{(n)}}{k^r(n-k+1)}-\frac{\ln n}{n^r}\right|\right]+O\left(\frac{1}{n^r}\right) & \text{when } r \text{ is odd},\\ O\left[\left|\sum_{k=1}^{n}\frac{1-\lambda_k^{(n)}}{k^r(n-k+1)}-\frac{\ln n}{n^r}\right|+\right. & \\ \qquad\left.+\left|\sum_{k=1}^{n}\frac{1-\lambda_k^{(n)}}{k^{r+1}}\right|\right]+O\left(\frac{1}{n^r}\right) & \text{when } r \text{ is even}.\end{cases} \tag{19}$$

8.5.5. We shall see that *if* 8.5(15) *or* 8.5(16) *hold for the matrix* λ *for any* n, *then the condition* 8.3(48) *is not only necessary* (see section 8.3.8) *but also sufficient for the corresponding process of approximation* 8.1(8) *to produce the best order of approximation by trigonometric polynomials on the class* $W_*^{(r)}M$ *when* r *is even, or on the class* $\tilde{W}_*^{(r)}M$ *when* r *is odd.*

This proposition remains valid for other integral values of r, subject to the condition that

$$\sum_{k=1}^{n} \frac{1-\lambda_k^{(n)}}{k^{r+1}} = O\left(\frac{1}{n^r}\right), \tag{20}$$

and in particular in all cases where $1-\lambda_1^{(n)} = O\left(\frac{1}{n^{r+1}}\right)$.

8.5.6. If the matrix λ, which satisfies the condition 8.5(15) or 8.5(16), also possesses the property that $\mu_1^{(n)} = O\left(\frac{1}{n^{r+1}}\right)$, then it is also possible to give asymptotic estimates for the upper bounds considered. The following theorem holds**.

If the bounded matrix λ *satisfies condition* 8.5(15) *or* 8.5(16) *for any* n *and* $\mu_1^{(n)} = O\left(\frac{1}{n^{r+1}}\right)$, *then, for any integer* $r \geqslant 0$ *and* $\alpha (0 \leqslant \alpha < 1)$, *the inequality*

$$\left.\begin{array}{l} \mathcal{E}_n(W_*^{(r)}H^{(\alpha)}M;x;\lambda) \\ \mathcal{E}_n(\tilde{W}_*^{(r)}H^{(\alpha)}M;x;\lambda) \end{array}\right\} = \frac{2^{\alpha+1}M}{\pi^2 n^{r+\alpha}} \int_0^{\frac{\pi}{2}} t^\alpha \sin t \, dt \left| \sum_{k=1}^{n} \frac{\lambda_k^{(n)}}{n-k+1} \right| + O\left(\frac{1}{n^{r+\alpha}}\right), \tag{21}$$

is satisfied for the upper bounds $\mathcal{E}_n(W^{(r)}H^{(\alpha)}M;x;\lambda)$ *and* $\mathcal{E}(\tilde{W}_*^{(r)}H^{(\alpha)}M;x;\lambda)$, where for $\alpha=0$ *the right-hand side must be doubled.*

By 8.3(36), after the application of the identity 8.5(17),*** this assertion follows from the following proposition.

8.5.7. For any integer $r \geqslant 0$ and for $0 \leqslant \alpha < 1$ the inequality****

$$\left.\begin{array}{l} \mathcal{E}_n(W_*^{(r)}H^{(\alpha)}M;x;\lambda) \\ \mathcal{E}_n(\tilde{W}_*^{(r)}H^{(\alpha)}M;x;\lambda) \end{array}\right\}$$

$$\leqslant \frac{2^{\alpha+1}M}{\pi^2 n^\alpha} \int_0^{\frac{\pi}{2}} t^\alpha \sin t \, dt \sum_{k=0}^{n-1} \left(\sum_{\nu=n-k}^{n} \frac{n-k}{\nu} \right) |\Delta^2 \mu_k^{(n)}| +$$

$$+O\left\{\frac{1}{n^\alpha} \sum_{k=1}^{n-1} (n-k)|\Delta^2 \mu_k^{(n)}| + n^{1-\alpha}\mu_1^{(n)}\right\} + O\left(\frac{1}{n^{r+\alpha}}\right) \tag{22}$$

is satisfied for any bounded matrix λ.

** A. F. Timan [12].

*** If we take into account that for the matrix λ considered here the relation

$$\sum_{k=1}^{n} \frac{\mu_k^{(n)}}{n-k+1} = \frac{1}{n^r} \sum_{k=1}^{n} \frac{\lambda_k^{(n)}}{n-k+1} + \frac{\ln n}{n^r} + O\left(\frac{1}{n^r}\right)$$

is satisfied.

**** See the footnote*** on page 548. In the case $\alpha = 0$ the right-hand side of 8.5(22) must be doubled.

In order to obtain the inequalities 8.5(22) it is necessary, as in section 8.5.1, to consider the four cases separately.

8.5.71. Let $r > 0$ be even and $0 \leqslant \alpha < 1$. In this case we commence with 8.5(4), and after the application of Abel's transformation we have

$$|f(0)-U_n^*(f;0;\lambda)| = \Big|-\sum_{k=0}^{n-1} \Delta^2\mu_{n-k-1}^{(n)} \sum_{\nu=0}^{k} S_{n-\nu}^*(\phi;0)-$$

$$-\mu_1^{(n)} \sum_{\nu=0}^{n} S_\nu^*(\phi,0)\Big| + O\left(\frac{1}{n^{r+\alpha}}\right). \qquad (23)$$

From this we obtain the inequality

$$|f(0)-U_n^*(f;0;\lambda)| \leqslant \sum_{k=0}^{n-1} (k+1)\,|\Delta^2\mu_{n-k-1}^{(n)}|\cdot|\sigma_n^*(\phi;k;0)| +$$

$$+(n+1)|\mu_1^{(n)}|\cdot|\sigma_n^*(\phi;n;0)| + O\left(\frac{1}{n^{r+\alpha}}\right),$$

where $\sigma_n^*(\phi;k;x)$ are the sums 8.4(4). Now applying theorem 8.4.4, we obtain the first of the estimates 8.5(22).

8.5.72. Let r be odd and $0 < \alpha < 1$. In this case we commence with 8.5(6) and after the application of Abel's transformation we have

$$|f(0)-U_n^*(f;0;\lambda)| \leqslant \sum_{k=0}^{n-1} (k+1)|\Delta^2\mu_{n-k-1}^{(n)}|\cdot|\tilde{\phi}(0)-\tilde{\sigma}_n^*(\phi;k;0)| +$$

$$+(n+1)|\mu_1^{(n)}|\cdot|\tilde{\phi}(0)-\tilde{\sigma}_n^*(\phi;n;0)| + O\left(\frac{1}{n^{r+\alpha}}\right),$$

where the $\tilde{\sigma}_n^*(\phi;k;x)$ are polynomials conjugate to 8.4(4). From this in virtue of theorem 8.4.4 we obtain the first of the inequalities of 8.5(22).

8.5.73. Let r be odd and $\alpha=0$. In this case we commence with 8.5(7), apply Abel's transformation and the estimate 8.4(58).

8.5.74. The second of the inequalities 8.5(22) in the case when $r > 0$, $\alpha=0$, follows from similar considerations, based on the integral representation 8.4(51).

8.5.75. In the case when $r=0 (0 < \alpha < 1)$, we apply Abel's transformation to 8.5(9) and 8.5(10), after which we use theorem 8.4.4.

8.5.8. We note separately the particular case of theorem 8.5.6 when $r = 0$. *If the system of numbers* $\lambda_k^{(n)}$ $(k = 0, 1, \ldots, n+1;\ \lambda_{n+1}^{(n)}=0;\ \lambda_0^{(n)} = 1)$

for any n is convex or concave and $1-\lambda_1^{(n)} = O\left(\frac{1}{n}\right)$, *then in the case when* $0 < \alpha < 1$

$$\left.\begin{array}{l}\mathcal{E}_n(H_*^{(\alpha)} M; x; \lambda)\\ \mathcal{E}_n(\tilde{H}_*^{(\alpha)} M; x; \lambda)\end{array}\right\}$$

$$= \frac{2^{\alpha+1} M}{\pi^2 n^\alpha} \int_0^{\frac{\pi}{2}} t^\alpha \sin t \, dt \sum_{k=1}^{n} \frac{\lambda_k^{(n)}}{n-k+1} + O\left(\frac{1}{n^\alpha}\right). \quad (24)$$

It is easily verified that the estimate 8.5(24) no longer applies in the general case if $\alpha=1$. It is sufficient, for example, to consider the matrix 8.4(6) for $k=n$. Since the sums $\sigma_n^*(f; n; x)$ (see 8.4(4) and 8.2(35)) can be represented in the form of the integral

$$\sigma_n^*(f; n; x) = \frac{1}{(n+1)\pi} \int_{-\pi}^{\pi} f(x+t) \frac{\sin^2 \frac{n+1}{2} t}{2 \sin^2 \frac{t}{2}} dt, \quad (25)$$

then for the periodic function $f(t) = M|t|$ of period 2π

$$\sigma_n^*(f; n; 0) - f(0) = \frac{M}{(n+1)\pi} \int_0^{\pi} t \frac{\sin^2 \frac{n+1}{2} t}{\sin^2 \frac{t}{2}} dt$$

$$> \frac{4M}{(n+1)\pi} \int_0^{\pi} \frac{\sin^2 \frac{n+1}{2} t}{t} dt = \frac{2 M \ln n}{(n+1)\pi} + O\left(\frac{1}{n}\right)$$

for $|t| \leqslant \pi$.

Therefore the upper bound $\mathcal{E}_n(H^{(1)} M; x; \lambda)$ corresponding to such a matrix λ cannot be of order $O(1/n)$ as $n \to \infty$, as is the right-hand side of 8.5(24) in this case.

8.5.9. It follows from theorem 8.5.6 that *if the matrix* λ *possesses the properties* 8.5(15) *and* 8.5(16) *for any n and* $\mu_1^{(n)} = O\left(\frac{1}{n^{r+1}}\right)$ *then the condition* 8.3(48) *is not only necessary* (see section 8.3.8) *but also sufficient for the corresponding process of approximation* 8.1(8) *to produce the best order of approximation by trigonometric polynomials on the classes* $W_*^{(r)} H^{(\alpha)} M (0 \leqslant \alpha < 1)$ *or* $\tilde{W}_*^{(r)} H^{(\alpha)} M$ $(0 \leqslant \alpha < 1;\ r+\alpha > 0)$.

8.5.10. We also note here that in virtue of proposition 8.4.6 the assertions of theorems 8.5.2, 8.5.3, 8.5.5 and 8.5.9, and also the relation 8.3(48) in theorem 8.3.8, remain valid after the replacement of r by any non-negative number $\varrho \leqslant r$. This is immediately obvious.

8.5.11. For the matrices $\lambda_k^{(n)}$ which satisfy the conditions of theorem 8.5.8, a relation similar to 8.2(56) holds between the upper bound $E_n(H_*^{(\alpha)}M; x; \lambda)$ and the upper bound

$$\tilde{\mathcal{E}}_n(H_*^{(\alpha)}M; x, \lambda) = \sup_{f \in H_*^{(\alpha)}M} |f(x) - V_n^*(f, n, \lambda)|, \tag{26}$$

where the $V_n^*(f, x, \lambda)$ are the polynomials 8.1(26).

The following theorem is valid**:

If the system of numbers $\lambda_k^{(n)} (k = 0, 1, \ldots, n+1;\ \lambda_0^{(n)} = 1;\ \lambda_{n+1}^{(n)} = 0)$ *is convex for any* n $(\Delta^2 \lambda_k^{(n)} \leqslant 0;\ k = 0, 1, \ldots, n-1)$ *and* $1 - \lambda_1^{(n)} = O\left(\frac{1}{n}\right)$, *then for the case* $0 < \alpha < 1$ *the relation*

$$\tilde{\mathcal{E}}_n(H_*^{(\alpha)}M; x, \lambda) = \left(\frac{\pi}{2}\right)^{\alpha+1} \frac{\left|\sin\left(n+\frac{1}{2}\right)x\right|}{\int\limits_0^{\frac{\pi}{2}} t^\alpha \sin t\, dt} \mathcal{E}_n(H_*^{(\alpha)}M; x, \lambda) + O\left(\frac{1}{n^\alpha}\right). \tag{27}$$

is satisfied.

The case $\alpha = 0$ corresponds to the relation 8.2(56).

We note that the upper bound 8.5(26) has a period equal to $\frac{2\pi}{N}$, and therefore, without loss of generality, it is possible to suppose that $0 \leqslant x < \frac{2\pi}{N}$ $(N = 2n+1)$.

Moreover, in calculating it we can put $M = 1$ and replace the class $H_*^{(\alpha)}1$ by its subclass $H_{*0}^{(\alpha)}$ of functions which vanish at the point $x = 0$.

8.5.111. As a preliminary we establish the following proposition.

For the upper bound $\tilde{\mathcal{E}}_n(k, x) = \sup\limits_{f \in H_{*0}^{(\alpha)}} \tau_n^*(f, k, x)$, *where the* $\tau_n^*(f, k, x)$ *are the polynomials* 8.2(57), *the asymptotic equality*

$$\tilde{\mathcal{E}}_n(k, x) = \frac{1}{\pi^{1-\alpha} n^\alpha} \left|\sin\left(n+\frac{1}{2}\right)x\right| \ln\frac{n}{k+1} + O\left(\frac{1}{n^\alpha}\right) \tag{28}$$

is satisfied as $n \to \infty$, *uniformly with respect to* x *and all* k $(0 \leqslant k \leqslant n)$.

** I. M. Ganzburg [3].

It follows from 8.2(59) that

$$\tau_n^*(f, k, x) = \frac{1}{N(k+1)}\left\{\sum_{\nu=1}^{n} f(y_\nu^{(n)}) \frac{\sin\frac{2n-k+1}{2}(x-y_\nu^{(n)})\sin\frac{k+1}{2}(x-y_\nu^{(n)})}{\sin^2\frac{x-y_\nu^{(n)}}{2}}\right.$$

$$\left.+\sum_{\nu=1}^{n} f(y_{-\nu}^{(n)}) \frac{\sin\frac{2n-k+1}{2}(x+y_\nu^{(n)})\sin\frac{k+1}{2}(x+y_\nu^{(2)})}{\sin^2\frac{x+y_\nu^{(n)}}{2}}\right\}+O\left(\frac{1}{n^\alpha}\right)$$

$$= \tau_{n_1}^*(f, k, x)+\tau_{n_2}^*(f, k, x)+O\left(\frac{1}{n^\alpha}\right) \quad \left(y_\nu^{(n)} = \frac{2\nu\pi}{N}\right).$$

Let us consider $\tau_{n_1}^*(f, k, x)$. Taking into account that for $f(x) \in H_{*0}^{(\alpha)}$ $(0 < \alpha < 1)$ the relations

$$\sum_{\nu=1}^{n} f(y_\nu^{(n)}) \frac{\left|\sin\frac{k+1}{2}(y_\nu^{(n)}-x)-\sin\frac{k+1}{2}y_\nu^{(n)}\right|}{\nu^2} = O\left(\frac{k+1}{n^{\alpha+1}}\right),$$

$$\frac{1}{k+1}\sum_{\nu=1}^{n} f(y_\nu^{(n)}) \sin\left(\frac{k+1}{2}y_\nu^{(n)}\right)\times$$

$$\times\left\{\frac{1}{\sin^2\frac{x-y_\nu^{(n)}}{2}}-\frac{1}{\sin^2\frac{y_\nu^{(n)}}{2}}\right\} = O(n^{1-\alpha}),$$

when $0 \leqslant x < \frac{2\pi}{N}$

are satisfied, we find

$$\tau_{n_1}^*(f, k, x) = S_1+S_2+O\left(\frac{1}{n^\alpha}\right),$$

where

$$S_1 = \frac{\sin\left(n+\frac{1}{2}\right)x}{N(k+1)} \sum_{\nu=1}^{n} (-1)^{\nu+1} f(y_\nu^{(n)})\times$$

$$\times\frac{\cos\frac{k}{2}\left(\frac{2\nu\pi}{N}-x\right)\sin\left(\frac{k+1}{N}\nu\pi\right)}{\sin^2\frac{\nu\pi}{N}},$$

$$S_2 = \frac{\cos\left(n+\frac{1}{2}\right)x}{N(k+1)} \sum_{\nu=1}^{n} (-1)^{\nu+1} f(y_\nu^{(n)}) \times$$

$$\times \frac{\sin\frac{k}{2}\left(\frac{2\nu\pi}{N}-x\right)\sin\left(\frac{k+1}{N}\nu\pi\right)}{\sin^2\frac{\nu\pi}{N}},$$

We prove that $S_2 = O\left(\frac{1}{n^\alpha}\right)$. In fact,

$$|S_2| \leqslant \frac{\left|\cos\left(n+\frac{1}{2}\right)x\right|}{N(k+1)} \sum_{\nu=1}^{\left[\frac{n}{2}\right]} |f(y_{2\nu}^{(n)}) - f(y_{2\nu-1}^{(n)})| \left| \frac{\sin\frac{2\nu k\pi}{N}\sin\frac{2\nu(k+1)\pi}{N}}{\sin^2\frac{2\nu\pi}{N}} \right|$$

$$+ \frac{\left|\cos\left(n+\frac{1}{2}\right)x\right|}{N(k+1)} \sum_{\nu=1}^{\left[\frac{n}{2}\right]} |f(y_{2\nu-1}^{(n)})| \left| \frac{\sin\frac{(2\nu-1)k\pi}{N}\sin\frac{(2\nu-1)(k+1)\pi}{N}}{\sin^2\frac{2\nu-1}{N}\pi} \right.$$

$$\left. - \frac{\sin\frac{2\nu k\pi}{N}\sin\frac{2\nu(k+1)\pi}{N}}{\sin^2\frac{2\nu\pi}{N}} \right| + O\left(\frac{1}{n^\alpha}\right) = S_2' + S_2'' + O\left(\frac{1}{n^\alpha}\right).$$

Since $\sum\limits_{\nu=\left[\frac{n}{k+1}\right]}^{\infty} \frac{1}{\nu^2} = O\left(\frac{k+1}{N}\right)$, it follows that

$$S_2' \leqslant \frac{1}{N(k+1)} \sum_{\nu=1}^{\left[\frac{n}{k+1}\right]} O\left(\frac{1}{n^\alpha}\right) O(k^2) = O\left(\frac{1}{n^\alpha}\right),$$

$$S_2'' \leqslant \frac{1}{N(k+1)} \left\{ \sum_{\nu-1}^{\left[\frac{n}{2}\right]} \frac{|f(y_{2\nu-1}^{(n)})|}{\sin^2\frac{2\nu\pi}{N}} \times \right.$$

$$\times \left| \sin\frac{(2\nu-1)k\pi}{N}\sin\frac{(2\nu-1)(k+1)\pi}{N} - \sin\frac{2\nu k\pi}{N}\sin\frac{2\nu(k+1)\pi}{N} \right| +$$

$$+\sum_{\nu=1}^{\left[\frac{n}{2}\right]} |f(y_{2\nu-1}^{(n)})| \left| \sin\frac{(2\nu-1)k\pi}{N} \sin\frac{(2\nu-1)(k+1)\pi}{N} \right| \times$$

$$\times \left| \frac{1}{\sin^2 \frac{2\nu-1}{N}\pi} - \frac{1}{\sin^2 \frac{2\nu\pi}{N}} \right| \Bigg\} = O\left(\frac{1}{n^\alpha}\right).$$

Hence

$$S_2 = O\left(\frac{1}{n^\alpha}\right).$$

From the last estimate it follows that

$$\tau_{n_1}^*(f, k, x) = S_1 + O\left(\frac{1}{n^\alpha}\right)$$

$$= \frac{\sin\frac{2n+1}{2}x}{N(k+1)} \sum_{\nu=1}^{\left[\frac{n}{2}\right]} \left\{ f(y_{2\nu-1}^{(n)}) \frac{\cos\frac{(2\nu-1)k\pi}{N} \sin\frac{(2\nu-1)(k+1)\pi}{N}}{\sin^2\frac{2\nu-1}{N}\pi} - \right.$$

$$\left. - f(y_{2\nu}^{(n)}) \frac{\cos\frac{2\nu k\pi}{N} \sin\frac{2\nu(k+1)\pi}{N}}{\sin^2\frac{2\nu\pi}{N}} \right\} + O\left(\frac{1}{n^\alpha}\right).$$

Reasoning exactly as in the case of the estimation of S_2, we obtain

$$\tau_{n_1}^*(f, k, x) = \frac{\sin\left(n+\frac{1}{2}\right)x}{N(k+1)} \sum_{\nu=1}^{\left[\frac{n}{k+1}\right]} \{f(y_{2\nu-1}^{(n)}) - f(y_{2\nu}^{(n)})\} \times$$

$$\times \frac{\cos\frac{2\nu k\pi}{N} \sin\frac{2\nu(k+1)\pi}{N}}{\sin^2\frac{2\nu\pi}{N}} + O\left(\frac{1}{n^\alpha}\right)$$

$$= \frac{\sin\left(n+\frac{1}{2}\right)x}{2N(k+1)} \sum_{\nu=1}^{\left[\frac{n}{k+1}\right]} \{f(y_{2\nu-1}^{(n)}) - f(y_{2\nu}^{(n)})\} \times$$

$$\times \frac{\sin\frac{4\nu(k+1)\pi}{N}}{\sin^2\frac{2\nu\pi}{N}} + O\left(\frac{1}{n^\alpha}\right).$$

Hence, using the relations

$$\frac{1}{\sin^2\alpha} - \frac{1}{\alpha^2} = O(1)\left(|\alpha| \leqslant \frac{3\pi}{4}\right), \qquad \frac{\sin\alpha}{\alpha^2} - \frac{1}{\alpha} = O(1),$$

we have

$$\tau^*_{n_1}(f, k, x) = \frac{\sin\left(n+\frac{1}{2}\right)x}{2\pi} \sum_{\nu=1}^{\left[\frac{n}{k+1}\right]} \frac{f(y^{(n)}_{2\nu+1}) - f(y^{(n)}_{2\nu})}{\nu} + O\left(\frac{1}{n^\alpha}\right).$$

Similar estimates for $\tau^*_{n_2}(f, k, x)$ with the replacement of ν by $-\nu$ show that

$$\tau^*_{n_2}(f, k, x) = \frac{\sin\left(n+\frac{1}{2}\right)x}{2\pi} \sum_{\nu=1}^{\left[\frac{n}{k+1}\right]} \frac{f(y^{(n)}_{-2\nu}) - f(y^{(n)}_{-2\nu+1})}{\nu} + O\left(\frac{1}{n^\alpha}\right).$$

Consequently

$$\tau^*_n(f, k, x) = \frac{\sin\left(n+\frac{1}{2}\right)x}{2\pi} \times$$

$$\times \sum_{\nu=1}^{\left[\frac{n}{k+1}\right]} \frac{f(y^{(n)}_{2\nu-1}) - f(y^{(n)}_{2\nu}) + f(y^{(n)}_{-2\nu}) - f(y^{(n)}_{-2\nu+1})}{\nu} + O\left(\frac{1}{n^\alpha}\right). \tag{29}$$

Since $f(x) \in H^{(\alpha)}_{*0}$, we obtain the inequality

$$|\tau^*_n(f, k, x)| \leqslant \frac{\left|\sin\left(n+\frac{1}{2}\right)x\right|}{\pi^{1-\alpha} n^\alpha} \ln \frac{n}{k+1} + O\left(\frac{1}{n^\alpha}\right). \tag{30}$$

On the other hand, the periodic function of period 2π

$$f_n(t) = \begin{cases} 0, \quad 0 \leqslant t \leqslant \frac{2\pi}{N}, \\ -\frac{1}{2}\omega\left(t - \frac{2\pi}{N}\right), \quad \frac{2\pi}{N} \leqslant t \leqslant \frac{4\pi}{N}, \\ (-1)^{\mu+1}\left\{\frac{1}{2}\omega\left(\frac{2\pi}{N}\right) - \omega\left(t - \frac{2\mu\pi}{N}\right)\right\}, \\ \frac{2\mu\pi}{N} \leqslant t \leqslant \frac{2(\mu+1)\pi}{N} \quad (\mu = 2, 3, \ldots, 4), \\ -\frac{1}{2}\omega(-t), \quad -\frac{2\pi}{N} \leqslant t \leqslant 0, \\ (-1)^{\mu}\left\{\frac{1}{2}\omega\left(\frac{2\pi}{N}\right) - \omega\left(t - \frac{2\mu\pi}{N}\right)\right\}, \\ \frac{2\mu\pi}{N} \leqslant t \leqslant \frac{2(\mu+1)\pi}{N} \quad (\mu = -2, -3, \ldots, -n), \end{cases} \tag{31}$$

where $\omega(u) = u^\alpha$, belongs to $H_{*0}^{(\alpha)}$, and for this function the inequality 8.5(30) becomes an equality.

8.5.112. Applying Abel's transformation twice to $V_n^*(f, x, \lambda)$, we obtain

$$V_n^*(f, x, \lambda) = -\sum_{k=0}^{n-1} (k+1)\, \Delta^2\lambda_{n-k-1}\tau_n^*(f, k, x)+ \\ +(n+1)\, \Delta\lambda_0^{(n)}\tau_n^*(f, n, x). \quad (32)$$

Since for every trigonometric polynomial $R_n(x)$ of order n we have $\tau_n^*(R_n, n, x) = \sigma_n^*(R_n, n, x)$, where $\sigma_n^*(f, k, x)$ is the polynomial 8.4(4), we have in virtue of 5.1(10)

$$|\tau_n^*(f, n, x)-\sigma_n^*(f, n, x)| \\ \leqslant \{\|\tau_n^*\|_C+\|\sigma_n^*\|_C\}\, E_n^*(f) = 2E_n^*(f) = O\left(\frac{1}{n^\alpha}\right).$$

Thus from 8.4(5) it follows that**

$$\tau_n^*(f, n, x) = O\left(\frac{1}{n^\alpha}\right).$$

Returning to 8.5(32), and taking into consideration the conditions of the theorem and the relation 8.5(28), we obtain

$$|V_n^*(f, x, \lambda)| \leqslant -\frac{\left|\sin\left(n+\frac{1}{2}\right)x\right|}{\pi^{1-\alpha}n^\alpha}\sum_{k=0}^{n-1}(k+1)\Delta^2\lambda_{n-k-1}\ln\frac{n}{k+1}+ \\ +O\left(-\frac{1}{n^\alpha}\sum_{k=0}^{n-1}(k+1)\Delta^2\lambda_{n-k-1}^{(n)}\right)+O\left(\frac{1}{n^\alpha}\right) \\ = \frac{\left|\sin\left(n+\frac{1}{2}\right)x\right|}{\pi^{1-\alpha}n^\alpha}\sum_{k=1}^{n}\frac{\lambda_{n-k}^{(n)}}{k}+O\left(\frac{1}{n^\alpha}\right). \quad (33)$$

Since the function $f_n(x)$ (see 8.5(31)) does not depend on k, it turns the inequality 8.5(33) into an equality***.

Consequently, *the asymptotic expansion*

$$\tilde{\mathscr{E}}_n(H_*^{(\alpha)}M; x; \lambda) = \frac{M}{\pi^{1-\alpha}n^\alpha}\left|\sin\left(n+\frac{1}{2}\right)x\right| \cdot \left|\sum_{k=0}^{n}\frac{\lambda_k^{(n)}}{n-k+1}\right|+O\left(\frac{1}{n^\alpha}\right) \quad (34)$$

holds for the upper bound 8.5(26) *as* $n \to \infty$.

** S. M. Nikol'skii [2].

*** This circumstance also plays an important part in the reasoning of section 8.2.43.

From 8.5(34) and 8.5(24) the relation 8.5(27) follows.

8.5.12. If the system of numbers $\mu_k^{(n)} = \dfrac{1-\lambda_k^{(n)}}{k}$ is non-decreasing and convex, then a relation of the type 8.5(27) can be extended to the class H_ω^*, where $\omega(t)$ is an arbitrary convex modulus of continuity.

The following theorem holds**:

If the matrix $\lambda_k^{(n)}$ is such that for any n the system of numbers $\mu_k^{(n)} (k = 0, 1, \ldots, n+1)$ satisfies condition 8.5(16) *and is non-decreasing, then the relation*

$$\tilde{\mathcal{E}}_n(H_\omega^*, x, \lambda) = \frac{\pi}{2} \frac{\omega\left(\dfrac{2\pi}{2n+1}\right)\left|\sin\left(n+\dfrac{1}{2}\right)x\right|}{\displaystyle\int_0^{\frac{\pi}{2}} \omega\left(\frac{4t}{2n+1}\right)\sin t \, dt} \times$$

$$\times \mathcal{E}_n(H_\omega^*, x, \lambda) + O\left(\omega\left(\frac{1}{n}\right)\right) \qquad (35)$$

is satisfied, where

$$\tilde{\mathcal{E}}_n(H_\omega^*, x, \lambda) = \sup_{f \in H_\omega^*} |f(x) - \tilde{V}_n^*(f, x, \lambda)|. \qquad (36)$$

The relation 8.5(35) follows directly from the following proposition***:

For matrices $\lambda_k^{(n)}$ which satisfy the conditions of the preceding theorem, and for any modulus of continuity $\omega(t)$, the following asymptotic equality holds:

$$\tilde{\mathcal{E}}_n(H_\omega^*, x, \lambda) = \frac{1}{\pi}\omega\left(\frac{2\pi}{2n+1}\right)\left|\sin\left(n+\frac{1}{2}\right)x\right| \times$$

$$\times \left|\sum_{k=0}^{n} \frac{\lambda_k^{(n)}}{n-k+1}\right| + O\left(\omega\left(\frac{1}{n}\right)\right). \qquad (37)$$

If $\omega(t)$ is an arbitrary convex modulus of continuity, then the asymptotic relation

$$\mathcal{E}_n(H_\omega^*, x, \lambda) = \frac{2}{\pi^2}\int_0^{\frac{\pi}{2}} \omega\left(\frac{4t}{2n+1}\right)\sin t \, dt \times$$

$$\times \left|\sum_{k=0}^{n} \frac{\lambda_k^{(n)}}{n-k+1}\right| + O\left(\omega\left(\frac{1}{n}\right)\right) \qquad (38)$$

is satisfied.

** I. M. Ganzburg [3].

*** I. M. Ganzburg [3]. For the case $\lambda_k^{(n)} = 1$, see S. M. Nikol'skii [2], [14].

8.5.121. As in section 8.5.11, we shall suppose in the calculation of $\tilde{\mathcal{E}}_n(H^*, x, \lambda)$ that $0 \leqslant x < \dfrac{2\pi}{N}$ and $f(x) = 0$.

Let us consider

$$\tilde{I}_n \equiv \tilde{I}_n(x, \lambda) = T_n(x) - V_n^*(f, x, \lambda), \tag{39}$$

where the $T_n(x)$ are the polynomials 8.1(22). It is obvious that

$$\tilde{I}_n = \frac{2}{N} \sum_{k=1}^{n} (1-\lambda_k^{(n)}) \sum_{\nu=0}^{2n} f(y_\nu^{(n)}) \cos k(x - y_\nu^{(n)}).$$

Applying Abel's transformation to the inner sum and taking account of the fact that $\cos ky_\nu^{(n)} = \cos \nu y_k^{(n)}$, we obtain

$$\tilde{I}_n = \frac{1}{N} \sum_{\nu=0}^{2n-1} \Delta f(y_\nu^{(n)}) \sum_{k=1}^{n} \frac{1-\lambda_k^{(n)}}{\sin \dfrac{k\pi}{N}} \sin k\left(\frac{2\nu+1}{N}\pi - x\right) + O\left(\omega\left(\frac{1}{n}\right)\right)$$

$$= \frac{1}{N} \sum_{\nu=0}^{2n-1} \Delta f(y_\nu^{(n)}) \sigma_\nu^{(n)} + O\left(\omega\left(\frac{1}{n}\right)\right),$$

where

$$\Delta f(y_\nu^{(n)}) = f(y_\nu^{(n)}) - f(y_{\nu+1}^{(n)}).$$

Carrying out Abel's transformation once more, we have

$$\sigma_\nu^{(n)} = \sum_{k=0}^{n-1} \Delta^2 y_{n-k-1} \frac{\cos \dfrac{2n-k+1}{2}\left(\dfrac{2\nu+1}{N}\pi - x\right) \sin \dfrac{k+1}{2}\left(\dfrac{2\nu+1}{N}\pi - x\right)}{2\sin^2 \dfrac{1}{2}\left(\dfrac{2\nu+1}{N}\pi - x\right)} +$$

$$+ \gamma_1^{(n)} \frac{\sin (n+1)\left(\dfrac{2\nu+1}{N}\pi - x\right)}{4\sin^2 \dfrac{1}{2}\left(\dfrac{2\nu+1}{N}\pi - x\right)} + (-1)^{\nu+1} \gamma_{n+1}^{(n)} \frac{\sin \dfrac{2n+1}{2} x}{2\sin \dfrac{1}{2}\left(\dfrac{2\nu+1}{N}\pi - x\right)},$$

where

$$\gamma_k^{(n)} = \frac{1-\lambda_k^{(n)}}{\sin \dfrac{k\pi}{N}} \quad (k = 1, 2, \ldots, n+1); \qquad \gamma_0^{(n)} = 0.$$

Since in virtue of the conditions of the theorem we have $\gamma_1^{(n)} = O\left(\frac{1}{n}\right)$, it follows that

$$\tilde{I}_n = \sum_{k=0}^{n-1} (k+1)\Delta^2\gamma_{n-k-1}V_{n,k}(f, x) + W_n(f, x) + O\left(\omega\left(\frac{1}{n}\right)\right), \tag{40}$$

where

$$V_{n,k}(f, x) = \frac{1}{N(k+1)} \sum_{\nu=0}^{2n-1} \Delta f(y_\nu^{(n)}) \times$$

$$\times \frac{\cos\frac{2n-k+1}{2}\left(\frac{2\nu+1}{N}\pi - x\right)\sin\frac{k+1}{2}\left(\frac{2\nu+1}{N}\pi - x\right)}{2\sin^2\frac{1}{2}\left(\frac{2\nu+1}{N}\pi - x\right)}, \tag{41}$$

$$W_n(f; x) = \frac{\gamma_{n+1}^{(n)}}{N}\sin\left(n+\frac{1}{2}\right)x \times$$

$$\times \sum_{\nu=0}^{2n-1}{}' (-1)^{\nu+1}\Delta f(y_\nu^{(n)}) \frac{1}{2\sin\frac{1}{2}\left(\frac{2\nu+1}{N}\pi - x\right)}. \tag{42}$$

We will show that $W_n(f, x) = T_n(x) + O\left(\omega\left(\frac{1}{n}\right)\right)$.

In fact, it follows from 8.5(42) and 8.1(22) that

$$W_n(f, x) = \frac{\sin\left(n+\frac{1}{2}\right)x}{2N}\left\{\sum_{\nu=1}^{n} \frac{(-1)^{\nu+1}\Delta f(y_\nu^{(n)})}{\sin\frac{\nu\pi}{N}} + \right.$$

$$\left. + \sum_{\nu=1}^{n} \frac{(-1)^{\nu}\Delta f(y_{-\nu}^{(n)})}{\sin\frac{\nu\pi}{N}}\right\} + O\left(\omega\left(\frac{1}{n}\right)\right)$$

$$= W_n'(f, x) + W_n''(f, x) + O\left(\omega\left(\frac{1}{n}\right)\right);$$

$$T_n(x) = \frac{\sin\left(n+\frac{1}{2}\right)x}{N}\left\{\sum_{\nu=1}^{n}\frac{(-1)^{\nu+1}f(y_\nu^{(n)})}{\sin\frac{\nu\pi}{N}}+\right.$$

$$\left.+\sum_{\nu=1}^{n}\frac{(-1)^{\nu}f(y_{-\nu}^{(n)})}{\sin\frac{\nu\pi}{N}}\right\}+O\left(\omega\left(\frac{1}{n}\right)\right)$$

$$= T_n'(x)+T_n''(x)+O\left(\omega\left(\frac{1}{n}\right)\right);$$

$$W_n'(f, x)-T_n'(x) = \frac{\sin\left(n+\frac{1}{2}\right)x}{2N}\sum_{\nu=2}^{n}f(y_\nu^{(n)})\alpha_\nu^{(n)},$$

where

$$\alpha_\nu^{(n)} = (-1)^{\nu+1}\left\{\frac{1}{\sin\frac{\nu-1}{N}\pi}-\frac{1}{\sin\frac{\nu\pi}{N}}\right\}.$$

Considering that $|\alpha_\nu^{(n)}| \geqslant |\alpha_{\nu+1}^{(n)}|$, $\operatorname{sign}\alpha_\nu^{(n)} = (-1)^{\nu+1}$ $(\nu = 2, 3, \ldots, n)$ we obtain after Abel's transformation

$$W_n'(f, x)-T_n'(x) = -\frac{\sin\left(n+\frac{1}{2}\right)x}{2N}\sum_{\nu=3}^{n}\Delta f(y_{\nu-1}^{(n)})\,G_\nu+f(x_2)\,G_2,$$

where $G_\nu = \sum_{i=\nu}^{n}\alpha_\nu^{(n)}$.

It follows from the inequalities $|G_\nu| \leqslant |\alpha_\nu^{(n)}|$ $(\nu = 2, 3, \ldots, n)$, and $f(x_2) = O\left(\omega\left(\frac{1}{n}\right)\right)$, that

$$|W_n'(f; x)-T_n'(x)| \leqslant \frac{1}{n}\,\omega\left(\frac{1}{n}\right)\sum_{\nu=3}^{n}|\alpha_\nu^{(n)}|+O\left(\omega\left(\frac{1}{n}\right)\right) = O\left(\omega\left(\frac{1}{n}\right)\right).$$

Similarly

$$W_n''(f, x)-T_n''(x) = O\left(\omega\left(\frac{1}{n}\right)\right).$$

Consequently

$$W_n(f, x) = T_n(x)+O\left(\omega\left(\frac{1}{n}\right)\right). \tag{43}$$

We now consider $V_{n,k}(f, x)$. From 8.5(41) we have

$$V_{n,k}(f, x) = V'_{n,k}(f, x) - V''_{n,k}(f, x), \tag{44}$$

where

$$V'_{n,k}(f, x) = \frac{1}{N(k+1)} \sum_{\nu=1}^{n} \Delta f(y_\nu^{(n)}) \times$$
$$\times \frac{\cos \frac{2n-k+1}{2}\left(\frac{2\nu+1}{N}\pi - x\right) \sin \frac{k+1}{2}\left(\frac{2\nu+1}{N}\pi - x\right)}{2\sin^2 \frac{1}{2}\left(\frac{2\nu+1}{N}\pi - x\right)};$$

$$V''_{n,k}(f, x) = \frac{1}{N(k+1)} \sum_{\nu=1}^{n} \Delta f(y_{-\nu}^{(n)}) \times$$
$$\times \frac{\cos \frac{2n-k+1}{2}\left(\frac{2\nu+1}{N}\pi + x\right) \sin \frac{k+1}{2}\left(\frac{2\nu-1}{N}\pi + x\right)}{2\sin^2 \frac{1}{2}\left(\frac{2\nu-1}{N}\pi + x\right)}.$$

It is not difficult to see that

$$V'_{n,k}(f, x) = \frac{N}{2\pi^2(k+1)} \sum_{\nu=1}^{n} \Delta f(y_\nu^{(n)}) \times$$
$$\times \frac{\cos \frac{2n-k+1}{2}\left(\frac{2\nu+1}{N}\pi - x\right) \sin\left(\frac{k+1}{N}\nu\pi\right)}{\nu^2} + O\left(\omega\left(\frac{1}{n}\right)\right)$$
$$= \frac{\sin\left(n+\frac{1}{2}\right)x}{2\pi} \sum_{\nu=1}^{\left[\frac{n}{k+1}\right]} \frac{(-1)^\nu \Delta f(y_\nu^{(n)})}{\nu} + O\left(\omega\left(\frac{1}{n}\right)\right).$$

Similarly

$$V''_{n,k}(f, x) = \frac{\sin\left(n+\frac{1}{2}\right)x}{2\pi} \sum_{\nu=1}^{\left[\frac{n}{k+1}\right]} \frac{(-1)^{\nu+1} \Delta f(y_{-\nu}^{(n)})}{\nu}.$$

From the last two equations and the relations 8.5(39) 8.5(40) 8.5(43), 8.5(44) for the functions $f(x)$ considered, the inequality

$$|V_n^*(f, x, \lambda)|$$
$$\leqslant \frac{1}{\pi}\omega\left(\frac{2\pi}{N}\right)\left|\sin\left(n+\frac{1}{2}\right)x\right| \sum_{k=0}^{n-1} (k+1)\,|\Delta^2\gamma_{n-k-1}| \ln\frac{n}{k+1} +$$
$$+ O\left(\omega\left(\frac{1}{n}\right)\sum_{k=0}^{n-1} (k+1)\,|\Delta^2\gamma_{n-k-1}|\right) + O\left(\omega\left(\frac{1}{n}\right)\right) \tag{45}$$

follows. We note that the system of numbers $\gamma_k^{(n)}$, like the $\mu_k^{(n)}$, satisfy the conditions

$$\Delta^2\gamma_k^{(n)} \geqslant 0; \qquad \gamma_k^{(n)} \leqslant \gamma_{k+1}^{(n)}.$$

Hence it follows from the equation

$$\sum_{i=1}^{k} (i+1)\Delta^2\gamma_{n-i-1} = \gamma_{n+1}^{(n)} - \gamma_{n-k}^{(n)} + (k+1)\Delta\gamma_{n-k+1}^{(n)}$$

and the inequality 8.5(45) that

$$|V_n^*(f, x, \lambda)| \leqslant \frac{1}{\pi}\omega\left(\frac{2\pi}{N}\right)\left|\sin\left(n+\frac{1}{2}\right)x\right| \times$$

$$\times \sum_{k=0}^{n-1} (k+1)\Delta^2\gamma_{n-k-1} \sum_{i=k+1}^{r} \frac{1}{i} + O\left(\omega\left(\frac{1}{n}\right)\right)$$

$$= \frac{1}{\pi}\omega\left(\frac{2\pi}{N}\right)\left|\sin\left(n+\frac{1}{2}\right)x\right| \times$$

$$\times \left|\sum_{k=1}^{n} \frac{1-\lambda_k^{(n)}}{(n-k+1)\sin\dfrac{k\pi}{N}} - \frac{\ln n}{\sin\dfrac{(n+1)\pi}{N}}\right| + O\left(\omega\left(\frac{1}{n}\right)\right). \tag{46}$$

The function $f_n(x)$ (see 8.5(31)), which belongs to the class considered, turns the inequality 8.5(46) into an equality. Consequently we have the asymptotic equality

$$\tilde{\mathcal{E}}_n(H_\omega^*, x, \lambda) = \frac{1}{\pi}\omega\left(\frac{2\pi}{N}\right)\left|\sin\left(n+\frac{1}{2}\right)x\right| \times$$

$$\times \left|\sum_{k=1}^{n} \frac{1-\lambda_k^{(n)}}{(n-k+1)\sin\dfrac{k\pi}{N}} - \frac{\ln n}{\sin\dfrac{(n+1)\pi}{N}}\right| + O\left(\omega\left(\frac{1}{n}\right)\right). \tag{47}$$

It follows from the properties of the sequence $\gamma_k^{(n)}$ that

$$\sum_{k=1}^{n} \frac{1-\lambda_k^{(n)}}{(n-k+1)\sin\dfrac{k\pi}{N}} - \frac{\ln n}{\sin\dfrac{(n+1)\pi}{N}} = -\frac{1}{\sin\dfrac{(n+1)\pi}{N}} \sum_{k=1}^{n} \frac{\lambda_k^{(n)}}{n-k+1} +$$

$$+ \frac{1}{n+1} \sum_{k=1}^{n} \frac{1-\lambda_k^{(n)}}{n-k+1} \left\{\frac{k}{\sin\dfrac{k\pi}{N}} - \frac{n+1}{\sin\dfrac{(n+1)\pi}{N}}\right\} + O(1).$$

Taking account of this equation, and also of 8.5(47), we obtain 8.5(37).

8.5.122. Without loss of generality the upper bound

$$\mathcal{E}_n(H_\omega^*, x, \lambda) = \sup_{f\in H_\omega^*} |f(x)-U_n^*(f, x, \lambda)|$$

can be calculated assuming that $x=0$ and $f(0)=0$.
Let us represent $U_n^*(f, 0, \lambda)$ in the form

$$U_n^*(f, 0, \lambda) = U_n^*(f, 0, \lambda)-S_n^*(f, 0)+S_n^*(f, 0),$$

where $S_n^*(f, x)$ is the Fourier sum of order n of the function $f(x)$.
Consider the difference

$$I_n \equiv I_n(f, \lambda) = U_n^*(f, 0, \lambda)-S_n^*(f, 0)$$
$$= \frac{1}{\pi}\sum_{k=1}^{n}(1-\lambda_k^{(n)})\sum_{\nu=0}^{2n}\int_0^{\frac{2\pi}{N}} f(t+y_\nu^{(n)})\cos k(t+y_\nu^{(n)})\,dt,$$

which is the integral analogue of $\tilde{I}_n$ from 8.5.121.

Continuing the reasoning for I_n as for $\tilde{I}_n$ in 8.5.121, we obtain the equation 8.5(38).

We note only that as the extremal function in this case we have the function

$$f_n(x) = \begin{cases} 0, & 0\leqslant x\leqslant \dfrac{2\pi}{N}, \\ \dfrac{1}{2}\omega[2(x-y_\nu^{(n)})](-1)^\nu, & \dfrac{2\nu\pi}{N}\leqslant x\leqslant \dfrac{2\nu+1}{N}\pi, \\ \dfrac{1}{2}\omega[2(y_{\nu+1}^{(n)}-x)](-1)^\nu, & \dfrac{(2\nu+1)\pi}{N}\leqslant x\leqslant \dfrac{2(\nu+1)\pi}{N} \\ & (\nu = 1, 2, \dots, n-1) \\ 0, & -\dfrac{2\pi}{N}\leqslant x\leqslant 0, \\ \dfrac{1}{2}\omega[2(x-y_\nu^{(n)})](-1)^{\nu+1}, & \dfrac{2\nu\pi}{N}\leqslant x\leqslant \dfrac{(2\nu+1)\pi}{N}, \\ \dfrac{1}{2}\omega[2(y_{\nu+1}^{(n)}-x)](-1)^{\nu+1}, & \dfrac{(2\nu+1)\pi}{N}\leqslant x\leqslant \dfrac{2(\nu+1)\pi}{N} \\ & (\nu = -2, -3, \dots, -n-1) \end{cases}$$

of period 2π, which belongs to H_ω^* if $\omega(t)$ is convex, and satisfies the condition $f_n(0) = 0$.

8.6. The approximative properties of orthogonal expansions

The methods of approximation $U_n^*(f; x; \lambda)$ considered in this chapter are based on an expansion in a Fourier series in the trigonometric system of functions $\sin kx, \cos kx\,(k=0, 1, 2, \ldots)$.

We have seen that the best order of approximation by the polynomials $U_n^*(f; x; \lambda)$ which can be attained for functions which have an r-th continuous derivative, when n increases without limit, is equal to $O\left[\frac{1}{n^r}\omega\left(f^{(r)}; \frac{1}{n}\right)\right]$.

It turns out that it is impossible to improve on this order of approximation in the general case, not only of the summation of a trigonometric Fourier series, but also in the class of all other orthogonal expansions.

8.6.1. Let $C^{(r)}$ denote the set of all functions $f(x)$ defined on $[0, 1]$ and having there an r-th ($r \geqslant 0$ an integer) continuous derivative, let $\{\phi_k(x)\}$ $(k=1, 2, \ldots)$ be an arbitrary system of functions orthonormal on $[0, 1]$, and

$$W_n(f; x; \lambda) = \sum_{k=1}^{n} \lambda_k^{(n)} c_k \phi_k(x), \tag{1}$$

where

$$c_k = \int_0^1 f(t)\,\phi_k(t)\,dt. \tag{2}$$

*For any system $\{\phi_k(x)\}$ orthonormal on $[0, 1]$, and for any bounded matrix λ subject to the condition that $|\lambda_k^{(n)}| \leqslant 1$, the inequality***

$$\sup_{f \in C^{(r)}} \frac{\int_0^1 |f(x) - W_n(f; x; \lambda)|\,dx}{\omega\left(f^{(r)}; \frac{1}{n}\right)} \geqslant \frac{B_r}{n^r} \qquad (n = 1, 2, \ldots), \tag{3}$$

in which the constant B_r depends only on r, is always satisfied.

Since

$$\int_0^1 |f(x) - W_n(f; x; \lambda)|\,dx \leqslant \operatorname*{vrai\,sup}_{0 \leqslant x \leqslant 1} |f(x) - W_n(f; x; \lambda)|,$$

and for the periodic function $f(x)$ of period 2π bounded matrices λ exist for which (see sections 8.4 and 5.1.4)

$$|f(x) - U_n^*(f; x; \lambda)| \leqslant \frac{B_r^*}{n^r}\,\omega\left(f^{(r)}; \frac{1}{n}\right),$$

** B. Sz. Nagy [4].

the relation 8.6(3) shows that the trigonometric system 1, $\sqrt{2}\cos k\pi x$ ($k=1, 2, \ldots$), among all other orthonormal systems of functions, is in a certain sense minimising on the classes $C^{(r)}$.

The inequality 8.6(3) follows from the following assertion:

8.6.11. *If* $\{\psi_k(x)\}$ *and* $\{\phi_k(x)\}$ *are any two systems of functions orthonormal on* $[0,1]$, *and* $\{M_k\}$ $(k=1,2,\ldots)$ *is an arbitrary sequence of positive numbers, then, whatever the natural numbers m and n may be, the following inequality holds***:

$$\max_{1\leq k\leq m} \frac{1}{M_k}\int_0^1 |\psi_k(x)-W_n(\psi_k;x;\lambda)|\,dx$$
$$\geq \frac{1}{\sum_{k=1}^m M_k}\left\{\sum_{k=1}^m \int_0^1 |\psi_k(x)|\,dx - \sqrt{mn}\max_{1\leq\nu\leq n}|\lambda_\nu^{(n)}|\right\}. \tag{4}$$

Put

$$K_n(x,t)=\sum_{\nu=1}^n \lambda_\nu^{(n)}\phi_\nu(x)\phi_\nu(t), \qquad c_j^{(k)}=\int_0^1 \psi_k(t)\phi_j(t)\,dt.$$

It is then obvious that

$$W_n(\psi_k;x;\lambda)=\int_0^1 K_n(x,t)\psi_k(t)\,dt$$

and

$$\sum_{j=1}^m \int_0^1 |W_n(\psi_j;x;\lambda)|\,dx=\sum_{j=1}^m \int_0^1\left|\int_0^1 K_n(x,t)\psi_j(t)\,dt\right|dx.$$

Since

$$\sum_{j=1}^m \int_0^1\left|\int_0^1 K_n(x,t)\psi_j(t)\,dt\right|dx \leq \sum_{j=1}^m\left\{\int_0^1\left|\int_0^1 K_n(x,t)\psi_j(t)\,dt\right|^2dx\right\}^{\frac{1}{2}}$$
$$\leq\sqrt{m}\left\{\sum_{j=1}^m \int_0^1\left|\int_0^1 K_n(x,t)\psi_j(t)\,dt\right|^2dx\right\}^{\frac{1}{2}},$$

$$\int_0^1\left|\int_0^1 K_n(x,t)\psi_j(t)\,dt\right|^2dx=\int_0^1\left|\sum_{\nu=1}^n \lambda_\nu^{(n)}c_\nu^{(j)}\phi_\nu(x)\right|^2dx=\sum_{\nu=1}^n|\lambda_\nu^{(n)}c_\nu^{(j)}|^2,$$

** B. Sz. Nagy [4].

and

$$\sum_{j=1}^{m} |c_\nu^{(j)}|^2 \leqslant \int_0^1 |\phi_\nu(x)|^2 \mathrm{d}x = 1,$$

then

$$\sum_{j=1}^{m} \int_0^1 |W_n(\psi_j; x; \lambda)| \mathrm{d}x$$

$$\leqslant \sqrt{m} \left\{ \sum_{j=1}^{m} \sum_{\nu=1}^{n} |\lambda_\nu^{(n)} c_\nu^{(j)}|^2 \right\}^{\frac{1}{2}} \leqslant \sqrt{mn \max_{1 \leqslant \nu \leqslant n} |\lambda_\nu^{(n)}|}.$$

Thus

$$\sum_{j=1}^{m} \int_0^1 |\psi_j(x)| \mathrm{d}x$$

$$\leqslant \sum_{j=1}^{m} \int_0^1 |\psi_j(x) - W_n(\psi_j; x; \lambda)| \mathrm{d}x + \sum_{j=1}^{m} \int_0^1 |W_n(\psi_j; x; \lambda)\ \mathrm{d}x$$

$$\leqslant \max_{1 \leqslant k \leqslant m} \frac{1}{M_k} \int_0^1 |\psi_k(x) - W_n(\psi_k; x; \lambda)| \mathrm{d}x \sum_{j=1}^{m} M_j + \sqrt{mn \max_{1 \leqslant \nu \leqslant n} |\lambda_\nu^{(n)}|},$$

i.e. 8.6(4) is satisfied.

8.6.12. In order to obtain theorem 8.6.1 from this, we consider the system of functions

$$\psi_k(x) = \sqrt{2} \cos k\pi x \qquad (k = 1, 2, 3, \ldots) \tag{5}$$

and put

$$M_k = \omega\left(\psi_k^{(r)}; \frac{1}{n}\right) \leqslant \frac{\sqrt{2}\ (k\pi)^{r+1}}{n}.$$

For any natural number m,

$$\sup_{f \in C^{(r)}} \frac{\int_0^1 |f(x) - W_n(f; x; \lambda)| \mathrm{d}x}{\omega\left(f^{(r)}; \frac{1}{n}\right)}$$

$$\geqslant \max_{1 \leqslant k \leqslant m} \frac{1}{M_k} \int_0^1 |\psi_k(x) - W_n(\psi_k; x; \lambda)| \mathrm{d}x. \tag{6}$$

If $m=2n$, then in virtue of 8.6(4)

$$\max_{1\leqslant k\leqslant m}\frac{1}{M_k}\int_0^1|\psi_k(x)-W_n(\psi_k;x;\lambda)|\,dx$$

$$\geqslant\frac{1}{\sum\limits_{k=1}^{m}M_k}\left\{\sum_{k=1}^{m}\int_0^1|\psi_k(x)|\,dx-\sqrt{mn}\right\}$$

$$\geqslant\frac{n}{\sqrt{2}\pi^{r+1}\sum\limits_{k=1}^{m}k^{r+1}}\left\{\frac{2m\sqrt{2}}{\pi}-\sqrt{mn}\right\}\geqslant\frac{B_r}{n^r},\tag{7}$$

where

$$B_r=\frac{\left(\frac{4}{\pi}-1\right)(r+2)}{\pi^{r+1}3^{r+2}}.$$

From 8.6(6) and 8.6(7) 8.6(3) follows.

8.6.13. Using proposition 8.6.11, it is possible to obtain inequalities of the type 8.6(3) for other classes of functions also. Consider, for example, the class $W^{(r)}H^{(\alpha)}$ of functions $f(x)$ having on [0, 1] an r-th ($r\geqslant 0$ an integer) derivative $f^{(r)}(x)$ which satisfies the Lipschitz condition

$$\sup_{x_1\neq x_2}\frac{|f^{(r)}(x_1)-f^{(r)}(x_2)|}{|x_1-x_2|^{\alpha}}=M_{r,\alpha}(f)<\infty\qquad(0<\alpha\leqslant 1).\tag{8}$$

For any system $\{\phi_k(x)\}$ *orthonormal on* [0, 1], *and for any bounded matrix* λ *subject to the condition that* $|\lambda_k^{(n)}|\leqslant 1$, *the inequality***

$$\sup_{f\in W^{(r)}H^{(\alpha)}}\frac{\int_0^1|f(x)-W_n(f;x;\lambda)|\,dx}{M_{r,\alpha}(f)}\geqslant\frac{B_{r,\alpha}}{n^{r+\alpha}}\tag{9}$$

is satisfied, in which the constant $B_{r,\alpha}$ *depends only on* r *and* α.

Since

$$\sup_{f\in W^{(r)}H^{(\alpha)}}\frac{\int_0^1|f(x)-W_n(f;x;\lambda)|\,dx}{M_{r,\alpha}(f)}$$

$$\geqslant\max_{1\leqslant k\leqslant m}\frac{\int_0^1|\psi_k(x)-W_n(\psi_k;x;\lambda)|\,dx}{M_{r,\alpha}(\psi_k)},$$

** B. Sz. Nagy [4].

where $\{\psi_k(x)\}$ is the system of functions 8.6(5), this estimate follows directly from 8.6(4) for $m=2n$ and

$$M_k = M_{r,\alpha}(\psi_k) \leqslant 2^{\frac{3}{2}-\alpha}(k\pi)^{r+\alpha}.$$

8.6.2. We note that the inequality 8.6(3) or 8.6(9) entails a similar inequality for the deviation $\|f(x)-W_n(f;\,x;\,\lambda)\|_{L_q}$ in any metric $L_q(1 \leqslant q \leqslant \infty)$ and, in particular, for $q=2$ leads to the estimate

$$\sup_{f\in C^{(r)}} \frac{\left\{\int_0^1 |f(x)-\sum_{k=1}^{n} c_k\phi_k(x)|^2 dx\right\}^{\frac{1}{2}}}{\omega\left((f^{(r)};\frac{1}{n}\right)} \geqslant \frac{B_r}{n^r} \tag{10}$$

or

$$\sup_{f\in W^{(r)}H^{(\alpha)}} \frac{\left\{\int_0^1 |f(x)-\sum_{k=1}^{n} c_k\phi_k(x)|^2 dx\right\}^{\frac{1}{2}}}{M_{r,\alpha}(f)} \geqslant \frac{B_{r,\alpha}}{n^{r+\alpha}}. \tag{11}$$

8.6.3. If we take account of the fact that the numerator of the left-hand side of 8.6(10) represents the best quadratic approximation of the function $f(x)$ by the polynomials $\sum_{k=1}^{n} a_k\phi_k(x)$, then 8.6(10) and 8.6(11) can be written in the form

$$\inf_{\phi_k}\sup_{f\in C^r} \frac{\inf_{a_k}\left\{\int_0^1 |f(x)-\sum_{k=1}^{n} a_k\phi_k(x)|^2 dx\right\}^{\frac{1}{2}}}{\omega\left(f^{(r)};\frac{1}{n}\right)} \geqslant \frac{B_r}{n^r},$$

$$\inf_{\phi_k}\ \sup_{f\in W^{(r)}H^{(\alpha)}} \frac{\inf_{a_k}\left\{\int_0^1 |f(x)-\sum_{k=1}^{n} a_k\phi_k(x)|^2 dx\right\}^{\frac{1}{2}}}{M_{r,\alpha}(f)} \geqslant \frac{B_{r,\alpha}}{n^{r+\alpha}}. \tag{12}$$

In this form the result is related to the following general problem formulated by A. N. Kolmogorov [3]. Let F be some linear normed space of functions f and let $W \subset F$. Any linearly independent system of functions $\phi_1, \ldots, \phi_n$ of F determines the upper bound of the best approximations $E_{n,\phi}(f) = \inf_{a_k}\|f-\sum_{k=1}^{n} a_k\phi_k\|_F$ on the class W, i.e. the quantity $E_{n,\phi}(W) = \sup_{f\in W} E_{n,\phi}(f)$.

The problem consists in the investigation of the lower bound $D_n(W) = \inf_{\phi_k} E_{n,\phi}(W)$ for all possible linearly independent systems $\phi_1, \ldots, \phi_n$.

In certain cases A. N. Kolmogorov obtained its exact solution. For example, if F is the space L_2, and W is the class $W_*^{(r)}(L_2)$, which consists of functions differentiable r times for which

$$\int_0^1 |f^{(r)}(x)|^2 dx \leqslant 1, \quad f^{(k)}(0) = f^{(k)}(1) \quad (k = 0, 1, \ldots, r-1),$$

then

$$D_{2n-1}[W_*^{(r)}(L_2)] = D_{2n}[W_*^{(r)}(L_2)] = \left(\frac{1}{2\pi n}\right)^r \quad (n = 1, 2, \ldots),$$

and this minimum is realised by the system $1, \sqrt{2}\sin 2k\pi x, \sqrt{2}\cos 2k\pi x$ $(k = 1, \ldots, n-1)$**.

8.6.4. Using the inequality 6.8(1), it is possible to obtain estimates for the order of decrease of the quantity $D_n(W)$ for certain other classes of functions. Thus, for example, from this inequality and the remark 6.8.23, it follows in virtue of 5.11(1) (see section 5.1.32) that for the set $K\Lambda_\alpha M(L_q^{*(1)})(1 \leqslant q \leqslant \infty)$

$$\frac{A}{(\ln n)^\alpha} \leqslant D_n(K\Lambda_\alpha M(L_q^{*(1)})) \leqslant \frac{B}{(\ln n)^\alpha} \quad (n > 1),$$

where A and B are certain positive constants independent of n.

8.7. Various problems and theorems

1. If $f(x)$ is a continuous function of period 2π, then the trigonometric polynomials

$$T_{n-1}(f; x) = \frac{1}{2n\pi}\int_0^{2\pi} f(t)\left[\frac{\sin\dfrac{n(x-t)}{2}}{\sin\dfrac{x-t}{2}}\right]^2 dt$$

$$= \frac{1}{\pi}\int_0^{2\pi} f(t)\left\{\frac{1}{2} + \sum_{k=1}^{n-1}\frac{n-k}{n}\cos k(x-t)\right\} dt, \qquad (\alpha)$$

$$T_n(f; x) = \frac{1}{2\pi}\,\frac{2.4\ldots 2n}{1.3\ldots(2n-1)}\int_0^{2\pi} f(t)\cos^{2n}\frac{x-t}{2}\,dt, \qquad (\beta)$$

possess the property that $\max_x |f(x) - T_n(f; x)| \to 0$ as $n \to \infty$.

** V. M. Tikhomirov proved in 1959 that in the space C^* of all continuous functions of period 2π,

$$D_n[W_*^{(r)}M] = \frac{MK_r}{(n+1)^r} \quad (n = 0, 1, 2, \ldots)$$

for any natural number r.

Fejér [1], De la Vallée-Poussin [1]. (The non-negative trigonometric polynomials $K_n(x-t)$, which are multiplied by $f(t)$ in the integrals 8.7,1(α) and 8.7,1(β), are such that $\int_0^{2\pi} K_n(t)\,dt = 1$ and for any $\delta > 0$ we have $\max_{\delta \leq |t| \leq \pi} K_n(t) \to 0$.) See also section 8.1.3.

2. If $f(x)$ is a continuous function of period 2π, then the trigonometric polynomials

$$T_n(f;\, x) = \frac{1}{p\pi} \int_0^{2\pi} f(t) \sum_{k=1}^{p} D_n(x-t+\alpha_n^{(k)})\,dt, \tag{γ}$$

where $D_n(t)$ is Dirichlet's kernel (see 4.1(10)), and

$$\text{(a)}\quad \alpha_n^{(k)} = O\left(\frac{1}{n}\right); \qquad \text{(b)}\quad \sum_{i,\,k=1}^{p} \cos\frac{2n+1}{2}(\alpha_n^{(i)} - \alpha_n^{(k)}) = O\left(\frac{1}{\ln^2 n}\right),$$

possess the property $\max_x |f(x) - T_n(f;\, x)| \to 0$ as $n \to \infty$. For the uniform convergence of the sequence of polynomials 8.7.2 (γ) on the class of all the continuous functions $f(x)$ of period 2π, when condition (a) is satisfied, condition (b) is not only sufficient but also necessary.

A. F. Timan and I. M. Ganzburg [1]. For the case $p = 2$, $\alpha_n^{(1)} = \frac{\pi}{2n} + o\left(\frac{1}{n \ln n}\right)$, $\alpha_n^{(2)} = -\frac{\pi}{2n} + o\left(\frac{1}{n \ln n}\right)$, see S. N. Bernstein [10], W. Rogosinski [1]. (The kernel $K_n(t) = \frac{1}{p\pi}\sum_{k=1}^{p} D_n(t+\alpha_n^{(k)})$, in addition to the properties indicated in the preceding theorem, satisfies the relation $\int_0^{2\pi} |K_n(t)|\,dt = O(1)$.)

3. If

$$T_n(f;\, x) = \frac{1}{2\pi}\int_0^{2\pi} f(t)[D_n(x-t+\alpha_n) + D_n(x-t-\alpha_n)]\,dt,$$

then in the case when $\alpha_n = \frac{2q+1}{2n}\pi + O\left(\frac{1}{n \ln n}\right)$, where $q = q(n)$ is a bounded integer-valued function of n,

$$\max_x |f(x) - T_n(f;\, x)| \to 0$$

as $n \to \infty$ for all continuous functions $f(x)$ of period 2π.

A. F. Timan [5]. See section 8.7.2.

4. If

$$T_n(f;\, x) = \frac{1}{3\pi}\int_0^{2\pi} f(t)[D_n(x-t+\beta_n) + D_n(x-t) + D_n(x-t+\gamma_n)]\,dt,$$

then in the case when

$$\beta_n = \frac{2(3r-1)}{3n}\pi + O\left(\frac{1}{n \ln n}\right), \qquad \gamma_n = \frac{2(3s+1)}{3n}\pi + O\left(\frac{1}{n \ln n}\right),$$

where $r = r(n)$ and $s = s(n)$ are bounded integer-valued functions of n,

$$\max_x |f(x) - T_n(f; x)| \to 0 \quad \text{as} \quad n \to \infty.$$

I. M. Ganzburg [1]. See section 8.7.2.

5. Let

$$l_k^{(n)}(x) = \frac{1}{n^2}(1 - x x_k^{(n)}) \left[\frac{\cos n \arccos x}{x - x_k^{(n)}}\right]^2, \qquad x_k^{(n)} = \cos \frac{2k-1}{2n}\pi.$$

If $f(x)$ is continuous on $[-1, 1]$, then the polynomials

$$P_{2n-1}(x) = P_{2n-1}(f; x) = \sum_{k=1}^{n} f(x_k^{(n)}) l_k^{(n)}(x) \tag{1}$$

interpolating it at the points $x_k^{(n)}$ $(k = 1, 2, \ldots, n)$ $\left(P_{2n-1}(x_k^{(n)}) = f(x_k^{(n)})\right)$ and having a zero derivative $(P'_{2n-1}(x_k^{(n)}) = 0)$, possess the property

$$\max_{-1 \leqslant x \leqslant 1} |f(x) - P_{2n-1}(f; x)| \to 0 \quad \text{as} \quad n \to \infty.$$

Fejér [3]. See V. L. Goncharov [1], I. P. Natanson [2]. (The condition $P'_{2n-1}(x_k^{(n)}) = 0$ $(k = 1, 2, \ldots, n)$ can be weakened: see Fejér [5], I. P. Natanson [2].

6. The ratio of the degree of the polynomials 8.7(1) to the number of points at which they interpolate the function $f(x)$ is equal to $\frac{2n-1}{n} \to 2$ as $n \to \infty$. For any $\lambda > 1$ it is possible to find a sequence of polynomials $P_m(x) = P_m(f; x)$ of degree $m < \lambda n$, which interpolate the function $f(x)$ continuous on $[-1, 1]$ at the n points $x_k^{(n)} = \cos \frac{2k-1}{2n}\pi$ and possess the property

$$\max_{-1 \leqslant x \leqslant 1} |f(x) - P_m(f; x)| \to 0 \quad \text{as} \quad m \to \infty.$$

S. N. Bernstein [12]. See I. P. Natanson [2].

7. Whatever the matrix of numbers $\lambda_k^{(n)}$ may be, the inequality

$$A \max_x M_n^*(x) \leqslant L_n \leqslant B \max M_n^*(x)$$

always holds between the Lebesgue constants 8.1(4) and the Lebesgue functions 8.1(30), where B and A are certain positive constants.

See 4.9(3). Marcinkiewicz and Zygmund [1]. See also S. M. Lozinskii [1].

8. If the sequence of polynomials 8.1(8) converges uniformly for any continuous function $f(x)$ of period 2π, then the sequence 8.1(28) possesses the same property, and conversely.

S. M. Lozinskii [1]. See section 8.7.7.

9. If for every n and for $k = 0, 1, \ldots, n-2$ we have $\Delta^2 \lambda_k^{(n)} \geqslant 0$ or $\Delta^2 \lambda_k^{(n)} \leqslant 0$ and condition 8.1(21) for $\nu = n$, condition 8.1(16), condition 8.2(9) and the condition

$$\nu_{n-1}^{(n)} - 2\lambda_n^{(n)} = O(1/\ln n) \tag{2}$$

are satisfied then the sequence of polynomials 8.1(8) converges uniformly on the class of all continuous functions of period 2π.

Karamata and Tomić [1].

10. Whatever the system of points $x_k = x_k^{(n)}$ $(k = 0, 1, \ldots, n)$, situated on $[-1, 1]$, the inequality

$$\max_{-1 \leqslant x \leqslant 1} \sum_{k=0}^{n} |l_{k,n}(x)| \geqslant \frac{2}{\pi}[1+o(1)] \ln n, \tag{3}$$

always holds, as $n \to \infty$ where $l_{k,n}(x)$ are the polynomials 4.1(2).

S. N. Bernstein [13]. In this connection see sections 4.1.2 and 4.12.10.

11. If the points $x_k = x_k^{(n)}$ $(k = 0, 1, \ldots, n;\ x_0^{n} = -1, x_n^{(n)} = 1)$ are equally spaced, then for arbitrarily large values of n

$$\max_{-1 \leqslant x \leqslant 1} \sum_{k=0}^{n} |l_{k,n}(x)| > Ca^n, \tag{4}$$

where C and a are positive constants and $a > 1$.

12. Whatever the system of points $x_k = x_k^{(n)}$ $(k = 0, 1, \ldots, 2n)$ situated on $[0, 2\pi]$,

$$\max_{x} \sum_{k=0}^{2n} |l_{k,n}^{*}(x)| \geqslant \frac{2}{\pi}[1+o(1)] \ln n \tag{5}$$

as $n \to \infty$, where $l_{k,n}^{*}(x)$ are the polynomials 4.2(2).

S. N. Bernstein [13]. In this connection see sections 4.2.1 and 4.12.8.

13. If the function $\lambda(u)$, defined on the segment $0 \leqslant x \leqslant 1$, is decreasing and convex, then for $\lambda_k^{(n)} = \lambda\left(\frac{k}{n+1}\right)$ $(0 \leqslant k \leqslant n)$

$$\tilde{L}_n = \sup_{|f(x)| \leqslant 1} \tilde{U}_n^{*}(f; x; \lambda) = \frac{4}{\pi} \sum_{\nu=0}^{\left[\frac{n-1}{2}\right]} \frac{1}{2\nu+1} \lambda\left(\frac{2\nu+1}{n}\right), \tag{6}$$

where $\tilde{U}_n^{*}(f; x; \lambda)$ are the trigonometric polynomials conjugate to 8.1(8). For $\lambda(u) = (1-u^{\beta})^{\delta}$ $(\beta > 0,\ \delta > 0)$ the sequence of Lebesgue constants L_n (see 8.1(14)) is always bounded.

B. Sz. Nagy [2]. For the case $\beta = 1$ see also Cramér [1].

14. The sequence of algebraic polynomials $P_n(x)$ of degree n which interpolate the function $|x|$ at the equidistant points $x_k^{(n+1)}$ $(k = 0, 1, \ldots, n;\ x_0^{(n+1)} = -1, x_n^{(n+1)} = 1)$ on $[-1, 1]$, diverges at every point x other than the ends or middle of the given segment.

S. N. Bernstein [4]. See I. P. Natanson [2].

15. There exists a function $f(x)$, continuous on $[-1, 1]$, for which the sequence of algebraic polynomials $P_n(x)$ of degree $\leqslant n$, which interpolate it at the Chebyshev points $x_k^{(n+1)} = \cos \frac{2k+1}{2(n+1)} \pi$ $(k = 0, 1, \ldots, n)$, diverges at every point $x \in [-1, 1]$.

Marcinkiewicz [1], Grünwald [1]. See I. P. Natanson [2].

16. There exists a continuous function $f(x)$ of period 2π for which the sequence of trigonometric polynomials $T_n(x)$ of order $\leqslant n$, which interpolate it at the equidistant points $y_k^{(n)} = \frac{2k\pi}{2n+1}$ $(k = 0, 1, 2, \ldots, 2n)$, diverges at any point x.

17. The relation 8.2(56) holds in all cases when $\lambda_k^{(n)}$ is the matrix 8.2(50), where the system of numbers $l_k^{(n)}$ is convex or concave and $a_n = O\left(\frac{1}{n}\right)$.

N. A. Pogodicheva [1].

18. For any integer-valued function $p = p(n)$, the relation 8.2(56) holds between the interpolation functions of Lebesgue $M_n^*(x)$ which correspond to the matrix of numbers $\lambda_k^{(n)} = \cos\frac{2kp\pi}{2n+1}$ $(k = 0, 1, \ldots, n)$ and the Lebesgue constants L_n.

A. F. Timan [1].

19. If $\sigma \geqslant 1$ and $0 < \theta \leqslant 1$, then

$$\frac{1}{\sigma}\int_{-\infty}^{\infty}\frac{|\cos\sigma(1-\theta)t-\cos\sigma t|}{t^2}\,dt = \frac{4\theta}{\pi}\ln\frac{1}{\theta}+O(\theta), \tag{7}$$

uniformly with respect to θ.

A. F. Timan [4].

20. In order that for all continuous functions of period 2π the method of approximation 8.1(8) should produce an approximation which satisfies the inequality

$$\max_x |f(x)-U_n^*(f;x;\lambda)| \leqslant B_\theta(f)E_{[\theta n]}^*(f), \tag{8}$$

where $\theta(0 < \theta < 1)$ is a given number and $B_\theta(f)$ is a constant depending only on f and θ, it is necessary that condition 8.2(9) should be satisfied and that for any n we should have $\lambda_k^{(n)} = 1$ when $0 \leqslant k \leqslant [\theta n]$. In the case when the system of numbers $\lambda_k^{(n)}$ $(k = 0, 1, \ldots, n+1;\ \lambda_{n+1}^{(n)} = 0)$ is convex $(\Delta^2\lambda_k^{(n)} \leqslant 0)$ for every n, these two conditions are also sufficient.

21. There exists a constant $c_0 > 1$ such that for all natural numbers n

$$\sup_f \int_0^{2\pi} f(t)\cos nt\,dt = \frac{4}{n}c_0, \tag{9}$$

where the upper bound extends to all continuous functions of period 2π for which $\omega_2(f;t) \leqslant 2t$.

A. V. Efimov [2].

22. Let $x_k^{(n)}$ $(k = 0, 1, \ldots, n;\ n = 0, 1, 2, \ldots)$ be a given system of points on [-1, 1 and for some β $(0 < \beta < 1)$, for any arbitrarily small $\varepsilon > 0$,

$$\overline{\lim_{n\to\infty}} \max_{-1\leqslant x\leqslant 1} \frac{1}{n^{\beta+\varepsilon}}\sum_{k=0}^{n} |l_{k,n}(x)| < c_1(\varepsilon) < \infty, \tag{10}$$

$$\underline{\lim_{n\to\infty}} \max_{-1\leqslant x\leqslant 1} \frac{1}{n^{\beta-\varepsilon}}\sum_{k=0}^{n} |l_{k,n}(x)| > c_2(\varepsilon) > 0, \tag{11}$$

where $l_{k,n}(x)$ are the polynomials 4.1(2). If $\alpha < \frac{\beta}{\beta+2}$, then there exists a function $f(x)$ continuous on $[-1, 1]$ which satisfies the Lipschitz condition $\omega(f;t) = O(t^\alpha)$ and is such that the sequence of algebraic polynomials $P_n(x)$ which interpolate it at the points $x_k^{(n)}$ is not bounded. If $\alpha > \beta$, then for any function $f(x)$ which satisfies the Lipschitz condition $\omega(f;t) = O(t^\alpha)$ on $[-1, 1]$, the sequence of interpolation polynomials $P_n(x)$ converges uniformly. If $\alpha > \frac{\beta}{\beta+2}$, then there exists a system of points for which conditions 8.7(10) and 8.7(11) are simultaneously satisfied and is such that on the class of all functions $f(x)$ possessing the property $\omega(f;t) = O(t^\alpha)$, the corresponding sequence of interpolation polynomials $P_n(x)$ converges uniformly. If $\alpha < \beta$, then there exists a system of points which satisfies conditions 8.7(10), 8.7(11), and the function $f(x)$ for which $\omega(f; t) = O(t^{\alpha})$ is such that the sequence of polynomials $P_n(x)$ which interpolate $f(x)$ at these points is not bounded.

P. Erdös and P. Turán [1].

23. If $f(x)$ is a function bounded on $(-\infty, \infty)$, then for $\sigma \geqslant 1$ and $0 < \theta \leqslant 1$

$$\sup_{-\infty<x<\infty} \left| f(x) - \frac{1}{\sigma\theta\pi} \int_{-\infty}^{\infty} f(t) \frac{\cos\sigma(1-\theta)(x-t) - \cos\sigma(x-t)}{(x-t)^2} dt \right| \leqslant A_{\sigma(1-\theta)}(f)\left[\frac{4}{\pi^2}\ln\frac{1}{\theta} + O(1)\right], \quad (12)$$

uniformly with respect to θ and σ.

See sections 4.12.11 and 8.7.19. See also N. I. Akhiezer and B. M. Levitan [1].

24. If $\int_{-\infty}^{\infty} |f(t)|\, dt < \infty$, then for $\sigma \geqslant 1$ and $0 < \theta \leqslant 1$

$$\int_{-\infty}^{\infty} \left| f(x) - \frac{1}{\sigma\theta\pi} \int_{-\infty}^{\infty} f(t) \frac{\cos\sigma(1-\theta)(x-t) - \cos\sigma(x-t)}{(x-t)^2} dt \right| dx \leqslant A_{\sigma(1-\theta)}(f)_L \left[\frac{4}{\pi^2}\ln\frac{1}{\theta} + O(1)\right].$$

25. If $H^{(\alpha)}M$ is the class of all functions $f(x)$ defined on $(-\infty, \infty)$ and satisfying the condition $|f(x_1)-f(x_2)| \leqslant M|x_1-x_2|^{\alpha}$, then for any α $(0 < \alpha < 1)$ and as $\sigma \to \infty$

$$\sup_{f\in H^{(\alpha)}M} \left| f(x) - \frac{1}{\sigma\theta\pi} \int_{-\infty}^{\infty} f(t) \frac{\cos\sigma(1-\theta)(x-t) - \cos\sigma(x-t)}{(x-t)^2} dt \right| = \frac{2^{\alpha+1}M}{\pi^2\sigma^{\alpha}} \ln\frac{1}{\theta} \int_0^{\frac{\pi}{2}} t^{\alpha} \sin t\, dt + O\left(\frac{1}{\sigma^{\alpha}}\right), \quad (13)$$

uniformly with respect to all θ $(0 < \theta \leqslant 1)$.

A. F. Timan [12]. See theorem 8.4.4.

26. If $f(x_1, \ldots, x_n)$ is a function which is continuous and of period 2π in each of the variables, then for $0 \leqslant p_\nu \leqslant n_\nu$

$$E^*_{n_1,\ldots,n_m}(f) \leqslant C\min\Big[\{E^*_{p_{\nu_1},\ldots,p_{\nu_i},\infty}(f) + E_{n_{\nu_{i+1}},\ldots,n_{\nu_m},\infty}(f)\} \ln\frac{n_{\nu_1}+2}{p_{\nu_1}+1} \cdots \ln\frac{n_{\nu_i}+2}{p_{\nu_i}+1}\Big] \quad (14)$$

$$\left(\nu_r = 1, \ldots, m;\ r = 1, \ldots, i;\ i \leqslant \left[\frac{m}{2}\right]\right),$$

where C is an absolute constant.

M. F. Timan [4]. A similar relation holds between the total and partial best approximations in the metric L if

$$\int_0^{2\pi} \cdots \int_0^{2\pi} |f(x_1, \ldots, x_m)|\, dx_1 \ldots dx_m < \infty,$$

and also in the case of best approximations of functions, bounded or integrable on the whole space, by arbitrary integral functions of exponential type.

27. Let $\lambda_k^{(n)} = 1-\left(\frac{k}{n+1}\right)^p$ $(0 \leqslant k \leqslant n)$. If the number $r > 0$ is even and $r \leqslant p$, then the relation

$$\sup_{f\in W_*^{(r)}M} \max_x |f(x)-U_n^*(f; x; \lambda)| = O\left(\frac{1}{n^r}\right) \tag{15}$$

holds for the polynomials 8.1(8), and in the case when $r \leqslant p-1$

$$\sup_{f\in W_*^{(r)}H_\omega} \max_x |f(x)-U_n^*(f; x; \lambda)| = O\left[\frac{1}{n^r}\omega\left(\frac{1}{n}\right)\right]. \tag{16}$$

The relation 8.7(15) remains valid also for odd values of r provided that $r \leqslant p-1$. For odd $r = p$,

$$\sup_{f\in W_*^{(r)}M} \max_x |f(x)-U_n^*(f; x; \lambda)| = O\left(\frac{\ln n}{n^r}\right) \tag{17}$$

when $n \to \infty$. In the case when r is odd and $r \leqslant p$,

$$\sup_{f\in \tilde{W}_*^{(r)}M} \max_x |f(x)-U_n^*(f; x; \lambda)| = O\left(\frac{1}{n^r}\right), \tag{18}$$

and for $r \leqslant p-1$

$$\sup_{f\in \tilde{W}_*^{(r)}H_\omega} \max_x |f(x)-U_n^*(f; x; \lambda)| = O\left[\frac{1}{n^r}\omega\left(\frac{1}{n^r}\right)\right]. \tag{19}$$

A. Zygmund [3]. The estimates 8.7(15), 8.7(17) and 8.7(18) (correspondingly 8.7(16) and 8.7(19) for $\omega(t) = Mt^\alpha (0 < \alpha < 1)$, follow from theorem 8.5.4 (correspondingly 8.5.6).

28. If $\lambda_k^{(n)} = 1-\frac{k}{n+1}$, then for $0 < \alpha < 1$

$$\mathcal{E}_n(H_*^{(\alpha)}M; x; \lambda) = \frac{2M\Gamma(\alpha)}{\pi(1-\alpha)n^\alpha}\sin\frac{\alpha\pi}{2}+o\left(\frac{1}{n^\alpha}\right), \tag{20}$$

and for $\alpha = 1$

$$\mathcal{E}_n(H_*^{(1)}M; x; \lambda) = \frac{2M}{\pi}\frac{\ln n}{n}+O\left(\frac{1}{n}\right). \tag{21}$$

Also

$$\mathcal{E}_n(W_*^{(1)}H^{(\alpha)}M; x; \lambda) = O\left(\frac{1}{n}\right). \tag{22}$$

S. M. Nikol'skii [2]. In the estimate 8.7(22) it is impossible to replace the order O by o (see section 8.3.1).

29. If $f(x)$ is a periodic function of period 2π and $\omega_k(f; t) \leqslant t$, then

$$\tilde{f}(x)-\tilde{\sigma}_{n-1}^*(f; n-1; n) = f\left(x-\frac{1}{2n}\right)-f\left(x+\frac{1}{2n}\right)+O\left(\frac{1}{n}\right), \tag{23}$$

where $\sigma_n^*(f; k; x)$ are the polynomials 8.4(4).

A. V. Efimov [3].

30. If $\sigma_n(f; k; x)$ are the algebraic polynomials 8.2(30), and $H^{(\alpha)}M$ is the class of all functions $f(x)$ which satisfy the Lipschitz condition $\omega(f; t) \leqslant Mt^{\alpha}$ on $[-1, 1]$, then for $0 < \alpha < 1$

$$\sup_{f \in H^{(\alpha)}M} |f(x) - \sigma_n(f; n; x)| = \frac{2M\Gamma(\alpha)}{\pi(1-\alpha)n^{\alpha}}(1-x^2)^{\frac{\alpha}{2}} \sin\frac{\alpha\pi}{2} + o\left[\frac{(1-x^2)^{\frac{\alpha}{2}}}{n^{\alpha}}\right] + \delta_n^{(\alpha)}(x) \tag{24}$$

uniformly with respect to $x \in [-1, 1]$, where

$$\delta_n^{(\alpha)}(x) = \begin{cases} O\left(\dfrac{|x|^{\alpha}}{n^{2\alpha}}\right) & \text{if} \quad 0 < \alpha < \dfrac{1}{2}, \\ O\left(\dfrac{\ln n}{n}\sqrt{|x|}\right) & \text{if} \quad \alpha = \dfrac{1}{2}, \\ O\left(\dfrac{|x|^{\alpha}}{n}\right) & \text{if} \quad \alpha > \dfrac{1}{2}. \end{cases} \tag{25}$$

A. F. Timan [6, 10].

31. At every point $x \in [-1, 1]$

$$\sup_{f \in H^{(1)}M} |f(x) - \sigma_n(f; n; x)| = M\sqrt{1-x^2}\, R_{2n-2}(x) + \frac{Mx}{n\pi}(\pi - 2\arccos x) \tag{26}$$

for $n = 1, 2, \ldots$ where

$$R_{2n-2}(x) = \frac{2}{n\pi}\left\{\sum_{k=1}^{n-1} \frac{1}{k}(1 - \cos 2k \arccos x) + \right.$$
$$\left. + \frac{\sin n \arccos x}{\sqrt{1-x^2}} \cos(n-1)\arccos x\right\}. \tag{27}$$

In particular, at the ends of the segment

$$\sup_{f \in H^{(1)}M} |f(\pm 1) - \sigma_n(f; n; \pm 1)| = \frac{M}{n}, \tag{28}$$

and at its centre

$$\sup_{f \in H^{(1)}M} |f(0) - \sigma_n(f; n; 0)| = \begin{cases} \dfrac{2}{n\pi}\displaystyle\sum_{k=1}^{n-1} \frac{1+(-1)^{k+1}}{k} & \text{if } n \text{ is even}, \\ \dfrac{2}{n\pi}\left[\displaystyle\sum_{k=1}^{n-1} \frac{1+(-1)^{k+1}}{k} + \frac{1}{n}\right] & \text{if } n \text{ is odd}. \end{cases} \tag{29}$$

A. F. Timan [6, 10].

32. If the function $f(x)$, defined on $[0, 1]$, is continuous, and $B_n(x)$ are the polynomials 1.1(3), then

$$\max_{0 \leqslant x \leqslant 1} |f(x) - B_n(x)| = O\left[\omega\left(f; \frac{1}{\sqrt{n}}\right)\right]. \tag{30}$$

Popoviciu [1]. The theorem that the left-hand side of 8.7(30) tends to zero as $n \to \infty$ is due to S. N. Bernstein [2].

33. If the function $f(x)$, bounded on $[0, 1]$, has a finite second derivative at the point x, then as $n \to \infty$

$$B_n(x) - f(x) = \frac{f''(x)}{2n} x(1-x) + o\left(\frac{1}{n}\right). \tag{31}$$

E. V. Voronovskaya [1].

34. The polynomials 1.1(3) possess the property that when $f(x)$ has a continuous derivative of the k-th order ($k \geqslant 0$) on $[0, 1]$, then $\max\limits_{0 \leqslant x \leqslant 1} |f^{(k)}(x) - B_n^{(k)}(x)| \to 0$ as $n \to \infty$.

I. N. Khlodovskii [1]. (See V. L. Goncharov [1].)

35. If $f(x)$ is an integral function, then the sequence of polynomials $B_n(x)$ converges to it at any point x on $(-\infty, \infty)$.

L. V. Kantorovich [2].

36. If $T_n(f; x)$ is the sequence of trigonometric polynomials 8.7.1(β), then for any continuous function $f(x)$ of period 2π

$$\max_x |f(x) - T_n(f; x)| = O\left[\omega\left(f; \frac{1}{\sqrt{n}}\right)\right]. \tag{32}$$

I. P. Natanson [1].

37. If the periodic function $f(x)$ of period 2π has a finite second derivative at the point x, then its deviations from the polynomials 8.7.1(β) as $n \to \infty$ are given by the equation

$$T_n(f; x) - f(x) = \frac{f''(x)}{n} + o\left(\frac{1}{n}\right). \tag{33}$$

I. P. Natanson [1].

38. If

$$\frac{\lambda_0^{(n)}}{2} + \sum_{k=1}^{n} \lambda_k^{(n)} \cos kt \geqslant 0 \qquad (n = 0, 1, 2, \ldots),$$

then there exist analytic functions $f(x)$ for which the rapidity of convergence of the approximation processes 8.1(8), 8.1(10), 8.1(26) and 8.1(27) as $n \to \infty$ is not higher than $\frac{1}{n^2}$.

See P. P. Korovkin [1, 2]. In the algebraic case consider the functions 1, x and x^2.

39. If $\lambda(u) = (1-u^\beta)^\delta$ ($\beta > 0$, $\delta > 0$) and $\lambda_k^{(n)} = \lambda\left(\frac{k}{n+1}\right)$, then

$$\mathcal{E}_n(H_*^{(\alpha)} M; x; \lambda) = \begin{cases} O\left(\dfrac{1}{n^\alpha}\right) & \text{when } \gamma = \min(\beta, \delta) > \alpha, \\[2ex] O\left(\dfrac{\ln n}{n^\alpha}\right) & \text{when } \gamma = \alpha, \\[2ex] O\left(\dfrac{1}{n^\gamma}\right) & \text{when } \gamma < \alpha. \end{cases} \tag{34}$$

B. Sz. Nagy [2]. For the case $\beta = 1, \delta \geqslant \alpha$, and also $\beta = 2$, $\delta \geqslant \alpha$ see M. Yakob [1].

40. For periodic functions the trigonometric polynomials 8.7.1(β) possess certain properties similar to the properties of the polynomials 1.1(3) in the non-periodic case. In particular, if the periodic function $f(x)$ of period 2π has a continuous derivative of the k-th order ($k \geqslant 0$), then $\max_x |f^{(k)}(x) - T_n^{(k)}(f; x)| \to 0$ as $n \to \infty$.

See section 8.7.34. I. P. Natanson [2].

41. If the function $f(x, y)$ is continuous in the closed region $\bar{G}$ belonging to the square $0 \leqslant x,\ y \leqslant 1$, then for any closed region $\bar{D} \subset G$ the polynomials

$$P_n(x, y) = P_n(f; x, y) = \left[\frac{3.5\ldots(2n+1)}{2.4\ldots 2n}\right]^2 \times$$
$$\times \frac{1}{4} \iint\limits_G f(u, v)\,[1-(u-x)^2]^n [1-(v-y)^2]^n \,du\,dv$$

possess the property that $\max_{x, y \in \bar{D}} |f(x, y) - P_n(f; x, y)| \to 0$ as $n \to \infty$. If, in addition, $f(x, y)$ has in G a continuous partial derivative

$$\frac{\partial^{r+s} f(x, y)}{\partial x^r \partial y^s}, \text{ then } \max_{x, y \in \bar{D}} \left| \frac{\partial^{r+s} f(x, y)}{\partial x^r \partial y^s} - \frac{\partial^{r+s} P_n(x, y)}{\partial x^r \partial y^s} \right| \to 0 \text{ as } n \to \infty.$$

See De la Vallée Poussin [4].

42. Let $k = 2^\nu + j$ ($\nu = 0, 1, 2, \ldots;\ 0 \leqslant j < 2^\nu$).

The system of Haar functions

$$\psi_k(x) = \begin{cases} (\sqrt{2})^\nu & \text{if } \dfrac{2j}{2^{\nu+1}} \leqslant x < \dfrac{2j+1}{2^{\nu+1}} \\ -(\sqrt{2})^\nu & \text{if } \dfrac{2j+1}{2^{\nu+1}} < x \leqslant \dfrac{2j+1}{2^{\nu+1}} \\ 0 & \text{for the remaining values of } x \end{cases} \tag{35}$$

is orthonormal on $[0, 1]$ and possesses the property that for any function $f(x)$ continuous on $[0, 1]$ the Fourier series

$$\sum_{k=1}^{\infty} c_k \psi_k(x) \quad \left(c_k = \int_0^1 f(t) \psi_k(t)\,dt\right)$$

converges uniformly to $f(x)$, and if $\omega(f; t) = O(t^\alpha)$ $(0 < \alpha \leqslant 1)$ then

$$\max_{0 \leqslant x < 1} \left| f(x) - \sum_{k=1}^{n} c_k \psi_k(x) \right| = O\left(\frac{1}{n^\alpha}\right). \tag{36}$$

Haar [1]. See S. Kaczmarz and H. Steinhaus [1].

43. If V is the class of all functions $f(x)$ which have bounded variation $V(f)$ on $[0, 1]$, then for any system of functions $\varphi_k(x)$ $(k = 1, 2, \ldots)$ orthonormal on $[0, 1]$

$$\sup_{f \in V} \frac{\inf_{\alpha_k} \left\{ \int_0^1 \left| f(x) - \sum_{k=1}^{n} \alpha_k \varphi_k(x) \right|^2 dx \right\}^{\frac{1}{2}}}{V(f)} \geqslant \frac{B}{n}, \tag{37}$$

where B is an absolute constant.

Rudin [1].

44. If the periodic function $f(x)$ of period 2π has a derivative $f^{(r)}(x)$ of order r of bounded variation with jumps $\sigma_1, \sigma_2, \sigma_3, \ldots$, then as $n \to \infty$

$$\left\{ \int_0^{2\pi} |f(x) - S_n^*(f; x)|^q \, dx \right\}^{\frac{1}{q}} = [1 + o(1)] \frac{1}{\pi} \left\{ \sum_{\nu=1}^{\infty} |\sigma_\nu|^q \right\}^{\frac{1}{q}} \frac{C_{r,q}}{n^{r+\frac{1}{q}}}, \tag{38}$$

where $1 < q \leqslant \infty$, $S_n^*(f; x)$ is the partial sum of order n of the Fourier series of the function $f(x)$, and $C_{r,q}$ is a constant depending only on r and q.

S. M. Nikol'skii [12]. To obtain this result it is convenient to use the formula

$$\sum_{k=n+1}^{\infty} \frac{\cos(kx+a)}{k^{r+1}} = \frac{1}{\Gamma(r+1)} \int_0^1 t^n \left(\ln \frac{1}{t}\right)^r \frac{\cos[(n+1)x+a] - t\cos(nx+a)}{t^2 - 2t\cos x + 1} \, dt,$$

for which see also A. N. Tveritin [1].

SOME RESULTS FROM THE THEORY OF FUNCTIONS AND FUNCTIONAL ANALYSIS

9.1. The binomial series

The series

$$1+mt+\frac{m(m-1)}{2!}t^2+\ldots+\frac{m(m-1)\ldots(m-n+1)}{n!}t^n+\ldots \tag{1}$$

for any real value of $m > 0$ converges uniformly on the segment $-1 \leqslant t \leqslant 1$ to the function $(1+t)^m$. (See G. M. Fikhtengol'ts [1], v. II.)

9.2 Abel's transformation

The identity

$$\sum_{k=m}^{n} u_k v_k = \sum_{k=m}^{n-1} U_k(v_k - v_{k+1}) - U_{m-1}v_m + U_n v_n, \tag{2}$$

holds, where $0 \leqslant m \leqslant n$, $U_k = u_0+u_1+\ldots+u_k$ $(k \geqslant 0)$, $U_{-1} = 0$. It is similar to the formula for integration by parts.

9.3. The theorem of Descartes on the positive roots of the equation with real coefficients

$$c_n x^n + c_{n-1}x^{n-1} + \ldots + c_1 x + c_0 = 0 \tag{3}$$

The number of positive roots of the equation (3) *does not exceed the number of changes of sign in the sequence of its coefficients* $c_n, c_{n+1}, \ldots, c_0$. (See A. G. Kurosh [1].)

9.4. The Bolzano–Weierstrass principle of compactness

Every bounded infinite set of points (on a straight line or on a plane) has at least one limit point, i.e. from any inifinite sequence of points of such a set it is possible to select a convergent subsequence. (See P. S. Aleksandrov and A. N. Kolmogorov [1].)

9.5. The diagonal process of Cantor

This is the name sometimes given to the following process of selection of a subsequence from a given sequence $A(a_1, a_2, a_3, \ldots)$. In accordance with some previously specified rule we select from A the part $A^{(1)}(a_{n_1^{(1)}}, a_{n_2^{(1)}}, \ldots)$. Then from $A^{(1)}$ we select the part $A^{(2)}(a_{n_1^{(2)}}, a_{n_2^{(2)}}, a_{n_3^{(2)}}, \ldots)$, from $A^{(2)}$ we select the part $A^{(3)}(a_{n_1^{(3)}}, a_{n_2^{(3)}}, a_{n_3^{(3)}}, \ldots)$, and so on. Finally, we construct the sequence $(a_{n_1^{(1)}}, a_{n_2^{(2)}}, a_{n_3^{(3)}}, \ldots)$, formed from the elements situated on the diagonal of the resulting infinite table $\{a_{n_i^{(k)}}\}$.

9.6. Compact sets in metric spaces

The set R of elements $x, y, z, \ldots$ of an arbitrary nature is said to be a metric space if for any pair x, y of "points" belonging to it there is defined a non-negative number $\varrho(x, y)$, called the distance between x and y and satisfying the following conditions:

(1) $\varrho(x, y) = 0$ when and only when x is identical with y;

(2) $\varrho(x, y) = \varrho(y, x)$;

(3) for any three points x, y and z of R the inequality $\varrho(x, y) \leqslant \varrho(x, z) + \varrho(z, y)$ is satisfied.

A metric space R is by definition complete, if any infinite sequence $\{x_n\}$ $(n = 1, 2, \ldots)$, for which $\varrho(x_n, x_m) \to 0$ when n and m simultaneously tend to infinity, converges to some point $x \in R$ i.e. $\varrho(x_n, x) \to 0$ when $n \to \infty$.

The set M in a metric space R is said to be compact** if from any infinite sequence of points $\{x_n\}$ $(n = 1, 2, \ldots)$ of M it is possible to select a subsequence $\{x_{n_k}\}$ $(k = 1, 2, 3, \ldots)$ which converges to some point $x \in R$. The set $M \subset R$ is said to be closed if it contains all its limit points, and open in R if $R - M$ is closed.

The set Q_ε is said to be an ε-net for M if for any $x \in M$ there exists an element $a \in Q_\varepsilon$, whose distance from x is not greater than ε.

Hausdorff's theorem

The set M in the complete metric space R is compact when and only when for any $\varepsilon > 0$ there exists in R a finite ε-net for R, i.e. an ε-net consisting of a finite number of elements.

In an n-dimensional Euclidean space every infinite bounded set is compact. (See L. A. Lyusternik and V. I. Sobolev [1].)

9.7. The Borel–Lebesgue lemma

From every infinite system of intervals covering a given finite segment $[a, b]$ it is possible to select a finite subset of the intervals possessing the same property. (See P. S. Aleksandrov and A. N. Kolmogorov [1].)

9.8. A bicompact topological space

By a topological space is understood a set R of elements of an arbitrary nature (called its points) in which there is distinguished a certain system $\mathfrak{M}$ of subsets, called open, possessing the following properties:

(1) $\mathfrak{M}$ includes the empty set and the whole of the set R;

(2) The sum or union of any (finite or infinite) number and the intersection of any finite number of sets of $\mathfrak{M}$ is a set belonging to the system $\mathfrak{M}$.

An arbitrary open set, which is in the system $\mathfrak{M}$ and contains the given point $x \in R$ serves, by definition, as a neighbourhood of this point.

The real function $f(x)$, defined at the points x of a topological space R, is continuous at the point x_0 if for any $\varepsilon > 0$ it is possible to find a neighbourhood of the point x_0 such that $|f(x) - f(x_0)| < \varepsilon$ provided only that x belongs to this neighbourhood.

The topological space R is said to be bicompact if from every infinite system of open sets covering R it is possible to select a finite subsystem possessing the same property.

** It should be noted that a different definition of compactness of a set in a metric space is given by some writers. (Translation editor).

The simplest example of a topological space is given by any metric space. A metric space is bicompact when and only when it is compact. (See P. S. Aleksandrov and A. N. Kolmogorov [1].)

9.9. A theorem on the continuation of continuous functions defined on closed sets

If at the points of a closed set Q of a metric space R there is defined a bounded continuous function, then it can be continued in a continuous manner to the whole space R without increasing the upper bound of the modulus. This means that there exists a function $F(x)$, defined on the whole space R, which is continuous at every point $x \in R$, is identical with $f(x)$ on the set Q and is such that $\sup_{x \in R} |F(x)| = \sup_{x \in Q} |f(x)|$. (See P. S. Aleksandrov and A. N. Kolmogorov [1].

9.10. Open and closed sets in an n-dimensional euclidean space

If Q is a closed set in the n-dimensional euclidean space R_n of points $x(\xi_1, \ldots, \xi_n)$, then its complement, that is $R_n - Q$, which is an open set, can be represented as the sum of an, at most, denumerable number of half-open n-dimensional parallelepipeds defined by $a_k \leqslant \xi_k < b_k$ $(k = 1, \ldots, n)$. (See V. I. Smirnov [1], Vol. V.)

9.11. Functions of bounded variation

A function $f(x)$ defined on the segment $[a, b]$ is a function of bounded variation if the upper bound

$$V_a^b(f) = \sup_{x_k} \sum_{k=0}^{n-1} |f(x_{k+1}) - f(x_k)|, \tag{4}$$

for all possible dissections of $[a, b]$ by points $a = x_0 < x_1 < \ldots < x_n = b$, is finite. This upper bound is called the total variation of $f(x)$ on $[a, b]$. Similarly the total variation of the function $f(x)$, defined on any set M of the real axis, is defined as the upper bound of the sum 9(4) for all possible systems of points $x_0 < x_1 < \ldots < x_n$, belonging to M.

Any function of finite total variation is the difference of two monotonically increasing bounded functions.

Functions of bounded variation on the segment $[a, b]$ can have on this segment only discontinuities of the first kind, the number of these being at most denumerable. At almost every point any such function $f(x)$ has a finite derivative $f^1(x)$, which is Lebesgue-integrable on the given segment. (See I. P. Natanson [3].)

9.12. The C-property of measurable functions

A function defined on a measureable set E of points x of an n-dimensional euclidean space is said to be measurable if, for any real value a, the set of all x of E where $f(x) > a$ is measurable.

A theorem of N. N. Luzin

If the function $f(x)$ is measurable on the set E of finite measure, and almost everywhere on E assumes finite values, then for any $\varepsilon > 0$ it is possible to find a closed set $F \subset E$, the measure of which differs from the measure of the set E by less than ε, and on which $f(x)$ is continuous.

In this definition of a measurable function the measure of the corresponding sets can be understood in the sense of Lebesgue or in a more general sense. Thus, for example, in the one dimensional case, when the sets considered are located on the real axis, we may specify a function $\varrho(x)$ which is arbitrary, bounded, non-decreasing and continuous on the left at every point, and define the generalised length of every half-open interval $[a, b]$ by the number $\varrho(b+0)-\varrho(a+0)$. Then we may, as usual, define the generalised measure of any open set on the real axis, the exterior measure of sets, and the class of all ϱ-measurable sets as the class of those sets which it is possible to cover by open sets, which differ from them on a set of arbitrarily small exterior measure.

The case $\varrho(x) = x$ corresponds to the ordinary Lebesgue measure. (See V. I. Smirnov [1], vol. V.)

9.13. The absolute continuity of a Lebesgue integral

If the function $f(x)$ is integrable in the sense of Lebesgue on the measurable set E, then for any $\varepsilon > 0$ it is possible to find a $\delta > 0$ such that for every set $M \subset E$ of measure $< \delta$ we have $\int\limits_M |f(x)|\,dx < \varepsilon$. This property of Lebesgue integrals remains valid also for Lebesgue–Stieltjes integrals $\int\limits_E f(x)\,d\varrho(x)$ if measurability and measure are understood in the generalised sense. Similarly for functions of many variables. (See V. I. Smirnov [1], Vol. V.)

9.14. The integral representation of absolutely continuous functions

The function $F(x)$ is, by definition, absolutely continuous if for any $\varepsilon > 0$ it is possible to find a $\delta > 0$ such that, for every finite system of disjoint segments $[a_k, b_k]$ $(k = 1, 2, \ldots, n)$ of total length less than δ, we have

$$\sum_{k=1}^{n} |F(b_k)-F(a_k)| < \varepsilon.$$

If the function $f(x)$ is integrable in the sense of Lebesgue on $[a, b]$, then the function $F(x) = \int\limits_a^x f(t)\,dt$ is absolutely continuous and for almost all x has the derivative $F'(x) = f(x)$.

Conversely, every function $F(x)$ absolutely continuous on $[a, b]$ is representable in the form $F(x) = \int\limits_a^x f(t)\,dt$, where $f(t)$ is Lebesgue-integrable on $[a, b]$. In this case $V_a^b(F) = \int\limits_a^b |f(t)|\,dt$. (See I. P. Natanson [3].)

9.15. Functions satisfying a Lipschitz condition of the first order

If the function $F(x)$, absolutely continuous on $[a, b]$, has almost everywhere a derivative $F'(x)$ for which $|F'(x)| \leqslant K$, then it satisfies the condition

$$|F(x_1)-F(x_2)| \leqslant K\,|x_1-x_2|, \tag{5}$$

where x_1 and x_2 are any two points of the segment $[a, b]$. Conversely, if the function $F(x)$ satisfies the condition 9(5), then the derivative $F'(x)$ exists almost everywhere, and possesses the property $|F'(x)| \leqslant K$. (See I. P. Natanson [3].)

9.16. Weierstrass's function

This is the name given to the sum of the series

$$f(x) = \sum_{k=1}^{\infty} a^k \cos b^k x, \tag{6}$$

where $0 < a < 1$. The function $f(x)$ is continuous and in the case when $b = \frac{1}{a}$, as was shown by Hardy [2], is not differentiable anywhere. In this case $\omega_2(f; t) = O(t)$, and $\omega(f; t) > C(t)|\ln t|$, where C is some positive constant. If $b > 1$ and $\frac{\ln 1/a}{\ln b} = \alpha < 1$, then $\omega(f; t) = O(t^\alpha)$. (See A. Zygmund [1], and also N. I. Akhiezer [3], where the case in which $b = 4m+1$ is studied in detail m being a natural number). The sum of the series $\sum_{k=1}^{\infty} a^k \sin b^k x$ possesses similar properties.

9.17. Fatou's lemma. Passage to the limit under the Lebesgue integral sign

If the sequence of functions $f_n(x)$ $(n = 1, 2, \ldots)$, measurable on the set E, converges to a function $f(x)$ almost everywhere on this set, then

$$\int_E |f(x)|\,dx \leqslant \varliminf_{n\to\infty} \int_E |f_n(x)|\,dx.$$

In particular, if $\int_E |f_n(x)|\,dx \leqslant A < \infty$, then the limit function $f(x)$ is Lebesgue-integrable on E and hence $\int_E |f(x)|\,dx \leqslant A$. In the case when $|f_n(x)| \leqslant \phi(x)$ $(n = 1, 2, \ldots)$, where the function $\phi(x)$ is Lebesgue-integrable on E, the equality $\lim_{n\to\infty} \int_E f_n(x)\,dx = \int_E f(x)\,dx$ is satisfied. (See I. P. Natanson [3].)

9.18. Fubini's theorem on the multiple Lebesgue integral

If the function $f(x_1, \ldots, x_m)$, measurable in the m-dimensional parallelepiped $G_m(a_k \leqslant x_k \leqslant b_k; k = 1, \ldots, m)$, is Lebesgue-integrable in G_m, then for all $x_{r+1}, \ldots, x_m(a_k \leqslant x_k \leqslant b_k;\ k = r+1, \ldots, m)$, with the exception perhaps of a set of measure zero in the $(m-r)$-dimensional space of these variables, f is integrable in the parallelepiped $G_r\ (a_k \leqslant x_k \leqslant b_k;\ k = 1, \ldots, r)$. Also

$$\int \cdots \int_{G_m} f(x_1, \ldots, x_m)\,dx_1 \ldots dx_m = \int_{a_m}^{b_m} \cdots \int_{a_1}^{b_1} f(x_1, \ldots, x_m)\,dx_1 \ldots dx_m, \tag{7}$$

where the repeated integral on the right-hand side is independent of the order of integration. In the case when the function $f(x_1, \ldots, x_m)$ is non-negative everywhere on G_m, it is sufficient for its integrability on G_m for the repeated integral on the right-hand side to be finite.

9.19. Hölder's inequality

The inequality

$$\sum_{k=1}^{\infty} |a_k b_k| \leqslant \Big(\sum_{k=1}^{\infty} |a_k|^p\Big)^{\frac{1}{p}} \Big(\sum_{k=1}^{\infty} |b_k|^q\Big)^{\frac{1}{q}}, \tag{8}$$

where $p > 1$ and $q = \frac{p}{p-1}$ $\left(\frac{1}{p} + \frac{1}{q} = 1\right)$, holds for any two sequences of numbers $\{a_k\}, \{b_k\}$ $(k = 1, 2, \ldots)$.

A similar inequality holds for Lebesgue integrals. If the functions $f(x)$ and $\phi(x)$ are measurable on the set E, then

$$\int_E |f(x)\phi(x)|\,dx \leqslant \left(\int_E |f(x)|^p\,dx\right)^{\frac{1}{p}} \left(\int_E |\phi(x)|^q\,dx\right)^{\frac{1}{q}}. \tag{9}$$

(See I. P. Natanson [3].)

9.20. Minkowski's inequality

For any two sequences of numbers $\{a_k\}$, $\{b_k\}$ $(k = 1, 2, \ldots)$ the following inequality is satisfied for $q \geqslant 1$:

$$\left(\sum_{k=1}^{\infty} |a_k + b_k|^q\right)^{\frac{1}{q}} \leqslant \left(\sum_{k=1}^{\infty} |a_k|^q\right)^{\frac{1}{q}} + \left(\sum_{k=1}^{\infty} |b_k|^q\right)^{\frac{1}{q}}. \tag{10}$$

If each of the two terms on the right-hand side of 9(10) is finite and $1 < q < \infty$, then this inequality can become an equality only in the case when $b_k = \lambda a_k$ $(\lambda \geqslant 0)$ for all values of k.

A similar inequality holds for Lebesgue integrals. For any two functions $f(x)$ and $\phi(x)$, measurable on the set E,

$$\left(\int_E |f(x)+\phi(x)|^q\,dx\right)^{\frac{1}{q}} \leqslant \left(\int_E |f(x)|^q\,dx\right)^{\frac{1}{q}} + \left(\int_E |\phi(x)|^q\,dx\right)^{\frac{1}{q}}, \tag{11}$$

where $q \geqslant 1$. A similar inequality holds for Lebesgue–Stieltjes integrals and for multiple Lebesgue integrals.

If each of the two terms on the right-hand side of 9(11) is finite, $\operatorname{mes} E > 0$ and $1 < q < \infty$, then this inequality can become an equality only in the case when $\phi(x) = \lambda f(x)$ $(\lambda \geqslant 0)$ almost everywhere. (See I. P. Natanson [3], V. I. Smirnov [1], vol. V.)

9.21. The generalised inequality of Minkowski

If the function $f(x, y)$ is measurable on the rectangle $a \leqslant x \leqslant b$, $c \leqslant y \leqslant d$ and for some $q \geqslant 1$ we have $\int_c^d \left\{\int_a^b |f(x, y)|^q\,dx\right\}^{\frac{1}{q}} dy < \infty$, then

$$\left\{\int_a^b \left|\int_c^d f(x, y)\,dy\right|^q dx\right\}^{\frac{1}{q}} \leqslant \int_c^d \left\{\int_a^b |f(x, y)|^q\,dx\right\}^{\frac{1}{q}} dy. \tag{12}$$

A similar inequality holds for functions of a greater number of variables, if the integrals along the segments $[a, b]$ and $[c, d]$ are replaced by the corresponding multiple integrals.

Jensen's inequality

If $F(t)$ is integrable on $[a, b]$, $p(t) \geqslant 0$,

$$\int_a^b p(t)\,dt = 1, \; q \geqslant 1, \quad \text{then} \quad \left\{\int_a^b |F(t)|\,p(t)\,dt\right\}^q \leqslant \int_a^b |F(t)|^q\,p(t)\,dt.$$

(See Hardy, Littlewood and Pólya [1].)

9.22. A linear normed space

The set F of elements $x, y, z, \ldots$ is called a linear normed space, if

(a) In F the operation of the addition of elements is defined, which puts into correspondence with every two elements $x, y \in F$ their sum $x+y$, which is an element of F;

(b) In F the operation of multiplication by a number is defined, which establishes a correspondence between each $x \in F$ and arbitrary number λ, and the product λx, which is an element of F;

(c) The given two operations satisfy the following requirements: (1) $x+y = y+x$; (2) $(x+y)+z = x+(y+z)$; (3) from the equality $x+y = x+z$ it follows that $y = z$; (4) $(\lambda+\mu)x = \lambda x+\mu x$; (5) $\lambda(x+y) = \lambda x+\lambda y$; (6) $\lambda(\mu x) = (\lambda\mu)x$; (7) $1 \cdot x = x$;

(d) A correspondence is established between every element $x \in F$ and a certain non-negative number $\| x\|$, called its norm and satisfying the following requirements: (1) $\| x\| = 0$ when and only when $x = \theta$, where θ is the null element of the space, i.e. the element possessing the property that $z+\theta = z$ for all $z \in F$; (2) $\|\lambda x\| = |\lambda| \, \| x\|$; (3) $\| x+y\| \leqslant \| x \|+\| y\|$.

The system of elements $x_1, \ldots, x_n$ of the linear normed space F is said to be linearly independent if from the inequality $\sum\limits_{k=1}^{n} \lambda_k x_k = \theta$ it follows that $\lambda_k = 0$ for all $k = 1, 2, \ldots, n$.

The space F is said to be n-dimensional, if there exist in it n linearly independent elements and no system of $n+1$ elements possesses the property of linear independence.

The part of a linear normed space F which contains together with any two elements x and y their sum $x+y$ and the element λx, where λ is an arbitrary number, is said to be a linear subspace of this space. Classical examples of linear normed spaces are:

1. The n-dimensional space $l_2^{(n)}$ of the vectors $x(\xi_1, \ldots, \xi_n)$, where $\|x\| = \left(\sum\limits_{k=1}^{n} |\xi_k|^2\right)^{\frac{1}{2}}$.

2. The space C of all the functions $x = x(t)$ continuous on the finite segment $[a, b]$ with the ordinary operations of addition of functions and multiplication of a function by a number, where $\|x\| = \max\limits_{a\leqslant t\leqslant b} |x(t)|$.

3. The space L_q $(q \geqslant 1)$ of all functions $x = x(t)$, measurable on a given set E (of positive measure), for which $\int\limits_E |x(t)|^q \, dt < \infty$**, with the ordinary operations of addition of functions and multiplication of a function by a number (subject to the condition that two functions $x(t)$ and $y(t)$ which assume identical values almost everywhere are regarded as identical), where $\|x\| = \left(\int\limits_E |x(t)|^q \, dt\right)^{\frac{1}{q}}$.

For $q = 2$ this space is called the Hilbert functional space.

In the general case a Hilbert space is defined as a linear normed space in which, for any two elements x and y, there is defined an operation of scalar multiplication which establishes a correspondence between a pair of elements and some complex number (x, y) and satisfies the conditions: $(x, x) = \|x\|^2$, $(y, x) = \overline{(x, y)}$, $(\lambda_1 x_1+\lambda_2 x_2, y) = \lambda_1(x_1, y)+\lambda_2(x_2, y)$. Let us note that here the inequality $|(x, y)| \leqslant \sqrt{(x, x)(y, y)}$ is always satisfied.

** It follows from the inequality 9 (9) that, when the measure of the set E is finite, any such function is Lebesgue-integrable on E. Let us note also that every such function $x(t)$, if we put $x(t) = 0$, when $t \,\overline{\in}\, E$, possesses the property of integral continuity, that is, $\int\limits_{-\infty}^{\infty} |x(t+h)-x(t)|^q \, dt \to 0$ as $h \to 0$.

4. The space M of all functions $x = x(t)$ measurable on the set E (of positive measure), for which $\operatorname{vrai\,sup}_{t\in E} |x(t)| < \infty$ ($\operatorname{vrai\,sup}_{t\in E} |x(t)|$ is the essential upper bound of $|x(t)|$ on E, i.e. the lower bound of all numbers C for which $|x(t)| > C$ on a set of measure zero), with the usual operations of addition of functions and multiplication of a function by a number (with the proviso that two functions $x(t)$ and $y(t)$ which assume identical values almost everywhere are identical), where $||x|| = \operatorname{vrai\,sup}_{t\in E} |x(t)|$.

A linear normed space F is said to be *strictly normed*, if the equality $||x+y|| = ||x|| + ||y||$ is possible only when $y = \lambda x$ ($\lambda \geqslant 0$).

The simplest example of a strictly normed space is the euclidean space $l_2^{(n)}$.

If $1 < q < \infty$, then the space L_q is strictly normed (see section 9.20). The spaces C, L_1 and M are not strictly normed.

The linear normed space F is said to be *complete* (a Banach space) if any infinite sequence of elements $\{x_n\}$ ($n = 1, 2, \ldots$) for which $||x_n - x_m|| \to 0$ as n and m increase without limit converges to an element $x \in F$ that is, $||x_n - x|| \to 0$ when $n \to \infty$.

The spaces $l_2^{(n)}$, C, L_q and M are complete.

Any linear normed space can be regarded as a metric space, where $\varrho(x, y) = ||x-y||$ (See L. A. Lyusternik and V. I. Sobolev [1].)

9.23. Convergent sequences of functions in the space L_q

If the sequence of functions $x_n(t) \in L_q$ ($n = 1, 2, \ldots$) converges in the sense of the metric of the space L_q to the function $x(t)$, i.e. $\left(\int_E |x_n(t) - x(t)|^q \, dt\right)^{\frac{1}{q}} \to 0$ when $n \to \infty$, then corresponding to this sequence there exists a subsequence $\{x_{n_k}(t)\}$ ($k = 1, 2, \ldots$) converging to $x(t)$ for almost all values of $t \in E$. A similar theorem holds for functions of many variables. (See I. P. Natanson [3].)

9.24. Linear operators and linear functionals

If F and G are two linear normed spaces and in virtue of some law each element $x \in F$ is put into correspondence with an element $y \in G$, it is said that the operator, $y = U(x)$ operates on F, transforming F into G.

In the case, when the operator $U(x)$ is additive, i.e. $U(x_1 + x_2) = U(x_1) + U(x_2)$ for any pair x_1, x_2 of F, and continuous, i.e. $U(x_n) \to U(x)$ as $x_n \to x$, it is said to be linear. An additive operator is linear when and only when it is bounded, i.e. for all $x \in F$

$$||U(x)|| \leqslant C\,||x||, \tag{13}$$

where C is a positive constant independent of x. The least possible value of the constant C in the inequality 9(13), said to be *the norm of the operator* U, is identical with the upper bound $||U|| = \sup_{||x|| \leqslant 1} ||U(x)||$.

The linear operator $U(x)$ is said to be finite-dimensional if it transforms the space F into a finite-dimensional space G.

A linear operator $U(x)$ which transforms F into the real axis, where the norm of a number is equal to its absolute value, is called a linear functional.

The Hahn–Banach theorem

If a linear functional $\Phi(x)$ is defined on the linear subspace E of the space F, then it can be extended to the whole space F with preservation of the norm. In other words, there exists a linear functional $\Phi(x)$ defined on the whole of F, which assumes at the points of E the same value as does $\phi(x)$, the norm of Φ on the whole space F being identical with the norm of ϕ on E.

9.25. Linear functionals in the spaces C and L_q

In the space C of functions $x = x(t)$ continuous on the finite segment $[a, b]$ (see section 9.22), every linear functional has the form

$$\Phi(x) = \int_a^b x(t)\, d\varrho(t), \tag{14}$$

where $\varrho(t)$ is a function of finite total variation $V_a^b(\varrho)$ on $[a, b]$ (see section 9.1.1). Here

$$||\Phi|| = \sup_{|x(t)| \leqslant 1} |\Phi(x)| = V_a^b(\varrho). \tag{15}$$

In the space $L_q(1 < q < \infty)$ of functions $x = x(t)$ defined on $[a, b]$ (see section 9.22), every linear functional $\Phi(x)$ is of the form

$$\Phi(x) = \int_a^b x(t)\, y(t)\, dt, \tag{16}$$

where $\int_a^b |y(t)|^{\frac{q}{q-1}} dt < \infty$. Here**

$$||\Phi|| = \left(\int_a^b |y(t)|^p\, dt\right)^{\frac{1}{p}} \quad \left(\frac{1}{p} + \frac{1}{q} = 1\right). \tag{17}$$

In the space L_1 of functions $x = x(t)$ defined on $[a, b]$, every linear functional $\Phi(x)$ is of the form

$$\Phi(x) = \int_a^b x(t)\, y(t)\, dt, \tag{18}$$

where $\operatorname{vrai\,sup}_{a \leqslant t \leqslant b} |y(t)| < \infty$. Here

$$||\Phi|| = \operatorname{vrai\,sup}_{a \leqslant t \leqslant b} |y(t)|. \tag{19}$$

(See L.A. Lyusternik and V. I. Sobolev [1].)

9.26. Sequences of linear operators on Banach spaces

Banach's theorem

If the sequence of linear operators $U_n(x)$ $(n = 1, 2, \ldots)$, which transforms the complete linear normed space F into the linear normed space G, possesses the property that $\overline{\lim_{n\to\infty}}\, ||U_n(x)|| < \infty$ *for any $x \in F$, then the sequence of norms of these operators is bounded.*

The sequence of linear functionals $\Phi_n(x)$ $(n = 1, 2, \ldots)$, defined in the space F, is said to be *weakly convergent* if for any $x \in F$ the sequence of numbers $\Phi_n(x)$ converges.

** The case when $q = 2$ is contained in the following theorem. In any complete Hilbert space H every linear functional $\Phi(x)$ is of the form $\Phi(x) = (x, y)$, where $y \in H$. Here $||\Phi|| = ||y||$.

The sequence of linear functionals $\Phi_n(x)$, defined in the Banach space F, converges weakly when and only when the sequence of norms $\|\Phi_n\|$ is bounded and the sequence of values of $\Phi_n(x)$ converges for all x of some set everywhere dense in F. (See L. A. Lyusternik and V. I. Sobolev [1].)

9.27. The integration and differentiation of Fourier series

If $f(x)$ is a periodic function of period 2π, integrable in the sense of Lebesgue on $[0, 2\pi]$, and

$$\frac{a_0}{2}+\sum_{k=1}^{\infty}(a_k\cos kx+b_k\sin kx) \tag{20}$$

is its Fourier series, i.e.

$$\begin{aligned} a_k &= \frac{1}{\pi}\int_0^{2\pi} f(t)\cos k{+}\mathrm{d}t \qquad (k=0,1,2,\ldots),\\ b_k &= \frac{1}{\pi}\int_0^{2\pi} f(t)\sin kt\,\mathrm{d}t \qquad (k=1,2,\ldots), \end{aligned} \tag{21}$$

then $|a_k|+|b_k|\to 0$ as $k\to\infty$ (the Riemann-Lebesgue theorem) and the series

$$\sum_{k=1}^{\infty}\frac{a_k\sin kx-b_k\cos kx}{k} \tag{22}$$

converges uniformly. Moreover, for all x

$$\int_0^x f(t)\,\mathrm{d}t=\sum_{k=1}^{\infty}\frac{b_k}{k}+\frac{a_0}{2}x+\sum_{k=1}^{\infty}\frac{a_k\sin kx-b_k\cos kx}{k}. \tag{23}$$

If the function $f(x)$ is absolutely continuous, then the Fourier series of its derivative $f'(x)$ is identical with the series obtained by term-by-term differentiation of the series 9(20). (See A. Zygmund [1].)

9.28. The convergence of Fourier series

The Jordan and Dini-Lipschitz tests

If the periodic function $f(x)$ of period 2π is continuous and of finite total variation on $[0, 2\pi]$ or satisfies the Dini-Lipschitz condition $\omega(f; t)\ln t\to 0$ as $t\to 0$, then its Fourier series converges uniformly.

A theorem of M. Riesz

If the periodic function $f(x)$ of period 2π is integrable on $[0,2\pi]$ and for some finite $q>1$

$$\int_0^{2\pi}|f(x)|^q\,\mathrm{d}x<\infty$$

then the partial sums of its Fourier series satisfy the relation

$$\lim_{n\to\infty}\int_0^{2\pi}\left|f(x)-\left\{\frac{a_0}{2}+\sum_{\nu=1}^{m}(a_\nu\cos\nu x+b_\nu\sin\nu x)\right\}\right|^q\mathrm{d}x=0.$$

Hence in virtue of section 9.23 it follows that for any given function it is possible to choose a subsequence $\{n_k\}$ $(k = 1, 2, \ldots)$ such that as $n_k \to \infty$

$$\frac{a_0}{2} + \sum_{\nu=1}^{n_k} (a_\nu \cos \nu x + b_\nu \sin \nu x) \to f(x)$$

almost everywhere.

The last assertion remains valid for any function integrable on $[0, 2\pi]$.

9.29. The arithmetic mean of the partial sums of a Fourier series

If $f(x)$ is a periodic function of period 2π, integrable in the sense of Lebesgue on $[0, 2\pi]$, and

$$S_n(f; x) = \frac{a_0}{2} + \sum_{k=1}^{m} (a_k \cos kx + b_k \sin kx)$$

is the partial sum of order n of its Fourier series, then the sequence of means

$$\sigma_n(f; x) = \frac{1}{n+1} \sum_{k=0}^{n} S_k(f; x) \tag{24}$$

converges to $f(x)$ almost everywhere as $n \to \infty$. (See A. Zygmund [1].)

9.30. Parseval's equality

If the periodic function $f(x)$ of period 2π is integrable on $[0, 2\pi]$ in the sense of Lebesgue and belongs to the space L_2, i.e.

$$\int_0^{2\pi} |f(x)|^2 dx < \infty, \tag{25}$$

then the coefficients of its Fourier series satisfy the equality

$$\frac{1}{2} a_0^2 + \sum_{k=1}^{\infty} (a_k^2 + b_k^2) = \frac{1}{\pi} \int_0^{2\pi} |f(x)|^2 dx \tag{26}$$

A similar equality holds for periodic functions of many variables. (See A. Zygmund [1].)

9.31. The Fischer-Riesz theorem

For any two sequences of real numbers $\{a_k\}$ $(k = 0, 1, 2, \ldots)$ *and* $\{b_k\}$ $(k = 1, 2, \ldots)$ *satisfying the condition*

$$\sum_{k=1}^{\infty} (a_k^2 + b_k^2) < \infty, \tag{27}$$

there always exists a measurable function $f(x)$ *of period* 2π *belonging to the space* L_2 *on* $[0, 2\pi]$ *for which these numbers are the corresponding coefficients of the Fourier series* (See A. Zygmund [1].)

9.32. The uniqueness theorem in the theory of trigonometrical series

If the trigonometric series

$$\frac{a_0}{2} + \sum_{k=1}^{\infty} (a_k \cos kx + b_k \sin kx)$$

converges everywhere with the exception, possibly, of an enumerable set of points, to a function $f(x)$ Lebesgue-integrable on $[0, 2\pi]$ then it is the Fourier series of this function. (See A. Zygmund [1].)

9.33. The Taylor series

If $\sum_{k=0}^{\infty} c_k(z-a)^k$ is the Taylor series of the function $f(z)$, analytic in the neighbourhood of the point a, then the radius of convergence of this series is identical with the distance from the point a of the nearest singularity of $f(z)$.

If everywhere in the circle $|z-a| < R$ the inequality $|f(z)| \leqslant M$ is satisfied, then

$$|c_n| \leqslant \frac{M}{R^n} \qquad (n = 0, 1, 2, \ldots). \tag{28}$$

If the Taylor series of the function $f(z)$ converges at a certain point $z_0 = a+Re^{i\phi_0}$ on the boundary of its circle of convergence to the value w_0, then $f(a+re^{i\phi_0}) \to w_0$ when $r \to R$ $(r < R)$.

The Laurent series

If the function $f(z)$ is analytic in the circular annulus $r < |z| < R$, then

$$f(z) = \sum_{k=-\infty}^{\infty} c_k z^k,$$

where

$$c_k = \frac{1}{2\pi i} \int_{\Gamma_\varrho} f(\zeta)\, \zeta^{-k-1} \mathrm{d}\zeta$$

and Γ_ϱ is the circumference $|z| = \varrho$ $(r < \varrho < R)$. See I. I. Privalov [2].)

9.34. Sequences of analytic functions

Weierstrass's theorem

If the sequence of functions $f_n(z) (n = 1, 2, \ldots)$, analytic in the domain G, converges to $f(z)$ uniformly in every closed bounded region $\overline{Q} \subset G$, then the function $f(z)$ is analytic in G and for any natural number k we have $f_n^{(k)}(z) \to f(z)$ when $n \to \infty$.

Vitali's theorem

If the sequence of functions $f_n(z)$, analytic in the domain G, possesses the property $\sup_{z \in G} |f_n(z)| \leqslant M < \infty$ and converges at the points z of some set D which possesses at least one limit point within G, then it converges uniformly in every closed bounded region $\overline{Q} \subset G$ and, consequently has as its limit some function $f(z)$ analytic in G.

The functions $f_n(z)$ $(n = 1,2,\ldots)$, defined on some set E in the complex plane, are said to be uniformly continuous on E, if for every $\varepsilon > 0$ it is possible to find a $\delta > 0$ such that for any pair of points z_1 and z_2 of E, at a distance apart of less than δ, the inequality $|f_n(z_1)-f_n(z_2)| < \varepsilon$ is satisfied for $n = 1, 2, \ldots$. (See I. I. Privalov [2] and E. C. Titchmarsh [1].)

9.35. The uniqueness theorem in the theory of analytic functions

If two functions $f(z)$ and $g(z)$, analytic in the domain G, assume identical values at the points z of some set D which has at least one limit point within G, then $f(z) \equiv g(z)$ everywhere in G. (See I. I. Privalov [2].)

9.36. Cauchy's theorem for analytic functions and some corollaries from it

If Γ is a piecewise smooth closed Jordan curve and the function $f(z)$ is analytic on Γ and in the domain G bounded by the contour Γ, then $\int_\Gamma f(z)dz = 0$. The values of the function $f(z)$ at the points z within G are connected with the values of this function on the boundary Γ by Cauchy's formula

$$f(z) = \frac{1}{2\pi i}\int_\Gamma \frac{f(\zeta)}{\zeta - z}\,d\zeta.$$

In the case when $f(z) \neq 0$ on Γ, the number of zeros of the function $f(z)$ within G is equal to

$$\frac{1}{2\pi i}\int_\Gamma \frac{f'(z)}{f(z)}\,dz.$$

Schwarz's formula.

If the function $f(z) = u(x, y) + iv(x, y)$ is analytic in the closed circle of radius R with centre at the origin of coordinates and at $z = 0$ assumes a real value, then at every point z within this circle

$$f(z) = \frac{1}{2\pi}\int_{-\pi}^{\pi} U(R\cos\phi, R\sin\phi)\frac{Re^{i\phi}+z}{Re^{i\phi}-z}\,d\phi. \tag{29}$$

Let $\Gamma_0, \Gamma_1, \ldots, \Gamma_n$ be closed piecewise smooth Jordan curves such that each of the curves $\Gamma_1, \ldots, \Gamma_n$ lies outside the remainder and all of them are situated within Γ_n. If the function $f(z)$ is analytic on $\Gamma_0, \Gamma_1, \ldots, \Gamma_n$, and also in the domain G which lies within Γ_0 and outside $\Gamma_1, \ldots, \Gamma_n$, then

$$\int_{\Gamma_0} f(z)\,dz = \int_{\Gamma_1} f(z)\,dz + \ldots + \int_{\Gamma_n} f(z)\,dz. \tag{30}$$

(See I. I. Privalov [2] and V. I. Smirnov [1], vol. III.)

9.37. Integral functions. Liouville's theorem

If the integral function $f(z)$ i.e. function which is analytic for any finite value of $z = x+iy$, is uniformly bounded throughout the whole complex plane, then $f(z) \equiv$ constant. If the integral function $f(z)$ possesses the property that $|f(z)| = O(|z|^k)$, then

$$f(z) = c_0 + c_1 z + \ldots + c_n z^n,$$

where $n \leqslant k$. (See I. I. Privalov [2].)

9.38. The order and type of an integral function

Every integral function $f(z)$ is the sum of a power series

$$f(z) = \sum_{k=0}^{\infty} c_k z^k \tag{31}$$

which converges for any z. The integral function 9(31) is said to be a function of finite order if

$$\overline{\lim_{n\to\infty}}\,\frac{\ln n}{\ln \dfrac{1}{\sqrt[n]{|c_n|}}} = \varrho < \infty. \tag{32}$$

Here the number ϱ is called the order of the function $f(z)$.

The integral function $f(z)$ of finite order $\varrho > 0$ is by definition a function of finite type, if

$$\overline{\lim_{n\to\infty}} \frac{1}{\varrho} \sqrt[n]{n!\,|c_n|} = \sigma < \infty. \tag{33}$$

The number σ is called the type of the function $f(z)$.

If $M(r) = \max_{|z|=r} |f(z)|$, then

$$\varrho = \overline{\lim_{r\to\infty}} \frac{\ln \ln M(r)}{\ln r} \tag{34}$$

and

$$\sigma = \overline{\lim_{r\to\infty}} \frac{\ln M(r)}{r^{\varrho}} \tag{35}$$

Conversely, 9(34) and 9(35) imply 9(32) and 9(33).

Integral functions of order $\varrho = 1$ and of finite type, and also integral functions of order $\varrho < 1$, are frequently called integral functions of exponential type or of finite degree.

(See E. C. Titchmarsh [1].)

9.39. The principle of the maximum modulus and the Phragmén–Lindelöf theorem

If the function $f(z)$ is analytic in the domain G, then $|f(z)|$ does not attain the value $\sup_{z\in G} |f(z)|$ anywhere within G except when $f(z) \equiv \text{const}$. In particular, if $|f(z)| \leqslant M$ at all points z on the boundary of G, then $\sup_{z\in G} |f(z)| \leqslant M$.

If the function $f(z)$, analytic in the region G between two straight lines making an angle $\frac{\pi}{\alpha}$ at the origin, and on the lines themselves, possesses the property that for some $\varrho < \alpha$ we have

$$|f(z)| = O\left(e^{|z|^{\varrho}}\right)$$

uniformly for all $z \in G$, and if on the boundary of this angular region $|f(z)| \leqslant M$, then $\sup_{z\in G} |f(z)| \leqslant M$. (See E. C. Titchmarsh [1].)

9.40. The bilinear function. The Riemann–Schwarz principle of symmetry

The bilinear function $w = \dfrac{az+b}{cz+d}$ maps every circle of the z-plane into a circle or straight line of the w-plane. Here any two points symmetrical (in the sense of inversion) with respect to the first circle are mapped into two points symmetrical with respect to its image.

The bilinear function which maps the unit circle of the z-plane into the upper w-half-plane is of the form $z = e^{i\theta} \dfrac{w-\beta}{w-\bar{\beta}}$, where $\operatorname{Im} \beta > 0$ and θ is a real number.

The mapping produced by an analytic function $w = f(z)$ possesses the property of preservation of angles at all points where the derivative $f'(z)$ is different from zero.

If the function $w = f(z)$ is analytic on one side of some arc of a circle C, is continuous up to this arc and assumes on it real values, then $f(z)$ may be continued analytically through this arc, and at points symmetrical (in the sense of inversion) with respect to the circle C it assumes conjugate values. (See I. I. Privalov [2].)

BIBLIOGRAPHY

OF MEMOIRS AND BOOKS REFERRED TO IN THE TEXT

ABRAMOV, L. M.

1. On the asymptotic behaviour of the Lebesgue functions of some methods of summation of Chebyshev series. *Dokl. Akad. Nauk. SSSR* **98**, No. 2, 173–176 (1954).

AKHIEZER, N. I.

1. Über die Funktionen die in gegebenen Intervallen am wenigsten von Null abweichen. *Izv. Kaz. fiz. mat. ob-va* **3** (1928).

2. On the best approximation of analytic functions. *Dokl. Akad. Nauk. SSSR* **18**, No. 4–5, 241–244 (1938).

3. Lectures on the theory of approximation. (Lektsii po teorii approksimatsii). Gostekhizdat (1947). English translation published as Theory of approximation, Ungar, New York; Constable, London (1956).

4. On the theory of integral functions of finite degree. *Dokl. Akad. Nauk. SSSR* **63** No. 5, 475–478 (1948).

5. On integral functions of finite degree which deviate least from zero. *Mat. sborn.* **31**, (73): 2, 415–438 (1952).

6. On weighted approximation of continuous functions by polynomials on the whole real axis. *Usp. mat. Nauk* **11**, No. 4. (70), 3–43 (1956).

AKHIEZER, N. I. and K. I. BABENKO

1. On weighted approximation by polynomials of functions continuous on the whole real axis. *Dokl. Akad. Nauk SSSR* **57**, No. 4, 315–318 (1947).

AKHIEZER, N. I. and M. G. KREIN

1. On the best approximation of periodic functions. *Dokl. Akad. Nauk. SSSR* **15**, (1937).

AKHIEZER, N. I. and B. M. LEVITAN

1. On an application of inequalities of H. Bohr and J. Favard. *Dokl. Akad. Nauk SSSR* **14**, 419–422 (1937).

AKHIEZER, N. I. and V. A. MARCHENKO

1. On some questions of approximation on the whole real axis (II). *Zapiski N. I. In-ta matem. i mekh. Khar'k. gos. universiteta im. A. M. Gor'kogo i Kkar'k. matem. ob-va*, **21**, 5-9 (1949).

ALEKSANDROV, P. S. and A. N. KOLMOGOROV

1. *Introduction to the general theory of sets and functions.* (Vvedenie v obshchuyu teoriyu mnozhestv i funktsii), Part I, Gostekhizdat, (1948).

APARISIO, B. E.

1. On certain properties of polynomials with integral coefficients and mean approximation of functions by polynomials with integral coefficients. *Izv. Akad. Nauk SSSR (seriya mat.)* **19**, 303–318 (1955).

BANACH, S.

1. *Course of functional analysis*. (Kurs funkstional'nogo analizu). Radyanska Shkola, Kiev (1948). (In Ukrainian).

BARI, N. K.

1. On the best approximation by trigonometric polynomials of two conjugate functions. *Izv. Akad. Nauk SSSR* (*seriya mat.*) **19**, 285–302 (1955).

2. Generalisation of the equations of S. N. Bernstein and A. A. Markov. *Izv. Akad. Nauk SSSR* (*seryia mat.*) **18**, No. 2, 159–176 (1954).

BARI, N. K. and S. B. STECHKIN

1. The best approximation and differential properties of two conjugate functions. *Trudy Mosk. matem. ob-va* **5**, 483–522 (1956).

BERMAN, D. L.

1. Approximation of periodic functions by linear trigonometric polynomial operations. *Dokl. Akad. Nauk SSSR* **91**, No. 6, 1249–1252 (1956).

2. On translational linear trigonometric polynomial operations. *Dokl. Akad. Nauk SSSR* **92**, No. 4, 693–694 (1953).

3. On a class of linear operations. *Dokl. Akad. Nauk SSSR* **85**, No. 1, 13–16 (1952).

BERNSTEIN, S. N.

1. On the best approximation of continuous functions by polynomials of given degree. (O nailuchshem priblizhenii nepreryvnykh funktsii posredstvom mnogochlenov dannoi stepeni). *Sobraniye sochinenii*, Vol. I, 11–104 (1912), *Izd. Akad. Nauk SSSR*, Vol. I (1952), Vol. II (1954).

2. Proof of a theorem of Weierstrass, based on the theory of probability. (Dokazatel'stvo teoremy Veiershtrassa, osnovannoyenna teorii veroyatnostei). *Sochineniya*, Vol. I, 105 (1912).

3. On the definition and properties of analytic functions of a real variable. (Ob opredelenii i svoistvakh analiticheskikh funktsii veshchestvennoi peremennoi). *Sochineniya*, Vol. I, 231–250 (1914).

4. Some remarks on interpolation. (Neskol'ko zamechanii ob interpolirovanii). *Sochineniya*, Vol. I, 253–263 (1916).

5. The problem of approximation of continuous functions on the whole real axis. (Zadacha priblizheniya nepreryvnykh funktsii na vsei veshchestvennoi osi). *Sochineniya*, Vol. I, 277–284 (1924).

6. On a property of integral functions of type O. (Ob odnom svoistve tselykh funktsii roda O). *Sochineniya*, Vol. I, 321–329 (1926).

7. Analytic functions of a real variable, their origin and means of generalisation. (Analiticheskie funktsii veshchestvennoi peremennoi, ikh vozniknovenniye i pyti obobshchenii). *Sochineniya*, Vol. I, 285–320.

8. Absolutely monotonic functions (Absolyutno monotonnyye funktsii). 1–66, *Sochineniya*, Vol. I, 370–425 (1928).

9. Some remarks on polynomials of least deviation with integral coefficients. (Neskol'ko zamechanii o mnogochlenakh naimen'shego ukloneniya s tselymi koeffitsiyentami). *Socheniniya*, Vol. I, 468–471 (1930).

10. On a method of summation of trigonometric series. (Ob odnom metode summirovaniya trigonometricheskikh ryadov). *Sochineniya*, Vol. I, 523–525 (1930).

11. On estimates of the derivatives of polynomials. (Ob otsenkakh proizvodnykh mnogochlenov). *Sochineniya*, Vol. I, 497–499 (1930).

12. On a modification of the interpolation formula of Lagrange. (Ob odnom vidoizmenenii interpolyatsionnoi formuly Lagranzha). *Sochineniya*, Vol. II,130–140 (1931).

13. On the limitation of the values of a polynomial $P_n(x)$ of degree n on the whole of a segment by its values at $n+1$ points of the segment. (Ob ogranichenii znachenii mnogochlena $P_n(x)$ stepeni n na vsem otrezke po ego znacheniyam v $n+1$ tochkakh otrezka). *Sochineniya*, Vol. II, 107–126 (1931).

14. On some extremal properties of sequences of integrals. (O nekotorykh ekstremal'nykh svoistvakh posledovatel'nykh integralov). *Sochineniya*, Vol. II, 170–172 (1935).

15. Extremal properties of polynomials and the best approximation of continuous functions of a single real variable. (Ekstremal'nyye svoistva polinomov i nailuchsheye priblizheniye nepreryvnykh funktsii odnoi veshchestvennoi peremennoi). State United Scientific and Technical Publishing House (1937).

16. On polynomials orthogonal on a finite segment. (O mnogochlenakh, ortogonal'nykh na konechnom otrezke). *Sochineniya*, Vol. II, 7–106 (1930–1931).

17. On the inverse problem of the theory of the best approximation of continuous functions. (Ob obratnoi zadache teorii nailuchshego priblizheniya nepreryvnykh funktsii). *Sochineniya*, Vol. II. 292–294 (1938).

18. On the best approximation of $|x|^p$ by means of polynomials of extremely high degree. (O nailuchshem priblizhenii $|x|^p$ pri pomoshchi mnogochlenov ves'ma vysokoi stepeni). *Sochineniya*, Vol. II, 262–272 (1938).

19. On a theorem of V. A. Markov. (O teoreme V. A. Markova). *Sochineniya*, Vol. II, 281–286 (1938).

20. On the best approximation of $|x-c|^p$. (O nailuchshem priblizhenii $|x-c|^p$). *Sochineniya*, Vol. II, 273–280 (1938).

21. On a question of local best approximation. (K voprosu o lokal'nom nailuchshem priblizhenii). *Sochineniya*, Vol. II, 306–309 (1940).

22. The constructive theory of functions as an extension of an idea of Chebyshev. (Konstruktivnaya teoriya funktsii kak razvitiye idei Chebysheva). *Sochineniya*, Vol. II, 349–360 (1945).

23. On the best approximation of continuous functions on the whole real axis by means of integral functions of given degree I. (O nailuchshem priblizhenii nepreryvnykh funktsii na vsei veshchestvennoi osi pri pomoshchi tselykh funktsii dannoi stepeni I). *Sochineniya*, Vol. II, 371–375 (1946).

24. On the best approximation of continuous functions on the whole real axis by means of integral functions of given degree II. (O nailuchshem priblizhenii nepreryvnykh funktsii na vsei veshchestvennoi osi pri pomoshchi tselykh funktsii dannoi stepeni II). *Sochineniya*, Vol. II, 375–379 (1946).

25. On the best approximation of continuous functions on the whole real axis by means of integral functions of given degree III. (O nailuchshem priblizhenii nepreryvnykh funktsii na vsei veshchestvennoi osi pri pomoshchi tselykh funktsii dannoi stepeni III). *Sochineniya*, Vol. II, 379–383 (1946).

26. On the best approximation of continuous functions on the whole real axis by means of integral functions of given degree V. (O nailuchshem priblizhenii nepreryvnykh funktsii na vsei veshchestvennoi osi pri pomoshchi tselykh funktsii dannoi stepeni V). *Sochineniya*, Vol. II, 390–395 (1946).

27. Limiting laws of the theory of best approximations. (Predel'nyye zakony teorii nailuchshikh priblizhenii). *Sochineniya*, Vol. II, 416–420 (1947).

28. On the upper bound of the maximum modulus of the derivative of a monotonic function of finite degree (O verkhnei granitse maksimuma modulya proizvodnoi monotonnoi funktsii konechnoi stepeni). *Sochineniya*, Vol. II, 396–398 (1946).

29. A new result and some formulae for best approximation (Novyi vyvod i obobshcheniye nekotorykh formul nailuchshego priblizheniya). *Sochineniya*, Vol. II, 402–404 (1946).

30. Generalisation of a result of S. M. Nikol'skii (Obobshcheniye odnogo rezul'tata S. M. Nikol'skii). *Sochineniya*, Vol. II, 399–401 (1946).

31. Some remarks on an inequality of Vladimir Markov (Neskol'ko zamechanii k neravenstvu Vladimira Markova). *Sochineniya*, Vol. I, 151–156 (1913).

32. On the best approximation of the function $\int_0^\infty |y|^s d\psi(s)$ on the segment [—1,1]. (O nailuchshem priblizhenii funktsii $\int_0^\infty |y|^s d\psi(s)$ na otrezke $[-1,1]$). *Sochineniya*. Vol. II, 361–370 (1946).

33. A supplement to the work of I. I. Ibragimov. (On the asymptotic value of the best approximation of a function which has a real singularity). (Dobavleniye k rabote I. I. Ibragimov "Ob asimptoticheskim znachenii nailuchshego priblizheniya funktsii, imeyushchei veshchestvennuyu osobuyu tockhy"). *Sochineniya*, Vol. II, 405–407 (1946).

34. On a theorem of Szegö (Ob odnoi teoreme Sege). *Sochineniya*, Vol, II, 173–177 (1935).

35. On limiting relations between the constants of the theory of best approximation. (O predel'nykh zavisimostyakh mezhdu konstantami teorii nailuchshego priblizheniya). *Sochineniya*, Vol. II, 413–415 (1947).

36. On the properties of homogeneous functional classes I. (O svoistvakh odnorodnykh funktional'nykh klassov I). *Sochineniya*, Vol. II, 421–432 (1947).

37. On integral functions of finite degree of many real variables. (O tselykh funktsiyakh konechnoi stepeni mnogikh veshchestvennykh peremennykh). *Sochineniya*, Vol. II, 437–441 (1948).

38. An extension of S. B. Stechkin's inequalities to integral functions of finite degree. (Rasprostraneniye neravenstva S. B. Stechkina na tselyye funktsii konechnoi stepeni). *Sochineniya*, Vol. II, 442–445 (1948).

39. An extension of properties of trigonometric polynomials to integral functions of finite degree. (Pereneseniye svoistv trigonometricheskikh polynomov na tselyye funktsii konechnoi stepeni). *Sochineniya*, Vol. II, 446–467 (1949).

40. On some properties of cyclically monotonic functions. (O nekotorykh svoistvakh tsiklicheski monotonnykh funktsii). *Sochineniya*, Vol. II, 493–516 (1950).

41. On weight functions. (O vesovykh funktsiyakh). *Sochineniya*. Vol. II 526–529 (1951).

42. On the best approximation of functions of several variables by means of polynomials or trigonometric sums. (O nailuchshem priblizhenii funktsii neskol'kikh peremennykh posredstvom mnogochlenov ili trigonometricheskikh summ). *Sochineniya*, Vol. II, 540–545 (1951).

43. On the connection between quasi-analytic functions and weight functions. (O svyazi kvazianaliticheskikh funktsii s vesovymi funktsiyami). *Sochineniya*, Vol. II, 530–534 (1951).

44. An application of a limit law to the theory of best approximations .(Odno primeneniye predel'nogo zakona teorii nailuchshikh priblizhenii). *Dokl. Akad. Nauk SSSR* **102**, No. 3, 435–436 (1955).

45. On the best approximation of analytic functions by means of integral functions of finite degree. (O nailuchshem priblizhenii analiticheskikh funktsii pri pomoshchi tselykh funktsii konechnoi stepeni). *Sochineniya*, Vol. II, 408 (1947).

BOAS, R. P. JR.

1. Interference phenomena for entire functions. *Mich. Math. J.* **3**, No. 2, 123–132 (1956).

2. Quelques généralisations d'un théorème de S. Bernstein sur la dérivée d'un polynôme trigonométrique. *Comptes Rendus* **227**, 618–619 (1956).

3. *Entire functions*, New York (1954).

BOCHNER, S.

1. *Vorlesungen über Fouriersche Integrale*, Leipzig (1952). English translation published as *lectures on Fourier integrals*, Princetown University Press (1959).

BOHR, H.

1. Un théorème générale sur l'intégration d'un polynôme trigonométrique. *Comptes Rendus* **200** (1935).

BRUDNYI, YU. A.

1. On the maximum modulus of a quasi-smooth function. *Usp. mat. Nauk* **12**, No. 4 (1957).

2. Approximation by integral functions on the exterior of a segment and semi-axis. *Dokl. Akad. Nauk SSSR* **124**, No. 4 (1959).

BRUDNYI, YU. A. and I. YE. GOPENGAUZ

1. On a problem of N. N. Luzin. *Dokl. Akad. Nauk SSSR* **113**, No. 1, 12–15 (1957).

BRUDNYI, YU. A. and A. F. TIMAN

1. Constructive characteristics of compact sets in Banach spaces and ε-entropy. *Dokl. Akad. Nauk SSSR* **126**, No. 5 (1959).

CARLEMAN, T.

1. *Les fonctions quasi-analytiques*, Paris (1926).

CARTAN, H.

1. *Sur les classes de fonctions définies par des inégalités portant sur leurs dérivées successives, Actualités scientifiques et industrielles*, No. 867 (1940).

CARTWRIGHT, M.

1. On certain integral functions of order 1. *Quart. J. Math.* (*Oxford*) Ser. 7, 46–65 (1936).

CHEBYSHEV, P. L.

1. Theory of mechanisms known by the title of parallelograms. (Teoriya mekhanizmov, izvestnykh pod nazvaniyem parallelogrammov (1854), Poln. sobr. sochinenii. (Complete collected works), Vol. II, 23–51. *Izd. Akad. Nauk SSSR* (1947).

2. Problems on the least values connected with the approximate representation of functions. (Voprosy naimen'shikh velichinakh, svyazannye s priblizhennym predstavleniyem funktsii). *Sochineniya*, Vol. II, 151–235 (1859).

CIVIN, P.

1. Inequalities for trigonometric integrals. *Duke math. J.* **8**, 656–665 (1941).

CRAMÉR, G.

1. *Études sur la sommation des séries de Fourier, Arkiv. för Matematik, Astronomi och fisik* **13**, 1–21 (1919).

DU BOIS REYMOND, P.

1. Untersuchungen über die Convergenz und Divergenz der Fourierschen Darstellungsformeln. *Abhand. Acad. München* **12**, 1–103 (1876).

DZHRBASHYAN, M. M.

1. Some questions of the theory of weighted-polynomial approximations in the complex domain. *Mat. sborn.* **36** (78), No. 3, 353–440 (1955).

DZYADYK, V. K.

1. On the best mean approximation of periodic functions with singularities. *Dokl. Akad. Nauk SSSR* **77**, No. 6, 949–952 (1951).

2. On the best approximation on a class of periodic functions having a bounded s-th derivative ($0 < s < 1$). *Izv. Akad. Nauk SSSR* (*seriya mat.*) **17**, 135–162 (1953).

3. On a constructive characteristic of functions satisfying the Lipschitz condition $\alpha (0 < \alpha < 1)$ on a finite segment of the real axis. *Izv. Akad. Nauk SSSR* (*seriya mat.*) **20**, 623–642 (1956).

4. On approximation of functions by ordinary polynomials on a finite segment of the real axis. *Izv. Akad. Nauk SSSR* (*seriya mat.*) **22**, No. 3, 337–354 (1958).

ERDÖS, P.

1. Note on some elementary properties of polynomials. *Bull. Amer. math. Soc.* **46**, 954–958 (1940).

2. Extremal properties of derivatives of polynomials. *Ann. Math.* **41**, 310–313 (1940).

ERDÖS, P. and P. TURÁN

1. On the role of the Lebesgue functions in the theory of Lagrange interpolation. *Acta Math. Acad. Sci. Hung.* **6**, No. 1–2, 47–60 (1955).

EVGRAFOV, M. A.

1. *The Abel-Goncharov interpolation problem* (Interpolyatsionnaya zadacha Abelya-Goncharova) Gostekhizdat, (1954).

FABER, G.

1. Über die interpolatorische Darstellung stetiger Funktionen. *Jber. D. Math. Vereinigung* **23** (1914).

FAVARD, J.

1. Sur les meilleures procédés d'approximation de certaines classes des fonctions par des polynômes trigonométriques. *Bull. Sci. math.* **61**, 209–224, 243–256 (1937).

FEJÉR, L.

1. Untersuchungen über Fouriersche Reihen. *Math. Ann.* **58**, 51–69 (1904).

2. Lebesguesche Konstanten und divergente Fourierreihen. *J. reine angew. Math.* **138**, 22–23 (1910).

3. Über Interpolation, Gött. Nachr., 66–91 (1916).

4. Über trigonometrische Polynome. *J. reine angew. Math.* **146,** 53–82 (1915).

5. Die Abschätzung eines Polynoms in einem Intervalle wenn Schranken für seine Werte und ersten Ableitungswerte in einzelnen Punkten des Intervalles gegeben sind und ihre Anwendung auf die Konvergenzfrage Hermitescher Interpolationsreihen. *Math.* Z. **32,** 426–457 (1930).

FEJES, L.

1. Two inequalities concerning trigonometric polynomials. *J. Lond. Math. Soc.* **14,** 44–46 (1939).

FEKETE, M.

1. Über die Verteilung der Wurzeln bei gewissen algebraischen Gleichungen mit ganzzahligen Koeffizienten. *Math. Z.* **17,** 228–249 (1923).

FIKHTENGOL'TS, G. M.

1. Course of differential and integral calculus (Kurs differentsial'nogo i integral'nogo ischisleniya). Gostekhizdat, Vol. I (1947), Vol. II (1948), Vol. III (1949).

FREUD, G.

1. Über gleichzeitige Approximation einer Funktion und ihrer Derivierten. *Int. Math. Nachr. Wien,* No. **47–48,** 36–37 (1957).

2. Über die Approximation reeller stetigen Funktionen durch gewöhnliche Polynome, *Math. Ann.* **137,** 1, 17–25 (1959).

GANZBURG, I. M.

1. On a method of approximation of continuous functions by trigonometric popolynomials. *Dokl. Akad. Nauk SSSR* **64,** No. 1, 13–16 (1949).

2. Generalization of some results of S. M. Nikol'skii and A. F. Timan. *Dokl. Akad. Nauk SSSR* **116,** No. 5, 727–730 (1957).

3. Extension of an asymptotic formula of A. F. Timan to a class of functions with a given modulus of continuity. *Dokl. Akad. Nauk SSSR,* **128,** No. 3 (1959).

GARVAKI, A. L.

1. On a combined approximation of a periodic function and its derivatives by trigonometric polynomials. *Izv. Akad. Nauk SSSR* (*seriya mat.*) **23** (1959).

GEL'FAND, I. M., D. A. RAIKOV and G. JE. SHILOV

1. Commutative normed rings. *Usp. mat. nauk* **1,** No. 2 (1946).

GEL'FOND, A. O.

1. Problems of representation and uniqueness for integral analytic function of the first order. *Usp. mat. nauk,* No. III, 144–174 (1937).

2. *The calculus of finite differences* (Ischisleniye konechnykh raznostei). Gostekhizdat (1952).

3. On uniform approximations by polynomials with integral coefficients. *Usp. mat. nauk* **10,** No. 1, 11–65 (1955).

GONCHAROV, V. L.

1. *The theory of interpolation and approximation of functions.* (Teoriya interpolirovaniya i priblizheniya funktsii), Gostekhizdat (1954).

GOPENGAUZ, I. YE. and A. F. TIMAN

1. On the question of the modules of continuity of periodic functions with given modulus of smoothness. *Usp. mat. nauk* **12,** No. 3 (75), 291–294 (1957).

GORNY, A.

1. Contribution à l'étude des fonctions dérivables d'une variable réelle. *Acta Math.* **71**, 317 (1939).

GRÜNWALD, G.

1. Über Divergenzerscheinungen der Lagrangeschen Interpolationspolynome stetiger Funktionen. *Annals of Math.* **37**, 908–918 (1936).

GUKEVICH, V. I.

1. The best mean approximation by polynomials of the function ln $(a - x)$. *Dokl. Akad. Nauk SSSR* **77**, No. 5, 785–788 (1951).

GUREVICH, V. B.

1. On some cases of coincidence of the best trigonometric polynomial approximation with approximations by polynomials of various degrees. *Izv. Akad. Nauk SSSR (seriya mat.)* **10**, 469–486 (1946).

HAAR, A.

1. *Zur Theorie der orthogonalen Funktionensysteme.* Inauguraldissertation, Göttingen (1909).

2. Die Minkowskische Geometrie und die Annäherung an stetige Funktionen. *Math. Ann.* **78** (1918).

HADAMARD, J.

1. Sur la généralisation de la notion de fonction analytique. *C. R. Séances Soc. Math. Fr.* **42**, 28 (1912).

HARDY, G. H.

1. Notes on some points in the integral calculus. *Messeng. Math.* (1922).

2. Weierstrass's non-differentiable function. *Trans. Amer. math. Soc.* **17**, 301–325 (1916).

HARDY, G. H. and J. E. LITTLEWOOD

1. Some properties of fractional integrals I, II. *Math. Zeitschr.* **27**, 565–606 (1928); *Math. Zeitschr.* **34**, 403–439 (1932).

HARDY, G. H., J. E. LITTLEWOOD and G. POLYA

1. *Inequalities.* Cambridge (1934).

HAUSDORFF, F.

1. Eine Ausdehnung des Parsevalschen Satzes über Fourierreihen. *Math. Zeitschr.* **16**, 163–169 (1923).

HELLY, E.

1. Über lineare Funktionaloperationen. *Sitzungsberichte der Math-Naturwiss. Klasse der Akad. der Wiss. zu Wien* **121**, 265–267 (1912).

HILLE, E.

1. *Functional analysis and semi-groups.* New York (1948).

IBRAGIMOV, I. I.

1. On the convergence of the Abel-Goncharov interpolation series. *Mat. sborn.* **21** (63), 49–52 (1947).

2. On the best approximation by polynomials of the function $[ax + b|x|]|x|^s$ on the segment $[-1,1]$. *Izv. Akad. Nauk SSSR (seriya mat.)* **14**, 405–412 (1950).

3. Extremal functions in the class of integral functions of exponential type. *Usp. mat. nauk* **12**, No. 3, 323–328 (1957).

4. On the best mean approximation of a function with s-th derivative of bounded variation on a segment. *Dokl. Akad. Nauk SSSR*, **90**, No. 1, 13–16 (1953).

5. Extremal problems in a class of integral functions of finite degree. *Izv. Akad. Nauk SSSR (seriya mat.)* **23**, No. 2, 423–256 (1959).

6. On the best approximation of functions, the s-th derivative of which has a discontinuity of the first kind. *Dokl. Akad. Nauk SSSR*, **89**, No. 6, 973–975 (1953).

IBRAGIMOV, I. I. and M. V. KELDYSH

1. On interpolation of integral functions. *Mat. sborn.* **20** (64), 283–291 (1947).

JACKSON, D.

1. Über Genauigkeit der Annäherung stetiger Funktionen durch ganze rationale Funktionen gegebenen Grades und trigonometrische Summen gegebener Ordnung, Diss. Göttingen (1911).

2. A general class of problems in approximation. *Amer. J. Math.* **46**, 215–234 (1924).

3. *The theory of approximation*, New York (1930).

4. Certain problems of closest approximation. *Bull. Amer. math. Soc.* **39**, 889–906 (1930).

KACZMARZ, S. and H. STEINHAUS

1. *Theorie der Orthogonalreihen.* Warsaw (1935).

KANTOROVICH, L. V.

1. Some remarks on approximation of functions by means of polynomials with integral coefficients. *Izv. Akad. Nauk SSSR (seriya mat.)*, 1163–1168 (1931).

2. On the convergence of a sequence of the polynomials of S. N. Bernstein outside the fundamental interval. *Izv. Akad. Nauk SSSR (seriya fiz.-mat.)*, 1103–1115 (1931).

KARAMATA, J. and M. TOMIĆ

1. On the summation of Fourier series. (O zbirlivosti Fourier-ovikh redova). *Glas Srpske akad. Nauka* (1953), **206**, *Od. prir.-mat. nauka* **5**, 89–126. (In Serbo-Croat).

KHAVINSON, S. YA.

1. On the dimensions of the polyhedron of best approximation in the metric L_1, Mosk. Inzh.-str. in-t im. V. V. Kuibysheva. *Sb. trudov kafedr mat. i mekhaniki*, No. 19, 18–19 (1957).

2. On the uniqueness of the function of the best approximation in the metric of the space L_1. *Izv. Akad. Nauk SSSR (seriya mat.)* **22**, No. 2, 243–270 (1957).

KHLODOVSKII, I. N.

1. On some properties of the polynomials of S. N. Bernstein. (O nekotorykh svoistvakh polinomov S. N. Bernsteina). *Trudy 1-go Vsesoyuznogo matematicheskogo s''ezda, Kharkov*, (1930).

KOLMOGOROV, A. N.

1. Über Kampaktheit der Funktionenmengen bei der Konvergenz im Mittel. *Gött. Nachr.*, 60–63 (1931).

2. Zur Grössenordnung des Restglieds Fourierischer Reihen differentzierbarer Funktionen, *Annals of Math.* **36**, 521–526 (1935).

3. Über die beste Annäherung von Funktionen einer gegebenen Funktionenklasse. *Annals of Math.* **37**, 107–110 (1936).

4. On inequalities between the upper bounds of sequences of the derivatives of a function on an infinite interval. *Uchenye zapiski MGU* **30**, *Matematika*, 3–16 (1939).

5. A note on the polynomials of P. L. Chebyshev deviating least from a given function. *Usp. mat. nauk* **3**, No. 1, 216–221 (1948).

6. On some asymptotic characteristics of completely bounded metric spaces. *Dokl. Akad. Nauk SSSR* **108**, No. 3, 385–388 (1956).

7. On the linear dimensions of vector topological spaces. *Dokl. Akad. Nauk SSSR* **120**, No. 2, 239–241 (1958).

KORENBLYUM, B. I.

1. On two theorems in the theory of absolutely monotonic functions. *Usp. mat. nauk* **6**, No. 4 (44), 172–175 (1951).

KOROVKIN, P. P.

1. On the order of approximation of functions by linear positive operators. *Dokl. Akad. Nauk SSSR* **114**, No. 6, 1158–1161 (1957).

2. *Linear operators and the theory of approximation* (Lineinyye operatory i teoriya priblizhenii) Fizmatgiz, (1959). English translation in preparation by Hindustan Publishing Company.

KRALIK, D.

1. Untersuchungen der Integrale und Derivierten gebrochener Ordnung mit den Metoden der konstruktiven Funktionentheorie. *Acta math. hung.* **7**, 49–64 (1956).

KREIN, M. G.

1. *L-problems in an abstract linear normed space, Collection of papers by N. I. Akhiyezer and M. G. Krein, On some questions of the theory of moments.* (L-problema v abstraktnom lineinom normirovannom prostranstve, Sbornik statei N. I. Akhiyezera i M. G. Kreina, O nekotorykh voprosakh teorii momentov). DNTVU (1938).

2. On the best approximation of continuous differentiable functions on the whole real axis. *Dokl. Akad. Nauk SSSR* **18**, 615–624 (1938).

3. On the representation of functions by Fourier-Stieltjes integrals. *Uchenyye zapiski Kuibyshevskogo ped. in-ta*, No. 7 (1943).

KREIN, M. G., D. P. MIL'MAN and M. A. RUTMAN

1. On a property of the basis in Banach space. *Zap. Khar'k. mat. ob-va* (4) **16**, 106–110 (1940).

KUROSH, A. G.

1. *Course of higher algebra.* (Kurs vysshei algebry). Gostekhizdat (1955).

LAVRENT'EV, M. A.

1. On the theory of conformal representation. (K teorii konformnykh otobrazhenii). *Trudy Fiz.-mat. in-ta im. V. A. Steklova, otdel mat.* **V**, 159–245 (1934).

LEBESGUE, H.

1. Sur les intégrales singulières. *Ann. de Toulouse* **1**, 25–117 (1909).

2. Sur l'approximation des fonctions. *Bull. Sc. Math.* **22**, (1898).

3. Sur la représentation trigonométrique approchée des fonctions satisfaisant à une condition de Lipschitz. *Bull. Soc. Math. de France* **38**, 184–210 (1910).

LEVIN, B. YA.

1. *The distribution of the zeros of integral functions.* (Raspredelenye kornei tselykh funktsii). Gostekhizdat (1956).

LEVITAN, B. M.

1. On a generalisation of inequalities of S. N. Bernstein and N. Bohr. *Dokl. Akad. Nauk SSSR* **15** (1937).

2. *Almost periodic functions.* (Pochti-periodicheskiye funktsii). Gostekhizdat (1953).

LIOUVILLE, J.

1. Sur le calcul des différentialles à indices quelconques. *J. Éc. polyt., Paris* **13**, 1–69 (1832).

LOZINSKI, S. M.

1. On the convergence and summability of Fourier series and interpolation processes. *Mat. sborn.* **14** (56), 175–268 (1944).

2. On a class of linear operations. *Dokl. Akad. Nauk SSSR* **61**, No. 2, 193–196 (1948).

3. The converse of Jackson's theorems. *Dokl. Akad. Nauk SSSR* **83**, No. 5, 645–647 (1952).

4. On the rapidity of convergence of a sequence of linear trigonometric polynomial operations. *Dokl. Akad. Nauk SSSR* **89**, No. 5. 785–787 (1953).

5. On the rapidity of convergence of a sequence of linear operations. *Dokl. Akad. Nauk SSSR* **89**, No. 4, 609–612 (1953).

LUSIN, N. N. and I. I. PRIVALOV

1. Sur l'unicité et la multiplicité des fonctions analytiques. *Ann. Sci. Ec. norm. Sup., série* 3, **42**, No. 6, 143–191 (1925).

LYUSTERNIK, L. A. and V. I. SOBOLEV

1. *Elements of functional analysis.* (Elementy funktsional'nogo analiza). Gostekhizdat (1951).

MANDELBROJT, S.

1. Sur les fonctions indéfiniment dérivables. *Acta Math.* **72**, 15 (1940).

2. *Sèries adherentes, regularisation des suites, applications.* Gauthier–Villars, Paris (1952).

MARCHOUD, A.

1. Sur les derivées et sur les différences des fonctions de variables réelles. *J. Math. pures appl.* **6**, 337–425 (1927).

MARCINKIEWICZ, J.

1. Sur la divergence des polynômes d'interpolation. *Acta Litterarum ac Scientiarum Szeged* **8**, 131–135 (1937).

MARCINKIEWICZ, J. and A. ZYGMUND

1. Mean values of trigonometrical polynomials. *Fund. Math.* **28** (1937).

MARKOV, A. A.

1. *On the limiting values of integrals connected with interpolation.* (O predel'nykh velichinakh integralov v svyazi s interpolirovaniyem (1898)). Izbr. trudy, (Collected Works). Gostekhizdat (1948).

MARKOV, V. A.

1. *On functions which deviate least from zero.* (O funktsiyakh, naimeneye uklonyayushchikhsya ot nulya). SPb (1892).

MARKUSHEVICH, A. N.

1. On best approximation. *Dokl. Akad. Nauk SSSR* **44**, No. 7, 290–292 (1944).

MATVEYEV, I. V.

1. On methods of summation of double Fourier series for functions of two variables. *Mat. sborn.* **28**, No. 1 (1951).

MERGELYAN, S. N.

1. On a theorem of M. A. Lavrent'ev. *Dokl. Akad. Nauk SSSR* **77**, No. 4, 565–568 (1951).

2. Weighted approximation by polynomials. *Usp. mat. nauk* **11**, No. 5, 107–152 (1956).

MITYAGIN, B. S.

1. On the second mixed derivative. *Dokl. Akad. Nauk SSSR* **123**, No. 4, 606–609 (1958).

MONTEL, P.

1. Sur les polynômes d'approximation. *Bull. Soc. math. Fr.* (1919).

MÜNTZ, C.

1. Über den Approximationssatz von Weierstrass. *Schwarz Festschrift* 303–312 (1944).

NAGY, B. SZ.

1. Über gewisse Extremalfragen bei transformierten trigonometrischen Entwicklungen. *Berichte Akad. d. Wiss. Leipzig* **90** (1938).

2. Sur une classe generale des procédés de sommation de séries de Fourier. *Hungarica Acta Mathematica* **1**, No. 3 (1948).

3. Méthodes de sommation de séries de Fourier I. *Acta Sc. Math. Szeged.* **12,** 204–210 (1950).

4. Approximation properties of orthogonal expansions. *Acta Sc. Math.* **15,** No. 1, 31–37 (1953).

NATANSON, I. P.

1. Some estimates connected with the singular integral of De la Vallée Poussin. *Dokl. Akad. Nauk SSSR* **45**, 290–293 (1944).

2. *The constructive theory of functions.* (Konstruktivnaya teoriya funktsii). Gostekhizdat (1949).

3. *The theory of functions of a real variable.* (Teoriya funktsii veshchestvennoi peremennoi). Gostekhizdat (1957). English translation published as *Theory of functions of a real variable,* Ungar, New York, Constable, London, Vol. I, 1955, Vol. II, 1960.

NEMYTSKII, V. V.

1. The method of fixed points in analysis. *Usp. mat. nauk*, No. 1 (1936).

NIKOL'SKII, S. M.

1. On some methods of approximation by trigonometric sums. *Izv. Akad. Nauk SSSR (seriya mat.)* **4**, 509–520 (1940).

2. Approximation of periodic functions by trigonometric polynomials. *Trudy Mat. in-ta im. V. A. Steklova Akad. Nauk SSSR* **15** (1945).

3. Mean approximation of functions by trigonometric polynomials. *Izv. Akad. Nauk SSSR (seriya mat.)* **10**, 207–256 (1940).

4. On the best approximation by polynomials of functions which satisfy a Lipschitz condition. *Izv. Akad. Nauk. SSSR (seriya mat.)* **10**, 295–318 (1946).

5. On the best approximation of functions, the s-th derivative of which has discontinuities of the first kind. *Dokl. Akad. Nauk SSSR* **55**, No. 2, 99–102 (1947).

6. On the best mean approximation by polynomials of functions with singularities of the form $|a-x|^s$. *Dokl. Akad. Nauk SSSR* **55**, No. 3, 195–198 (1947).

7. On the best mean approximation by polynomials of the function $|a-x|^s$. *Izv. Akad. Nauk SSSR (seriya mat.)* **11**, 139–180 (1947).

8. On the best linear method of mean approximation of differentiable functions by polynomials. *Dokl. Akad. Nauk SSSR* **58**, No. 2, 185–188 (1947).

9. On linear methods of summation of Fourier series. *Izv. Akad. Nauk SSSR (seriya mat.)* **12**, 259–278 (1948).

10. Generalisation of an inequality of S. N. Bernstein. *Dokl. Akad. Nauk SSSR* **60**, No. 9, 1507–1510 (1948).

11. Generalisation of a proposition of S. N. Bernstein on differentiable functions of many variables. *Dokl. Akad. Nauk SSSR* **59**, 1535–1536 (1948).

12. Fourier series of functions which have a derivative of bounded variation. *Izv. Akad. Nauk SSSR (seriya mat.)* **13**, No. 6, 513–532 (1949).

13. Inequalities for integral functions of finite degree and their application in the theory of differentiable functions of many variables. *Trudy Mat. in-ta im. V. A. Steklova* 38, 244–278 (1951).

14. Fourier series of functions with a given modulus of continuity. *Dokl. Akad. Nauk SSSR* **52**, 191–194 (1946).

PALEY, R. E. A. C. and N. WIENER

1. *Fourier transforms in the complex domain.* New York (1934).

PETROVSKII, I. G.

1. *Lectures on the theory of ordinary differential equations.* (Lektsii po teorii obyknovennykh differentsialnykh uravnenii). Gostekhizdat (1947).

PINKEVICH, V. T.

1. On the order of the remainder term of the Fourier series of functions differentiable in the sense of Weyl. *Izv. Akad. Nauk SSSR (seriya mat.)* **4**, 521–578 (1940).

PLANCHEREL, M.

1. Contribution à l'étude de la représentation d'une fonction arbitraire par des ntégrales définies. Rend. Palermo (1910).

PLANCHEREL, M. and G. PÓLYA

1. Fonctions entières et intégrales de Fourier multiples. *Comm. Helv.* **9**, 224–248 (1937).

POGODICHEVA, N. A.

1. *Investigation of a class of linear processes of approximation by interpolation polynomials.* (Issledovaniye odnogo klassa lineinykh protsessov priblizheniya interpolyatsionnymi polinomami). Dissertation, Dnepropetrovsk (1956).

POGODICHEVA, N. A. and A. F. TIMAN.

1. On a relation in the theory of the summation of interpolation polynomials and Fourier series. *Dokl. Akad. Nauk SSSR* **111**, No. 3, 542–543 (1956).

PÓLYA, G.

1. Sur un algorithme toujours convergent pour obtenir les polynômes de meilleure approximation de Tchebycheff pour une fonction continue quelconque. *Comptes Rendus* **157** (1913).

PÒLYA, G. and G. SZEGÖ

1. *Aufgaben und Lehrsätze aus der Analysis*, Vols. I, II. Springer Verlag, Berlin (1925).

PONOMARENKO, V. G.

1. *On linear processes of approximation of continuous periodic functions of two variables.* (O lineinykh protsessakh priblizheniya nepreryvnykh periodicheskikh funktsii dvukh peremennykh). Dissertation, Dnepropetrovsk (1956).

PONTRYAGIN, L. S. and L. G. SHNIREL'MAN

1. On a metric property of dimension. (Ob odnom metricheskom svoistve razmernosti). *Annals of Maths.* **33**, 156–162 (1932). This same paper is given as an application in the book: W. Hurewicz and H. Wallman, *Dimension Theory*, 2nd ed., Princetown Univ. Press (1948).

POPOVICIU, T.

1. Sur l'approximation des fonctions convexes d'ordre supérieur. *Mathematica* **10**. 49–54 (1934).

PRIVALOV, I. I.

1. Sur les fonctions conjuguées. *Bull. Soc. Math. Fr.* **44** (1916).

2. *Introduction to the theory of functions of a complex variable.* (Vvedeniye v teoriyu funktsii kompleksnogo peremennogo). Gostekhizdat (1948).

PRINGSHEIM, A.

1. Über die notwendigen und hinreichenden Bedingungen des Taylor'schen Lehrsatzes für Funktionen einer reellen variablen. *Math. Ann.* **44**, 57–82 (1897).

QUADE, E.

1. Trigonometric approximation in the mean, *Duke Math. Journal* **3** (1937).

RABINOVICH, S. N.

1. On some estimates connected with linear methods of approximation by algebraic polynomials. (O nekotorykh otsenkakh, svyazannykh s lineinymi metodami priblizheniya algebraicheskimi mnogochlenami). *Nauch. dokl. vyssh. shkoly, fiz-mat. nauki,* No. 3, 104–106 (1958).

RAIKOV, D. A.

1. On local approximation of differentiable functions. *Dokl. Akad. Nauk SSSR* **24** 652–655 (1939).

REMEZ, E. YA.

1. *On methods for the best approximate representation of functions according to P. L. Chebyshev.* (Pro metodi naikrashchogo v rozyminni P. L. Chebysheva nablizhenogo predstavleniya funktsii). Kiev, (1935). (In Ukrainian).

2. *General computational methods for Chebyshev approximation.* (Obshchie vychislitel' nye metody Chebyshevskogo priblizheniya). Kiev, (1957).

RIEMANN, B.

1. *An experiment in the generalisation of the operations of integration and differentiation,* Collected works *(Sochineniya)*, Gostekhizdat, 262–277 (1948).

RIESZ, M.

1. Sur les fonctions conjuguées. *Math. Zeitschr.* **27**, 218–244 (1927).

2. Formule d'interpolation pour la dérivée d'un polynôme trigonométrique. *Comptes Rendus* **158**, 1152 (1914).

ROGOSINSKI, W.

1. Über die Abschnitte trigonometrischer Reihen. *Math. Ann.* **95** (1925).

2. Extremum problems for polynomials and trigonometrical polynomials. *Lond. Math. Soc.* **29**, No. 3, 259–275 (1944)

RUBINSHTEIN, G. Sh.

1. On a method of investigation of convex sets. *Dokl. Akad. Nauk SSSR* **102**, No. 3, 451–454 (1955).

RUDIN, W.

1. *L*-approximation by partial sums of orthogonal developments. *Duke Math. Journ.* **19**, 1–4 (1952).

SAKOVICH, G. N.

1. Some properties of generalised quasi-smooth functions. (Nekotorye svoistva oboshchennogo-kvasigladkikh funktsii). *Kievskii universitet stud. nauch. raboty* **16** (1955).

SCHAEFFER, A. C. and R. J. DUFFIN

1. On some inequalities of S. Bernstein and W. Markoff for derivatives of polynomials. *Bull. Am. Math. Soc.* **44**, No. 4, 289–297 (1938).

2. A refinement of an inequality of the brothers Markoff. *Trans. Am. Math. Soc.* **50**, No. 3, 517–522 (1941).

SCHAUDER, J.

1. Zur Theorie stetiger Abbildungen in Funktionalräumen. *Math. Zeitschr.* **26**, 47–65 (1927).

SCHUR, I.

1. Über das Maximum des absoluten Betrages eines Polynoms in einem gegebenen Intervalle. *Math. Zeitschr.* **4**, 271–287 (1919).

SELIVANOVA, S. G.

1. Asymptotic estimates of approximations of a periodic functions by Chebyshev sums. *Dokl. Akad. Nauk SSSR* **105**, No. 4, 648–651 (1955).

SHCHERBINA, A. D.

1. On a method of summation of series conjugate to Fourier series. *Mat. sborn.* **27** (69): 2, 157–170 (1950).

SHOHAT, J. A. and J. D. TAMARKIN

1. *The problem of moments.* New York (1943).

SMIRNOV, V. I.

1. *Course of higher mathematics.* (Kurs vysshei matematiki). Vols. I–V. (English translation in preparation by Pergamon Press).

SOBLE, A.

1. Majorants of polynomial derivatives. *Amer. Math. Monthly* **64**, No. 9, 639–643 (1957).

STECHKIN, V. A.

1. Generalisation of some inequalities of S. N. Bernstein. *Dokl. Akad. Nauk SSSR* **60**, No. 9, 1511–1514 (1948).

2. On the order of the best approximations of continuous functions. *Izv. Akad. Nauk SSSR (seriya mat.)* **15**, 219–242 (1951).

3. *The methods of summation of S. N. Bernstein and W. Rogosiński.* (Metody summirovaniya S. N. Bernsteina i W. Rogosińskogo) (appendix to the book: G. H. Hardy, Divergent series, Oxford, 1949).

4. The best approximation of functions representable by trigonometric series with gaps. *Dokl. Akad. Nauk SSSR*, **76**, No. 1, 33–36 (1951).

5. On the best approximation of conjugate functions by trigonometric polynomials. *Izv. Akad. Nauk SSSR (seriya mat.).* **20**, 197–206 (1956).

6. On the absolute convergence of orthogonal series. *Mat. sborn.* **29**, (70): 1, 225–231 (1951).

STEIN, E.

1. Functions of exponential type. *Ann. Math.* **65**, No. 3, 585–592 (1957).

STEKLOV, V. A.

1. Théorème de fermeture pour les polynômes de Laplace-Hermite-Tchebychef. *Izv. Akad. Nauk* **10**, 403–416 (1916).

2. Theoreme de fermeture pour les polynomes de Tchébychef-Laguerre. *Izv. Akad. Nauk* **10**, 633–642 (1916).

STONE, M. H.

1. Applications of the theory of Boolean rings to general topology. *Trans. Amer. math. Soc.* **41**, 375–481 (1937).

SUN YUN-SHEN.

1. On the best approximation of periodic differentiable functions by trigonometric polynomials. *Izv. Akad. Nauk. SSSR (seriya mat.)* **23**, No. 1, 67–72 (1959).

SZEGÖ, G.

1. *Über einen Satz des Herrn Bernstein.* Schriften Königsb. Ges. (1928).

2. *Orthogonal polynomials.* New York (1939).

3. On an inequality of P. Turán concerning Legendre polynomials. *Bull. Amer. math. Soc.* **54**, 401–405 (1948).

SZIDON, S.

1. Verallgemeinerung eines Satzes über die absolute Konvergenz von Fourierreihen mit Lücken. *Math. Ann.* **97**, 675–676 (1927).

TIMAN, A. F.

1. On a method of approximation of continuous functions by trigonometric polynomials. *Izv. Akad. Nauk SSSR (seriya mat.)* **11**, 263–282 (1947).

2. On the Lebesgue constants for some methods of summation. *Dokl. Akad. Nauk SSSR* **61**, No. 6, 989–992 (1948).

3. Some extremal properties of periodic functions. (Nekotoryye ekstremal'nyye svoistva periodicheskikh funktsii). Dnepropetr. universitet, *Nauch. Zap.* **34**, 153–162 (1948).

4. Some asymptotic estimates for the polynomials of N. I. Akhiyezer and B. M. Levitan. *Dokl. Akad. Nauk SSSR* **64**, No. 2, 175–178 (1949).

5. On some methods for the summation of Fourier series. *Izv. Akad. Nauk SSSR (seriya mat.)* **14**, No. 1, 85–94 (1950).

6. Approximation by ordinary polynomials of functions satisfying a Lipschitz condition. *Dokl. Akad. Nauk SSSR* **77**, No. 6, 969–972 (1951).

7. Strengthening of Jackson's theorem on the best approximation of continuous functions on a finite segment of the real axis. *Dokl. Akad. Nauk SSSR* **78**, No. 1, 17–20 (1951).

8. Generalisation of some results of A. N. Kolmogorov and S. M. Nikol'skii. *Dokl. Akad. Nauk SSSR* **81**, No. 4, 509–511 (1951).

9. On quasi-smooth functions. *Izv. Akad. Nauk SSSR (seriya mat.)* **15**, No. 3. 243–254 (1951).

10. *Investigations in the theory of approximation of functions.* (Issledovaniya po teorii priblizheniya funktsii). Dissertation, Khar'kov (1951).

11. On interference phenomena in the behaviour of integral functions of finite degree. *Dokl. Akad. Nauk SSSR* **89**, No. 1, 17–20 (1953).

12. Approximation properties of linear methods of summation of Fourier series. *Izv. Akad. Nauk SSSR (seriya mat.)* **17**, 99–134 (1953).

13. On linear processes of approximation by algebraic polynomials, Lebesgue functions and some applications to Fourier series. *Dokl. Akad. Nauk SSSR* **101**, No. 2, 221–224 (1955).

14. Generalisation of a theorem of Stone. *Dokl. Akad. Nauk SSSR* **111**, No. 5, 955–958 (1956).

15. Converse theorems of the constructive theory of functions defined on a finite segment of the real axis. *Dokl. Akad. Nauk SSSR* **116**, No. 5, 762–765 (1957).

16. A supplement to the work of A. B. Efimov "An estimate of the modulus of continuity of functions of the class $H\frac{1}{2}$, *Izv. Akad. Nauk SSSR (seriya mat.)* **21**, No. 4, 595–598 (1957).

17. Some remarks on trigonometric polynomials and Fourier-Stieltjes series. *Usp. mat. Nauk* **12**, No. 2 (74), 175–183 (1957).

18. On the best approximation of differentiable functions by algebraic polynomials on a finite segment of the real axis. *Izv. Akad. Nauk SSSR (seriya mat.)* **22**, 355–360 (1958).

TIMAN, A. F. and V. K. DZYADYK.

1. On the best approximation of quasi-smooth functions by ordinary polynomials. *Dokl. Akad. Nauk SSSR* **75**, 499–502 (1950).

TIMAN, A. F. and I. M. GANZBURG.

1. On the convergence of some processes of summation of Fourier series. *Dokl. Akad. Nauk SSSR* **63**, No. 6, 619–622 (1948).

TIMAN, A. F. and M. F. TIMAN.

1. The generalised modulus of continuity and best mean approximation. *Dokl. Akad. Nauk SSSR* **71**, No. 1, 17–20 (1950).

2. On the relation between the moduli of smoothness of functions defined on the whole real axis. *Dokl. Akad. Nauk SSSR* **113**, No. 5, 995–997 (1957).

TIMAN, A. F. and L. I. TUCHINSKII.

1. Approximation by algebraic polynomials of differentiable functions defined on a finite segment. *Dokl. Akad. Nauk SSSR* **111**, No. 4, 771–772 (1956).

TIMAN, M. F.

1. On the relationship between total and partial best mean approximations of functions of many variables. *Dokl. Akad. Nauk SSSR* **112**, No. 1, 24–26 (1957).

2. Converse theorems of the constructive theory of functions in the spaces L_p. *Mat. sborn.* **46** (88): 1, 125–132 (1958).

3. Converse theorems of the constructive theory of functions of many variables. *Dokl. Akad. Nauk SSSR* **120**, No. 6, 1207–1209 (1958).

4. On the relationship between total and partial best approximations of functions of many variables. *Dokl. Akad. Nauk SSSR* **124**, No. 3, 527–528 (1959).

TITCHMARSH, E. C.

1. *The theory of functions*. Oxford.

2. *Introduction to the theory of Fourier integrals*. Oxford.

TRICOMI, F. G.

1. *Vorlesungen über Orthogonalreihen*. Berlin (1955).

TSENOV, I. V.

1. Some questions in the theory of approximation of functions. *Mat. sborn.* **21**, 435–439 (1947).

TULAIKOV, A. N.

1. Zur Kompaktheit im Raum L_p für $p = 1$. *Gött. Nachr.* 167–170 (1933).

TURÁN, P.

1. Über die Ableitung von Polynomen. *Compos. Math.* 7, 89–95 (1939–40).

TVERITIN, A. N.

1. An application of the theory of moments in the theory of trigonometric series. *Dokl. Akad. Nauk SSSR* **61**, No. 6, 985–988 (1948).

VALLÉE-POUSSIN, CH. J. de la,

1. Sur l'approximation des fonctions d'une variable réelle et de leurs dérivées par des polonômes et des suites finies de Fourier. *Bull. Acad. Sc. Belgique*, 193–254 (1908).

2. Sur les polonômes d'approximation et la représentation approchée d'un angle. *Bull. Acad. Sc. Belgique*, 808 (1910).

3. *Leçons sur l'approximation des fonctions d'une variable réelle*. Paris (1919).

4. *Cours d'analyse infinitesimale*, Vols. 1–2.

VIDENSKII, V. S.

1. On inequalities relating to the derivatives of a polynomial. *Dokl. Akad. Nauk SSSR* **67**, No. 5, 777–780 (1949).

2. On the estimation of the derivatives of a polynomial. *Dokl. Akad. Nauk SSSR* **73**, No. 2, 258–259 (1950).

3. On estimates of the derivatives of a polynomial. *Izv. Akad. Nauk SSSR (seriya mat.)* **15**, 401–420 (1951).

4. Generalisation of the inequalities of V. A. Markov. *Dokl. Akad. Nauk SSSR* **120**, No. 3, 447–450 (1958).

5. Generalisation of a theorem of A. A. Markov on the estimation of the derivative of a polynomial. *Dokl. Akad. Nauk SSSR* **125**, No. 1. 15–18 (1959).

6. On uniform approximation in the complex plane. *Usp. mat. nauk* **11**, No. 5 (71), 169–175 (1956).

VINTUSHKII, A G.

1. The absolute ε-entropy of metric spaces. *Dokl. Akad. Nauk SSSR* **117**, No. 5, 745–747 (1957).

VORONOVSKAYA, E. V.

1. Determination of the asymptotic form of approximation of functions by the polynomials of S. N. Bernstein. *Dokl. Akad. Nauk SSSR, A*, 79–85 (1932).

WEIERSTRASS, K.

1. Über die analytische Darstellbarkeit sogennanter willkürlicher Funktionen reeller Argumente. *Sitzungsberichte der Acad. Berlin*, 633–639; 789–805 (1885).

WEYL, H.

1. Bemerkungen zum Begriff der Differentialquotienten gebrochener Ordnung, *Vierteljahrsschr Naturf. Ges. Zürich*, 62, 296–302 (1917).

WIDDER, D. V.

1. Necessary and sufficient conditions for the representation of a function as a Laplace integral. *Trans. Amer. Math. Soc.* **33**, 851–892 (1931).

WIENER, N.

1. *The Fourier integral and certain of its applications.* Cambridge (1933).

YAKOB, M.

1. Application of the Laplace transformation to the summation of Fourier series and trigonometric interpolation polynomials. *Dokl. Akad. Nauk SSSR* **32**, 390–394 (1941).

YEFIMOV, A. V.

1. Bounds of the modulus of continuity for functions of the class $H\frac{1}{2}$. *Izv. Akad. Nauk SSSR (seriya mat.)* **21**, No. 2, 283–288 (1957).

2. On the Fourier coefficients of functions of the class $\tilde{H}_2$. *Usp. mat. Nauk*, **12**, No. 3, 303–311 (1957).

3. Approximation of conjugate functions by Fejer sums. *Usp. mat. Nauk* **14**, No. 1, 183–188 (1959).

YOUNG, W. H.

1. On the integration of Fourier series. *Proc. Lond. Math. Soc.* (1911).

ZHIRNOVA, G. A.

1. On approximation of functions by polynomials with integral coefficients. *Izvestiya vysshikh uchebnykh zavedenii, Matematika* **4**, 80–88 (1958).

ZINGER, I.

1. Caractérisation des éléments de meilleure approximation dans un espace de Banach quelconque. *Acta Sci. Math. Szeged.* **17**, No. 3–4, 181–194 (1956).

ZOLOTAREV, E. I.

1. *On a question of least values* (Ob odnom voprose o naimen'shikh velichinakh). Dissertation, Complete collected works, Vol. II. (1868).

ZUKHOVITSKII, S. I.

1. On approximation of real functions in the sense of P. L. Chebyshev. *Usp. mat. nauk* **11**, No. 2 (68), 125–159 (1956).

2. On the problem of Chebyshev approximation in Hilbert space. *Dopovidi Akad. Nauk URSR*, No. 1, 7 (1955). (In Ukrainian).

3. Some theorems of the theory of Chebyshev approximations in Hilbert space. *Mat. sborn.* **37** (79), No. 1, 3–20 (1955).

4. An algorithm for the construction of the Chebyshev approximation of a continuous function by a polynomial. *Dokl. Akad. Nauk SSSR* **120**, No. 4, 693–697 (1958).

ZUKHOVITSKII, S. I. and M. G. KREIN.

1. Remark on a possible generalisation of theorems of A. Haar and A. N. Kolmogorov. *Usp. mat. nauk* **5**, No. 1 (35), 217–229 (1950).

ZUKHOVITSKII, S. I. and S. B. STECHKIN.

1. On approximation of abstract functions. *Usp. mat. nauk* **12**, No. 1 (73), 187–191 (1957).

ZYGMUND, A.

1. *Trigonometrical series, Warsaw*, (1935). Revised edition in two volumes published as Trigonometric series, Cambridge University Press. (1959).

2. Smooth functions. *Duke Math. Journal* **12**, 47–76 (1945).

3. The approximation of functions by typical means of their Fourier series. *Duke Math. Journal.* **12**, 4, 685–754 (1945).

4. A remark on the integral modulus of continuity. Univ. Nac., Tucuman Revista A7, 259 (1950).

INDEX

A CATALOG OF SELECTED

DOVER BOOKS

IN SCIENCE AND MATHEMATICS

A CATALOG OF SELECTED

DOVER BOOKS

IN SCIENCE AND MATHEMATICS

RELATIVITY, THERMODYNAMICS AND COSMOLOGY, Richard C. Tolman. Landmark study extends thermodynamics to special, general relativity; also applications of relativistic mechanics, thermodynamics to cosmological models. 501pp. 5⅜ × 8½. 65383-8 Pa. $12.95

APPLIED ANALYSIS, Cornelius Lanczos. Classic work on analysis and design of finite processes for approximating solution of analytical problems. Algebraic equations, matrices, harmonic analysis, quadrature methods, much more. 559pp. 5⅜ × 8½. 65656-X Pa. $12.95

SPECIAL RELATIVITY FOR PHYSICISTS, G. Stephenson and C.W. Kilmister. Concise elegant account for nonspecialists. Lorentz transformation, optical and dynamical applications, more. Bibliography. 108pp. 5⅜ × 8½. 65519-9 Pa. $4.95

INTRODUCTION TO ANALYSIS, Maxwell Rosenlicht. Unusually clear, accessible coverage of set theory, real number system, metric spaces, continuous functions, Riemann integration, multiple integrals, more. Wide range of problems. Undergraduate level. Bibliography. 254pp. 5⅜ × 8½. 65038-3 Pa. $7.95

INTRODUCTION TO QUANTUM MECHANICS With Applications to Chemistry, Linus Pauling & E. Bright Wilson, Jr. Classic undergraduate text by Nobel Prize winner applies quantum mechanics to chemical and physical problems. Numerous tables and figures enhance the text. Chapter bibliographies. Appendices. Index. 468pp. 5⅜ × 8½. 64871-0 Pa. $11.95

ASYMPTOTIC EXPANSIONS OF INTEGRALS, Norman Bleistein & Richard A. Handelsman. Best introduction to important field with applications in a variety of scientific disciplines. New preface. Problems. Diagrams. Tables. Bibliography. Index. 448pp. 5⅜ × 8½. 65082-0 Pa. $12.95

MATHEMATICS APPLIED TO CONTINUUM MECHANICS, Lee A. Segel. Analyzes models of fluid flow and solid deformation. For upper-level math, science and engineering students. 608pp. 5⅜ × 8½. 65369-2 Pa. $13.95

ELEMENTS OF REAL ANALYSIS, David A. Sprecher. Classic text covers fundamental concepts, real number system, point sets, functions of a real variable, Fourier series, much more. Over 500 exercises. 352pp. 5⅜ × 8½. 65385-4 Pa. $10.95

PHYSICAL PRINCIPLES OF THE QUANTUM THEORY, Werner Heisenberg. Nobel Laureate discusses quantum theory, uncertainty, wave mechanics, work of Dirac, Schroedinger, Compton, Wilson, Einstein, etc. 184pp. 5⅜ × 8½. 60113-7 Pa. $5.95

INTRODUCTORY REAL ANALYSIS, A.N. Kolmogorov, S.V. Fomin. Translated by Richard A. Silverman. Self-contained, evenly paced introduction to real and functional analysis. Some 350 problems. 403pp. 5⅜ × 8½. 61226-0 Pa. $9.95

PROBLEMS AND SOLUTIONS IN QUANTUM CHEMISTRY AND PHYSICS, Charles S. Johnson, Jr. and Lee G. Pedersen. Unusually varied problems, detailed solutions in coverage of quantum mechanics, wave mechanics, angular momentum, molecular spectroscopy, scattering theory, more. 280 problems plus 139 supplementary exercises. 430pp. 6½ × 9¼. 65236-X Pa. $12.95

ASYMPTOTIC METHODS IN ANALYSIS, N.G. de Bruijn. An inexpensive, comprehensive guide to asymptotic methods—the pioneering work that teaches by explaining worked examples in detail. Index. 224pp. 5⅜ × 8½. 64221-6 Pa. $6.95

OPTICAL RESONANCE AND TWO-LEVEL ATOMS, L. Allen and J.H. Eberly. Clear, comprehensive introduction to basic principles behind all quantum optical resonance phenomena. 53 illustrations. Preface. Index. 256pp. 5⅜ × 8½. 65533-4 Pa. $7.95

COMPLEX VARIABLES, Francis J. Flanigan. Unusual approach, delaying complex algebra till harmonic functions have been analyzed from real variable viewpoint. Includes problems with answers. 364pp. 5⅜ × 8½. 61388-7 Pa. $8.95

ATOMIC SPECTRA AND ATOMIC STRUCTURE, Gerhard Herzberg. One of best introductions; especially for specialist in other fields. Treatment is physical rather than mathematical. 80 illustrations. 257pp. 5⅜ × 8½. 60115-3 Pa. $5.95

APPLIED COMPLEX VARIABLES, John W. Dettman. Step-by-step coverage of fundamentals of analytic function theory—plus lucid exposition of five important applications: Potential Theory; Ordinary Differential Equations; Fourier Transforms; Laplace Transforms; Asymptotic Expansions. 66 figures. Exercises at chapter ends. 512pp. 5⅜ × 8½. 64670-X Pa. $11.95

ULTRASONIC ABSORPTION: An Introduction to the Theory of Sound Absorption and Dispersion in Gases, Liquids and Solids, A.B. Bhatia. Standard reference in the field provides a clear, systematically organized introductory review of fundamental concepts for advanced graduate students, research workers. Numerous diagrams. Bibliography. 440pp. 5⅜ × 8½. 64917-2 Pa. $11.95

UNBOUNDED LINEAR OPERATORS: Theory and Applications, Seymour Goldberg. Classic presents systematic treatment of the theory of unbounded linear operators in normed linear spaces with applications to differential equations. Bibliography. 199pp. 5⅜ × 8½. 64830-3 Pa. $7.95

LIGHT SCATTERING BY SMALL PARTICLES, H.C. van de Hulst. Comprehensive treatment including full range of useful approximation methods for researchers in chemistry, meteorology and astronomy. 44 illustrations. 470pp. 5⅜ × 8½. 64228-3 Pa. $10.95

CONFORMAL MAPPING ON RIEMANN SURFACES, Harvey Cohn. Lucid, insightful book presents ideal coverage of subject. 334 exercises make book perfect for self-study. 55 figures. 352pp. 5⅜ × 8¼. 64025-6 Pa. $9.95

OPTICKS, Sir Isaac Newton. Newton's own experiments with spectroscopy, colors, lenses, reflection, refraction, etc., in language the layman can follow. Foreword by Albert Einstein. 532pp. 5⅜ × 8½. 60205-2 Pa. $9.95

GENERALIZED INTEGRAL TRANSFORMATIONS, A.H. Zemanian. Graduate-level study of recent generalizations of the Laplace, Mellin, Hankel, K. Weierstrass, convolution and other simple transformations. Bibliography. 320pp. 5⅜ × 8½. 65375-7 Pa. $8.95

NUMERICAL METHODS FOR SCIENTISTS AND ENGINEERS, Richard Hamming. Classic text stresses frequency approach in coverage of algorithms, polynomial approximation, Fourier approximation, exponential approximation, other topics. Revised and enlarged 2nd edition. 721pp. 5⅜ × 8½.
65241-6 Pa. $14.95

THEORETICAL SOLID STATE PHYSICS, Vol. I: Perfect Lattices in Equilibrium; Vol. II: Non-Equilibrium and Disorder, William Jones and Norman H. March. Monumental reference work covers fundamental theory of equilibrium properties of perfect crystalline solids, non-equilibrium properties, defects and disordered systems. Appendices. Problems. Preface. Diagrams. Index. Bibliography. Total of 1,301pp. 5⅜ × 8½. Two volumes. Vol. I 65015-4 Pa. $14.95
Vol. II 65016-2 Pa. $14.95

OPTIMIZATION THEORY WITH APPLICATIONS, Donald A. Pierre. Broad-spectrum approach to important topic. Classical theory of minima and maxima, calculus of variations, simplex technique and linear programming, more. Many problems, examples. 640pp. 5⅜ × 8½. 65205-X Pa. $14.95

THE MODERN THEORY OF SOLIDS, Frederick Seitz. First inexpensive edition of classic work on theory of ionic crystals, free-electron theory of metals and semiconductors, molecular binding, much more. 736pp. 5⅜ × 8½.
65482-6 Pa. $15.95

ESSAYS ON THE THEORY OF NUMBERS, Richard Dedekind. Two classic essays by great German mathematician: on the theory of irrational numbers; and on transfinite numbers and properties of natural numbers. 115pp. 5⅜ × 8½.
21010-3 Pa. $4.95

THE FUNCTIONS OF MATHEMATICAL PHYSICS, Harry Hochstadt. Comprehensive treatment of orthogonal polynomials, hypergeometric functions, Hill's equation, much more. Bibliography. Index. 322pp. 5⅜ × 8½. 65214-9 Pa. $9.95

NUMBER THEORY AND ITS HISTORY, Oystein Ore. Unusually clear, accessible introduction covers counting, properties of numbers, prime numbers, much more. Bibliography. 380pp. 5⅜ × 8½. 65620-9 Pa. $9.95

THE VARIATIONAL PRINCIPLES OF MECHANICS, Cornelius Lanczos. Graduate level coverage of calculus of variations, equations of motion, relativistic mechanics, more. First inexpensive paperbound edition of classic treatise. Index. Bibliography. 418pp. 5⅜ × 8½. 65067-7 Pa. $11.95

MATHEMATICAL TABLES AND FORMULAS, Robert D. Carmichael and Edwin R. Smith. Logarithms, sines, tangents, trig functions, powers, roots, reciprocals, exponential and hyperbolic functions, formulas and theorems. 269pp. 5⅜ × 8½. 60111-0 Pa. $6.95

THEORETICAL PHYSICS, Georg Joos, with Ira M. Freeman. Classic overview covers essential math, mechanics, electromagnetic theory, thermodynamics, quantum mechanics, nuclear physics, other topics. First paperback edition. xxiii + 885pp. 5⅜ × 8½. 65227-0 Pa. $19.95

SPECIAL FUNCTIONS, N.N. Lebedev. Translated by Richard Silverman. Famous Russian work treating more important special functions, with applications to specific problems of physics and engineering. 38 figures. 308pp. 5⅜ × 8½. 60624-4 Pa. $8.95

OBSERVATIONAL ASTRONOMY FOR AMATEURS, J.B. Sidgwick. Mine of useful data for observation of sun, moon, planets, asteroids, aurorae, meteors, comets, variables, binaries, etc. 39 illustrations. 384pp. 5⅜ × 8¼. (Available in U.S. only) 24033-9 Pa. $8.95

INTEGRAL EQUATIONS, F.G. Tricomi. Authoritative, well-written treatment of extremely useful mathematical tool with wide applications. Volterra Equations, Fredholm Equations, much more. Advanced undergraduate to graduate level. Exercises. Bibliography. 238pp. 5⅜ × 8½. 64828-1 Pa. $7.95

POPULAR LECTURES ON MATHEMATICAL LOGIC, Hao Wang. Noted logician's lucid treatment of historical developments, set theory, model theory, recursion theory and constructivism, proof theory, more. 3 appendixes. Bibliography. 1981 edition. ix + 283pp. 5⅜ × 8½. 67632-3 Pa. $8.95

MODERN NONLINEAR EQUATIONS, Thomas L. Saaty. Emphasizes practical solution of problems; covers seven types of equations. ". . . a welcome contribution to the existing literature. . . ."—*Math Reviews*. 490pp. 5⅜ × 8½. 64232-1 Pa. $11.95

FUNDAMENTALS OF ASTRODYNAMICS, Roger Bate et al. Modern approach developed by U.S. Air Force Academy. Designed as a first course. Problems, exercises. Numerous illustrations. 455pp. 5⅜ × 8½. 60061-0 Pa. $9.95

INTRODUCTION TO LINEAR ALGEBRA AND DIFFERENTIAL EQUATIONS, John W. Dettman. Excellent text covers complex numbers, determinants, orthonormal bases, Laplace transforms, much more. Exercises with solutions. Undergraduate level. 416pp. 5⅜ × 8½. 65191-6 Pa. $9.95

INCOMPRESSIBLE AERODYNAMICS, edited by Bryan Thwaites. Covers theoretical and experimental treatment of the uniform flow of air and viscous fluids past two-dimensional aerofoils and three-dimensional wings; many other topics. 654pp. 5⅜ × 8½. 65465-6 Pa. $16.95

INTRODUCTION TO DIFFERENCE EQUATIONS, Samuel Goldberg. Exceptionally clear exposition of important discipline with applications to sociology, psychology, economics. Many illustrative examples; over 250 problems. 260pp. 5⅜ × 8½. 65084-7 Pa. $7.95

LAMINAR BOUNDARY LAYERS, edited by L. Rosenhead. Engineering classic covers steady boundary layers in two- and three-dimensional flow, unsteady boundary layers, stability, observational techniques, much more. 708pp. 5⅜ × 8½. 65646-2 Pa. $18.95

LECTURES ON CLASSICAL DIFFERENTIAL GEOMETRY, Second Edition, Dirk J. Struik. Excellent brief introduction covers curves, theory of surfaces, fundamental equations, geometry on a surface, conformal mapping, other topics. Problems. 240pp. 5⅜ × 8½. 65609-8 Pa. $7.95

TENSOR CALCULUS, J.L. Synge and A. Schild. Widely used introductory text covers spaces and tensors, basic operations in Riemannian space, non-Riemannian spaces, etc. 324pp. 5⅜ × 8¼. 63612-7 Pa. $8.95

A CONCISE HISTORY OF MATHEMATICS, Dirk J. Struik. The best brief history of mathematics. Stresses origins and covers every major figure from ancient Near East to 19th century. 41 illustrations. 195pp. 5⅜ × 8½. 60255-9 Pa. $7.95

A SHORT ACCOUNT OF THE HISTORY OF MATHEMATICS, W.W. Rouse Ball. One of clearest, most authoritative surveys from the Egyptians and Phoenicians through 19th-century figures such as Grassman, Galois, Riemann. Fourth edition. 522pp. 5⅜ × 8½. 20630-0 Pa. $10.95

HISTORY OF MATHEMATICS, David E. Smith. Nontechnical survey from ancient Greece and Orient to late 19th century; evolution of arithmetic, geometry, trigonometry, calculating devices, algebra, the calculus. 362 illustrations. 1,355pp. 5⅜ × 8½. 20429-4, 20430-8 Pa., Two-vol. set $23.90

THE GEOMETRY OF RENÉ DESCARTES, René Descartes. The great work founded analytical geometry. Original French text, Descartes' own diagrams, together with definitive Smith-Latham translation. 244pp. 5⅜ × 8½. 60068-8 Pa. $6.95

THE ORIGINS OF THE INFINITESIMAL CALCULUS, Margaret E. Baron. Only fully detailed and documented account of crucial discipline: origins; development by Galileo, Kepler, Cavalieri; contributions of Newton, Leibniz, more. 304pp. 5⅜ × 8½. (Available in U.S. and Canada only) 65371-4 Pa. $9.95

THE HISTORY OF THE CALCULUS AND ITS CONCEPTUAL DEVELOPMENT, Carl B. Boyer. Origins in antiquity, medieval contributions, work of Newton, Leibniz, rigorous formulation. Treatment is verbal. 346pp. 5⅜ × 8½. 60509-4 Pa. $8.95

THE THIRTEEN BOOKS OF EUCLID'S ELEMENTS, translated with introduction and commentary by Sir Thomas L. Heath. Definitive edition. Textual and linguistic notes, mathematical analysis. 2,500 years of critical commentary. Not abridged. 1,414pp. 5⅜ × 8½. 60088-2, 60089-0, 60090-4 Pa., Three-vol. set $29.85

GAMES AND DECISIONS: Introduction and Critical Survey, R. Duncan Luce and Howard Raiffa. Superb nontechnical introduction to game theory, primarily applied to social sciences. Utility theory, zero-sum games, n-person games, decision-making, much more. Bibliography. 509pp. 5⅜ × 8½. 65943-7 Pa. $12.95

THE HISTORICAL ROOTS OF ELEMENTARY MATHEMATICS, Lucas N.H. Bunt, Phillip S. Jones, and Jack D. Bedient. Fundamental underpinnings of modern arithmetic, algebra, geometry and number systems derived from ancient civilizations. 320pp. 5⅜ × 8½. 25563-8 Pa. $8.95

CALCULUS REFRESHER FOR TECHNICAL PEOPLE, A. Albert Klaf. Covers important aspects of integral and differential calculus via 756 questions. 566 problems, most answered. 431pp. 5⅜ × 8½. 20370-0 Pa. $8.95